A New Island Biogeography of the Sea of Cortés

A New

ISLAND BIOGEOGRAPHY OF THE SEA OF CORTÉS

Edited by

Ted J. Case
Martin L. Cody
Exequiel Ezcurra

2002

UNIVERSITY PRESS

Oxford New York
Auckland Bangkok Buenos Aires Cape Town Chennai
Dar es Salaam Delhi Hong Kong Istanbul Karachi Kolkata
Kuala Lumpur Madrid Melbourne Mexico City Mumbai Nairobi
São Paulo Shanghai Singapore Taipei Tokyo Toronto

Copyright © 2002 by Oxford University Press

Published by Oxford University Press, Inc.
198 Madison Avenue, New York, New York 10016

www.oup.com

Oxford is a registered trademark of Oxford University Press

All rights reserved. No part of this publication may be reproduced,
stored in a retrieval system, or transmitted, in any form or by any means,
electronic, mechanical, photocopying, recording, or otherwise,
without the prior permission of Oxford University Press.

Library of Congress Cataloging-in-Publication Data
A new Island biogeography of the sea of Cortés / edited by Ted. J. Case, Martin L. Cody,
and Exequiel Ezcurra.
   p. cm.
  Includes bibliographical references (p.).
  ISBN 978-0-19-513346-2
  1. Island ecology—Mexico—California, Gulf of—Congresses. 2.
Biogeography—Mexico—California, Gulf of—Congresses. 3. California, Gulf of
(Mexico)—Congresses. I. Case, Ted J. II. Cody, Martin L., 1941– III. Ezcurra, Exequiel.

QH107 .I85 2002
577.5'2'091641—dc21   2001050007

To Gary Polis

Preface

A quarter-century ago a group of researchers convened at UCLA to share interests and results from biogeographical studies in the Sea of Cortés (or Gulf of California), the seaway that separates the Lower California peninsula from mainland Mexico; almost 20 years ago, the first *Island Biogeography in the Sea of Cortez* was published by the University of California Press. The book is long out of print, and since its publication field work in the region has expanded vigorously, adding studies on several new taxonomic groups (e.g., arthropods), as well as many new findings on taxa previously covered. Research on the geology of the gulf and peninsula carried out over the last several decades indicates a need to reevaluate previous historical scenarios of the timing of possible colonization events and of vicariant origins of island species. Beyond the scores of new species records for islands and a constantly shifting nomenclature that required accommodation, important new insights and trends arose that clearly justified a new synthesis. For example, DNA sequence analysis has illuminated the origins and phylogenetic relationships within several taxa, most notably the mammals and reptiles. Expanded databases support a more precise evaluation of island endemics, of colonization and extinction events, and enhanced precision of our species–area curves and island incidence functions. These developments, taken together, argued strongly for an updated and expanded new *Biogeography*, to integrate new and broader studies encompassing more taxa and more complete island coverage. We hope the present synthesis will provide a basis for further research and exploration in upcoming years of the biologically fascinating Sea of Cortés region.

Among the new results presented here are longer term studies of the population dynamics of species, including endemic lizards and gulls that live on the gulf islands. These populations exhibit strong pulses in response to El Niño events, which also dramatically affect island mouse populations and likely other terrestrial animals. During relatively wet El Niño years, terrestrial ecosystems show higher plant productivity, leading to strong recruitment in the animal populations, as well as presumably in the plants. On the other hand, the same El Niño events signal reduced productivity in the marine environment, and some seabirds experience reproductive failure as their food supplies are diminished. Recent research on these relationships has raised appreciation of the role of marine inputs as important energy sources for terrestrial plants and animals on small islands, where productivity is otherwise minimal. As these factors were recognized relatively recently, they received no attention in the first synthesis.

An especially satisfying new element in this edition is the inclusion of insular distributions of some arthropod groups, which are now quite well known. Ants and tenebrionid beetles, both taxa representing important, conspicuous, and diverse inhabitants of deserts, are included in the new book, and their distributions on the islands may now be compared to that of the vertebrates. The untimely death of Dr. Gary Polis precluded even a larger presence here of the island arthropod studies that he initiated and pursued assiduously for most of his productive career.

Most of the contributors to the original *Biogeography* play a role in this second version, with new syntheses and expanded and updated treatments of their taxonomic specialties. But whereas the first book did not include a single contribution from the Mexican academic community, nearly half of our current authors are Mexican scientists, a dramatic turnaround in just two short decades and a tribute to the progress and accomplishment of Mexican science in this area and discipline. In the early 1980s, when the first edition of this book was edited, scientific research in Mexico was mostly concentrated in Mexico City, and regional research centers were just starting to develop. The Centro de Investigación Científica y Educación Superior de Ensenada (CICESE) had been founded a few years before and initially concentrated its research in physical oceanography. Likewise, the Centro de Investigaciones Biológicas del Noroeste (CIBNOR) had existed for just a few years and initially concentrated its research on experimental biochemistry. At the time *Island Biogeography in the Sea of Cortez* was published, both centers had cautiously started lines of ecological research. Similarly, in the state of Sonora the Centro Ecológico de Sonora and the Centro de Investigación y Desarrollo de los Recursos Naturales de Sonora (CIDESON) were founded in the early 1980s, with the enthusiastic support of the state authorities. The regional universities (Baja California, Sonora, Baja California Sur, and Sinaloa) also started ambitious projects around that time to study and understand the environment of the Gulf of California. Finally, many young Mexican researchers that started their professional lives doing studies in the Sea of Cortés later became conservationists and founded nongovernmental conservation groups that are very active in the region. It is hoped that the continued growth and unfailing enthusiasm that has been the most distinctive trait of these organizations continues in the future.

The gulf region is increasingly being exploited, for its natural resources by way of marine fisheries in particular, and for its stunning natural beauty by way of a burgeoning tourism industry. Further, the region's human population is increasing apace. It is appropriate, therefore, that this volume discusses these evolving circumstances and the efforts of the Mexican government to regulate and manage them. Accordingly, we include a chapter on the conservation issues in the Sea of Cortés, past accomplishments, and conservation needs as yet outstanding. The Sea of Cortés is facing a large number of ever-growing threats, but the number of people that are confronting these threats in Mexico is also growing exponentially. There is reason to be cautiously optimistic about the long-term preservation of this invaluable region.

We thank those who helped us in the preparation and construction of the book, in particular, Kirk Jensen of Oxford University Press for his encouragement and useful suggestions. We thank Lee Grismer, Bill Newmark, Bruce Patterson, Andy Suarez, Ken Petren, and the many anonymous reviewers of chapters in the new *Biogeography*. Particular thanks also go to Fulvio Eccardi and Patricio Robles-Gil for kindly allowing use of their stunning photographs.

Ted J. Case
Martin L. Cody
Exequiel Ezcurra

Contents

Contributors xv

I. THE PHYSICAL SCENE 1

1. History of Scientific Exploration in the Sea of Cortés 3
   George E. Lindsay and Iris Engstrand

2. Geology and Ages of the Islands 14
   Ana Luisa Carreño and Javier Helenes

3. Physical Oceanography 41
   Saul Álvarez-Borrego

II. THE BIOLOGICAL SCENE 61

4. Plants 63
   Martin L. Cody, Jon Rebman, Reid Moran, and Henry J. Thompson

5. Ants 112
   April M. Boulton and Philip S. Ward

6. Tenebrionid Beetles 129
   Francisco Sanchez Piñero and Rolf L. Aalbu

7. Rocky-Shore Fishes   154
   Donald A. Thomson and Matthew R. Gilligan

8. Nonavian Reptiles: Origins and Evolution   181
   Robert W. Murphy and Gustavo Aguirre-Léon

9. Reptiles: Ecology   221
   Ted J. Case

10. Land Birds   271
    Martin L. Cody and Enriqueta Velarde

11. Breeding Dynamics of Heermann's Gulls   313
    Enriqueta Velarde and Exequiel Ezcurra

12. Mammals   326
    *Timothy E. Lawlor, David J. Hafner, Paul T. Stapp,
       Brett R. Riddle, and Sergio Ticul Alvarez-Castaneda*

13. Island Food Webs   362
    *Gary A. Polis, Michael D. Rose, Francisco Sanchez-Piñero,
       Paul T. Stapp, and Wendy B. Anderson*

III. THE HUMAN SCENE   381

14. Human Impact in the Midriff Islands   383
    Conrad J. Bahre and Luis Bourillón

15. Cultural Dispersal of Plants and Reptiles   407
    Gary P. Nabhan

16. Ecological Conservation   417
    *Exequiel Ezcurra, Luis Bourillón, Antonio Cantú, María Elena
       Martínez, and Alejandro Robles*

IV. APPENDIXES   445

1.1. New Measurements of Area and Distance for Islands in the Sea of Cortés   447

4.1. Vascular Plants of the Gulf Islands   465

4.2. The Vascular Flora of Cerralvo Island   512

4.3. Plants on Some Small Gulf Islands   527

4.4. Plants of Small Islands in Bahía de Los Ángeles   535

4.5. Plants Endemic to the Gulf Islands   540

5.1. Checklist of the Ants of the Gulf of California Islands   545

6.1. Tenebrionid Beetle Species on Islands in the Sea of Cortés and the Pacific   554

6.2. List of Tenebrionidae on Islands in the Sea of Cortés and the Pacific   562

7.1. Location of Icthyocide Collections of Rocky-Shore Fishes in the Sea of Cortés   565

7.2. List of Data of Island and Mainland Fish Collection Stations in the Gulf of California (1973–1976)   566

7.3. Physical Data of Mainland Areas Sampled in the Gulf of California   568

7.4. Physical Data of Islands Sampled in the Gulf of California   569

7.5. List of Rocky-Shore Fishes Used in the Analyses, Including Their Residency Status, Biogeographic Affinity, Type of Egg, and Food Habits   571

8.1. Updated mtDNA Phylogeny for *Sauromalus* and Implications for the Evolution of Gigantism   574

8.2. Distributional Checklist of Nonavian Reptiles and Amphibians on the Islands in the Sea of Cortés   580

8.3. Distribution of Nonavian Reptiles and Amphibians on Major Islands in the Sea of Cortés   586

8.4. Distribution of Nonavian Reptiles on Minor Islands in the Sea of Cortés   592

10.1. Bird Census Data from Ten Mainland and Peninsular Sites in Similar Sonora Desert Habitat   595

10.2. Analysis of Bird Species Counts in Twelve Mainland and Peninsular Sites   600

10.3. Factors Affecting γ-Diversity (Species Turnover) among the Birds of Desert Sites in Southwestern North America   602

10.4. Bird Census Data from Eleven Mainland and Peninsular Sites in Similar Thorn-Scrub Habitats   604

10.5. Explanation of Island Numbers and References   611

10.6. Distribution of Shoreline Species over Islands   612

10.7. Raptors, Owls, and Goatsuckers   615

10.8. Land Bird Records from Northern Islands   620

10.9. Land Birds Records from Southern Islands   626

10.10. Wintering, Migrant, and other Land Birds of Casual Occurrence   630

10.11. Breeding Bird Densities Recorded at Sonoran Desert Sites on Peninsula, Mainland, and Gulf Islands   640
12.1. Native Terrestrial Mammals Recorded from Islands in the Sea of Cortés   642
12.2. Records of Occurrence and Likely Origins of Terrestrial Mammals from Islands in the Sea of Cortés   649
12.3. Distribution of Bats on Islands in the Sea of Cortés   652

Index   655

Contributors

Rolf L. Aalbu
Department of Entomology
California Academy of Sciences
Golden Gate Park
San Francisco, CA 94118

Gustavo Aguirre-Léon
Instituto de Ecologia
Antigua Carretera a Coatepec km 2.5,
Xalapa, Veracruz, Mexico
aguirreg@sun.ieco.conacyt.mx

Sergio Ticul Alvarez-Castañeda
Centro de Investigaciones Biologicas del
 Noroeste, S.C.,
Mar Bermejo, 195
Playa Palo Santa Rita
La Paz, Baja California Sur, Mexico 23090

Saul Alvarez-Borrego
CICESE Departamento de. Ecologia
km 107 Carretera Tijuana-Ensenada
Ensenada, Baja California 22860, México
alvarezb@arrecife.cicese.mx

Wendy B. Anderson
Department of Biology
Drury University
900 N. Benton
Springfield, MO 65802
wanderso@lib.drury.edu

Conrad J. Bahre
Department of Land, Air and Water
 Resources
University of California at Davis
Davis, CA 95616
cjbahre@ucdavis.edu

April M. Boulton
Department of Environmental Science
 and Policy
University of California at Davis
Davis, CA 95616
amboulton@ucdavis.edu

Luis Bourillón
ITESM-Campus Guaymas
Terminación Bahía de Bacohibampo
Guaymas, Sonora, 85450 México
cobiac@campus.gym.itesm.mx

Antonio Cantú
Conservacion del Territorio Insular
  Mexicano, A.C. (ISLA)
Paseo Álvaro Obregón 735
Colonia El Esterito
La Paz, Baja California Sur, 23020 México
isla1@Prodigy.net.mx

Ana Luisa Carreño
Instituto de Geología
Circuito Exterior C.U.
Delegación de Coyoacán
04510 D.F., México
anacar@servidor.unam.mx

Ted J. Case
Section of Ecology, Behavior and Evolution, 0116
University of California at San Diego
La Jolla, CA 92093
case@biomail.ucsd.edu

Martin L. Cody
Department of Organismic Biology, Ecology
  and Evolution
University of California at Los Angeles
Los Angeles, CA 90095-1606
mlcody@ucla.edu

Iris Engstrand
Department of History
University of San Diego
5998 Alcala Park
San Diego, CA 92110-2492
iris@acusd.edu

Exequiel Ezcurra
Instituto Nacional de Ecología
  and San Diego Natural History Museum
Av. Periférico Sur 5000
Delegación Coyoaćan
México, D.F. 04530
eezcurra@ine.gob.mx

Richard S. Felger
Drylands Institute
2509 North Campbell Avenue
Tuscon, AZ 85719

Matthew R. Gilligan
Biology Department
Savannah State University
Savannah, GA 31404
gillganm@savstate.edu

David J. Hafner
New Mexico Museum of Natural History
1801 Mountain Road NW
Albuquerque, NM 87104

Javier Helenes
CICESE, Departamento de Geología
km 107 Carretera Tijuana-Ensenada
Ensenada, Baja California, 22860 México
jhelenes@cicese.mx

L. David Humphrey
Department of Rangeland Resources
Utah State University
Logan, UT 84322-5230
ldavidhu@yahoo.com

Timothy E. Lawlor
Department of Biological Sciences
Humboldt State University
Arcata, CA 95521
tel1@humboldt.edu

George E. Lindsay
Director Emeritus
California Academy of Science
Golden Gate Park
San Francisco, CA 94118

Maria Elena Martînez
Conservacion del Territorio Insular Mexicano
  (ISLA), A.C.
Paseo Álvaro Obregón 735
Colonia El Esterito
La Paz, Baja California Sur, 23020 México
isla1@Prodigy.net.mx

Reid Moran
2316 Valley West Dr.
Santa Rosa, CA 95401-5737

Robert W. Murphy
Royal Ontario Museum,
100 Queen's Park, Toronto M5S 2C6,
  Canada
drbob@zoo.toronto.edu

Gary P. Nabhan
Center for Sustainable Environments
Northern Arizona University
Flagstaff, AZ 86001
Gary.Nabhan@nau.edu

Kenneth Petren
Department of Biological Sciences
University of Cincinnati
Cincinnati, OH 45221-0006
ken.petren@uc.edu

Gary Polis (deceased)
Department of Environmental Science
and Policy
University of California at Davis
Davis, CA 95616

Jon Rebman
San Diego Natural History Museum
P.O. Box 121390
San Diego, CA 92112-1390
jrebman@sdnhm.org

Brett R. Riddle
Department of Biological Sciences
University of Nevada Las Vegas
4505 Maryland Parkway
Las Vegas, NV 89154

Alejandro Robles
Conservation International México, A.C.
Camino al Ajusco 124-1°piso
Fracc. Jardines de la Montaña
México, D.F. 14210
CI-Mexico@conservation.org

Michael D. Rose (deceased)
Department of Environmental Science
and Policy
University of California at Davis
Davis, CA 95616

Francisco Sanchez Piñero
Dpto. Biologia Animal y Ecologia
Facultad de Ciencias
Universidad de Granada
18071 Granada, Spain
fspinero@ugr.es

Paul Stapp
Department of Environmental Science
and Policy
University of California at Davis
Davis, CA 95616

Henry Thompson
842 B Avenue
Coronado, CA 92118

Donald A. Thomson
Department of Ecology and Evolutionary
Biology
University of Arizona
Tucson, AZ 85721
dat@u.Arizona.edu

Enriqueta Velarde
Centro de Ecologia y Pesquerias
Unidad de Investigaciones, Universidad
Veracruzana
Avenida Dos Vistas s/n
Xalapa, Veracruz, CP 91190, Mexico
evelarde@bugs.invest.uv.mx

Philip S. Ward
Department of Entomology
University of California at Davis
Davis, CA 95616
psward@ucdavis.edu

Patricia West
Center for Sustainable Environments
Northern Arizona University
P.O. Box 5765
Flagstaff, Arizona 86011-5765
paw3333@yahoo.com

PART I

THE PHYSICAL SCENE

1

History of Scientific Exploration in the Sea of Cortés

GEORGE E. LINDSAY
IRIS H. W. ENGSTRAND

    The Sea of Cortés (*el Mar de Cortés*), also known as the Gulf of California, is the body of water that separates the California peninsula from the mainland of Mexico. It extends in a northwest–southeast axis for 1070 km, varying in width from 100 to 200 km. The gulf was formerly much longer, but sediments carried by the Colorado River created a delta and dammed off its upper end, forming what is now the Imperial Valley. The western side of the gulf is dotted with islands (figs. 1.1.–1.3, 13.1), the longest of which is Ángel de la Guarda, measuring 67 km long, up to 16 km wide, and 1315 m high (see app. 1.1 for a list of island names and measurements). Most of the islands are geological remnants of the peninsula's separation from the mainland, a continuing process that started 4 or more million years ago (see chap. 2). One central gulf island, Tortuga, is an emerged volcano, whereas San Marcos Island to its west is largely gypsum, possibly precipitated from an ancient lake. The largest island in the gulf is Tiburón, with an area of approximately 1000 km^2. It is barely separated from the mainland to the east and has a curiously mixed biota of peninsular and mainland species. One tiny island, San Pedro Nolasco, is only 13 km off shore in San Pedro Bay, Sonora, but has an unusual flora that includes a high percentage of endemics.

    The isolation of organisms that colonized or were established previously on the Sea of Cortés islands provided an opportunity for genetic and ecological change. In one plastic and rapidly evolving plant family, the Cactaceae, about one-half of the 120 species found on the islands are endemic. Similarly, populations isolated by climate on peninsular mountains are well differentiated. Because of the topographical diversity of the area and its effect on the disruption and integration of populations, the Sea of Cortés and its islands have been called a natural laboratory for the investigation of

4 The Physical Scene

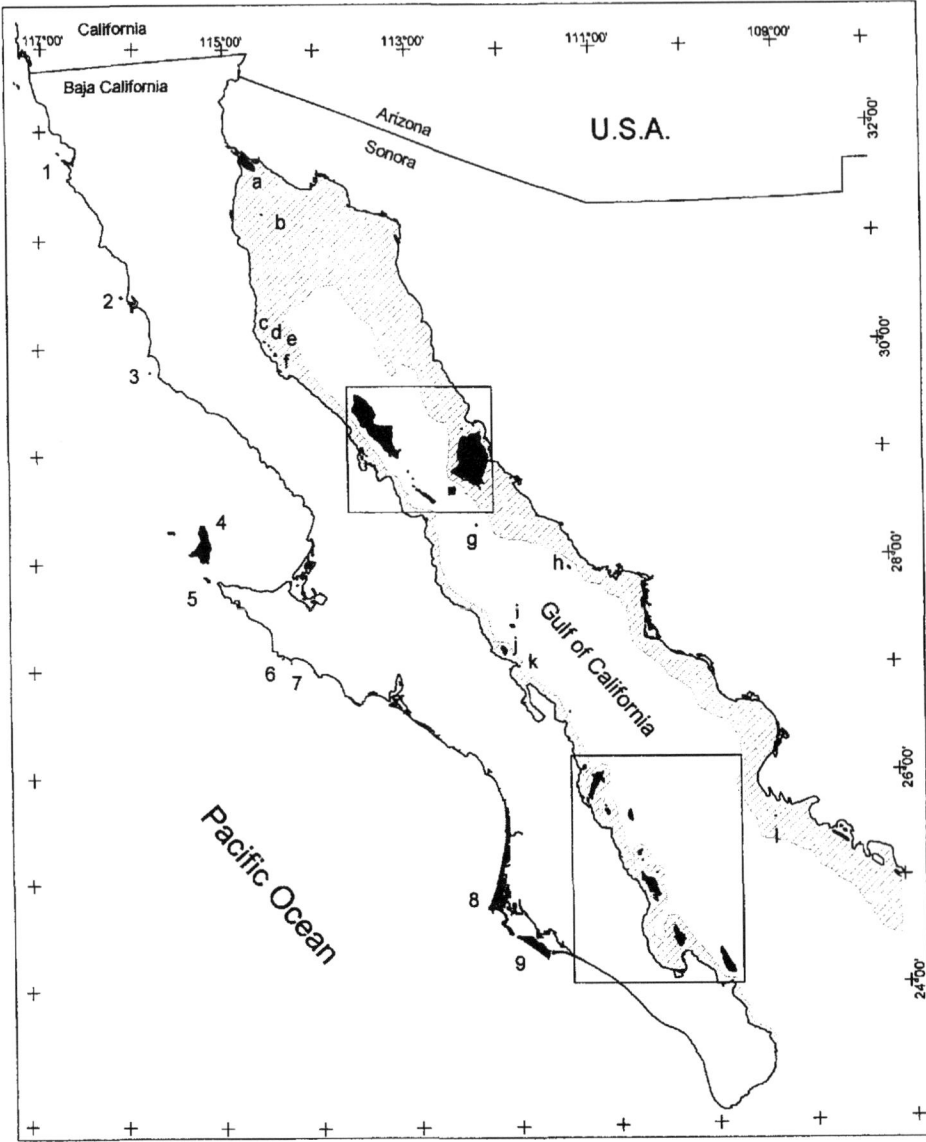

Figure 1.1 Location of islands that are cited in the text and in figures throughout the volume. The shaded contour marks the 100 m bathymetric line, separating landbridge islands from deep-ocean islands. The islands in the rectangles are shown in figures 1.2 and 1.3. The figure was developed by Dr. Hugo Riemann using the geographic information system at the *Colegio de la Frontera Norte* in Tijuana. The identity of the minor islands is as follows: islands in the Pacific Ocean: 1, Todos Santos; 2, San Martín; 3, San Jerónimo; 4, Cedros; 5, Natividad; 6, San Roque; 7, Asunción; 8, Magdalena; 9, Santa Margarita. Islands in the Sea of Cortés: a, Montague; b, Roca Consag; c, El Muerto; d, Encantada; e, San Luis (Encantada Grande); f, Willard; g, San Pedro Mártir; h, San Pedro Nolasco; i, Tortuga; j, San Marcos; k, Santa Inés; l, Farallón.

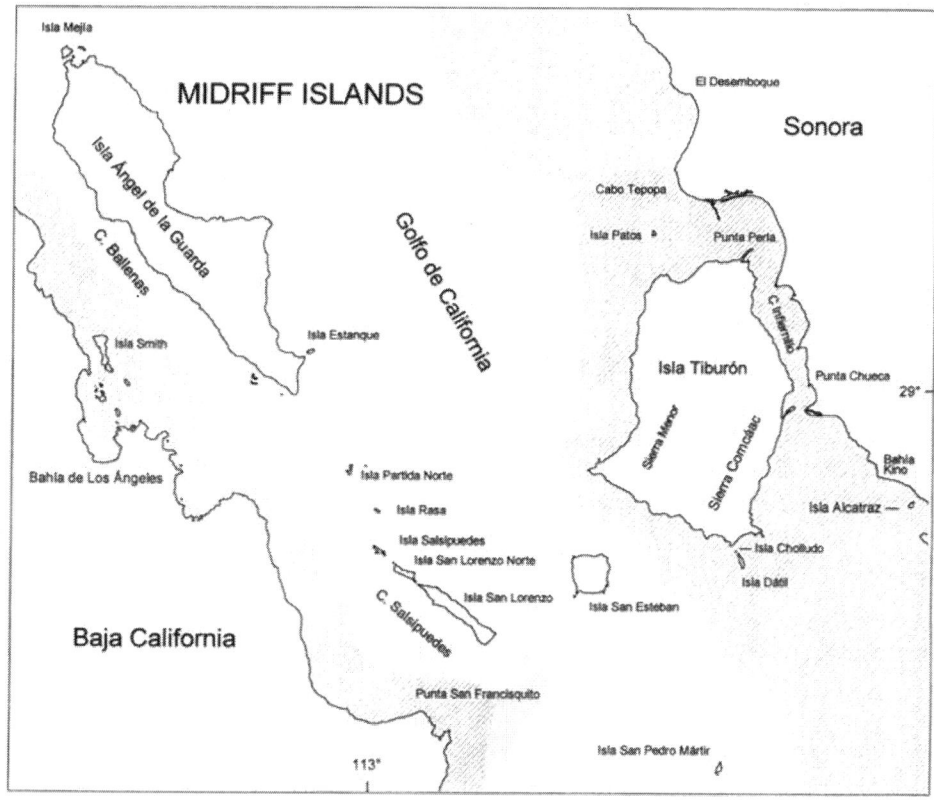

Figure 1.2 The islands of the Midriff region in the Sea of Cortés. The shaded contour marks the 100 m bathymetric line. The figure was developed by Dr. Hugo Riemann using the geographic information system at the *Colegio de la Frontera Norte* in Tijuana.

speciation. The area is still incompletely explored biologically, and its full research potential is as yet unrealized. Even the basic inventory of the terrestrial life of the islands is incomplete, and the waters of the Sea of Cortés are likewise little known.

Human activities have altered the natural resources of the Sea of Cortés over the past four centuries, particularly so during the last six decades. Recently some fisheries' resources have been exploited to near extinction, and the diversion of the fresh water that formerly flowed from the Colorado River has changed the salinity and apparently the biota of at least the northern portion of the gulf. Fortunately, there is now concern for the natural values of the Sea of Cortés, its islands, and the California peninsula. On 30 May 1964, President Adolfo López Mateos declared Rasa Island a migratory waterfowl preserve, and in August 1978, President José López Portillo decreed that all of the islands were to have similar protection and that their plants and animals were to remain undisturbed.

The structural history of the Sea of Cortés is thought by some to be as short as 4 million years, and there are various theories concerning the time and manner of its origin (see chap. 2). The political history has a more precise starting date, with the

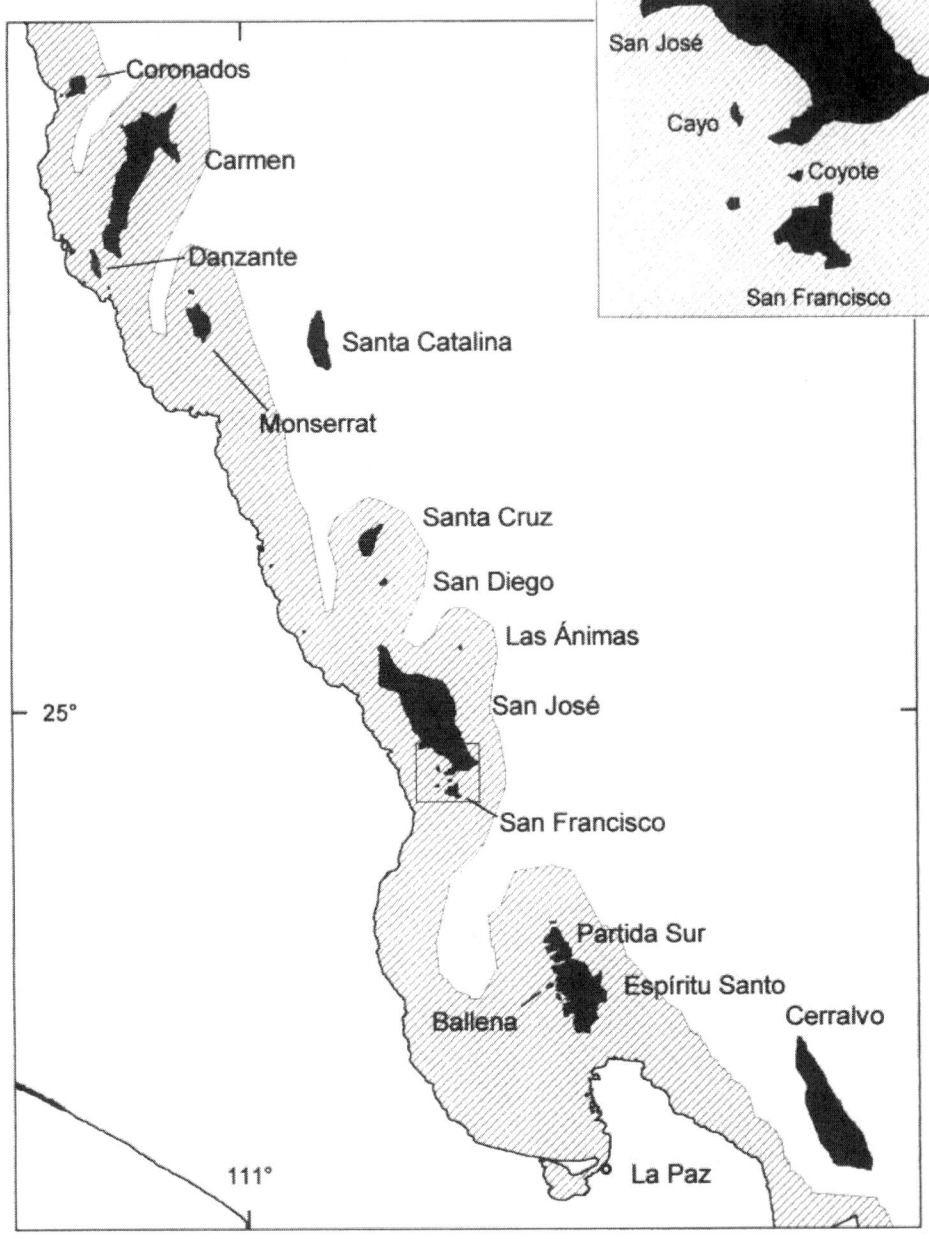

Figure 1.3 The southern islands of the Sea of Cortés. The shaded contour marks the 100 m bathymetric line. The figure was developed by Dr. Hugo Riemann using the geographic information system at the *Colegio de la Frontera Norte* in Tijuana.

discovery of the peninsula by mutineers who murdered their captain and took over the exploring vessel *La Concepción*, which Hernán Cortés had dispatched in late 1533. Fortún Jiménez was the leader, and he, along with 20 of his companions, were killed by natives at a place that was to be called Santa Cruz, eventually La Paz.

Survivors told tales of pearls, and Cortés established a short-lived colony at Santa Cruz on 3 May 1535. It lasted only 2 years, but Cortés persisted in his exploration efforts by sending Francisco de Ulloa northward in 1539. Ulloa reached the head of the Gulf, which he called *Mar de Cortés*, proving Baja California to be a peninsula. He continued down its eastern shore, around Cabo San Lucas, and northward at least as far as Cedros Island.

The Viceroy of México, Antonio de Mendoza, was also interested in these new northwestern islands. First he sent an overland expedition under Fray Marcos de Niza and then another under Francisco Vásquez de Coronado, beyond the headwaters of the Gulf, in 1540. In support of Coronado, Captain Hernándo de Alarcón sailed up the Sea of Cortés to the mouth of the Colorado River, and then traveled upriver in longboats possibly as far as the Gila River, reconfirming the peninsular character of Baja California. In 1542 the viceroy sent Juan Rodríguez Cabrillo to explore the western coast of California and to continue north until he found the end of the American continent or Northwest Passage. Cabrillo discovered San Diego harbor on 28 September 1542, and may have reached 44°N. The Sea of Cortés and the California peninsula thus became known to cartographers and geographers.

No further attempts were made to explore or colonize California for half a century. Then in 1579 Francis Drake passed through the Strait of Magellan to the Pacific, and Spain's navigational monopoly and security were challenged. Thomas Cavendish captured the Spanish galleon *Santa Ana*, returning from Manila, near Cabo San Lucas in 1587. During the wars of the Reformation, both Dutch and English freebooters touched the Sea of Cortés, trading with the Indians for pearls while waiting in the Bay of Ventana and Pichilingue harbor for the Manila galleons.

These threats induced the Spanish crown to send Sebastián Vizcaíno in 1596 to settle and fortify the ports of California. His attempt to colonize Santa Cruz, which he renamed La Paz, failed, but in 1602 he explored the outer coast as far as Bahía de Todos Santos, where Ensenada is now located. His voyage continued northward as he explored the Pacific coast of California and selected Monterey as a possible port for Spain's Manila galleon. This voyage ended the exploration of the West Coast for a century and a half, while seventeenth-century expeditions sought settlement of California on the gulf rather than facing the stormy outer coast. In 1615 Nicolás Cardona and Juan de Iturbe explored the gulf coast for pearls and, after 9 years of attempted voyages by Cardona, failed to establish a colony. Others searched the gulf for pearls between 1629 and 1668, but were unsuccessful in acquiring a sufficient quantity to justify the expenses involved. An English privateer, Captain Edward Cooke, described the natives and natural resources of Puerto Seguro near Cabo San Lucas in 1712.

The California peninsula was successfully settled by the Jesuits under Juan María Salvatierra, who disembarked near Loreto on 25 October 1697. The Jesuits, who eventually established 17 missions among the native inhabitants, remained until 1768. They supplied the first written accounts of California and the Sea of Cortés. As early as 1721, Father Juan de Ugarte, S.J., sailed northward to confirm once again Ulloa's finding of 1539 that Baja California was a peninsula. His expedition visited the Sonoran coast, the island of Tiburón, and the mouth of the Colorado. His report, however, fell on deaf ears, and maps continued to feature the "Island of California."

Of primary importance among Jesuit documents is a 683-page manuscript (a copy of which is in the Bancroft Library at the University of California, Berkeley) written by Miguel Venegas, a Jesuit priest living in Mexico City. In 1739, he completed an epic work entitled *Empresas Apostólicas* based on original reports and letters written by the Jesuits working on the peninsula. Eventually sent to Spain, it was edited by Andres Marcos Burriel, S.J., and published in Madrid in 1757 as *La Noticia de la California y de su conquista temporal y espiritual hasta el tiempo presente* by Miguel Venegas, S.J. An English translation, including a section about the fauna, flora, and indigenous people called *A Natural and Civil History of California*, was published in London in 1759. Neither Venegas nor Burriel had been in California, and they depended on the reports of their colleagues for their information. On pages 42–43 the following passage appears:

> But among the plants and shrubs which most abound in California, the principal is the pitahaya [*Stenocereus thurberi*], a kind of beech, the fruit of which forms a great harvest for the poor inhabitants here. This tree is not known in Europe, and differs from all other trees in the world; its branches are fused and rise vertically from the stem, so as to form a very beautiful top; they are without leaves, the fruit growing to the boughs. The fruit is like a horse chestnut and full of prickles; but the pulp resembles that of a fig, only more soft and luscious.... California has also great plenty of red junas, called in New Spain, *junas japonas* and a particular species of fig. Father Ascensión says, "That the Bay of San Barnabe abounds with various trees, as fig trees, lentisks [pitahayas, an infinite number of plum trees], which, instead of resin or gum, yield a very fine and fragrant incense in great quantity."

Miguel del Barco, S.J., who served for 30 years as a California missionary until the expulsion of the Jesuits in 1768, wrote a comprehensive natural history and historic chronology to augment and correct the account of Venegas and Burriel. Barco's manuscript was an important source for Francisco Javier Clavijero, S.J., whose *Historia de la California* was published posthumously in 1789. Clavijero, like Venegas, did not have personal knowledge of California and depended on the reports by refugees among his fellow Jesuits for much of his information. Unfortunately, Barco's manuscript remained unpublished in Roman archives for two centuries, but at last appeared in 1973 as *Historia Natural y Crónica de la Antigua California*, beautifully edited by Miguel León-Portilla. Selections in English are found in Miguel del Barco, *The Natural History of Baja California*, translated by Froylán Tiscareño (1980).

The task of exploring the Sea of Cortés fell in 1746 to Fernando Consag, a Croatian Jesuit priest at the Misión San Ignacio from 1734 to 1759. His flotilla of four canoes, manned by six soldiers and a crew of mission Indians, worked its way north from Santa Rosalía for a month until entering the mouth of the Colorado. Convinced he had reached the head of the Gulf, Consag again turned south and recommended the Bahía de los Angeles as a base of supply for the missions. As a corollary to Consag's effort, the Jesuits attempted to open a land route between Sonora and California. Another Jesuit, Father Francisco Inama of Vienna, had studied natural science and had an interest in zoology, especially rattlesnakes. He recorded his observations at the Misión San José de Comondú between 1751 and 1768. The Bohemian-born Wenceslao Linck, who served at the Jesuit Mission of San Borja from 1762 to 1768, explored the lands to the north and along the Gulf, hoping to find suitable land for a new mission. Finally, an Alsatian missionary, Johann Jakob Baegert, S.J., who returned to his homeland after the expulsion, published in 1771 a colorful book, *Nachrichten von der Amerikanischen Halbinsel Californien* (translated as *Observations in Lower California*, by Brandenburg and Baumann, 1952) relating his personal experiences while serving at Misión de San Luis Gonzaga from 1751 until 1768. Baegert's scientific observations were brief but interesting, and include this information on the barrel cactus:

> This is a single, green, soft shoot, without branches or arms, four to six spans high, three to four spans thick, grooved from top to bottom, and studded all over with red thorns. These become longer and more curved towards the upper part of the plant, and at the top center, they are as long as a finger and curved like a firehook. These thorns make excellent toothpicks, and one of them will serve many years without repointing or resharpening. (pp. 33–34)

Following a pattern set in other areas of the New World, the Jesuits from California were expelled and were replaced by Franciscans, who relinquished the peninsula to Dominicans 5 years later. Although these latter orders did not publish natural history observations, they hosted the first scientific expedition through Baja and Alta California. José Longinos Martínez, member of the Royal Scientific Expedition to New Spain (1785–1803), undertook the overland journey from Cabo San Lucas to Monterey in 1792. He had a military escort but was without scientific colleagues. A true naturalist of the time, he studied fauna and flora as well as the geology and mineralogy of the peninsula. He gave an interesting and detailed report about the natives and their living habits, changes introduced by Europeans, and of the animals and plants, particularly those with a medicinal use. Longinos sent scientific descriptions and specimens to Spain, but these seem not to have been utilized, and his work remained unpublished until the twentieth century (Simpson 1961).

The biological exploration of the peninsula and Sea of Cortés really started in the mid-nineteenth century when plant collections were made by I.G. Voznesenskii. The Russian colony at Sitka, Alaska, depended on salt purchased at Carmen Island for the preservation of furs, and it sent a ship to replenish its supply every 3 years. In 1841 Voznesenskii, a preparator in the Zoological Museum of the Imperial Academy of Sciences, Saint Petersburg, was aboard the ship *Naslednik Aleksandr* and landed on Carmen Island on Christmas Day. He was able to visit Loreto, Puerto Escondido, and other places. Before leaving the Gulf on 4 February 1842, he collected 360 plant specimens representing 113 species; presumably he had made other large natural history collections. The plants are in the Komarov Botanical Institute of the Academy of Sciences, Leningrad. Voznesenskii's total expedition to the New World lasted 10 years. While at Fort Ross, Alta California, he made extensive botanical, zoological, and geological collections, and climbed and named Mount Saint Helens.

The British were concerned about the situation in western North America when Russia and the United States exhibited obvious interest in expanding at the expense of both the young Republic of Mexico and British-claimed territory. England dispatched two major mapping and charting expeditions toward the middle of the nineteenth century, probably to evidence her own interest in the area. The survey ship *Sulphur* touched several spots along the Pacific side of the peninsula in October and

November 1839, and the ship's surgeon, Richard Brinsley Hinds, made botanical collections which were published in *The Botany of H.M.S. Sulphur* in 1844. It was at Magdalena Bay that Hinds collected the type material of the strange *copalquín* or elephant tree, *Pachycormus discolor*, and wrote, "I have also seen some attempts at trees; imagine what the bones and muscles of a giant would be distorted into three feet, such looked these trees. They twisted and twirled, but could not assume the erect position. Their diameters were far from inconsiderable" (p. 5).

H.M.S. *Herald* surveyed the Gulf of California in early 1850, but the expedition's botanist, Berthold Seeman, left the ship in Mazatlán to cross the Sierra Madre to Durango. Seeman was particularly interested in cacti for Kew Gardens, and it was his misfortune to miss the collections he could have made on the islands of the Gulf.

From 1859 to 1861 a controversial Hungarian naturalist, John (János) Xántus, was employed by the U.S. Coast Survey and the Smithsonian Institution to make tidal observations and natural history collections in the cape region of Baja California, primarily at Cabo San Lucas. The flora and fauna of the cape were not well known, and the Xántus collections furnished many type specimens, drawing attention to the unique biology of that area. His name is commemorated in the Spencer F. Baird report on the birds collected and Asa Gray's report on the plants.

In 1867 William M. Gabb made the overland trip from Cabo San Lucas to San Diego on what was to be known as the J. Ross Browne Expedition, even though Browne returned to La Paz from Magdalena Bay, while Gabb, with another geologist, F. von Lohr, proceeded northward. They were primarily interested in mines and minerals, but Gabb collected some natural history specimens, of which the cacti were reported by Dr. George Engelmann. Browne described the trip in his *Resources of the Pacific Slope* (1869).

Between 1870 and 1900 Edward Palmer made extensive botanical collections on several expeditions, the first of which was to Carmen Island. He was also responsible for the best collection of plants and birds from Guadalupe Island in the Pacific, in 1875 and 1879, being the first naturalist to work there.

The pace of biological exploration of Baja California and neighboring islands accelerated toward the end of the century. Lyman Belding collected birds for the United States National Museum starting in 1881. In 1888 the California Academy of Sciences began a series of expeditions that involved Walter Bryant, Charles D. Haines, T. S. Brandegee, Gustav Isen, Rollo H. Beck, E. W. Gifford, and several others. Extensive collections resulting from their work were lost in the San Francisco earthquake and fire of 1906. The Academy resumed its role in 1921, with a major expedition to all of the main Sea of Cortés islands between April 16 and July 10; its interest in the area has continued to the present, and its publications on the natural history amount to several thousand pages.

A French chemical engineer, Leon Diguet, became intrigued by the ethnology and natural history of Baja California while employed by the Boleo mines in Santa Rosalia between 1888 and 1894. He made some collections which so interested the Museum d'Histoire Naturelle that he was commissioned to return as a field collector and investigator. Two conspicuous plants that bear his name are the giant barrel cactus of Catalina Island (*Ferocactus diguetii*) and the bushy ocotillo of the lower peninsula and gulf, *Fouquieria diguetii*. The French pearling industry at La Paz provided facilities and the assistance of pearl divers for marine fauna studies at several locations in the Gulf. Diguet also studied the rock paintings of the central peninsula, made extensive botanical collections, and eventually went on to central Mexico, where he spent several years in fieldwork. He returned to Baja California in 1900, 1904, and 1913.

In 1905 and 1906 the Bureau of Biological Survey of the United States Department of Agriculture sent Edward W. Nelson, assisted by E. A. Goldman, to make detailed studies of Baja California, the distribution of birds and mammals, and their environmental relationships. On this mission they traveled more than 2000 miles by mule, traversing the length of the peninsula and crossing it 8 times. They made general collections but were handicapped by problems of transportation. Goldman prepared an important paper that was published by the U.S. National Herbarium in 1916, *Plant Records of an Expedition to Lower California*; Nelson wrote *Lower California and Its Natural Resources*, which was a Memoir of the National Academy of Sciences in 1921 and remains one of the best descriptions of the area.

In early 1911 the United States Bureau of Fisheries vessel *Albatross* made a 2-month cruise from San Diego to the upper gulf under the sponsorship of the bureau, the American Museum of Natural History, the New York Zoological Society, the New York Botanical Garden, and the United States National Museum. Collections were made in several branches of science, particularly by J. N. Rose, botany; Paul Bartsch, invertebrates; H. E. Anthony, mammals; and P. I. Osburn, birds.

To continue to review the biological exploration of the Gulf of California through the past 85 years would be arduous indeed. There have been too many expeditions to list, and the California Academy of Sciences alone has been represented there every year. So this introduction will include only a brief mention of some modern investigations, particularly botanical, and apologies for those we have neglected.

Naturalist Laurence M. Huey, Curator of Vertebrates for the San Diego Museum of Natural History, began to amass a significant collection of birds and mammals from southern California and northern Baja California during the 1920s. In July 1923 Huey accompanied an expedition jointly sponsored by several American institutions and the Mexican government for scientists to explore the coasts and islands of Baja California. Accompanying Huey on the expedition were, among others, José Gallegos from Mexico City and Joseph Slevin from the California Academy of Sciences.

Ira L. Wiggins of Stanford University began studies of Baja California natural history, particularly its plants, in 1929. Some of his early work was in collaboration with Forrest Shreve, who published *Vegetation of the Sonoran Desert*. Wiggins made several trips covering the length of the peninsula and to the islands of the gulf. His monumental *Flora of the Sonoran Desert* was published in 1964, accompanied by a reprint of Shreve's *Vegetation*, and was followed by his *Flora of Baja California* (1980).

Howard Scott Gentry started his Baja California fieldwork in 1938, and it continues to the present. His *Agaves of Baja California* was published in December 1978. Some of Gentry's early work was with the Alan Hancock Foundation of the University of Southern California, which sponsored the numerous expeditions of a series of oceanographic vessels named *Velero*. Perhaps the best-known marine exploration in the Gulf was that conducted by E.F. "Doc" Ricketts and John Steinbeck aboard the *Western Flyer* in 1940. The log and scientific discoveries of this trip were written as the classic *Sea of Cortez* (1941/1971), an unparalleled description of the marine biology, and thoroughly good reading besides.

In 1952 the Sefton Foundation of San Diego and Stanford University sponsored a major expedition to the Gulf of California on the foundation's research vessel, *Orca*, with botanist Reid Moran from the San Diego Natural History Museum on board. A 1953 expedition aboard the *Orca* directed by researchers from Stanford University was shared with faculty and students of the University of California, Los Angeles.

Between 1958 and 1972 the Belvedere Scientific Fund of San Francisco sponsored many investigations, cruises, air-supported land expeditions, and publications on the Gulf of California, as well as the production of two 1-hour conservation motion pictures by the California Academy of Sciences, *Sea of Cortez* and *Baja California*. The personal interest of the late Kenneth K. Bechtel, sponsor of the Belvedere Scientific Fund, resulted in the establishment of the Isla Rasa preserve. His generous but anonymous support was responsible for much of the scientific work in the Sea of Cortés.

Reid Moran, curator of botany at the San Diego Natural History Museum from 1956 to 1982, spent much of his research time collecting and caring for plants from Baja California and the islands of the Gulf and the Pacific. His contributions in the first edition of this book evidence his thorough work. Laurence Huey accumulated the largest collection of mammals. Alan J. Sloan, Michael Soulé, and Charles Shaw collected reptiles on museum expeditions. From 1960 to 1969 the museum operated the Vermilion Sea Field Station at Bahía de los Angeles, with support from the National Science Foundation and the Belvedere Scientific Fund. The museum's 1962 Belvedere Expedition was its major single expedition to the Gulf. Collected specimens included botany (2000), entomology (10,270), birds and mammals (174), reptiles (100), and invertebrates (fossils from 35 locations).

A major highway provides overland access along the length of Baja California, and resort hotels have removed any discomfort from travel from the northern border to Cabo San Lucas. Airline service to Loreto, La Paz, and Los Cabos at the end of the peninsula is readily available. Whale-watching trips from San Diego, by boat and by land vehicles, penetrate the lagoons of the west coast. The isolation that formerly protected the unique biota and its natural setting on the peninsula is virtually gone. Transportation by water in the Sea of Cortés and its islands is still limited, and these areas remain essentially unchanged. It is our responsibility, however, to conserve this unique area for the future.

References

Baegert, Johann Jakob S.J. 1952. *Observations in Lower California*, trans. from the original 1771 German edition *Nachricten von der Amerikanischen Halbinsel Californien* by M. M. Brandenburg and Carl I. Baumann. Berkeley: University of California Press.

Barco, Miguel del. 1973. *Historia Natural y Crónica de la Antigua California, adiciones y correcciones a la noticia de Miguel Venegas*, edited by Miguel Leon Portilla. Mexico: Universidad Autónoma de México.

Barco, Miguel del. *The Natural History of Baja California.* 1980. Trans. by Froylan Tiscareño. Los Angeles: Dawson's Book Shop.

Browne, J. Ross. 1869. *Resources of the Pacific Slope: A Statistical and Descriptive Summary . . . with a sketch of the settlement of Lower California*. New York: D. Appleton and Company.

Clavijero, Francisco Javier, S.J. 1852. *Historia de la Antigua o Baja California.* Mexico: J.R. Navarro. 1852 (original edition Madrid, 1789).

Hinds, Richard Brinsley. 1844. *The Botany of H.M.S. Sulphur*. The Botanical Descriptions by George Bentham. London: Smith, Elder and Co.

Gentry, Howard Scott. *Agaves of Baja California*. 1978. Occasional Paper No. 130. California Academy of Sciences, December.

Goldman, Edward A. 1916. *Plant Records of an Expedition to Lower California*. Washington, D.C.: U.S. National Herbarium.

Nelson, Edward W. 1921. *Lower California and Its Natural Resources*. Memoir of the National Academy of Sciences. Washington, D.C.: Government Printing Office.

Ricketts, Edward. F. and John Steinbeck. 1941. *Sea of Cortez: a leisurely journal of travel and research*. New York: Viking Press.

Simpson, Lesley Byrd, trans. and ed. 1961. *Journal of José Longinos Martínez*. San Francisco: Published in original Spanish in Salvador Bernabéu, *"Diario de la Expediciones a las Californias" de José Longinos Martínez*. Madrid: Doce Calles, 1994.

Venégas, Miguel, S. J. *La Noticia de la California y de su conquista temporal y espiritual hasta el tiempo presente*, edited by Andres Marcos Burriel, S.J., Madrid, 1757; *A Natural and Civil History of California*. English trans. London, 1759.

Wiggins, Ira L. *Flora of Baja California*. Stanford: Stanford University Press, 1980, 1993.

Wiggins, Ira L. and Forrest Shreve. 1964. *Vegetation and Flora of the Sonoran Desert*. Stanford: Stanford University Press.

Xántus, John. 1986. *The Letters of John Xantus to Spencer Fullerton Baird from San Francisco and Cabo San Lucas, 1859–1861*, ed. by Ann H. Zwinger. Los Angeles: Dawson's Book Shop.

2

Geology and Ages of the Islands

ANA LUISA CARREÑO
JAVIER HELENES

Before middle Miocene times, Baja California was attached to the rest of the North American continent. Consequently, most of the terrestrial fauna and flora of the peninsula had its origins in mainland Mexico. However, the separation of the peninsula and its northwestward displacement resulted in a variety of distribution patterns, isolations, extinctions, origins and ultimate evolution of fauna and flora in several ways.

The islands in the Gulf of California have been colonized by species from Baja California and mainland Mexico. Some workers (Soulé and Sloan 1966; Wilcox 1978) consider that many of these islands originated as landbridges. Geographically, most of the islands are closer to the peninsula than to the mainland. Therefore, it has been assumed that the Baja California Peninsula was the origin of most of the organisms inhabiting them (Murphy 1983). Islands separated by depths of 110 m or less from the peninsula or mainland Mexico apparently owe their current insular existence to a rise in sea level during the current interglacial period (Soulé and Sloan 1966). In contrast, little information exists for deep-water islands.

Any complete analysis of the distribution and origin of several organic groups inhabiting the Gulf of California islands should involve the consideration of several contrasting models arguing in favor of or against the equilibrium theory (MacArthur and Wilson 1967). In any model, one of the most important features to consider is the relationship between the species inhabiting the gulf islands and the physical and geological processes of formation of the islands, as well as their age, size, and distance from either the peninsula or the mainland. Understanding colonization, migration, and distribution, particularly in some groups, requires information on whether a particular island was ever connected to a continental source. For example, to explain some characteristics of the populations of any island, which presumably had a recent

(<10,000–15,000 years) connection to a continental source, it is necessary to evaluate the coastal erosion or the relative rise in the sea level. These factors might contribute to effectively isolating an insular habit or to forming landbridges.

Previous Geological Studies in the Islands

There are more than 100 islands and islets in the Gulf of California (see the figures in chap. 1). Detailed geological reconnaissance of many islands is still missing or incomplete, but a few studies present detailed descriptions for some important islands. Johnson's (1924) report of the California Academy of Science expedition to the gulf in 1921 contains paleontological information on the Late Tertiary formations of several of the islands visited during this expedition. Durham (1950) and Anderson (1950) made geological and paleontological reconnaissance of Roca Consagrada, Ángel de la Guarda, San Lorenzo, Tortuga, Tiburón, San Pedro Nolasco, San Marcos, Coronado, Carmen, Danzante, Monserrat, Santa Catalina, Santa Cruz, San Diego, San José, Espíritu Santo, and Cerralvo. Phillips (1964) presented data concerning the geology of Ángel de la Guarda, Coronado (Smith), Ventana, Cabeza de Caballo, Estanque (Pond), Rasa, San Lorenzo, Patos, Turner (Dátil), and San Esteban.

Phillips (1966a,b) made a geological reconnaissance of some of the northwestern islands in the Gulf of California during the California Academy of Sciences expedition of 1966. Work on Tiburón includes a geological reconnaissance by Gastil and Krummenacher (1977) and a detailed geological study including mapping by Gastil et al. (1999). Aranda-Gómez and Pérez-Venzor (1986) made accurate lithostratigraphic descriptions on Espíritu Santo. Gastil et al. (1975) published a 1:250:000 geologic map of Ángel de la Guarda, and Escalona-Alcázar and Delgado-Argote (1998) produced a detailed geological reconnaissance map of the same island. Desonie (1992) and T. Calmus (pers. commun., 1988) made a comprehensive study on Isla San Esteban.

A doctoral dissertation evaluated Tortuga (Batiza 1978). A master's and a bachelor's thesis have been made on La Encantada (Rossetter 1974) and on San José (Puy-Alquiza 1992). The volcanic evolution of San Luis has been described in two papers (Rossetter and Gastil 1971; Paz-Moreno and Demant 1999) and detailed geological mappings of the San Lorenzo and Las Ánimas islands have been recently completed (Escalona-Alcazar 1999; Escalona-Alcázar and Delgado-Argote 2000; Escalona-Alcázar et al., 2001).

Age of the Gulf of California

In a broad geological sense, there are two stages in the evolution of the Gulf of California. One is a Miocene stage (fig. 2.1a), which is well developed beneath the eastern waters of the present gulf, and beneath the Pleistocene alluvium of western Sonora. The other is a Pliocene stage (fig. 2.1b) and has a similar, but narrower, shape than the one we observe today (Gastil and Fenby 1991), and with a shallower bathymetry, particularly in the south.

Before plate tectonics, some models (Rusnak and Fisher 1964) suggested that the gulf was formed by gravity sliding of the peninsula off a gently sloping East Pacific

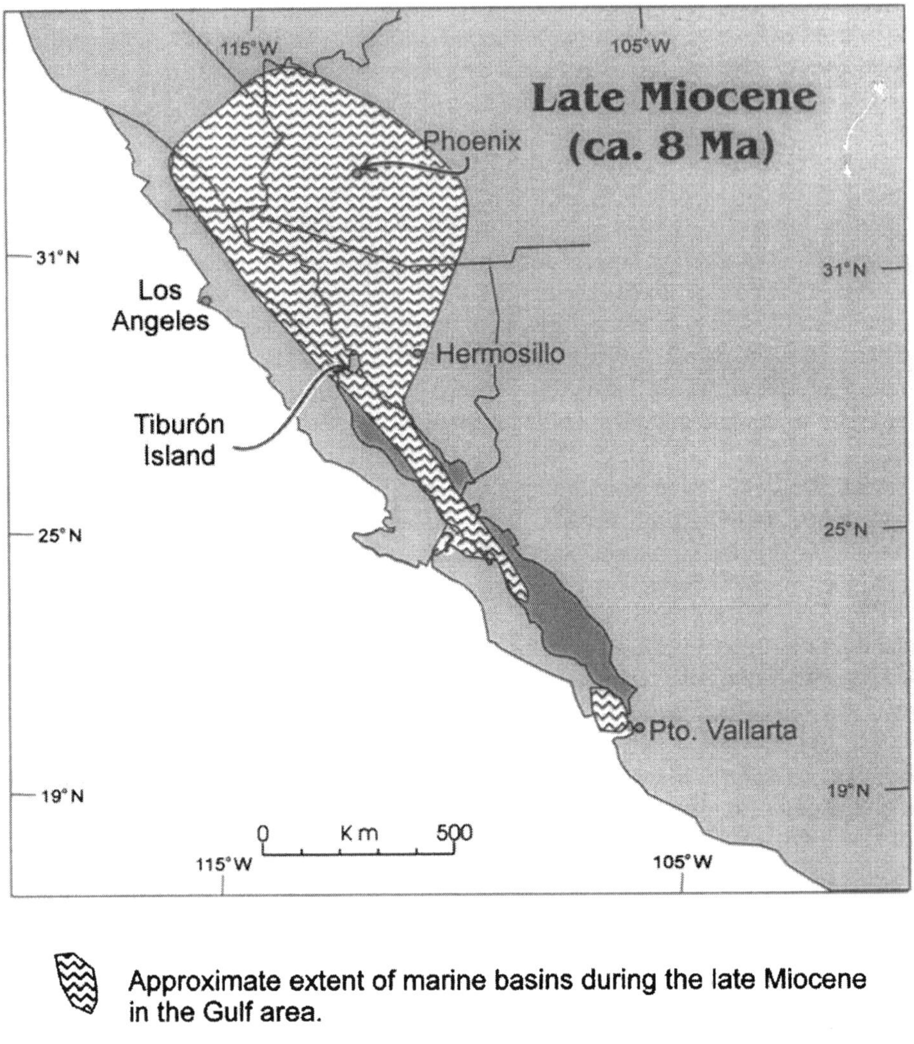

16 The Physical Scene

 Approximate extent of marine basins during the late Miocene in the Gulf area.

Figure 2.1 Evolution of the marine basins throughout Neogene times. (a) Late Miocene proto-Gulf.

Rise, and that the gulf began opening in middle Miocene times. Later, considering plate tectonics concepts, a set of plate-driven rifting models was developed (Moore and Buffinton 1968; Karig and Jensky 1972; Larson 1972) to explain the evolution of the gulf. Mammerickx (1980) was the first to suggest that the proto-gulf be formed as part of a continuous rifting and drifting process in which it became oceanic in character. Later, Curray et al. (1982) and Moore and Curray (1982) argued that the gulf opened a maximum of 300 km by rifting, which started some 5.5 Ma ago, and true sea-floor spreading began 3.5 Ma.

Some of the earlier models (Atwater 1970; Karig and Jensky 1972) propose that even when the gulf opened 4 Ma ago, there was east–west normal faulting in the area

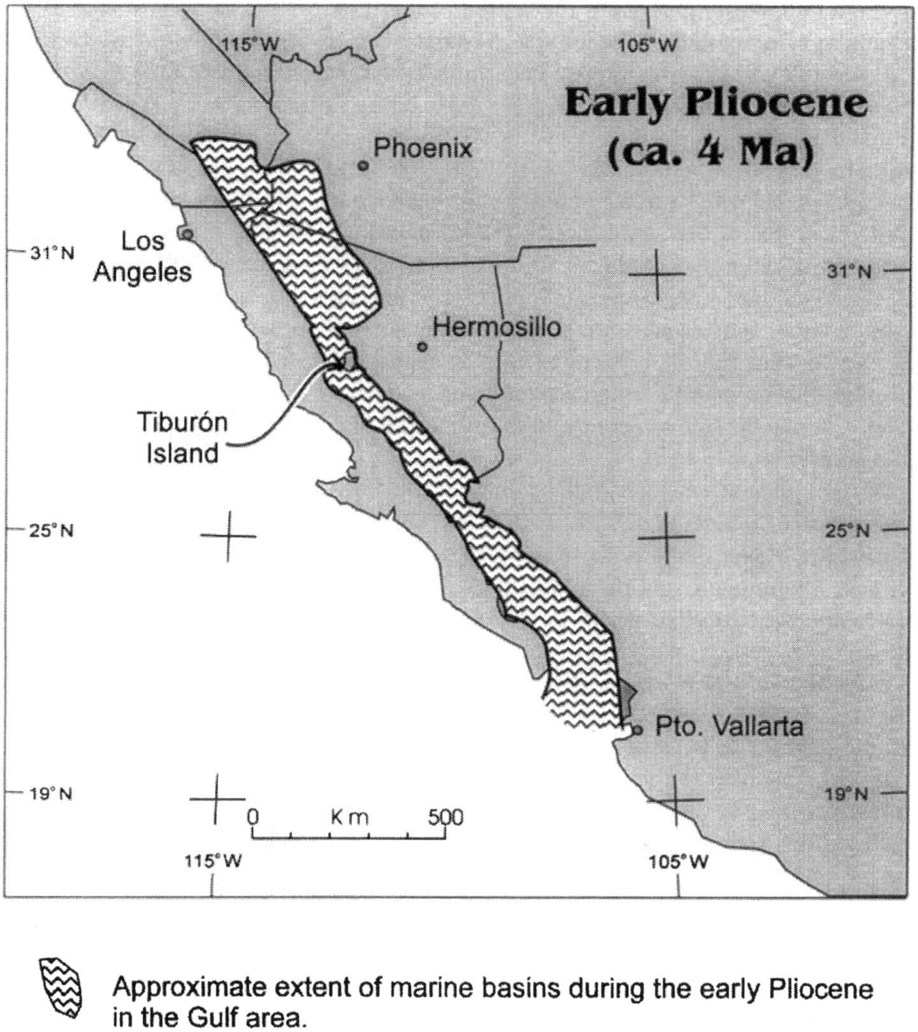

Approximate extent of marine basins during the early Pliocene in the Gulf area.

Figure 2.1 (*Continued*) Evolution of the marine basins throughout Neogene times. (b) Late Pliocene Gulf of California.

much earlier, before 10 Ma. This normal faulting was probably related to Basin and Range extension (Dokka and Merriam 1982). Models based on magnetic anomalies on the sea floor constrain the age of the opening of the southern mouth of the gulf to less than 5 Ma.

Considering the width of the gulf, the magnetic anomaly patterns, the position of the Rivera Rise in the gulf, and the time of deposition on the Magdalena Fan, Lyle and Ness (1991) determined the timing and rates of the initial rift. They proposed three events related to the opening. An initial rifting of the continental crust in the area at approximately 14 Ma; exposure of the oceanic crust before 8.3 Ma, and a total of 450–600 km of right-lateral translation of the Baja California Peninsula since the beginning of the middle Miocene (ca. 14 Ma).

Paleontologic and stratigraphic data indicate that before the late Oligocene to late Miocene (from approximately 30–5 Ma), the Baja California Peninsula was attached to mainland Mexico. Also, as early as 12 Ma ago, a marine basin developed in the northern part of the gulf (Ingle 1974; Smith 1991a; Gastil et al. 1999). On the other hand, paleomagnetic data from the sea floor indicate that the southern part of the gulf opened less than 4 Ma ago (Atwater 1989; Londsdale 1991).

In the southern part of Baja California, vertical motions near the tip of the peninsula involved both subsidence and uplift since the mid-Miocene. At the San José del Cabo Trough, subsidence formed an onshore, nonmarine basin in the mid-Miocene. By late Miocene times this event resulted in deposits of relatively deep-water (approximately 200 m) marine diatomaceous mud, between 6.5 and 3.2 Ma ago (McCloy 1984; Carreño 1992). Afterward, in the Pliocene, the area was uplifted and it is now above sea level.

Several lines of evidence indicate that the Gulf of California began forming in the mid-Miocene, before the oldest (~3.2 Ma) oceanic crust between the tip of Baja California and the East Pacific Rise (Curray and Moore 1984; Londsdale 1991) was formed. Stratigraphic and paleontological data from uplifted marine strata along the gulf margins and on the islands suggest the presence of shallow marine deposits in the northern gulf, as early as about 18 Ma on Isla Ángel de la Guarda (Escalona-Alcázar and Delgado-Argote 1998) and 12.5 Ma on Isla Tiburón (Gastil et al. 1999). Oskin et al. (2000) restudied the Tiburón sequence and reported an early Pliocene age for the marine rocks associated with the proto-gulf. However, the reworked marine mollusks contained in these sediments are undoubtedly of middle Miocene age (J.T. Smith, personal communication, 2000), thus indicating mid-Miocene marine environments in the area. To the north of Tiburón, marine sedimentary sequences indicate the presence of marine basins as deep as 200 m before 6 Ma in the vicinity of San Felipe (Boehm 1984) and in the Salton Sea area (McDougall et al. 1999).

During the late Miocene to Pliocene (8–2 Ma), a marine basin on the northwestern side of the gulf, in the areas near San Felipe and Santa Rosalía, subsided to an approximate depth of 200 m (Carreño 1981, 1982; Boehm 1984). During this interval, the shorelines migrated west from Tiburón, following the escarpment of the rising Sierra San Pedro Mártir from Puertecitos in the south, to San Felipe, to the west of the Sierra Cucapás, and to the Salton Trough, California, in the north. At the same time, the sea in the Loreto area and near San José was shallower and deposition changed between marine and nonmarine environments (Carmen, Cerralvo, and San Marcos islands).

Near the town of San Ignacio, in the central part of the peninsula, marine sediments and a series of tholeiitic basalt vents have been used as evidence to suggest a seaway from the Pacific to the proto-gulf during late Miocene times (Helenes and Carreño 1999).

Based on stratigraphic position, some authors report a middle Miocene (16–10 Ma) to Pliocene (< 5 Ma) age range for the oldest marine sedimentary rocks in the Cabo Trough (McCloy 1984; Carreño 1985; Carreño and Segura-Vernis 1992; Rodríguez-Quintana and Segura-Vernis 1992; Smith 1992). However, most microfossil evidence indicates that the earliest seawater entered this area in late Miocene times (10–5 Ma). Probably by the end of the middle Miocene, and surely by late Miocene times, the sea invaded the area of the Cabo Trough. The exact configuration of the embayment is hard to determine. However, a significant change in the direction of deposition, from a western to an eastern direction, coincides with the uplift of the Sierra La Victoria in the cape region.

Later, during the late Pliocene to Quaternary interval, tectonism and volcanism influenced local sedimentary settings. North of Santa Rosalía, uplift of the Caldera la Reforma domed up the sedimentary sequences and tilted the strata. Young volcanoes such as those on Isla Tortuga, Isla Coronado, and El Púlpito created new habitats for dispersing organisms. Pleistocene terraces are reported throughout the gulf, particularly in the northern gulf, in the islands, and in the Loreto and Santa Rosalía basins (Ortlieb 1991).

The Origin of Islands

An island is a landmass completely surrounded by water. The term has been loosely applied to land-tied, submerged areas and to landmasses cut off on two or more sides by water. As Gastil et al. (1983) noted, several major geologic events are ultimately related to the origins of islands, and these include subsidence, uplift, erosion, and volcanism. The origin and evolution of most of the islands is related to these structural events (table 2.1; fig. 2.2). The numerous islands and islets in the gulf can be separated into three different regions with general characteristics, which unite them: northern, central midriff, and southern gulfs areas.

Northern Gulf Area or North Midriff Islands

In the northern part of the Gulf of California, much of the tectonic relief has been obliterated by deposition of sediments delivered by the Colorado River, which has been building its delta for about 5 Ma. Most of the islands in the northern part of the gulf were formed by deltaic sedimentation. The delta has extensive barrier-islands, such as Montague, Pelícano, Gore, and probably the El Bajito. These islands were formed during the Holocene (0–10 ka) mainly by accumulation of exotic detritus, swampy terrenes, and coastal dunes (Colletta and Angelier 1984). The remaining islands in the area were formed by volcanic activity in the last 2 Ma.

According to the chronostratigraphic scale based on oxygen-isotope curves of the V-28-238 and V-28-239 deep sea cores (Shackleton and Opdyke 1973, 1976), in this chapter the Pleistocene interglacial stages corresponding to the highest sea level episodes are referred to by their odd number in the V-28-238 isotopic curve. Isotopic stage I.S.5 corresponds to the Sangamon Interglacial (~80–130 ka). Thus, according to Ortlieb (1991), in general one may assume that coastal regions showing a last interglacial (I.S. 5e) shoreline a few meters above present mean sea level, and with no highly elevated middle Pleistocene terraces, have not been subject to recent vertical motion. Conversely, a staircase arrangement of marine terraces, with several interglacial marine platforms lying significantly higher than +10 m, may be taken as evidence of uplift motions during the last few hundred thousand years.

Elevations indicate that no large vertical motion has occurred recently and that the relative motion observed for the Late Quaternary is limited to a narrow area northeast

Table 2.1 Geologic features of the islands of the Gulf of California

Island Name and Synonym[a]	Main Rocks	Probable Origin	Probable Age of First Isolation	Reference[b]
Gore, El Pelícano, Montague	Quaternary deltaic sediments	Accumulation of detritus by river discharge	Holocene (0–10 ka)	11, 16
Roca Consagrada	Cretaceous granitic	Block faulting	Pliocene–Pleistocene	1
El Huerfanito, Miramar (El Muerto, Link), Coloradito (Los Lobos, Lobos, Salvatierra), El Cholludo	Miocene volcanic	Block faulting	Pleistocene–Holocene	11, 22
Pómez (El Pomo, El Puma)	Miocene volcanic	Volcanism	Pleistocene–Holocene	11, 22
San Luis (Encantada Grande)	Obsidian dome	Volcanism	Pleistocene–Holocene	10, 17, 18, 22, 23
San Luis Gonzága (Willard)	Dacite domes and breccias	Volcanic	Holocene	11, 14, 15, 22
Mejía	Paleozoic metamorphic, Cretaceous granitic, middle to late Miocene volcanic	Block faulting	Pleistocene	11, 19
Ángel de la Guarda	Cretaceous tonalite and granitic (includes biotite schist enclaves) Plio-Quaternary marine and non marine deposits	Block faulting	Pleistocene	1, 7, 11, 19
Estanque (Pond)	Miocene volcanic	Volcanic	Miocene–Pliocene	11, 19
Smith (Coronado)	Paleozoic basement and Pliocene dacitic volcanic	Volcanic, submergence (subsidence?)	Pleistocene–Holocene	11, 19
Cabeza de Caballo	Miocene basalt flows	Volcanic	Pleistocene	11, 19
La Ventana	Paleozoic metamorphic, Cretaceous granitic	Submergence (subsidence?)	Pleistocene	11, 19
Partida	Miocene volcanic	Volcanic	Pleistocene	11
Rasa	Tholeiitic basalts	Volcanic	Pleistocene–Holocene	11
Salsipuedes	Miocene volcanic	Block faulting	Pliocene	11

Island	Geology	Process	Age	References[b]
San Lorenzo Norte (Las Ánimas)	Pliocene andesitic basalt flows and marine rocks	Block faulting	Pliocene	9, 11, 19
San Lorenzo (San Lorenzo Sur)	Batholithic basement, Miocene volcanic	Block faulting	Pliocene	9, 11, 19
Patos, Tiburón, San Esteban, Turners (Dátil)	Basement, Miocene volcanic, and Miocene and Pliocene marine	Faulting, uplift, and erosion	Pleistocene–Holocene	4, 11, 12, 19
San Pedro Nolasco	Cretaceous granodiorite	Faulting	Pleistocene?	14
Tortuga	Tholeiitic and andesitic flows	Eruption (volcanic)	Pleistocene–Holocene	3, 11
Santa Marcos	Miocene volcanic, Pliocene marine lava flows, and Pleistocene terraces	Uplift	Pleistocene	5, 9, 11
San Ildefonso	Upper Cretaceous granitic, Tertiary volcanic and marine terraces	Uplift	Pleistocene–Holocene	9, 11
Carmen, Danzante	Miocene volcanic, Miocene and Pliocene marine rocks and andesite flows	Uplift and volcanic	Pleistocene–Holocene	5, 9, 11, 16
Coronado (Coronados)	Miocene volcanic, Pliocene and Pleistocene marine and Holocene andesite volcano	Uplift, erosion, and volcanic	Pleistocene?–Holocene	9, 11
Monserrat	Miocene volcanic and Pliocene and Pleistocene marine	Faulting and uplift	Late? Pleistocene	9, 11
Santa Catalina, Santa Cruz & San Diego	Granitic basement	Faulting and erosion		11
San José	Miocene volcanic and Pliocene marine	Block faulting and uplift	Pliocene–Pleistocene	11, 21
San Francisco	Miocene volcanic	Block faulting	Pliocene	11
La Partida, Espíritu Santo	Cretaceous basement, Miocene volcanic, and Pleistocene marine	Block faulting and uplift	Pleistocene	2, 11, 13
Cerralvo	Basement, Miocene volcanic, Pleistocene marine	Uplift	Pleistocene	11, 13

[a] Island names and synonyms from Secretaría de Gobernación and Secretaría de Marina (1987) and L. A. Delgado-Argote (pers. commun., 28 April 1999).
[b] References: 1. Anderson 1950; 2. Aranda-Gómez and Pérez-Venzor 1986; 3. Batiza 1978; 4. Desonie 1992; 5. Durham, 1950; 6. Escalona-Alcázar 1999; 7. Escalona-Alcázar and Delgado-Argote 1998; 8. Escalona-Alcázar and Delgado-Argote 2000; 9. Gastil and Krummenacher 1977; 10. Gastil et al. 1975; 11. Gastil et al. 1983; 12. Gastil et al. 1999; 13. Hausback 1984; 14. Londsdale 1991; 15. Martín-Bavajas et al. 2000; 16. Oertlieb 1991; 17. Paz-Moreno and Demant 1995; 18. Paz-Moreno and Demant 1999; 19. Phillips 1964; 20. Phillips 1966; 21. Puy-Alquiza 1992; 22. Rossetter 1974; 23. Rosseter and Gastil 1975.

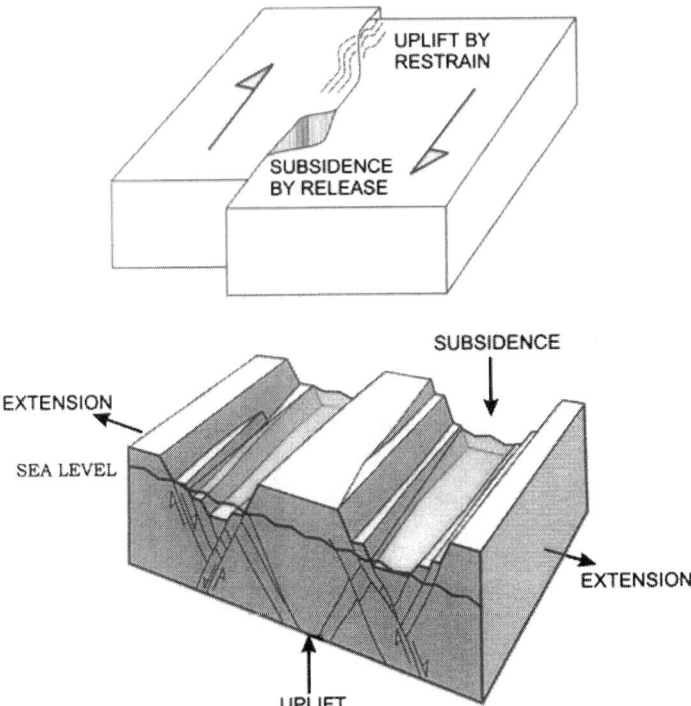

Figure 2.2 Diagrams illustrating uplift and subsidence. (a) Right-lateral strike slip fault can cause subsidence in a bend releasing stress, and uplift in a restraining bend. (b) Block diagram of uplift and subsidence caused by thermal extension.

of the trace of the Cerro Prieto Fault. According to Ortlieb (1991), a partly emerged deltaic sequence correlative with isotopic stage 5e (I.S.5e; see Shackleton 1987), that is approximately 125 ka old, presents a coquina (detrital limestone composed of fossil debris) at the top. This sequence is associated with a shoreline observed at a height of +7 m to the northwest of Bahía Adair, at +12 m to the southwest of the same bay, and at +25 m at Punta Gorda. Strata contemporaneous with the coquina crop out approximately +10 m on the west bank of the actual deltaic plains (on the eastern side of Sierra El Mayor), and at +8 m in the San Felipe area.

According to Anderson (1950), Isla Roca Consagrada, at the extreme northern end of the gulf, is granitic in composition. It is undoubtedly related to the granitic rocks of Baja California and is probably an eastward extension of the Peninsular Batholith. El Huerfanito, Miramar (El Muerto or Link), Coloradito (Los Lobos, Lobos, or Salvatierra), and El Cholludo islands are formed mainly by Miocene volcanic rocks. According to Rossetter (1974) and Gastil et al. (1983), they were separated from the peninsula by block faulting during the Pleistocene. In contrast, Isla Pómez (El Pomo or El Puma) was formed by volcanic eruptions during the Pleistocene or Holocene times. Similarly, San Luis (Encantada Grande) is formed by an obsidian dome, which rises up on the continental shelf (<100 m below sea level). The well-preserved morphology of the volcanic structures clearly supports a late Quaternary–Holocene age for all the eruptive activity (Londsdale 1989; Paz-Moreno and Demant 1995, 1999). San Luis is linked to Encantada by a reef, which appears during low tidal cycles.

Central Gulf Area or Midriff Islands

The central gulf area includes the largest islands in the gulf and is more complex than the northern gulf area. Its structural history is closely related to the northwestward movement of the peninsula, and its main trend is dominated by the Guaymas Fault system, whose northwestern part includes three *en echelon* fault zones. These faults share a complex rift valley between Baja California and the midriff islands (Ángel de la Guarda, Partida Norte, San Lorenzo, Tiburón, San Pedro Mártir, and San Esteban), as well as a number of small islands and islets. Many of the islands in this area contain evidence of a middle (15–10 Ma) or mostly late Miocene (10–5 Ma) proto-gulf, together with contemporaneous volcanic rocks. Later, in Pleistocene to Holocene (5–<2 Ma) times, movement along the many faults uplifted the islands.

Ángel de la Guarda is the second largest island in the gulf. It is located 33 km northeast of Bahía de los Ángeles, and the deep and narrow Ballenas Channel (minimum width of 13 km off Punta Remedios) separates it from the peninsula. This island contains a lithologic sequence similar to the one described at Bahía de Los Ángeles and the San Luis Gonzága area (Gastil et al. 1975; Delgado-Argote and García-Abdeslem 1999). Here, Paleozoic metamorphic and Cretaceous granitic rocks constitute the basement and are overlain by middle-to-late Miocene volcanic (rhyolithic and andesitic) rocks, which interdigitate with marine rocks.

Ángel de la Guarda has been related to sea floor spreading during the Pleistocene (Gastil et al. 1983). However, Escalona-Alcázar and Delgado-Argote (1998) point out that even when the structure of the island followed the evolutionary pattern of the Gulf of California after the Pliocene, it contains 18 Ma dacites interfingering with marine sedimentary rocks of a probably middle-to-late Miocene age. Based on seismic reflection images from the bottom of Ballenas Channel, Delgado-Argote et al. (1999) concluded that similar volcanic rocks form part of its walls and bottom. These data suggest that Ángel de la Guarda was formed by a leaky transform fault around 2–3 Ma, when apparently the location of this fault changed from east to west of the island (Lonsdale,1989).

Isla San Luis Gonzága is a very recent volcanic island composed by dacitic domes and breccias, as well as fossiliferous beach deposits interbedded with volcanic material of a supposedly younger age that those cropping out in the San Luis Gonzága volcanic field (Martín-Barajas et al. 2000).

The structural style of Isla Mejía, a landbridge with respect to Ángel de la Guarda (table 2.2), is similar to that of Ángel de la Guarda. The latter has a rectilinear pattern at almost all its margins. The occidental and oriental coasts at Mejía are almost straight and similar to the peninsular coast between Bahía San Luis Gonzága and Bahía de los Ángeles. They have an orientation of N40°W, parallel to the Ballenas transform fault zone. The southeastern coast of Ángel de la Guarda has an north–south orientation, parallel to the structural pattern of the continental crust subsidence zones in the rift margins of the inactive Tiburón Basin. According to radiometric dates in Bahía de los Ángeles and San Luis Gonzága, these structural trends seem to have

Table 2.2 Geographic parameters characterizing the islands in the Gulf of California

Island Name and Synonym	Area (km^2)	Distance (km)	Measured to	Minimum Channel Depth (m)	Between Island and
Gore, El Pelícano, Montague	—	—	—	< 20	Among them, peninsula
Roca Consagrada	—	—	—	≈ 30	Peninsula
El Huerfanito,	0.08	0.48	Peninsula		
Miramar (El Muerto, Link),	1.33	3.39	Peninsula		
Coloradito (Los Lobos, Lobos, Salvatierra),	0.36	7.45	Peninsula	≤ 40	Among them
		5.21	Miramar	≈ 40	Peninsula
		2.97	Encantada		
El Cholludo	0.05	0.55	Encantada		
Pómez (El Pomo, El Puma)	0.22	1.33	San Luis	≤ 40	Peninsula
San Luis (Encantada Grande)	6.85	5.45	Peninsula	≈ 100	Peninsula
		5.76	San Luis		
San Luis Gonzága (Willard)		1	Peninsula	< 40	Peninsula
Mejia	2.26	0.18	Ángel de la Guarda	< 10	Ángel de la Guarda
				≈ 600	Peninsula
Ángel de la Guarda	936.04	12.12	Peninsula	≈ 200	Partida
				< 10	Estanque
Estanque (Pond)	1.03	0.61	Ángel de la Guarda	< 10	Ángel de la Guarda
				≈ 400	Peninsula
Coronado (Smith)	9.13	2.18	Peninsula	< 20	Peninsula
Cabeza de Caballo	0.77	2.24	La Ventana	< 20	Peninsula
La Ventana	1.41	3.15	Peninsula	< 20	Peninsula
		8.30	Rasa	≈ 200	Ángel de la Guarda
Partida Norte (Cardonazo)	1.38	17.88	Peninsula	< 200	Rasa
		12.18	Ángel de la Guarda	≈ 400	Peninsula
		20.79	Peninsula	≈ 200	Salsipuedes
Rasa	0.68	9.58	Salsipuedes	≈ 1000	Peninsula
		9.52	Partida Norte		
Salsipuedes	1.16	1.52	San Lorenzo Norte	< 200	San Lorenzo Norte
		19.21	Peninsula	< 1000	Peninsula
San Lorenzo Norte (Las Ánimas)	4.26	0.12	San Lorenzo Sur	< 200	San Lorenzo Sur
				≈ 1500	Peninsula
San Lorenzo (San Lorenzo Sur)	33.03	16.36	Peninsula	400	San Esteban
				≈ 1500	Peninsula
Patos	0.45	7.45	Tiburón	< 80	Tiburón, peninsula
Tiburón	1223.53	1.70	Sonora	< 50	Peninsula
				< 600	San Lorenzo
San Esteban	40.72	11.64	Tiburón	≈ 400	Tiburón
Turners (Dátil, Lobos)	1.25	1.94	Tiburón	≈ 80	Sonora
San Pedro Nolasco	3.45	14.61	Sonora	< 200	Sonora
Tortuga	11.36	36.30	Peninsula	≈ 600	Peninsula
		26.65	San Marcos		

Table 2.2 (*Continued*)

Island Name and Synonym	Area (km²)	Distance (km)	Measured to	Minimum Channel Depth (m)	Between Island and
Santa Marcos	30.07	4.01	Peninsula	< 200	Peninsula
San Ildefonso	1.33	9.94	Peninsula	< 100	Peninsula
Carmen	143.03	6.03	Peninsula	250	Peninsula
				250	Danzante
Coronado (Coronados)	7.59	2.61	Peninsula	200	Carmen
				50	Peninsula
Danzante	4.64	2.61	Peninsula	< 100	Peninsula
Monserrat	19.86	13.70	Peninsula	≈ 200	Peninsula
Santa Catalina	40.99	25.15	Peninsula	400	Peninsula and Danzante
		20.73	Monserrat		
Santa Cruz	13.06	19.82	Peninsula	> 400	Peninsula
		16.91	San José	≈ 200	San José
San Diego	0.60	6.36	Santa Cruz	> 400	Peninsula
		9.28	San José	≈ 200	San José
San José	187.16	4.61	Peninsula	60	Peninsula
San Francisco	4.49	2.36	San José	80	Peninsula
		7.39	Peninsula	40	San José
La Partida,	19.29	0.17	Espíritu Santo	40	Espíritu Santo
Espíritu Santo	87.55	6.15	Peninsula	200	Peninsula
Cerralvo	10.46	8.73	Peninsula	800	Peninsula

From Shepard (1950), Gastil et al. (1983), Londsdale (1989), Dauphin and Ness (1991). Island areas and distance measurements were kindly given by R.W. Murphy (Royal Ontario Museum).

occurred during the Miocene, between 18 and 14 Ma, probably before the island emerged. The small satellite islands in the area, Granito (or La Lobera) and Estanque (or Pond) are landbridges with respect to Ángel de la Guarda.

The area between Bahía de los Remedios and Punta Quemada includes the Miocene volcanic Isla Coronado (Smith), where the Paleozoic basement is overlain by a dacitic volcano of probable Pliocene age. Terraces with Pliocene–Pleistocene coquinas are similar to those cropping out in the Don Juan area (Delgado-Argote 2000). This agrees with the observation of Phillips (1964), who interprets the Pleistocene isolation of this island as caused by submergence, or subsidence.

Isla Ventana, with a granitic and metamorphic basement, probably followed a similar evolution to that of Coronado, whereas Isla Cabeza de Caballo is formed by 12 Ma andesitic basalt flows similar to those at Sierra de Las Ánimas. Coronado, Piojo, Cabeza de Caballo, and the small islands located off shore between Punta Gringa and Punta Arena are landbridges with respect to the peninsula.

Salsipuedes, Las Ánimas, and San Lorenzo have an orientation parallel to the peninsular coast and are separated from it by the Salsipuedes Channel. This channel is the continuation to the south of the Ballenas Channel. Las Ánimas is south of Salsipuedes and separated from it by a channel 3.5 km wide.

According to Anderson (1950), the San Lorenzo islands represent an emerged portion of the submarine ridge extending from Ángel de la Guarda. Calc-alkalik rocks older than 12 Ma compose San Lorenzo Norte and Sur, Salsipuedes and Partida Norte islands, which were probably separated by normal faulting from the peninsula sometime during the Pliocene and were left behind as Baja California moved to the northwest (Moore 1973; Desonie 1992). On San Lorenzo Norte, andesitic flows dated 5 Ma (Delgado-Argote 2000) overlie marine rocks with abundant mollusks. According to Escalona-Alcázar (1999), the San Lorenzo and Las Ánimas islands represent a faulted, uplifted block, suggesting that both islands were a rigid block uplifted before that they migrated southeast along the Ballenas, Partida, and San Lorenzo transform faults, which are active faults since 1 Ma (Londsdale 1989).

Isla Estanque appears to have had a volcanic origin sometime around 5–3 Ma, based on its structural relationship with San Esteban, San Lorenzo, and the southern part of Ángel de la Guarda islands. A palinspastic reconstruction of the Gulf of California situates Lorenzo Sur (formed by Cretaceous batholithic rocks) facing the Bahía and Sierra de Las Ánimas, where continental sandstones are capped by a 7.8 Ma basalt.

Isla Rasa is composed of Pleistocene–Holocene olivine tholeiitic basalts. It formed on top of what appears to be a leaky transform fault (Rossetter 1974). Marine rocks on San Esteban (Carreño, work in progress) and on Tiburón (Gastil et al. 1999) have a probable age of middle–late Miocene to early Pliocene. These ages require the presence of a shallow sea in the north-central Gulf of California as early as 12.7 Ma. Ash layers with marine shell impressions are evidence of marine deposition and indicate that the area was submerged at the beginning of the pyroclastic phase (9.5 Ma). The volcano depositing these ash layers was emergent during most of its eruptive history. After cessation of the volcanic activity, at least the eastern side of Isla San Esteban became submerged as indicated by the presence of Pliocene marine invertebrates above the 2.5–2.9 Ma volcanic sequence (Desonie 1992). The steep cliffs surrounding San Esteban and the height of the fossiliferous tuff (a few hundred meters above sea level) indicate that the island's present height is at least partially the result of uplift along submerged faults.

At Tiburón, Patos, and Turner (Dátil) islands (the last two are landbridges with respect to Tiburón), the Miocene marine sedimentary rocks have been dated as 12 Ma. The invasion of the East Pacific Rise with sea-floor spreading, oblique rifting, and formation of the modern Gulf of California started at the southern end of the gulf after 5 Ma. Strontium isotope dates on calcareous megafossils in a sedimentary sequence, as well as K-Ar ages of volcanic rocks overlying these sedimentary rocks, indicate that marine water entered the gulf as early as 13–12 Ma. This marine transgression took place during the extensional Basin and Range phase of gulf history. However, microfossils occurring on marine beds cropping out on Tiburón indicate an age younger than 6.5 Ma. This dating is in apparent conflict with an 11.2 Ma radiometric date on an ash flow tuff that unconformably overlies the microfossil-bearing sedimentary rocks. However, the precise structural relationship between the two sets of rocks is still uncertain. In any case, strata older than 9 Ma are tilted and separated by an angular unconformity from younger, flat-lying rhyolites and sediments; the composition of the volcanic rocks older than 12 Ma points to their origin within an arc, whereas the rocks dated as 6 Ma and younger display compositions reflecting the transtensional opening of the modern Gulf of California (Gastil et al. 1999). Finally, San Pedro Nolasco is mainly composed of granodiorite, apparently of the same petrographic type as that cropping out near Guaymas, and may be related to the granitic rocks of Cretaceous age in Baja California (Anderson 1950).

Southern Midriff Islands or Southern Gulf Area

Islands in the southern gulf are composed of either volcanic or granitic rocks related to the peninsula, and some of them contain Pliocene marine sediments. In general, the islands in this area were uplifted when the southern end of the gulf opened and seafloor spreading started on the eastern side of the Baja California peninsula.

Batiza (1978) suggested that Tortuga, which is formed by tholeiitic and andesitic flows, was constructed on 1.7 Ma-old oceanic crust, over a leaky fracture zone. Batiza also suggested that the oldest activity on the island started on its southeastern portion and migrated northwest to its present position, and that the normal faulting, which is perpendicular to the fracture zone, was originated by tectonic subsidence. Fabriol et al. (1999) interpreted this fracture zone as a volcanic ridge, noting that the normal faults are parallel to the trend of the Guaymas spreading center. They also interpreted the alignment of gravimetric and bathymetric highs as evidence that the islands are part of the 290° trending Tortuga Volcanic Ridge, which intersects the Rosalía Volcanic Ridge in the Vírgenes Heights.

On Isla San Marcos, the oldest formations exposed correspond to volcanic rocks overlain by marine sedimentary rocks of Pliocene age, capped by lava flows and Pleistocene terraces in the same attitude and elevation as on neighboring islands. Undeformed, low terraces found at +9 m are visible all around San Marcos, indicating that this island is structurally different from the Santa Rosalía Basin where the terraces are deformed.

A 10-km wide, deep channel separates Isla San Ildefonso from nearby Bahía San Nicolás. The island is composed of Upper Cretaceous granitic rocks and of Tertiary volcanic rocks, similar to those cropping out in Punta San Antonio and Península Concepción (Staines-Urias 1996). The terraces have elevations of +6 and +12 m (Ledesma-Vázquez and Johnson 1997) and are associated with the Pleistocene isotopic stage I.S.5e (~120/ ~135 ka; Shackleton 1987).[1]

Isla Carmen is elongated in a north–northeast direction, in contrast to the north–south trend of the peninsular Baja California coast. It is composed of Miocene volcanic rocks overlain by Miocene–Pliocene marine rocks (San Marcos, Carmen, and Marquer formations). A small shield or volcanic cone of andesitic flows is found at the northeastern coast of the island, and a mid-Pliocene limestone located at the higher elevations of the island covers it. No indications of submarine eruptions are found, as would be expected if the andesite erupted during sedimentation in the mid-Pliocene. Thus, uplift of the limestone must have preceded the eruption. A Pleistocene andesitic volcano was probably built on Coronado, and it presents a similar degree of erosion as the cone on Carmen Island (Anderson 1950; C. Siebe, pers. commun., June 1999).

Remnants of the I.S.5a/c isotopic stage (~85/ ~105 ka; Shackleton 1987), have not been positively identified outside La Reforma-Santa Rosalía area. Nevertheless, littoral notches and narrow benches have been observed at elevations between +3 and +6 m—that is, below the well-developed I.S.5e terrace in Isla del Carmen (~125 ka).

Monserrat is composed of Miocene volcanic rocks covered by a fossiliferous limestone (supposedly middle Pliocene in age), which is in turn covered by Pleistocene gravels. All these units are faulted into several terraces or mesas. Anderson (1950) states that after the Miocene volcanic accumulation, the rocks were tilted eastward, and by the middle Pliocene the area subsided and marine calcareous rocks accumulated. These beds were tilted slightly to the east, and during the early Pleistocene a transgression of the sea left a thin veneer of fossiliferous gravels on the limestone and the volcanic rocks. Later on, this surface was broken into fault blocks, the faults striking to the north, and in the central part of the island there was considerable elevation of the fault blocks. After faulting, during the latest Pleistocene, a marine terrace was formed which is now some few meters above the sea level.

According to Anderson (1950), Miocene volcanic rocks similar to those exposed on Carmen cover Isla Danzante. Johnson (1924) believed that the island represents less elevated fragments than that mass with similar structures observed in the Sierra de La Giganta.

Isla San José is located to the north of Bahía de La Paz. Its eastern side is formed by scarps produced by the tectonic subsidence of the sea floor toward the El Farallón Basin, which was affected by northeast–southwest faulting (Cromwell 1962). According to Puy-Alquiza (1992), the San José Channel, located between Isla San José and Punta Larga and Punta Mechudo on the peninsula, was formed by late Miocene activity of the La Giganta Fault. During the early Pliocene, San José subsided and received marine sediments, which were then uplifted in the late Pliocene. The combination of these events and the mentioned faulting ultimately formed the San José Channel and San Francisco island.

Espíritu Santo and Partida islands are 20 km to the north–northwest of La Paz and are separated from the peninsula by the San Lorenzo Channel. The islets of Ballena and Gallina form a landbridge between both islands. Espíritu Santo and Partida are probably composed of a Cretaceous basement and a thick sequence of Miocene volcanic and volcano-clastic rocks similar to those cropping out in Punta Coyote and nearby La Paz. The combined system of high-angle normal faults in the eastern part of both islands is related to the opening of the Gulf of California (late Pliocene), and they represent the continuation of faults mapped in the Punta Coyote area (Aranda-Gómez and Pérez-Venzor 1986).

Santa Catalina, Santa Cruz, San Diego, and part of the San José and Cerralvo islands are granitic (Johnson 1924). These granitic rocks might be related to those forming the southern tip of the peninsula (Anderson 1950). Isla Cerralvo also contains exposures of a sequence of probable Miocene age, Comondú equivalent ash-flow tuffs, one of which is densely welded, together with sedimentary rocks (Hausback 1984).

Use of Fossil Deposits in Dating Insular Isolation

Many of the gulf islands, as well as the Baja California Peninsula, have sedimentary rocks containing numerous Paleozoic, Mesozoic and Cenozoic fossils that are mostly marine. These fossils are a valuable tool for reconstructing ancient coastal lines and for interpreting transgressive-regressive cycles. Their presence on islands permits us to infer when these areas were submerged or when were they isolated, or when terraces formed on them.

Paleobiologists have tried to relate species diversity and area of habitat to the fossil record in a number of ways. However, effective isolation is difficult to determine exclusively by conventional paleontological methods. It is clear that the study of fossil and living marine organisms offers little, if any, information to support the equilibrium theory of island biogeography. Nevertheless, the fossil record provides a unique, albeit imperfect, opportunity to test hypotheses about whether provinces of different sizes, and thus different frequency distributions of geographic ranges among their component species, exhibit different speciation and extinction rates.

In this respect, Lawlor et al. (chap. 12 of this volume) state that several mammalian taxa support the existence of a mid-peninsular seaway across the Vizcaíno Desert, approximately 1 Ma ago, an idea also supported by the combined data from mitochondrial DNA phylogeography and geological events published by Upton and Murphy (1977). The Tertiary sedimentary record in the northwestern part of the Sebastián Vizcaíno Peninsula is characterized by the lack of Paleogene rocks; the Neogene formations tilting and/or faulting are bounded by angular unconformities. Volcanic material and deep-water diatomaceous mudstones and laminated diatomites dominated the upper parts of the middle-to-late Miocene formations, whereas the uppermost Miocene–Pliocene formations consist of arenaceous, shallow sedimentary rocks. In contrast, the northeastern part of the Sebastián Vizcaíno Peninsula contains an almost continuous marine sedimentary sequence from Albian to middle Eocene, with a hiatus encompassing from middle Eocene to Pliocene, receiving marine sediments until the Pleistocene. According to Ortlieb (1991), the Bahía Asunción area in the Vizcaíno is the only region where the amino acid racemization stratigraphic data can support chronologic interpretation of middle Pleistocene marine deposits probably I.S.5e (~125 ka). Helenes and Carreño (1999) proposed, on the basis of the mixed cool-tropical characteristics of the fossil assemblages mentioned above and the distribution of marine deposits in the north and nonmarine deposits in the central part and south of the Gulf of California, the hypothetical seaway crossing San Ignacio, at least until Miocene–Pliocene. Any geological data permit us to confirm if the seaway was present in the area 1 Ma.

Fragmentation of landmasses through plate-tectonic processes results in addition, deletion, spread, and contractions of provinces, but its treatment has been largely theoretical (Flessa and Imbrie 1973). Concerning terrestrial organisms, the most quantitative study is that of the Great American Interchange of land mammals between North and South America faunas and floras. Of particular interest is the Baja California Tertiary record of terrestrial fossils, which are scarce but significant (table 2.3). For example, the Pliocene–Pleistocene record of the reptilian *Crocodylus* cf. *C. moreletti* at the tip of Baja California (Miller 1980) suggests that the southern part of Baja California was adjacent to the Mexican mainland during or just after late Pliocene times. According to W. Langston and W. Miller (pers. commun., May 2000), this species was a crocodile from the eastern portion of Central America that did not tolerated salt water and could have lived in the west and became extinct in this region.

Other evidence that southernmost Baja has not been separated for a long time is the similarity of the Pliocene faunas from that area and from central Mexico. Mammalian species can change in as little as 300,000 years, although many can live much longer without appreciable change. However, many of the Pliocene species could have moved into Baja California through Arizona or California. A better fossil record of

Table 2.3 Tertiary terrestrial fossils documented throughout the peninsula of Baja California

Locality and Age	Reference	Fossils
Early Paleocene–Eocene Punta Prieta, Baja California	Novacek et al. (1991)	Marsupialia, Creodonta, Condylarthra, Acreodi, Pantodonta, Tillodontia, Perissodactyla, Artiodactyla, Rodentia, Eutheria *incertae sedis*, Squamata, Sepentes, and Archosauromorpha
Middle (?) Miocene at La Misión, Baja California	Minch et al. (1970)	Isolated record of a Camellidae, birds, and undetermined mammals
Middle (?) Miocene at La Purísima, Baja California Sur	Ferrusquía-Villafranca and Torres-Roldán (1980)	Isolated record of a Canidae
Pliocene and Pleistocene (Blancan) at Las Tunas	Miller (1980)	Canidae, Felidae, Perissodactyla, Artiodactyla, Antilocapridae, Rodentia, Lagomorpha, Amphibia, Chelonia, Squamata, Crocodylia, and Accipitriformes
Pliocene and Pleistocene (Blancan) at Rancho Algodones Area, Baja California Sur	Ferrusquía and Torres-Roldán (1980)	Canidae, Rodentia, Lagomorpha
Pleistocene (?) at Isla San José, Baja California Sur	Applegate and Carranza-Castañeda (1983)	Isolated record of Rhyncotherium
Pleistocene (Rancholabrean) at Rancho El Carrizal, Baja California Sur	Ferrusquía-Villafranca and Torres-Roldán (1980)	Perissodactyla and Artiodactyla
Pleistocene (Rancholabrean) at Santa Rita	Ferrusquía-Villafranca and Torres-Roldán (1980)	Proboscidea, Perissodactyla, Artiodactyla, Lagomorpha, and Quelonia

rodents and other taxa with species characteristic of geographically small areas would provide important data helping to understand the isolation of the cape region of Baja California. The close relationship of the Baja fauna with other western North American faunas does not support the theory of an isolated southern cape region for a long interval.

Data presented by Lawlor et al. in chapter 12 suggest the separation of Baja California, the northern extent of the gulf to the Salton Trough, the hypothesized seaway across the Vizcaíno Desert (discussed earlier), and the Isthmus of La Paz (see discussion below) were geological events that contributed effectively to isolate the peninsular and the Sonoran mammalian faunas, whereas the isolation of the population of mammals on landbridge islands from those on the nearest mainland probably occurred during the close of the Wisconsinian glaciation or earlier.

Some researchers suggest that the Los Cabos block migrated northeastward along the La Paz fault, approximately 50 km to its present location relative to the rest of Baja California (Anderson 1971; Londsdale 1989), and therefore this block was largely isolated. Several lines of investigation suggest that the tip of Baja California did not separate from the mainland as an isolated landmass but remained connected to the rest of the peninsula during the entire process.

Geochemical and isotopic data suggest an independent paleogeographical pre-Miocene evolution of the Los Cabos block, in relation to the rest of Baja California Peninsula. Thus, the La Paz fault could possibly be the accretional structure of the Los Cabos block with the rest of the peninsula (Schaaf et al. 2000). However, geologic data do not confirm the existence of the La Paz fault as defined by Hausback (1984), and fission-track data suggest contrasting cooling histories for the western and eastern margins of the Los Cabos block (Fletcher et al. 2000). Field relations and thermal modeling indicate that the rocks exposed at the eastern margin were recently exhumed from a significantly deeper crustal level, where the San José del Cabo fault accommodated 5.2–6.5 km of exhumation at rates as high as 1.5–2 mm/year (Fletcher et al. 2000). In contrast, the presence of early Miocene marine strata at the western side of Sierra El Novillo (Schwennicke et al. 1996), overlying crystalline basement, suggest that the basement rocks of the western margin reached the surface by the early Miocene (Fletcher et al. 2000). According to these ideas, the western margin of the Los Cabos block was inundated during the early–middle Miocene, by a shallow seaway (Schwennicke et al. 1996) that separated the geographic range of several species into vicariant cape region and northern populations. After its exhumation, the eastern margin received bathyal deposits in early Miocene times (McCloy 1984; Carreño 1992). A post-Pliocene regressive event is evidenced by late Pliocene to early Pleistocene shallower marine and terrestrial deposits (Miller 1980). Considering a scenario similar to the one discussed here, Grismer (1994) interpreted the evolution of xerophilic reptile taxa in southern Baja California as a vicariant complex.

In spite of the abundance of marine fossils throughout the peninsula and islands, fossil use is rather limited for the purpose of this chapter, but nevertheless, fossils can help approach some important questions. For example, it has been assumed that some species became extinct in the Pacific following the stepwise closure of the Panamanian seaway, combined with cooling of the climate and local paleooceanographic changes (Keigwin 1978; Crouch and Poag 1979). Possibly these taxa underwent speciation and dispersion before and during the time when the boundary between the North American and Pacific plates changed from the western side of the Baja California Peninsula to the Gulf of California. This is difficult to assess because, in general, faunas from the Caribbean and western Atlantic are better known than those from the eastern Pacific.

A good example of biostratigraphic distributional patterns is the report of Durham and Allison (1960) on the invertebrate fossil record throughout the Caribbean and Panamanian Provinces. Another example is the cognate molluscan genera lists presented by Woodring (1966), as well as the 279 pairs of Holocene molluscan species and subspecies that Vermeij (1978) tabulated. Smith (1991b) pointed out the role of Cenozoic giant pectinids from California and the Tertiary Caribbean Province as a point of departure for more detailed paleo-biogeographic studies.

Communities on seamounts, which reflect a widespread and more or less homogenous distribution (Wilson and Kauffmann 1987), represent benthic invertebrate faunas with pelagic larval stages. However, some organisms such as the podocopid Ostracoda do not possess a pelagic larval stage, and only the Cypridacea can swim short distances. The remaining species are incapable of swimming. Thus, dispersal of shallow-water species of Ostracoda across deep water relies on passive distribution mechanisms for the colonization of islands. Titterton and Whatley (1988) discussed possible shallow-water mechanisms for dispersal and colonization. They showed that passive dispersal of benthonic Ostracoda is driven primarily by prevalent surface ocean currents. For these reasons, Ostracoda are a valuable tool in understanding the evolution of species' distributional areas and the factors affecting biotic diversity through time.

Ostracoda from the Caribbean and western Atlantic are better known than those from the eastern Pacific. When explaining similarities between the Atlantic and Pacific assemblages, most authors prefer to either describe new species or postulate possible faunal interchange after closing of the Central American isthmus. Most refer to those populations that have a high degree of similarity as conspecific.

Morphological and paleo-zoogeographic characteristics (Cronin 1985) of Cenozoic marine Ostracods from the Atlantic, Caribbean, and Pacific areas suggest that the closing of the Isthmus of Panama represents a classic dumbbell allopatric split of large populations of tropical species. Before and during the formation of the Isthmus of Panama (and also during the opening of the Gulf of California in the early to mid-Pliocene), large, widespread populations of Ostracoda were divided. But the resulting geographical isolation had little if any effect on carapace morphology during the following 3.0 Ma. Pre- and post-isthmus specimens seem to represent environmentally controlled species, or ecophenotypic variations, with significant interpopulation polymorphism in size and carapace ornamentation. However, some of these species are recognized as endemic to the Gulf of California and the Pacific side of Baja California. These endemic forms split from ancestral populations located near the northern part of their range in the Pacific region. Rather than being the result of the formation of the Isthmus of Panama or a tectonic event related to the opening of the Gulf of California, this speciation was most probably related to the appearance of peripheral isolates, unrelated to the main populations in the Pacific and in the Caribbean. Speciation in Ostracoda occurs most frequently when populations are isolated in the periphery as a result of climatic changes. However, speciation also occurs in the middle of ranges, when small populations are isolated on island margins by passive mechanisms. Isolation of large populations by land barriers did not result in speciation. Speciation occurs by means of cladogenesis or lineage splitting, and the duration of most species varies within 5 and 8 Ma of morphologic stability (Cronin 1988).

Even when biogeographic data are integrated with geophysical and geological evidence, the controls of biogeographic and evolutionary patterns in the oceanic biota are still poorly understood. The relationship between geographic range, provinciality, environmental variation, and the paleobiological properties of organisms are very complex and are still being actively explored (Prothero 1997).

Species related to Quaternary marine fossils are still living in the Gulf of California. Some assumptions might be made about changes in the paleo-sea level only when absolute or numerical ages are available. Only then can we evaluate the absolute vertical motion experienced by either local or regional uplift, warping, and/or block faulting in any given area during the Quaternary–Holocene.

Ortlieb (1991) conducted a regional survey of emerged Pleistocene marine terraces in the Gulf of California. He used morphology, sedimentology, chronostratigraphic data, and the faunal content of marine deposits. Among his conclusions is that the peninsula of Baja California experienced relatively slow and uniform vertical motions in the Quaternary at a mean rate of 100 ± 50 mm/10^3 years over the last million years. Only the Pliocene–Quaternary La Reforma Caldera, north of Santa Rosalía, shows significantly more uplift (maximum rate of about 200 to 300 mm/10^3 years). However, uplift rates seem to have decreased, at least locally, based on the relative stability of the late Pleistocene terraces (ca. 125,000 years BP) in the northeastern (eastern side of the Colorado River mouth), southern (Bahía de La Paz and Punta San Telmo shores) and west-central (central Sonora and Isla Tiburón shores) parts of the peninsula.

No other information on Quaternary fossils is available. In fact, the fossils encountered in several key localities around the Gulf of California, as well as on the islands, are much older and have helped us to understand much of the geological evolution of the Gulf of California. Undoubtedly, the information necessary to evaluate the natural history of the islands in the Gulf of California awaits further research.

Summary of Island Age and Origins

The islands in the gulf can be subdivided into three main areas according to their age and style of formation: the northern gulf, the central or Midriff, and the southern gulf.

In the northern part of the Gulf of California, the islands are very young; most were formed by deltaic sedimentation of the Colorado River, and some were formed by volcanic activity in the last 2 Ma.

The central gulf area is structurally the most complex. Many of the islands in this area received marine sediments in middle Miocene times (14–12 Ma) and were uplifted in Pleistocene or Holocene times (5–<2 Ma) along the many faults present.

In the southern gulf area, the islands are composed of either volcanic or granitic rocks related to the peninsula, and some of them contain Pliocene marine sediments. In general, the islands in this area were uplifted when the southern end of the gulf opened (<4 Ma) and sea-floor spreading started on the eastern side of the Baja California Peninsula.

The Gulf of California represents an extensive and classical area where biologists attempt to explain the distribution of animals and plants in terms of present-day climatic and ecological conditions. Additionally, the paleontologists attempt to do something similar with the past distributions of organisms and use them to explain many present-day occurrences. The Gulf of California and surrounding areas (mainland, peninsula, and gulf islands) also represent a unique area where biogeographers may test the relationship between the land areas of islands and the diversity of life they support. Nevertheless, to test hypotheses concerning origin, migration, evolution, and extinction of organisms, it is important to know the geologic evolution, of a given area. In this case, a crucial factor in hypothesis testing is the knowledge of the origin and age of the gulf islands.

Since 1983, when Gastil et al. wrote *The Geology and Ages of Islands* in the first edition of this book, only a small amount of information regarding the origin and ages of the gulf islands has been added. The main goal of many research projects in the area during these two last decades has been the characterization of the two main phases in the geological evolution of the Gulf of California. The first phase is formed by a complex history of continental-margin sedimentation, tectonic accretion and magmatism. It started in the Miocene, with an andesitic arc overlying lithosphere that underwent a process of subduction along the western margin of Baja California. After subduction ceased, about 12 Ma, an east–west belt of extension marked by alkali-basalt volcanism and Basin and Range normal faulting that produced the "proto-gulf" occupied the site of the future gulf.

Marine geophysical data and drilling in the gulf, even though restricted, permits us to infer some general trends in the origin and evolution of some of the gulf islands. Nevertheless, many questions, most of them postulated by Gastil et al. (1983), remain unsolved. Biogeographers must be wary of the possible problems and pitfalls in the data presented here.

Acknowledgments We would like to recognize and express our indebtedness to the landmark work of Gordon Gastil, a pioneer in the geological research in Baja California whose studies have been fundamental for other researchers. We would especially like to acknowledge E. Ezcurra, T. Case, and M. Cody for inviting us to rewrite Gastil and coworkers' original paper. Thanks are also extended to L.A. Delgado-Argote (Department of Geology at CICESE) for discussions and for releasing unpublished information for our study. This chapter benefited from the thoughtful revisions of two anonymous reviewers. We are particularly thankful for the detailed critical revision and the insightful comments of one of them. This chapter was partly supported by UNAM-PAPIID Grant Num. IN-I02995.

References

Anderson, C.A. 1950. Geology of islands and neighboring land areas (Pt. 1 of the 1940 E. W. Scripps cruise to the Gulf of California). *Geological Society of America Memoir* 43:1–53.
Anderson, D.L. 1971. The San Andreas Fault. *Scientific American* 225:58–68.
Applegate, S.P., and O. Carranza-Castañeda. 1983. El primer reporte de un mamífero fósil de las Islas del Golfo de California. II International Meeting on the Geology of the Baja California Peninsula: Universidad Autónoma de Baja California and Sociedad de Geológica Peninsular, Abstracts, p. 4.
Aranda-Gómez, J.J., and J.A. Pérez-Venzor. 1986. Reconocimiento geológico de las islas Espíritu Santo y La Partida, Baja California Sur. *Revista* (Instituto de Geología, Universidad Nacional Autónoma de México) 6:103–116.
Atwater, T. 1970. Implications of plate tectonics for the Cenozoic tectonic evolution of Western North America. *Geological Society of America Bulletin* 81:3513–3536.
Atwater, T. 1989. Plate tectonic history of the Northeast Pacific and Western North America. In: E.L. Winterer, D.M. Hussong, and R.W. Decker (eds), *The Eastern Pacific Ocean and Hawaii. The Geology of North America*. N: Geological Society of America, Boulder, Colorado, pp. 21–72.
Batiza, R. 1978. Geology, petrology, and geochemistry of Isla Tortuga, a recently formed tholeiitic island in the Gulf of California. Special Issue of the *Geological Society of America Bulletin* 89:1309–1324.
Boehm, M.C. 1984. An overview of lithostratigraphy, biostratigraphy, and paleoenvironments of the late Neogene San Felipe marine sequence, Baja California, Mexico. In: V.A. Frizzell, Jr. (ed), *Geology of the Baja California Peninsula*. Society of Economic Paleontologists and Mineralogists, Los Angeles, California: pp. 253–265.

Carreño, A.L. 1981. Ostrácodos y foraminíferos planctónicos de la localidad Loma del Tirabuzón, Santa Rosalía, Baja California Sur e implicaciones bioestratigráficas y paleoecológicas. *Revista* (Instituto de Geología, Universidad Nacional Autónoma de México) 5:55–64.

Carreño, A.L. 1982. Biostratigraphy at the Loma del Tirabuzón (Corkscrew Hill) Santa Rosalía, Baja California Sur, Mexico. In: B. Mamet and M.J. Copeland (eds), *III North American Paleontological Convention*, 5–7 August, Montreal, Canada. Business & Economic Services Limited, Toronto, Canada, pp. 67–69.

Carreño, A.L. 1985. Biostratigraphy of late Miocene to Pliocene on the Pacific Island María Madre, Mexico. *Micropaleontology* 31:139–166.

Carreño, A.L. 1992. Neogene microfossils from the Santiago Diatomite, Baja California Sur, Mexico. In: M. Alcayde-Orraca and A. Gómez-Caballero (eds), *Calcareous Neogene Microfossils of Baja California Sur, Mexico. Paleontología Mexicana* (Instituto de Geología, Universidad Nacional Autónoma de México) 59:1–37.

Carreño, A.L., and L.R. Segura-Vernis. 1992. Ostrácodos de la Formación Trinidad, Baja California Sur, México. In: A. Carrillo-Chávez and A. Álvarez-Arellano (eds), *Memoir of the First International Meeting on the Geology of Baja California*. Sociedad Geológica Peninsular, Universidad Autónoma de Baja California Sur, pp. 101–110.

Colletta, B., and J. Angelier. 1984. Deformations of middle and late Pleistocene deltaic deposits at the mouth of the Río Colorado, northwestern Gulf of California. In: V.M. Malpica, et al. (eds), *Neotectonics and Sea Level Variations in the Gulf of California Area. A Symposium*. Hermosillo, México, 1984. Instituto de Geología, Universidad Nacional Autónoma de México, pp. 31–53.

Cromwell, J.C. 1962. Displacement along the San Andreas fault, California. *Geological Society of America, Memoir* 71:1–61.

Cronin, T.M. 1985. Speciation and stasis in marine Ostracoda: climatic modulation of evolution. *Science* 227:60–63.

Cronin, T.M. 1988. Geographical isolation in marine species: evolution and speciation in Ostracoda. In: T. Hanai, et al. (eds), *Evolutionary Biology of Ostracoda*. Proceedings of the Ninth International Symposium on Ostracoda. Kodansha Ltd., Tokyo, Japan, pp. 871–889.

Crouch, R.W., and C.W. Poag. 1979. *Amphistegina gibbosa* D'Orbigny from the California borderlands: the Caribbean connection. *Journal of Foraminiferal Research* 9:85–105.

Curray, J.R., and D.G. Moore. 1984. Geologic history of the mouth of the Gulf of California. In: J.K. Crouch and Bachman, S.B. (eds), *Tectonics and Sedimentation along the California Margin*. Pacific Section, Society of Economic Paleontologists and Mineralogists, pp.17–36.

Curray, J.R., D.G Moore, K. Kelts, and G. Einsele. 1982. Tectonics and geological history of the passive continental margin at the tip of Baja California. In: J.R. Curray, et al. (eds), *Initial Reports of the Deep Sea Drilling Project Leg 64*. U.S. Government Printing Office, Washington, D.C.: 1089–1116.

Dauphin, J.P., and G.E. Ness. 1991. Bathymetry of the Gulf and Peninsular Province of the Californias. In: J.P. Dauphin and B.R.T. Simoneit (eds), *The Gulf and Peninsular Province of the Californias. American Association of Petroleum Geologists Memoir* 47:21–24.

Delgado-Argote, L.A. 2000. Evolución Tectónica y magmatismo neógeno de la margen oriental de Baja California Central. Ph.D. dissertation. Universidad Nacional Autonoma de México, México, D.F.

Delgado-Argote L.A., and J. García-Abdeslem. 1999. Shallow Miocene Basaltic magma reservoirs in the Bahía de los Ángeles Basin, Baja California, Mexico. *Journal of Volcanology and Geothermal Research* 88:29–46.

Delgado-Argote, L.A., M. López-Martínez, and M.C. Perrilliat. 1999. Geologic reconnaissance and Miocene age of volcanism and associated fauna from sediments of Bahía de los Ángeles, Baja California. In: H. Delgado-Granados et al. (eds), *Cenozoic Volcanism and Tectonics of Mexico*. Special Paper 335. Geological Society of America, Boulder, Colorado.

Desonie, D.L. 1992. Geological and geochemical reconnaissance of Isla San Esteban: postsubduction orogenic volcanism in the Gulf of California. *Journal of Volcanology and Geothermal Research* 52:123–140.

Dokka, R.K., and R.H. Merriam. 1982. Late Cenozoic extension of Northeastern Baja California, Mexico. *Geological Society of America Bulletin* 93:371–378.

Durham, J.W. 1950. Megascopic paleontology and marine stratigraphy (Pt. 2 of the *E. W. Scripps cruise to the Gulf of California*). *Geological Society of America Memoir* 43:1–216.

Durham, J.W., and E.C. Allison. 1960. The geologic history of Baja California and its marine faunas. *Systematic Zoology* 9:47–91.

Escalona-Alcázar, F.J. 1999. Reconocimiento geológico de las islas San Lorenzo y Las Ánimas y la margen nororiental de la Sierra Las Ánimas, Baja California Central. Undergraduate thesis, CICESE, Ensenada, Baja California, México.

Escalona-Alcázar F.J., and L.A Delgado-Argote.1998. Descripción estratigráfica de la zona El Paladar y litológica de la isla Ángel de la Guarda, Golfo de California. *GEOS* 18:197–205.

Escalona-Alcázar, F.J., and L.A. Delgado-Argote. 2000. Estudio de la deformación en las islas San Lorenzo y Las Ánimas, Golfo de California: implicaciones sobre su desplazamiento como bloque rígido desde el Plioceno tardío. *GEOS* 20:8–20.

Escalona-Alcázar, F.J., L.A. Delgado-Argote, M. López-Martínez, and G. Rendón-Márquez. *Revicta Mexicana de Ciencias Geológicas* (Instituto de Geologia, Universidad Nacional Autónoma de México) 18: 111–128.

Fabriol, H., L.A. Delgado-Argote, J.J. Dañobeitia, D. Córdoba, A. González, J. García-Abdeslem, R. Bartolomé, and B. Martín-Atienza. 1999. Backscattering and geophysical features of volcanic rifts offshore Santa Rosalia, Baja California Sur, Gulf of California, Mexico. In: *Rift-related Volcanism: Geology, Geochemistry and Geophysics*. Special Issue Journal of Volcanology and Geothermal Research 93:75–92.

Ferrusquía-Villafranca, I., and V. Torres-Roldán. 1980. El registro de mamíferos terrestres del Mesozoico y Cenozoico de Baja California. *Revista* (Instituto de Geología, Universidad Nacional Autónoma de México) 4:56–62.

Flessa, K. W., and J. Imbrie. 1973. Evolutionary pulsations: Evidence from Phanerozoic diversity patterns. In: D.H. Tarling and S.K. Runcorn (eds), *Implications of Continental Drift to the Earth Sciences*. Academic Press, London, pp. 245–284.

Fletcher, J.M., B.P. Kohn, D.A. Foster, and A.J.W. Gleadow. 2000. Heterogeneous Neogene cooling and exhumation of the los Cabos block, southern Baja California: evidence from fission-track thermochronology. *Geology* 28:107–110.

Gastil, R.G., and S.S. Fenby. 1991. Detachment faulting as a mechanism for tectonically filling the Gulf of California during dilatation. In: J.P. Dauphin and B.R.T. Simoneit (eds), *The Gulf and Peninsular Province of the Californias*. Special Issue of the *American Association of Petroleum Geologists Memoir* 47:371–375.

Gastil, R.G., and D. Krummenacher. 1977. Reconnaisance geology of coastal Sonora between Puerto Lobos and Bahía Kino. *Geological Society of America Bulletin* 88:189–198.

Gastil, G., J. Minch, and R.P. Phillips. 1983. The geology and ages of islands. In: T.J. Case and M.L. Cody (eds), *Island Biogeography in the Sea of Cortéz*. University of California Press, Berkeley; pp. 13–15.

Gastil, J., J. Neuhaus, M. Cassidy, J. Smith, J. Ingle, and D. Krummenacker. 1999. Geology and paleontology of Southwestern Isla Tiburón, Mexico. *Revista Mexicana de Ciencias Geológicas* 16:1–34.

Gastil, R.G., R.P. Phillips, and E.C. Allison. 1975. Reconnaissance geology of the state of Baja California (with a geologic map, scale 1:250,000). *Geological Society of America Memoir* 140:1–70.

Grismer, L.L. 1994. The origin and evolution of the Peninsular herpetofauna of Baja California, Mexico. *Herpetological Natural History* 2:51–106.

Hausback, B.P. 1984. Cenozoic volcanic and tectonic evolution of Baja California Sur, Mexico. In: V.A. Frizzell, Jr. (ed). *Geology of the Baja California Peninsula*. Society of Economic Paleontologists and Mineralogists, Los Angeles California; pp. 219–236.

Helenes, J., and A.L. Carreño. 1999. Neogene sedimentary evolution of Baja California in relation to regional tectonics. *Journal of South American Earth Sciences* 12:589–605.

Ingle J.C., Jr. 1974. Paleobathymetric history of Neogene marine sediments, Northern Gulf of California. In: G. Gastil and J. Lillegraven (eds), *The Geology of Peninsular California*. Guidebook 49 of the Pacific Section, American Association of Petroleum Geologists & Society of Economic Paleontologists and Mineralogists; pp. 121–138.

Johnson, I.M. 1924. Expedition of the California Academy of Science to the Gulf of California in 1921. *California Academy of Science, Proceedings* (Geology and Paleontology, 4th ser) 16:137–157.

Karig, D.E., and W. Jensky. 1972. The Proto-Gulf of California. *Earth & Planetary Science Letters* 17:169–174.

Keigwin, L.D. 1978. Pliocene closing of the Isthmus of Panama based on biostratigraphic evidence from nearby Pacific and Caribbean cores. *Geology* 6:630–634.

Larson, R.L. 1972. Bathymetry, magnetic anomalies, and plate tectonics history of the mouth of the Gulf of California. *Geological Society of America Bulletin* 83:3345–3360.

Ledesma-Vázquez, J., and M.E. Johnson. 1997. Evolución Pliocénica-Pleistocénica de Bahía Concepción, Golfo de California, B.C.S. In: *Reunión Anual de la Unión Geofísica Mexicana*, Puerto Vallarta, Jalisco, México, *GEOS* 17:115–122.

Londsdale, P. 1989. Geology and tectonic history of the Gulf of California. In Winterer, E.L. et al. (eds), *The Eastern Pacific Ocean and Hawaii*. Geological Society of America, The Geology of North America Series, Boulder, Colorado, pp. 499–521.

Londsdale, P. 1991. Structural patterns on the Pacific floor offshore of Peninsular California. In: J.P. Dauphin and B.R.T. Simoneit (eds), *The Gulf and Peninsular Province of the Californias*. American Association of Petroleum Geologists Memoir 47:87–125

Lyle, M., and G.E. Ness. 1991.The opening of the southern Gulf of California. In: J.P. Dauphin and B.R.T. Simoneit (eds), *The Gulf and Peninsular Province of the Californias. American Association of Petroleum Geologists Memoir* 47:403–423.

MacArthur, R.H., and E.O. Wilson. 1967. *The Theory of Island Biogeography*. Princeton University Press, Princeton, New Jersey.

Mammerickx, J. 1980. Neogene reorganization of spreading between the Tamayo and the Rivera fracture zone. *Marine Geophysical Research* 4:305–318.

Martín-Barajas, A., J.M. Fletcher, M. López-Martínez, and R. Mendoza-Borunda. 2000. Waning Miocene subduction and arc volcanism in Baja California: the San Luis Gonzaga volcanic field. *Tectonophysics* 318: 27–51.

McCloy, C. 1984. Stratigraphy and depositional history of the San José del Cabo trough, Baja California Sur, Mexico. In: V.A. Frizzell, Jr. (ed), *Geology of the Baja California Peninsula*. Society of Economic Paleontologists and Mineralogists, Pacific Section, Los Angeles, California; pp. 267–273.

McDougall, K., R.Z. Poore, and J. Matti. 1999. Age and paleoenvironment of the Imperial Formation near San Gorgonio Pass, southern California. *Journal of Foraminiferal Research* 29:4–25.

Miller, W.A. 1980. The late Pliocene Las Tunas local fauna from southernmost Baja California. *Journal of Paleontology* 54: 762–805.

Minch, J.A., K.C. Schulte, and G. Hofman. 1970. A middle Miocene age for the Rosarito Beach Formation in northwestern Baja California, Mexico. *Geological Society of America Bulletin* 81:3149–3154.

Moore, D.G. 1973. Plate-edge deformation and crustal growth, Gulf of California structural province. *Geological Society of America Bulletin* 84:1883–1906.

Moore, D.G., and E.C. Buffinton.1968. Transform faulting and growth of the Gulf of California since the late Pliocene. *Science* 161:1238–1241.

Moore. D.G., and J.R. Curray. 1982. Geologic and tectonic history of the Gulf of California. In: J.R. Curray et al. (eds), *Initial Reports of the Deep Sea Drilling Project*. Government Printing Office, Washington, DC, pp. 1279–1294.

Murphy, R.W. 1983. The reptiles: origins and evolution. In: T.J. Case and M.L. Cody (eds), *Island Biogeography in the Sea of Cortéz*. University of California Press, Berkeley; pp. 130–158.

Novacek, M.J., I. Ferrusquía-Villafranca, J.J. Flynn, A.R. Wyss, and M. Norell. 1991. Wasatchian (early Eocene) mammals and other vertebrates from Baja California, México; the Las Lomas-Las Tetas de Cabra Formation. *Bulletin of the American Museum of Natural History* 208:1–88.

Ortlieb, L. 1991. Quaternary vertical movements along the coasts of Baja California and Sonora. In: J.P. Dauphin and B.R.T. Simoncit (eds), *The Gulf and Peninsular Province of the Californias*. American Association of Petroleum Geologists Memoir 47:447480.

Oskin, M., J. Stock, A. Martín-Barajas, and C. Lewis. 2000. New constraint on timing and total offset across the Gulf, and an early Pliocene age of the Isla Tiburón Proto-Gulf section. In: *V International Meeting on Geology of the Baja California Peninsula*. Loreto, Baja California Sur, April 25-May 1, 2000. Sociedad Geológica Peninsular, Universidad Autónoma de Baja California Sur, La Paz; pp. 9–10.

Paz-Moreno, F.A., and A. Demant. 1995. Isla San Luis: a Holocene eruptive centre of tholeiitic affinity in the Gulf of California, Mexico. In: *III International Meeting on Geology of the Baja California Peninsula*. La Paz, Baja California Sur, 17–21 April 1995. Sociedad Geológica Peninsular, Universidad Autónoma de Baja California Sur, La Paz; pp.138–140.

Paz-Moreno, F.A., and A. Demant. 1999. The Recent Isla San Luis volcanic centre: petrology of a rift-related volcanic suite in the northern Gulf of California, Mexico. *Journal of Volcanology and Geothermal Research* 93:31–52.

Phillips, R.P. 1964. Geophysical investigations of the Gulf of California. Ph.D. dissertation, University of California, San Diego.

Phillips, R.P. 1966a. Reconnaissance geology of some of the northwestern islands in the Gulf of California. In: *Program with Abstracts*. Geological Society America, Cordilleran Section. p. 59.

Phillips, R.P. 1966b. Geological framework of the Gulf islands. In: Lindsay, G.E. (ed). *The Gulf Islands Expedition of 1966*. Special issue of the *Proceedings of the California Academy Sciences* 30:341–342.

Prothero, D.R. 1997. *Bringing Fossils to Life. An Introduction to Paleobiology*. McGraw-Hill, Boston, Massachusetts.

Puy-Alquiza, M.I. 1992. *Geología de la isla San José, Canal San José y su posible correspondencia con el Macizo Peninsular, Baja California Sur, México*. Bch. thesis, Universidad Autónoma de Baja California Sur, Área Interdiciplianria de Ciencias del Mar, Departamento de Geología Marina, La Paz, B.C.S., México.

Rodríguez-Quintana, R., and L.R. Segura-Vernis. 1992. Gasterópodos de la Formación Trinidad, Baja California Sur. In: A. Carrillo-Chávez and A. Álvarez-Arellano (eds), *Memoir of the First International Meeting on the Geology of Baja California.* Universidad Autónoma de Baja California Sur, Sociedad Geológica Peninsular; pp. 111–133.

Rossetter, R.J. 1974. *Geology of La Encantada Island, Baja California, Mexico.* Undergraduate research report, California Sate College, San Diego.

Rossetter, R.G., and G. Gastil. 1971. Isla San Luis, a rift volcano in the Gulf of California. *Geological Society of America* (Abstracts and Program) 3:187–188.

Rusnak, G.A., and R.L. Fisher. 1964. Structural history and evolution of Gulf of California. In: Tj. H. van Andel and G.G. Shor (eds), *Marine Geology of the Gulf of California. American Association of Petroleum Geologists Memoir* 3:144–155.

Schaaf, P., H. Böhnl, and J.A. Pérez-Venzor. 2000. Pre-Miocene paleogeography of the Los Cabos Block, Baja California Sur: geochronological and palaeomagnetic constrainsts. *Tectonophysics* 318:53–69.

Schwennicke, T., G. González-Barba, and N. DeAnda-Franco. 1996. Lower Miocene marine and fluvial beds at Rancho La Palma, Baja California Sur. *Boletín del Departamento Geológico de la Universidad de Sonora* 13:1–14.

Secretaría de Gobernación and Secretaría de Marina. 1987. *Islas mexicanas, régimen jurídico y catálogo.* Talleres Gráficos de la Nación, México, D.F.

Shackleton, N.J. 1987. Oxygen isotope, ice volume and sea level. *Quaternary Science Reviews* 6:183–190.

Shackleton, N.J., and N.D. Opdyke. 1973. Oxygen isotope and paleomagnetic stratigraphy of equatorial Pacific core V-28-238: oxygen isotope temperature and ice volumes on a 105 Year and 106 Year time scale. *Quaternary Research* 3:39–55

Shackleton, N.J., and N.D. Opdyke. 1976. Oxygen isotope and paleomagnetic stratigraphy of Pacific core V28-239, late Pliocene to latest Pleistocene. *Geological Society of America Memoirs* 145: 449–464

Shepard, F.P. 1950. Part III. Submarine topography of the Gulf of California. In: *1940 E.W. Scripps Cruise to the Gulf of California. Geological Society of America Memoir* 43:1–32.

Smith, J.T. 1991a. Cenozoic marine mollusks and paleogeography of the Gulf of California. In: Dauphin, J.P., and B.R.T. Simoneit (eds), *The Gulf and Peninsular Province of the Californias. American Association of Petroleum Geologists, Memoir* 47:637–666.

Smith, J.T.1991b. Cenozoic giant pectinids from California and the Tertiary Caribbean Province: *Lyropecten, "Macrochlamis", Vertipecten and Nodipecten species. U.S. Geological Survey Professional Paper* 1391:1–155.

Smith, J.T. 1992. The Salada Formation of Baja California Sur, Mexico. In Carrillo-Chávez, A. and A. Álvarez-Arellano (eds), *Memoir of the First International Meeting on the Geology of Baja California.* Universidad Autónoma de Baja California Sur, Sociedad Geológica Peninsular, La Paz, Baja California Sur, México; pp. 23–32.

Soulé, M., and A.J. Sloan. 1966. Biogeography distribution of the reptiles and amphibians on islands in the Gulf of California, Mexico. *Transactions, San Diego Society Natural History* 14:137–156.

Staines-Urias, F. 1996. *Neotectónica del área San Nicolás, B.C.S.* Unpublished undergraduate thesis, Facultad de Ciencias Marinas, Universidad Autónoma de Baja California, México.

Titterton, R., and R.C. Whatley. 1988. The provincial distribution of shallow water Indo-Pacific marine Ostracoda: origins, antiquity, dispersal routes and mechanisms. In: Hannai, T. et al. (eds), *Evolutionary Biology of Ostracoda.* Kodansha/Elsevier, Tokyo, pp. 759–786.

Upton, D.E., and R.W. Murphy. 1997. Phylogeny of the side-blotched lizards (Phrynosomatidae: *Uta*) based on mtDNA sequences: support for a midpeninsular seaway in Baja California. *Molecular Phylogenetics and Evolution* 8:104–113.

Vermeij, G.J. 1978. *Biogeography and adaptation: patterns of marine life.* Harvard University Press, Cambridge, Massachusetts.

Wilcox, B.A. 1978. Supersaturated island faunas: a species-age relationship for lizzard faunas on post-Pleistocene land-bridge islands. *Science* 199:996–998.

Wilson, R.R., and S. Kauffmann. 1987. Seamount biota and biogeography. In: Keating B.H. et al. (eds), *Seamounts, Islands and Atolls.* Special issue of *Geophysical Monographs* (American Geophysical Union) 43:355–379.

Woodring, W.P. 1966. The Panama land bridge as a sea barrier. *American Philosophical Society Proceedings* 110:425–433.

3

Physical Oceanography

SAÚL ÁLVAREZ-BORREGO

The nature of the relationships between physical and biological processes in the ocean is subtle and complex. Not only do the physical phenomena create a structure, such as a shallow, mixed layer or a front, within which biological processes may proceed, but they also influence the rates of biological processes in many indirect ways. In the ocean, physical phenomena control the distribution of nutrients necessary for phytoplankton photosynthesis. Places with higher kinetic energy have higher concentrations of planktonic organisms, and that makes the whole food web richer (Mann and Lazier 1996). For example, in the midriff region of the Sea of Cortés (Tiburón and Ángel de la Guarda; fig. 1.2), tidal currents are strong, and intense mixing occurs, creating a situation similar to constant upwelling. Thus, primary productivity is high, and this area supports large numbers of sea birds and marine mammals (Maluf 1983).

The Gulf of California is a dynamic marginal sea of the Pacific Ocean and has been described as an area of great fertility since the time of early explorers. Gilbert and Allen (1943) described it as fabulously rich in marine life, with waters fairly teeming with multitudes of fish, and to maintain these large numbers, there must be correspondingly huge crops of their ultimate food, the phytoplankton. Topographically the gulf is divided into a series of basins and trenches, deepening to the south and separated from each other by transverse ridges (Shepard 1950; fig. 3.1). Input of nutrients into the gulf from rivers is low and has only local coastal effects. The Baja peninsula has only one, very small river, near 27°N; rivers in mainland Mexico and the Colorado River have dams that divert most of the water for agricultural and urban use.

The gulf has three main natural fertilization mechanisms: wind-induced upwelling, tidal mixing, and thermohaline circulation. Upwelling occurs off the eastern coast

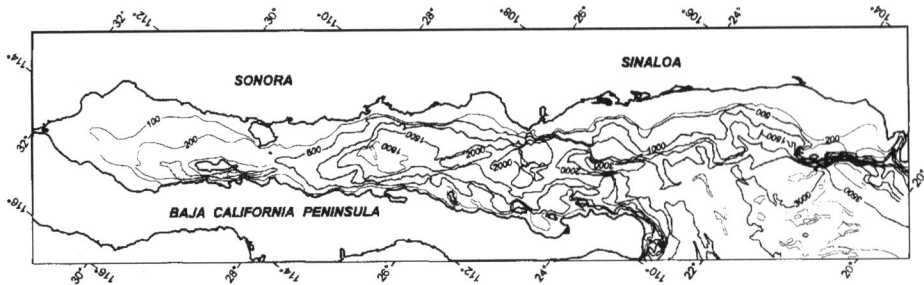

Figure 3.1 Bathymetry of the Gulf of California in meters (based on Dauphin and Ness 1991).

with northwesterly winds (winter conditions from December through May) and off the Baja California coast with southeasterly winds (summer conditions from July through October), with June and November as transition periods (Álvarez-Borrego and Lara-Lara 1991). Upwelling events last for few days, and then relaxation allows for stabilization of the water column and blooming of phytoplankton communities.

Tidal energy dissipation is strongest in the upper gulf (up to >0.5 W/m^2) and in the Midriff (>0.3 W/m^2; Argote et al. 1995). Tidal mixing between the islands of San Lorenzo and San Esteban produces vigorous stirring of the water column down to >500 m depth, with the net effect of carrying colder, nutrient-rich water to the surface. Tidal mixing has a fortnightly modulation; it is strongest with post-spring tides and weakest with post-neap tides (Simpson et al. 1994). In the gulf the annual mean of the net water–atmosphere heat flux is into the sea (Castro et al. 1994). This requires an oceanic export of heat and salt out of the gulf to achieve a balance and implies that the annual mean thermohaline circulation must have an inflowing component at depth (Bray and Robles 1991). This has a profound ecological implication because inflowing, deep water has higher inorganic nutrient concentrations than outflowing surface water (Álvarez-Borrego and Lara-Lara 1991).

In the southern Gulf, in contrast, El Niño events have a suppressing effect on primary productivity. El Niño events may have an enhancing effect on phytoplankton production in places of large turbulence, such as in Ballenas Channel between Ángel de la Guarda and the peninsula (Santamaría-Del-Angel et al. 1994a,b). These events can cause the reproductive failure of higher organisms in the water and on the islands (Jiménez-Castro 1989; Velarde and Ezcurra, chap. 11, this volume) due to suppression of primary production or changes in the planktonic community structure, or both.

Meteorological Aspects

The Gulf's Climate

Everywhere but at its southern end, the Gulf of California is surrounded by elevated topography. The moderating effect of the Pacific Ocean on the climate of the gulf is greatly reduced by an almost uninterrupted chain of mountains, 1–3 km high, in Baja

California. To the east and north, elevations in the Sierra Madre Occidental typically exceed 1500 m. To the northwest, the gulf region extends into the lowlands of the greater Sonoran desert, which is divided by the Colorado River valley and includes portions of northwestern Mexico, southeastern California, and southwestern Arizona. Significant air flow below 800 m is usually channeled along the gulf and is open to direct oceanic influence only from the south. The Gulf of California is thus a semienclosed basin in a meteorological as well as in an oceanic sense (Badan-Dangon et al. 1991). The climate of the gulf is therefore more continental than oceanic, a fact which contributes to the wide annual and diurnal temperature ranges observed there (Hernández 1923; cited by Roden 1964).

The climate over the gulf is typically divided into two seasons: a mid-latitude winter and a subtropical summer (Mosiño and García 1974). In winter the air temperatures decrease toward the interior of the gulf. Winter air temperature differences between the gulf and the Pacific coasts of Baja California are small, but at the same latitude the air temperatures of Mexico mainland's coast are higher by about 2°C than those of Baja California (fig. 3.2a). In summer the air temperature increases toward the interior of the gulf, and the temperature differences between the east and west coasts of Baja California become large, sometimes exceeding 10°C (fig. 3.2b). There is more precipitation on the east than on the west side of the gulf. The northern half of the gulf is dry and desertlike, with annual rainfall less than 100 mm. In the southeast, rainfall along the coast increases to about 1000 mm per year (fig. 3.2c). South of Ángel de la Guarda and Tiburón, most of the rain falls between June and October, but in the north most of the rain falls during winter (fig. 3.2d). The mean annual air temperature range increases from about 6°C at Cape Corrientes to 18°C at the northern end of the gulf.

The number of rainy days per year decreases from about 60 at Cape Corrientes to about 5 along the central Baja California coast. Skies are generally clear. The year-to-year variation of rainfall in the gulf is large and strongly depends on the incidence of tropical storms and hurricanes. Although summer tropical storms along the west coast of Mexico seldom enter the gulf region (Serra 1971), those that do enter dissipate at different places in the states of Sinaloa and Sonora, and a few travel all the way to Arizona (Harris 1969). Occasionally during the summer, tropical Pacific air surges up the gulf and over the United States, and a portion of the large pool of tropical moisture present off western Mexico is channeled up the gulf and over the southwestern United States (Reyes and Cadet 1986). These are 5-day pulses of relatively strong southeasterly winds, blowing at speeds of 10 m/s or more (Badan-Dangon et al. 1991).

A marine layer is well defined over the gulf (100–300 m thick) during both seasons but dissipates within a few kilometers inland. The summer marine layer is on the order of 200–300 m thick, with dew point temperatures of 26°–28°C (a moisture content of 21–24 g/kg). Immediately above the layer is a dry zone. The atmosphere over the gulf is much drier in the winter. The winter marine layer is thinner, on the order of 100–200 m, and it is defined by a dew point temperature of 6°–11°C (a moisture content of 6–8 g/kg). Perturbations to the marine layer occur in the vicinity of islands, since even the narrowest have a significant wake that extends several island diameters downstream. The marine layer is also significantly eroded in Ballenas Chan-

44   The Physical Scene

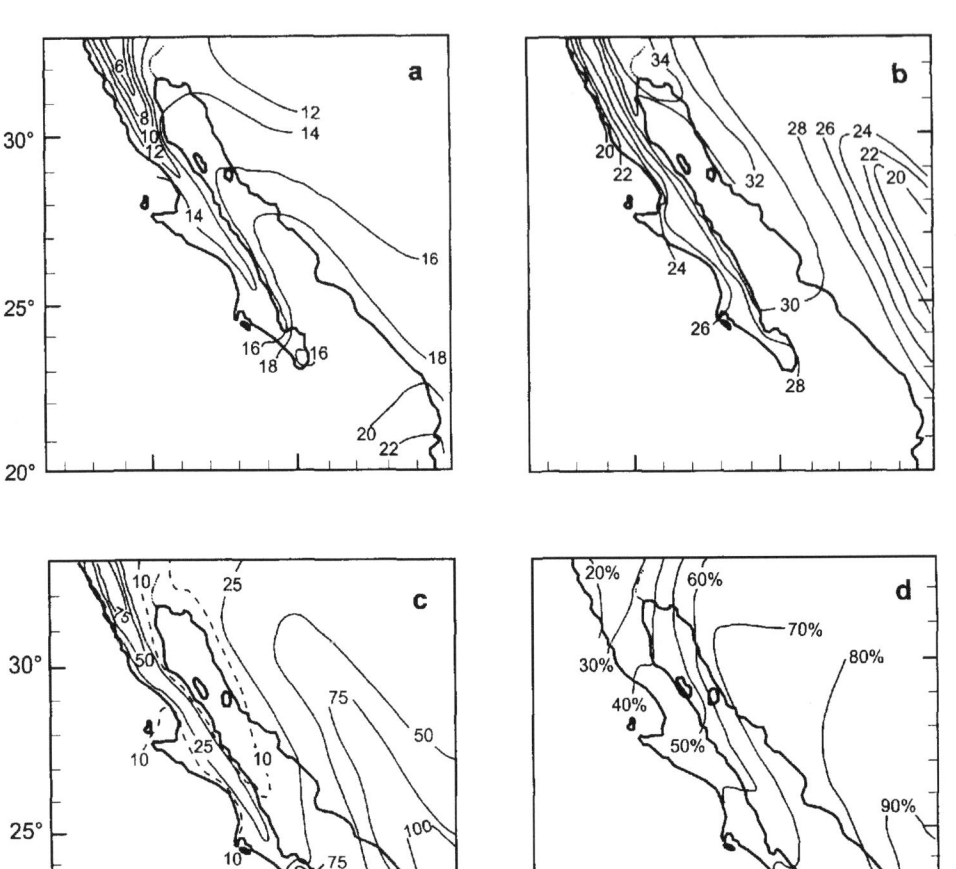

Figure 3.2 (a) Mean air temperature for January (°C); (b) mean air temperature for July (°C); (c) mean annual rainfall (cm); and (d) mean percentage of summer (May–October) rain (after Page 1930 and Ward and Brooks 1936; cited by Roden 1964).

nel, possibly because the passage is narrow enough to perceive the cross-channel circulation from both the steep slopes of Isla Ángel de la Guarda and from Baja California (Badan-Dangon et al. 1991).

Winds and Upwelling

The wind field over the gulf is nearly monsoonal in nature and is markedly coherent both along and across the gulf (Bray and Robles 1991). The seasonal signal of the monsoon-type winds tends to be modified by the nearshore seabreeze. Moderate northwest gales that last two or three days at a time are frequently experienced in the upper gulf between December and February. These winds are particularly strong in Ballenas Channel; they may on occasion raise such a heavy sea that navigation becomes impossible (Roden 1964).

Wind blowing across the sea surface sets up a stress that causes water to begin moving in the same direction as the wind. The effect of the rotation of the earth (Coriolis effect) causes the current to veer to the right in the Northern Hemisphere. When the situation has persisted for some time, the net movement of the water is at right angle to the direction of the wind (Ekman transport). When the direction of this transport is away from the coast, surface waters move offshore and their place is taken by deeper water that upwells close to shore. If this deeper water is nutrient-enriched, upwelling serves as a stimulant of primary production. During winter, when northwesterly winds generate strong upwellings off the eastern coast of the gulf, nutrient enrichment has a marked effect on phytoplankton communities (the concentration of chlorophyll a reaches values >10 mg/m^3), and due to eddy circulation nutrient enrichment increases the phytoplankton biomass across the gulf (Santamaría-Del-Angel et al. 1994a). It is interesting to note that, because isograms such as 1 ml/l of dissolved oxygen, 2.5 µM phosphate, and 25 µM nitrate are very shallow (~100 m), winter upwelling areas in the gulf have some of the highest surface nutrient concentrations in any of the oceans of the world (Álvarez-Borrego et al. 1978).

However, because of strong stratification during summer, upwelling off the Baja coast with southeasterly winds has a weak effect on phytoplankton biomass, causing chlorophyll a to increase only to values around 0.5 mg/m^3, in spite of similar wind magnitudes as those in winter (Santamaría-Del-Angel et al. 1999). In the gulf, 5–15 times the energy to move water from 100 m to the surface is needed during summer compared to that during winter. An index of stratification of the near surface water column (0–100 m) is only 15–60 J/m^3 for winter and up to 260–310 J/m^3 for summer (Cortés-Lara et al. 1999). The effect of these differences does not show when calculating Ekman transport for the two seasons because Ekman oversimplified the transfer of momentum from the air to the water and assumed, among other things, that the water column is homogeneous (Pond and Pickard 1983). Another factor that summer upwelling does not have a large positive effect on the productivity of the gulf is that high mixed-layer temperatures (often >29°C) reduce the phytoplankton photosynthetic capacity (Santamaría-Del-Angel et al. 1999).

Satellite-derived composite images showing the surface distribution of chlorophyll a concentration for the weeks of 11–17 April 1980 and 26–31 August 1979 are representative of the extreme conditions in the gulf: winter (plate 1A) and summer (plate 1B). During winter, pigment concentrations are high in most of the gulf. The effect of upwelling is evident off the coast of mainland Mexico, and very high pigment values are also evident in the northern gulf. Plumes of high pigment concentration may be seen elongating from the eastern side of the gulf toward the Baja California coast. Pigment concentrations are low only at the entrance to the gulf, south from 25°N (plate 1A; Santamaría-Del-Angel et al. 1994a). During summer, most of the gulf has low pigment concentrations (plate 1B). With winds from the southeast, the equatorial surface water enters into the gulf with high temperatures and low nutrients (Álvarez-Borrego and Schwartzlose 1979). Pigment concentrations were higher only in the waters around the midriff islands, in the upper gulf adjacent to the Colorado

River mouth, and in a narrow strip off mainland Mexico (plate 1B; Santamaría-Del-Angel et al. 1994a).

Air–Water Heat Exchange

The location of the gulf between two arid land masses results in a net flux of moisture from the ocean to the atmosphere, making the gulf the only evaporative basin of the Pacific (Roden 1964). Evaporation increases to the north. Maximum evaporation occurs during late summer or early fall due to the very high sea-surface temperatures in late summer (Bray and Robles 1991). In spite of evaporation, there is an annual mean net heat flux into the sea of >100 W/m^2. On the average across seasons, a net 18×10^{12} W enter the gulf through the surface. This heat has to be exported to the Pacific somehow; otherwise the gulf's temperature would be increasing (Lavín et al. 1997). Less dense, warmer surface water flows out from the gulf into the Pacific. To balance this flow, relatively deep water flows into the gulf. This is an inverse circulation with respect to the circulation between the Mediterranean and the Atlantic through the Gibraltar strait.

Gilbert and Allen (1943) studied the phytoplankton communities of the gulf with samples mainly collected during February–March. They postulated that an inflow of deep water and an outflow of surface water characterize the general circulation of the gulf. As a consequence of this circulation, deep water, rich in plant nutrients, is brought to the euphotic zone where the nutrients can be used by the phytoplankton. According to Gilbert and Allen, the main factor affecting this circulation is the wind, which brings about upwelling and an outward transport of surface water.

Bray (1988a) proposed an oversimplified three-layer circulation. Water is generally outflowing between 50 and 250 m, inflowing between 250 and 500 m, and variable in the top 50 m. According to Bray (1988a), the surface layer shows some tendency to reverse direction with the seasonal winds.

Water Masses, Tides, and Circulation

Water Masses

South of Ángel de la Guarda and Tiburón islands, the gulf has basically the same thermohaline structure as that of the Eastern Tropical Pacific, with modifications at the surface due to excess evaporation (Sverdrup 1941; Roden 1964). Different authors have assigned diverse names to the water masses of the Gulf of California. The most accepted nomenclature is that proposed by Torres-Orozco (1993), based on the characteristics of these waters in the gulf, and not on their origin, with the exception of the Gulf of California water mass (GCW; $S \geq 35.0$‰ and $T > 12°C$; S is salinity and T is temperature). According to Torres-Orozco (1993), besides GCW, there are California current water (CCW; $S \leq 34.50$‰ and $12.0 \leq T < 18.0°C$); equatorial surface water (ESW; $S < 35.0$ and $T \geq 18.0°C$); subtropical subsurface water (SSW; $34.5 < S < 35.0$‰ and $9.0 \leq T < 18.0°C$); Pacific intermediate water (PIW; $34.5 \leq S < 34.8$‰ and $4.0 \leq T < 9.0°C$); and Pacific bottom water (PBW; $S > 34.5$‰ and $T < 4.0°C$).

Oceanographically, the southernmost limit of the Gulf of California is a line connecting Cape San Lucas and Cape Corrientes (Roden 1964). The region of the en-

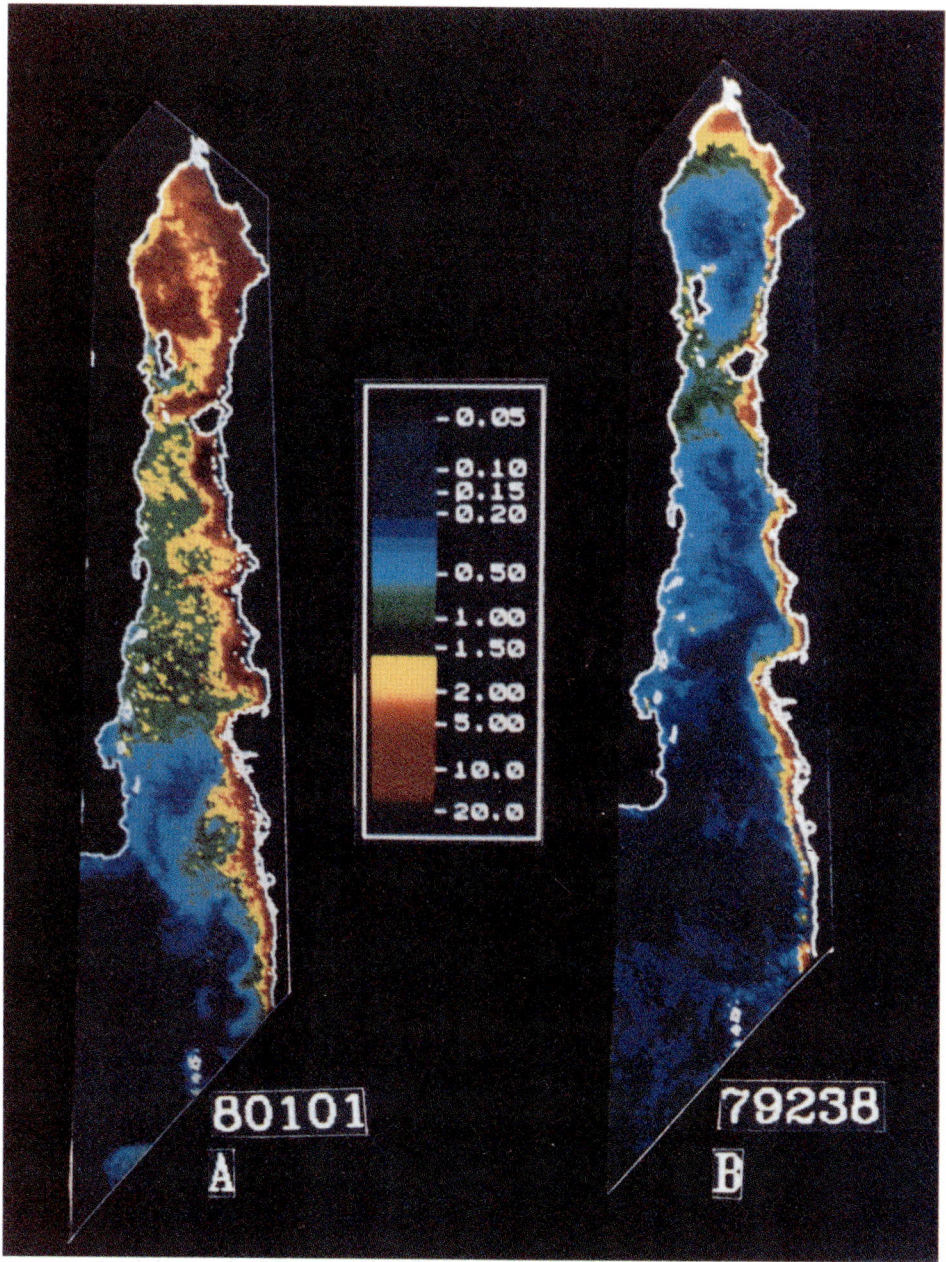

Plate 1. Gulf of California weekly composites of the CACS scenes for (a) early spring (April 11–17, 1980) and (b) summer (August 26–31). Numbers in the color scale are pigment concentrations in mg m^{-3}. Numbers at the bottom are the year (first two digits) and the first Julian day of the week. (Adapted from Santamaria-Del-Angel et al. 1994a.)

trance to the gulf is considered roughly as that between this line and one connecting La Paz with Topolobampo. This is a transitional zone that has a complicated and dynamic oceanographic structure. The influence of the gulf on the adjacent Pacific Ocean is small. At the entrance to the Gulf of California there are three kinds of surface water: CCW, which flows southward along the west coast of Baja California; ESW, of intermediate salinity, which flows into the area from the southeast (Costa Rica Current); and warm, highly saline GCW (Griffiths 1968). Beneath these three water masses, successively with depth, are SSW, characterized by a salinity maximum; PIW, characterized by a salinity minimum; and PBW, below the intermediate water, characterized by an increase in salinity to about 34.68 parts per thousand.

Torres-Orozco (1993) calculated the volume of these water masses in the gulf and their seasonal fluctuations. For that purpose, he defined the mouth of the gulf as a line connecting Punta Arenas, Baja California (23°28′N, 109°27′W), with Altata, Sinaloa (24°31′N, 107°43′W). Thus, Torres-Orozco did not consider most of the region of the entrance to the gulf. As an average, the PBW occupies 41% of the gulf's water volume, the PIW 33%, the SSW 19%, the GCW 6%, and the ESW 1% (fig. 3.3). The CCW was not detected in the volumetric analysis; its volume is too small north of the line between Punta Arenas and Altata. The lowest proportion of GCW and ESW occurs in winter–spring (6% and <1%, respectively), and that of SSW during fall

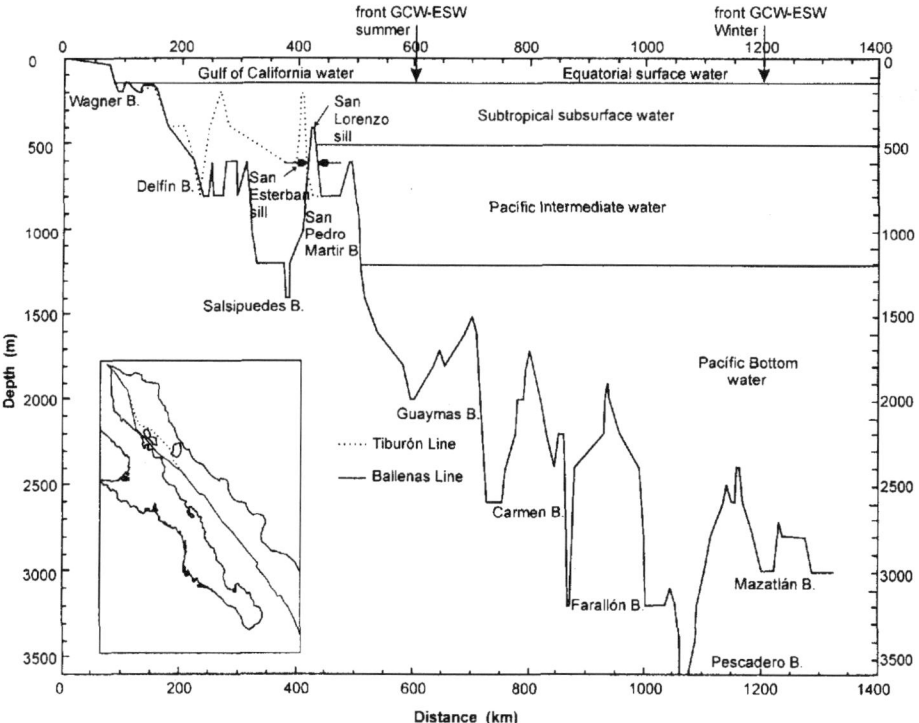

Figure 3.3 Distribution of water masses in the Gulf of California (adapted from Lavín et al. 1997).

(18%); the highest proportion of GCW and SSW occurs during summer (7% and 20%, respectively), and the highest for ESW during fall (2%).

The formation of fronts at the entrance to the gulf is a feature that has attracted the attention of those interested in commercial fisheries, such as tuna fisheries. Griffiths (1968) noted that the most important oceanographic feature of the entrance to the gulf is a strong front between the CCW and gulf surface waters. At Cape San Lucas this front is roughly straight, but to the south and west it becomes more sinuous and much weaker and is formed more and more by CCW and subtropical surface waters. According to Griffiths (1968), during spring of 1960, to the south of Cape San Lucas the CCW penetrated gulf surface waters at depths between 50 and 100 m, spreading horizontally or affecting in some way the entire gulf entrance, often in a complicated manner. Álvarez-Sánchez et al. (1978) detected CCW at the western part of the entrance to the gulf in March–April 1970, from the surface to 150 m, and as far into the gulf as 25°N (off San José island). These latter authors reported that CCW was associated with a current entering the gulf at the western part of the gulf's mouth, while at the same time GCW was associated with a current outflowing at the eastern side. Castro et al. (2000) carried out eight cruises to study the water masses present at the Punta Arenas–Altata line and reported that the CCW did not present a clear seasonal change, but it was present throughout the whole year with increasing volume from August to October–November.

The GCW is either ESW or SSW that has been transformed by evaporation (Roden and Groves 1959; Bray 1988b). Formation of GCW occurs during the whole year, in the upper gulf and different shallow regions, mainly off Sonora north of Tiburón, and even on the wide shelf off southern Sonora (Torres-Orozco 1993). Based on the salinity minimum and maximum that characterize the ESW and SSW, respectively, Álvarez-Borrego and Schwartzlose (1979) suggested that these two water masses are present only at the mouth of the gulf during winter and the beginning of spring. At the end of spring and during summer and fall they invade all the portion of the gulf south of Ángel de la Guarda. However, with Torres-Orozco's (1993) criteria for defining the water masses, it is now clear that the SSW invades the whole gulf, including most of the northern region. The ESW has a clear seasonal pulse penetrating the gulf up to the Guaymas Basin during summer and only near the entrance during winter (fig. 3.3), in accordance with seasonal fluctuations of the wind.

Álvarez-Borrego and Schwartzlose (1979) were able to confirm Sverdrup's (1941) hypothesis concerning the winter convective water movement at the head of the gulf. Cold and highly saline water sinks from the surface at the northernmost extreme of the gulf and moves southeastward along the bottom. North of Ángel de la Guarda, this water forms a distinct salinity maximum at 150–200 m, with values around 35.5 parts per thousand. Due to this convective movement, in the northern gulf vertical distributions of salinity and oxygen show minima, and those of nutrients show maxima, during winter and spring which are not present in summer and fall.

The upper gulf, near the Colorado River delta, is a shallow area with depths <30 m and has the greatest seasonal hydrographic changes. After construction of Glen Canyon Dam was completed in 1963, regular input of Colorado River water to the upper gulf was stopped. During 1979–87, water releases became necessary due to abnormal snowmelts in the upper basin of the river (Lavín and Sánchez 1999). Water releases also occurred in 1993 and 1998–99. In the upper gulf, the surface horizontal temperature gradient reverses at the beginning of spring and fall due to the annual cycle of atmospheric temperature. Surface temperature increases from the southeast to the northwest in summer, whereas the opposite trend occurs in winter. Minimum and maximum surface temperatures have been recorded west of Montague island: 8.25°C in December and 32.58°C in August. In the deeper waters of the southeast region, the seasonal range is 17°–30.75°C.

A monthly hydrographic sampling during 1972–73 (years of no water release from the Colorado River) showed that salinity in general maintained a surface gradient with values increasing northwestward. It ranged from a minimum of 35.28 in October to a maximum of more than 38.50 in July (Álvarez-Borrego et al. 1973). With data obtained between December 1993 and June 1996, this hydrographic behavior has been shown to persist at present essentially the same (Lavín et al. 1998) when there is no freshwater release from the Colorado River. Lavín and Sánchez (1999) used the opportunity provided by the Colorado River freshwater release of March–April 1993 to observe the effect of the positive estuary behavior on the upper gulf. In spite of the 1993 discharge being only about 25% of the mean discharge before 1935 (in m^3/s), and of very short duration, a clear hydrographic impact was recorded. In opposition to the normal inverse estuarine situation, salinity and density decreased toward the head. Surface salinity decreased from 35.4 near San Felipe to 32 at about 10 km south of Montague island. The surface temperature was normal for the time of the year.

Tides, Tidal Currents, and Tidal Mixing

The northern Gulf of California exhibits spectacular tidal phenomena. With an amplitude of >7 m during spring tides in the upper gulf and >4 m in the Midriff, tidal energy dissipation rates are great. Strong tidal currents in the channels between the midriff islands create a great turbulence that mixes the water column to great depths. This also has the effect of making the areas around the islands of the gulf a source of CO_2 to the atmosphere (Hidalgo-González et al. 1997).

Tides have different components or harmonics, and 20 or more constituents may be required to predict the tidal height accurately. In general, the most important components of tides are those with semidiurnal, diurnal, and fortnightly periods. Semidiurnal components are originated by the gravitational attraction of the moon and sun on the ocean water. The greatest effect is the one corresponding to the moon's attraction. This is known as the principal lunar semidiurnal component (M_2), with a period of 12 h 25 min (the 25 min are due to the moon's movement around the earth). The attraction of the sun (which is not as strong as that of the moon because of distance) also creates a semidiurnal component (S_2) with a period of 12 h. Diurnal components are originated by the declination of the moon's orbit with respect to the equator (the moon is displaced to the north and south of the equator in 27.2 days). This declination introduces an asymmetry of the water protuberance (principal lunar diurnal constituent, O_1). The small period difference between the M_2 and S_2 components produces the fortnightly variation of tides, the one caused by the moon's phases. When the sun, moon, and earth are on the same line (full and new moon) the combined gravitational forces produce maximum tides (spring tides). When they are in quadrature, the forces of the sun and moon oppose each other, and tidal amplitudes are minimum (neap tides).

Tides in the gulf are produced by co-oscillation with the adjacent Pacific. They are due mainly to variations of sea level at the entrance, not to gravitational attraction of the moon and sun on the gulf's waters (Ripa and Velázquez 1993). In the gulf the tidal wave is progressive. The time of high or low water is progressively later traveling north up the gulf. The time difference between the entrance and the vicinity of the Colorado River is approximately 5.5 h for high water and 6 h for low water. The result is that low water at one end of the gulf occurs at about the same time as high water at the other end (Roden 1964).

The difference between semidiurnal and diurnal tides is striking. Semidiurnal components dominate at the entrance and at the head of the gulf. The semidiurnal tide (M_2) enters the gulf with moderate amplitude of 30 cm. The speed of the wave decreases within the gulf, and the amplitude decreases to one third of its initial value near the middle of the gulf. It then accelerates, and its amplitude increases at the head to 165 cm. The semidiurnal components have a virtual amphydromic point near Santa Rosalia, in the central gulf, and diurnal components dominate there (Marinone and Lavín 1997). In contrast, the amplitude of the diurnal tide increases slowly and monotonically to about twice its amplitude at the mouth.

Stock (1976) computed tidal currents for different points in the gulf and found that the current ellipses for the M_2 constituent have the major axis increasing from the mouth (3 cm/s) to the head (60 cm/s; fig. 3.4). Total tidal currents in the narrows between the islands and the coast and in the passages connecting semi-enclosed lagoons with the gulf are strong. The speed of these currents is variable and depends on the stage of the moon and the prevailing winds, but they have been reported up to 3 m/s for Ballenas Channel (6 knots) (Roden 1964; Álvarez-Sanchez et al. 1984). Ballenas Channel is isolated from the central part of the gulf by a submarine ridge. The sill depth of this ridge is approximately 450 m (Rusnak et al. 1964), so that exchange between the northern and southern parts of the gulf is limited to relatively shallow waters. Maximum depth of Ballenas Channel is 1600 m. The water in the basin comes from mixing between the surface and sill depths. The lowest surface temperatures (Robinson 1973; Soto-Mardones et al. 1999) and the highest surface nutrient and CO_2 concentrations (Álvarez-Borrego et al. 1978; Hidalgo-González et al. 1997) in the entire gulf are persistently found here. However, in spite of strong stratification during summer, turbulence at the sill carries >30% of the phytoplankton below the euphotic zone, down to >200 m, decreasing photosynthesis (Cortés-Lara et al. 1999).

Turbulence created by these strong tidal currents acts as a physical pump that carries CO_2 from deep waters to the surface and the atmosphere in the Midriff. This is a mechanism opposite to the biological pump that carries CO_2 from surface to deep waters (photosynthesis consumes CO_2 at the surface, and then biogenic particles sink to deep waters). This region of the gulf acts as a continuous source of CO_2 to the atmosphere, with greatest CO_2 fluxes with spring tides. Largest surface partial pressure of CO_2 (pCO_2) value calculated for July 1990 in this area was >400 µatm (~39.2 Pa), compared to an atmospheric pCO_2 of 350 µatm (~34.3 Pa; corresponding largest CO_2 flux was 23 mM/m^2·d^{-1}). With a much less stratified water column during winter, larger CO_2 fluxes to the atmosphere than those of summer are to be expected (Hidalgo-González et al. 1997). Events of strong upwelling at the eastern coast, during

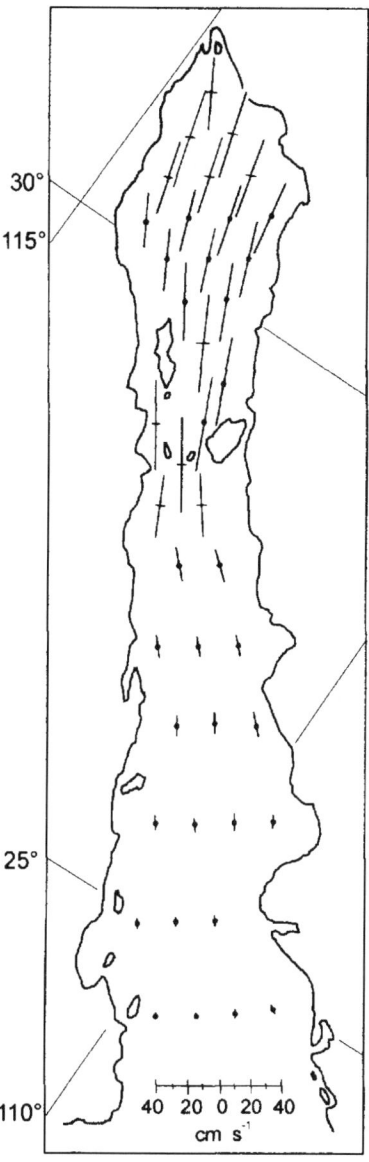

Figure 3.4 Model current ellipses of the principal lunar semidiurnal tidal component (M_2) (after Stock 1976).

winter should also create a temporary CO_2 flux from the water to the atmosphere. In spite of strong tidal currents north from 30°N, during winter there are no high CO_2 values in the surface waters because the water column is relatively shallow. CO_2 behaves as a nutrient; it increases with depth. In the northern gulf there is not enough depth for waters to have high CO_2 concentrations to mix with surface waters. In this region, pCO_2 of surface waters is practically in equilibrium with that of the atmosphere (Montes-Hugo et al. 1998).

Circulation

Circulation is the system of currents that do not contain the tidal currents; that is, circulation is formed by the residual currents. Tidal currents produce a forward and backward motion. Circulation is of interest because it is responsible for transporting materials (e.g., pollutants, eggs, and larvae) from one place to another. Many of the characteristics of the Gulf of California circulation mentioned in the literature have not been observed directly through measurements, or they are based on very weak evidence. In particular, it is important to make clear that winds are not the main generator of currents in the gulf at the seasonal scale (Lavín et al. 1997).

Roden (1958) used 1947 charts of ship-drifting prepared by the U.S. Hydrographic Office. He concluded that in winter southward currents north of Cape Corrientes characterize the surface circulation of the gulf. During summer a current flows northward along the coast of Mexico and enters the gulf at the eastern and central regions of the mouth, with a southward flow near Baja California. This is in agreement with the results obtained by Granados-Gallegos and Schwartzlose (1974), who used drift bottles. Lepley et al. (1975) used satellite photographs (color, infrared, and multispectral imagery) to detect what they interpreted as a surface counterclockwise (cyclonic) gyre during summer in the northern gulf, reversing during winter to form an anticyclonic gyre.

A large fraction of the seasonal circulation in the gulf results from the influence of the Pacific Ocean through the mouth. Ripa (1990, 1997) proposed the hypothesis that this influence is in the form of an internal wave of annual period. Beier (1997) developed a linear, bidimensional, baroclinic numerical model with two layers and bathymetry. This model is forced with the same seasonal signals used by Ripa (1997): the effect of Pacific Ocean (PO), the wind (W), and the air–water heat flux (Q). The effect of PO is the main component of the seasonal signal of stored heat (Castro et al. 1994; Ripa 1997). It is simulated with a baroclinic (velocity changes with depth, contrary to barotropic where velocity does not change with depth) perturbation trapped to the first 30 km adjacent to the eastern coast. The structure of an internal Kelvin wave is specified at the entrance to the gulf. The wave is then distorted by the topography as it propagates to the interior at the continental coast, up to the head. Then it returns southward, trapped to the peninsula's coast, and exits the gulf with a little lower amplitude due to loss of energy by friction. The effects of PO and W are in phase; both produce the same kind of circulation during a certain season. In Beier's (1997) model, the effect of Q on the circulation is smaller than the effects of PO and W.

The results of Beier's (1997) model for the surface layer (0–70 m) show an anticyclonic circulation during winter (fig. 3.5a) and cyclonic circulation during summer (fig. 3.5b). However, there are continuous changes throughout the year. Circulation is not always clearly cyclonic or anticyclonic. In general, the results shown in figure 3.6 are in agreement with observations reported by previous authors. One striking result is the gyre that forms in the northern gulf. This gyre is in agreement with the one postulated by Lepley et al. (1975) and with the one described by Lavín et al. (1997) using direct observations with ARGOS drifters. Lavín et al. (1997) described the presence of a cyclonic gyre during summer, and an anticyclonic gyre during winter, in the northern gulf (fig. 3.6). Both gyres had mean speeds of approximately 30 cm/s near the edge. However, the character of the gyre is different in the two seasons. In

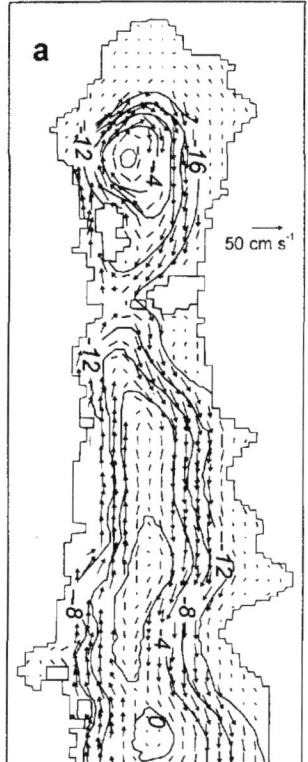 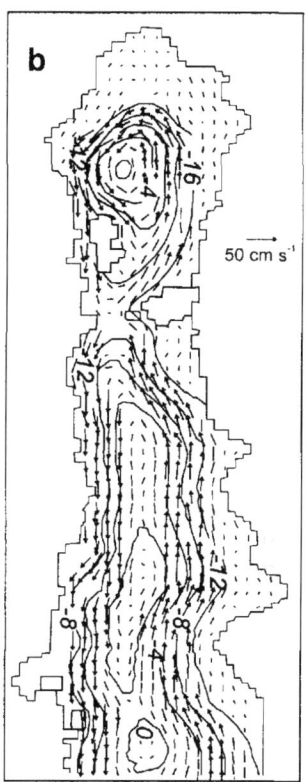

Figure 3.5 Contours of sea level (cm) and circulation of the Gulf of California (a) in February and (b) in August (after Lavín et al. 1997).

summer, the gyre is persistent and clearly baroclinic. In winter, the situation is less clear: The gyre appears to be a mixture of baroclinic and barotropic and to be highly variable from year to year. This kind of circulation suggests that neutrally buoyant substances and passive organisms may get trapped for extended periods in the northern gulf.

There are evidences of other gyres in the central and southern gulf not predicted by Beier's (1997) model, but in agreement with the circulation predicted for areas near the two coasts. Emilsson and Alatorre (1997) also used drifters to describe a baroclinic cyclonic gyre at the entrance to the gulf in August 1978, with maximum speeds of >55 cm/s at its western edge. The center of this gyre was in a line between La Paz Bay and Topolobampo. Badan-Dangon et al. (1985) used March–April infrared satellite imagery to describe the evolution of an upwelling plume, which originated north of Guaymas, and eventually reached most of the way across the gulf. The movement of this plume was associated with a basinwide, anticyclonic circulation. Santamaría-Del-Angel et al. (1994b) used satellite-derived photosynthetic pigment concentration data to describe the temporal and spatial variability within the gulf. Their results show a clear seasonal pattern. Pigment maxima occurred on both sides of the

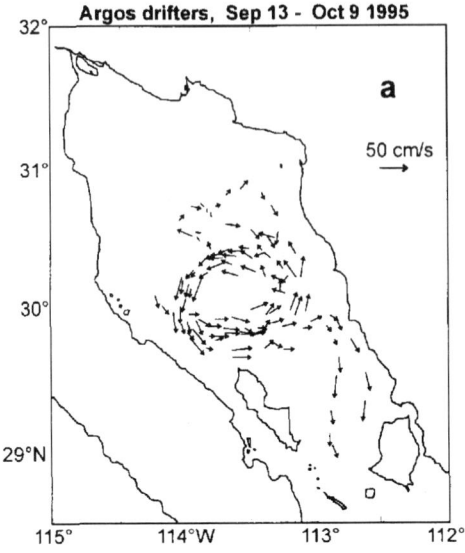

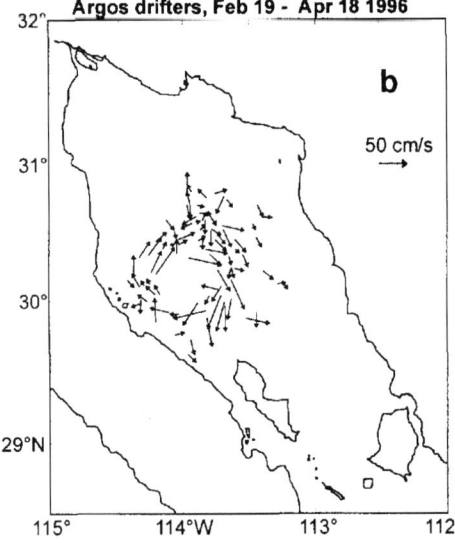

Figure 3.6 Surface velocity fields for (a) summer and (b) winter (after Lavín et al. 1997).

gulf during winter, a fact that they attributed to surface circulation that brings upwelled water from the eastern to the western coast.

Interannual and Interdecadal Changes of Hydrographic Conditions

Álvarez-Borrego and Schwartzlose (1979) noticed that during 1957, an El Niño year, the ESW and SSW showed a much stronger invasion into the gulf than during 1974, a non–El Niño year. Baumgartner and Christensen (1985) used time series of atmospheric and oceanic indices from 1950 through 1974, and concluded that interannual ocean climate in the Gulf of California is related only to the equatorial mode of variability commonly known as the El Niño-Southern Oscillation (ENSO). They found no discernible relationship between the gulf and independent modes of variability in the North Pacific gyre. Interannual variability of the gulf is associated with the cyclonic north equatorial circulation composed of the north equatorial countercurrent, the north equatorial current, and the Costa Rica current.

Torres-Orozco (1993) indicated that in the period 1939–1986, the greatest ESW volume was estimated for March 1983 due to the 1982–83 ENSO event, while the ESW volume for March 1984 was lower than the long-term spring average. During March 1983, the ESW invaded the gulf up to the Guaymas Basin, while in March 1984 it only reached the Cape San Lucas region (Lavín et al. 1997). During an El Niño event the warmer ESW causes an increase in sea level. Tidal records show an El Niño event with a clear positive anomaly of mean sea level. Baumgartner and Christensen (1985) concluded that the principal source of interannual variability in the sea level of the gulf is the changing intensity of the equatorial circulation associated with the El Niño phenomenon.

In situ, sedimentological, and satellite data from the Gulf of California indicate that phytoplankton communities here respond to El Niño events differently from the communities off Peru and the Pacific coast of Baja California. Based on satellite-derived photosynthetic pigment data, Santamaría-Del-Angel et al. (1994b) concluded that El Niño events have different effects on distinct regions of the gulf. Places with low kinetic energy, such as the southern gulf, have a strong suppression of phytoplankton biomass. Places with high kinetic energy, such as the upper gulf or Ballenas Channel, have no effect or even show an increase of primary production due to reduced mixing. Tershy et al. (1991) reported the greatest numbers of cetaceans and seabirds in Ballenas Channel during 1983, with a decline during 1984 and 1985. At the same time, pelicans had zero reproduction at La Paz Bay in 1983 (Jiménez-Castro 1989). Green-Ruiz and Hinojosa-Corona (1997) studied the spawning area of anchovy (*Engraulis mordax*) in the gulf during 1990–1994. They found that in February 1992, an El Niño year, anchovies spawned only in the big islands region, and the highest abundance of eggs coincided with the lowest water temperatures (17°–18°C). This was 2°–3°C higher than the lowest temperatures observed there in 1990–1991.

The stronger winter invasion of the ESW during an El Niño event changes the phytoplankton community structure, and this may have a profound effect in the food web. Baumgartner et al. (1985) studied the phytoplankton size fraction >24 µm preserved in the laminated sediments of Guaymas Basin, composed of the heavily silicified diatoms and silicoflagellates. A 20-year record, from 1953 through 1972, of year-to-year variability of these siliceous phytoplankton assemblages shows a strong correlation with interannual sea-level anomalies. El Niño periods are generally marked by increases in preserved abundances of the total siliceous assemblage, and in particular by greater numbers of individuals within species whose distribution is limited to tropical and subtropical waters.

Based on their high numbers of adult cetaceans and seabirds in Ballenas Channel during 1983, Tershy et al. (1991) suggested that this area may serve as a refugium of high productivity and prey abundance for these highly mobile marine animals during

El Niño years. According to Velarde and Ezcurra (chap. 11, this volume), the results of 21 years of systematic censuses of birds on Isla Rasa, situated in the Midriff region, show that Heermann's Gulls (*Larus heermanni*) have had small fluctuations around 250,000 individuals. Even during El Niño years (e.g., 1998) most adult individuals of this and other seabird species were present in their nesting territories. However, Velarde and Ezcurra report that during winter–spring of El Niño years 1992 and 1998, the reproductive success of Heermann's Gulls, measured both as nesting success and breeding success, was at its lowest. During El Niño events, female gulls lose weight, and the survival of chicks drops almost to zero. Satellite-derived ocean-color data show that phytoplankton biomass, estimated as photosynthetic pigment concentration, was not significantly lower during winter–spring of 1998 than those of non-El Niño years in the Midriff region (not illustrated). Thus, normal levels of phytoplankton biomass around Isla Rasa do not guarantee a structure of the food web to support successful reproduction of seabirds. Also, Heermann's Gulls start arriving to the Midriff in mid-February, and by late March a couple of hundred thousand have reached the area, possibly with an anomalous low weight already. More data are needed to understand the geographic variation of the biological impacts of El Niño events in the gulf.

Temporal changes in relative abundance of pelagic fishes have been characterized by reading the paleo-ecological records in the laminated sediments of Guaymas Basin. Dramatic interdecadal changes of fish biomass are due to unknown physical variations, and this affects organisms around and on the islands. Holmgren-Urba and Baumgartner (1993) reconstructed time series of scale-deposition rates for Pacific sardine, northern anchovy, Pacific mackerel, Pacific hake, and an undifferentiated group of myctophids. The time series are resolved into 10-year sample blocks and extend from approximately 1730 to nearly 1980. This reconstruction shows a strongly negative association between the presence of sardines and anchovies, with anchovies dominating throughout the nineteenth century, and with only two important peaks of sardine scale deposition—one in the twentieth century and one at the end of the eighteenth century. Both the mackerel and the myctophids group vary more like sardines than like anchovies (with the hake intermediate between the two). This suggests an overall coherent pattern in changing ecosystem structure that operates over a period of about 120–140 years. According to Holmgren-Urba and Baumgartner, the relation to climate is not clear yet, but altogether, the information suggests that climate does mediate population sizes but that this process is still subject to strong filtering through biological interactions among species.

References

Álvarez-Borrego, S., L.A. Galindo-Bect, and B.P. Flores-Baez. 1973. Hidrología. In: *Estudio Químico sobre la Contaminación por Insecticidas en la Desembocadura del Río Colorado*, Tomo I, Reporte a la Dirección de Acuacultura de la Secretaría de Recursos Hidráulicos, Univ. Aut. Baja California, Ensenada, pp. 6–177.

Álvarez-Borrego, S., and J.R. Lara-Lara. 1991. The physical environment and primary productivity of the Gulf of California. In: J.P. Dauphin and B.R. Simoneit (eds), *The Gulf of California and Peninsular Province of the Californias*. American Association of Petroleum Geologists Memoir 47; pp. 555–567.

Álvarez-Borrego, S., J.A. Rivera, G. Gaxiola-Castro, M.J. Acosta-Ruiz, and R.A. Schwartzlose. 1978. Nutrientes en el Golfo de California. *Ciencias Marinas* **5**:21–36.

Álvarez-Borrego, S., and R.A. Schwartzlose. 1979. Water masses of the Gulf of California. *Cien. Mar.* **6**:43–63.

Álvarez-Sánchez, L.G., A. Badan-Dangon, and J.M. Robles. 1984. Lagrangian observations of near-surface currents in Canal de Ballenas. *CalCOFI Rep.* **25**:35–42.

Álvarez-Sánchez, L.G., M.R. Stevenson, and B. Wyatt. 1978. Circulación y masas de agua en la región de la boca del Golfo de California en la primavera de 1970. *Cien. Mar.* **5**:57–69.

Argote, M.L., A. Amador, M.F. Lavín, and J.R. Hunter. 1995. Tidal dissipation and stratification in the Gulf of California. *J. Geophys. Res.* **100**:16103–16118.

Badan-Dangon, A., C.E. Dorman, M.A. Merrifield, and C.D. Winant. 1991. The lower atmosphere over the Gulf of California. *J. Geophys. Res.* **96**:16877–16896.

Badan-Dangon, A., C.J. Koblinsky, and T. Baumgartner. 1985. Surface thermal patterns in the Gulf of California. *Oceanologica Acta* **8**:13–22.

Baumgartner, T.R., and N. Christensen Jr. 1985. Coupling of the Gulf of California to large-scale interannual climatic variability. *J. Marine Res.* **43**:825–848.

Baumgartner, T.R., V. Ferreira-Bartrina, H. Schrader, and A. Soutar. 1985. A 20-year varve of siliceous phytoplankton variability in the central Gulf of California. *Mar Geol.* **64**: 113–129.

Beier, E. 1997. A numerical investigation of the annual variability in the Gulf of California. *J. Phys. Oceanogr.* 27:615–632.

Bray, N.A. 1988a. Thermohaline circulation in the Gulf of California. *J. Geophys. Res.* **93**: 4993–5020.

Bray, N.A. 1988b. Water mass formation in the Gulf of California. *J. Geophys. Res.* **93**:9223–9240.

Bray, N.A., and J.M. Robles. 1991. Physical oceanography of the Gulf of California. In: Dauphin, J.P., and B.R. Simoneit (eds), *The Gulf of California and Peninsular Province of the Californias*. American Association Petroleum Geologist Memoir 47; pp. 511–553.

Castro R., M.F. Lavín, and P. Ripa. 1994. Seasonal heat balance in the Gulf of California. *J. Geophys. Res.* **99**:3249–3261.

Castro, R., A.S. Mascarenhas, R. Durazo, and C. Collins. 2000. Variación estacional de la temperatura y salinidad en la entrada del Golfo de California, México. *Cien. Mar.* **26**: 561–583.

Cortés-Lara, M.C., S. Álvarez-Borrego, and A.D. Giles-Guzmán. 1999. Efecto de la mezcla vertical sobre la distribución de nutrientes y fitoplancton en dos regiones del Golfo de California, en verano. *Rev. Soc. Mex. Hist. Nat.* **49**:193–206.

Dauphin, J.P., and G.E. Ness. 1991. Bathymetry of the Gulf and Peninsular Province of the Californias. In: J.P. Dauphin and B.R. Simoneit (eds), *The Gulf of California and Peninsular Province of the Californias*. American Association Petroleum Geologists Memoir 47; pp. 21–23.

Emilsson, I., and M.A. Alatorre. 1997. Evidencias de un remolino ciclónico de mesoescala en la parte sur del Golfo de California. In: M.F. Lavín (ed), *Contribuciones a la Oceanografía Física en México*. Unión Geofís. Mex., Monografía 3; pp. 173–182.

Gilbert, J.Y., and W.E. Allen. 1943. The phytoplankton of the Gulf of California obtained by the "E.W. Scripps" in 1939 and 1940. *J. Marine Res.* **5**:89–110.

Granados-Gallegos, J.L., and R.A. Schwartzlose. 1974. Corrientes superficiales en el Golfo de California. In: F.A. Manrrique (ed), *Memorias del V Congreso Nacional de Oceanografía*. Escuela de Ciencias Marítimas del Instituto Tecnológico de Monterrey, Guaymas, pp. 271–285.

Green-Ruiz, Y.A., and A. Hinojosa-Corona. 1997. Study of the spawning area of the Northern anchovy in the Gulf of California from 1990 to 1994, using satellite images of sea surface temperatures. *J. Plankton Res.* **19**:957–968.

Griffiths, R.C. 1968. Physical, chemical, and biological oceanography of the entrance to the Gulf of California, spring of 1960. *U.S. Fish Wildlife Service Special Scientific Report Fisheries*, no. 573.

Harris, M.F. 1969. *Effects of tropical cyclones upon Southern California.* Unpublished Ph.D. thesis, San Fernando Valley State College, San Fernando, CA.

Hidalgo-González, R.M, S. Álvarez-Borrego, and A. Zirino. 1997. Mixing in the region of the Midriff islands of the Gulf of California: Effect on surface pCO2. *Cien. Mar.* **23**:317–327.

Holmgren-Urba, D., and T.R. Baumgartner. 1993. A 250-year history of pelagic fish abundance from the anaerobic sediments of the central Gulf of California. *CalCOFI Rep.* **34**:60–68.

Jiménez-Castro, C.M. 1989. Hábitos alimenticios, requerimiento energético y consumo alimenticio del pelícano café (Pelecanus occidentalis) en la Bahía de La Paz, B.C.S. Thesis, Univ. Aut. Baja California Sur, La Paz.

Lavín, M.F., E. Beier, and A. Badan. 1997. Estructura hidrográfica y circulación del Golfo de California: escalas estacional e interanual. In: M.F. Lavín (ed), *Contribuciones a la Oceanografía Física en México*, Unión Geofís. Mex., Monografía 3, pp. 141–171.

Lavín, M.F., R. Durazo, E. Palacios, M.L. Argote, and L. Carrillo. 1997. Lagrangian observations of the circulation in the northern Gulf of California. *J. Physical Oceanogr.* **27**:2298–2305.

Lavín, M.F., V.M. Godínez, and L.G. Álvarez. 1998. Inverse-estuarine features of the Upper Gulf of California. *Est. Coastal Shelf Sci.* **47**:769–795.

Lavín, M.F., and S. Sánchez. 1999. On how the Colorado River affected the hydrography of the Upper Gulf of California. *Cont. Shelf Res.* **19**:1545–1560.

Lepley, L.K., S.P. Vonder-Haar, J.R. Hendrickson, and G. Calderon-Riveroll. 1975. Circulation in the northern Gulf of California from orbital photographs and ship investigations. *Cien. Mar.* **2**:86–93.

Maluf, L.Y. 1983. Physical oceanography. In: Case, T.J., and M.L. Cody (eds), *Island Biogeography in the Sea of Cortez*, University of California Press, Berkeley; pp. 26–45.

Mann, K.H., and J.R.N. Lazier. 1996. *Dynamics of Marine Ecosystems. Biological-Physical Interactions in the Oceans.* Blackwell Science, Oxford.

Marinone, S.G., and M.F. Lavín. 1997. Mareas y corrientes residuales en el Golfo de California. In: M.F. Lavín (ed), *Contribuciones a la Oceanografía Física en México.* Unión Geofis. Mex., Monografía 3; pp. 113–139.

Montes-Hugo, M.A., S. Álvarez-Borrego, and A. Zirino. 1998. The winter air-water CO_2 net flux is not significant in the Gulf of California to the north of 30°N. *Cien. Mar.* **24**: 483–490.

Mosiño, P., and E. García. 1974. The climate of Mexico. In: Bryson, R.A., and F.K. Hare (eds), *Climates of North America.* Elsevier, New York; pp. 345–404.

Pond, S., and G.L. Pickard. 1983. *Introductory Dynamical Oceanography.* Pergamon Press, New York.

Reyes, S., and D.L. Cadet. 1986. Atmospheric water vapor and surface flow patterns over the tropical Americas during May-August 1979. *Mon. Weather Rev.* **114**:582–593.

Ripa, P. 1990. Seasonal circulation in the Gulf of California. *Ann. Geophys.* **8**:559–564.

Ripa, P. 1997. Towards a physical explanation of the seasonal dynamics and thermodynamics of the Gulf of California. *J. Physical Oceanogr.* **27**:597–614.

Ripa, P., and S.G. Marinone. 1989. Seasonal variability of temperature, salinity, velocity, vorticity and sea level in the central Gulf of California, as inferred from historical data. *Quat. J. R. Meteorol. Soc.* **115**:887–913.

Ripa, P., and G. Velázquez. 1993. Modelo unidimensional de la marea en el Golfo de California. *Geofís. Intern.* **32**:41–56.

Robinson, M.K. 1973. *Atlas of monthly mean sea surface and subsurface temperatures in the Gulf of California, Mexico.* San Diego Society of Natural History Memoir 5.

Roden, G.I. 1958. Oceanographical and meteorological aspects of the Gulf of California. *Pacific Sci.* **12**:21–45.

Roden, G.I. 1964. Oceanographic aspects of the Gulf of California. In: Van Andel, Tj. H, and G.G. Shor, Jr. (eds), *Marine Geology of the Gulf of California: A Symposium*, American Association of Petroleum Geologists Memoir 3; pp. 30–58.

Roden, G.I., and G.W. Groves. 1959. Recent oceanographic investigations in the Gulf of California. *J. Mar. Res.* **18**:10–35.

Rusnak, G.A., R.L. Fisher, and F.P. Shepard. 1964. Bathymetry and faults of the Gulf of California. In: Van Andel, Tj. H., and G.G. Shor, Jr. (eds), *Marine Geology of the Gulf of California: A Symposium*, American Association of Petroleum Geologists Memoir 3; pp. 59–75.

Santamaría-Del-Angel, E., S. Álvarez-Borrego, R. Millán-Nuñez, and F.E. Muller-Karger. 1999. Sobre el efecto de las surgencias de verano en la biomasa fitoplanctónica del Golfo de California. *Rev. Soc. Mex. Hist. Nat.* **49**:207–212.

Santamaría-Del-Angel, E., S. Álvarez-Borrego, and F.E. Muller-Karger. 1994a. Gulf of California biogeographic regions based on coastal zone color scanner imagery. *J. Geophys. Res.* **99**:7411–7421.

Santamaría-Del-Angel, E., S. Álvarez-Borrego, and F.E. Muller-Karger. 1994b. The 1982–1984 El Niño in the Gulf of California as seen in coastal zone color scanner imagery. *J. Geophys. Res.* **99**:7423–7431.

Serra, C.S. 1971. Hurricanes and tropical storms of the west coast of Mexico. *Mon. Weather Rev.* **99**:302–308.

Shepard, F.P. 1950. Submarine topography of the Gulf of California (P. 3 of the 1940 *E.W. Scripps* cruise to the Gulf of California). Geological Society of America Memoir 43.

Simpson, J.H., A.J. Souza, and M.F. Lavín. 1994. Tidal mixing in the Gulf of California. In: Beven, K.J., P.C. Chatwin, and J.H. Millbank (eds), *Mixing and Transport in the Environment*. John Wiley & Sons, London; pp. 169–182.

Stock, G. 1976. Modelling of tides and tidal dissipation in the Gulf of California. Ph.D. thesis, University of California, San Diego.

Sverdrup, H.U. 1941. The Gulf of California: preliminary discussion on the cruise of the E.W. Scripps in February and March 1939. Sixth Pacific Sci. Congr. Proc. **3**:161–166.

Tershy, B.R., D. Breese, and S. Álvarez-Borrego. 1991. Increase in cetacean and seabird numbers in the Canal de Ballenas during an El Niño-Southern Oscillation event. *Mar. Ecol. Prog. Ser.* **69**:299–302.

Torres-Orozco, E. 1993. Análisis volumétrico de las masas de agua del Golfo de California. M.Sc. thesis, CICESE, Ensenada, Mexico.

PART II
THE BIOLOGICAL SCENE

4

Plants

MARTIN CODY
REID MORAN
JON REBMAN
HENRY THOMPSON

   This chapter deals with the general features of plant diversity and distribution on the Sea of Cortés islands and to a lesser extent with adaptive features of plants such as morphology and phenology. This review is based mainly on the plant lists in appendixes 4.1–4.5 describing the island floras, endemics, and relicts. In our interpretations of these lists we draw from various floristic, systematic, and distributional works that pertain to the Gulf of California region and beyond. After an introductory section, we work from broader biogeographical questions to matters that have more local, specific, or taxonomically restricted perspectives.

A Visitor's View

There must be few experiences in the biologist's world to compare with approaching a "new" island by boat. The ingredients are adventure and suspense, mystery and perhaps even a little danger. There are feelings of discoveries to be made, knowledge to be extended, curiosity to be both piqued and satisfied. Such feelings are shared not only among natural historians but by any adventurous and curious traveler; the more difficult the island is to reach, the keener the excitement of the visit.
   The islands in the Sea of Cortés would seem ideally qualified to generate this sort of bioadventurism. They are mostly uninhabited and have been little explored biologically, and the mounting of small-scale expeditions to successfully reach (and leave) the islands is not always a trivial matter. First appearing as blurry, near-colorless breaks between the unrelenting blue of gulf and sky, the islands leave early impressions of abrupt topography and a seeming lack of vegetation. The dull-green

smudge of plant life on foothill outwash fans and in the arroyos eventually becomes apparent, but almost up to the point of a landing the islands preserve the impression of rock masses broken only occasionally by bajadas on which a few cardons (*Pachycereus pringlei*) are conspicuous (see fig. 16.8).

Islands larger than a few square kilometers have well-developed drainage courses that reach the coasts as dry arroyos and provide breaks in the generally steep coastal cliffs (see fig. 16.7). The beaches where the larger arroyos reach the coast are logical landing points, and here the visitor gets the first close look at the vegetation. By and large, its constituents are the same beach shrubs (such as the chenopods *Allenrolfea, Salicornia, Suaeda*) as on the Baja California peninsula, and the same leguminous trees (mesquites *Prosopis*, ironwood *Olneya*, paloverdes *Cercidium*) and columnar cacti (cardon; senita *Lophocereus*; pitaya agria *Stenocereus* [*Machaerocereus*]) in the arroyos and on the flatter alluvial plains. Here early impressions of desolation are dispelled; although the steeper, rocky slopes are indeed sparsely vegetated and unappealing, the vicinity of the arroyos has the high diversity of plant species and growth forms and the almost gardenlike order typical of the Sonoran Desert in general. The arroyos and bajadas on the larger islands appear similar to those on the peninsula. In general, the relative luxuriance and diversity of the vegetation increase with the size of the arroyo and its drainage basin, and so vary both between peninsula and islands and within and among islands.

Smaller islands lack the surface area prerequisite for the formation of arroyos and alluvium; these islands support only the extremely sparse, low, and open vegetation of rocky slopes with minimal soil and lack much of the vegetative and floristic diversity of the larger islands.

In the southern gulf, the larger islands have the distinctly brushier aspect characteristic of the southern gulf coastal area of the peninsula, which receives higher total and more summer rainfall than the north. In addition, the southern islands are in warmer water and sheltered bays and lagoons are fringed with red mangroves (*Rhizophora mangle*) and their associates. Although these shallow, silty lagoons make for easier landings than the rocky coves to the north, one contends instead with the coral heads (especially threatening to rubber boats), the infuriating sand flies (Phlebotomidae), and distinctly higher humidity and temperature that favor none but the lightest of sleeping bags.

The Physical Environment

As might be expected of islands located mostly in the coastal parts of a gulf averaging less than 100 miles wide, the physical environment is not much different from that of the nearby or adjacent mainland or peninsular localities. This is particularly true because the climate on the two sides of the gulf is essentially the same, especially in its harsher aspects of low overall precipitation and extremely hot summers. By interpolation, the islands between these coasts cannot diverge much in these same aspects. Any differences in climate of the islands must result from their small sizes and low elevations and to the buffering effects of the surrounding water, particularly in respect to temperature maxima and minima.

Unfortunately, almost no climatic records are available from the islands, and our knowledge of climate in the gulf region comes largely from Hastings and associates (Hastings and Humphrey, 1969a,b; Hastings et al., 1972) for the areas around the gulf. From these sources we have compiled charts of yearly precipitation and average temperature (figs. 4.1–4.3), exercising some license in the extrapolation of isotherms and isohyets across the gulf. The large northern islands of Ángel de la Guarda and Tiburón appear similar in climate to the nearest mainland (including the peninsula),

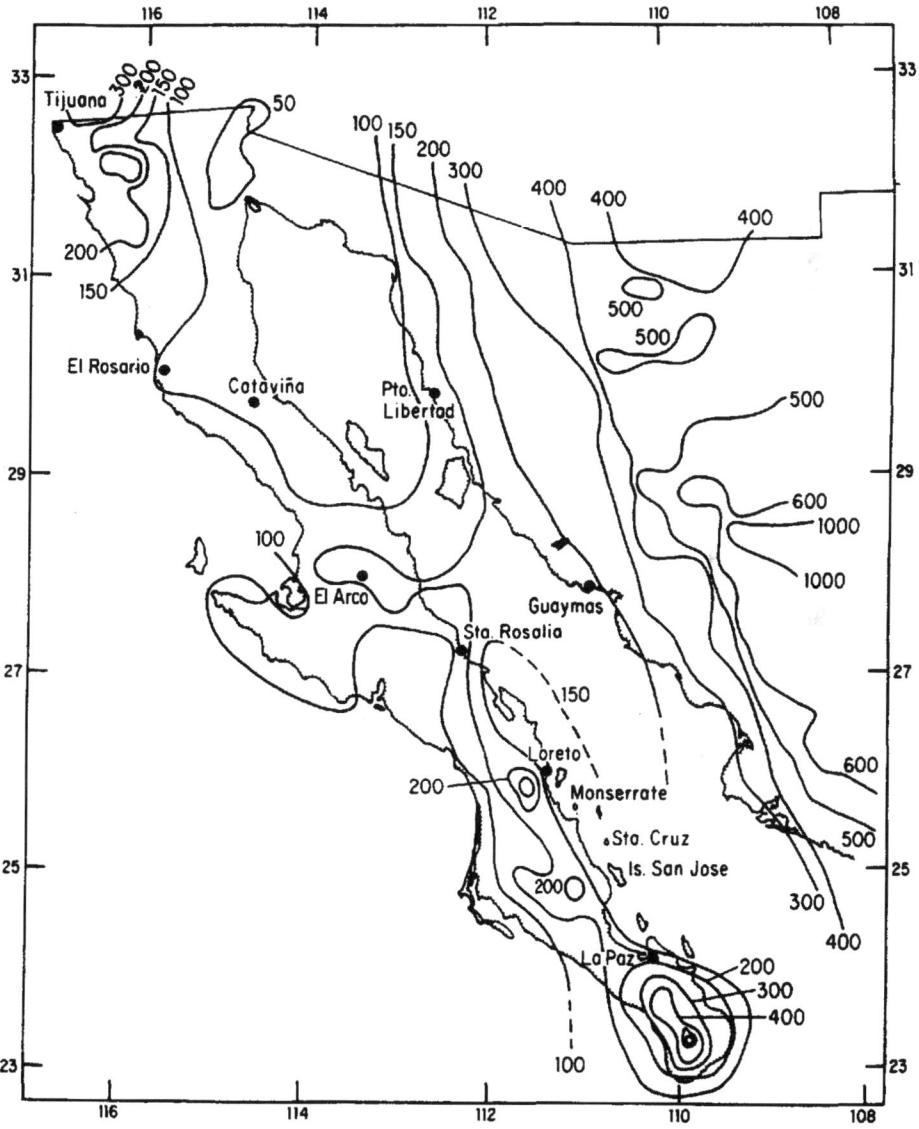

Figure 4.1 Isohyets of annual precipitation (mm), around the Gulf of California.

66 The Biological Scene

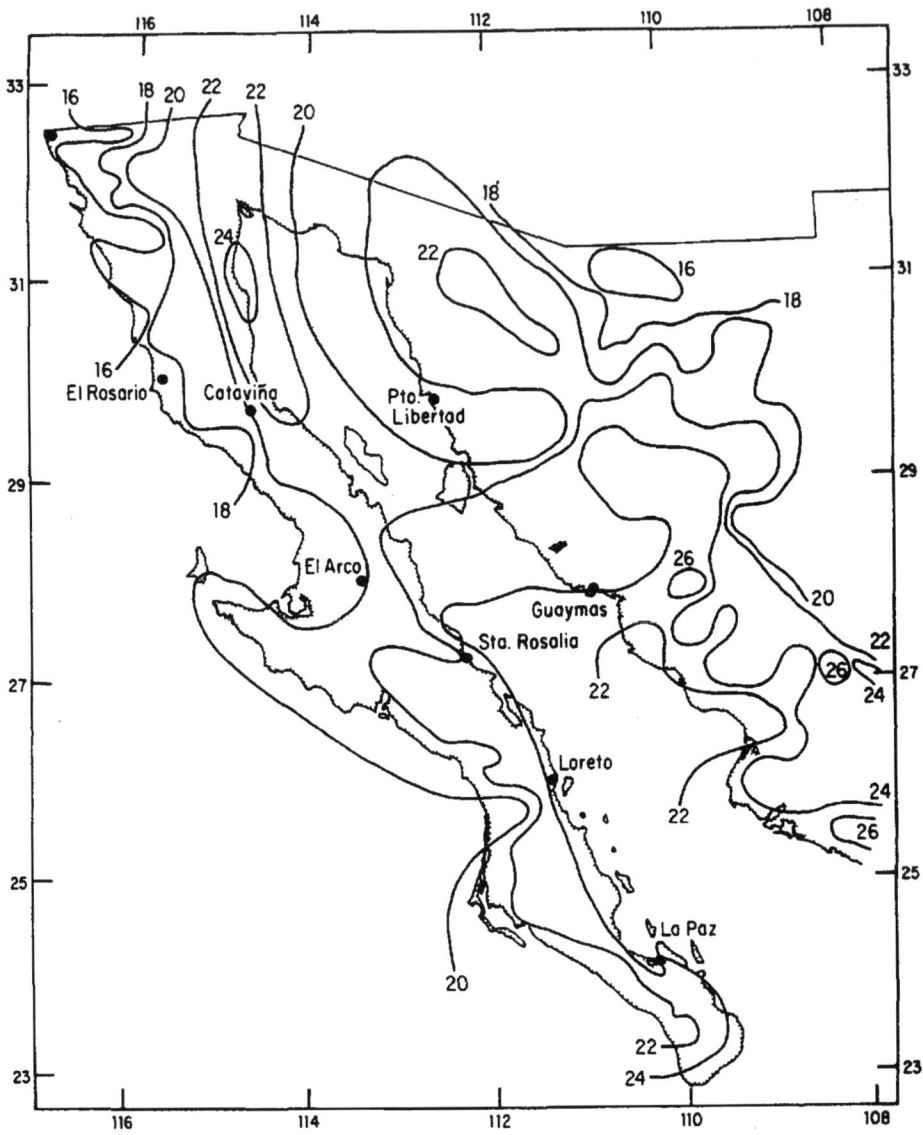

Figure 4.2 Isotherms of mean annual temperature (°C), around the Gulf of California.

and the same can be said for the other midriff islands, which all lie between the same isohyets as the peninsular coast (El Barril) and the Sonoran coast at Tiburón and north to nearly 31°N.

The southern islands, from Carmen south to Cerralvo, lie off the peninsula where isohyets parallel the coast and increase inland. This implies that these islands receive rather less rainfall than the adjacent peninsula, but because the opposite gulf coast in southern Sonora and northern Sinaloa receives appreciably higher rainfall, the precipitation on these islands cannot be accurately estimated.

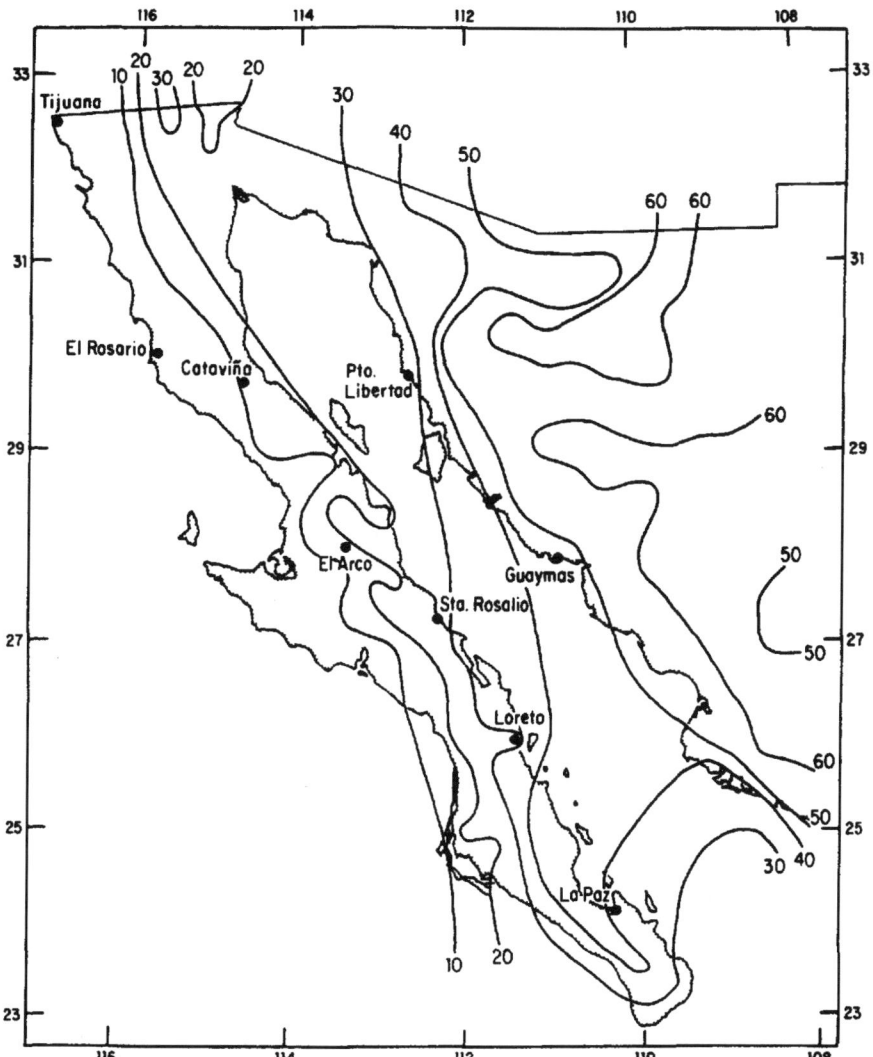

Figure 4.3 Contours of the percentage of the mean annual precipitation that falls as summer rainfall during June, July, and August. Based partly on Hastings and Turner (1965).

It is accurate, however, and for most purposes adequate, to characterize all the gulf islands as hot and dry. Most of them probably receive between 100 and 150 mm of rainfall per year. The usefulness of the rainfall is diminished by a low year-to-year predictability of precipitation (with zero rainfall in some years) and by the uncompromisingly high temperatures in summer, mostly averaging around 30°C in July, with annual means of 21°–23°C. In the north, Ángel de la Guarda apparently receives less rainfall than some other islands, but its higher elevation and larger collection area might help counter reduced annual precipitation. It seems just as useful to estimate island precipitation from the vegetation it supports as to estimate it from patchy and distant rainfall records. Using vegetation of comparable habitats as an indicator, we estimate that rainfall on the gulf islands is generally similar to that on the adjacent peninsula. A possible exception is Isla Cerralvo, where the vegetation is less tall and dense compared to that either on the peninsula near El Sargento or on Isla San José farther north, possibly signaling reduced precipitation.

A further aspect of climate especially relevant to plant life is the seasonal distribution of rainfall. In the northern gulf the rainfall patterns are more like those of the Mediterranean-climate zone to the northwest, with a high preponderance of winter rains. Both southward down the peninsula and gulf and eastward across the gulf to Sonora, the incidence of summer rains increases. Less than 20% of the precipitation falls in 3 summer months in the northern half of the peninsula, but this increases to 40% in parts of the southern peninsula and to 50% in Sonora (fig. 4.3). Thorn-scrub areas receive even more summer rain, generally around 60% in Sonora but less in the cape region, where the thorn scrub is generally less well developed. Summer rainfall in the Sonoran Desert is important not only to a large group of annual plants, the "summer ephemerals," but also to the many shrubs of tropical and subtropical derivation (e.g., *Bursera, Jatropha*), which leaf out and flower after summer rains.

Another major physical determinant of vegetation and plant distribution is substrate. The gulf islands' underlying rock types are characterized in chapter 2. In general, the range of parent material is more restricted on the gulf islands than on the peninsula or on the Sonoran mainland. But perhaps more important for plants than the parent rock are the physical conditions provided by the derived substrates. There are significant changes in plant species with soil depth and its particle size and water-holding capacity—changes readily apparent from the most cursory examination of vegetation from deep alluvial, sandy soils to coarser gravel to rocky slopes. The range of substrate conditions on an island is thus more a function of precipitation and drainage patterns than of parental rock type.

Vegetation Types on the Islands

The islands in the Sea of Cortés all support vegetation typical of the Sonoran Desert, and a general classification of this vegetation into plant community types is given by Shreve (1951) and summarized by Wiggins (1980). The comprehensive monograph by Felger and Lowe (1976) describes in detail the vegetation of the Sonoran islands in the northeastern gulf. Plant communities are characterized by the dominant plants and their growth forms that largely follow variations in the physical environment. Thus, plant communities change geographically as the climate changes and change locally with environmental factors such as moisture availability.

All the principal gulf islands and the entire gulf coast closest to them (except west of Ángel de la Guarda) are within the area that Shreve (1951) called "sarcocaulescent desert." Among the dominant plants are trees of *Cercidium, Bursera hindsiana*, and *Jatropha cuneata*, and cacti of the genus *Opuntia* (subgenus Cylindropuntia), and especially the large columnar *Pachycereus pringlei*. The peninsular coast west of Ángel de la Guarda and northward on the eastern side of the Sierra San Pedro Mártir and Sierra Juárez, and thence around the head of the gulf to the northern interior of Sonora, is within Shreve's "microphyllous desert." It is a drier desert type in general and characterized by a higher incidence of trees and shrubs with small leaves and by fewer with swollen trunks. Among the plants of this association, the creosotebush *Larrea tridentata* is common; *Larrea* is conspicuous in its scarcity on most of the gulf islands. But, notably, several of the characteristic dominants, *Ambrosia* spp., *Olneya tesota*, and *Bursera microphylla*, are common on the islands as well as throughout the sarcocaulescent desert to the south.

Similarly, plants that dominate Shreve's sarcophyllous desert of the mid-peninsular interior and the Magdalena region to its south have wide distributions that include the gulf coast and the islands. Note that the identification of certain geographic regions as particular desert types is strictly one of convenience and of course overrides both the wide occurrence of certain plant assemblages throughout the Sonora Desert and the changes of plant assemblages within local areas of varied topography.

At the local level, the terminology of Felger and Lowe (1976) is useful, being both simple and descriptive. They distinguish (1) two kinds of littoral scrub, one of mangroves and one of other halophytic (salt-tolerant) shrubs; (2) six kinds of desert scrub, including creosotebush, mixed microphyllous shrubs, a scrub dominated by columnar cacti, and a riparian version of desert scrub; and (3) three kinds of thorn scrub, ranked roughly by stature from plains to riparian sites but all composed mainly of drought-deciduous plants in a dense and often spiny vegetation. Large islands such as Carmen, San José, or Cerralvo have all these plant community types, often within a 1-km transect. Figure 4.4 summarizes on a large scale the variation in vegetation types around the gulf, and figure 4.5 shows stylistically the appearance and disposition of plant communities on a typical southern gulf island, naming some of the more conspicuous vegetation types.

Plant Distributions

Phytogeographic Regions Around the Sea of Cortés

Plant geographers find it convenient to divide the world into phytogeographic regions that reflect two distinct and independent notions. First, within regions the range of physical environments is limited and characterized by some set of climatic features. Second, many plants within regions share an evolutionary past, perhaps originating in and adapting to abiotic features and coadapting to each other within the same area. Clearly, phytogeographic regions change over time, shifting over the map with changing climate patterns and changing land surfaces and varying in plant makeup as different species wax and wane over evolutionary time and advance into and retreat from particular vegetation types. First we describe the current picture around the Gulf of California, and then in the following section we discuss the historical perspective.

The occurrence of plant species on the gulf islands is best interpreted in relation to the plant distribution over the surrounding land areas that so nearly encircle the gulf: the peninsula to the west and the mainland states of Sonora and Sinaloa to the east. In Lower California, the high-elevation, temperate coniferous forests of the northern mountains, the Mediterranean-climate region between these mountains and the northwest Pacific coast, and the cape region in the south are outside the Sonoran Desert region (fig. 4.4). The southern end of the Californian floristic province extends

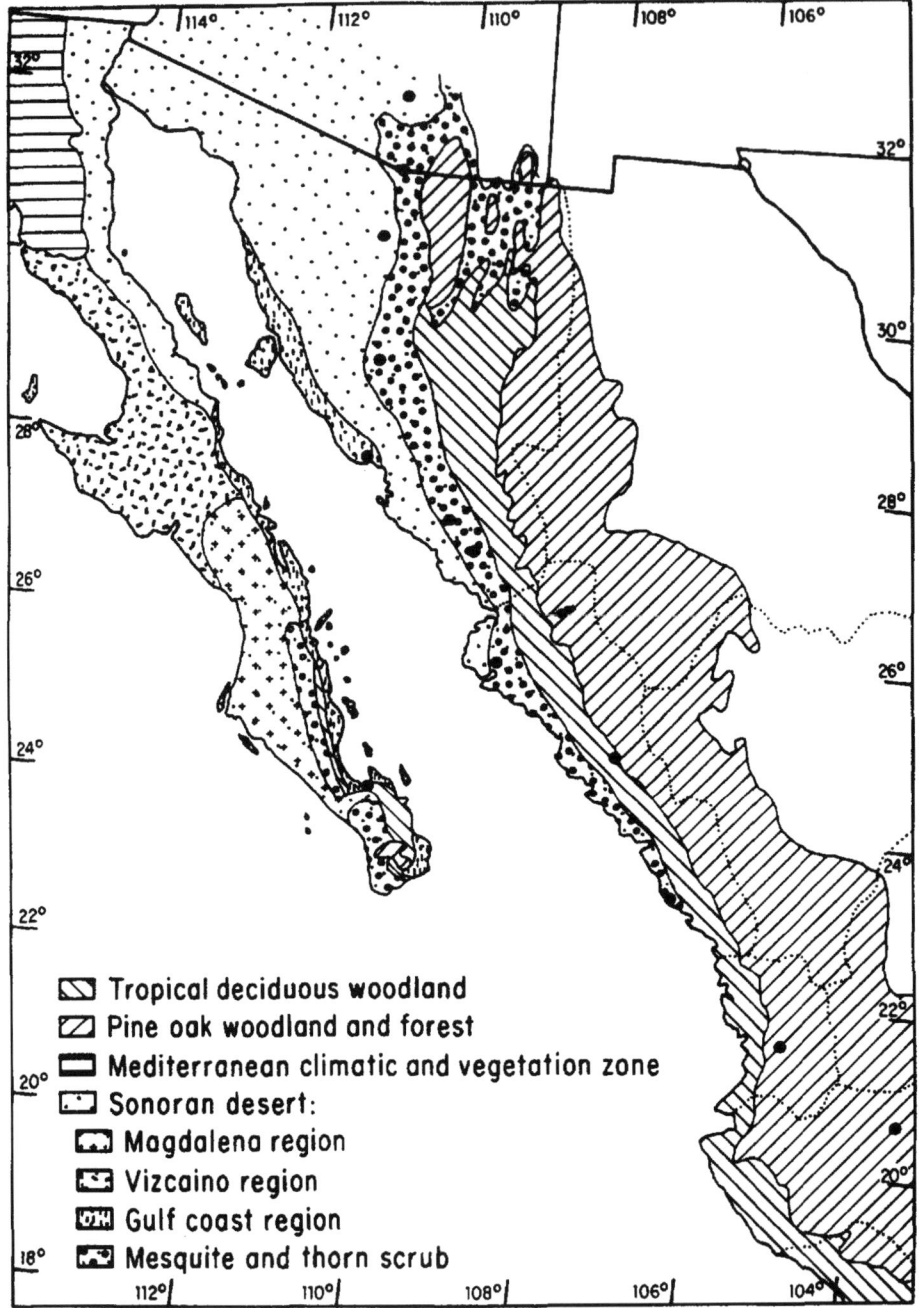

Figure 4.4 Distribution of the major vegetation types around the Gulf of California. The Sonoran Desert occurs around the head of the gulf and over much of Sonora, Mexico; it is represented by an additional two subdivisions on the Lower California peninsular (Vizcaíno, Magdalena regions), and a third (gulf coast) along both gulf coasts and on islands in the gulf.

Figure 4.5 Schematic representation of different types of vegetation on a moderate-sized gulf island.

into the peninsula as far as the Mediterranean-climatic zone dominates the Pacific coast (roughly as far south as El Rosário). This region includes the montane areas of the Sierra San Pedro Mártir and the Sierra Juárez, but not the area between these mountains and the gulf coast to the east. Its vegetation and flora are much like those of coastal southern Alta California and its peninsular ranges (Laguna, Santa Ana, San Jacinto Mountains), though they gradually change as one moves southward. In elevational succession from the Pacific coast, the vegetation includes coastal sage scrub, chaparral, evergreen oak woodland, pinyon pine, and montane coniferous forest (Munz 1974). A large proportion of the plant species does not extend much south of El Rosario, and thus the northwest has little bearing on plant distributions in the gulf (but see below).

The second phytogeographic zone on the peninsula distinct from the deserts is the cape region. This zone includes the cape block southeast of the low-lying land from La Paz south and west to the Pacific coast, and it extends north through the Sierra de la Giganta and along the gulf coast to around Loreto and farther north in a much attenuated form to perhaps Santa Rosalía. The region has somewhat higher rainfall and is dominated by summer rains from tropical anticyclones. The vegetation at lowest elevations is thorn scrub, with elements of tropical deciduous woodland, and at higher elevations in the Sierra de la Laguna and the Sierra Victoria passes first to a chaparral-like vegetation and thence to evergreen oak and pine forest. Pines do not occur in the Sierra de la Giganta, and oaks are less prevalent there. Many of the plants do not extend northward into the desert areas of the peninsula. At higher elevations, this vegetation reflects the origin of the cape landmass adjacent to the mainland Mexican coast at Colima and Guerrero; a review by Wiggins (1960) emphasized the floristic uniqueness of this part of the peninsula and its connection with the Pacific Coast region of southern Mexico.

A more diverse version of thorn scrub and tropical deciduous woodland occurs in northern Sinaloa and Sonora, where it lies between the Sonoran Desert and the western escarpment of the Sierra Madre Occidental (fig. 4.4). The thorn scrub reaches the coast from Mazatlan south and to the north occurs at increasing distances inland, following the lower slopes of the northerly trending cordillera (Sierra Madre Occidental). This results in a broader expanse of desert lowlands northward between the scrub-covered foothills and the eastern gulf coast. The correspondence between the distribution of habitat and the distribution of climate is readily seen: compare the total precipitation figures, the seasonal distribution of this precipitation, and the temperature maxima for the regions around the gulf (figs. 4.1–4.3) with the habitat distributions in figure 4.4.

Because no gulf island is high enough to support pine–oak vegetation, their main plant associations are Sonoran Desert and, in the south only, thorn scrub. For good analyses of these vegetation types, see Shreve (1951) for the Sonoran Desert in general, Felger and Lowe (1976) for the northeastern gulf coast facies in particular, and Gentry (1942) and van Devender and Friedman (2000) for thorn scrub and tropical deciduous woodland.

Plant Distributions on the Gulf Islands

By most criteria the region around the Gulf of California is one of moderately high plant diversity; the Lower California peninsula alone supports nearly 4000 plant species (Wiggins 1960), and Arizona nearly 3500 (Kearney and Peebles 1951). But because the gulf islands are within a relatively homogeneous climatic area, and because their elevations are modest, much of this diversity is precluded from a potential island distribution. The tally of island plant species in our earlier edition (Cody et al. 1983) was about 570, when we predicted the list would reach perhaps 650 species after more thorough exploration. Appendix 4.1 now lists 649 species, and clearly there are more to be recorded or described. It is difficult to escape the impression that the islands are not at all unique in their plant species lists; rather they appear to be simply samples of the nearest mainland, in terms of both numbers and kinds of species. These samples vary as the islands vary, of course, being more comprehensive with increased island size and topographic diversity, and they vary geographically with the position of the island, just as mainland plant lists vary with the size of the area surveyed and its geographical position. A superficial assessment gives the impression that one might just as well be looking at a comparably circumscribed section of the nearest mainland area: island-specific characteristics show up only with more careful scrutiny. In the next section we treat aspects of this nonuniqueness that pertain to diversity; here we will pursue distributional matters.

The occurrence of a given plant on an island can be interpreted most readily in terms of two sorts of factors: geographic and climatic. The position of the island relative to mainland determines, at least in part, the available plant species pool; the climatic regime on the island influences its suitability for potential colonists; and the dispersal capacity together with the resident flora determine whether a plant can reach an island and survive there. The Sonoran islands of Tiburón, San Esteban, and San

Pedro Mártir (and a few much smaller satellites) have plants clearly derived from Sonora. This is particularly true of Tiburón, a large landbridge island separated from the mainland coast by a narrow and shallow channel: in terms of its plant species the island is hardly distinguishable from a chunk of mainland. Of the 298 plant species on Tiburón, 75 occur on no other gulf island, and 9 more are otherwise found only on San Esteban. The second largest Sonoran island, San Esteban, lies in the mid-gulf (the Midriff area) with large islands to the east (Tiburón) and west (Ángel de la Guarda, San Lorenzo) that decrease its isolation from the Sonoran and peninsular coasts. Of the 123 plants recorded for San Esteban, only 3 are found exclusively there, 8 span the Ángel-San Esteban-Tiburón islands (only), and others restricted to this east–west bridge are found only to the east (7) or only to the west (3) of San Esteban.

Similarly, at the other end of the gulf, the plant list for Cerralvo reflects its geographic position as the only large island of appreciable topographic diversity close to the cape region. Some 34 of its 228 species occur on no other gulf island, and a further 15 are shared only with the landbridge island of Espíritu Santo north of La Paz. On this latter, the corresponding figures are 30 of 242 taxa. Species restricted to one or both of these islands include such cape forms as *Gochnatia arborescens, Acacia goldmanii* and *A. pacensis, Mimosa xanti, Erythrina flabelliformis, Ipomoea pescaprae, Phyllanthus galeottianus, Fouquieria burragei, Russellia retrorsa, Schoepfia californica*, and *Aloysia barbata*. Most of these are typical of the scrub and woodland that dominate the cape vegetation.

Rainfall is higher in the cape region and on the Sonoran coast than on the northwestern gulf islands and, in particular, climate is similar between Cerralvo in the southwestern gulf and Tiburón and San Pedro Nolasco in the northeast. Perhaps corresponding to this climatic similarity, several shrubs (e.g., the sage *Salvia similis*, the cheesebush *Hymenoclea monogyra*, hopbush *Dodonaea viscosa*) share a distribution limited to these gulf islands. Numerous other shrubs (see range maps in Hastings et al. 1972) have distributions in the cape and central or eastern Sonora, to the east of longitude 110°30′ in areas with precipitation greater than 300 mm.

In an attempt to define island-specific characteristics of plant distributions, we can search for groups that are unusually well or poorly represented there, but nothing dramatic emerges. A case could be made for the success of the family Cactaceae on gulf islands, but then most gulf islands are drier than all but the coastal areas of the peninsula and thus are predisposed to support cacti in abundance and diversity. Perhaps surprisingly, though, their island prominence suggests little or no dispersal limitation in cacti. Certainly, plants of riparian and more mesic habitats are scarce on the islands, but so are the conditions necessary for their maintenance. The large family Asteraceae, with diverse growth forms and habitat preferences, is well represented in the gulf island flora, but only in the same proportion as on the peninsular coast. The apparent dearth of annuals on islands may be as much a result of infrequent late-summer plant collections as a response to possibly less frequent late-summer rains on the islands.

A measure of uniqueness of an island's flora is percentage of endemic species (that is, forms that occur only on that island.) Endemics might evolve in response either to unique physical conditions on islands or to the unique biotic conditions that result from different and usually less diverse sets of coexisting species. Neither the abiotic nor the biotic conditions on the gulf islands are obviously much different from those of adjacent mainland areas. Not surprisingly, the list of endemic species is short (see app. 4.5): 23 taxa are now recognized as endemic, 15 at the full species level. Nearly half of them are cacti (in the genera *Echinocereus, Mammillaria,* and *Ferocactus*; see fig. 16.9. Several other near-endemics occur along 100–200 km stretches of adjacent coast, such as *Cryptantha angelica* and *Xylorhiza frutescens* on Ángel and the adjacent peninsular coast, *Passiflora arida* var. *cerralbensis* on Cerralvo and the opposite cape coast, and *Agave chrysoglossa* on San Pedro Nolasco, Tiburón, and the nearby Sonoran coast. Two-thirds of the endemics occur on single islands, and half the remaining endemics occur on just two islands: *Echinocereus grandis* on San Lorenzo and San Esteban and *Euphorbia polycarpa* var. *johnstonii* on Carmen and Monserrat are examples. Although endemics occur both on landbridge and on non-landbridge islands, half of them are on three large non-landbridge islands: Ángel de la Guarda, San Esteban, and Cerralvo.

Again, it is by no means certain that these endemics are products of the insular situation per se. Just as narrowly circumscribed areas of the mainland can support endemic species, the peculiarities of local conditions wherever these occur might favor the evolution of local differentiates. Restricted gene flow or reproductive isolation can partition patchily distributed populations on contiguous land areas as well as on islands.

Plant species now found on the gulf islands might be long-term residents that have survived *in situ* since the island was created, either by rifting off the trailing edge of the peninsula or by rising sea levels that submerged a previous mainland or peninsula connection. Alternatively, some plants might have colonized over water since the island was created. Clearly this must be the case on islands such as Rasa, formed *de novo* as products of sea-floor spreading and never connected to larger land areas. The evidence from present plant distributions supports the notion that not only do plant populations survive extremely well on islands, as the similarity in species richness between islands and island-sized mainland areas attests, but also plants colonize islands readily enough. This notion is supported by the lack of any distinct dichotomy between very old islands and new islands, whether newly isolated or newly formed in the sea, but also by the fact that island species numbers do decline, regardless of island age, with increasing isolation from peninsular or mainland sources (see below).

Historical Perspective on the Plant Distributions

History of the Sonoran Desert

The land areas now referred to as the Sonoran Desert region have not always supported desert vegetation. Indeed, in geological time this desert is a recent phenomenon. The history of the area and its plants was pieced together from studies of geology, paleoclimates, and paleobotany; a comprehensive version of this story was published in 1979 by the chief proponent of such studies, Daniel Axelrod (for earlier related studies, see Axelrod 1950, 1958; Raven and Axelrod 1978). The salient features of Sonoran Desert history according to Axelrod are as follows. First, the peninsula of Baja California has rifted some 500 km northwest to its present position, beginning with active plate subduction and volcanic action in the mid-Cretaceous. Most of the shift occurred in the last 5 million years (also see chap. 2). The cape region block, originating to the south near the state of Jalisco, brought with it a legacy of the old Madro-Tertiary flora, to mingle later with Arcto-Tertiary plants entering the peninsula from the north. In addition, the climate of these landmasses has undergone dramatic changes during this 30-million-year period, considerably affected by the retreat of the mid-continental seas in the early Tertiary; further, the uplift of the peninsular and Transverse mountain ranges created rainshadows to the east, contributing to an increasingly arid climate. Many plants now typical of dry semidesert were widely distributed in late Miocene but were then more restricted during the wetter climates of the Pleistocene.

The sequence of vegetation types now observed from desert to higher rainfall areas along longitudinal, latitudinal, or elevational gradients is seen in the fossil record with increasing age at increasing distances north of present desert areas. Thus, plant species of the temperate forests of the Mexican highlands and the short-tree woodlands below are found in Paleocene to lower Oligocene deposits as far north as Washington state and Wyoming and as far east as Georgia. Madrean woodland remains are found in Miocene and Pliocene deposits as far north as southern Oregon and Idaho, and thornscrub elements are present in fossil beds of the same age in southern California. The region's climate became drier as high-pressure areas stabilized over the lower-middle latitudes, favoring the development of a higher diversity of plants resistant to drought and high temperatures. The main sources of these plants were tropical, via the vegetational sequence of tropical savannah, short-tree woodland, and thorn scrub. Thus, more than half the Sonoran Desert plant genera occur also in southern Sonoran short-tree woodland, and nearly a third live in similar vegetation farther south in Jalisco. A minor source of the present Sonoran Desert flora was pine–oak evergreen woodland.

The areal extent of the Sonoran Desert has waxed and waned with dry and wet phases of climate from the late Pliocene through the Pleistocene to the present day. The desert probably has never dominated such great land areas as it does today, but it was constrained to relatively tiny areas during the pluvial periods of the Pleistocene. The locations of such refugia, where deserts persisted during unfavorable times, are still debated; they were constrained by areas which, although desert in the present climatic regime, in fact supported more mesic vegetation types 15,000–30,000 years ago. The hard evidence comes from relictual plant remains in ancient woodrat middens; the softer inferences derive from topographical considerations of where desert environments might have survived the pluvials and from present-day disjunctions and turnovers in both animal and plant species and races.

For islands in the Gulf of California, the climatic and vegetation shifts over the last million years or so have two main implications. First, arid-climate plants with expanding ranges might find it more difficult to reach a suitable habitat on the islands if such a habitat became available only after the island was formed. Conversely, mesic-adapted plants that expanded during pluvial periods would find it difficult to reach the island and supplant the preexisting plant communities, which might therefore survive the wetter periods more successfully.

The Landbridge Islands

The occurrence of a given plant on a given island today is potentially affected by the history of the island landmass—by the position and size of the island, which determine those plant species that might be available as colonists as well as which of these might be able to survive on the island with its particular physical properties. Historical attributes include the age of the land that now constitutes an island, its age as an island, and its possible past connections to mainland or peninsula. Given the stochastic nature of both colonization and extinction, only a large sample of islands in each of many categories, size by age by position, can isolate the effects of these factors on plant species distribution; the limited number of islands in the gulf prevents definitive conclusions.

Islands can be grouped into size categories and further classified as landbridge or not (i.e., with or without Pleistocene connections to the mainland and peninsula). The small islands of San Francisco and San Diego seem to be roughly comparable in topography, as are the medium-sized islands of Monserrat and Santa Cruz. The first-mentioned in each case is the more recent island, and in each case has the larger flora (by 27% and 45%, respectively). This effect, if real, further dissipates on larger islands: Cerralvo and Ángel de la Guarda, older, non-landbridge islands, have floras similar in size to those of Espíritu Santo, San José, and Carmen, which are landbridge islands. This fits our conception of extinction rates, which are expected to diminish with increasing population or island size. Further questions on plant diversity on old versus new, close versus distant, large versus small islands are discussed later in this chapter.

A landbridge effect may help explain distributional anomalies with specific plant taxa. For example, the mesquites (*Prosopis*; three island species, nine island records) occur collectively on only landbridge islands with the single exception of San Esteban.

Mid-Peninsular Relicts of Mediterranean-Climate Vegetation

The cooler and wetter intervals of the Pleistocene permitted the chaparral and oak woodlands now occurring north of El Rosario and west of the Sierra San Pedro Mártir crest to extend south perhaps as far as the cape region. Relicts of this vegetation, reflecting its former southerly extent on the Lower California peninsula, may still be encountered on isolated mountaintops southward. In the Sierra San Borja, at the latitude of the Bahía de los Ángeles, are such characteristic chaparral shrubs as *Adenostoma fasciculatum, Arctostaphylos glauca, A. peninsularis, Ceanothus greggii* var. *perplexans, Cercocarpus betuloides, Garrya grisea, Keckiella antirrhinoides, Prunus ilicifolia, Rhus ovata,* and *Xylococcus bicolor.* Other northern plants occur there too: *Berberis higginsiae Diplacus aridus, Juniperus californica, Quercus peninsularis,* and *Ribes quercetorum* are examples. Five of these species, and other northern ones including *Ceanothus oliganthus,* reach Volcán las Tres Vírgenes, northwest of Santa Rosalía. A few other species, such as *Malosma [Rhus] laurina* and *Heteromeles arbutifolia,* are patchily distributed all the way to the cape region mountains.

There are also interesting connections between the mid-peninsular deserts and southeastern Arizona and Sonora. *Rhus kearneyi,* for example, extends (with three subspecies) from southern Yuma County, Arizona, to the Sierra de la Giganta, with a population on Ángel de la Guarda. From the isolated occurrences of this and the foregoing species on the southern edges of their ranges, a wider, more continuous, and more southerly distribution in more mesic times may be inferred.

The Mojave Desert and Northwestern Link

It appears likely that, during the Pleistocene glacial maxima, not only did the chaparral and oak woodland vegetation extend farther south on the western and central parts of the peninsula, but so did other northern vegetation types. The Mojave Desert vegetation extended south between the Sierra Juárez and the northwestern gulf coast. Thus, this vegetation would then have been a potential source of colonists for Ángel de la Guarda in the northwestern gulf; correspondingly, the climate on Ángel de la Guarda would have been more conducive to their survival there. Further, Ángel de la Guarda appears to have originated as a block of the peninsula north of the peninsular coast to which it now lies closest, and in trailing southeast as the peninsula moved northwest may have retained in its flora relicts of plants with current distributions to the north. This argument is supported by the presence on Ángel de la Guarda of *Acacia greggii* and *Trichoptilium incisum*, which are found farther north on the eastern side of the peninsula, and of *Plagiobothrys jonesii* and *Gutierrezia microcephala*, a typical Mojave Desert annual and shrub, respectively, that have not yet been found on the peninsula itself. In addition, *Xylorhiza [Machaeranthera] frutescens* in the Ángel de la Guarda area is related to the northerly *X [M.] orcuttii*; *Ferocactus gatesii* on islets in Bahía de los Ángeles is most closely related to the northern *F. cylindraceus*. All of these forms appear to be relictual on islands considerably south of the ranges of the closest relatives in the northeastern peninsula, supporting the hypothesis that the Bahía de los Ángeles area once had vegetation now typical of cooler deserts to the north.

The Midriff Chain and the Guaymas Connection

In the northern gulf at 29°N, not only is the seaway at its narrowest but islands of the midriff group form a partial bridge between the peninsula and the Sonoran coast. The ends of this chain are Ángel de la Guarda and San Lorenzo to the west and Tiburón to the east, with Partida Norte, San Pedro Mártir, and San Esteban more or less between. The largest water gap between these stepping stones is the 15 km between San Lorenzo and San Esteban. The midriff islands act as a filter barrier to the floras at both western and eastern ends of the island bridge, components of which extend part or all of the way across it. The peninsular and Sonoran mainland floras are in general similar but do include species that are ecological counterparts or vicariants. Many of these are closely related congeners; pairs of allopatric species, with one on the peninsula and the other in Sonora, are found in the genera *Esenbeckia*, *Schoepfia*, *Acacia*, *Randia*, *Ambrosia*, *Fouquieria*, and many others. Genera such as chollas and prickly-pears *Opuntia* and copal trees *Bursera* contain some species shared across the gulf, while others are represented by close counterparts, each restricted to the western or eastern sides. Yet other species pairs might play similar roles but are less closely related. An example is the columnar cacti *Carnegiea gigantea*, absent from the peninsula where the morphologically similar *Pachycereus pringlei* is widespread; the latter is common on the Sonoran side only south of 29°N.

A sizeable group of plant species is quite broadly distributed over Baja California, from there onto the midriff islands, and thence a limited distance onto the Sonoran coastal area. A number of such species are recorded on the mainland only near Tiburón (e.g., *Stenocereus gummosus, Pithecelobium confine, Desmanthus*); others range wider, as far north as Puerto Libertad (e.g., *Euphorbia xanti, Errazurizia megacarpa*) or south to Guaymas (e.g., *Coreocarpus parthenioides parthenioides, Viscainoa geniculata*). Yet others occupy the Sonoran coast between Guaymas and Libertad (*Euphorbia tomentulosa, Bursera hindsiana*). Other species crossing the Midriff do not quite reach the mainland coast, and some 13 species extend from the peninsula onto the midriff islands as far as Tiburón: *Perityle aurea, Pelucha trifida*, and *Porophyllum crassifolium* are examples.

Another group of species has a peninsular origin, judging from its wider distribution there, and reached Sonora by the midriff chain. But with increasing aridity in the region, these species are now observed in various stages of retreat to the south. Several of these species are concentrated on the mainland near Guaymas where, with 200 mm of precipitation, it is much less arid than Tiburón, with 130 mm. Examples are *Lysiloma candida, Fouquieria diguetii*, and *Euphorbia magdalenae*. The southerly retreat leaves the most arid location, Ángel de la Guarda, vacant first, and the least xeric of these species have pulled south on both sides of the gulf. *Bourreria sonorae*, for example, which now occurs near Guaymas and down the gulf coast of the peninsula from Bahia Concepción to the cape, might have reached and crossed the midriff islands during more mesic times. *Lysiloma candida* is now absent from all midriff islands, but begins a continguous peninsular distribution just to the south (28°30′) and survives also at Guaymas.

Conspicuously few species have a wide distribution in Sonora and extend across the Midriff to the peninsula. The rare exception is *Opuntia leptocaulis*, extending from Oklahoma to San Luis Potosí, Arizona, and Sonora, including Tiburón; *O. lindsayi* occurs locally in central Lower California, and *O. tesajo* is a morphological counterpart in the northern peninsula. Other cacti get part way: *O. fulgida* crosses the Midriff to San Lorenzo, *Echinocereus scopulorum* is found on Tiburón and San Esteban, and a derived endemic, *E. grandis*, gets as far west as San Lorenzo, while *O. versicolor* reaches west to San Esteban. Why has traffic across the Midriff been largely one way? An explanation could be that in the 15,000 years since the last pluvial period, aridity has spread eastward, from the peninsula to Sonora. Such a chronology might be accounted for by the rainshadow effect of the Sierra San Pedro Mártir and enhanced by a possible refuge for arid climate plants in the same rainshadow region.

Finally, why are peninsular plants that have reached Sonora so limited in distribution there? It may be because in Sonora they meet ecological counterparts that also were expanding from Pleistocene refuges, though at a later time. For example, the continued spread of *Lysiloma candida* south or east to the foothills might be precluded by the two-species combination of *L. divaricata* and *L. watsonii*, that of *Fouquieria diguetii* by its close relative and counterpart *F. macdougallii*, and *Ambrosia chenopodifolia* by its Sonoran equivalent *A. deltoidea*.

Note that, in contrast, distinct monotypic taxa such as *Forchammeria watsonii* that likely used the midriff stepping stones now have a comparably wide distribution on both sides of the gulf (and on large southern islands), although now centered somewhat south of the putative crossing point.

The Cape Region

As described above, the floristic legacy of the cape region is that of the mainland Mexican coast at Jalisco and Nayarit, to which the cape block was once attached. Evidence of this legacy is abundant, especially in the oaks and pines of the higher areas of the cape, all of which show affiliations with the mainland coastal region to the south. The same association will show up later with the birds and reptiles (see chaps. 6, 10). The gulf island that supports the southerly origins of the cape region is Cerralvo, a trailing block off the wake of the rifting peninsula. Further, since Cerralvo has had no recent connections with the peninsula (or mainland), it has not received the northerly-originating immigrants that a landbridge island might have had during pluvial periods of lowered sea levels.

Some connections with the southern mainland survive on Cerralvo only; *Acacia filicifolia* occurs on the island but nowhere on the peninsula and is encountered on the mainland from Guerrero to Veracruz (Standley 1920–26). In the Asteraceae, *Vernonia triflosculosa* var. *palmeri* is the only member of its tribe of tropical affinities on the peninsula; Standley (1926) lists 33 species in the genus in Mexico, but only *V. triflosculosa* extends north to Sonora and to Cerralvo. *V. littoralis* is recorded on Socorro Island farther south of the cape region. *Sclerocarpus divaricatus* has a mainland distribution from southern Sinaloa to Central America and also occurs in the cape region (but on no islands).

Some plant taxa based in the cape region extend farther north on gulf islands than on the peninsula. For example, *Coulterella capitata* apparently is restricted on the peninsula to the vicinity of La Paz, but it extends north on islands to Monserrat; *Machaeranthera (Haplopappus) arenaria* occurs on the peninsula south of La Paz but on the gulf islands reaches Coronado and Carmen. The anacard tree ciruelo (*Cyrtocarpa edulis*) is common in the cape region and makes its northernmost appearance on Isla Carmen. There are two explanations for this phenomenon. Possibly these plants were able to expand north on islands but not on the peninsula, encountering less ecological resistance from islands with a somewhat lower plant-species richness than mainland communities. Alternatively, the plants may have expanded north on the peninsula and reached the islands as well (perhaps later). But thereafter, with changing climate, they became extinct at higher latitudes on the peninsula where invading competitors had ready access, while remaining extant as relictual populations on northern islands that were more difficult for invaders to access.

Exotics, Introductions, and Recent Invasions

In accordance with our contention that the islands in the Gulf of California are nearly pristine, records of non-native plants are few. A few European weeds occur locally, such as *Chenopodium murale* on Ángel de la Guarda and Tiburón, *Plantago ovata* on Ángel, Tiburón, and San José, and *Sisymbrium irio* on Carmen. Such weedy exotics are far commoner in the Mediterranean-climate region of the peninsula and extend into the desert only in the most disturbed areas, which generally exclude the islands. *Nicotiana glauca*, for example, is a common roadside weed over much of southern California and throughout the peninsula, but is recorded only on San José among the gulf islands. On Santa Catalina, the palms *Brahea brandegeei* and *Phoenix dactylifera* were almost certainly both introductions; apparently only the latter survives there. A lone palm (*Phoenix?*) grows also near the rancho on the northeastern side of San José. An interesting experiment in progress is the advent into the cape region of the mainland Mexican *Acacia cymbispina*. Although currently restricted to a few sites in the cape (e.g., a roadside population in the San Antonio area), it might be expected to spread throughout the region of thorn scrub and perhaps ultimately to some southern gulf islands.

Plant Species Diversity

Regional Diversity

The islands in the Gulf of California are part of the Sonoran Desert, which includes both coasts of the gulf and, according to our estimates of updates of Shreve and Wiggins' (1964) flora, supports about 3500 plant species. To put this total in perspective, the Californian floristic province has more than 4100 species, California has more than 5000 native species, and other regions have much higher totals in smaller areas. For example, in South Africa the Cape flora has some 8850 species in 2/7 the area of the Sonoran Desert, and the South African Cape Peninsula flora has 2256 in 1/550 the area. Thus the flora of the Sonoran Desert is not especially diverse; viewed as a relatively new flora recently derived from largely subtropical roots, it presumably may be still expanding. About 650 plant species have been identified on the islands, with the highest species total (298) on the largest island, Tiburón. Undoubtedly there are additional species that remain to be found.

The island floras represent a subset of the Sonoran Desert flora, and, given that some islands have a predominantly Sonoran influence, while others are affiliated to Baja California, this subset is unique. As described above, a few species on the islands are relicts surviving a southerly expansion of the northern Mojave Desert or are northern outliers of central Mexican thorn scrub. Others are vicariant counterparts from the mainland and the peninsula; the Cactaceae provide many examples. Thus the overall total of 650 species collectively represents several different aspects or components of diversity, some of which we discuss next.

Species–Area Relations

The Species-Area Curve

One approach to species diversity is by way of the species-area curve, usually plotted with logarithmic axes. Although such plots contain limited information (May 1975; cf. Cody 1975), they are useful in a comparative way between regions and among taxa and can indicate the extent to which islands are depauperate in species relative to mainland sites. Figure 4.6 shows species–area plots from the 31 gulf islands for which we have reliable data, and for comparison includes small peninsular sites and the regional totals from larger areas around the Gulf of California. The peninsular sites are both nested and non-nested areas between 1 m^2 and 100 ha in size, the bulk

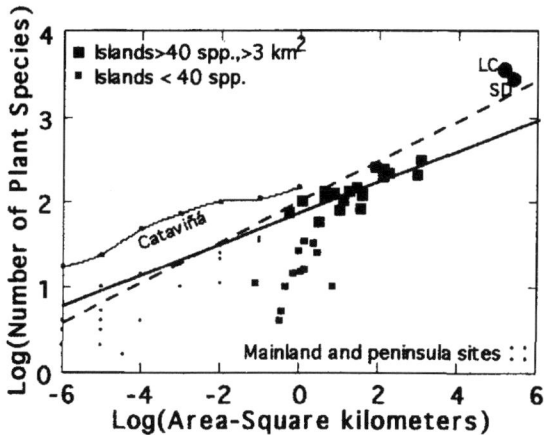

Figure 4.6 Species–area plot for the gulf islands (squares) and select peninsular sites or areas (circles). A series of nested census plots of Cataviñá, north-central peninsula, is identified. The solid line is the regression based on islands >3 km^2 in area (larger squares; the dashed regression line is based on all peninsular sites). Regional totals are labeled LC (Lower California) and SD (Sonoran Desert). See text for discussion.

of which lie between Bahía de los Ángeles and El Arco, with one series from the more diverse Cataviñá area.

With the notable exception of the smaller islands, points in figure 4.6 fall on a roughly linear plot of slope ±1/4. This value is close to Phillip Darlington's (1957) rule of thumb that a 10-fold increase in area produces a doubling (here a 1.9-fold) of species numbers. However, a linear relation is apparent only for the larger islands, and species numbers fall rather precipitously on smaller islands (<3 km^2) with fewer than about 40 species. The regression line for the subset of 19 larger islands [logSPP = 1.855 + 0.182 × logAREA; $R^2 = .67$, $p < .001$] has a significantly lower slope (0.182) than does the regression line based on the small peninsular sites [$n = 33$; logSPP = 1.996 + 0.240 × logAREA; $R^2 = .63$, $p < .001$]. Although the small islands are clearly species poor compared to the mainland (where as many or more species occur in much smaller areas), it is not clear whether the larger islands share this feature. In fact, given the scatter in the mainland data, the mainland regression does not differ significantly, in either parameter or in combination, from the island regression line ($p > .05$).

The larger gulf islands form a reasonable bridge between the peninsular sites and the regional totals, and support species numbers insignificantly different from those predicted by the mainland regression. Thus the effects of insularity show up convincingly only on the smaller gulf islands. This is presumably related to the lower range of physical conditions on islands too small to have well-defined drainage channels, well-developed arroyos, and deeper alluvial fans, and which thus lack all but the shallow-rooted xeric plants of rocky slopes. For comparison, Harner and Harper (1976) showed the effects of local topographic and edaphic diversity on plant species richness in pine–oak–juniper woodlands.

Note that if the small peninsular samples are used to predict the number of plant species in Lower California (LC) or the Sonoran Desert (SD), they underestimate the regional species totals by a factor of 2, but the large-island regression is notably worse in this regard (fig. 4.6).

Species-area data for various mainland regions of the Mediterranean-climate vegetation of central and southern California along with the California Channel Islands are shown for comparative purposes in figure 4.7. With a broader array of mainland sites and areas in this data set but fewer islands ($n = 31$ and 9, respectively), a subtle but significant island effect is detected. The mainland regression line [logSPP = 2.12 + 0.263 × logAREA; $R^2 = .83$] is steeper than that of the channel islands ($b = 0.240$), and its intercept is higher ($a = 1.95$ for the islands), differences that, in combination, are statistically differentiated ($p = .04$). Note that the larger islands in figure 4.7 do not appear very different from mainland sites of comparable area, but it is again the smaller islands that are relatively species poor and contribute to the island–mainland difference. The gulf island regression has a slope significantly lower than either California mainland or islands, and in general the gulf islands hold a relatively smaller percentage of the mainland flora than do the California islands. Individual Channel Islands with 1/100 and 1/1000 the area of the California Floristic Province have 14% and 8% of its species respectively; in the gulf, islands with similar fractional areas support 8.5% and 5.5% of the regional flora.

Another comparison is made in figure 4.8 with the coastal coniferous forest vegetation of British Columbia, featuring Vancouver Island and 200 small to large (up to around 1 km²) islands in Barkley Sound on the outer coast of Vancouver Island (Cody 1999). All of these islands lie on the continental shelf. We again draw the regression line through the larger islands (SPP > 40; $n = 30$), in comparison with which the smaller islands once more fall off with steeply declining species numbers. This line [logSPP = 2.05 + 0.174 × logAREA; $R^2 = .82$] once more underestimates substantially the flora of Vancouver Island and of British Columbia (ca. 3750 spp.). But when Barkley Sound islands are grouped into archipelagoes and their areas thereby nested to produce collective species totals, a new regression line that includes the archipelagoes is steeper [logSPP = 2.17 + 0.234 × logAREA; $R^2 = .82$, $n = 37$] and predicts mainland and regional counts with great accuracy (see fig. 4.8). Note that species counts on these far wetter islands are much higher than on the gulf islands, but the slope of the species–area curve is much the same. The degeneration of species richness on smaller islands occurs in Barkley Sound at a much smaller threshold of island size (beginning around 1/100 km²). This effect that was already apparent on the desert islands in the gulf where areas declined to several square kilometers.

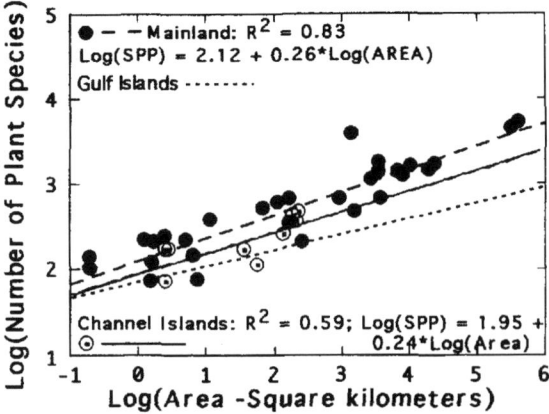

Figure 4.7 Species–area plots for sites in the Mediterranean-type climate zone of Southern California: mainland (solid circles) and its channel islands (open circles). The island (solid) and mainland (dashed) regression lines are significantly different from each other and from the Gulf of California line (dotted), where there are fewer plant species per area.

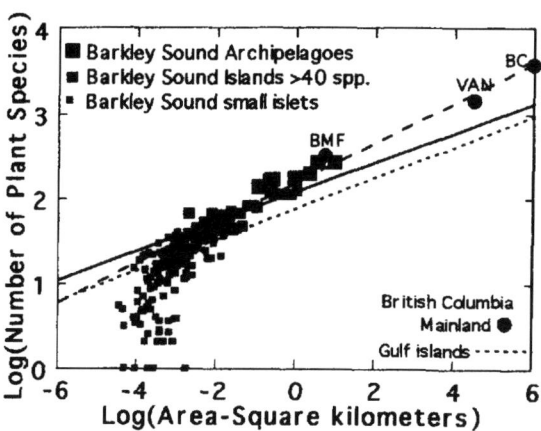

Figure 4.8 Species–area plots for islands in Barkley Sound, Vancouver Island, British Columbia, and adjacent mainland plus regional totals. Larger islands in the sound (>40 plant spp.) provide a regression line (solid line) parallel to but above that of the Gulf of California islands (dotted line). When plant counts are grouped over archipelagoes in the sound, the new regression line (dashed) accurately predicts regions species totals. BC, British Columbia; VAN, Vancouver Island.

Not surprisingly, area has been identified in most island studies as the dominant factor in determining plant species richness. A vast amount of data was summarized by Williams (1964), including mainland and islands counts, and there also nonlinearities in the relationship below and above island sizes around 1 km^2 were observed. Area was found to be the best predictor of species numbers by Hamilton et al. (1963) for the Galapagos Islands, and by Johnson et al. (1968) and Johnson and Raven (1970) for several island sets, including the California Channel Islands and the islands around northern Britain (see also Cody et al. 1983). The areas of both modern and glacial islands of paramo vegetation in the high Andes best determine flora size, as described by Simpson (1974). For more discussion of species–area curves on islands, see Williamson (1981) and Rosenzweig (1995).

Incidence Functions

In a comprehensive study of island species diversity, Diamond (1975) introduced the descriptive technique of incidence functions. Incidence functions are plots of the proportion of the islands in each size category on which a given species occurs. They show at a glance whether a species is represented only on the largest islands or occupies a broad range of island sizes, and whether incidence abruptly falls off at a critical island size or tails off gradually. Reasons for species-specific differences in incidence are an interesting subject for speculation and raise questions about the biology of dispersal and persistence, minimum area requirements, and minimum viable population sizes.

Incidence functions can be used to describe the occurrence of plant species over gulf island sizes by maximum likelihood estimates of logistic regression curves. Figure 4.9a shows that cactus species persist on islands of decreasing size to different extents. Cardons (*Pachycereus pringlei*) occur on virtually all islands, regardless of size, whereas pitahaya (*Stenocereus gummosus*) and nipple cacti (*Mammillaria* spp.) are found on all but some of the smallest islands. Two species of cylindropuntias, one thick-stemmed (*Opuntia cholla*) and the other thinner (*O. alcahes*), have similar and more restricted distributions, and the barrel cacti (*Ferocactus* spp.) are even more

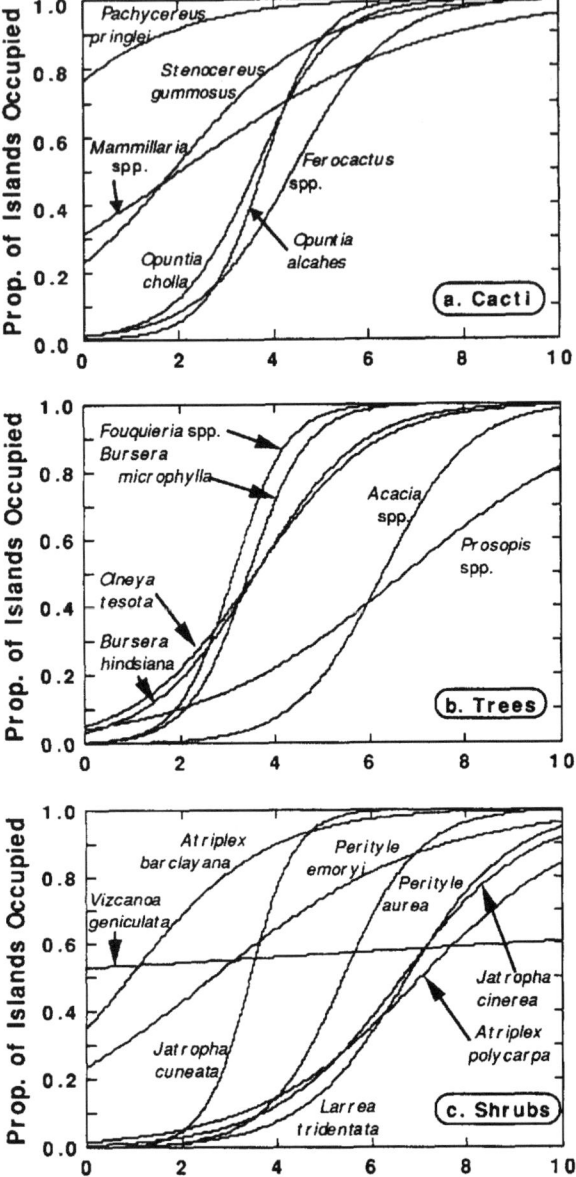

Figure 4.9 Incidence functions for plant species in three different growth forms: (a) cacti, (b) trees, and (c) shrubs. The curves are fitted (maximum likelihood fit) logistic regression plots showing how incidence of a given plant species falls from high levels (right; most islands occupied) to low levels (right) as island size decreases. For shrubs, congeneric or otherwise related species are compared to show how close relatives often differ strikingly in incidence. Gulf islands are grouped into eight size classes for this analysis; see text.

limited to the larger islands. Figure 4.9b gives incidence functions for several of the more conspicuous small trees (mesquites *Prosopis* spp., acacias *Acacia* spp., ironwood *Olneya tesota*, torote *Bursera* spp.), and the bushy *Fouquieria* species. Of the leguminous trees, the ironwood enjoys a wide island distribution, and mesquites are in contrast quite restricted. The smaller-sized and smaller-leaved torote colorado (*Bursera microphylla*) is more successful on medium-sized islands than is its congener, the larger sized and larger leaved torote prieto (*B. hindsiana*).

The contrast in incidence among pairs of related species, including some congeners, is illustrated with shrubs in figure 4.9c. Among the saltbushes (Chenopodiaceae), *Atriplex barclayana* is one of the most widely distributed shrubs on the islands, whereas the congeneric *A. polycarpa* is by contrast restricted to the largest islands. The congeneric matacora *Jatropha cuneata* and torotillo *J. cinerea* (Euphorbiaceae) show corresponding distributional differences, as do the rock daisies *Perityle emoryi* and *P. aurea*, with the former widely distributed and the latter restricted in both instances.

The incidence of two conspicuous shrubs in the family Zygophyllaceae, creosotebush (*Larrea tridentata*) and guayacán (*Viscainoa geniculata*) is shown in figure 4.9c. The former has a distribution surprisingly restricted to six islands, including the two largest, Tiburón and Ángel. The latter occurs on 11 islands (app. 4.1), also including the 2 largest, and over the size range of islands on which guayacán is found, its incidence is fairly constant at around 50%. Neither species appears to be affected by landbridge status (or otherwise) of islands, as each occurs on both sorts of islands, though neither is found on Cerralvo (large, isolated, and non-landbridge). Most provocative is the fact that the two zygophyll species co-occur on no islands other than the largest two, even though the size range of demonstrably occupiable islands overlaps widely between the species. Creosotebush, for example, occurs on San Francisco but not on adjacent and much larger San José, with respect to which San Francisco is a landbridge satellite; guayacán occurs on San José but not on San Francisco. It is difficult to avoid the conclusion that their checkerboard distributions are influenced each by the other species.

Components of Plant Species Diversity Within and Between Habitats

Diversity Components

Universally, the numbers of plant species and the numbers of higher taxa, increase with increasing area sampled. The biological significance of this relationship is moderated by other factors that vary simultaneously with area, such as habitat type and range and the number of living opportunities thereby provided. In a nested series of species counts over increasing area in uniform habitat, such as on the Magdalena Plain, the species count rapidly approaches a maximum as area approaches just a few hectares. This maximum, the number of species that can coexist in the uniform habitat, is the α-component of diversity. The species total will increase only when increasing sample area includes different habitat types, such as different substrates, slopes, aspects, soil depths, or moisture conditions. The additional species that are accumulated as a result of changing habitat is the β-component of diversity.

Still larger areal samples will encompass the ranges of species that are geographical replacements of some of those counted earlier, independently of habitat change. This increase has been termed γ-diversity (Cody et al. 1977; Cody 1981, 1986, 1993; cf. Whittaker 1972, who reserved γ-diversity for a landscape species inventory and used β-diversity for turnover between geographic areas). Thus, our γ-diversity is composed of ecological equivalents or counterparts, the "vicariant species" of current biogeography jargon. The point at which first β-diversity and next γ-diversity components are added to the initial α-diversity of plant communities in uniform habitat depends, of course, on the topography, local as well as regional, and the components cannot be easily separated in species–area curves.

The peninsular samples in the 1–10 ha range represent largely α-diversity; no comparable counts have been made on the gulf islands, but we would expect comparable α-diversities, at least on the larger islands. Likewise, we expect plant species to respond to changing habitat in the same way on the islands as on the peninsula and thus expect β-diversities to be comparable. Because plants are commonly supposed to be more strongly associated than animals with specific habitats due to their rigid physiology and limited ability to avoid or moderate unfavorable conditions, niche shifts and density compensation are likely to have minor roles in island plant communities. Such effects would tend to conserve α-diversity. In Sonoran Desert habitats in general, the α-diversities of perennial plants are low, around 20–40 species in a hectare of uniform alluvial fan. The diversity of annual plants is less easily determined because they are unpredictable in both space and time. Because any change in substrate or slope drastically affects moisture availability, however, β-diversity in deserts is very high. In moving from an arroyo onto a bajada, thence onto a stony slope, and thence to a rocky outcrop, one usually loses at least 50% of the species at each habitat shift, but these are replaced by new species appropriate to the new habitat. Thus, β-diversity is high, notwithstanding the wide distribution over habitat of some of the more conspicuous plants such as cardons and organpipe cacti.

Species Turnover Among Islands

Next we consider changes in species lists from island to island and ask, for islands of comparable size, whether the species lists are similar or whether each island appears to be an independent sample of the regional flora. The species turnover from one island to another can be expressed by the formula SPTURN = $100\{1 - [C(T_1 + T_2)/2T_1T_2]\}$, where T_1 and T_2 are the species totals on the two islands, and C is the number of species in common. First consider six large islands: Tiburón, Ángel de la Guarda, Carmen, San José, Espíritu Santo, and Cerralvo, with between 191 (Carmen) and 298 (Tiburón) plant species recorded. Species turnover among pairs of these islands generally varies from 40–70%. The flora of Ángel de la Guarda averages a two-thirds turnover in species with the other five islands, but with the closer Tiburón is just 47% different, and with the distant Cerralvo there is a 72% turnover. Likewise, Carmen is more distinct in plant species from more distant Tiburón (52% turnover) and Cerralvo (50%) than from the closer San José (39%). Among these large islands, species turnover is quite closely related to and increases with geographic position and separation distance DIST (fig. 4.10). The relationship, SPTURN = $39.1 + 0.045 \times$ DIST, implies that close islands will differ in their floras by nearly 40%. This can be interpreted either as a sampling effect (of the source flora by the islands), as a historical effect of some sort, or as a consequence of interisland differences in abiotic factors such as climate, substrate, or geographical position. Thence turnover increases by 4.5% for each 100 km of interisland separation, and factors related to isolation or distance include those just mentioned.

Next consider 6 smaller islands between 0.6 and 11.4 km^2 in area, with 29–75 plant species recorded: Tortuga, San Pedro Nolasco, San Pedro Mártir, San Ildefonso,

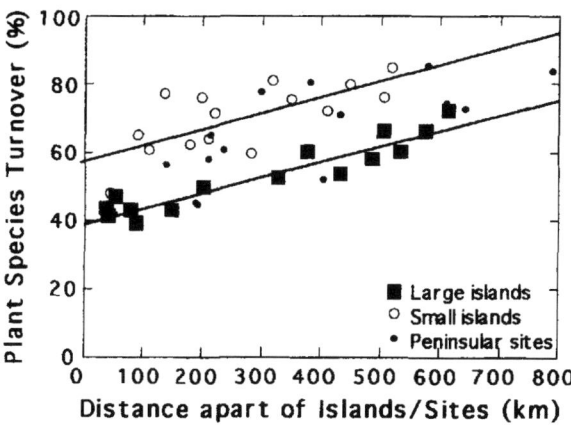

Figure 4.10 The floras of pairs of gulf islands are compared and differences are reflected in a percent species turnover value (ordinate). Turnover increases with distance between the islands (abscissa) but is lower on large islands, around 20% higher on small islands, and generally intermediate between peninsular sites.

San Diego, and San Francisco (see table 4.1). Turnover rates among these islands is given as SPTURN = 57.4 + 0.047 × DIST (R^2 = .50, p = .003). In comparison to the larger islands, there is a nearly 20% higher turnover (i.e., less predictability in species composition, per isolation), and although the slopes of the curves are very similar, the influence of distance is less critical (lower R^2 value). Among even smaller islands (SPP = 10–28; n = 6), the turnover at zero distance (intercept) is yet higher (60%), but the slope is about the same.

These relations are shown in figure 4.10. In summary, small islands support less predictable species sets that are more a random sample of the species pool, the species that occur on small islands show higher γ-diversity than do the species sets of larger islands, and, while turnover between islands is correlated with island separation, this influence is less predictable on the smaller islands.

Are the high turnover rates among the small islands strictly an island phenomenon, or do similar rates obtain on the peninsula? This question is answered using six 10-ha plots from a range of peninsular sites: at (a) Cataviñá, (b) 20 km north of Punta Prieta, (c) 22 km from San Borja, (d) 31 km south of Santa Rosalía, (e) 6 km south of Loreto, and (f) 53 km northwest of La Paz (unpublished data of R. M. Turner and J. K. Hastings, pers. commun.). Table 4.1 shows turnover numbers and percentages for these species lists, and the data are included in figure 4.10. The peninsular plots (SPTURN = 45.3 + 0.053 × DIST; R^2 = .59, p = .001) in general lie between those of small and large islands; the small-island line approximates an upper bound to the mainland points, and the large-island line approximates a lower bound.

Finally, we ask whether the species sets of larger islands are simply built up by adding to the species sets of smaller islands (i.e., are they nested). To answer this question, we take a series of nine islands in the northern gulf, from Tiburón across the Midriff to Ángel de la Guarda—9 islands spanning more than 3 orders of magnitude in size, about 2.5 is species richness. In comparing each island to the next larger in the size sequence, a maximum bound to their combined floras is given by the sum of the two island floras (if they hold no species in common); a minimum bound is given by the size of the larger flora (if the smaller flora is entirely a subset of the larger). Figure 4.11a gives the results of this analysis. Turnover appears to be rela-

Table 4.1 Species numbers (diagonal), species in common (above diagonal), and percent species turnover (below diagonal) between large, small, and tiny islands and peninsular 10-ha sites

Large Islands

	Ángel	Tiburon	Carmen	San Jose	Espíritu Santo	Cerralvo
Ángel	205	130	80	72	76	61
Tiburon	46.5	298	112	117	114	104
Carmen	59.5	51.9	191	124	122	106
San Jose	65.8	53.3	38.8	216	135	127
Espíritu Santo	65.8	57.3	42.9	40.9	242	133
Cerralvo	71.7	59.7	49.0	42.8	43.4	228

Small Islands

	San Francisco	San Ildefonso	San Pedro Martir	San Diego	San Pedro Nolasco	Tortuga
San Francisco	78	14	8	35	16	29
San Ildefonso	71.6	36	11	17	10	20
San Pedro Martir	79.5	54.9	26	10	14	19
San Diego	47.4	69.8	75.8	58	11	24
S. Pedro Nolasco	75.5	78.7	62.4	74.8	56	19
Tortuga	62.6	59.7	52.3	68.0	72.2	79

Tiny Islands

	Salsipuedes	Partida Norte	Rasa	Cholludo	Las Animas	Pato
Salsipuedes	15	7	7	7	3	4
Partida N.	54.8	16	5	7	2	5
Rasa	51.7	66.5	14	6	3	5
Cholludo	63.2	64.7	67.0	26	4	5
Las Animas	76.4	84.7	75.6	74.1	11	3
Pato	66.7	59.4	57.1	65.4	71.4	10

Mainland

	Cataviñá	Punta Prieta	San Borja	Santa Rosalia	Loreto	La Paz
Cataviñá	39	17	25	5	3	6
Punta Prieta	55.8	38	26	10	6	10
San Borja	44.8	41.7	54	12	5	11
Santa Rosalia	80.4	60.5	57.3	19	9	12
Loreto	85.4	70.7	77.5	44.2	14	7
La Paz	83.7	72.6	74.1	51.3	65.0	35

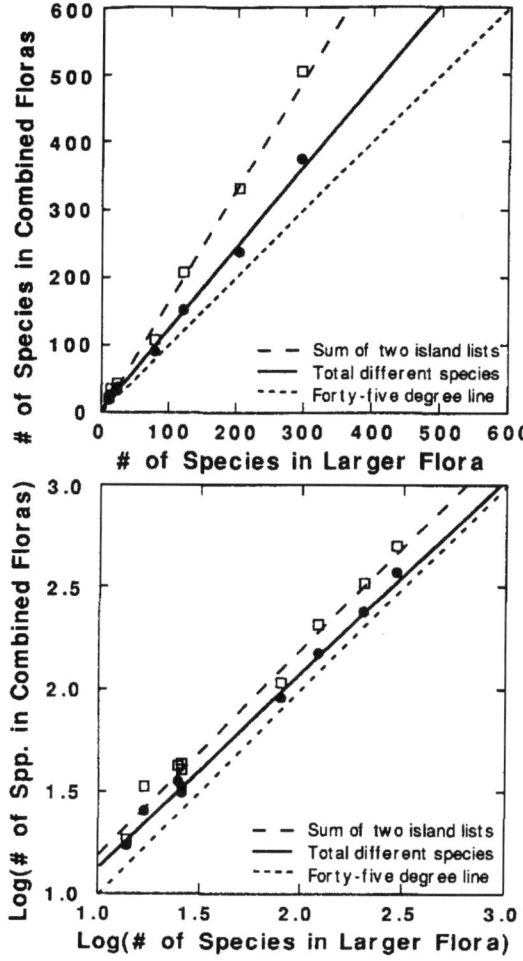

Figure 4.11 For a series of islands of different sizes (Midriff of the gulf), each island pair of adjacent sizes has a combined species total (solids circles and regression line) intermediate between a minimum bound (dotted 45° line: species on smaller island are a subset of species on larger island) and a maximum bound (open symbols and dashed regression line: sum of the two islands' floras). Turnover between islands is rather constant over the large range of island areas; see text for further discussion.

tively independent of island size, in that about the same proportion of add-ons and drop-outs is observed over a wide range of island sizes. With log-transformed data this is clearer (fig. 4.11b); the actual number of species held by island pairs is intermediate between the natural upper and lower bounds [logTOTAL = $0.162 + 0.961 \times$ logSP1; $R^2 = .995$, $p < .001$). While a regression coefficient of 0.961 indicates a somewhat lower species turnover rate between larger islands than smaller islands, this slope is not significantly different from unity. Presumably, chance effects may be somewhat more prevalent in the composition of the smaller island floras, but countering this effect may be a tendency for smaller islands to support only more generalized, widely dispersing species. Likely the larger islands are more easily reached and, with their broader habitat range, capable of supporting a more predictable species set with less precarious and larger populations, in which there is less room for chance omissions.

Historical and Other Factors

The regression line (species–area curve) that relates plant species number to island area, for all 19 islands larger than 3 km² with >40 species, is $S = 71.6(A^{0.182})$, a significant relation that accounts for 67% of the variance in species numbers. Most of the remaining one-third of this variance is attributable to an isolation factor (ISOL), the distance of the island from mainland or peninsula. In figure 4.12 the residuals from the species–area regression equation are plotted against distance from mainland. There is a strong trend for those islands that have positive residuals (RESID), which are above the regression line and species-rich, to be close to the mainland or peninsula (RESID = 0.112 − 0.012 ISOL; $R^2 = .58$, $p < .001$, $n = 19$). The two islands that are relatively species-poor per area and per isolation of around 15 km are San Lorenzo and San Pedro Nolasco.

When residuals from this last regression line are partitioned between landbridge (LB) and non-landbridge (NLB) islands ($n = 10$, 9 respectively), there is a tendency for landbridge islands to show higher species richness (mean LB residuals = 0.045 ± 0.103 SD; mean NLB residuals = −0.50 ± 0.112 SD; t test, $.05 < p < 0.1$ that there is a significant historical effect associated with landbridge status). The test is somewhat equivocal because most landbridge islands are also those close to the peninsula or mainland, and so these two factors cannot easily be separated in their effects on species numbers. Islands to the left (low isolation) in figure 4.12 are species rich either because they are closer to potential sources of colonists or because they are generally landbridge islands with relatively recent connections (≤15,000 YBP) to peninsula or mainland.

In summary, plant species diversity on the gulf islands shows several conspicuous patterns, the most obvious being the increase in species count with island size. Probably this results from the broader range of habitats for plants, in terms of substrates, slopes, elevations, and moisture conditions, on larger islands. Although most islands <3 km² in area are obviously depauperate compared to the peninsula, larger islands are not obviously so. Species numbers are significantly enhanced by proximity to mainland or peninsula, implying that these putative sources of colonists are a necessary feature for the maintenance of the island species counts. However, Holocene

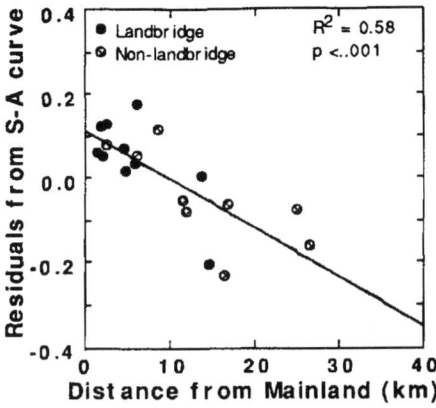

Figure 4.12 Residuals from the gulf island species–area curve of figure 4.6 (ordinate) show the effects of island isolation (abscissa), with negative residuals on distant islands with relatively smaller floras. There is no conspicuous effect of island landbridge status.

landbridge connections appear to be insignificant in their contribution to island species counts, implying that short water gaps are as easily crossed as are short overland distances by dispersing propagules.

Species turnover occurs largely as a function of interisland separation, and although it slows somewhat as species become more predictable on larger islands, it is a constant feature over a wide range of island sizes, and nestedness is not prominent. Some species show an orderly response to island size, producing smooth incidence functions and narrowly defined lower limits to the islands on which they can persist. Others do not, indicating that a substantial stochastic component contributes to the island floras. All the gulf islands together have more than double the number of species on the largest island alone (Tiburón) This means that there is a substantial turnover of species beyond that controlled by habitat, range, island size, and distributional limits, and a lack of nestedness of smaller island floras within those of larger islands.

Ecological Considerations

Adaptation to Island Life

The bland facts of plant distribution and diversity tell only whether a certain species is present on an island, and clearly there is a lot more to island biology than this. Unfortunately, few island studies have progressed beyond this stage with plants, and there is much about the lives of gulf island plants that we do not know. The missing data might enhance, in particular, our knowledge of adaptation in circumscribed populations of limited size in sites of decreased diversity and most likely increased intraspecific density. Given that a plant species occurs on an island, does it maintain the same reproductive system there, with the same reproductive rates, pollinators, and seed dispersal mechanisms, and the same phenology of these events? Has selection kept the island population phenotypically similar to source populations in aspects of vegetative structure such as stature, leaf size and shape, and branching patterns? If an island species lacks close competitors that co-occur with it on the peninsula or mainland, does this allow wider phenotypic variation on the island, increased population density, the occupancy of a wider range of habitats, a changed physiognomy? Are communities of plants structured in the same way on islands as elsewhere in similar environments, with similar architecture and relative abundances, the same patterns of species segregation and species packing?

These are questions no one has yet attacked on the gulf islands, nor are generalities available from other island systems. We can guess that gulf islands will not show dramatic differences at the community ecology level, since by and large the plant communities there appear to be relatively complete. Yet the first impressions from many islands suggest that changes have taken place and that the islands are not merely replicate samples of the peninsular vegetation. We notice, perhaps, the unusual abundance of *Ruellia* near the beach on Danzante, the different branching patterns of the cardons on Partida Norte or Santa Cruz, the apparent scarcity of *Acacia*, the abundance of *Ibervillea* on the bajadas of Cerralvo, and the surprisingly dense thickets of vegetation on San José. Many more subtle changes might appear only after measurements of, for instance, leaf sizes of *Jatropha*, flower production in *Echinocereus*, or the habitat range of *Trixis*. The remainder of this chapter discusses in as much detail as possible aspects of the life histories and roles of some of the more important plant groups on the islands, though admittedly this treatment is a poor substitute for comprehensive studies on island plant biology.

The Cacti

Cacti are among the most conspicuous and dominant plants in the island vegetation; many genera and species are represented, and some island populations are dramatically different from their closest mainland or peninsular counterparts. The family Cactaceae is taxonomically confusing for professional and amateur botanists alike, but nevertheless can be used to illustrate several aspects of island biogeography and ecology in the gulf. Some of the proliferation of and confusion in cactus nomenclature is due to the attraction of the group for nonscientists and the value they place on naming more taxa, but a more interesting aspect of the taxonomic confusion has, perhaps, a biological basis. Undoubtedly, more detailed studies of anatomy, morphology, and chemical systematics will reduce uncertainties, but there are basic reasons for believing that cactus classification is intrinsically difficult. The reasons for this, and its consequences, are worth discussing.

Cacti show extremes of adaptation to arid environments. In most cacti, leaves are ephemeral, vestigial, or quite absent, and photosynthesis takes place in the stems. The stems not only contain and display the photosynthetically active tissues, but they are often modified for water storage and in addition act as the supportive structure of the plant. Photosynthesis involves the crassulacean acid metabolism (CAM) system common in xerophytic plants, with gas exchange taking place at night when water loss from open stomata is reduced (see, e.g., Kluge and Ting 1978). Further, the stems, being relatively nonligneous and a great potential source of nutrients and water for herbivores, are strongly defended by spines, thick (but transparent) cuticles, and toxic chemicals. Because cacti are morphological extremes, there must be a close correspondence, within a given niche, between morphology and environment. This means that the plant's morphology may be expected to track changes in the environment and to vary with both physical and biotic selective forces. Thus one reason for classification difficulties is variability in many cactus species over habitat gradients and throughout their geographic ranges.

Reproductive isolation is poorly developed between species within and even among genera of cacti; it is based mainly on habitat difference, on phenological (temporal) segregation of flowering, and on differences in flower morphology that restrict pollinator access. These mechanisms are in contrast to the genetic or chemical regulation of compatibility found in perhaps the majority of plant species. As a consequence, hybrids may be quite common, especially among species with generalized flower morphologies that occur adjacent on habitat gradients. Such hybrids might not be at any great disadvantage, however, especially if they have an intermediate morphology in an intermediate environment. And in some groups, such as cylindropuntias, vegetative reproduction is far more important than sexual reproduction; their dispersal units are the distal joints that are shed at drier periods or seasons (K. Niessen, unpublished data).

It is possible that related species differentiated in different refuges when desert environments were more restricted, without development of reproductive barriers other than those of population isolation. Thus ecological counterparts in *Echinocereus* and likewise in *Opuntia* in different parts of the Sonoran Desert and in the Mojave Desert freely exchange genes where they meet. Even species as grossly different as *Bergerocactus emoryi* and *Pachycereus pringlei* can hybridize where they meet (in the northwestern peninsula; Moran 1962). Although taxonomic difficulties limit the utility of the cacti for biogeographic studies, these plants are nevertheless of great ecological and evolutionary interest.

Island Distributions

The cacti are extremely successful at reaching and surviving on the gulf islands, and indeed on most eastern Pacific islands at the desert latitudes. Their radiation is most spectacular in the Galapagos Islands (see Carlquist 1965). Their success in the gulf seems not to depend on seed vectors or dispersal mode, as their fruits are extremely variable, from large to small, colorful to cryptic, fleshy to dry, dehiscent to entire, and variously spiny to spineless. Different dispersal modes are indicated by different fruit types, ranging from zoochory (animal dispersal) in fruits large, colorful, and dehiscent to small, bright, and berrylike, to anemochory (wind dispersal) in fruits that are dry and fall to the ground as entire, spiny spheres, to others that propagate vegetatively with zoochorous stems.

Of the largest columnar cacti, saguaro *Carnegiea gigantea* extends south in Sonora to latitude 26° and reaches Tiburón and a few smaller Sonoran islands such as Alcatraz and Cholludo. Its "replacement" in Baja California, cardon *Pachycereus pringlei*, occurs on all the islands, small and large, near and far, landbridge or not. The congeneric *P. pecten-aboriginum* might be expected in the rather sparse thorn-scrub vegetation on Cerralvo and San José, but it is not recorded there. Organpipe cactus, *Stenocereus thurberi*, occurs widely on gulf islands, but it is absent from several in the northwestern gulf (probably Ángel de la Guarda, to San Lorenzo), as well as the small, rocky, and isolated San Pedro Mártir. Its absence from the former group and from the peninsula north of 29° coincides with the drop of annual precipitation below 100 mm, and the same factor may explain its absence from the Vizcaíno and Magdalena deserts in the southwestern peninsula and from the head of the gulf west of 113°.

Stenocereus gummosus is a near-endemic to Baja California south of 32° and extends across the midriff islands to Punta Sargento on the Sonoran coast. It reaches all major islands except San Pedro Mártir and San Pedro Nolasco. This species is a presumed derivative of *S. frickii* in southwestern Mexico (see Gibson and Horak 1978 for systematic revisions in the columnar cacti). A second peninsular species, *S. eruca*, likely differentiated on the Pacific islands of Santa Margarita and Magdalena and subsequently reinvaded the peninsula in the Magdalena Plains region, to which its peninsular range is now restricted. Last in this group is senita or garambullo, *Lophocereus schottii*, widespread all around the Gulf of California and related to the Pachycereii via the central Mexican *S. marginatus*. *L. schottii* is absent from the same northwestern gulf islands that lack *S. thurberi*, most notably from Ángel de la Guarda and San Lorenzo, and also from Espíritu Santo and Partida in the south. There appears to be no reason except insularity for its absence from Ángel de la Guarda. Its absence from the southern landbridge islands might be related to the adaptation of *L. schottii* in the cape region to the densely vegetated thorn-scrub habitat, via a taller, more branched, and arborescent habit (Cody 1984, 1986b), and this may compromise its inhabiting the dry, open flats of these southern islands.

Differentiation among the columnar cacti on gulf islands is most noticeably apparent in growth-form variations; in particular, the cardon, *Pachycereus pringlei*, can occur on islands in a remarkably short and almost trunkless form that is a dramatic contrast to the usual tall peninsular forms that branch 2–3 m above ground. Figure 4.13 shows how the branching patterns of the columnar cacti vary among sites from the peninsula near La Paz, the large landbridge island of San José, the questionable landbridge island of Monserrat, and the definitely oceanic Isla Santa Cruz. Among the possible determinants of these growth-form differences is climate, yet the sites are arranged along precipitation isopleths of about 150 mm. Alternatively, the difference might be due to topography, yet the extreme habitats of steep, rocky slopes and deep, sandy flats on San José produce only minimal shifts in the branching of *P. pringlei*.

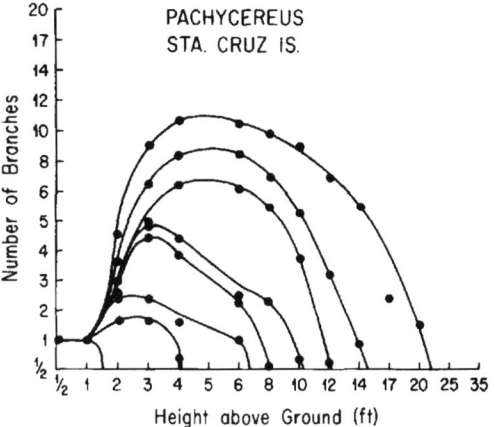

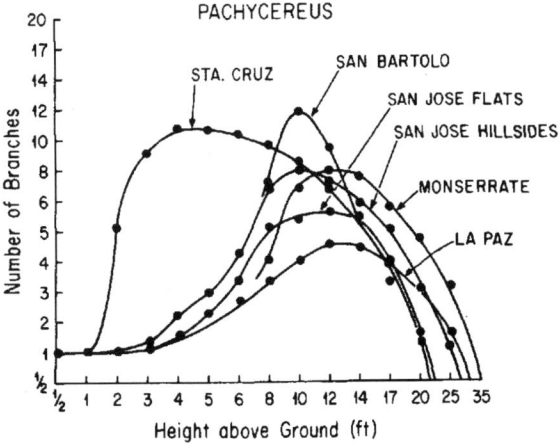

Figure 4.13 Variation in branch height and branch numbers of cardón, *Pachycereus pringlei*, among mainland and gulf island sites. Upper graph: On Santa Cruz island, where there are few other columnar cacti, branching begins low to the ground, and the largest individuals have >10 branches; different curves are summary envelopes of different size classes of the cactus. Bottom graph: Branching number is also high at San Bartolo on the mainland in the dense thorn scrub, but occurs much higher off the ground. Typical mainland branching pattern (few branches high off the ground) is shown at the La Paz site; other island variations are represented by Monserrat and San José, the latter showing the influence of habitat on the pattern.

This leaves the species composition of associated cacti at the site as the strongest correlate of branching pattern, with the implication that coexisting species segregate in branching pattern to maintain relative interspecific differences. Thus, species vary in branching pattern according to the identities and morphologies of coexisting cacti at the site, and morphologies that are species-specific constants are of no advantage (Cody 1984, 2002).

Barrel cacti (*Ferocactus* spp.) reach the gulf islands from several sources. From Sonora *F. wislizenii* gets to Tiburón, and the endemic *F. gatesii* on islets in Bahía de los Ángeles appears to be relictual from cooler times when *F. cylindraceus* reached farther south in the northern peninsula. In the central and southern peninsula, the group *townsendianus-peninsulae-rectispinus* replaces *F. cylindraceus* and reaches San José and Ildefonso. A well-differentiated gulf island group of barrels is that of *F. johnstonianus* (on Ángel de la Guarda) and related *F. diguetii* on islands from Coronado and Carmen to Cerralvo, excluding only the southerly landbridge islands. *Ferocactus diguetii* is unusually large on Santa Catalina (fig. 16.9) and unusually small on Carmen, the most and least isolated of these islands, respectively.

The hedgehog cacti (*Echinocereus* spp.) are represented on most of the gulf islands. *E. scopulorum* of the Sonoran coast reaches Tiburón and is called *E. websterianus* on San Pedro Nolasco. The spectacular island endemic *E. grandis*, derived from *E. scopulorum*, occurs on San Esteban and San Lorenzo, interestingly the only large midriff islands without barrel cacti. *E. grandis* is broader and taller than congeners and occasionally grows as a single stem; thus it makes a reasonable morphological counterpart for absent barrel cacti. *E. ferrerianus* occurs in the vicinity of Bahía de los Ángeles and on a few islets there and as far south as the Tres Vírgenes, and suggests a northern derivation. The most widespread of the genus on gulf islands is *E. brandegeei*, which occurs on the peninsula from the central part to the cape and on almost all islands from San Marcos to Espíritu Santo.

Among the diminutive and variable nipple cacti (*Mammillaria* and *Cochemiea* spp.), the distinctive subgenus Bartschella occurs in the cape region and on the landbridge Espíritu Santo (as *M. schumannii*). Genus *Cochemiea* is distinguished by its long, red tubular flowers, presumably attractive to hummingbird pollinators. *C. setispina* occurs on Ángel de la Guarda and the adjacent peninsula, and *C. poselgeri* occurs on all islands between San Ildefonso and Cerralvo and on the adjacent southern peninsula as well. The remaining mammillarias segregate into two groups, one with milky and the other with clear juice. The milky-juiced *M. evermanniana* occurs on three southern islands, and milky *M. tayloriorum* occurs only on San Pedro Nolasco, but clear-juiced species are more widespread. The *dioica* group (including *M. milleri* on Tiburón, together with *M. insularis*, *M. estebanensis*, *M. armillata*, and *M. fraileana*) occurs on most islands from San José north to Ángel de la Guarda, and the related *M. albicans* group is found on islands from Santa Cruz south to Cerralvo.

Even more difficult to handle than the mammillarias, both taxonomically and practically, are chollas and prickly pears of the genus *Opuntia*. Only two platyopuntias occur on the gulf islands, *O. tapona*, from Coronados south to Espíritu Santo, and *O. bravoana* on San Pedro Nolasco and Cerralvo. However, the cylindropuntias are far more diverse. Typical mainland and peninsular sites have three or four species, differing in joint diameter and degree of arborescence, characters in turn associated with substrate rockiness and slope angle. At Cataviñá in the upper peninsula, for example, the three common chollas are *O. cholla*, *O. molesta* and *O. alcahes* (M.L. Cody, unpublished data); those near the Arizona/Sonora border are *O. fulgida*, *O. acanthocarpa*, *O. arbuscula*, and *O. leptocaulis* (Yeaton and Cody 1979); those of the Mojave Desert are *O. acanthocarpa*, *O. echinocarpa*, and *O. ramosissima*. At each site the cholla species tend to segregate ecologically and morphologically, the more branched species with narrower joints on flats with deep, sandy soils, the less branched and thicker jointed species on steeper and rockier slopes. The segregation appears to be related to the ratio of surface area to volume of the plants, which is inversely proportional to the joint diameter, r. On the flats, photosynthetic area per unit volume is maximized by small r, but on the slopes where water storage is of more importance, large r might be selectively advantageous.

Despite some segregation of Cylindropuntia species over habitat type, at any given site several species characteristically occur together within a few hectares. These are often from different parts of the genus. Cylindropuntia (*sensu* Gibson 1977) species are grouped into different series. In series Bigelovianae, *O. prolifera* occurs mainly in the northern peninsula and on its Pacific side, and *O. fulgida* is widely distributed over southern Arizona and Sonora and extends across the midriff islands from Tiburón and San Pedro Mártir to Ángel de la Guarda. Three other taxa belong to the series: *O. bigelovii* on the Sonoran coast and across the midriff islands and onto the peninsula, albeit sporadically; *O. ciribe* is recorded regularly over the middle sections of the peninsula and on most gulf islands; and *O. burrageana* occurs on Espíritu Santo.

Another two series are the Echinocarpae and the Ramosissimae, which are common in the more northerly and cooler deserts. A fourth, of little importance on gulf islands, is the Leptocaules, a narrow-stemmed species group represented on Tiburón and at a few mid-peninsular localities by *O. lindsayi*, and on the peninsula exclusively by *O. tesajo*, with no island records. The last series is large and important, the Imbricatae, which includes two species endemic to the peninsula, *O. cholla*, widely distributed from 31°30′N to the cape, and *O. molesta*, common over the central third of lower California. A third species in the series is *O. versicolor* from Tiburón and San Esteban. While the last two have very restricted island distributions, *O. cholla* is reported from every larger gulf island except San Pedro Nolasco (see app. 4.1), and from many smaller islands also. To complete the picture, we mention subgenus *Corynopuntia* (club chollas); *O. invicta* is spottily distributed from Bahía de los Ángeles over the southern half of the peninsula and shows up on the islands of San Marcos and Carmen.

We can conclude that chollas, like cardons and other wide-stemmed cacti discussed above, are excellent island colonists, in that almost every gulf island has a representative of the Imbricatae and the Bigelovianae. Which cacti, then, are poorly represented on the islands? Rather less successful is the remaining widespread mainland and peninsular cactus genus with tuberous roots, *Peniocereus*. *Peniocereus striatus* occurs from southern Arizona to southern Lower California and to central Sonora and is reported on every landbridge island in the southern gulf plus Tortuga and Coronado but no others. *P. johnstonii* is endemic to southeastern Lower California and occurs on Monserrat, San José, and Cerralvo. Three cactus genera on the gulf side of the peninsula fail to reach the gulf islands; these are *Myrtillocactus*, *Pereskiopsis*, and *Morangaya*. *Myrtillocactus cochal* is common in the northwest part of the peninsula and extends south to the cape region but is never common near the gulf coast. *Pereskiopsis porteri* is a primitive, large-leaved, scrambling bush in thorn scrub and woodland of the cape region and might conceivably occur on such southern and shrubby islands as Cerralvo and San José. *Morangaya pensilis* is endemic between 950 and 1,900 m in the mountains of the cape region.

The Family Agavaceae

The family Agavaceae includes *Agave*, *Yucca*, and their relatives. The two large northern islands, Ángel de la Guarda and Tiburón, support the distantly related *Nolina bigelovii* and *Dasylirion wheeleri*, respectively, now placed in the family Nolinaceae and here near the southern end of their ranges. *Agave* species occur on these two large islands and on most of the larger islands to the south. Both *Yucca valida* and *Y. whipplei* occur on the peninsula opposite the Midriff, but neither reaches the islands. H. S. Gentry (1972, 1978) has monographed agaves both in Sonora and on the peninsula, and from these accounts the biogeography of *Agave* spp. in the gulf can be summarized. Because these plants were important to the Native Americans, who derived up to 50% of their food from the young shoots in spring, the current distributions may have been considerably altered by harvesting and husbandry. Agaves are especially restricted on Tiburón, where the Seri tribe is still strong (R. S. Felger, pers. commun.), but abundant on San Esteban, which was only occasionally visited by the Seri.

Gentry (1978) divided the genus *Agave* into four groups on the peninsula, three of which occur on no gulf islands: group Umbelliflorae, with *A. shawii* and *A. sebastiana* in mostly coastal areas in the northwest of the peninsula and on Pacific islands near 28°N; group Datyliones with only *A. datylio* in the south-central peninsula south of 26°30′N, but related to *A. aktites* of a similarly restricted distribution in coastal southern Sonora and northern Sinaloa; group Campaniflorae with two species restricted to the southern and western cape region (*A. capensis* and *A. promontorii*) and a third species, *A. aurea*, in the cape and also in Sierra de la Giganta. The largest group, members of which are found on the gulf islands, is the Deserticolae.

In the Deserticolae is the group of small, narrow-leaved xerophytes related to *A. deserti*, which ranges from southern California, southwestern Arizona, and northwestern Sonora down the peninsula to perhaps 30°N. It is replaced to the south on the peninsula by *A. cerulata*, with the shorter- and broader-leaved *A. c. nelsonii* toward the Mediterranean-climate Pacific coast area near 30° (i.e., with a leaf morphology that approaches *A. shawii* in that region), and *A. c. subcerulata* at the southern end of the species range near 27°N and reaching the landbridge island of San Marcos. The mid-peninsular *A. c. cerulata* is common on Ángel de la Guarda, and its derivative *A. c. dentiens* is endemic to San Esteban. Across the gulf, *A. deserti* is replaced by *A. subsimplex*, which occurs on the northern Sonoran coast and also on Tiburón and its satellites; probably as a result of convergent evolution in response to similar abiotic factors, rather than by close phylogeny, as it looks much like *A. c. subcerulata* at about the same longitude but west of the gulf.

The same group of agaves includes two species of the coast and islands off the Pacific side of the peninsula, *A. vizcainoensis* and *A. margaritae*, and a related trio of sierran species, *A. moranii* (north), *A. avellanidens* (central), and *A. gigantensis* (Sierra de la Giganta), none of which reaches the gulf islands. Last of the group is *A. sobria*, broadly distributed along the southern gulf coast and reaching Carmen and Danzante (ssp. *sobria*), with ssp. *roseana* on Espíritu Santo and the adjacent peninsula near La Paz. Thus the genus as a whole has a spotty gulf distribution, with major landbridge islands (e.g., San José) and large, diverse islands (e.g., Cerralvo) lacking any agaves. The island forms *A. cerulata dentiens* and *A. sobria roseana* differ from peninsular counterparts in similar ways, with long, narrow, and thin leaves in a more open rosette, and with marginal teeth reduced or absent. Conceivably, these features might be related to reduced browsing pressure on the islands, which lack ungulates and other potential herbivores.

The Leguminous and Other Trees

Trees are conspicuous and important if only for their size in both desert and thorn scrub; most of these trees are leguminous (family Fabaceae; Leguminosae), but first we will mention those in non-leguminous groups.

The two mangrove species, *Rhizophora mangle* and *Avicennia germinans* (in different families), are found in sheltered lagoons on islands from Coronado (*Avicennia*) and Carmen (both species) south, and *Rhizophora* occurs on Tiburón, with an outlying population in Bahía de los Ángeles. The mangroves occur on mainland and peninsular coasts from Guaymas and Mulege south, where winter sea-surface temperatures exceed 18°C; they avoid islands with steep, rocky coasts, such as Santa Cruz and San Diego, and apparently have not gained a foothold on Cerralvo either. The fig *Ficus palmeri* (Moraceae) is also conspicuous and a common feature of rocky coastlines, being found on nearly every gulf island. The handsome evergray *Sideroxylon leucophyllum* (Sapotaceae), otherwise restricted to the east side of Baja California, reaches Islas Ángel de la Guarda and San Esteban.

Most nonleguminous trees in desert regions are restricted to canyons and arroyos, but *Forchammeria watsonii* (Capparidaceae) grows on open flats; it occurs from San Ignacio south on the peninsula and reaches Islas San José, Espíritu Santo, and Cerralvo. Two trees in the Anacardiaceae reach gulf islands: *Pachycormus discolor* of rocky slopes is patchily distributed both on islands (Ángel de la Guarda and Carmen) and on the peninsula. *Cyrtocarpa edulis* is found on sandy flats and foothill slopes in the cape region, and occurs on all islands from Cerralvo to Santa Cruz except San Francisco and San Diego, with a northerly station on Carmen well beyond its peninsular range. Palms (family Arecaceae), generally restricted to the deeper canyons and washes, are even less successful at reaching the islands; *Brahea armata* is distributed throughout the northern part of the peninsula from the Sierra Juárez to the Sierra San Borja and is replaced in the Sierra de la Giganta south to the cape by *B. brandegeei*. *B. armata* occurs only on Ángel de la Guarda and *B. brandegeei* just on Santa Cruz, to which it may have been introduced and where it seems now extinct. Other canyon and arroyo trees include two in the Bignoniaceae: desert willow *Chilopsis linearis* occurs in deserts of the northern third of the peninsula but is found on none of the gulf islands, whereas *Tecoma stans*, distributed from Loreto south to the cape and beyond to Central America, occurs on Danzante, Carmen, San José, and Cerralvo.

Within the thorn scrub and tropical deciduous vegetation of the cape region are several trees that might be considered candidates for island life but are absent (e.g., *Ceiba acuminata*, Bombacaceae; *Sapindus saponaria*, Sapindaceae), but most of the woodland vegetation of oaks (*Quercus*) and pines (*Pinus*) occurs at higher elevations on the peninsula and is therefore precluded from the islands (maximum elevation of landbridge islands: Tiburón, 875 m, and of nonlandbridge islands: Ángel de la Guarda 1100 m).

Leguminous trees of low-elevation desert and arroyo habitats are successful colonists of and/or survivors on the gulf islands, where they are common and conspicuous. The pattern emerges that the monospecific or species-poor genera such as *Olneya* and *Prosopis* (in the gulf) are best represented on islands, whereas the speciose genera such as *Acacia* are less successful there. The explanation appears to be that genera with many species include taxa that are locality or habitat specific (with high β- and γ-diversity components) and thus represent more specialized species with smaller source populations, possibly with more limited dispersal potential, and therefore suitable conditions for island survival are more limited. Monospecific genera are more broadly distributed, and such species are apparently tolerant of a wider range of physical conditions; therefore they have the advantage of larger and better positioned source populations and likely find a variety of island conditions acceptable for growth and survival.

The genus *Olneya* is monotypic (ironwood, *O. tesota*) and extends from southeastern California and southwestern Arizona to southern Sonora and southern Lower California. It occurs on all the landbridge islands except San Ildefonso (the smallest, at 1.3 km^2) and on all of the larger non-landbridge islands down to Tortuga (11.4 km^2) and, remarkably, San Diego (0.6 km^2). It is absent only from islands that either are very small (San Pedro Mártir, Rasa, Salsipuedes), or low and dry (San Lorenzo), or very isolated (Santa Catalina). Mesquites (*Prosopis* spp.) have a range similar to that of *Olneya* on the peninsula and mainland, but occur only on landbridge islands larger than 11 km^2 with the single exception of San Esteban, which supports two species. A note of the occurrence of mesquite on Ángel de la Guarda by Gentry (1949) has not been substantiated. The peninsular endemic *P. palmeri*, though it lives near both landbridge islands Carmen and San José, has been found on no gulf islands. This situation, at first appearance, is reversed with the genus *Lysiloma*, since the widely distributed *L. divaricatum* (Sonora, Baja California, south to Costa Rica) reaches no islands except Tiburón, and the peninsular endemic *L. candidum* reaches all islands in its latitudinal range (south of 28°30′N) except the isolated Santa Catalina and the small San Diego. However, *L. candidum* is much more generally distributed on the peninsula than is *L. divaricatum*, which is common only in the cape region.

The palo verdes, *Cercidium* spp., are well represented on the islands. The northerly *C. floridum floridum* occurs in Sonora and reaches Tiburón; its peninsular counterpart *C. f. peninsulare* occurs in more mesic sites from 27°30′N south to the cape and is found on all islands from Coronado to Cerralvo except the smaller Monserrat, San Diego, and San Francisco. *Cercidium praecox* has a peninsular and mainland distribution similar to that of *C. floridum*, but occurs only on three mid-gulf islands, Carmen, Ildefonso, and Tortuga. *Cercidium microphyllum* extends south in the peninsula as far as the southern end of the Sierra de la Giganta and to all islands larger than 4 km^2 north of San José. Thus, islands generally have one species of *Cercidium* if they lie south of the range of *C. microphyllum* or north of the range of *C. f. peninsulare*; two species if they exceed 4 km^2 and are within both species ranges; three species at central latitudes where *C. praecox* is added (Tortuga), or even four species where the hybrid taxon *C. Xsonorae* occurs (Carmen only).

The most diverse genus of leguminous trees and shrubs in our range is *Acacia*, with at least a dozen species on the peninsula and nearly as many again in mainland Sonora; there are more than 1000 species in Australia alone, and Standley (1920, 1926) listed 64 for Mexico. Many of the gulf region species are geographic (γ) replacements; some of these are closely related and intergrade, such as the two species pairs *A. greggii* and *A. wrightii* and *A. mcmurphyi* and *A. goldmanii*. Others are close morphological counterparts, such as *A. coulteri* in Sonora for the peninsular pair *A. goldmanii* and *A. mcmurphyi*. Leaf morphology of *Acacia* species is shown graphically in figure 4.14; the relation shows a trade-off of smaller number of leaflets for larger leaflet size but, in general, species with fewer and larger leaflets, N, have a larger total leaf size, A: $[N(A^{1/2}) = 70$, or $\log N = 1.85 - (\log A)/2]$. A second axis of diversification is in the stature of the plant: those most similar in leaf morphology are least similar in stature (cf. *A. constricta* and *A. occidentalis*, *A. californica* and *A. crinita*).

This range of morphological features contributes to both α- and β-diversity components of species richness. In dry desert such as coastal Sonora or the northern part of the peninsula, the range of leaf morphology is restricted (see fig. 4.14); in higher rainfall areas southward in the peninsula and eastward in Sonora, the range of leaf morphotypes steadily increases. These morphotypes cover between them a range of habitats and roles within habitats. At the northern end of the Sierra de la Giganta, for example, *A. farnesiana* occupies the open flood plains and lower arroyos (intermediate leaf morphology, broad distribution worldwide); *A. brandegeei* ranges somewhat higher (to 850 m) in arroyos and on mesas; *A. mcmurphyi* is still higher (350–1300

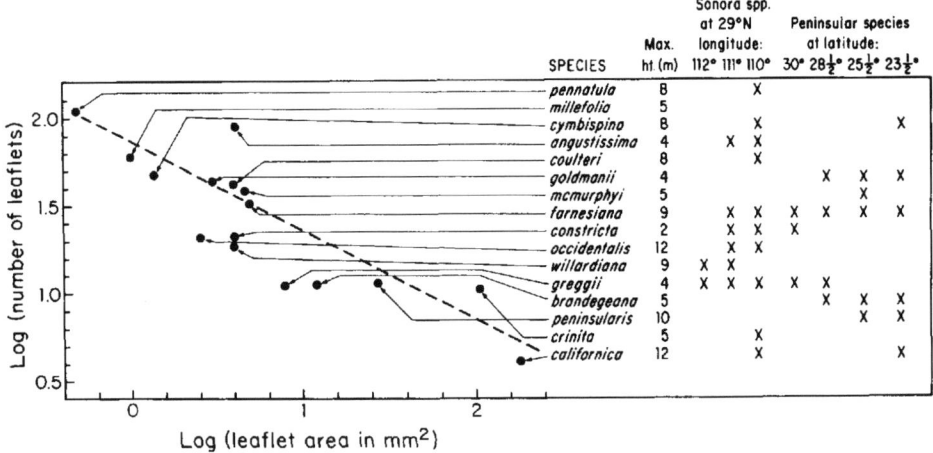

Figure 4.14 Leaf and leaflet morphology in the genus *Acacia*. The graph identifies *Acacia* species on the inverse relationship between number and area of leaflets. To the right, plant height is indicated, with a chart showing which species are represented in Sonora at latitude 29° and at various latitudes on the Lower California peninsula. Note that species numbers decrease with longitude in Sonora in moving from thorn scrub to desert vegetation, decrease with latitude on the peninsula, and the more extreme *Acacia* morphotypes are represented only in the most diverse associations.

m); and *A. goldmanii* is highest (700–1700 m). *A. peninsularis* is common between 275 m and 375 m, a newly described *A. kelloggii* is yet higher, between 690 m and 1600 m on north slopes, and *A. constricta* occurs on plains of the Pacific drainage only (A. Carter, pers. commun.).

The island occurrences of these habitat-specific *Acacia* species are extremely sparse: 7 species have a combined total of 12 island records. In order of island occurrences, they are: *A. goldmanii* on four islands from Santa Cruz to Cerralvo; the northerly *A. greggii* on both Ángel de la Guarda and Tiburón, *A. willardiana* on the Sonoran islands of Tiburón and San Pedro Nolasco; and single island records for *A. filicioides* (Cerralvo), *A. mcmurphyi* (Santa Cruz), and both *A. constricta* and a presumed derivative, *A. pacensis*, on Espíritu Santo (V. Rudd, pers. commun.). These last two species might represent a modified case of speciation via repeated invasion, if the latter evolved in isolation and the former recolonized possibly at a time of a landbridge connection. Despite the ability of *Acacia* seeds to pass through browsers such as cows and goats (and giraffes and elephants) and still be viable, and despite distributional aids (to species such as *A. farnesiana*) such as the use and transportation of their twigs and pods as forage for livestock, the group appears to be singularly incapable of reaching or surviving in insular habitats.

The Family Burseraceae

The shrubs and trees of *Bursera* make an interesting lesson in biogeography in and around the Gulf of California, at the northwesterly limit of this mainly subtropical group. The genus occurs from Peru to the southern United States, with about 100 species at least, 60 of them in Mexico with a concentration of species in the tropical deciduous woodlands of the middle Pacific states (e.g., 48 species in Guerrero). The basic taxonomy is given by McVaugh and Rzedowski (1965), with more recent updates (including reduction of taxonomic entities) and ecological notes by Rzedowski (1968), Porter (1974), and Rzedowski and Kruse (1979).

The genus has two distinct sections: *Bullockia* (exfoliating bark, flowers 3,4,5-merous, ovary 3-celled, fruit 3-valved), and *Bursera* (non-exfoliating bark, flowers 4-merous, ovary 24-celled, fruit 2-valved). Each section is represented in the gulf region by a species in drier, desert conditions and a species of more southerly thorn scrub or tropical deciduous woodland affinities. In *Bullockia*, *B. hindsiana* (including *B. epinnata* and *B. cerasifolia*; A. Gibson, pers. commun.) extends throughout the peninsula to 30°N and up the gulf coast to its head and thence on the Sonoran coast to between 28° and 30°N. *B. hindsiana* occurs on virtually all larger gulf islands (see fig. 4.9). The related species of thorn scrub rather than desert affinities is *B. laxiflora* (including *B. filicifolia*). The *laxiflora* group occurs throughout Sonora from the Sierra Madre foothills to near the gulf coast and in the cape region and Sierra de la Giganta of Baja California. Like many taxa of thorn scrub and deciduous woodland, it occurs on none of the gulf islands except Tiburón, but it seems a candidate for island life at least on Cerralvo, San José, and Carmen.

Sect. *Bursera* has a northwesterly desert species in *Bursera microphylla*, which extends from southern California and southwestern Arizona through the peninsula to the cape and in Sonora to 27°30′N. This species, like *B. hindsiana*, occurs on virtually all gulf islands, whereas its relative with thorn scrub and woodland affinities, *B. fagar-*

oides (including *B. confusa* and *B. odorata*), has a distribution similar to that of the *B. laxiflora* group in Sonora and the peninsula and occurs on no gulf islands except Tiburón.

Throughout most of the gulf region we have a two-species *Bursera* system, with one species per site from each section of the genus, their identities varying with habitat—desert or thorn scrub/woodland (fig. 4.15). Much of the taxonomic confusion stems from variable leaf characteristics, with leaflet numbers variable in, for example, the *hindsiana* group from one to five. Rzedowski and Kruse (1979) concluded that multipinnate leaves are primitive in *Bursera* and reduction to few or single leaflets is an adaptation to different, perhaps drier, habitats. The adaptive significance of leaf morphology in *Bursera*, however, remains to be elucidated.

The Family Asteraceae (Compositae)

The sunflower family is the largest family of flowering plants; it is also celebrated for its ability to colonize remote islands. The Asteraceae is very well represented on oceanic and volcanic islands around the world, often by endemic species and genera of arboreal forms whose closest mainland relatives are herbaceous plants. Examples are *Darwinothamnus* and *Scalesia* on the Galapagos Islands and *Dendroseris*, *Phoenicoseris*, and *Centaurodendron* on the Juan Fernandez Islands in the southeastern Pacific. Saint Helena, in the southern Atlantic, has seven tree composites in the genera *Commidendron*, *Melanodendron*, *Senecio*, and *Aster* (see Carlquist 1965). Typically, herbaceous genera such as *Senecio*, *Sonchus*, and *Solidago* have undergone impressive adaptive radiations in larger archipelagoes, such as the Canary Islands, where they produce large, shrublike species. A more local success story is that of the tarweeds (tribe Madiinae), with an endemic genus (*Adenothamnus*) on the northwest coast of Lower California and several shrubby endemic *Hemizonia* species there and on its offshore islands. But the group also has extended westward to Hawaii and radiated into such spectacular island endemics as the silverswords *Argyroxyphium* (Baldwin et al. 1991).

Asteraceae represents 15% of the Baja California flora and 14% of the gulf islands flora. Half of the endemic genera of Lower California are Asteraceae, as are just less than 20% of the endemic species (13% of the island endemics are in this family). Thus there are some fundamentally distinct composites in the gulf area, but there has not been any exceptional degree of speciation there. Just two species, *Hofmeisteria filifolia* on Ángel de la Guarda and *Coreocarpus sanpedroensis* on San Pedro Nolasco, are endemic to gulf islands, as is a subspecies of *Machaeranthera* (*M. pinnatifida incisifolia*) on some seven islands (see app. 4.1).

Of 13 tribes of the Asteraceae represented in Wiggins' (1980) Baja California flora, 3 occur on no gulf island (e.g., the northerly distributed mayweeds Anthemideae), and others are scarce (e.g., the chicories, Cichorieae, with 1 insular species out of 17 peninsular genera and 37 species); these species are scarce not just on the islands, but in desert habitats in general. The peninsular species list includes only 3 trees but almost 100 shrub species, rather more herbaceous perennials, and even larger numbers of annuals (see table 4.2). We consider those species occurring near the gulf coast south of about 29°30′N as potential candidates for island life in the gulf; thus the peninsular list does not include the many species that occur only in the north and

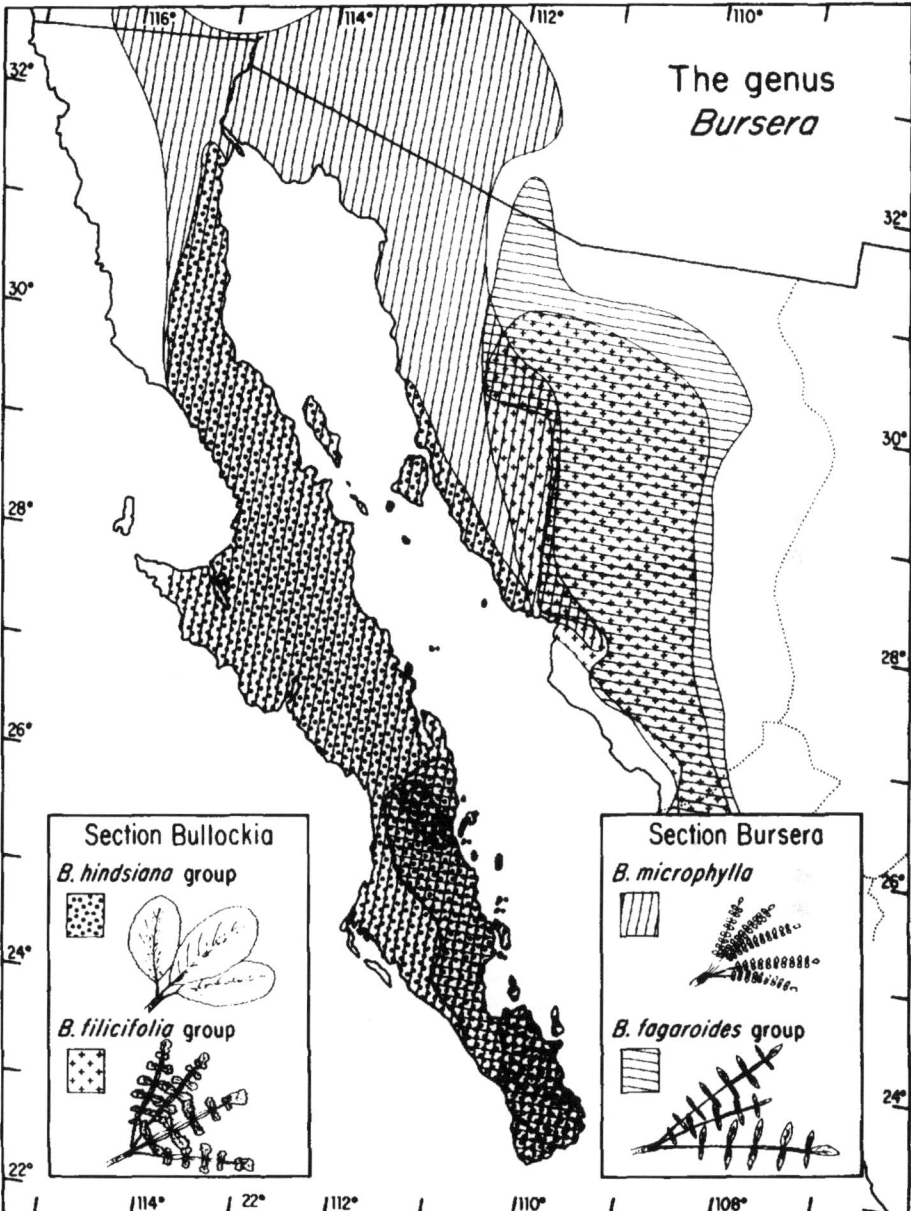

Figure 4.15 Distribution of *Bursera* species around the Gulf of California. The two distinct sections of the genus, *Bullockia* and *Bursera*, each have two species or species complexes, one northern (*Bursera hindsiana* and *Bursera microphylla*) and one southern (*Bursera filicifolia* and *Bursera fagaroides*).

104 The Biological Scene

Table 4.2 Peninsular and gulf coast source flora of composite plant species for potential island distribution and island incidence by growth form on gulf islands

Growth Form	No. of Species on Peninsula	% Ranging to Gulf Coast	No. of Species Near Coast	% Ranging to Gulf Islands	No. of Species on Islands	No. of Islands Occupied per Species
Trees	3	100	3	100	3	1.0
Shrubs	98	59	57.5	83	49	4.7
Perennial forbs	118	37	43.5	34	15	
Annual herbs	125	40	50	28	14	3.5
Total/average	344	45	154	53	81	4.1

especially in the northwestern Mediterranean-climate region, that are restricted to higher elevations, and others that occur only down the Pacific side of the peninsula. Among many others, this excludes, for example, many high-elevation herb species of *Bidens*, many temperate-climate weeds of the northern peninsula (many introduced from Europe), several *Baccharis* and all *Artemisia* species of the northwest, and many of the small annual *Chaenactis* that prefer the broad flats and plains of the Pacific drainage. This selectively removes annuals and perennial forbs, then shrubs and trees, a selective process that is continued from the gulf coast out to the islands, with the same bias against nonwoody species. Thus two small trees in the family Asteraceae are plants of the wetter thorn-scrub woodland of the southern Gulf; both are common in the cape region, and both reach a single island, Cerralvo. One is *Gochnatia arborescens*, a member of the Mutisieae, a typically southern tribe more common in Central America. The second species is *Ambrosia carduacea*, the most arborescent of its large genus in the region.

The particular success on the gulf islands of species with shrubby growth forms, relative to the herbs and annuals, is notable. Although collection records for annuals are relatively patchy (late July and August, when the summer annuals bloom, is an unpopular time for botanists to drag plant presses across these deserts), further work is unlikely to change the picture much. Not only are the shrubs of the peninsula better represented on the islands (with 49 of 98 peninsular species, or 50%, versus 29 of 243 peninsular nonwoody composites, or <12%), but they reach more islands per species (4.7 islands per species, versus 3.5 islands for herbaceous perennials and annuals combined; see table 4.2).

The number of island species of Asteraceae is related to island size as expected ($S = 3.77\ (A^{0.371})$; $R^2 = .62$), with a steeper slope than the all-plants species–area curve, indicating relatively fewer composites on smaller islands. However, the residuals are significantly and negatively correlated with latitude ($r = -.525, n = 31, p < .01$), showing relatively higher composite diversity on the southern islands, where herbaceous species perhaps benefit from the higher summer rainfall.

Three of the extremely successful shrub genera on the gulf islands are *Viguiera*, *Bebbia*, and *Encelia* of the tribe Heliantheae; all are widely distributed at least north to the Mojave Desert. *Viguiera deltoidea* is a complex recently partitioned into four species-level taxa that reach the islands (*V. deltoidea*, *V. chenopodina*, *V. parishii*, and *V. tomentosa*) with others on the mainland and peninsula. These, together with *Bebbia juncea* and *Encelia farinosa*, are all quite variable in habit, as indicated by the many named subspecies. Variation in *Viguiera* includes larger flower heads in *V. d. tastensis* of cape region, and in leaf morphology of the other peninsular and island forms. *Viguiera chenopodina*, common on the Pacific plains of peninsula and on islands across the gulf to San Pedro Nolasco, has leaves cinereous underneath and soft rather than harsh as in congeneric relatives from drier, rockier, and higher sites. Both *Viguiera* and *Encelia* include a few shrubby species with restricted ranges in the peninsula, but none of these reaches the islands or shows morphological variability.

Also in Heliantheae is the endemic peninsular genus *Alvordia*, with southern affinities. Two species, *A. brandegeei* and *A. fruticosa*, have narrow allopatric ranges in the cape region (chromosome numbers of $n = 15$ and $n = 30$, respectively), and the former occurs on San José. The polyploid *A. glomerata* ($n = 60$) extends north through the Sierra de la Giganta to Volcan Tres Vírgenes (*ssp. glomerata*), including only Espíritu Santa among the islands, whereas ssp. *insularis* is found on Islas Carmen, San Marcos, and Santa Catalina, as well as on San José. This strictly insular subspecies has shorter leaves and smaller achenes, with a relatively larger pappus (Carter 1964). This might reflect a relatively recent invasion from peninsular stock, as generally the pappus becomes smaller and the achene larger in insular anemochores (Cody and Overton 1996).

Another interesting shrub of the same tribe is *Coulterella capitata*, which likewise is based in the cape region and extends north of its peninsular range only on the landbridge islands of Espíritu Santo, San Francisco, San José, and Monserrat.

The aforementioned southern tribe Mutisieae (besides the small tree *Gochnatia* and three herbaceous *Perezia* species, which reach no islands from the peninsula) includes the widely distributed *Trixis californica*. This species extends from Texas to central Mexico and is found throughout the Sonoran Desert to the southern Mojave Desert (Sheephole Mountains, southern California). The taxon occurs on virtually all islands from Carmen north. Its distribution in the southern peninsula is spotty, and south of Carmen it reaches only Isla San Francisco. Apparently it is replaced in the southern peninsula by *T. angustifolia* and *T. peninsularis*, but like the more local members of *Viguiera* and *Encelia*, these other *Trixis* species do not reach gulf islands. *Trixis californica* covers its broad geographical range without any of the morphological variation of the three good-colonist Heliantheae shrubs discussed above.

A close morphological counterpart of *Trixis*, in the northern tribe Senecioneae, is *Peucephyllum schottii*. This monotypic genus extends farther north, to Nevada and Utah, and its southern limits are along the northwestern gulf coast as far south as San Marcos island. It is recorded at several peninsular sites south to Bahía de los Ángeles, and it reaches the islands of Ángel de la Guarda, San Esteban, and San Lorenzo. This distribution might date from the time when present Mojave Desert taxa extended much farther south down the eastern peninsula during wetter and especially cooler climates. The only other member of Senecioneae on a gulf island is the annual *Senecio mohavensis* on Ángel de la Guarda.

Other genera of composite shrubs achieve wide distribution through many partly allopatric species rather than with single variable species or single uniform species. A prime example is *Ambrosia*, of the tribe Ambrosieae, with 17 species in Wiggins' (1980) peninsular flora. This total includes two northern weedy species and a further two herbaceous species that occur over the northern half of the peninsula but reach no islands. The remaining 13 are shrub species that cover the peninsula and Sonora very well among the various species but are relatively poor island colonists (or survivors) in comparison to other composite shrubs of broader distribution. Two species, *A. dumosa* and *A. ilicifolia*, are common only as far south as the midriff islands, throughout which, from Ángel de la Guarda to Tiburón, their coverage is complete. *A. bryantii* extends over the southern two-thirds of the peninsula and onto the three largest southern landbridge islands. Four other species are less predictable within the gulf: *A. camphorata* only on Santa Catalina, *A. chenopodifolia* on Ángel de la Guarda and in a restricted area of the Sonoran coast (its Sonoran replacement is *A. deltoidea*), *A. divaricata* in mid-peninsula, on Tiburón and at a couple of Sonoran coastal sites, and *A. magdalenae* again on Ángel de la Guarda and in Sonora. Most of these species are a priori candidates for wider island distributions. Yet other species, both in Sonora or on the peninsula (*A. ambrosioides, A. deltoidea*, and *A. cordifolia*) have been found on no islands. A second shrubby genus in this tribe, *Hymenoclea*, is of generally northern associations, but one species (*H. salsola* ssp. *pentalepis*) reaches mid-peninsula, and a second (*H. monogyra*) reaches the cape region. Tiburón has both species, Ángel de la Guarda just the former, and Cerralvo just the latter.

One of the largest tribes of Asteraceae in the gulf area is the Astereae, with 67 species on the peninsula well allocated among annuals, herbaceous perennials, and shrubs. Characteristically, it is the shrubs that succeed on the islands: *Baccharis* (two species, five northern islands), *Gutierrezia* (one species on Ángel de la Guarda), and especially the taxon until recently known as *Haplopappus* (and now subdivided into *Ericameria, Machaeranthera, Xylothamnia*, and others). In this group, a herbaceous perennial reaches the islands (*M. pinnatifida*), but its great diversity is in woody shrubs, with 21 species (in the old genus) on the peninsula. Many have restricted ranges and habitats, with just four appearing likely candidates for islands ranges, and two of these fulfilling the expectation. *M. arenaria* does not extend farther north than La Paz on the peninsula, but it is on most major islands from Carmen and Coronado south. *Xylothamnia diffusa* (= *H. sononensis*) is more broadly distributed on both sides of the gulf, on all islands larger than 25 km^2 except Santa Catalina, and on all smaller landbridge islands. The only other island record for the tribe is *Xylorhiza* [ex. *Machaeranthera*] *frutescens*, a derivative of the northerly X. [*M*]. *orcuttii*, endemic to Ángel de la Guarda and its vicinity.

In the tribe Eupatorieae, a conspicuous and speciose genus of suffruticose or fruticose plants is *Brickellia*. Of 14 peninsular species, 6 are northerly or of high elevations, and 2 are on the Pacific coast and reach Pacific islands, with 1 (*B. hastata*) on Cerralvo. A further two species are endemic in the Sierra de la Giganta and reach no islands, and a further two extend from these mountains to the cape. One of these is *B. brandegeei*, which is recorded on Espíritu Santo. *B. coulteri* ranges all around the gulf and onto Tiburón and San Esteban; *B. glabrata* covers the southern half of the peninsula and the central gulf islands from San Marcos to Carmen and its neighbors and thence to Cerralvo.

More remarkable in this tribe is the genus *Hofmeisteria* and its closely related (and recently segregated) genus *Pleurocornis*. These taxa perhaps come closer than any other plant group to an adaptive radiation on gulf islands. The complex includes the endemic *H. filifolia* on Ángel de la Guarda, a local peninsular endemic (*H. gentryi* in the Sierra de la Giganta), two broadly distributed taxa, *H. laphamioides* and *H. fasciculata*, both very variable and both occurring in one form or another on most gulf islands. A fifth species, *H. crassifolia*, has a Sonoran range that includes San Pedro Nolasco. All peninsular taxa except *H. gentryi* occur on the gulf islands, and these subshrubs appear to be in the process of further differentiation.

A similar adaptive radiation seems to be underway in the gulf region in a second genus of the tribe Helenieae, the rock-daisies *Perityle*, which includes both annual and shrubby forms. Three peninsular species are not recorded from islands (*P. cuneata, P. incompta,* and *P. lobata*), and another one (*P. incana*) is endemic to Isla Guadalupe on the Pacific side. Five species occur on gulf islands (see fig. 4.9 for two of these) *P. aurea, P. californica, P. crassifolia, P. emoryi,* and *P. microglossa*, although none is restricted to islands. Certainly the habit in these daisies of clinging to the most desolate rock stands these plants in good stead for insular life, and they reach many of the smallest and most isolated islands in the gulf. The number of taxa on the islands peaks at five, on Cerralvo, which lacks only *P. aurea*, is the only island with *P. microglossa*, and like many other southern islands supports both forms of *P. crassifolia*. Species numbers decline with both island size and increasing latitude [$\log(\text{SPP}+1) = 2.41 + 0.145*\log\text{AREA} - 0.08$ LAT; $R^2 = .79$, $p < .001$], but is unaffected by island isolation or landbridge status. Most of the northern islands, despite large size or landbridge status, have just the two widespread species, *P. aurea* and *P. emoryi*. In figure 4.16 (upper), contours show these joint effects on *Perityle* species richness, and the 75% confidence ellipses of each taxon's distribution are shown in the lower part of the figure. Aspects of incipient speciation in this plant group, as well as the morphological features in which they diverge and their adaptive significance, would be a rewarding subject of future studies.

This analysis of the family Asteraceae permits the following conclusions: As with leguminous trees, the species most likely to be widely distributed on gulf islands are those with broad distributions around and beyond the gulf rather than restricted distributions within the region. Plants that are morphologically variable show up on gulf islands frequently, but other taxa are equally successful with a uniform morphology. Endemism on gulf islands is notable only because it is so low, presumably because the islands are not sufficiently isolated to have severely diminished species counts, and, further, the physical habitats on islands are much the same as those on parts of the adjacent peninsula or mainland. These desert habitats selectively favor shrubs over herbaceous plants generally, but it is within several groups of herbaceous or suffrutescent plants that any ongoing adaptive radiation might be sought.

Because this family is well known for its wind-dispersed achenes, we consider last dispersal potential and its possible constraints on insular distributions. Within Asteraceae, dispersal mechanisms and potential are diverse. At one extreme are light achenes extended above to form a beak and topped with a pappus like a parachute— the dandelion model (see Cody and Overton 1996 for a study of island evolution in achene/pappus morphology). At the other extreme are large, heavy, and smooth achenes—the sunflower model. Between these extremes are achenes with a copious plumose pappus, a sparser pappus of bristles, a pappus of a few awns, or just a few scales. All these fruit types are represented in taxa broadly distributed on the gulf islands. The dandelion-type parachute is particularly common among annual chicories, species underrepresented on the islands. Conversely, many shrubby Heliantheae are

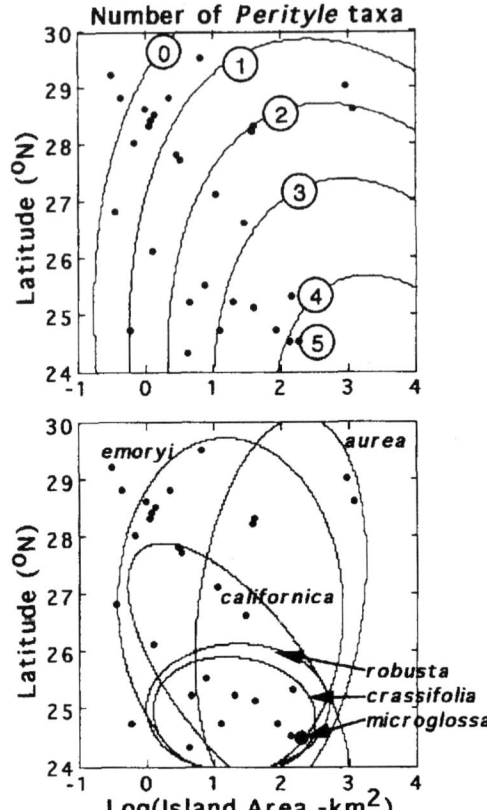

Figure 4.16 Some degree of adaptive radiation is underway in the gulf region and on islands in the Rock-daisies *Perityle*. Upper graph: The number of taxa on islands peaks at five (Cerralvo) and decreases with both diminishing island size and increasing latitude. Lower graph: The ranges of the six island taxa are indicated, with just two taxa reaching the northern islands and just one of these (*P. emoryi*) common on very small islands.

common and diverse on the islands, and these have achenes with little in the way of adornments predisposing them to wind dispersal. We conclude that the structure of the achene/pappus dispersal unit has very little predictive value for likely island colonists or island survivors.

References

Axelrod, D. 1950. The evolution of desert vegetation in Western North America. Carnegie Institute of Washington Publication **590**:215–306.

Axelrod, D. 1958. Evolution of the Madro-Tertiary geoflora. *Botanical Review* **24**:433–509.

Axelrod, D. 1979. Age and origin of Sonoran Desert vegetation. *California Academy of Science Occasional Papers* **132**:1–74.

Baldwin, B., D.W. Kyhos, J. Dvorak and G.D. Carr. 1991. Chloroplast DNA: evidence for the North American origin of the Hawaiian silversword alliance (Asteraceae). *Proceedings of the National Academy of Science USA* **88**:1840–1843.

Carlquist, S. 1965. *Island Life*. New York: Natural History Press.

Carter, A. 1964. The genus *Alvordia* (Compositae) of Baja California, Mexico. *Proceedings of the California Academy of Science* **30**:157–174.

Cody, M.L. 1975. Towards a theory of continental species diversities: bird distributions over Mediterranean habitat gradients. In: M.L. Cody and J.M. Diamond (eds), *Ecology and Evolution of Communities*. Belknap Press, Cambridge, MA; pp. 214–257.

Cody, M.L. 1984. Branching patterns in columnar cacti. In: N. Margaris, M. Arianoutsou-Farragitaki, and W. Oechel (eds), *Being Alive on Land. Trends in Vegetation Science*, vol. 14. Junk, The Hague; pp. 201–236.

Cody, M.L. 1986a. Diversity, rarity, and conservation in Mediterranean-Climate regions. In: M. Soulé (ed), *Conservation Biology: The Science of Scarcity and Diversity*. Sinauer and Associates, Sunderland, MA.

Cody, M.L. 1986b. Distribution and morphology of columnar cacti in tropical deciduous woodland, Jalisco, Mexico. *Vegetatio* **66**:137–145.

Cody, M.L. 1993. Bird diversity patterns and components across Australia. In: R. Ricklefs and D. Schluter (eds), *Species Diversity in Ecological Communities*. University of Chicago Press, Chicago. pp. 147–158.

Cody, M.L. 1999. Assembly rules at different scales. In: E. Weiher and P. Keddy (eds), *Ecological Assembly Rules: Perspectives, Advances, Retreats*. Cambridge University Press, Cambridge. pp. 165–205.

Cody, M.L. 2002. Growth form variations in columnar cacti (Cactaceae: Pachycereeae) within and between North American habitats. In: T. Fleming and A. Valiente-Banuet (eds), *Biology of Columnar Cacti and Their Mutualists*. University of Arizona Press, Tucson; pp. 204–235.

Cody, M.L., E.R. Fuentes, W. Glanz, J.W. Hunt, and A.R. Moldenke. 1977. Convergent evolution in the consumer organisms of Mediterranean Chile and California. In: H. Mooney (ed), *Convergent Evolution in Chile and California*. Dowden, Hutchinson & Ross, New York. pp. 144–192.

Cody, M.L., R. Moran and H.J. Thompson. 1983. The Plants. In: T. Case and M.L. Cody (eds), Island Biogeography in the Sea of Cortez. University of California Press; pp. 49–97.

Cody, M.L. and J. Mcl. Overton. 1996. Short-term evolution of reduced dispersal in island plant populations. *Journal of Ecology* **84**:53–61.

Darlington, P.J. Jr. 1957. *Zoogeography*. John Wiley & Sons, NY. Rept. 1982, R.F. Krieger Publ., Malabar, FL.

Diamond, J.M. 1975. Assembly of species communities. In: M.L Cody and J.M. Diamond (eds), *Ecology and Evolution of Communities*. Belknap Press, Cambridge, MA; pp. 342–444.

Felger, R.S., and C.H. Lowe. 1976. The island and coastal vegetation and flora of the northern part of the Gulf of California. *Los Angeles County Natural History Museum Contributions in Science* **285**:1–59.

Gentry, H. 1942. Rio Mayo plants. *Carnegie Institute of Washington Publications* **527**:1–328.

Gentry, H. 1949. Land Plants collected by the Velero III, Allan Hancock Pacific Expeditions 1937–1941. Allan Hancock Pac. Exped. 13(2): 3–245.

Gentry, H. 1972. *The Agave Family in Sonora*. U.S. Department of Agriculture Handbook 399. Washington, DC: Government Printing Office.

Gentry, H. 1978. The agaves of Baja California. *California Academy of Sciences Occasional Papers* **130**:1–119.

Gibson, A.C. 1977. Wood anatomy of *Opuntias* with cylindrical to globular stems. *Botanical Gazette* **138**:334–351.

Gibson, A.C., and K.E. Horak. 1978. Systematic anatomy and phylogeny of Mexican cacti. *Annals of the Missouri Botanical Garden* **65**:999–1057.

Hamilton, T.H., I. Rubinoff, R.H. Barth, and G. Bush. 1963. Species abundance: natural regulation of insular variation. *Science* **142**: 1575–1577.

Harner, T.H., and K.T. Harper. 1976. The role of area, heterogeneity, and favorability in plant species diversity of pinyon-juniper woodlands. *Ecology* **57**:271–277.

Hastings, J.R. 1964a. Climatological data for Baja California. Technical Report, Meteorology and Climatology of Arid Regions 14. University of Arizona Institute of Atmospheric Physics, Tucson.

Hastings, J.R. 1964b. Climatological data for Sonora and northern Sinaloa. Climatological data for Baja California. Technical Report, Meteorology and Climatology of Arid Regions 15. University of Arizona Institute of Atmospheric Physics, Tucson.

Hastlngs, J.R., and R.R. Humphrey. 1969a. Climatological data and statistics for Baja California. Climatological data for Baja California. Technical Report, Meteorology and Climatology of Arid Regions 18. University of Arizona Institute of Atmospheric Physics, Tucson.

Hastlngs, J.R., and R.R. Humphrey. 1969b. Climatological data and statistics for Sonora and northern Sinaloa. Technical Report, Meteorology and Climatology of Arid Regions 19. University of Arizona Institute of Atmospheric Physics, Tucson.

Hastings, J.K., and R.M. Turner. 1965. Seasonal precipitation regimes in Baja California, Mexico. *Geografisker Annaler* **47**:204–223.

Hastings, J.K., R.M. Turner, and D.K. Warren. 1972. An atlas of some plant distributions in the Sonoran Desert. Technical Report, Meteorology and Climatology of Arid Regions 19. University of Arizona Institute of Atmospheric Physics, Tucson.

Johnson, M.P., L.G. Mason, and P.H. Raven. 1968. Ecological parameters and plant species diversity. *American Naturalist* **102**:297–306.

Johnson, M.P., and P. Raven. 1970. Natural regulation of plant species diversity. *Evolutionary Biology* **4**:127–162.

Kearney, T.H., and R.H. Peebles. 1951. *Flora of Arizona*. University of California Press, Berkeley.

Kluge, M., and I.P. Ting. 1978. *Crassulacean Acid Metabolism: Analysis of an Ecological Adaptation*. Springer-Verlag, Berlin.

May, R.M. 1975. Patterns of species abundance and diversity. In: M.L. Cody and J.M. Diamond (eds), *Ecology and Evolution of Communities*. Belknap Press, Cambridge, MA; pp. 81–120.

McVaugh, R., and J. Rzedowski. 1965. Synopsis of the genus *Bursera* L. in western Mexico, with notes on the material of *Bursera* collected by Sesse and Mociño. *Kew Bulletin* **18**:317–382.

Moran, R.1962. *Pachycereus orcuttli*—a puzzle solved. *Cactus and Succulents Journal* (Los Angeles) **34**:88–94.

Munz, P.A. 1974. *A Flora of Southern California*. University of California Press, Berkeley.

Porter, D.M. 1974. The Burseraceae in North America north of Mexico. *Madroño* **22**:273–276.

Raven, P.H., and D.I. Axelrod. 1978. Origin and relationships of the California flora. *University of California Publications in Botany* **72**:1–134.

Rosenzweig, M.L. 1995. *Species Diversity in Space and Time*. Cambridge University Press, Cambridge.

Rzedowski, J. 1968. Notas sobre el genero *Bursera* (Burseraceae) en el Estado de Guerrero, (Mexico). *Annales de la Escuela Nacional de Ciencias Biologicas Mexico* **7**:17–36.

Rzedowski, J., and H. Kruse. 1979. Algunas tendencias evolutivas en Bursera (Burseraceae). *Taxon* **28**:103–116.

Shreve, F. 1951. The vegetation of the Sonoran Desert. *Carnegie Institute of Washington Publications* **591**:1–1192.

Shreve, F., and I.L. Wiggins. 1964. *Vegetation and Flora of the Sonoran Desert*, 2 vols. Stanford University Press, Stanford, CA.

Simpson, B.B. 1974. Glacial migrations of plants; island biogeographical evidence. *Science* **185**:698–700.

Standley, P.C. 1920–26. Trees and shrubs of Mexico. *US National Herbarium Contributions* **23:**1–1721.
Van Devender, T.R., and S.L. Friedman. 2000. From tropical forest to the Sonoran Desert: gradients in space and time. In: R.H. Robichaux and D.A. Yeton (eds), *The Tropical Deciduous Forest of Alamos, Sonora: Ecology and Conservation of a Threatened Ecosystem.* University of Arizona Press, Tucson; pp. 38–101.
Whittaker, R.H. 1972. Evolution and measurement of species diversity. *Taxon* **21:**213–251.
Wiggins, I.L. 1960. The origin and relationships of the land flora. *Systematic Zoology* **9:**148–165.
Wiggins, I.L. 1980. *Flora of Baja California.* Stanford University Press, Stanford, CA.
Williams, C.B. 1964. *Patterns in the Balance of Nature.* Academic Press, New York.
Williamson, M. 1981. *Island Populations.* Oxford Univ. Press, Oxford.
Yeaton, R.I., and M.L. Cody. 1979. The distribution of cacti along environmental gradients in the Sonoran and Mojave Deserts. *Journal of Ecology* **67:**529–541.

5

Ants

APRIL M. BOULTON
PHILIP S. WARD

The distribution and abundance of ants on islands has attracted considerable attention from ecologists and biogeographers, especially since the classic studies by Wilson on the ants of Melanesia and the Pacific islands (Wilson 1961; Wilson and Taylor 1967a,b; see also updates by Morrison 1996, 1997). The species–area curve for Polynesian ants was an important contribution in the development of island biogeography theory (MacArthur and Wilson 1967). Subsequent studies of other island ant faunas, such as those of the Caribbean (Levins et al. 1973; Wilson 1988; Morrison 1998a,b), Japan (Terayama 1982a,b, 1983, 1992), Korea (Choi and Bang 1993; Choi et al. 1993), and island archipelagos in Europe (Baroni Urbani 1971, 1978; Pisarski et al. 1982; Vepsäläinen and Pisarski 1982; Ranta et al. 1983; Boomsma et al. 1987) and North America (Goldstein 1976; Cole 1983a,b), have confirmed the general features of this relationship, although the underlying causative agents and the relative contribution of stochastic and deterministic processes to ant community composition remain points of controversy.

The islands in the Sea of Cortés are particularly interesting from a biogeographic standpoint because they vary considerably in size, topography, and isolation. In addition, both oceanic and landbridge islands occur in the gulf, allowing comparisons between faunas that resulted from colonization (assembly) versus relaxation. Nevertheless, the ant assemblages of the gulf islands have received little study. There are a few scattered island records in taxonomic and faunistic papers (Smith 1943; Cole 1968; MacKay et al. 1985). Bernstein (1979) listed 16 ant species from a total of nine Gulf of California islands, but a number of evident misidentifications occur in her list (see below). To the best of our knowledge, no other publications have appeared on the ant communities of these islands.

In this chapter, we document the ant species known from islands in the Sea of Cortés and analyze species composition in a selected subset of the better sampled islands. Most of the data come from recent collections made within the last two decades. After describing in qualitative terms the ant fauna of the islands and adjacent mainland areas, we address the following questions: (1) How does ant species richness vary as a function of island area and isolation? (2) Is ant species richness correlated with other abiotic and biotic factors? (3) Is there evidence that competitive interactions affect species composition of Baja-island ant communities? We compare our results with those of other studies of ants in insular habitats and discuss the implications for community structure.

Collections and Data Analysis

The database for this study was compiled from several sources. P. S. Ward examined ant specimens in the following collections, searching for records from both the Sea of Cortés islands and the adjacent mainland areas of Baja California and Sonora: California Academy of Sciences, San Francisco (CASC), California Department of Food and Agriculture, Sacramento (CDAE), Natural History Museum of Los Angeles County (LACM), and Bohart Museum of Entomology, University of California at Davis (UCDC). In addition, recent ant collections from the region made by Rolf Aalbu, April Boulton, Ted Case, Brian Fisher, Robert Johnson, Andy Suarez, and others were processed and identified (material deposited in UCDC).

The literature records of Smith (1943), Cole (1968), and MacKay et al. (1985) were added to the database. This involved only three ant species, all of which were also known to be island inhabitants from the above-mentioned museum material. Bernstein (1979) recorded 16 species of ants from the gulf islands, attributing identification of the specimens to R. E. Gregg. We have not examined this material, but a careful evaluation of the records in light of current knowledge suggests the following taxonomic updates. We judge that the record of *Monomorium minimum* refers to *Monomorium ergatogyna*; that *Solenopsis aurea* is actually *Solenopsis xyloni*; and that *Aphaenogaster mutica* is a misidentification of *Aphaenogaster boulderensis*. The reports of *Myrmecocystus comatus* and *Myrmecocystus melliger* from the islands are almost certainly incorrect (see Snelling 1976), but we cannot be certain which species are involved. They might be *M. placodops* and *M. semirufus*, respectively, but we have omitted these records for now. The 10 other species recorded from various islands are plausible identifications and have been provisionally accepted.

We list all ant species recorded from islands in the Sea of Cortés in appendix 5.1. In an attempt to assess sampling intensity, we also tallied the number of collections from each island. For our own field work we have used accession numbers, where a different number is assigned to each collection from a given micro-site, such as a nest series, a series of foragers from one location, a collection from spider or ant-nest middens, or a series of pitfall trap samples from one transect. Each accession number counts as one collection. For other recently collected material, sent to us by other field workers, accession numbers were generally not used, so we count as one collection all the ants from a given vial. Finally, for the older point-mounted museum material and for literature records, a "collection" refers to a record of a species from a given locality by a given collector.

Considering the variable number and types of collections from each island, we confined our quantitative analyses to two data sets (table 5.1), termed "high confidence" and "low confidence," and we excluded the least well-surveyed islands. Our low confidence (LC) data set includes all those islands (1) for which we have four or more collection records, regardless of collection type, or (2) which were sampled with pitfall traps. The LC data set covers 37 islands (table 5.1) and was used in traditional biogeographic analyses of species richness, island area, and isolation (after MacArthur and Wilson 1967).

The high confidence (HC) data set is restricted to 13 islands in Bahía de los Ángeles that were systematically surveyed by the senior author from 1998 to 1999 using pitfall traps, opportunistic collecting, and a large set of ant samples from spider middens. An aerial view of islands in this region is shown in fig. 16.7. Middens were collected both opportunistically and along a 100-m transect by locating rocks within 1 m of the line and cleaning them of all spider-midden remains. Because insect exoskeletons can remain intact for long periods of time, midden samples are excellent sources for ant remains. Pitfall trapping was implemented at least twice every season (1998–1999) and consisted of five traps/transect spaced at 2-m intervals. Three to five transects were used per island, depending on island size. The HC data set was examined for island biogeographic trends, dominance patterns, species-association trends, and relationships with plant-species richness. Plant-species richness was surveyed on the HC islands (see Polis et al., chap. 13, this volume) in 1998–present. Other plant records were taken from Cody et al. (1983). We also used estimates of ocean channel depth between each island and any point of the adjacent mainland areas (both Sonora and the peninsula) to examine the effect of previous land connections to mainland source areas (as described in chaps. 2, 6, 9; app. 1.1).

To assess the efficacy of sampling, we calculated species-accumulation curves for the three largest islands in our HC data set (i.e., Coronado, Cabeza de Caballo, and La Ventana). These show a substantial decline in slope with increasing sample size (fig. 5.1), indicating that our sampling regime is reasonably effective in revealing the composition of the ground-foraging ant community on these small islands.

The theory of island biogeography was developed by MacArthur and Wilson (1967) as a means of explaining differences among islands in the number of species. They proposed that immigration rate was a function of island distance from the mainland or source pool, such that islands closer to the source should have greater immigration rates than isolated islands. They also asserted that extinction was a function of island size: larger islands should experience lower rates of extinction than smaller islands. Following the logic of MacArthur and Wilson (1967), we examined the dependence of ant species number (log scale) on both log island area and log isolation. Isolation was computed as the distance to the closest mainland, either in Sonora or on the Baja California peninsula. In addition to assessing these relationships, we performed multiple regressions of ant species richness on island area, elevation, and isolation. Elevation was represented by the highest point on each island.

Data were analyzed with parametric statistics using the statistical computer package SPSS (version 8.0 for Windows). Species-accumulation curves were calculated using

Table 5.1 Data for islands considered to be sampled with high confidence (HC) and low confidence (LC)

Island	Type	No. of Collections	No. of Ant spp.	Area (km^2)	Isolation (km)	Maximum Elevation	Ocean Depth (m)	Sea Bird Status	No. of Plant spp.
Coronadito	HC	6	5	0.072	2.913	15	35	2	21
Bota	HC	9	6	0.093	2.919	40	35	1	29
Cabeza de Caballo	HC	9	10	0.704	2.015	140	40	1	38
Cerraja	HC	6	4	0.037	2.957	12	35	2	22
Coronado [=Smith]	HC	24	16	8.684	2.187	465	35	1	74
Flecha	HC	6	2	0.129	2.907	50	35	2	18
Jorobado	HC	7	3	0.039	2.179	25	35	2	7
La Ventana	HC	12	14	1.275	3.15	120	35	1	63
Llave	HC	3	2	0.022	2.957	8	35	2	13
Los Gemelitos Este	HC	3	2	0.047	0.826	20	40	3	10
Los Gemelitos Oeste	HC	5	4	0.02	0.848	20	40	3	11
Mitlán	HC	5	9	0.156	1.96	40	35	1	35
Pata	HC	7	7	0.136	2.559	40	35	1	40
El Piojo	LC	2	5	0.533	6.323	60	35	3	26
El Pescador	LC	3	2	0.035	0.516	12	35	2	9
"Bahía Ánimas Blanca"	LC	2	4	0.029	0.643	12	10	2	6
"Bahía Ánimas Est"	LC	2	4	0.004	0.204	10	10	3	6
"Bahía Ánimas Oeste"	LC	2	4	0.004	0.121	10	10	3	14
Ángel de la Guarda	LC	49	25	924.1	12.12	1316	244	1	199
Partida	LC	2	2	0.931	17.392	122	400	3	
Islote east of Partida	LC	1	1	0.14	19.392	15	400	3	
Rasa	LC	3	3	0.377	20.253	31	400	3	
Salsipuedes	LC	2	3	0.998	18.054	114	400	3	
San Lorenzo Norte	LC	2	3	4.192	19.305	200	400	3	
San Esteban	LC	25	10	41.534	34.5	431	300	1	117
San Pedro Mártir	LC	5	4	2.9	47.25	320	275	3	24
Tiburón	LC	35	37	1173.4	1.745	875	10	1	298
Carmen	LC	5	14	138.65	5.76	479	150	1	163
Mestiza	LC	4	8	0.087	0.239	20	15	1	
Islote Blanco	LC	4	7	0.03	0.63	20	10	2	
Danzante	LC	5	14	4.49	2.679	320	30	1	114
Las Tijeras	LC	4	5	0.025	1.787	20	20	2	
Pardo	LC	4	10	0.038	0.457	10	20	1	
San José	LC	8	7	162.26	4.696	633	61	1	138
San Francisco	LC	5	3	3.868	7.163	210	63	1	76
Espíritu Santo	LC	4	2	80.599	6.269	576	12	1	
Cerralvo	LC	5	5	139.87	9.189	767	235	1	143

LC analyses included both LC and HC islands. Plant species richness data for HC islands are from appendix 4.4; plant data for the LC islands are from Cody et al. (1983). Sea bird status codes follow Sánchez-Piñero and Polis (2000): 1 = nonbird; 2 = roosting; 3 = nesting.

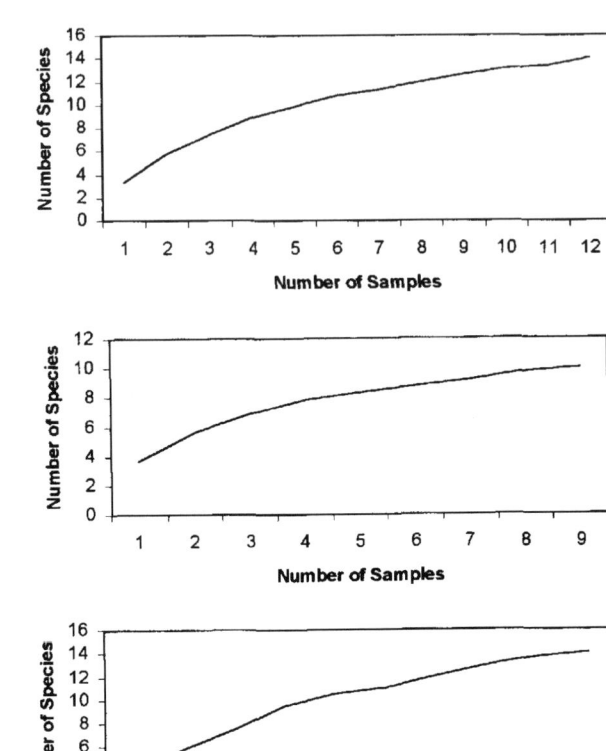

Figure 5.1 Species-accumulation curves for (a) Coronado (= Smith); (b) Cabeza de Caballo; and (c) La Ventana.

the program EstimateS (R. K. Colwell, unpublished software, version 5.0). This program randomizes sample order 100 times and computes means for each succeeding sample.

Results and Discussion

Historical Background

The earliest known ant collections from the islands in the Gulf of California were made by E. P. Van Duzee and J. C. Chamberlin in 1921, during an expedition sent by the California Academy of Sciences (Slevin 1923). Based on material in the CASC, these two individuals collected eight ant species from a total of eight islands. Later island collections (1953–1979) by D. W. Anderson, P. H. Arnaud, E. V. Dawson, J. T. Doyen, J. Nyhan, and S. C. Williams added nine more species (material in CASC, LACM, and UCDC). More recent field work on the islands (1986–present) by R. Aalbu, A. M. Boulton, T. J. Case, B. L. Fisher, F. Sanchez-Piñero, and P. S.

Ward has greatly expanded our knowledge of island ant composition, bringing the total number of specimen-based records to 63 species.

For two of the species collected by Van Duzee and Chamberlin there are no additional island records, and several of the species subsequently collected on the islands are known only from single collections. It is clear that more extensive surveys are needed for many of the islands.

Ant Fauna of Adjacent Regions

The ant faunas occurring on either side of the Sea of Cortés represent the source pools for the gulf islands. The ants of Sonora have not been intensively studied, but those occurring on the Baja California peninsula are now reasonably well known, at least to the point where we can circumscribe, to a first approximation, the parent fauna for the islands on the western side of the gulf.

To date, 176 species of ants belonging to 36 genera and 7 subfamilies have been recorded from Baja California, of which 47 species (27%) appear to be endemic to the peninsula (Johnson and Ward, in press). Only six species are non-native, making the ant fauna of Baja California one of the most pristine (i.e., least invaded by exotics) of any warm temperate/subtropical region in the world. About 55 ant species are confined to the relatively mesic northwestern corner of Baja California (the California floristic province), nine more species are either shared with or endemic to the Sierra de la Laguna of the cape region, and three species are restricted to other peripheral areas at the northern limit of the peninsula. This leaves about 100 indigenous ant species that occupy the arid heartland of the peninsula and that can be thought of as potential colonists of the desert islands (or as contributing to the fauna that was cut off from the mainland, in the case of landbridge islands).

Most of these ants are ground-nesting species, and the workers are generalized scavengers and predators, feeding on dead and living arthropods, vertebrate carrion, homopteran honeydew, floral and extrafloral nectar, and seeds. The ants in certain genera, such as *Messor*, *Pogonomyrmex*, and *Pheidole*, are generally considered seed-harvesting specialists, but even they remain opportunistically omnivorous (Snelling and George 1979). The fungus-growing ants (tribe Attini) are represented in the Baja California deserts by at least four species, belonging to the genera *Acromyrmex*, *Cyphomyrmex*, and *Trachymyrmex*. Specialized arthropod predators include several species of army ants (genus *Neivamyrmex*) and the endemic ponerine ant *Leptogenys peninsularis*.

Ant Fauna of the Islands

In the present study we have accumulated records of 65 ant species from islands in the Sea of Cortés (app. 5.1), of which 45 are found in the western islands. (We use the term "western islands" to refer to those adjacent to the Baja California peninsula—i.e., the islands lying within the states of Baja California and Baja California Sur.) Nearly all of these ant species are known from adjacent mainland areas in Baja California, Sonora, or both. Only four species appear to be island endemics (one *Pheidole* species, two *Leptothorax* species, and one *Forelius* species), and more intensive mainland collecting is likely to reduce this number even further. In the ant list [app. 5.1] species which could not be associated with any known taxon have been assigned code numbers (e.g., *Forelius* sp. BCA-1).

The island ant fauna is representative of the adjacent mainland regions. The 65 species are apportioned among 19 genera and 6 subfamilies. The largest genera in the deserts of the Baja California peninsula (*Pheidole*, 18 species; *Myrmecocystus*, 15 species; *Leptothorax*, 9 species) are also the most species-rich on the islands (*Pheidole*, 13 species; *Myrmecocystus*, 6 species; *Leptothorax*, 5 species), although the relative richness of *Pheidole* and *Myrmecocystus* suggests that species in the latter genus are less successful colonists. Fungus-growing ants (tribe Attini) are represented on the islands by several species of the relatively primitive genus *Cyphomyrmex*, but the more specialized leaf-cutting ant, *Acromyrmex versicolor*, appears to be absent from all islands except Tiburón, despite being common in desert washes on the Baja California mainland. Ponerine ants, represented on the Baja peninsula by the genera *Hypoponera* and *Leptogenys*, also appear to be rare or absent from the islands.

One species of non-native ant, *Paratrechina longicornis*, has colonized the gulf islands. It occurs sporadically on islands in Bahía de los Ángeles and off Loreto. This species is common around human settlements in parts of the Baja mainland (e.g., in the town of Bahía de los Ángeles), and has presumably been introduced to the islands by human transport.

Army ants (genus *Neivamyrmex*) are known only from Isla Tiburón and Isla San José. About six species of *Neivamyrmex* occur on the Baja California peninsula (the exact number is uncertain because males are often collected away from the colonies, and male- and worker-based names have not been completely reconciled). Their absence from offshore islands presumably reflects their limited dispersal capabilities: *Neivamyrmex* queens are completely wingless, and reproduction occurs by colony fission (Gotwald 1995). In addition, we can expect them to have disappeared on small landbridge islands on which they were initially present, given that army ant populations in small habitat patches are especially extinction prone (Terborgh et al. 1997; Suarez et al. 1998). These ants are rapacious predators on other ant colonies, especially species in the genus *Pheidole*, and their absence from most islands may well have significant community-level effects.

The native ant species which inhabit the gulf islands are mostly ground-nesting, opportunistic scavengers, although the two most frequent island inhabitants, *Pogonomyrmex californicus* and *Solenopsis xyloni*, are also granivorous. Seed-harvesting species of *Messor*, prominent in mainland habitats, are absent from most islands. Arboreal ants, already uncommon on mainland areas, are scarce. They include one species of *Crematogaster* (*C. arizonensis*), one species of *Camponotus* (*C. mina*) and three *Pseudomyrmex* species. Not surprisingly most islands lack these arboreal species (app. 5.1).

Species Number

Abiotic Factors: Area, Isolation, and Sea Depth

For both LC and HC data sets, log area and log elevation account for the major portion of variance in ant species number (table 5.2). This pattern is stronger for the HC islands than for the LC data. Isolation explains none of the variance for either data set. We examined the independence of these three factors with a correlation

Table 5.2 The percent variance in log ant species number explained by single independent variables

Variable	LC Data Set (N = 37)	HC Data Set (N = 13)
Log area	24.1	65.9
Log isolation	0.0	0.0
Log elevation	23.6	58.9

LC, low confidence; HC, high confidence.

matrix. There was not a significant correlation between log isolation and log area ($R = .55$). When we correlated log area and log elevation, they were highly associated with each other ($R = .96$). Thus, elevation and area may represent a "single" variable on which ant species number depends.

There is a positive relationship between log number of ant species and log island area (fig 5.2). The z-value for the HC islands is 0.33, whereas $z = 0.11$ for the LC islands. The steeper z-value for the HC islands may be a result of increased

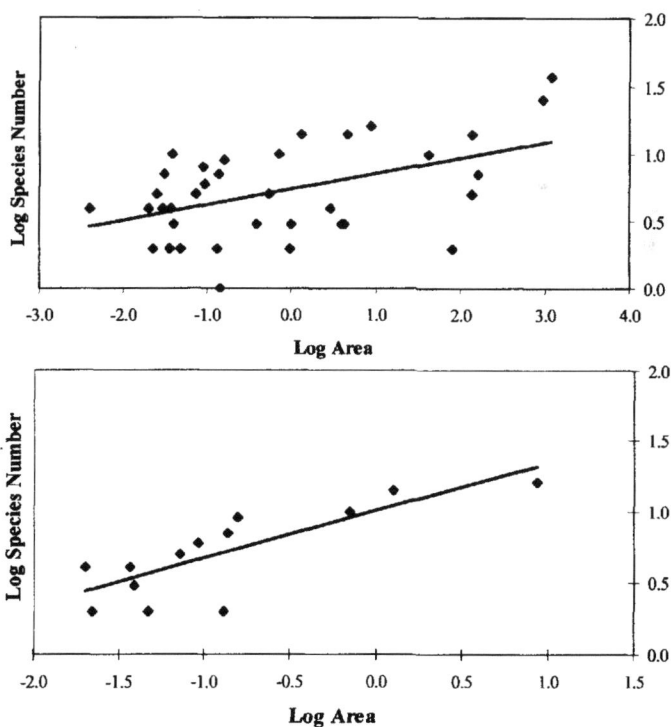

Figure 5.2 Log ant species plotted against log island area for (a) LC data set (z-value = 0.11; $y = 0.1135x + 0.7374$, $R^2 = .2408$) and (b) HC data set (z-value = 0.33; $y = 0.3342x + 1.0009$, $R^2 = .6591$).

Table 5.3 Multiple regression coefficients and the total amount of explained variance (R^2) in an analysis of log ant species.

Variable	LC Data Set ($N = 37$)	HC Data Set ($N = 13$)
Log area	0.2*	0.3*
Log isolation	−0.3*	0.0
Log elevation	0.04	−0.1
y-intercept	1.0*	1.0*
R^2	48.6%	65.9%

LC, low confidence; HC, high confidence.
*Variable has a significant t-value at $p < .05$.

sampling intensity. The lower z-value for the LC islands may be due to the fact that the two data sets span different ranges of island size (HC: small only; LC: small to large) and that the HC islands are included in the LC data set.

A stepwise multiple regression for the HC data set of log species, with log area, isolation, and elevation as independent variables, revealed that area explained most of the variation in ant species number; elevation and isolation were excluded from the model (table 5.3). For the LC data set, the model included both area and isolation, while elevation was excluded (table 5.3). Elevation explained almost as much of the variation in ant species number as area did in the independent linear regressions (table 5.2). The model may have excluded elevation in both the HC and LC multiple regressions because it detected collinearity between area and elevation (Tabachnick and Fidell 1996). It is not surprising that isolation was excluded in the HC multiple regression model because the islands in that data set are all relatively equidistant from the mainland. The LC islands, on the other hand, have a larger range of isolation values, and the regression coefficient shows that isolation has a slightly negative effect on ant species number. Overall, island area explains most of the variation in ant species number for both LC and HC data sets.

Finally, we examined another measure of isolation, shallowest sea-depth between islands and the nearby mainland. Sea depth between an island and mainland areas (both Sonoran and peninsular) did not appear to affect ant species number in either data set (HC: $F = 0.2$, $p = .7$; LC: $F = 1.4$, $p = .2$). Recall that ant species number was also not associated with island isolation as an independent factor (table 5.2). Because island isolation is often related to ocean depth (Brown and Lomolino 1998), we would expect ants to have a similar relationship (or lack thereof) with both variables. In fact, shallowest depth is significantly correlated with isolation for these islands ($R = .8$, $p < .01$).

Biotic Factors: Plants and Sea Birds

A number of studies (e.g., Terayama 1992; Quiroz-Robledo and Valenzuela-González 1995; Torres and Snelling 1997) have shown that plant diversity is often the best predictor of ant species richness. This could be due to the fact that seeds and plant

Table 5.4 Multiple regression coefficients and the total amount of explained variance (R^2) in an analysis of log ant species on log plant richness and log area

Variable	LC Data ($N = 27$)	HC Data ($N = 13$)
Island area	−0.1	0.2
Plant species richness	0.5*	0.9*
y-intercept	0.0	−0.5*
R^2	61.7%	78.0%

LC, low confidence; HC, high confidence. Plant species richness data were available for only 27 of the 37 LC islands.
*Variable has a significant t-value at $p < .05$.

tissue are heavily utilized by a number of ant species (e.g., Davidson 1977; Devall and Thien 1989; Crist and MacMahon 1992). However, the relationship between plant and ant richness may also reflect their correlated response to a third, underlying causative agent, such as habitat heterogeneity.

In this study, log plant richness explains 78% of the variation in ant species richness for the HC islands and 61.7% of the variation for the LC data. Area explained a significant proportion of the variance in the previous section. When we performed a stepwise multiple regression of log ant species on log plant species and log island area, plant richness explained significantly more variation in ant richness than island area (table 5.4). These analyses of ant species richness on plant number and island area indicate that plants may be the better predictor of ant diversity.

We tested whether or not sea bird presence affected ant species number. Polis and colleagues (Sánchez-Piñero and Polis 2000; Polis et al., chap 13, this volume) have shown that productivity and community dynamics on Bahía de los Ángeles islands are driven by allochthonous marine input in the form of sea bird guano and carcasses (Sánchez-Piñero and Polis 2000). This marine input combines with terrestrial productivity by plants to set productivity on an island, and it is these factors that dictate the abundance and diversity of many components of the terrestrial biota, such as spiders (Polis and Hurd 1996), lizards (G. Polis, unpublished data), rodents (Stapp et al. 1999), and beetles (Sánchez-Piñero and Polis 2000).

Table 5.5 Analysis of covariance of ant species number with resulting F-values

Source of Variation	LC Data Set ($N = 37$)	HC Data Set ($N = 13$)
Main effect (sea birds)	7.8*	3.7[a]
Covariate (island area)	0.7	3.6[a]

LC, low confidence; HC, high confidence.
*Variable has a significant F-value at $p < .05$.
[a]Values approached significance at $p < .09$.

Table 5.6 Mean ± SD for island size and plant and ant species richness per island type (based on sea bird activity) for high confidence (HC) islands

HC Data	Nonbird Islands (N = 6)	Roosting Islands (N = 5)	Nesting Islands (N = 2)
Mean island size	1.8 ± 3.4	0.06 ± 0.04	0.03 ± 0.02
Mean plant species	46.5 ± 17.8	16.2 ± 6.2	10.5 ± 0.7
Mean ant species	10.3 ± 4.0	3.2 ± 1.3	3.0 ± 1.4

Sánchez-Piñero and Polis (2000) have categorized these islands into three groups: nonbird (sea birds not present or restricted to a few gulls nesting on cliffs), roosting (sea birds frequently rest but do not breed in large numbers), and nesting (islands that support tens to thousands of active nests). A one-way analysis of variance of ant species richness, with sea bird status (nonbird, roosting, and nesting) as the main effect revealed that seabird presence has a significant, negative impact on ant richness in both data sets (LC: $F = 8.6$, $p < .05$; HC: $F = 9.7$, $p < .05$).

Recall that area was an important predictor of ant richness in the previous section. In an analysis of covariance of ant richness, with sea bird status as the main effect and island area as a covariate, seabird presence was still an important predictor of ant species number (table 5.5). For the LC islands, area was not a significant covariate. For HC islands, area was just as important as sea bird status in predicting ant richness. This result can probably be explained by the fact that all bird islands (both nesting and roosting) in the HC data set are small islands (analysis of variance of log area on sea bird presence $F = 5.7$, $p < .05$).

In summary, island area, sea bird presence, and plant species richness explain a significant amount of variation in ant species richness. How do these three variables interact? Table 5.6 shows that nonbird islands in the HC data set are the largest islands in the Bay and also have the most plant species, while small islands have reduced plant diversity and tend to be roosting/nesting islands. The same trend is seen in the LC data (table 5.7). In an analysis of covariance with sea bird status (nonbird, roosting, and nesting) as the main effect and plant richness as a covariate, only plant-species richness is significantly associated with ant richness (table 5.8). A one-way analysis of variance shows that plant richness significantly varies with sea bird pres-

Table 5.7 Mean ± SD for island size and plant and ant species richness per island type (based on seabird activity) for low confidence (LC) islands

LC Data	Nonbird Islands (N = 17)	Roosting Islands (N = 9)	Nesting Islands (N = 11)
Mean island size	157.6 ± 343.0	0.05 ± 0.03	0.9 ± 1.4
Mean plant species	109.1 ± 76.0	13.7 ± 6.7	15.2 ± 8.1
Mean ant species	11.6 ± 8.6	3.8 ± 1.7	3.2 ± 1.2

Plant species richness was not available for all the LC islands (nonbird, $N = 14$; roosting, $N = 7$; nesting, $N = 6$).

Table 5.8 Analysis of covariance of ant species number with resulting F-values

Source of Variation	LC Data Set ($N = 27$)	HC Data Set ($N = 13$)
Main effect (sea birds)	1.0	1.2
Covariate (log plant number)	5.3*	4.7*

LC, low confidence; HC, high confidence.
*Variable has a significant F-value at $p < .05$.

ence for both data sets (HC: $F = 9.6$, $p < .01$, LC $F = 9.6$, $p < .01$); like ants, there are fewer plant species on seabird islands. This suggests that sea birds reduce ant richness indirectly via decreasing plant species richness. Thus, out of all variables examined, plant species richness seems to be the best predictor of ant species richness for both data sets.

Species Associations

We performed chi-square tests on all possible pairwise combinations of species (Morrison 1996) found on the HC islands (except the six rare species that occurred only once in the data set). For each species pair that was significantly associated, we calculated an index of association value using the formula (observed number of joint occurrences)/(expected number of joint occurrences) (after Eickmeier 1988). This formula produces a value that is either <1 (signifying a negative association between two species) or >1 (representing a positive association). There were no significant negative co-occurrences between any two species, and there were 20 positive associations between pairs of ant species (table 5.9); a sign test shows that there were significantly more positive associations than predicted by a binomial distribution ($p = .002$).

The overall trend is that large islands (e.g., Smith and Ventana) contain the most positive interactions. *Forelius* sp. BCA-2 and *Camponotus festinatus* occur with moderate frequency on the HC islands (three and five islands, respectively), yet they are involved in the greatest number of positive associations ($n = 6$). Moreover, rare species on these islands (e.g., *Messor julianus, Cyphomyrmex* BCA-1) were involved in a disproportionately large number of significant associations, whereas the cosmopolitan species (e.g., *Pogonomyrmex californicus* and *Solenopsis xyloni*) were not involved in any significant associations (as would be expected).

Synthesis

The ant fauna of the islands in the Sea of Cortés essentially represents a subset of the ant species occurring in deserts of the Baja California peninsula and the state of Sonora. All of the genera and most of the species are shared with these neighboring regions. Moreover, most of the ant genera on the desert mainland are represented by at least one species on the islands. Some of these are infrequent island inhabitants, however, which are more or less confined to the larger islands. Examples include the

Table 5.9 Results of a chi-square analysis of the frequency of all possible pairwise combinations of ant species on the high confidence islands

	1	2	3	4	5	6	7	8	9	10	11	12	13	14	15
1. *Dorymyrmex* sp. BCA-1	■	0	+	0	0	++	0	0	0	++	0	+	0	0	0
2. *Forelius pruinosus*		■	0	+	0	0	0	0	0	0	0	0	0	++	0
3. *Forelius* sp. BCA-2			■	++	0	+	0	++	++	0	0	0	0	++	0
4. *Camponotus festinates*				■	0	0	0	+	+	0	++	0	0	+	0
5. *Paratrechina longicornis*					■	0	0	++	++	0	0	0	0	0	0
6. *Aphaenogaster megommata*						■	0	0	0	++	0	0	0	0	0
7. *Crematogaster depilis*							■	0	0	0	0	0	0	0	0
8. *Cyphomyrmex* sp. BCA-1								■	++	0	0	0	0	0	0
9. *Messor julianus*									■	0	0	0	0	0	0
10. *Pheidole cf. californica*										■	0	0	0	0	0
11. *Pheidole yaqui*											■	++	0	0	0
12. *Pheidole* sp. BCA-5												■	0	0	0
13. *Pogonomyrmex californicus*													■	0	0
14. *Solenopsis molesta*														■	0
15. *Solenopsis xyloni*															■

0 indicates a nonsignificant relationship; + indicates a positive association between two species at the $.1 > p > .05$ level; ++ indicates a positive association at the $p < .05$ level. Species numbers in the top row follow the order in column one.

army ants (*Neivamyrmex*), leaf-cutting ants (*Acromyrmex*), seed-harvesting ants in the genus *Messor*, twig-nesting ants in the genus *Pseudomyrmex*, and the ponerine genus *Odontomachus*.

Conversely, certain taxa appear to be particularly well represented on the islands. This includes species in the genera *Forelius*, *Leptothorax*, and *Pheidole*. The last two are small myrmicine ants that may benefit from the general absence of army ants (*Neivamyrmex*), for which the myrmicines are an important prey item. The presence on the islands of such rare mainland species as *Crematogaster rossii* and *Aphaenogaster boulderensis* (both known on the Baja peninsula from only one or two localities) also suggests that the islands may serve as refuges.

Ant species richness varies substantially among islands, and we have examined a number of biotic and abiotic factors previously demonstrated to influence ant species numbers in insular environments (e.g., Wilson and Taylor 1967; Boomsma et al. 1987; Terayama 1992). Our quantitative analyses use data taken from only a fraction of the islands in the gulf (37 out of 130+ islands), so it is worth stressing the need for additional field surveys. Nevertheless, analysis of our HC data set, based on 13 islands, consistently produced results similar to those of the LC set, indicating the robustness of the findings.

Three factors seem to influence ant species richness on the gulf islands: island area, plant species richness, and sea bird presence. More ant species are found on larger, nonbird islands and on islands with more plant species. The strong positive correlation between island area and island elevation makes it difficult to disentangle the effects of these two variables. For the LC data set, isolation (i.e., distance from mainland) has a significant negative effect on ant species richness in a multiple regression analysis that includes area, elevation, and isolation. Thus, our results are at least consistent with classic island biogeography theory in which species numbers are viewed as an equilibrium between area- and distance-dependent immigration and extinction rates. We recognize, however, that many of the islands in our data set (and all of the HC islands) are landbridge islands whose composition will have been affected by historical connections to the mainland. Future studies of the Sea of Cortés ant fauna, including better sampling of oceanic islands, should permit more exact comparisons between landbridge islands and those with no history of attachment to the mainland.

In our analyses, the effect of sea bird presence on ant species richness is striking, even when island area is taken into account (seabird islands tend to be small). Presumably the presence of sea birds alters the island environment in ways that make it much less suitable for most ant species. When sea birds either roost or nest on these islands, their impact is widely felt (Sánchez-Piñero and Polis 2000). An exaggerated example of the impact of sea birds is demonstrated on at least three islands in our data set variously named "Blanco" or "Blanca" due to their complete covering with white bird guano. This extreme guano effect has been shown to reduce plant and animal diversity (Sánchez-Piñero and Polis 2000), especially on the smaller islands. Thus, it is not surprising that ant species richness might be affected by these birds and/or their guano. Attesting to the difficulty in interpreting such a relationship between sea birds and ants, Duffy (1991) noted that sea-bird chicks are sometimes preyed upon by ants on islands off Peru. He then suggested that sea birds may be the ones excluded from an island by ants, not vice versa.

We have preliminary data from a medium-sized sea-bird island in Bahía de los Ángeles (El Piojo) where pelican colonies are scattered around the island and do not homogeneously dominate the landscape (as on smaller islands). Pitfall traps set in both bird-dominated and nonbird areas of El Piojo reveal higher abundance and species richness of ants in the absence of birds. In fact, only one ant species, *Solenopsis xyloni*, was collected in bird areas, while three additional species were encountered at other sites on the island. The causal aspects of this relationship need further investigation, but we provisionally conclude that there is a rather strong negative association between sea birds and ants.

In addition to the ants, sea bird presence negatively impacts plant richness. Throughout all of our analyses with abiotic and biotic factors, plant-species richness remained the best predictor of ant richness. As we suggested earlier, there is no a priori reason to believe that ant diversity is determined by plant diversity. The number of seeds, seed set, and/or plant biomass would be more likely to affect ant abundance and diversity than plant richness alone. Thus, there may be a third, underlying variable controlling both ant and plant richness. At this time, we can only conjecture that this third variable is habitat or landscape heterogeneity. However, there is some evidence suggesting that sea bird presence may be controlling ant and plant richness in that sea birds had a significant negative impact on both plants and ants, separately.

There were a number of positive associations between ant species. In this type of analysis, the chance of a type I error is no longer 0.05 because so many comparisons are being made simultaneously (Morrison 1996). Thus, it is not the total number of significant interactions that is of interest, but the ratio of positive to negative interactions. There were no negative interactions in the results. We do have preliminary data showing that the dominant ant on the Bahía de los Ángeles islands, *Solenopsis xyloni*, is also competitively dominant via interference competition in both mainland and island trials (A.M. Boulton, unpublished data). However, as is often the case, other species (e.g., *Pheidole* spp., *Aphaenogaster megommata*) compensate for this at the baits by being superior exploitative competitors (see also Holway 1999). Consequently, there appears to be no evidence for competitive exclusion on the HC islands.

Our results are concordant with previous findings for ants (and other insects) on islands, in that much of the variance in species number is explained by variation in island area (Wilson and Taylor 1967a,b, Simberloff and Wilson 1969, Polis et al., chap 13, this volume) and plant richness (e.g., Boomsma et al. 1987, Terayama 1992, Torres and Snelling 1997, but see also Davidson 1977, Catangui et al. 1996). The most novel result presented here is the reduced diversity of ants (and plants) on islands with nesting or roosting sea birds. At this stage we are uncertain about the directionality or causality of this negative association.

Finally, we note that our survey of the ant communities on islands in the Sea of Cortés is far from complete. While the islands in Bahía de los Ángeles are now reasonably well known, this is a biased subset with respect to island size and isolation. Understanding of the biotic and abiotic factors influencing these island assemblages of desert ants will be improved by more extensive surveys of the larger and more distant islands in the gulf and by better sampling of the adjacent mainland faunas.

References

Baroni Urbani, C. 1971. Studien zur Ameisenfauna Italiens XI. Die Ameisen des Toskanischen Archipels. Betrachtungen zur Herkunft der Inselfaunen. *Revue Suisse de Zoologie* **78**: 1037–1067.

Baroni Urbani, C. 1978. Analyse de quelques facteurs autécologiques influençant la microdistribution des fourmis dans les îles de l'archipel toscan. *Mitteilungen der Schweizerischen Gesellschaft* **51**:367–376.

Bernstein, R.A. 1979. Evolution of niche breadth in populations of ants. *American Naturalist* **114**:533–544.

Boomsma, J.J., A.A. Mabelis, M.G.M Verbeek, and E.C. Los. 1987. Insular biogeography and distribution ecology of ants on the Frisian islands. *Journal of Biogeography* **14**:21–37.

Brown, J.H., and M.V. Lomolino. 1998. *Biogeography.* 2nd ed. Sinauer Associates, Sunderland, MA.

Catangui, M.A., B.W. Fuller, A.W. Walz, M.A. Boetel, and M.A. Brinkman. 1996. Abundance, diversity and spatial distribution of ants (Hymenoptera: Formicidae) on mixed-grass rangelands treated with Diflubenzuron. *Environmental Entomology* **25**:757–766.

Choi, B.-M., and J.R. Bang. 1993. Studies on the distribution of ants (Formicidae) in Korea (12). The analysis of ant communities in 23 islands. [in Korean.] *Chongju Sabom Taehakkyo Nonmunjip [Journal of Chongju National University of Education]* **30**:317–330.

Choi, B.-M., K. Ogata, and M. Terayama. 1993. Comparative studies of ant faunas of Korea and Japan. I. Faunal comparison among islands of southern Korea and northern Kyushu, Japan. *Bulletin of the Biogeographical Society of Japan* **48**:37–49.

Cody, M., R. Moran, and H. Thompson. 1983. The plants. In: T.J. Case and M.L. Cody (eds), *Island biogeography in the Sea of Cortéz.* University of California Press, Berkeley; pp. 49–97.

Cole, A.C., Jr. 1968. *Pogonomyrmex Harvester Ants. A Study of the Genus in North America.* University of Tennessee Press, Knoxville.

Cole, B.J. 1983a. Assembly of mangrove ant communities: patterns of geographical distribution. *Journal of Animal Ecology* **52**:339–347.

Cole, B.J. 1983b. Assembly of mangrove ant communities: colonization abilities. *Journal of Animal Ecology* **52**:349–355.

Crist, T.O., and J.A. MacMahon. 1992. Harvester ant foraging and shrub-steppe seeds: interactions of seed resources and seed use. *Ecology* **73**:1768–1779.

Davidson, D.W. 1977. Species diversity and community organization in desert seed-eating ants. *Ecology* **58**:711–724.

Devall, M.S., and L.B. Thien. 1989. Factors influencing the reproductive success of *Ipomoea pes-caprae* (Convolvulaceae) around the Gulf of Mexico. *American Journal of Botany* **76**:1821–1831.

Duffy, D.C. 1991. Ants, ticks, and nesting seabirds. In: J.E. Loye and M. Zuk (eds), *Bird-Parasite Interactions*. Oxford University Press, Oxford; pp. 242–257.

Eickmeier, W.G. 1988. Ten years of forest dynamics at Radnor Lake, Tennessee. *Bulletin of the Torrey Botanical Club* **115**:100–107.

Goldstein, E.L. 1976. Island biogeography of ants. *Evolution* **29**:750–762.

Gotwald, W.H., Jr. 1995. *Army Ants: The Biology of Social Predation*. Cornell University Press, Ithaca, NY.

Holway, D.A. 1999. Competitive mechanisms underlying the displacement of native ants by the invasive Argentine ant. *Ecology* **80**:238–251.

Johnson, R.A., and P.S. Ward. in press. Biogeography and endemism of ants (Hymenoptera: Formicidae) in Baja California, Mexico: a first overview. *Journal of Biogeography*.

Levins, R., M.L. Pressick, and H. Heatwole. 1973. Coexistence patterns in insular ants. *American Scientist* **61**:463–472.

MacArthur, R.H., and E.O. Wilson. 1967. *The Theory of Island Biogeography*. Princeton University Press, Princeton, NJ.

MacKay, W.P., E.E. MacKay, J.F. Perez Dominguez, L.I. Valdez Sanchez, and P.V. Orozco. 1985. Las hormigas del estado de Chihuahua Mexico: el genero *Pogonomyrmex* (Hymenoptera: Formicidae). *Sociobiology* **11**:39–54.

Morrison, L.W. 1996. The ants (Hymenoptera: Formicidae) of Polynesia revisited: species numbers and the importance of sampling intensity. *Ecography* **19**:73–84.

Morrison, L.W. 1997. Polynesian ant (Hymenoptera: Formicidae) species richness and distribution: a regional survey. *Acta Oecologia* **18**:685–695.

Morrison, L.W. 1998a. A review of Bahamian ant (Hymenoptera: Formicidae) biogeography. *Journal of Biogeography* **25**:561–571.

Morrison, L.W. 1998b. The spatiotemporal dynamics of insular ant metapopulations. *Ecology* **79**:1135–1146.

Pisarski, B., K. Vepsäläinen, S. Ås, Y. Haila, and J. Tiainen. 1982. A comparison of two methods of sampling ant communities. *Annales Zoologici Fennici* **48**:75–80.

Polis, G.A., and S.D. Hurd. 1996. Linking marine and terrestrial food webs: allochthonous input from the ocean supports high secondary productivity on small islands and coastal land communities. *American Naturalist* **147**:396–423.

Quiroz-Robledo, L., and J. Valenzuela-González. 1995. A comparison of ground ant communities in a tropical rainforest and adjacent grassland in Los Tuxtlas, Veracruz, Mexico. *Southwestern Entomologist* **20**:203–213.

Ranta, E., K. Vepsäläinen, S. Ås, Y. Haila, B. Pisarski, and J. Tiainen. 1983. Island biogeography of ants (Hymenoptera, Formicidae) in four Fennoscandian archipelagoes. *Acta Entomologica Fennica* **42**:64.

Sánchez-Piñero, F., and G.A. Polis. 2000. Bottom-up dynamics of allochthonous input: direct and indirect effects of seabirds on islands. *Ecology* **81**:3117–3132.

Simberloff, D.S., and E.O. Wilson. 1969. Experimental zoogeography of islands: the colonization of empty islands. *Ecology* **50**:278–296.

Slevin, J.R. 1923. Expedition of the California Academy of Sciences to the Gulf of California in 1921. General account. *Proceedings of the California Academy of Sciences* (4)**12**:55–72.
Smith, M.R. 1943. *Pheidole (Macropheidole) rhea* Wheeler, a valid species (Hymenoptera: Formicidae). *Proceedings of the Entomological Society of Washington* **45**:5–9.
Snelling, R.R. 1976. A revision of the honey ants, genus *Myrmecocystus* (Hymenoptera: Formicidae). *Natural History Museum of Los Angeles County Science Bulletin* **24**:1–163.
Snelling, R.R., and C.D. George. 1979. The taxonomy, distribution and ecology of California desert ants. Report to California Desert Plan Program, Bureau of Land Management, U.S. Department of the Interior, Washington, DC.
Stapp, P.T., F. Sánchez-Piñero, and G.A. Polis. 1999. Stable isotopes reveal strong marine and El Niño effects on island food webs. *Nature* **401**:467–469.
Suarez, A.V., D.T. Bolger, and T.J. Case. 1998. Effects of fragmentation and invasion on native ant communities in coastal southern California. *Ecology* **79**:2041–2056.
Tabachnick, B.G., and L.S. Fidell. 1996. *Using Multivariate Statistics.* 3rd ed. Harper Collins, New York.
Terayama, M. 1982a. Regional differences of the ant fauna of the Nansei Archipelago based on the quantitative method. I. Analysis using Nomura-Simpson's coefficient. *Bulletin of the Biogeographical Society of Japan* **37**:1–5.
Terayama, M. 1982b. Regional differences of the ant fauna of the Nansei Archipelago based on the quantitative method. II. Analysis using harmonity index of taxa. *Bulletin of the Biogeographical Society of Japan* **37**:7–10.
Terayama, M. 1983. Biogeographic study of the ant fauna of the Izu and the Ogasawara Islands. *Bulletin of the Biogeographical Society of Japan* **38**:93–103.
Terayama, M. 1992. Structure of ant communities in east Asia. 1. Regional differences and species richness. [In Japanese.] *Bulletin of the Biogeographical Society of Japan* **47**:1–31.
Terborgh, J., L. Lopez, J. Tello, D. Yu, and A.R. Bruni. 1997. Transitory states in relaxing ecosystems of land bridge islands. In: W.F. Laurance and R.O. Bierregaard, Jr. (eds). *Tropical Forest Remnants. Ecology, Management, and Conservation of Fragmented Communities.* University of Chicago Press, Chicago; pp. 256–274.
Torres, J.A., and R.R. Snelling. 1997. Biogeography of Puerto Rican ants: a non-equilibrium case? *Biodiversity and Conservation* **6**:1103–1121.
Vepsäläinen, K., and B. Pisarski. 1982. Assembly of island ant communities. *Annales Zoologici Fennici* **19**:327–335.
Wilson, E.O. 1961. The nature of the taxon cycle in the Melanesian ant fauna. *American Naturalist* **95**:169–193.
Wilson, E.O. 1988. The biogeography of the West Indian ants (Hymenoptera: Formicidae). In J.K. Liebherr (ed), *Zoogeography of Caribbean Insects.* Cornell University Press, Ithaca, NY; pp. 214–230.
Wilson, E.O., and R.W. Taylor. 1967a. An estimate of the potential evolutionary increase in species density in the Polynesian ant fauna. *Evolution* **21**:1–10.
Wilson, E.O., and R.W. Taylor. 1967b. The ants of Polynesia (Hymenoptera: Formicidae). *Pacific Insects Monograph* **14**:1–109.

6

Tenebrionid Beetles

FRANCISCO SÁNCHEZ PIÑERO
ROLF L. AALBU

Desert islands in the Sea of Cortés are inhabited by a rich arthropod fauna. Although the seemingly barren landscape may appear devoid of arthropods at midday, at night one may often see members of one of the dominant groups of animals on the islands, the tenebrionid beetles (fig. 6.1). Highly variable in shape and size, the Tenebrionidae are one of the most highly evolved and diverse families of beetles. Tenebrionidae is perhaps the fifth largest beetle family (>2000 genera), and about half of these are uniquely adapted to arid environments and form a dominant group in desert ecosystems (Crawford 1991). For example, the adult biomass of one species (*Asbolus verrucosus*) at a site in the northern Mojave Desert was found to be greater (275 g/ha) than the combined biomass of all mammals, birds, and reptiles (a total of 263 g/ha) in the same area (Thomas 1979).

The major problem that insects encounter in desert environments is water loss; their relatively high surface–volume ratios cause rapid desiccation in dry air (Crawford 1981). Water is lost mainly through transpiration (combined cuticular and respiratory water loss) but also through defecation, defensive secretions, and oviposition. Adaptations to desert conditions can thus be categorized as either ecological adaptations (finding or creating moist conditions) or morpho-physiological adaptations (protection and resistance against desiccation) (Ghilarov 1964).

Adaptations which have made tenebrionids desert specialists all contribute to either reducing transpiratory or other water loss or allowing survival in harsh desert environments: body-shape diversity; the ability to seek refuge underground by either digging or using existing holes and crevices; omnivorous feeding habits; timing of both daily and seasonal activities to coincide with the most favorable environmental conditions; fused elytra composed of unique, straight-chain hydrocarbons (which allow tight mo-

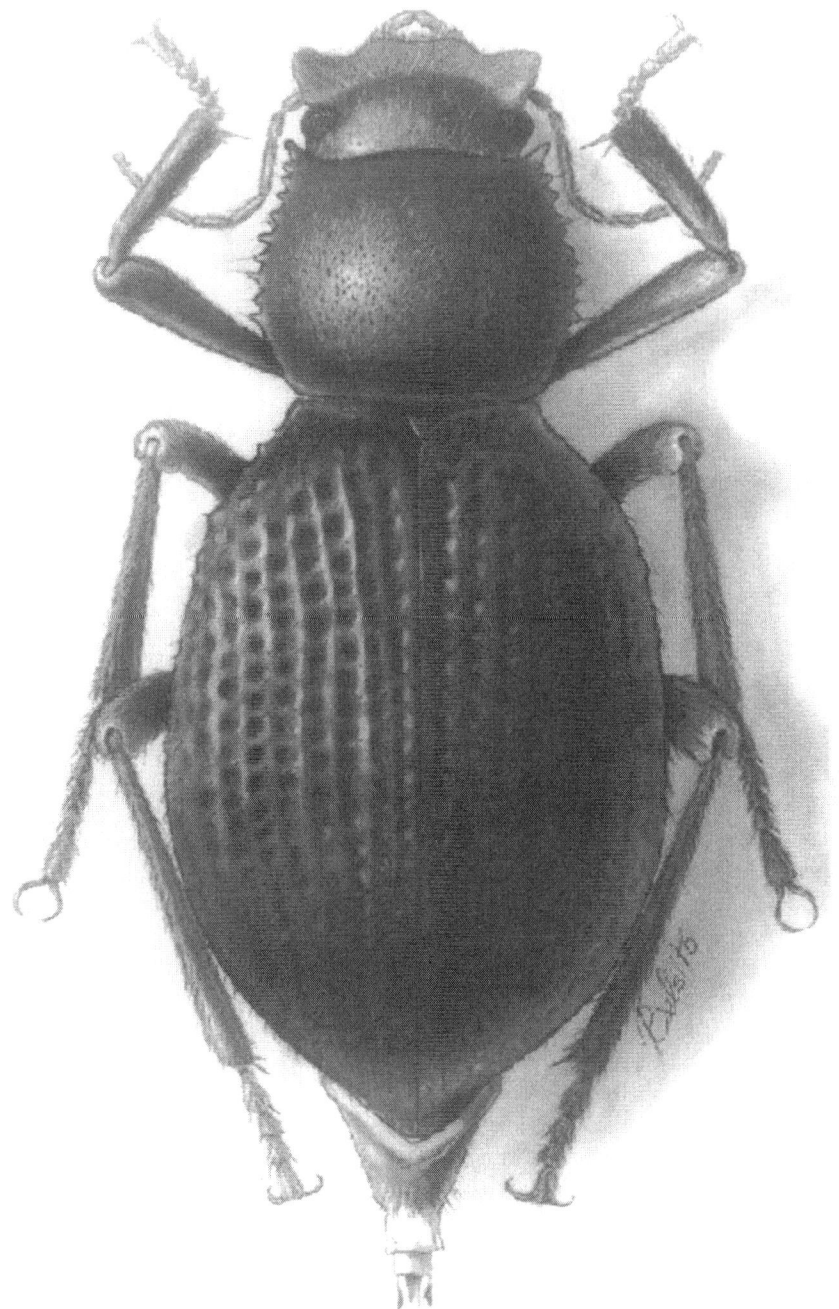

Figure 6.1 *Craniotus marecortezi* Aalbu et al. (*in litteris*), an endemic tenebrionid species from San Esteban island.

lecular packing), creating a protective, sealed subelytral cavity; the ability to secrete protective wax layers on the cuticular surface; and loss of defensive glands (for review, see Aalbu in press).

Tenebrionid species found in arid environments can be categorized into three groups according to their capacity to withstand harsh desert conditions. The tenebrionids most highly adapted to arid environments are characterized by flightlessness, fused elytra, subelytral cavities, and lack of defensive glands. An otherwise similarly adapted second group of tenebrionids have retained the use of defensive glands. The first group includes tenebrionids such as *Asbolus*, which are able to tolerate higher temperatures (>40°C) than species belonging to the second group (e.g., *Eleodes*, 30°C) without a significant increase in water loss (Ahearn and Hadley 1969) and with transition temperatures (at which cuticular permeability abruptly increases) of above 50°C for *Asbolus* and *Cryptoglossa* versus 40°C in *Eleodes* (Hadley 1978). A third group is composed of species that have developed strategies of avoidance rather than the ability to withstand harsh environmental conditions. These are usually winged and often have defensive glands (e.g., *Hymenorus*, *Phaleria*).

Flightless desert tenebrionids with fused elytra are very resistant to immersion in sea water and so are potentially able to survive rafting easily; the subelytral air cavities provide additional buoyancy (Peck and Kukalová-Peck 1990; Peck 1996). In addition, being omnivorous enables tenebrionids to more easily survive the harsh conditions and low resource availability on desert islands. These beetles are very diverse and abundant in Mediterranean, subtropical, and tropical environments and are a faunal component of islands in these regions. Tenebrionids have colonized isolated oceanic islands such as the Galápagos (Peck and Kukalová-Peck 1990; Peck 1996) and islands in the Pacific (Kaszab 1955a,b, 1982). Most island studies including Tenebrionidae (e.g., Wollaston 1854; van Dyke 1953) consist of species lists or descriptions without any biogeographical analysis, although Kaszab (1955a,b, 1982) analyzed the endemic genera and species of Tenebrionidae of major groups of Pacific islands. More recent studies have analyzed endemism, adaptive radiation, and estimations of rates and modes of colonization events of beetles (including tenebrionids) in the Galápagos Islands (Peck and Kukalová-Peck 1990; Peck 1996). Tenebrionids have also recently received attention as model organisms to investigate the genetic differentiation of island biotas (e.g., Finston and Peck 1995, 1997).

A large number of islands are found in the Sea of Cortés (>900 islands and islets). However, published data on the tenebrionid fauna in this region are sparse. Other than simple species descriptions (Horn 1894), the early descriptive phase resulted from collecting expeditions sponsored mainly by the California Academy of Sciences (Blaisdell 1923, 1925a,b, 1936, 1943). When the main road to the cape was paved in the 1970s, increased faunal surveys resulted in more extensive generic and tribal revisions (see below), with newer studies incorporating elements of cladistic biogeography (Doyen 1984, 1987; Aalbu in press).

In this chapter we make a first attempt to analyze the tenebrionid fauna of the Gulf of California. We are aware that additional collections and research are necessary to get a better understanding of the biogeographical patterns inherent to this fauna. Nevertheless, we hope that this chapter will provide insight into the biogeography of this dominant and speciose group of insects on islands in the Gulf of California.

Methods and Data Analysis

The tenebrionid fauna of the islands was compiled from recent collection efforts on the islands as well as from records in literature sources (Horn 1894; Blaisdell 1923, 1925a,b, 1936, 1943; Triplehorn 1965, 1996; Triplehorn and Brown 1971; Berry 1975, 1980; Doyen 1976, 1982, 1984, 1987; Triplehorn and Watrous 1979; Papp 1981; Aalbu and Triplehorn 1985; Thomas 1985; Triplehorn and Aalbu 1987; Brown and Doyen 1991; Aalbu et al. 1995; Aalbu in press). We also examined material from collections in the California Academy of Sciences (San Francisco, California), California Department of Food and Agriculture (Sacramento, California), College of Idaho (Caldwell, Idaho), University of California (Berkeley, California), Field Museum of Natural History (Chicago, Illinois), and the Kirby W. Brown Collection (Paradise, California), as well as beetles in our personal collections and voucher specimens deposited at University of California, Davis.

After examination of numerous specimens and types (Aalbu and Sánchez-Piñero, in preparation) we adopted the following synonymies: *Argoporis insularis* (Berry 1980) as synonymous with *Argoporis apicalis* (Blaisdell 1943); *Steriphanus torpidus* (Blaisdell 1923) and *S. mucronatus* (Blaisdell 1923) with *S. subopacus* (Horn 1870); *Stibia tortugensis* (Blaisdell 1936) with *Stibia sparsa* (Blaisdell 1923); and *Triphalopsis minor* (Blaisdell 1923) with *Triphalopsis partida* (Blaisdell 1923). In addition, a number of hitherto undescribed genera and species (Aalbu and Sánchez-Piñero, in preparation), are included in our data analysis; higher classification follows Aalbu et al. (in press).

Data from 382 tenebrionid species occurring on the peninsula and from 268 islands around Baja California were gathered and maintained in a relational database for rapid access and analysis. These data included records of 144 species occurring on 84 islands (66 islands in the Gulf of California and 18 Pacific islands). The principal analysis focused on specimen data from islands in the Gulf of California. However, due to the lack of physical information on some of these islands (see below), only 64 islands were used in analyses involving island area. Specimen records from 18 Pacific islands were also included for comparative purposes.

On 25 islands in Bahía de los Ángeles, Bahía de Loreto, and in the central Midriff area (see fig. 13.1), carrion-baited traps (Sánchez-Piñero and Polis 2000) were used to sample tenebrionids every year from 1995 to 1999. (An aerial photograph of the islands in Bahía de los Ángeles is shown in fig. 16.7.) On these 25 islands and 5 others (Bahía Ánimas northeast and southwest, Blanca, Calaveras, Pescador; see Polis et al., chap. 13), we sampled tenebrionids using additional methods including hand collecting, pitfall traps (both dry and containing propylene glycol antifreeze) and traps baited with peanut butter mixed with oatmeal. For some relatively large islands (Smith, Cabeza de Caballo, Ventana), traps were set in different areas of the island to include the main habitat types. However, for other large islands (Carmen, Danzante, San Lorenzo Norte, Salsipuedes, Partida) our collecting did not cover the entire island, and we consider the data from these islands to be incomplete compared with both smaller islands and those large islands with a higher collecting effort (see below). In addition, collections were made (by hand collecting and antifreeze, oatmeal, and carrion-baited pitfall traps) on the peninsula for comparison with the islands.

The islands were divided into three categories according to the relative reliability of the sampled data set: (1) occasional collecting (36 islands), based mainly on literature records and museum specimens; (2) a low-confidence data set, composed of 12 relatively large islands surveyed every year (1995–1999), but on which collections were not made over the entire island; and (3) a high-confidence data set, composed of 18 islands surveyed every year (1995–1999), on which sampling was carried out over the entire island area. For data analysis, we arranged the above categories into three groups: group A (all 66 islands), group B (only low- + high-confidence islands), and group C (exclusively high-confidence islands).

Physical attributes of the islands such as latitude, size (area), elevation, distances to mainland and/or closest source areas, isolation by minimum ocean depth (i.e., minimum channel depth separating the island from the mainland; Richman et al. 1988), and plant-species richness were gathered from different sources. Area, distance, and elevations (Due 1992) are provided in appendix 1.1. Latitude, ocean depths and additional information was taken from published maps (Topography International 1986; Sea of Cortéz Charts, North & South, by Fish-n-Map Co., Arvada, Colorado) and other available island information (University of California—Santa Cruz Island Conservation & Ecology Group online Island Conservation Database [macarthur.ucsc.edu] and Figueroa and Castrezana 1996). Plant-species richness for 18 islands (group C) was evaluated by P. West (unpublished data).

We defined five categories of island size [in km^2: < 0.1 ($n = 21$); ≥ 0.1 to < 1.0 ($n = 17$); ≥ 1.0 to < 10.0 ($n = 12$); ≥ 10.0 to < 100.0 ($n = 9$); ≥ 100.0 ($n = 5$)]. Major islands in island groups were considered to be faunal source areas for satellite islands within the group. Isolation was considered to be the distance from the mainland to the major islands (of an island group) plus the distance from the major island to the satellite island. Five categories of distance were defined [in km: < 1.0 ($n = 10$); ≥1.0 to <5.0 ($n = 22$); ≥5.0 to <10.0 ($n = 10$); ≥10.0 to <15.0 ($n = 10$); ≥15.0 ($n = 14$)]. Categories of minimum ocean depth between the mainland and island or island group corresponds to the lowering of sea level in the order of 130 m during the last glaciation (Lowe and Walker 1997), differentiating landbridge islands [< 130 m depth ($n = 44$)] and deep-sea islands [> 130 m depth ($n = 22$)]. Three categories of species endemism were recognized: single island, island group, and multiple-island endemics. Multiple-island endemics (where species are found on widely separate islands) as well as single specimens of New Genus 1 sp. 2 and New Genus 1 sp. 3, collected on Tiburón and San Marcos islands, respectively, were not used in the analysis on endemism because these are likely to be collected on mainland sites as more collecting efforts are carried out in the peninsula and Sonora.

Two software packages were used for data analysis: JMP IN (version 3.2.1; SAS Institute Inc.) and Statview 4.1 (Abacus Concepts, 1992–1994). Analyses included simple and multiple regressions (area, elevation, distance, ocean depth, plant richness); analysis of covariance (seabird use of islands); frequency distribution analysis (Chi-square tests) of the proportion of species occurrence in the defined categories of area, distance, and ocean depth. Association between pairs of species were analyzed using a Chi-square test on combinations of the most abundant species to test the frequency distribution of pairwise occurrences. For significantly associated species pairs, we calculated an index of association (observed/expected number of joint occurrences). For this index, values < 1 indicate a negative association between two species; values > 1 represent positive associations between species (Eickmeier 1988). In addition, to test whether a given pair of species were significantly related in terms of abundance, correlations in abundances between species pairs were performed.

The Tenebrionid Fauna of Baja California

The peninsula of Baja California and its islands have an extremely rich fauna of tenebrionid beetles: 9 subfamilies, 30 tribes, 116 genera, and 382 species are known from this region. For an area about 2% the size of the contiguous United States, Baja California has more than 60% the number of genera and 32% the number of species in the United States (10 subfamilies, 44 tribes, 191 genera, and 1190 species; Aalbu et al. in press). The tenebrionid fauna of Baja California has twice the number of taxa found on New Zealand (16 tribes, 36 genera, and 149 species; Watt 1992) in about one-half the area. The diversity of the tenebrionid fauna in Baja California is very high compared with other arthropod groups. Tenebrionidae is represented by more than 3 times the number of ant genera and nearly 2.4 times the number of ant species found in Baja California (Boulton and Ward, chap. 5, this volume).

Although the island fauna of the Sea of Cortés contains a few elements with affinities to mainland Sonora, comparatively little is known about the Sonoran fauna compared to that of the peninsula, and thus a corresponding mainland–island analysis cannot yet be made.

The Tenebrionid Fauna of the Islands of the Sea of Cortés

Forty-seven genera and 103 species are found on islands in the Sea of Cortés (i.e., 40.5% of genera and 27% of peninsular species are present on gulf islands). On the Pacific side, 21 genera (18% of the number of peninsular genera) and 42 species (11% of the number of the peninsular species) occur on islands. Overall, 60 genera and 144 species are recorded from islands around Baja California (i.e., 51.7% of the total peninsular genera and 37.6% of the total peninsular species). These are listed in appendixes 6.1 and 6.2. On gulf islands, the diversity of tenebrionids is also higher than ant diversity (with more than 2.4 times the number of genera and 1.5 times the number of species of ants; Boulton and Ward, chap. 5).

The tenebrionid fauna on the islands of the Gulf of California is similar in composition to that on the mainland, as indicated by the similar frequency distribution of the number of species/genus in mainland and insular faunas ($\chi^2 = 33.74$, $p = .91$, df = 46). In addition, in the pool of species in Baja California, the genera present on islands are generally more species-rich (mean number of species/genus = 4.26 ± 0.67) than genera absent from islands (mean number of species/genus = 2.18 ± 0.32; $F = 13.32$, $p < .0001$, df = 1, 106). The fauna on the islands is composed primarily of Pimeliinae (69% of species), Tenebrioninae (16% of species), and Opatrinae (7% of species). The Eurymetopini is the most species-rich tribe on the islands with 42 species. The most species-rich genera on the islands include *Eusattus* (nine species), *Argoporis, Eleodes, Telabis* (six species each), *Stibia, Cryptoglossa* (five species each), and *Steriphanus* (four species). In terms of abundance, the assemblages are dominated by *Triphalopsis* (*T. partida* and *T. californicus*), *Cryptoglossa* (*C. asperata* and *C. spiculifera*), and *Argoporis* (*A. apicalis*) species, which represent 60% of the abundance of tenebrionids on the islands. However, the abundance (and presence) of each species varies greatly among islands. Some species are dominant only on certain islands, such as *Cryptadius tarsalis* on Rasa, *Steriphanus obsoletus* on Partida, *Stibia sparsa* on both Gemelitos or New Genus 2 on Pardo. Other abundant species include *Batuliodes confluens, Conibius opacus* and *Tonibius sulcatus*.

Tenebrionid beetles can be divided into two major groups according to habitat use: species associated with soil or sand and species associated with trees. Among tenebrionids that are associated with trees, different areas are exploited both as larvae and adults including dead, rotten wood and its associated fungus. Few species found on gulf islands (eight species belonging to Diaperini, Belopini, Bolitophagini, Epitragini, Tenebrionini, and Alleculini) are associated with trees. In contrast, most tenebrionids inhabiting the islands are ground dwellers. These include most Pimeliinae (Asidini, Coniontini, Cryptoglossini, Eurymetopini, Stenosini), Opatrinae (Opatrini, Melanimimi), Tenebrioninae (Cerenopini, Eleodini), and some Diaperinae (Phaleriini; see app. 6.2).

The gulf island tenebrionid fauna is composed mainly of highly adapted desert species. We have distinguished three different groups of species according to their adaptation to arid conditions. One group, most highly adapted to arid conditions (see above) include most Pimeliinae. A second group, which have retained defensive glands, is composed of most Opatrinae and some Tenebrioninae. The last group (which on gulf islands include the Alleculinae, Diaperini, Bolitophagini, Lagriinae, Epitragini and Vacronini of Pimeliinae, and the Triboliini and Tenebrionini of Tenebrioninae) are characterized by mainly winged forms more susceptible to desiccation. We categorize these as groups 1 (most highly adapted), 2 (highly adapted), and 3 (least adapted; see app. 6.2).

In comparing these groups on the Pacific islands with those of the gulf, we find that in the Pacific islands (with generally less harsh environmental conditions due to the colder ocean temperatures, resulting in the frequent presence of a protective fog layer), the composition of the three groups differs. For instance, the ratio of groups 1–3 in the gulf islands is 68:20:12, while in the Pacific islands it is 38:57:05, the main difference being the ratio between groups 1 and 2. This same trend applies to endemic species with ratios of 31:69:00 (Pacific islands) and 92:08:00 (gulf islands; see app. 6.2).

Endemism

One genus (2%) and 25 species (24%) are endemic to the islands on the Sea of Cortés. This includes 14 single-island endemics and 11 multiple-island endemics (no island group endemics are present). Although the number of endemic species is higher than that found in other arthropod taxa on the islands (e.g., ants with five endemic species; Boulton and Ward, chap. 5, this volume), the proportion of endemic tenebrionids species in the gulf islands (23%) is lower than the proportion of endemic tenebrionids on the peninsular areas of Baja California (45% of the species). These figures can be compared to the Pacific side where 16 species (38%) of the island fauna are endemic (see below). Thus, for islands on both the gulf and Pacific sides of Baja California, endemics total 2 genera (3%) and 41 species (28%).

The proportion (%) of single-island endemics is similar on gulf islands (2.12% ± 0.83%; $n = 66$) and on Pacific islands (4.22% ± 2.82%; $n = 10$; $F = 0.83$, $p = .36$, df = 1, 74; ANOVA), removing the three farthest Pacific island groups (Revillagigedo, Guadalupe, and San Benitos). Similarly, the proportion of multiple-island endemics on Pacific islands (9.93% ± 3.94%) did not differ from gulf islands (12.56% ± 2.90%; $F = 0.02$, $p = .89$, df = 1, 74; ANOVA). However, the proportion of endemics on more remote island groups including the Revillagigedo, Guadalupe, and San Benitos groups is much higher, ranging from 25–40% in the San Benitos to 100% on the oceanic Guadalupe and Revillagigedo islands. We should caution that, in some cases, even genera described as endemic on some oceanic islands have been found to belong to more widespread genera after closer comparative studies and more mainland sampling (Doyen 1972; Aalbu and Triplehorn 1991).

Endemism on gulf islands appears to be related to isolation (ocean depth) and island area. Single-island endemics occur in a significantly larger proportion of deep-sea islands (22.7% of the islands) than landbridge islands (2.3%) (the two undescribed species of New genus 1, only known from single specimens collected on Tiburón and San Marcos islands, but very likely occurring in the poorly surveyed mainland areas of Sonora and Baja California, were not considered as endemics in this analysis). Deep-sea islands are likely to have been isolated from mainland sources for a longer time, thus facilitating the occurrence of endemic species via speciation or relictual persistence of isolated populations. Unfortunately, the lack of fossil records for tenebrionids on Baja California does not allow any conclusion to be made concerning the relative importance of speciation, relictual distribution patterns, or insufficient mainland sampling on gulf islands.

The occurrence of single-island endemics shows also the effect of island area on endemism: the frequency of islands with single-island endemics is higher as islands increase in size (fig. 6.2). This effect of area on endemism is corroborated by the number of endemic species on islands. The number of endemic species in the Gulf of California is related to area (table 6.1). Adler et al. (1995) indicated that the number of endemics increases with island area in Pacific and Antillean butterflies and in

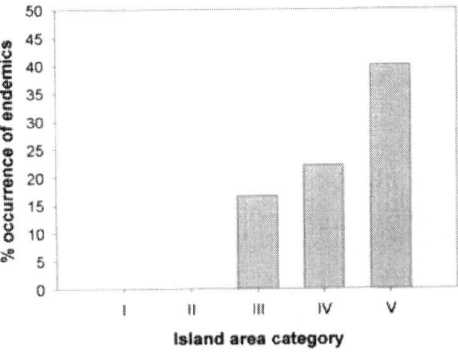

Figure 6.2 Proportion of Gulf of California islands on which single-island endemics occur according to island area categories (I: < 0.1 km^2; II: ≥ 0.1 < 1.0 km^2; III: ≥ 1.0 < 10.0 km^2; IV: ≥ 10.0 < 100 km^2; V: > 100 km^2; see Methods).

Table 6.1 Multiple regression coefficients and slopes for the effects of area, distance (from islands to the mainland), and ocean depth on the number of single-island endemics

	R	
Area	.04**	
Distance	.03	(ns)
Ocean depth	.08	(ns)
Intercept	−.06	(ns)
R^2	.21	

Only area is a significant factor affecting the number of endemic species.
**$p < .01$; ns, not significant.

Pacific birds. These authors suggested that a larger area provides the persistence and variety of habitats most conducive to speciation on islands. In fact, intra-island speciation is restricted to islands large enough to allow effective segregations of populations within the island (Whittaker 1998), as shown by *Stomion* species in the Galápagos (Finston and Peck 1995) and suggested for flightless tenebrionids in continental areas (Doyen and Slobodchikoff 1984). The higher probability of persistence of certain habitats on larger islands may also be a factor allowing the persistence of relictual species on larger islands.

The above results suggest that isolation may be a factor determining the occurrence of endemics on islands, with deep-sea islands, separated for longer time from the mainland, having a higher probability of occurrence of endemic species (via speciation or persistence of relictual species) than landbridge islands (more recently separated from the mainland). Because the number of endemic species on an island increases with island area, larger islands also have a larger probability of occurrence of endemic species. Thus, deep-water and large islands (e.g., Cerralvo, Santa Catalina, San Esteban) contain most of the single-island endemics in the gulf.

Disharmony

The fauna of the island in the Gulf of California is disharmonic with respect to the mainland source faunas. Disharmony occurs not only because the presence of endemic species on the islands but also because island species assemblages do not merely represent a random subset of species from the mainland sources.

Comparison of the mainland source faunas at Bahía de los Ángeles area with the nearby island faunas indicates that they are a subset of the surrounding mainland, the island fauna accounting for 40% of the species present in mainland assemblages. Only two of the species found in the islands have not been collected on the nearby mainland: *Eschatomoxys* sp. 1 and *Cheirodes californicus*. The absence of these species in mainland surveys indicates insufficient sampling. This is corroborated by additional collecting in new mainland areas in 1999, which resulted in the discovery of two new species previously collected only on islands (*Craniotus* new sp. 1, and New Genus 1, new sp.: Eurymetopini). Farther south, near Loreto, 12 species occur on nearby islands but are absent from the adjacent mainland surveys. Most of these locally absent species occur in other areas of the Baja California peninsula. Some of these (such as New Genus 2, new sp., also Eurymetopini) are known only from a few of the islands in Loreto Bay but most likely also occur on the mainland, again indicating that more thorough mainland surveys are necessary.

We also point out that although island faunas are for the most part subsets of adjacent mainland species pools, island assemblages lack some of the more abundant mainland species. Thus, *Craniotus pubescens*, a species relatively abundant on the mainland, is not present on islands around Bahía de los Ángeles (note: there is a record requiring confirmation from Partida Norte, the only putative island occurrence). On the other hand, *Craniotus* new sp. is a species less abundantly collected on the mainland but occurs on most of the islands. This pattern is repeated in species of the genus *Cryptoglossa* (*C. muricata* being more abundant in mainland areas, while on islands only *C. spiculifera* is found, except on Partida Norte, where *C. muricata* has been recorded historically, but not collected recently, again a record requiring confirmation) as well as in *Eleodes* (*E. mexicana* and *E. discinctus* are found on islands, while *E. loretensis*, the dominant species on the mainland, is absent on these islands).

The island fauna is disharmonic with the mainland species pool (i.e., does not merely represent a random subset either in terms of species composition or abundance of the potential mainland pool). On islands in Bahía de los Ángeles and central Midriff, *Argoporis apicalis*, *Triphalopsis californicus* (or *Stibia sparsa*), *Cryptoglossa spiculifera*, *Batuliodes confluens*, and *Tonibius sulcatus* are the dominant elements of the assemblage, a pattern which recurs on a number of islands. However, there are variations to this pattern of species composition, in that some of these species are not present on some islands, where other species become locally dominant (for example, *Cryptadius tarsalis* on Rasa and *Steriphanus subopacus* on Partida Norte). These variations in composition and abundance are probably related to specific habitat requirements (see below).

Factors Affecting Occurrence on Islands

A number of factors determine the presence of species on islands. These are related to habitat suitability, island size, distance to mainland sources, and species interactions. Previous studies have found a strong relationship between tenebrionid species and substrate type (Medvedev 1965; Andrews et al. 1979; Jenkins 1971; Thomas, 1983, 1985). Substrate appears to be an important factor determining the occurrence of tenebrionids on gulf islands. Certain species can be readily associated with habitat and soil characteristics: *Phaleria* (sandy beaches, including intertidal zones), *Edrotes*, *Cryptadius*, *Eusattus* (sand dunes, loose sand), *Cryptoglossa*, *Batulioides*, *Steriphanus* (sandy compact soils, flat areas), *Argoporis*, *Triphalopsis*, *Stibia* (rocky soils, hillsides and cliffs), and *Asbolus mexicanus*, *Typlusechus* (fine silt soils found under overhangs and crevices). Thus, the set of species present on an island is determined by its habitat and substrate characteristics. Differences in the faunistic composition between islands reflects differences in each island's physical characteristics.

In addition to habitat traits, island size appears to affect species occurrence on an island. Half of the species (12 out of the 24 analyzed) appear to be able to maintain populations on islands of almost any size category. Others show differences in the proportion of islands on which they occur depending on island size. In general, species occurrence on islands increases according to island size, although the slope and shape of the curve may differ (fig. 6.3). Some species occur on very small to very large islands (*Cryptoglossa asperata* and *C. spiculifera*), whereas other species inhabit the

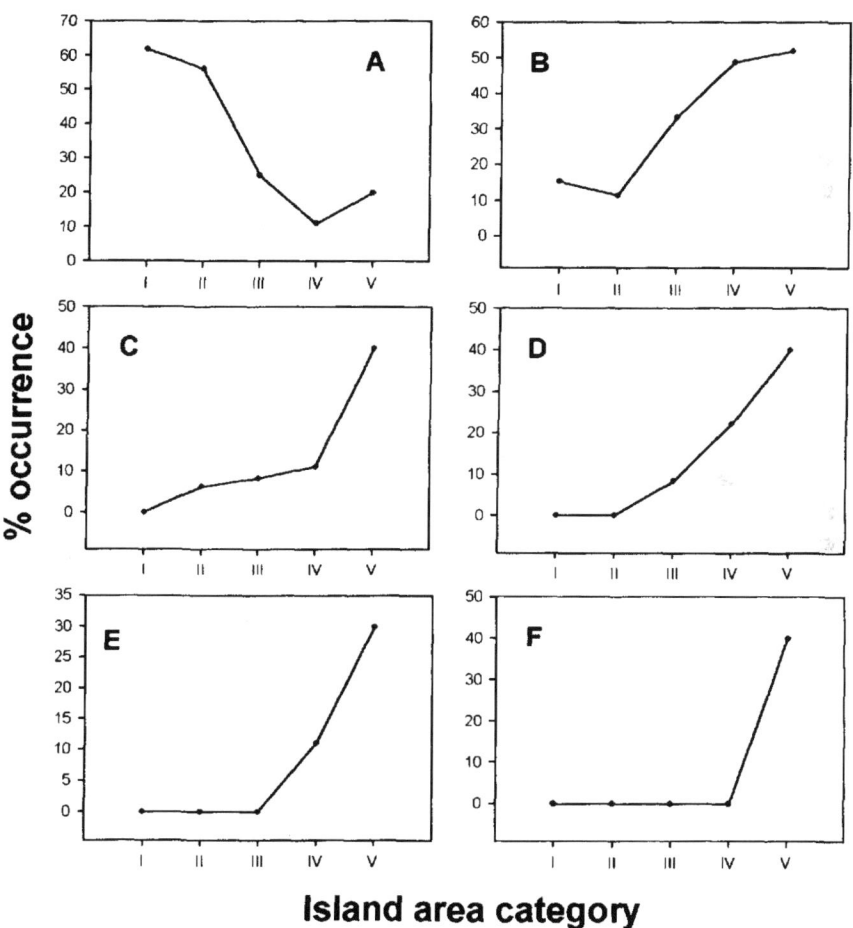

Island area category

Figure 6.3 Patterns of species with differences in occurrence on islands with respect to island size (I: < 0.1 km^2; II: ≥ 0.1 < 1.0 km^2; III: ≥ 1.0 < 10.0 km^2; IV: ≥ 10.0 < 100 km^2; V: > 100 km^2) in the Gulf of California. (A) *Argoporis apicalis*; (B) *Cryptoglossa spiculifera, C. asperata, Conibius opacus, Eusattus erosus,* and *Triphalopsis partida*; (C) *Asbolus mexicanus*; (D) *Cerenopus concolor*; (E) *Trimytis obtusa* and *Stibia puncticollis*; (D) *Eleodes eschscholtzi* and *E. insularis*.

largest islands exclusively (*Eleodes eschscholtzi* and *E. insularis*). These patterns show an interesting aspect: beetle size appears not to be a factor limiting occurrence on islands, as large *Cryptoglossa* species are able to maintain populations on some very small islands (for example, on Gemelitos West, where *C. spiculifera* is abundant). The fact that some of the smallest islands receive resources from the ocean (principally from seabirds, such as guano and carcasses, see below) may help explain the persistence of populations of large beetles on small islands.

The other pattern to consider which is related to area is represented by *Argoporis apicalis*. This species is more frequent on small islands rather than on larger islands (fig. 6.3A). Low competitive ability and high dispersal ability have been argued to explain this pattern of species occurrence in birds (Diamond 1974). Thus, although potentially excellent colonists, species with this distributional pattern are unable to maintain populations on islands where more effective competitors are present (a condition more likely found on larger islands). However, in the case of *Argoporis apicalis*, other factors may better explain this pattern of occurrence. Three independent components may be responsible for the observed pattern: habitat requirements, trophic niche width, and sampling biases.

Habitat requirements may be related to the strong association of *Argoporis apicalis* with rocky, often steep, hillsides, which tend to occupy relatively more area on small islands where beaches practically do not exist, and where cliffs constitute a major proportion of the island. Second, *Argoporis* may be better able to make use of seabird remains and marine detritus as part of its diet (Stapp et al. 1999) than other tenebrionid species as shown by higher abundances of this species on seabird islands, where *Argoporis* apicalis becomes the dominant species (F. Sánchez-Piñero, unpublished data). Food selection experiments indicate that this species is also able to eat the brown algae *Sargassum* spp. (Phaeophyta), a food item rejected by other species inhabiting the islands (e.g., *Cryptoglossa spiculifera;* F. Sánchez-Piñero and C. Harvey, unpublished data). The higher abundance levels of *Argoporis* may be a factor of both increased seabird effects and other marine subsidies (Stapp et al. 1999; Sánchez-Piñero and Polis 2000), resulting in a better record of this species' presence on these small islands rather than on larger islands, where beetles are generally much less abundant (especially on large islands without seabird colonies; see below).

Isolation (distance to mainland sources and ocean depth) does not appear to be an important factor affecting occurrence on islands for most tenebrionid species. Only a few species show differences in their occurrence on islands related with distance and ocean depth: *Craniotus* new sp. 1 ($\chi^2 = 26.4$, $p < .0001$), *Eschatomoxys* sp. 1 ($\chi^2 = 12.8$, $p = .01$), *Microschatia championi* ($\chi^2 = 15.13$, $p < .01$) and *Triphalopsis californicus* ($\chi^2 = 11.6$, $p = .02$; fig. 6.4). Distance and ocean depth show the same patterns of occurrence in these species. This is probably related to the correlation between these two variables ($r = .76$, $p < .0001$). In general, these species occur on nearer islands with a higher frequency than on distant islands, a pattern predicted for poor dispersers. However, islands closer than 1 km to the mainland show in general a lower occurrence of these species, which indicates that this may be a sampling artifact related to the fact that islands between 1 and 5 km from the mainland include those in Bahía de los Ángeles, a group of islands that have been better surveyed than the other islands. Differences in the occurrence of other species (*Steriphanus subopacus* and *Conibius opacus*) on islands related to distance and ocean depth simply corresponds

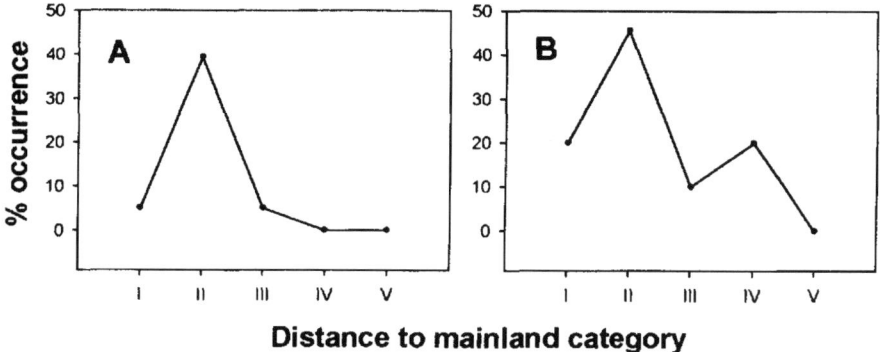

Figure 6.4 Patterns of species showing differences in occurrence on islands depending to distance from the mainland (I: < 1.0 km; II: ≥ 1.0 < 5 km; III: ≥ 5.0 < 10.0 km; IV: ≥ 10.0 < 15.0 km; V: ≥ 15.0 km). (A) *Craniotus* new sp. 1, *Eschatomoxys* new sp. 1, and *Microschatia championi*; (B) *Triphalopsis californicus*.

to the traits of the islands in the main region inhabited by both species (central Midriff and southern gulf, respectively).

The lack of significance from isolation on the proportion of species occurrence on islands for most tenebrionids, as well as the relatively low proportions of endemics present on the islands, suggest that the tenebrionid fauna present on gulf islands results mainly from recent isolation from mainland sources (in fact, most islands in the gulf are landbridge islands). However, the lack of effective isolation due to continuous arrivals to the islands (Williams 1996) may provide an alternative explanation to the observed pattern. This may also explain the similarity of species composition on oceanic islands (e.g., Rasa and Tortuga) with mainland sources.

Species Associations

The occurrence or absence of certain species on islands is also related to species interactions. Correlations between species abundances and analysis of species associations showed that only three taxa are significantly negatively associated: *Tonibius sulcatus* and *Conibius opacus* ($\chi^2 = 8.97$, $p < .01$, df = 1; these two species do not occur together on any of the 25 islands surveyed), *Triphalopsis partida* and *T. californicus* ($\chi^2 = 10.86$, $p < .01$, df = 1; only one of the two species occur on the surveyed islands) and *Stibia* spp. and *Triphalopsis* spp. ($\chi^2 = 4.17$, $p < .05$, df = 1; index of association = 0.63; these genera coexist on relatively large islands such as Partida Norte, San Lorenzo Norte [*S. sparsa* and *T. partida*], Ventana, Cabeza, Smith [*S. sparsa* and *T. californicus*] and one small island, Tijeras [*S. fallaciosa* and *T. partida*]). In these three cases, negative associations occur between species that are similar morphologically and which may perform similar roles in the community. Although the distributional patterns of these species overlap at least partially, the two species in a given pair only coexist on a few islands (the two *Triphalopsis* spp. occur on Ángel de la Guarda and Mejía, and *T. sulcatus* and *C. opacus* on San Lorenzo Sur). This pattern is particularly interesting in the case of *Stibia* and *Triphalopsis*: *Triphalopsis* is generally a dominant species on the islands, but on islands where *Triphalopsis* is absent, *Stibia* becomes the dominant species (e.g., both Gemelitos islands, Cardonosa). However, both these genera occur together on one small island (Tijeras) and five larger islands (Cabeza, Ventana, Smith, Partida Norte and San Lorenzo Norte), where neither of the two is abundant. These patterns may well be related to species competition, due to the fact that only species pairs with similar morphological characteristics show negative associations. Therefore, coexistence may only be possible in cases where island size (i.e., relatively large islands with greater habitat heterogeneity) and/or scarce, patchy resource availability prevents competitive exclusion and allows both species to coexist. Other gulf island species do not show any significant association (presence/absence of species) or correlation (abundance) with other species. These results indicate that, except for closely related taxa with probably similar niche requirements, most species on islands form independent assemblages.

Species Richness

Species richness on islands is usually considered a function of island area and isolation (MacArthur and Wilson 1967) although other factors, such as latitudinal gradients and effects of other organisms, may also alter richness (Huston 1994).

Abiotic Factors: Area, Elevation, Latitude, and Isolation Effects

Area affects the number of species on islands significantly, although such an effect is strongly affected by survey intensity: The slope of the regression line increased from $z = 0.15$ ($R^2 = .31$; $F = 27.69$, $p < .0001$, $n = 64$) to $z = 0.20$ ($R^2 = .56$; $F = 36.13$, $p < .0001$, $n = 30$) and $z = 0.29$ ($R^2 = .68$; $F = 33.44$, $p < .0001$, $n = 18$) for data groups A, B, and C, respectively (fig. 6.5). Island elevation, although related with island area ($r = .92$, $p < .0001$, $n = 59$), is a poorer predictor of beetle-species richness than area (group A: $R^2 = .22$, $F = 16.37$, $p = .0002$, $n = 61$; group B: $R^2 = .49$, $F = 26.74$, $p < .0001$, $n = 30$; group C: $R^2 = .44$, $F = 12.53$, $p = .0027$, $n = 18$).

A multiple regression considering the effects of area, latitude, distance (from each island to mainland) and ocean depth shows that only area had a significant effect on species richness on the islands (table 6.2). Although distance to mainland showed a significant effect when the 64 islands in group A were analyzed, this effect is probably due to the fact that islands in Bahía de los Ángeles, Bahía de Loreto, and the central Midriff region have been better surveyed than other, more remote islands. Thus, a multiple regression removing these islands (i.e., considering only those islands with occasional collecting) showed that only area is significantly related to species richness (table 6.2). We believe that the significant effect of ocean depth evidenced in group C (which includes islands in Bahía de los Ángeles and Bahía de Loreto) is spurious because ocean depth in all but one of these islands is 10–40 m, the exception being for Calaveras, a small island with one species, with 100 m ocean depth separating the island from the mainland. When this island is omitted from the analysis, the significance of ocean depth is negated. A multiple regression including different measures of distance (from satellite islands to main islands [source area] in island groups, or to

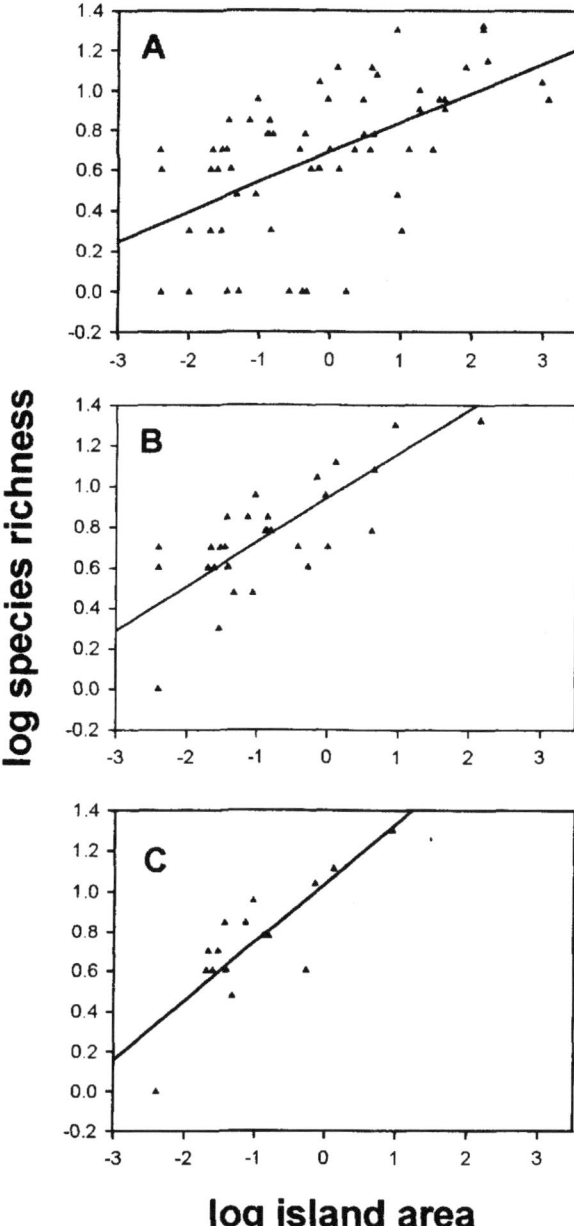

Figure 6.5 Linear regression of tenebrionid species richness on islands (log number of species) as a function of island area (log island area). The different graphs represent the fit for each group (A, B, and C) defined according to the islands considered in the data set (see Methods). A: islands with occasional collecting; B: low-confidence collecting; C: high-confidence collecting.

the mainland as the ultimate source area) provided similar results to those shown in table 6.2.

In addition, the lack of isolation effects are shown by the similar slope describing the species–area relationships on near (< 10 km from mainland; all islands: $R^2 = .32$, $F = 19.48$, $p < .0001$, $n = 43$; islands with occasional collecting only: $R^2 = .60$, $F = 23.56$, $p = .0002$, $n = 18$) and far islands (> 10 km from mainland; all islands: $R^2 =$

Table 6.2 Multiple regression coefficients and slopes for the effects of area, distance (from islands to the mainland), ocean depth, and latitude on tenebrionid species richness considering the three different survey types

	Group A ($n = 64$)	Group B ($n = 30$)	Group C ($n = 18$)	Occasional Collecting ($n = 34$)
Area	0.19†	0.24***	0.24**	0.25***
Distance	−0.26*	−0.004 (ns)	−0.05 (ns)	0.15 (ns)
Ocean depth	0.14 (ns)	−0.17 (ns)	−0.72*	0.004 (ns)
Latitude	1.85 (ns)	2.45 (ns)	5.48 (ns)	−2.03 (ns)
Intercept	−2.03***	−2.32 (ns)	−5.85 (ns)	3.22 (ns)
R^2	0.42	0.64	0.80	0.65

Area significantly affects beetle species richness in all cases, whereas distance and ocean depth have significant effects in groups A and C, respectively (see text for discussion). Analysis discarding islands with higher sampling efforts as those in Bahía de los Ángeles, Bahía de Loreto, and central Midriff (i.e., including islands with occasional collecting only), suggest that the distance effect found in group A is an artifact of greater sampling efforts on those islands.
***$p < .0001$; **$p < .001$; *$p < .05$; ns, not significant.

.38, $F = 11.56$, $p = .003$, $n = 21$; islands with occasional collecting only: $R^2 = .63$, $F = 24.05$, $p = .0002$, $n = 16$; fig. 6.6). In addition, an oceanic island in the Gulf of California such as Rasa has a similar number of species (five) as landbridge islands (Piojo: four species; Mitlan: six species) of similar size and collecting effort. These results corroborate the hypothesis that isolation is not an important factor influencing darkling beetle diversity in the islands of the Gulf of California.

Biotic Factors: Plant Species Richness and Sea Birds

Tenebrionid species richness is related to plant species richness ($R^2 = .72$, $F = 39.46$, $p < .0001$, $n = 17$). However, plant species richness is a function of island area ($R^2 = .65$, $F = 28.08$, $p < .0001$, $n = 17$). When a multiple regression is performed including both plant species richness and area as predictors for beetle diversity, only the number of plant species appears significantly related to tenebrionid species richness ($R^2 = .73$).

These results indicate that both plant and tenebrionid species richness are a function of island area in the gulf. However, plant species richness is a better predictor of beetle diversity rather than area per se. Thus, tenebrionid diversity may be a function of habitat heterogeneity, which is better reflected by plant species richness than by area. In fact, substrate and microclimatic conditions (which affect plant diversity) determine tenebrionid species distribution (e.g., Medvedev 1965, Wharton and Seely 1982; Sheldon and Rogers 1984), selection of oviposition sites (Hafez and Makky 1959; Brun 1975), and activity patterns (Kenagy and Stevenson 1982).

No significance was detected on tenebrionid species richness resulting from sea bird use of islands (categorized in terms of nesting, roosting, and no sea birds); numbers of sea birds nesting on islands (Velarde and Anderson 1994), or guano cover (as a measurement of sea bird presence on the islands). Although sea birds have a negative effect on plant diversity and plant cover on these islands (Sánchez-Piñero and

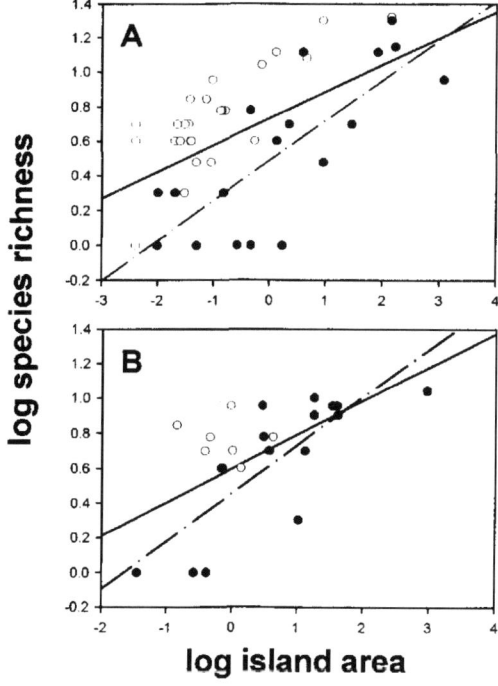

Figure 6.6 Linear regression of tenebrionid species richness on islands situated (A) less than 10 km, and (B) more than 10 km from the mainland. Filled circles and dashed lines represent the regression for the subset (occasional collecting) of the data comprising all islands.

Polis 2000; G. Polis et al., unpublished data) due to plant destruction by sea birds for nest construction and guano scorching of the vegetation, their impact on islands does not appear to significantly affect tenebrionid species richness. Further studies are needed to test this hypothesis.

Factors Influencing Abundance

Tenebrionids beetles account for as much as 70% of the ground arthropod fauna (except ants) on gulf islands (Sánchez-Piñero, unpublished data). Species aggregate at night in large numbers at sources of food on some islands (Triplehorn and Aalbu 1987; Sánchez-Piñero and Polis 2000). Our data show that beetle abundance varies by three orders of magnitude among islands. Although these variations are related to a number of factors, both biotic and abiotic, we focus here on the effects of resource availability and predation as determinants of tenebrionid abundance.

Terrestrial productivity on islands in the Gulf of California is low (Polis et al. 1997). However, these islands are surrounded by highly productive marine waters which provide additional resources that enter the islands via onshore drift and sea birds (guano and remains) (Polis and Hurd 1995, 1996; Polis et al., chap. 13, this volume). This marine input is a key component of the dynamics of terrestrial ecosystems on islands in the Gulf of California. The amount of marine input arriving on the islands via onshore drift is a function of the proportion that the coastal perimeter represents with respect to the total island area. Thus, this perimeter–area ratio is inversely related to island area, with smaller islands receiving a relatively larger contribution from the ocean productivity than larger islands. The higher prey availability in the supralittoral from onshore drift of algae and marine carrion leads to an increase in orb-web spider population densities, which is more appreciable on small islands (Polis and Hurd 1995, 1996; Polis et al., chap. 13, this volume).

Although marine productivity probably affects a number of animal groups (such as orb-web spiders, lizards, and rodents), tenebrionid abundance is not related to perimeter–area ratios ($R^2 = .10$, $F = 2.55$, $p = .12$). This is probably a result of life-history requirements of tenebrionids, whose larvae develop in specific substrates and are probably constrained by substrate physical traits, salinity, and submergence during tidal changes. In fact, most intertidal and supralittoral animals (including specialized *Phaleria* and *Cryptadius* beach dwellers) show physiological and behavioral adaptations to cope with the special conditions in these habitats (Chelazzi and Colombini 1989; Colombini et al. 1994; see also McLachlan 1991). However, adult beetles on islands may forage in supralittoral areas during dry years, when terrestrial resources are low (Stapp et al. 1999).

Marine input also enters the islands through a second conduit, the sea birds. Sea birds are diverse and abundant in the Gulf of California, where they maintain large breeding colonies on certain islands (e.g., Anderson 1983; Velarde and Anderson 1994). The birds feed in the ocean and roost and nest on islands, influencing island ecosystems with guano deposition, fish remains, and sea bird carcasses (Sánchez Piñero and Polis 2000). Guano enhances plant productivity and increases plant nutrient content (i.e., food quality) during wet El Niño years, when water is available on these islands (Anderson and Polis 1999). Sea bird remains are an abundant source of detrital food in nesting colonies, where carrion availability ranges from 5–100 g/m^2 fresh weight (Sánchez-Piñero and Polis 2000; Polis et al., chap. 13, this volume).

Tenebrionid abundance is related to seabird inputs on these islands (Sánchez-Piñero and Polis 2000; Polis et al., chap. 13, this volume). In a study on 25 islands in the Gulf of California differentially used by sea birds in terms of negligible use by sea birds, roosting islands, and nesting islands, beetles were more abundant on islands where sea birds either breed or roost than on islands without seabirds or on the mainland (fig. 6.7). This pattern of higher abundance on islands with sea birds was consistent during the 3 years in which the study was conducted. The effect of sea birds on beetles is still more evident within nesting islands: abundance (number of beetles per trap) inside colonies was up to six times higher than in areas outside the sea bird colonies in large colony islands (Rasa, San Pedro Mártir, and San Lorenzo Norte) (Sánchez-Piñero and Polis 2000).

The increase of beetle abundance on sea bird islands is a likely consequence of direct and indirect effects acting differentially on nesting and roosting islands, respectively. On roosting islands, sea birds affect beetle populations mainly indirectly, via guano and plants, and only after wet years. Guano is rich in nitrogen and phosphorous, increasing the nutritive content of the soil and therefore enhancing plant productivity as well as the nutrient content of plant tissues and detritus (Anderson and Polis 1999). On these islands tenebrionids primarily eat plant detritus from previous pulses of productivity, as indicated by diet records and stable isotope analysis (Stapp et al. 1999; Sánchez-Piñero and Polis 2000). Pulses of productivity during El Niño years produce stores of plant detritus, which are consumed by tenebrionids during subse-

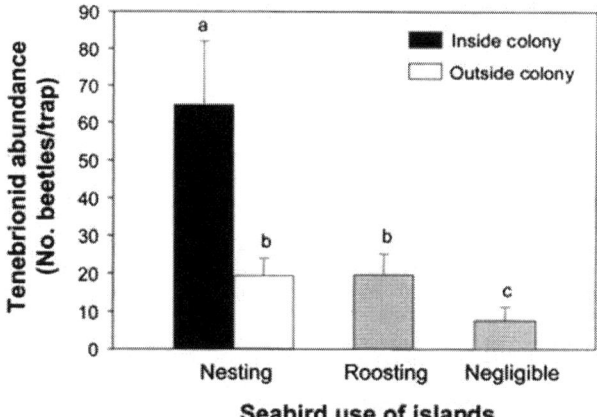

Figure 6.7 Tenebrionid abundance (mean number of beetles/trap) on islands where seabirds nest, roost, or have a negligible presence. Islands with seabirds have significantly higher beetle abundance than islands without seabirds. Also, within nesting islands, seabird colony areas support higher beetle abundances than areas outside the colonies.

quent dry years. Therefore, roosting islands, because of their enhanced guano productivity, maintain higher population levels of tenebrionids than islands without sea birds during and after pulses of productivity.

On nesting islands, beetle abundance is a function of sea bird density. Increased beetle numbers are probably related to the availability of carrion on these islands. In fact, most beetles (94%) were recorded eating sea bird carrion and dropped fish on large nesting islands. Although plant cover is reduced inside the nesting colonies by trampling and the large amount of guano (scorching vegetation), tenebrionid abundances in nesting areas are still up to six times higher than outside colony areas. Outside the colonies, beetle abundance is high through indirect effects (guano, etc.), as on roosting islands. In fact, beetle abundance does not differ between roosting islands and sites outside colonies on nesting islands.

In addition to productivity, predation may have effects in determining the abundance of tenebrionids on islands. Tenebrionid predators on gulf islands include ground spiders (principally Filistatidae, Theridiidae, and Agelenidae; J. Barnes, pers. commun., 1997), lizards (*Uta stansburiana* and *Cnemidophorus tigris*; Wilcox 1981; Hews 1990a,b), birds (e.g., *Corvus corax*; F. Sánchez-Piñero, pers. obs.), and rodents (*Peromyscus maniculatus*; see Parmenter and MacMahon 1988). Because of complex intraguild predation interactions among these groups of predators, the effect of each taxon on tenebrionid beetle abundance is difficult to assess. Our data and preliminary experiments indicate that tenebrionids are the largest proportion of the diet of ground spiders, but the impact on the beetle populations is probably low, and spider populations may simply track rather than determine the abundance of beetles (and other prey). However, spider populations may also be controlled by lizards (J. Sabo, pers. commun., 1999) and rodents.

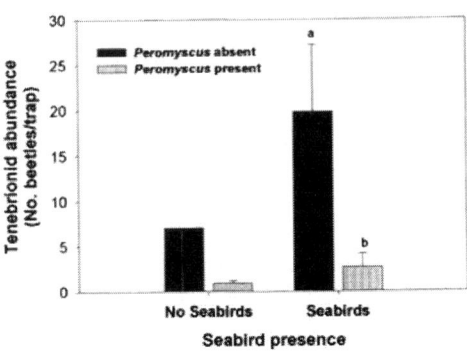

Figure 6.8 Beetle abundance (number of beetles/trap) on islands in Bahía de los Ángeles differentially used by seabirds according to the presence and absence of rodents. On seabird islands, beetle abundance is significantly higher on islands where the omnivorous mouse *Peromyscus maniculatus* is absent. Similarly, on islands not used by seabirds, the only island on which *Peromyscus* is absent (Cabeza de Caballo) shows a higher abundance of tenebrionid beetles than islands on which the mouse is present.

Lizards do not appear to influence beetle abundance, and correlational and experimental data suggest that, although tenebrionids can be eaten by lizards, the effects of lizard predation on tenebrionid abundance is not significant (F. Sánchez-Piñero, unpublished data). The fact that most tenebrionids are crepuscular or nocturnal, spending most of the day in sheltered areas or below ground to avoid extreme temperatures and desiccation, and are thus unavailable for lizards, would support these results.

However, the crepuscular or nocturnal pattern of activity of desert tenebrionids coincides with the activity of rodents, making tenebrionids potential prey for omnivorous rodents (e.g., Parmenter and MacMahon 1988). Our preliminary results indicate that islands (both used and not used by sea birds) where rodents are present support a lower abundance of tenebrionids than those islands where rodents are absent (fig. 6.8). Overall, the inclusion of rodent presence/absence in an analysis considering plant cover and sea bird presence on the 14 islands in Bahía de los Ángeles increases the predictive power of the variables from 41% to 79% of the variability of beetle populations (F. Sánchez-Piñero and P.T. Stapp, unpublished data), suggesting that rodent predation potentially has a strong effect on beetle abundance on these islands. However, more research is necessary to uncover the interactions among tenebrionids and their predators on the islands.

Summary

Due to their specialized adaptations to arid environments, the tenebrionid fauna of the islands in the Sea of Cortés are diverse, with 103 species known at present to inhabit 66 islands. Although most island species in the Sea of Cortés are represented on the mainland, a number of island endemics are also found (25 species). In general, the most species-rich taxa on the mainland are the most species-rich on the islands. However, island species do not form random subsets of mainland assemblages, but rather recurrent patterns of assemblages. This is indicated by the nested sets of species occurrence (as shown by incidence curves based on area) and by the fact that certain combinations of species (e.g., *Cryptoglossa*, *Triphalopsis*, *Argoporis*, also the most abundant species) often recur on different islands. Because tenebrionids are highly related to substrate, variability in the composition of the assemblages is likely due to differences in habitat availability on the different islands. In addition, species interactions (most likely competition) constrains the occurrence of similar species (*Stibia* spp. vs. *Triphalopsis* spp., *Conibius opacus* vs. *Tonibius sulcatus*, *Triphalopsis californicus* vs. *Triphalopsis partida*) on the islands.

The number of species on islands increases as a function of area; only plant species richness (a function of area as well) is a better predictor of tenebrionid species richness. This area effect suggests that tenebrionid species richness is related to habitat heterogeneity, as plant diversity is a better predictor of soil and habitat characteristics (and diversity) than area per se. The fact that incidence curves show that species occur in different proportions on islands depending on island size may reflect the proportion of islands in which certain habitat types occur (with larger islands increasing the probability of having a certain type of habitat). Area also influences the number of endemics found on islands. Thus, both the number of endemics and the proportion of islands with endemic species increase with area.

Isolation apparently has no effects on species richness, although it seems related to the presence of endemic species. Whether this character is due to recent landbridge connections to mainland areas during glaciations (see Murphy 1983; app. 1.1) or due to lack of effective isolation because of the continuous arrival of colonizers (Williams 1996) requires further analysis.

Abundance in tenebrionid populations on the islands is determined by sea bird input and probably also to some extent by predation. Sea birds subsidize beetles indirectly via guano deposition and plant growth and directly via carrion. The availability of these resources strongly affects the abundance of tenebrionid beetles, as evidenced by the differences in abundance within islands: seabird colony areas have up to six times higher tenebrionid abundance than areas outside colonies. Predators such as omnivorous rodents seem to negatively influence the abundance of tenebrionids, as suggested by the lower abundance of beetles on islands where rodents are present.

Acknowledgments This work would have never been possible without the support and help of Gary Polis and Mike Rose and their enthusiasm for the islands in the Sea of Cortés, and this chapter is dedicated to their memory. We thank April Boulton, Kirby Brown, and Fred Andrews for their help in collection of specimens and data. This study was partially funded by National Science Foundation grants DEB-9207855 and DEB-9527888 and the California Academy of Sciences. We thank the Mexican government for research permits DOO 700 (2)-2341 and DOO 750–1502.

References

Aalbu, R.L. In press. The Pimeliine tribe Cryptoglossini: classification, biology and inferred phylogeny (Coleoptera: Tenebrionidae). *Knull Series, Ohio Biol. Survey*.

Aalbu, R.L., T.J. Spilman, and K.W. Brown. 1995. The systematic status of *Amblycyphrus asperatus*, *Threnus niger*, *Pycnomorpha californica*, *Emmenastus rugosus*, and *Biomorphus tuberculatus* Motschulsky (Coleoptera: Tenebrionidae). *Proc. Entomol. Soc. Wash.* **97**:481–488.

Aalbu, R.L., and C.A. Triplehorn. 1985. Redefinition of the Opatrine tribes in North America with notes on some apterous genera (Coleoptera: Tenebrionidae: Tenebrioninae). *Coleop. Bull.* **39**:272–280.

Aalbu, R.L., and C.A. Triplehorn. 1991. *Pedonoeces* G.R. Waterhouse *Blapstinus* Sturm, relevant name changes for California and Galápagos Island species and new insular species from Mexico (Coleoptera: Tenebrionidae). *Coleop. Bull.* **45**:169–175.

Aalbu, R.L., C.A. Triplehorn, J.M. Campbell, K.W. Brown, R. Somerby, and D.B. Thomas. In press. The Tenebrionidae. In: M. Thomas (ed), *Arnett's The Beetles of the United States States (A manual for identification)*.

Adler, G.H., C.A. Austin, and Dudley, R. 1995. Dispersal and speciation of skinks among archiplelagos in the tropical Pacific Ocean. *Evolutionary Ecology* **9**:529–541.

Ahearn, G.A., and N.F. Hadley. 1969. The effects of temperature and humidity on water loss in two desert tenebrionid beetles, *Eleodes armatus* and *Cryptoglossa verrucosa*. *Comp. Biochem. Physiol.* **30**:739–749.

Anderson, D.W. 1983. The seabirds. In: T.J. Case and M.L. Cody (eds), *Island Biogeography in the Sea of Cortéz*. University of California Press, Berkeley; pp. 246–264.

Anderson, W.B., and Polis, G.A. 1999. Nutrient fluxes from water to land: seabird effects on plant quality in the Gulf of California islands. *Oecologia* **118**:324–332.

Andrews, F.G., A.R. Hardy, and D. Giuliani. 1979. The coleopterous fauna of selected California sand dunes. California Department of Food & Agriculture Report, Sacramento, CA.

Berry, R.L. 1975. A revision of the genus *Cerenopus* (Coleoptera: Tenebrionidae). *Ann. Entomol. Soc. Am.* **68**(6):925–934.

Berry, R.L. 1980. A revision of the North American genus *Argoporis* (Coleoptera: Tenebrionidae: Cerenopini). *Ohio Biol. Surv. Bull.* **6**:1–109.

Blaisdell, F.E. 1923. Expedition of the California Academy of Sciences to the Gulf of California in 1921. *Proc. Calif. Acad. Sci.* **12**:201–288.

Blaisdell, F.E. 1925a. Expedition to Guadalupe Island, Mexico in 1922 (Coleoptera). *Proc. Calif. Acad. Sci.* **14**:321–343.

Blaisdell, F.E. 1925b. Revised checklist of the species of *Eleodes* inhabiting America, north of Mexico, including Lower California and adjacent islands. *Pan-Pacific Entomol.* **2**:77–80.

Blaisdell, F.E. 1936. Studies in the tenebrionid tribe Triorophini: A monographic revision of the species belonging to the genus *Stibia* (Coleoptera). *Trans. Am. Entomol. Soc.* **62**:57–105.

Blaisdell, F.E. 1943. Contributions toward a knowledge of the insect fauna of Lower California. No. 7. Coleoptera; Tenebrionidae. *Proc. Calif. Acad. Sci., Ser.* **4**:171–287.

Brown, K.W., and J.T. Doyen. 1991. Review of the genus *Microschatia* (Solier) (Tenebrionidae: Coleoptera). *J. New York Entomol. Soc.* **99**:539–582.

Brun, G. 1975. Recherches sur l'ecologie de *Pimelia bipunctata* (Col., Tenebrionidae) des dunes du littoral de Camargue. *Bull. Ecol.* **6**:99–112.

Chelazzi, L., and I. Colombini. 1989. Zonation and activity patterns of two species of the genus *Phaleria* Latreille (Coleoptera, Tenebrionidae) inhabiting an equatorial and Mediterranean sandy beach. *Ethology, Ecology and Evolution* **1**:313–322.

Colombini, I., L. Chelazzi, M. Fallaci, and L. Palesse. 1994. Zonation and surface activity of some tenebrionid beetles living on a Mediterranean sandy beach. *Journal of Arid Environments* **28**:215–230.

Crawford, C.S. 1991. The community ecology of macroarthropod detritivores. In: G. A. Polis (ed), *The Ecology of Desert*. University of Arizona Press, Tucson; pp. 89–112.

Crawford, C.S. 1981. *Biology of Desert Invertebrates*. Springer-Verlag, New York.

Diamond, J.M. 1974. Colonization of exploded volcanic islands by birds: the supertramp strategy. *Science* **184**:803–806.

Doyen, J.T. 1972. Familial and subfamilial classification of the Tenebrionoidea (Coleoptera) and a revised generic classification of the Coniontini (Tentyriidae). *Quaest. Entomol.* **8**:357–376.

Doyen, J.T. 1976. Biology of the genus *Coelus* (Coleoptera: Tentyriidae). *Kansas Entomol. Soc.* **49**:595–624.

Doyen, J.T. 1982. New species of Tenebrionidae from western North America (Coleoptera). *Pan-Pac. Entomol.* **58**:81–91.

Doyen, J.T. 1984. Systematics of *Eusattus* and *Conisattus* (Coleoptera; Tenebrionidae; Coniontini; Eusatti). *Occ. Pap. Calif. Acad. Sci.* **141**:1–104.

Doyen, J.T. 1987. Review of the tenebrionid tribe Anepsiini (Coleoptera). *Proc. California Acad. Sci.* **44**:343–371.

Doyen, J.T., and C.N. Slobodchikoff. 1984. Evolution of microgeographic races without isolation in a coastal dune beetle. *J. Biogeogr.* **11**:13–25.

Due, A.D. 1992. *Biogeography of scorpions in Baja California, Mexico and an analysis of the insular scorpion fauna in the Gulf of California.* Ph.D. dissertation, Vanderbilt University, Nashville, TN.

Eickmeier, W.G. 1988. Ten years of forest dynamics at Radnor Lake, Tennessee. *Bulletin of the Torrey Botanical Club* **115**:100–107.

Figueroa, A.L., and B.J. Castrezana. 1996. Recommendations for conducting tours in the Gulf of California islands: Working together in the conservation of the Gulf of California Islands. Institute Nacional de Ecología, SEMARNAP, Mexico.

Finston, T.L., and S.B. Peck. 1995. Population structure and gene flow in *Stomion*: a species swarm of flightless beetles of the Galápagos Islands. *Heredity* **75**:390–397.

Finston, T.L., and S.B. Peck. 1997. Genetic differentiation and speciation in Stomion (Coleoptera: Tenebrionidae): flightless beetles of the Galápagos Islands, Ecuador. *Biol. J. Linnean Soc.* **61**:183–200.

Ghilarov, M.S. 1964. The main directions in insect adaptations to life in the desert. *Zool. Zh.* **43**:443–444.

Hadley, N.F. 1978. Cuticular permeability of desert tenebrionid beetles: correlations with epicuticular hydrocarbon composition. *Insect Biochem.* **8**:17–22.

Hafez, M., and A.M.M. Maky. 1959. Studies on desert insects in Egypt. III. On the bionomics of *Adesmia bicarinata* Klug. *Bull. Soc. Entomal. Egypt* **43**:89–113.

Hews, D.K. 1990a. Resource defense, sexual selection and sexual dimorphism in the lizard *Uta palmeri*. Ph.D. dissertation, University of Texas, Austin.

Hews, D.K. 1990b. Examining hypothesis generated by field measures of sexual selection of male lizards, *Uta palmeri*. *Evolution* **44**:1956–1966.

Horn, G.H. 1894. The Coleoptera of Baja California. *Proc. Acad. Natl. Sci. Phila., 2nd Ser.* **4**:302–437.

Huston, M.A. 1994. *Biological Diversity. The Coexistence of Species on Changing Landscapes.* Cambridge University Press, Cambridge.

Jenkins, S.L. 1971. An ecological survey of the Tenebrionidae and Curculionidae (Coleoptera) occurring in the Pleasant Valley area, Joshua Tree National Monument (JTNM). M. S. Thesis, California State University, Long Beach.

Kaszab, Z. 1955a. Die Tenebrioniden der Fiji-Inseln. *Proc. Hawaiian Entomol. Soc.* **15**:423–563.

Kaszab, Z. 1955b. Die Tenebrioniden der Samoa-Inseln. *Proc. Hawaiian Entomol. Soc.* **15**:639–671.

Kaszab, Z. 1982. Die Tenebrioniden Neukaledoniens und er Loyalté-Inseln (Coleoptera). *Folia Ent. Hung.* **43**:1–294.

Kenagy, G.J., and R.D. Stevenson. 1982. Role of the body temperature in the seasonality of daily activity in tenebrionid beetles of eastern Washington. *Ecology* **63**:1491–1503.

Lowe, J.J., and M.J.C. Walker. 1997. *Reconstructing Quaternary Environments*, 2nd ed. Longman, Essex.

MacArthur, R.H., and E.O. Wilson. 1967. *The Theory of Island Biogeography*. Princeton University Press, Princeton, NJ.

McLachlan, A. 1991. Ecology of coastal dune fauna. *Journal of Arid Environments* **21**:229–243.

Medvedev, G.S. 1965. Adaptations of leg structure in desert darkling beetles (Coleoptera, Tenebrionidae). *Entomol. Rev.* **44**:473–485.

Murphy, R.W. 1983. The reptiles: Origin and evolution. In: T.J. Case and M.L. Cody (eds.), *Island Biogeography in the Sea of Cortéz*. University of California Press, Berkeley; pp. 130–158.

Papp, C.S. 1981. Revision of the genus *Araeoschizus* LeConte (Coleoptera: Tenebrionidae). *Ent. Arb. Mus. Frey* **29**:273–420.

Parmenter, R.R., and J.A. MacMahon. 1988. Factors limiting populations of arid-land darkling beetles (Coleoptera: Tenebrionidae): predation by rodents. *Environ. Entomol.* **17**:280–286.

Peck, S.B. 1996. Origin and development of an insect fauna on a remote archipelago: The Galápagos Islands, Ecuador. In: A. Keast and S.E. Miller (eds), *The Origin and Evolution of Pacific Islands Biotas, New Guinea to Eastern Polynesia: Patterns and Processes*. SPB, Amsterdam; pp. 91–122.

Peck, S.B., and J. Kukalová-Peck. 1990. Origin and biogeography of beetles (Coleoptera) of the Galápagos Archipelago, Equador. *Can. J. Zool.* **68**:1617–1638.

Polis, G.A., and S.D. Hurd. 1995. Extraordinarily high spider densities on islands: flow of energy from the marine to terrestrial food webs and the absence of predation. *Proc. Nat. Acad. Sci. USA* **92**:4382–4386.

Polis, G.A., and S.D. Hurd. 1996. Linking marine and terrestrial food webs: allochthonous input from the ocean supports high secondary productivity on small islands and coastal communities. *American Naturalist* **147**:396–423.

Polis, G.A., S.D. Hurd, C.T. Jackson, and F. Sánchez-Piñero. 1997. El Niño effects on the dynamics and control of an island ecosystem in the Gulf of California. *Ecology* **78**:1884–1897.

Richman, A., T.J. Case, and T. Schwaner. 1988. Natural and unnatural extinction rates for island reptiles. *American Naturalist* **31**:611–630.

Sánchez-Piñero, F., and G.A. Polis. 2000. Bottom-up dynamics of allochthonous input: direct and indirect effects of seabirds on islands. *Ecology*.

Sheldon, J.K., and Rogers, L.E. 1984. Seasonal and habitat distribution of tenebrionid beetles in shrub-steppe communities of the Hanford Site in Eastern Washington. *Environ. Entomol.* **13**:214–220.

Stapp, P.T., G.A. Polis, and F. Sánchez-Piñero. 1999. Stable isotopes reveal strong marine and El Niño effects on island food webs. *Nature* **401**:467–469.

Topography International, Inc. 1986. *The Baja Topographic Atlas Directory*. TPI, San Clemente, CA.

Thomas, D.B. 1979. Patterns in the abundance of some tenebrionid beetles in the Mojave Desert. *Environ. Entomol.* **8**:568–574.

Thomas, D.B. 1983. Tenebrionid beetle diversity and habitat complexity in the eastern Mojave Desert. *Coleop. Bull.* **37**:135–147.

Thomas, D.B. 1985. A morphometric and revisionary study of the littoral beetle genus *Cryptadius* LeConte, 1852 (Tenebrionidae: Coleoptera) *Pan-Pacif. Entomol.* **61**:189–199.

Triplehorn, C.A. 1965. Revision of Diaperini of America north of Mexico with notes on extralimital species (Coleoptera: Tenebrionidae). *Proc. U.S. Natl. Mus.* **117**:349–458.

Triplehorn, C.A. 1996. *Eleodes* of Baja California (Coleoptera: Tenebrionidae). *Ohio Biol. Surv. Bull.* n.s. **10**:1–39.

Triplehorn, C.A., and R.L. Aalbu. 1987. *Eleodes blaisdelli* Doyen, a synonym of *E. caudatus* (Horn) (Coleoptera: Tenebrionidae). *Coleop. Bull.* **41**:370–372.

Triplehorn, C.A., and K.W. Brown. 1971. A synopsis of the species of *Asidina* in the United States with description of a new species from Arizona (Coleoptera: Tenebrionidae). *Coleop. Bull.* **25**:73–86.
Triplehorn, C.A., and L.E. Watrous. 1979. A synopsis of the genus *Phaleria* in the United States and Baja California (Coleoptera: Tenebrionidae). *Coleop. Bull.* **33**:275–295.
van Dyke, E.C. 1953. The Coleoptera of the Galápagos Islands. *Occas. Papers Calif. Acad. Sci.* **22**:1–181.
Velarde, E., and Anderson, D.W. 1994. Conservation and management of seabird islands in the Gulf of California: setbacks and successes. In: D.N. Nettleship, J. Burger, and M. Gochfeld (eds), *Seabirds on Islands. Threats, Case Studies and Action Plans*. BirdLife Conservation Series 1. BirdLife International; Cambridge, U.K. pp. 229–243.
Watt, J.C. 1992. *Fauna of New Zealand 26: Tenebrionidae*. DSIR Plant Protection, Auckland, New Zealand.
Wharton, R.A., and M.K. Seely. 1982. Species composition of and biological notes on Tenebrionidae of the Lower Kuiseb River and adjacent gravel plain. *Madoqua* **13**:5–25.
Whittaker, R.J. 1998. *Island Biogeography. Ecology, Evolution and Conservation*. Oxford University Press, Oxford.
Wilcox, B.A. 1981. *Aspects of the biogeography and evolutionary ecology of some island vertebrates*. Ph.D. dissertation, University of California, San Diego.
Williams, M.R. 1996. Species-area curves: the need to include zeroes. *Global Ecology and Biogeog. Lett.* **5**:91–93.
Wollaston, T. 1854. *Insecta Maderensia; Being an Account of the Insects of the Islands of the Madeiran Group*. London.

7

Rocky-Shore Fishes

DONALD A. THOMSON
MATTHEW R. GILLIGAN

    Marine systems have provided little empirical or theoretical support for the equilibrium theory of island biogeography introduced by MacArthur and Wilson (1967; hereafter referred to as MacArthur-Wilson equilibria). In particular, although marine islands represent isolated habitats for shoreline-restricted marine organisms, it is clear that they do not have impoverished biotas relative to adjacent mainland shores as do their terrestrial counterparts. Additionally, it is not clear that colonization rates based on distance from propagule sources, and extinction rates based on island size, play a substantial role in determining the number and kind of species that may exist here.

    In this chapter we ask whether the gulf islands are biogeographic islands to rocky-shore fishes as they are to terrestrial plants and animals. Although the adults and juveniles of most marine shore fishes cannot readily cross the deep waters separating landmasses, most marine fishes have pelagic eggs and larvae which are often found great distances from shore (Leis and Miller 1976; Leis 1991). Certain families of teleostean fishes (e.g., the blennioids and gobioids) have demersal eggs that are attached to a substrate, and only the larvae are dispersed by ocean currents. Some of these fishes have short-lived larvae that are normally found only close to shore (Brogan 1994). Considering such different types of dispersal mechanisms, one must conclude that distance over open water must be as formidable a barrier to dispersal in some fishes as it is to terrestrial organisms. In line with this conclusion, shore-fish faunas of oceanic islands show high degrees of endemism—for example, 23% in Galapagos shore fishes (Walker 1966), 23.1% and 22.2% in Hawaiian and Easter Island fishes, respectively (Randall 1998).

It is well known that the marine insular environment differs considerably from the mainland or continental environment (Robins 1971). Essentially, the former is characterized by a more stable, predictable physical regime with moderate fluctuations in physical factors such as sea temperature, salinity, and turbidity, whereas the latter usually has wider and more unpredictable fluctuations in physical parameters. Robins (1971) compared the difference in species richness between insular and continental fish faunas of the tropical western Atlantic to that between a tropical and a temperate forest, respectively. Similar differences in patterns of species richness in fossil assemblages of benthic invertebrates have been noted for offshore versus inshore marine environments (Valentine and Moores 1974; Bambach 1978).

Most aquatic studies of insular biogeography have examined small-scale insular habitats (Schoener 1974a,b). Results indicate that colonization processes through rather than over water may differ in fundamental ways for many taxa. Molles's (1978) experimental studies of model patch reefs in the Gulf of California, however, showed patterns of immigration and extinction of reef fishes that approximated prediction of the MacArthur-Wilson equilibrium model when species turnover was highest. A similar study of colonization of artificial reefs on the Great Barrier Reef of Australia (Talbot et al. 1978) showed high species turnover but no evidence of a persistent species equilibrium. This was attributed to high predation and the unpredictable seasonal recruitment of juveniles. Bohnsack's (1979) results of fish colonization on model and natural patch reefs in the Florida Keys also showed high species turnover and generally confirmed the predictions of a MacArthur-Wilson equilibrium.

Conversely, the recolonization of larger Gulf of Mexico patch reefs after the natural defaunation by a red tide showed that the reef-fish communities developed according to well-defined successional sequences and not by chance colonization processes (Smith 1979). A climax community formed rather than a dynamic species equilibrium effected by continual species turnover. Although certain features of reef-fish colonization were consistent with the MacArthur-Wilson model, the basic requirements of the model were not fulfilled as witnessed by the low, erratic decolonization rate, low species turnover rate, and the development of a compositionally stable community nearly identical to the original.

It should be emphasized that the colonization potentials of marine fishes differ greatly from those of terrestrial animals in that dispersal is achieved chiefly by planktonic larvae rather than by adults. Larvae grow into juveniles that often occupy habitats and niches very different from those of the adults. As the juveniles mature, a type of pseudo–species-turnover occurs as the propagules gradually assume adult niches. This kind of immigration, which involves a gradual accommodation rather than an abrupt confrontation with adults, may result in a significantly lower extinction rate, leading to a low species-turnover rate for marine shore fishes.

In this chapter we analyze the distribution patterns of some resident rocky-shore fishes on islands and mainlands of the Sea of Cortés to determine how well these distributions conform to existing models and hypotheses of island biogeography and community structure.

We can make two sets of predictions about the distribution patterns of Gulf of California resident reef fishes. If these distributions have resulted from a continuing series of MacArthur-Wilson dynamic equilibria, then: (1) within a biogeographic region there should be more or less stochastic differences in species composition and relative abundances between sites owing to steady-state species turnover; (2) larger, closer islands should have more species per collection than smaller, farther islands because of greater colonization and lower extinction rates; and (3) mainland shores should have more species per collection that any island owing to species–area effects.

If these distributions have resulted primarily from deterministic processes involving physical and biological accommodations that will depress the rate of local extinction and reduce species turnover then (1) within a biogeographic region, species composition and the relative abundance of species within habitats should be similar and predictable; (2) islands, regardless of size or distance, should have similar species diversity and composition and low species-turnover rates, ecological barriers to colonization should be more formidable than geographic barriers to dispersal within a biogeographical region; and (3) mainlands should show lower species diversity than islands owing to higher species-turnover rates, which result from catastrophic density-independent mortalities caused by unpredictable physical changes (silting, winter kills, etc.).

Note that MacArthur-Wilson equilibria require stochastic immigration and extinction with competitive interactions influencing the extinction rates as species number increases. Deterministic processes imply predictable recruitment, habitat selection, and competitive interactions that structure the community but do not necessarily lead to competitive exclusion. The stochastic and deterministic views differ with regard to where disturbance is most likely to have its effects on the community (i.e., embayments and mainlands vs. points and islands).

Although these predictions are extreme and simplified and may not be totally appropriate, they serve to guide our analysis of Gulf of California island–mainland distribution patterns of rocky-shore fishes within the context of current biogeographic and community theories.

Areas Studied and Methods

We sampled 28 islands and 22 mainland sites in the Gulf of California (apps. 7.1, 7.2) from June 1973 to July 1976, principally during the highest sea-temperature months of summer and fall. Quantitative collections of fishes were made along the shorelines of small, protected rocky coves of similar substrate complexity (e.g., crevices, ledges, loose rock) and depth profile (fig. 7.1). This type of habitat is commonly found along steeply sloping, rocky coastlines in the gulf. The collections were standardized by spreading 500 ml of Pronoxfish, a rotenone-base ichthyocide, over an area not smaller that 10 m^2, bounded on one side by the shoreline. In most cases the slope of the substrate below the sea surface was nearly 45°, and the maximum depth was 3 m. All fishes in the area that were partially or totally immobilized by the toxicant were collected using hand nets and a modified "slurp gun" device (Gilligan 1976). The more mobile residents that were usually able to escape the ichthyocide (e.g., damselfishes) and transient species (e.g., mullets, anchovies) were collected but not included in this analysis. Samples were preserved in 10% formalin, stored in 40% isopropyl alcohol, and counted and weighed by species lots on a Mettler K7T top-loading balance. All specimens were deposited in the University of Arizona Fish Collection,

Figure 7.1 View of a protected rocky cove and a near-shore island just north of the San Carlos/Guaymas region of the Sonoran coast typical of island/mainland rocky-shores fish collection sites.

where about 1000 collections of reef fishes made from 1966 to 1978 also provided ancillary distributional data on gulf shore fishes. We recorded sea-surface temperature and secchi disc depth (a measure of water transparency) at the time of the collection. Latitude, mean monthly sea surface temperatures, island area and perimeter, distance of the island to the nearest mainland, and volume of ocean within 20 km were determined for each site (apps. 7.3, 7.4). Numerical H' diversity (Shannon and Weaver 1949) and biomass H' diversity (Wilhm 1968) were calculated for each sample as follows:

$$H = \sum_{i=1}^{s} p_i \ln p_i,$$

where p_i is the numerical or weight proportion of the ith species and S is the number of species in the sample. Community samples were compared directly using several numerical methods. Horn's measure of similarity of overlap (Horn 1966) was calculated for all pairs of samples, producing in this case a 1225-element (50 taking 2 at a time) similarity matrix. This measure of similarity varies from 0, when no species are held in common between two collections, to 1, when the species' identities and their proportional contributions are identical.

In addition to the analyses of rotenone collections, visual censuses of reef fishes on islands in the upper and lower gulf were taken by Gilligan (1980a,b) and further visual censuses were taken during expeditions to the Gulf of California islands in the summer of 1993 and 1994 (Thomson and Mesnick, 1994; Thomson et al. 1996).

Physical Regime of Gulf Rocky Shores

The best predictor of shore-fish distribution in the gulf is sea-surface temperature (Lehner 1979). The offshore sea-surface temperature curves for the upper, central, and lower gulf are similar during the warmer months (April–November) but diverge considerably during the colder months (December–March). This thermal regime approximates the mean sea-surface temperature around gulf islands; inshore sea-surface temperatures along the mainland coast are more extreme. For example, the January mean monthly sea-surface temperature at Puerto Peñasco, Sonora, for the years 1965–1979 is 13.8°C, whereas the mean offshore sea-surface temperature is about 16°C for January, over 2°C warmer (fig. 7.2). Near-surface temperatures show a latitudinal gradient from 13° to 21°C during the winter but virtually no latitudinal gradient during the summer when temperatures are typically between 29° and 30°C. Note that the gulf waters are considerably warmer than Pacific waters along the western coast of Baja California in the summer but the isotherms are nearly parallel within and outside the gulf at the same latitudes in the winter.

The development of a rocky-shore community in the upper gulf is severely constrained by the periodic exposure of wide bands of shoreline. At Puerto Peñasco where the spring tidal amplitude is about 7 m, two-thirds of the area of the platform reefs bordering the shoreline are exposed during spring tides (Thomson and Lehner 1976). Tide pools become refuges for resident and transient reef fishes on the greatly sloping mainland platform reef. These tide pools are practically nonexistent along the steep, rocky shores of the northernmost gulf islands. The combination of the lowest tides occurring in the winter and periodic severe temperature depressions limits the distribution and species diversity of intertidal fishes and may even result in winter kills.

Sandy and muddy bottoms may act as barriers to the distribution of some rocky-bottom fishes because the juveniles and adults of most reef fishes are unwilling or unable to cross even moderate expanses of sand that are uninterrupted by patch reefs or rocky habitat. Although the Baja California gulf coast is nearly continuous rocky coastline, extensive sandy beaches and estuarine lagoons, interrupted by short sections of rocky headlands, are characteristic of the mainland gulf coast (Thomson et al. 2000). Nevertheless, Lehner (1979) has suggested that extensive stretches of uninterrupted sandy or muddy shores are not important barriers to the distribution of rocky-

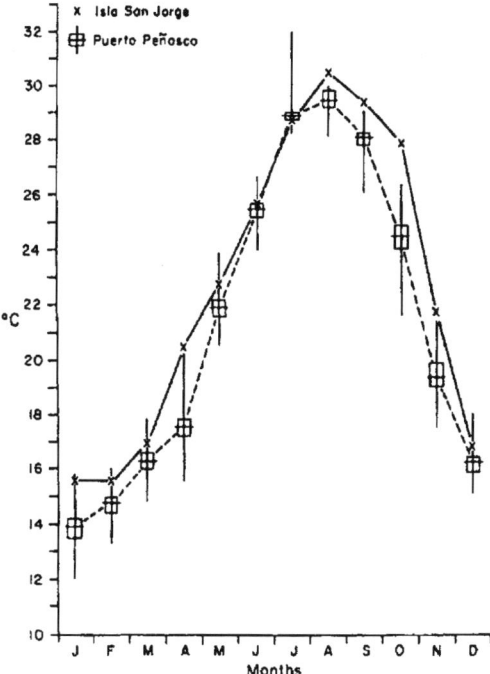

Figure 7.2 Mean monthly sea surface temperatures near Isla San Jorge in the upper Gulf of California (extrapolated from Robinson 1973) and mean, range, and standard error of monthly sea surface temperatures from 1964 to 1978 at Puerto Peñasco (corresponding mainland).

shore fishes. Evidently the pelagic dispersal of the early life stages of most reef fishes is adequate to traverse such barriers.

Biogeography

Walker (1960) analyzed the distribution patterns of 526 species of fishes limited to the area of the continental shelf, excluding all deep-water forms. He distinguished four faunal areas within the gulf, which he named the upper gulf, the central gulf, the Cabo San Lucas area, and the southeastern gulf. He concluded that the gulf fish fauna was clearly part of the Panamic fauna but had certain distinctive elements that gave it a special character of its own. Walker listed 17% of the gulf fish fauna as endemic species; 10% consisted of northern elements, and the remainder (73%) were Panamic.

Our latest tally of all Gulf of California fishes (Findley et al. 1999; Thomson et al. 2000) shows that 872 species have been recorded from this sea with about 10% (86 spp.) endemic to the gulf (see table 7.1). Overall, excluding deep-sea fishes, 92% of the gulf's fish fauna has tropical affinities and 8% temperate affinities. Reef fishes, in general, have more restricted geographic distributions than other marine teleosts. We have analyzed the distribution patterns of 232 species of gulf reef fishes (about 83% of the total number of rocky-shore fishes in the gulf) and found that 64% of these have Panamic affinities and 22% are either gulf or Mexican endemics (table 7.2).

Of the 100 species categorized as primary residents based on limited adult and larval mobility (demersal eggs and short-lived pelagic larvae), 47% are either re-

Table 7.1 Fish species diversity of the Sea of Cortés based on percentages of total number of species recorded (872 spp.)

Category	Percent of Total Species
Teleostean fishes (bony fishes)	90.0
Chondrichthyan fishes (sharks, rays, etc.)	9.5
Myxinoids (hagfishes)	0.5
Endemic species	10.0
Rocky-shore species	32.0

See Findley et al. (1999) and Thomson et al. (2000).

stricted to the gulf or range no farther than southern Mexico. The secondary residents, those species with greater dispersal potential (pelagic eggs and longer-lived larvae), show low endemicity. Leis and Miller (1976) showed that the larvae of Hawaiian fishes from families with demersal eggs tend to be found closer to shore than those with pelagic eggs. Their categorization of inshore reef fishes according to egg type essentially agrees with our residency classification (primary and secondary). The sole exception is the family Pomacentridae (damselfishes), in which we consider gulf species of the genera *Stegastes* and *Microspathon* as primary residents of the genera *Abudefduf*, and *Chromis* as secondary residents, chiefly because of their greater mobility as juveniles and adults.

The pattern we see emerging from studies on the ecology of gulf fishes is that the gulf and tropical eastern Pacific rocky-shore fish communities are strongly influenced by latitudinal physical gradients and marked seasonality in the northern and southern boundaries of the faunal distributions. The distribution of reef fishes within the gulf and throughout the tropical eastern Pacific can be characterized as much more parochial than the more homogeneous distribution pattern of tropical western Atlantic

Table 7.2 Zoogeographical affinities of Sea of Cortés reef fishes

				% Affinities					
	Families	Genera	Species	Panamic	Mexican[b]	Gulf[c]	San Diegan	Indo-west Pacific	World-wide
Resident reef fishes[a]	38	123	232	64	22	10	3	9	1
Primary residents	11	49	100	47	47	22	5	1	0
Secondary residents	28	74	132	77	4	1	2	15	2

[a]Distribution data from Thomson et al. (1979), excluding species with poorly known distributions.
[b]Mexican endemic species (distribution restricted to Mexican waters, including Gulf of California endemic species).
[c]Gulf of California endemic species (distribution restricted to within the gulf).

fishes, as pointed out earlier by Rosenblatt (1967). Besides the generally more rigorous physical environment which accounts for the scarcity of coral reefs in the eastern Pacific, contributors to the formation of distinct faunal subunits within the tropical eastern Pacific include the seasonal patterns of upwelling along the western coasts of North and South America, the narrowness of the continental shelf, fewer islands, and the long stretches of uninterrupted sandy shores. The Gulf of California with its three ichthyofaunal subunits (upper, central, and lower) has the most distinctive and most endemic fish fauna in this region.

Faunal Analysis

The 28 island and 22 mainland sites sampled in the Gulf of California, together with physical and community diversity data for each sample, are given in appendix 7.2. Island and mainland physical data are listed in appendixes 7.3 and 7.4, respectively.

From the 50 quantitative samples of the rocky-shoreline fish community, 68 species from 13 families were used in the analyses (app. 7.5). These species form an ecologically homogeneous group with regard to general habitat requirements. They can be additionally defined by their relatively low adult mobility and inability to avoid an ichthyocide. Eighty-four percent have demersal eggs or are viviparous (i.e., they are primary residents), and 16% have pelagic eggs (secondary residents). Endemism in this group is high in general, with 34% of the species having geographic distributions entirely or primarily within the Gulf of California. If only the 17 most abundant species in the 50 samples are considered, the level of endemism is 62%. These 17 species account for 90% of all individuals collected.

Collection Variables

The major trends among the physical variables for each fish collection site have been presented in a correlation matrix (see Gilligan 1980b). Lower mean monthly minimum sea-surface temperature, greater tidal range, and decreasing volume of ocean within 20 km of site are all positively correlated with latitude ($p < .005$). Although sea temperature at collection sites is not correlated with latitude, there is a positive relationship ($p < 0.05$) between water transparency and sea temperature at collection sites. An increase in water volume within 20 km of each site may reflect both a greater maximum depth and a larger area of ocean impinging on a coastline site. Both of these factors may moderate inshore sea temperatures and expose the sites to open ocean currents that will influence dispersal and colonization by planktonic propagules. Higher sea surface temperatures in the lower gulf suggest a buffering of temperature extremes by large volumes of Pacific water at the mouth of the gulf.

Species diversity (numerical and biomass) is positively correlated with water clarity ($p < .005$) and maximum sea temperature ($p < .05$). Species number is positively correlated with water clarity ($p < .005$), and all sea temperature variables are negatively correlated with tidal range ($p < .05$) and latitude ($p < .005$). The statistical analysis of all gulf samples demonstrates the marked latitudinal gradient within the gulf, which is essentially a relationship between sea temperature and species diversity.

162 The Biological Scene

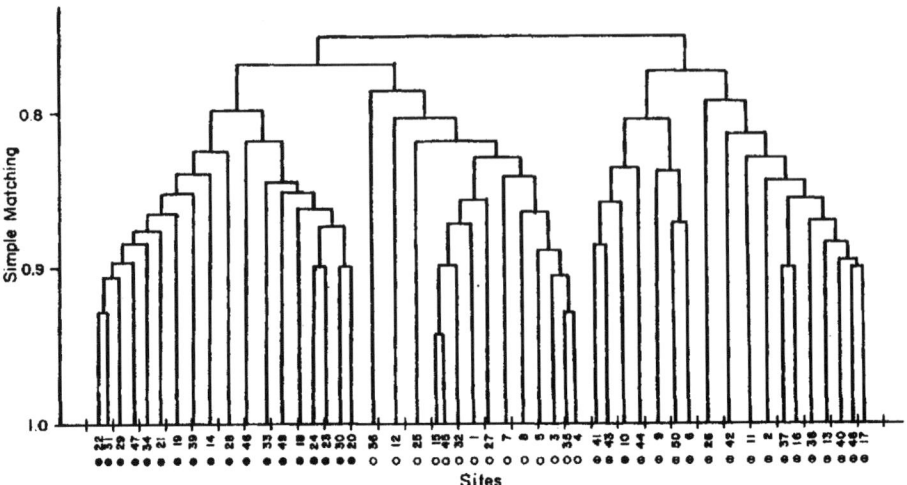

Figure 7.3 Dendrograph of rocky-shore fish collections at 50 sites in the Gulf of California based on similarity of species identity and proportional abundances (Horn's index). Filled circles, upper gulf; open and divided circles, central gulf.

Biogeographic Analysis

Fish collections at each site were compared directly and quantitatively. Horn's measure of similarity or Horn's index (Horn 1966) produces a dendrograph (McCammon and Weninger 1970) that graphically represents the association among sites in a hierarchical scheme based on the proportional abundance as well as the identity of species in each sample. The use of similarity measures sufficiently sensitive to differences in proportional abundance should provide better resolution of biogeographic areas for groups such as marine shore fishes that often have, because of the vagaries of larval dispersal, distributional ranges that extend beyond their principle or breeding ranges. Four groups of sites can be identified at the 0.6 level of similarity in the dendrograph (fig. 7.3), which supports the division of fish communities into upper, central, and lower gulf types (fig. 7.4).

Although the upper and lower gulf groups are rather distinct geographically, it appears that a large part of the central gulf is composed of two broadly overlapping community types. These poorly separated groups may reflect variance in local habitat-associated species abundance rather than regional differences in the community. If we amend Walker's (1960) biogeographic scheme based on these collections of rocky-shoreline residents, then the central gulf would include more of the Midriff area and Baja coastline to the north and more of the western gulf and Baja coastline south of La Paz. Even so, this scheme agrees remarkably well with Walker's biogeographic regions and those based on other marine taxa (Glassell 1934; Dawson 1960; Garth 1960; Soulé 1960; Yensen 1974).

An observation worth noting is that among the two central gulf groups, all but one of the nine mainland sites along the Sonoran coast near Guaymas are of one type and all but one of the five island sites in this area are of the other type (fig. 7.4). Sites 39

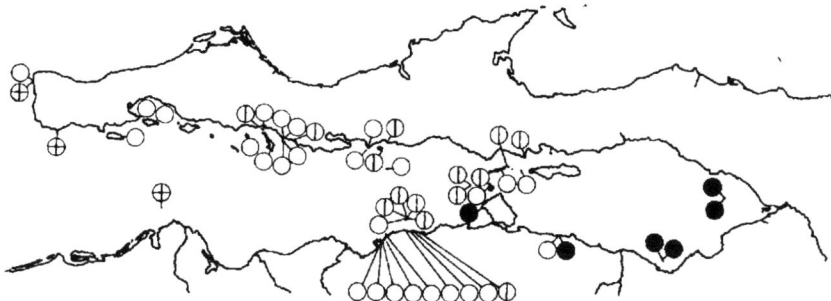

Figure 7.4 Site groups or clusters from the dendrograph (fig. 7.3) at the 0.6 level of similarity of 50 rocky-shore fish collections in the Gulf of California (Horn's index). Filled circles, upper gulf group; open and divided circles, central gulf groups; circles with cross, lower gulf group.

and 30 (app. 7.1), the northernmost Baja collections, were not well separated from the upper gulf group (fig. 7.3) and with only slight differences could have been included in that group. This suggests a pattern of southward extension of the upper gulf community on mainlands and northward extension of the central gulf community on islands (fig. 7.4). Although the upper and lower gulf sites are different physically, the diversity measures show stronger divergence between upper and central areas. Physical and diversity differences between central and lower gulf areas are not as great as either of the former comparisons. The boundary zone between the upper and central gulf is characterized by an abrupt change in the tidal range and tidal currents. This is a region of turbulence and upwelling and has a more pronounced geographical barrier (the larger midriff islands) than does the central-lower gulf region.

Community Composition in the Upper,
Central, and Lower Gulf

The numerical rankings of the 12 most abundant species on islands was compared to the ranking of the 12 most abundant primary resident tidepool fishes from Puerto Peñasco during comparable sea-temperature periods (Thomson and Lehner 1976). Using Spearmann's rank correlation coefficient, it was found that the ranks were quite dissimilar ($r_s = 0.207$) and in fact independent and uncorrelated ($p > .05$).

The composition of the large central gulf area (36 of the 50 total collections made) is perhaps most characteristic of the gulf in general. The 13 most abundant species in the central gulf (fig. 7.5) represent 85% of all individuals collected in the study. Additionally, not only are 8 of the 13 species endemic to the gulf (62%), but these 8 endemics represent 56% of all individuals collected in the entire gulf from the 50 sites samples. These 13 species are adapted for living in rocky surge zones. They compose a guild of crevice dwellers with similar refuge, food, and breeding-site requirements. A comparison of the abundance ranks of these 13 species from the 20 island and 16 mainland sites in the central gulf by Spearmann's rank correlation coefficient revealed that the rankings of the species do not differ significantly; there are certain key species whose ranking changes may be taken as an indication of the community differences

164 The Biological Scene

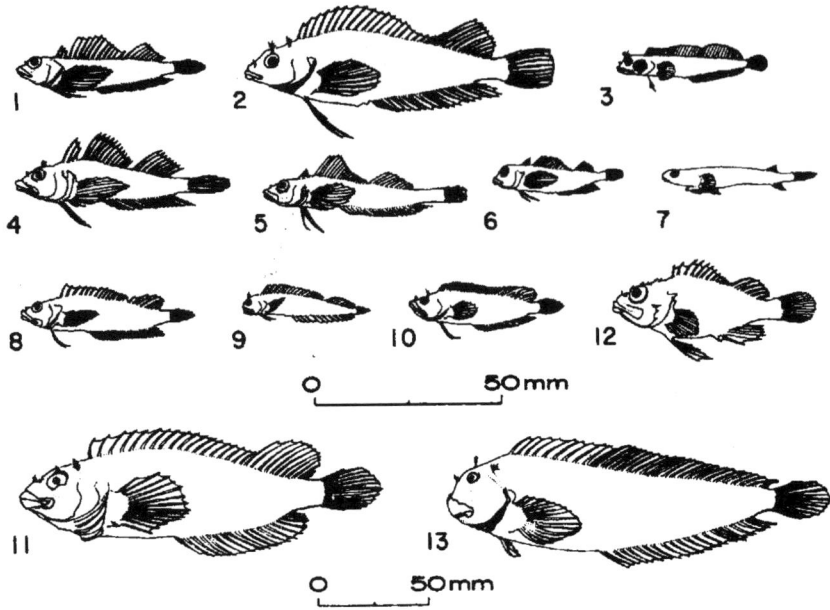

Figure 7.5 The 13 most abundant rocky-shore fish species collected by ichthyocide methods in the central Gulf of California (36 collections). Listed in decreasing order of abundance: 1, *Axoclinus nigricaudus*; 2, *Malacoctenus hubbsi*; 3. *Acanthemblemaria crockeri*; 4, *Enneanectes reticulatus*; 5, *Crocodilichthys gracilis*; 6, *Axoclinus carminalis*; 7, *Tomicodon boehlkei*; 8, *Xenomedea rhodopyga*; 9, *Coralliozetus micropes*; 10, *Paraclinus sini*; 11, *Labrisomus xanti*; 12, *Scorpaenodes xyris*; 13, *Ophioblennius steindachneri*.

between island and mainland sites. The fifteen most abundant species in central gulf island and mainland collections are listed in decreasing order of abundance in table 7.3. It can be seen from this table that *Acanthemblemaria crockeri*, *Enneanectes reticulatus*, and *Coralliozetus micropes* represent a proportionally larger part of the island community and that *Axoclinus carminalis* and *Xenomedea rhodopyga* represent a proportionally larger part of the mainland community. There were no species collected only on islands or mainlands that had abundances greater than or equal to one individual per collection. Although five species were found only in island samples and three species were found only in mainland samples in the central gulf, all were rare in collections, and nothing can be inferred from their presence or absence.

Island-Mainland Analysis

Island and mainland collections showed significant negative correlations of number of species with latitude. Islands had stronger correlation ($r = -.614$, $p < .001$, $n = 29$) than mainlands ($r = -.386$, $p < .05$, $n = 21$), which is attributed to the lack of comparable samples in the northernmost gulf, an area of generally low diversity. The gently sloping, rocky shores of the upper gulf mainland, along with the great spring tidal

Table 7.3 Rank and relative abundance of the 15 most common species in island and mainland collections in the central Gulf of California

	Island (n = 20 Collections)		Mainland (n = 16 Collections)	
Rank	Species	Mean No./ Collection	Mean No./ Collection	Species
1	*Axoclinus nigricaudus*	80	39	*Axoclinus nigricaudus*
2	*Malacoctenus hubbsi*	64	32	*Axoclinus carminalis*
3	*Acanthemblemaria crockeri*	62	29	*Malacoctenus hubbsi*
4	*Enneanectes reticulatus*	49	22	*Xenomedea rhodopyga*
5	*Crocodilichthys gracilis*	40	17	*Tomicodon boehlkei*
6	*Tomicodon boehlkei*	29	14	*Acanthemblemaria crockeri*
7	*Coralliozetus micropes*	26	14	*Crocodilichthys gracilis*
8	*Paraclinus sini*	21	12	*Paraclinus sini*
9	*Axoclinus carminalis*	18	10	*Enneanectes reticulatus*
10	*Xenomedea rhodopyga*	18	10	*Labrisomus xanti*
11	*Labrisomus xanti*	14	7	*Coralliozetus micropes*
12	*Ophioblennius steindachneri*	12	5	*Apogon retrosella*
13	*Scorpaenodes xyris*	11	4	*Ophioblennius steindachneri*
14	*Apogon retrosella*	10	4	*Tomicodon humeralis*
15	*Ogilbia* sp.	7	3	*Malacoctenus tetranemus*

range, prohibited sample sites comparable to the steeper rock surfaces of the two northern gulf islands.

The northernmost islands (Roca Consag and San Jorge) physically resemble rocky shorelines in the central gulf. To compare island and mainland sites in this region, we used data from fish collections analyzed by Thomson and Lehner (1976) at Puerto Peñasco, which approximate the kind of collections made in this study. Comparisons were made between four collections from one mainland tide pool gathered from 1967 to 1973 and two collections each from the northern gulf islands. Island and mainland diversities are similar; however, the tide-pool samples are considerably larger. The major difference between these island and mainland collections is the identity of the species found. The mainland collection included 3 species that were absent from the 4 island collections; however, 10 species were collected on the island that are extremely rare or absent on the adjacent mainland (Puerto Peñasco), where more than 130 collections from rocky habitats have been made from 1963 to 1978. These island species are all common in collections throughout the rest of the gulf. Eight other reef species (*Haemulon maculicauda, Chromis atrilobata, Alphestes immaculatus, Serranus psittacinus, Malacoctenus tetranemus, Ptereleotris* sp., *Lythrypnus dalli, Tomicodon boehlkei*) occur regularly on these islands but are absent or have been only rarely reported on the adjacent mainland (Puerto Peñasco).

In the central gulf, where a relatively large number of island and mainland sites were sampled in the same range of latitude, the island–mainland comparisons yielded the most surprising and significant results. Table 7.4 gives the island and mainland means and t tests for a number of physical and fish diversity measures. There is a

Table 7.4 Means and t-tests of physical and diversity measures for 20 island and 16 mainland sites in the central Gulf of California

	Mean Island	Mean Mainland	Probability
SD	9.9	5.8	< .05
Volume	50.0	21.7	< .001
S	20.6	17.1	< .05
N	507.9	245.2	< .001
B	594.1	361.4	< .05
H'_n	2.228	1.984	< .05
H'_b	1.967	1.586	< .001

SD = secchi disc depth (m); Volume = volume of ocean within 20-km radius from site (km^3); S = number of species; N = number of individuals; B = biomass (g); H'_n = numerical diversity; H'_b = biomass diversity.

significantly greater number of species, higher density, greater biomass, and higher species diversity (H_n and H_b) on islands. These differences parallel significant differences in water clarity and volumes of ocean near island sites.

Habitat area was estimated by perimeter for islands and contiguous rocky coastline length for mainlands (apps. 7.3, 7.4) and is loosely correlated to fish species number in the 50 ichthyocide samples ($r = .230$, $p = .054$). Island perimeter and rocky coastline length are better estimates of habitat area for shore fishes than total island area because suitable habitats of the submerged shelf of rocky islands assume a ring configuration for all but the smallest islands and patch reefs. It must be pointed out that these are only rough estimates of supposed contiguous habitat areas and that the decisions regarding the continuity of some areas were arbitrary. A log/log species–area curve for all island samples was significant ($r = .355$, $p = .032$, $n = 28$), but considering only the 20 central gulf islands it was not ($r = .135$, $p = 0.285$). We found no correlation between number of species in island collections and distance to the nearest mainland. In addition to collections, visual censuses are required to obtain reasonable estimates of the total species number of reef fishes on islands. Visual censuses of reef fishes on patch reefs and small near-shore islands (in the central gulf) have produced remarkably good species–area curves (fig. 7.6).

Species Turnover Within and Between Island and Mainland Sites in the Central Gulf

The rate of replacement of species along a gradient and the rate of replacement of species from time to time at a locality owing to colonization and extinction events have been termed β diversity and species turnover, respectively. Although formulation of β-diversity measures (Pielou 1975) may differ from measures of species turnover (Diamond 1969; Simberloff 1969), both are concerned with the comparison of presence or absence of species in two or more samples. If we view the samples of the resident rocky-shore fish community in the central gulf as successive samples at various localities from an island or the mainland, then we may observe indirectly the regularity or patchiness with which species are distributed on islands and mainlands by measuring this site to site compositional difference or species turnover.

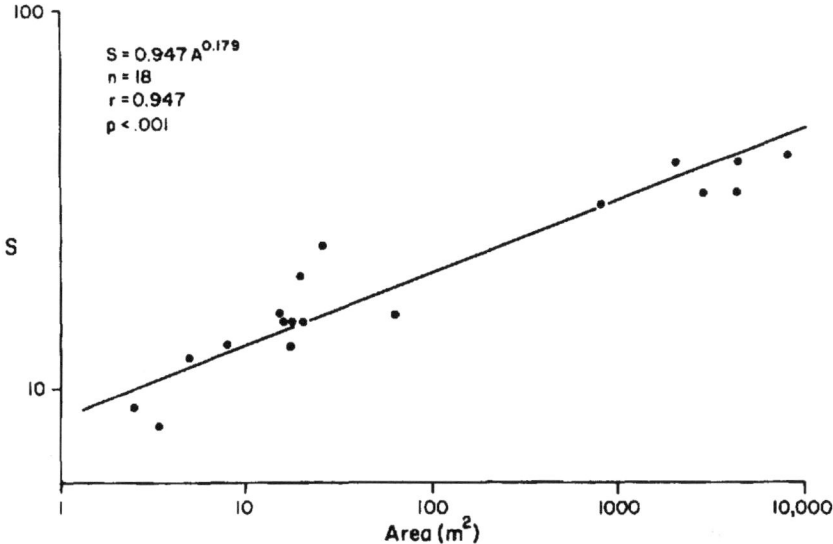

Figure 7.6 Log/log species–area curve for rocky-shore fish species on small, near-shore natural patch reefs (<100 m^2; Molles 1976) and small, near-shore islands (>100 m^2; Gilligan, unpublished data) in the Guaymas-San Carlos, Sonora vicinity of the central Gulf of California. Habitat area for islands was estimated by rocky shoreline area to the depth of observation (4 m) rather than by terrestrial island area.

A simple measure of species turnover is just the number of species involved in absence–presence events (colonizations) and presence–absence events (extinctions) per species in two samples; this can be calculated as 1 minus the Jaccard coefficient of similarity. The Jaccard coefficient is the number of co-occurrences (presence–presence species) per species in two samples (Clifford and Stephenson 1975). A Kruskal-Wallis one-way analysis of variance of the ranked turnover values for the 20 island and 16 mainland sites in the central gulf showed that the average species turnovers were not equal within and between islands and mainlands ($\chi^2 = 64.95$, $p < .001$) and that β diversity (site-to-site species turnover) is lowest within islands and highest within the mainland; that is, species composition among island sites is more predictable than among mainland sites. Not surprisingly, the same results were obtained when Isla San Pedro Nolasco (five collections) was compared with the corresponding mainland coastline (nine collections). Gilligan (1980a) also found, using visual census methods along a habitat gradient in the gulf, low species turnover at exposed points and high species turnover at protected, rocky shores.

Latitudinal and Island Body-Size Trends in the Gulf

We calculated mean weight per individual for each species in each collection by dividing total weight of all individuals of a species in a collection by the total number of individuals in that collection; this, in turn, was used as an estimate of body size. Among the 30 species collected at at least 10 sites, 5 (*Gobiesox pinniger*, *Tomicodon*

168 The Biological Scene

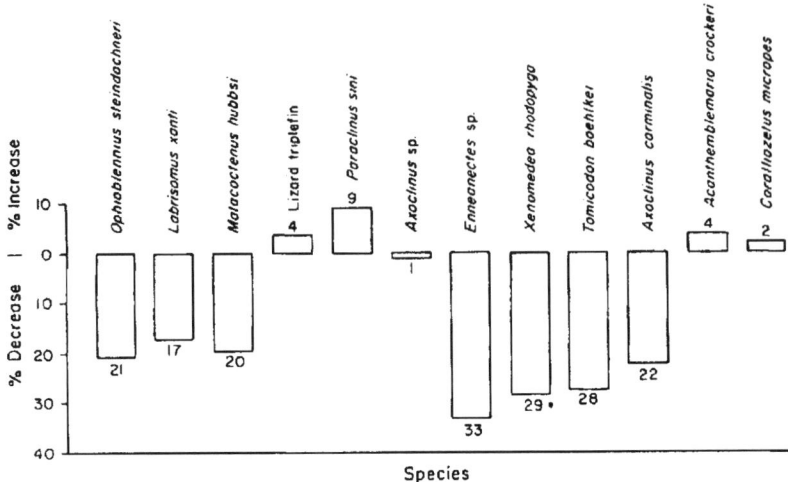

Figure 7.7 Island body-size trends of the most abundant central gulf rocky-shore fish species in ichthyocide collections. Percent change in body size on islands for each species is based on mean weight per individual of species lots from 20 island and 16 mainland collections in the central Gulf of California. Species appear left to right in order of decreasing overall body size. *Difference in mean body weight was significant at the .05 level.

boehlkei, *Hypsoblennius brevipinnis*, *Axoclinus carminalis*, *Gobiosoma chiquita*) had significant positive correlations ($p < .05$) of body size with latitude, and only 1 (*Hypsoblennius gentilis*) had a significant negative correlation. At a lower significance level ($.05 < p < .19$), another 9 of the species had positive correlations of body size with latitude (*Apogon retrosella*, *Crocodilichthys gracilis*, *Labrisomus striatus*, *L. xanti*, *Malacoctenus hubbsi*, *Xenomedea rhodopyga*, *Acanthemblemaria crockeri*, *Coralliozetus angelica*, *Barbulifer pantherinus*), and none showed a negative correlation. Although this kind of estimate of body size may be somewhat biased by the contribution of juvenile individuals in samples, there are two lines of evidence that suggest that the trend of increasing body size with latitude is a real one and that it may even be quite prevalent among small rocky-shore fishes in the gulf. Measurements of specimens in the University of Arizona Fish Collections have found that latitudinal size clines do exist in species in the families Labrisomidae and Tripterygiidae (measuring only the maximum length of individuals in hundreds of collections) in the Gulf of California, despite the fact that the trend was undetectable for some of these species in the present study. Second, unusually large individuals of species in the families Labrisomidae, Chaenopsidae, and Gobiidae have been collected in the midriff island area of the upper-central Gulf of California. For some of these species this area represents the northern limits of their geographic distribution.

Among the 13 most frequently collected species (pictured in fig. 7.5), there appears to be a trend toward decreasing body size on islands in the central gulf. Although the only species that had a statistically significant difference ($p < .05$) in mean body size between the 20 island and 16 mainland collections was *Xenomedea rhodopyga* (with a 29% decrease in islands specimens; fig. 7.7), 8 of the 13 species showed trends of decreasing body size on islands, with 7 of these showing decreases between 17% and 33%. Of the five species with trends of body-size increase on islands, only one increased more than 9%. An analysis using only the standard length of the largest individual in collections showed that, again, *Xenomedea rhodopyga* was smaller on islands ($p < .02$). However, no clear trend was shown for other species.

Patterns of Diversity and Community Structure

The biogeographic and ecological nature of islands in marine systems is quite different from that of terrestrial systems. Although open ocean and deep water are barriers to the movement of many marine shore species, they are not formidable barriers to the distribution of species with planktonic means of dispersal, which includes a great majority of the marine biota. Distance over open ocean must be great or environmental conditions very different, due to major ocean current differences, to effect major changes in the faunal composition of shore biotas. The Eastern Pacific Barrier (EPB) is a distance barrier that separates the Indo-west Pacific biota from that of the Eastern Tropical Pacific; it is about 90% effective going from west to east and nearly 100% effective from east to west (Briggs 1961; Rosenblatt 1967), although favorable ocean currents exist for both paths of immigration.

Besides distance and ocean current direction, it is clear that ecological factors such as species-specific dispersal potential become important determinants in successful colonization. Of the transpacific shore fishes that have established populations in the Tropical Eastern Pacific (Rosenblatt et al. 1972), most are secondary resident reef fishes with pelagic eggs and long-lived pelagic larvae. They also tend to be species that are common throughout their range, undoubtedly providing an abundant source of propagules. In the Eastern Tropical Pacific the transpacific immigrants are most commonly found on the far offshore islands (e.g., Isla Revillagigedo, Galapagos) and at the tip of Baja California, a region which at one time may have been an island (Durham and Allison 1960) and which now has what has been described as an insular fauna (Walker 1960). It is not clear, however, what an insular fauna is. In the eastern Pacific there is clearly no insular biogeographic region since, as Rosenblatt and Walker (1963) have noted, of the seven species listed by Snodgrass and Heller (1905) and "eastern Pacific insular species" (restricted to two or more islands), four have ranges that extend to the mainland. Yet many shore-fish species have restricted distributions, and the fauna of isolated oceanic tends to have a high percentage of endemic species (e.g., Hawaiian, Galapagos, and Easter islands). The endemics are usually derived from those taxa that have more limited means of dispersal (e.g., primary resident species with demersal eggs and short-lived pelagic larvae), a correlation between differentiation and vagility noted by Rosenblatt and Walker (1963). The relatively high endemism of tropical rocky-shore fishes (22%) along Mexico's Pacific coast is largely due to the primary resident fishes (47 spp.) rather than to the secondary residents (5 spp.) It appears that insular faunas at this level are characterized by fugitive species with wide distributions and dispersal capabilities or, on very isolated archipelagos, by endemics that are generally much more restricted in their dispersal capability.

Individual islands in the Gulf of California accommodate a high proportion of the available species pool in comparison to the rather low proportion of the total available species pool of the terrestrial fauna found on even large islands close to shore. Although this may be due primarily to the much greater effectiveness of open ocean as a barrier to terrestrial species, it may also indicate something about the nature of competition, resource utilization, and coexistence in terrestrial versus marine communities. Furthermore, processes of marine immigration and colonization are continuous and rates are high, perhaps leading to depression of extinction or to the rescue effect as proposed by Brown and Kodric-Brown (1977). Both island and mainland reefs receive propagules primarily from outside sources (Sale 1977) because the majority of the reproductive effort of reef-dwelling species is swept away by currents and becomes part of the plankton community (meroplankton). The planktonic early life stages settle and transform, either selectively (Marliave 1977) or randomly in response to hard or relatively fixed substrates, in habitats that are often quite different from the habitats of the adults. Considering the difference between marine and terrestrial systems in immigration, colonization, resource utilization, and complexity of life histories, one might not expect MacArthur-Wilson–type equilibrium to be appropriate for most marine insular habitats. However, in small-scale insular systems, Molles (1978) found that patterns of immigration and extinction of Gulf of California reef fishes on experimental model reefs approximated MacArthur-Wilson predictions when species turnover was highest. Nevertheless, fish species composition was similar on all of Molles's 12 model reefs, contrasting sharply with results of Sale and Dybdahl (1975), who showed random patterns of fish species composition on experimental coral heads in the Great Barrier Reef. This difference can be attributed mainly to the larger species pool and smaller size of reefs in Sale and Dybdahl's experiments. We contend that colonization will always appear stochastic or random if the scale of habitats or the sample areas are too small to contain some reasonable proportion of the species pool.

The Marine Environment of Islands and Mainlands

The physical environments of oceanic islands and continental coasts differ markedly (table 7.5) and may be associated with major differences in marine coastal faunas (Robins 1971; Gilbert 1972). The key to understanding these differences is recognizing the nature of the ecotone between their respective shoreline zones and the surrounding ocean. Offshore island-shore faunas form a rather sharp gradient with the pelagic ocean province, whereas mainland coasts form broader ecotones with the pelagic neritic province and the benthic sublittoral zone of the inner and outer continental shelf. Thus offshore islands surrounded by depths greater than 200 m tend to be influenced strongly by the open ocean and its characteristic biota. Gulf of California islands range from offshore (e.g., San Pedro Mártir and Tortuga) to inshore (e.g., Carmen and Espíritu Santo). The marine environment of gulf island and mainland coasts essentially reflects the differing physical regimes of the surrounding water masses. One can expect a moderation of sea-surface temperatures with distance offshore from the continental landmasses and shelf areas. Although data on sea-surface temperatures off islands are unavailable, offshore isotherm profiles produced by several hydrographic cruises indicate that the water around islands gets neither as cold in the winter nor as warm in the summer as near-shore mainland water (see fig. 7.2). This phenomenon alone may be important in preventing winter kills of inshore fishes

Table 7.5 Summary of island–mainland environmental differences in the Gulf of California

Physical and Biological Environments	Islands	Mainlands
Sea temperatures	Seasonal fluctuations moderated; lethal winter low temperatures unlikely to occur	Seasonal fluctuations more extreme; lethal winter low temperatures cause periodic winterkills in upper Gulf
Water clarity	Usually clear	Often turbid
Wave action and currents	Extent of unprotected (exposed) shores exceed protected (sheltered) shores; surge conditions common; current velocity consistently greater than inshore	Extent of protected (sheltered) shores exceeds unprotected (exposed) shores; surge conditions not usual; current velocity tends to be less than offshore
Tides	Narrow intertidal zone is exposed at low tide because of greater slope of island shoreline	Broad intertidal zone is exposed at low tide because of more gentle shoreline slope
Substrate and sediment	Cliff faces and consolidated boulders typical; unconsolidated sediments coarse-grained; high level of turbulence required to produce turbidity	Loose and consolidated boulders typical; fine-grained sediments common; low level of turbulence produces turbidity
Plankton	High diversity; low density; proportion of holoplankton exceeds proportion of meroplankton	High density; low diversity; proportion of meroplankton exceeds proportion of holoplankton
Benthic algae	Low profile forms (coralline reds) dominant	High profile forms (Sargassum) dominant
Benthic invertebrates	Stony corals, hydroids, sponges and sea stars dominant forms	Soft corals, anemones, polychaetes and sea cucumbers dominant forms

The same set of physical and biological conditions that characterize the differences between islands and mainlands in the gulf can also be found to a lesser degree along the rocky-shore habitat gradients of small peninsulas in the central gulf (Gilligan 1980a).

on islands. Winter kills from low sea temperatures occur periodically in the upper gulf (Thomson and Lehner 1976), and the gulf islands may act as refugia for warm-water species during these severe climatic events.

The greater water clarity around islands in contrast to continental shores is a function of lower plankton densities and smaller quantities of suspended sediments. Although the gulf as a whole is a protected sea, severe storms can cause great disturbances to mainland communities on account of greater abundance of unconsolidated sediments (boulders, cobble, sand), which may scour or completely cover portions of the reef when wave action is heavy. The silting over of reefs caused by the deposition of these fine sediments along the coast leads to the mortality of sessile invertebrates and algae.

Although current measurements are generally unavailable for gulf shoreline areas, observations show that currents are stronger around islands and exposed mainland points. This greater water circulation rate prevents silting of rocky substrates and efficiently carries away the pelagic eggs and larvae of reef fishes and invertebrates, although eddy formation may serve to retain some of these (Boden 1953; Emery 1972).

Tides show less effect on the shoreline communities of islands because of steeper slopes and greater wave action, which expose less of the overall reef area at low tide. The gentle slopes and great tides of the upper gulf platform reefs severely restrict the use of the reef area by marine fishes at low tides, and the intertidal fish communities are regulated by the availability of tide pools. Plankton communities have a strong influence on the reef community both as a food source for planktivores and by their propagule content. The ecological efficiencies of productivity of the open ocean (10%), coastal zones (15%), and upwelling areas (20%), and the nature of planktonic food chains (Ryther 1969) have implications for island and mainland reef communities. The offshore islands in the gulf would be expected to have a more oceanic-type plankton community characterized by a greater proportion of holoplanktonic species (entire life as plankton) than meroplanktonic forms (larvae of benthic invertebrates and fishes). In contrast, the neritic plankton community of the continental shelf and inshore areas would be expected to have a greater abundance of meroplankton. Preliminary analysis of numerous surface zooplankton samples in the gulf indicated that this is the case (D. Siegel-Causey, pers. commun.). The longer food chains of oceanic plankton communities would provide a greater diversity of food particle size for planktivores, which, if resource partitioning according to food item size is important, could allow higher diversity of filter feeders and other selective planktivores. There does appear to be more profuse growth (greater percentage of cover) of stony corals (branching types and sea fans) and hydroids on island reefs. The only extensive growths of coral in the eastern Pacific are along islands (Isla Jaltemba, Clipperton Island) and insular mainland shores (Bahias Pulmo and Los Frailes in the cape region of Baja California). This is consistent with Robins's (1971) conclusion about coral reefs around islands in the tropical western Atlantic.

Other trends among the non-fish biota are the dominance of encrusting red algae (Rhodophyta) on islands and the greater abundance of large brown algae (Phaeophyta, e.g., *Sargassum*) in sheltered bays along the mainland coasts of the gulf. Asteroids (sea stars) are more diverse on islands and holothuroids (sea cucumbers) on mainland coasts (L.Y. Maluf, pers. commun.), the former perhaps because of the greater variety and density of prey around islands and the latter because of the greater abundance of sediments along the mainland.

Rocky-Shore Fishes

The significantly greater density of rocky-shore fishes around islands indicates a higher carrying capacity for island reefs and perhaps a better or more productive environment for shore fishes (table 7.6). Island collections had a higher α diversity (see table 7.4) as measured by both number of species and species-diversity indices (H_n and H_b); greater richness (number of species), however, may be due partially to the higher densities of individuals (larger sample sizes) in island collections. Nevertheless, there appears to be a real trend in greater diversity of fishes along the gulf island shores, although the difference is not great in our samples. We predict a much greater difference in diversity if total species counts of reef species are used for island and adjacent mainland areas of comparable size.

Table 7.6 Summary of island–mainland differences in the rocky-shore fish communities of the Gulf of California

	Islands	Mainlands
Density and biomass	Greater number of individuals per collection and higher biomass diversity	Smaller number of individuals per collection and lower biomass diversity
Body size	Numerically dominant species tend to be smaller	Numerically dominant species tend to be larger
α-Diversity	Tend to have more species and greater evenness per collection	Tend to have fewer species and less evenness per collection
β-Diversity (species turnover)	Lower rate between (spatially) and within (temporally) habitats	Higher rate between and within habitats
Resident species composition	Triplefin blennies most dominant guild, followed by tube blennies	Triplefin blennies most dominant guild, followed by labrisomid blennies
Transient fishes	Tend to be oceanic pelagic and deep benthic species	Tend to be neritic pelagic and sandy shore-estuarine species
Competition	More intense because of higher density	Less intense because of low density
Predation	More pelagic piscivorous species; seabird and sea lion predation more severe on large reef fishes (secondary residents)	More benthic piscivorous species; fishing predation greater on larger reef fishes
Vertical zonation	Vertical zones more compressed because of steep shoreline; deeper-water fishes found closer to shoreline	Vertical zones broad because of gentle slopes; deeper-water fishes farther from shoreline
Latitudinal gradient	Northern islands have more southern elements than adjacent mainland; northern boundaries of distribution extended on islands	Northern mainlands show greater dominance by northern disjunct species

A trend of smaller adult body size in primary resident fishes on islands is indicated (fig. 7.7). We suggest that smaller body size is associated with precocious metamorphosis from larvae to juvenile and that this may be selected for on islands where, given the necessity of a temporary planktonic existence to reduce predation (Helfman 1978; Johannes 1978), a recolonization rather than a extended dispersal strategy may be more successful. Larvae that metamorphose into juveniles earlier and at a smaller size would have an advantage over longer-lived larvae in returning to their parent island.

Species turnover between sites was greatest for mainlands and lowest for islands. This may reflect the greater disturbance factor on mainland shores as compared to the more benign island environment. The species composition and relative abundance ranks of the 15 most common fishes on central Gulf islands and mainlands were similar (see table 7.3), and the dominant guild, triplefin blennies (tripterygiids: 41% island, 43% mainland), consisted of four species of active fishes that live on the surface of boulders along the shoreline. Two species of tube blennies (chaenopsids), small fishes that live in tubular dwellings vacated by various invertebrates (polychaetes, gastropods, barnacles), were relatively more numerous on islands (19% vs. 9%), whereas five species of labrisomid blennies were relatively more numerous on mainlands (34% vs. 25%). The most abundant species in both island and mainland collections was the small, cryptically colored Cortés triplefin, *Axoclinus nigricaudus*, endemic to the Gulf of California. There was remarkable similarity in the identity and abundance ranks of the 15 most abundant central gulf species (over 90% of all individuals collected) on islands and mainlands (see table 7.3). The few significant changes in ranking (e.g., *Axoclinus carminalis*, *Enneanectes reticulatus*, *Acanthemblemaria crockeri*) were not dramatic, and not a single species could be considered primarily insular or continental. The great similarity in species composition of the central gulf collections is in concordance with Molles's (1978) results of visual censuses of model and natural patch reefs in the Guaymas-San Carlos region of the central gulf and with Thomson and Lehner's (1976) repetitive defaunation of tide pools in the upper gulf. Such results lend no support to a lottery hypothesis of community structure but suggest that habitat requirements of reef fishes are not as generalized as Sale (1978) has maintained, and there may be more order than chaos in recruitment to a site in this system.

The roles of competition and predation are too complex to evaluate on the community level on account of such confounding factors as age- and size-dependent habitat shifts and flexible growth rates that allow a great deal of accommodation to occur. We do assume, however, that competition is more intense in island communities simply because of increased density. Piscivorous predation is different in the two environments. Pelagic predators such as yellowtails, black skipjacks, and sharks may be more common off islands where needlefishes, barracudas, and sierra mackerel might be more frequent along mainland shores. It is difficult, however, to assess the effects of this type of predation because these kinds of predators generally do not prey on reef fishes. Mainland rocky shores are often adjacent to shallow sandy areas where jacks, corvinas, snappers, flounders, small sharks, and stingrays may prey on juvenile reef fishes. We suggest that predation pressure may be more intense on mainland shores because of the presence of such rock–sand interface predators. This could also account for the lower density of fishes in mainland collections.

Although the compositions of island and mainland collections in the central gulf were similar, upper gulf island and mainland collections were markedly dissimilar. Many species of fishes that are rare or absent on the mainland coast were found in the northernmost island collections. The presence on northern islands of species commonly found farther south may be attributed to the more benign environment around islands and to the severe physical environment associated with the broad intertidal and shallow mainland coasts of the upper gulf.

Reef-fish diversity is strongly influenced by another, generally overlooked, factor: the vertical relief of the substrate. Molles (1978) found good correlations between species diversity of central gulf fishes and height of patch reefs but found no significant differences in species diversity when he varied the interspace size diversity of his model patch reefs. The rocky shorelines and narrow shelves surrounding gulf islands are typically steeper than the gently sloping, broad shelves of the mainlands. Thus there is greater vertical profile heterogeneity of nearshore habitats around islands; equal area samples should show greater diversity on islands than on mainlands. Our shoreline collections, however, may not reflect the compression effects of vertical zonation because the slopes of island–mainland collection sites were similar.

Recent Visual Censuses of Island Reef Fishes

In 1993 (Thomson and Mesnick 1994) and 1994 (Thomson et al. 1996) visual fish censuses were conducted by students of the senior author's marine ecology class along island sites throughout the gulf. In 1993 the following islands were censused: Puerto Refugio, San Esteban, San Pedro Mártir, San Ildefonso, Santa Catalina, and Las Ánimas (Sur). In 1994 the island censused included: San Pedro Mártir, Partida (Norte), Ventana, Salsipuedes, Ildefonso, Las Ánimas (Sur), San Diego, and Farallon. Although the censuses were qualitative (absent, rare, uncommon, common, and abundant) and about 60 species of easily observable and identifiable reef fishes were selected (secondary residents only, not including the primary resident species collected for this study), the pattern of relative abundance was clear. Compared to earlier collecting and censusing such as during Stanford's cruise 16 of the *Te Vega* (1967 unpublished cruise report) and Gilligan's (1980a,b) visual censuses in 1974, it was clear that there was little or no change in the relative abundance of common reef fishes. However, the commercially exploited fishes such as large groupers, snappers, and parrotfishes were considerably less abundant and rarely seen, even at the most remote island sites. The one exception to this was the heavily exploited leopard cabrilla, *Mycteroperca rosacea*, which was commonly observed and was often abundant at some island sites. Nevertheless, large individuals were uncommon. We were surprised to discover that considering that nearly all groupers are protogynous hermaphrodites, examination of the gonads of several individuals of the leopard grouper showed that several small males had ripe testes, indicating that they are gonochorists and not hermaphrodites (D.A. Thomson and Neff Nash, unpublished obs.). The question remains whether the leopard grouper's sexual strategy, unique among the groupers, was responsible for maintaining reasonably abundant populations in the face of heavy commercial exploitation.

Island Biogeography and Reef-Fish Community Structure Models Reevaluated

The results of our studies on the distribution patterns of rocky-shore fishes did not conform with our first set of predictions based on a MacArthur-Wilson equilibrium. The high degree of uniformity in the species composition and relative abundance ranks in island collections and in island versus mainland collections in the central gulf indicated a low degree of stochasticity and high degree of predictability in the dynamics controlling community structure. Between-site species turnover was not great enough to suggest that random events played a significant role in species composition; however, predictability was lower on mainland sites. Smith (1979) concluded that "eastern Gulf of Mexico reef-fish communities develop according to predictable rather than chance processes implicit in the M-W [MacArthur-Wilson] model" (p. 59). He found that colonization of patch reefs after natural defaunation by a red tide seemed to follow a successional sequence leading to a climax community consisting of a fairly constant assemblage of species that resisted further colonization by other species. Thomson and Lehner (1976) showed that repeated defaunation of tide-pool communities with rotenone ichthyocides and natural winter kill did not change the community composition of intertidal fishes in the upper Gulf of California, although colonization, in this case, was probably by juveniles and adults rather than by larvae. The reason that these marine systems behave differently from terrestrial systems may be that marine colonization processes do not necessarily fit a diffusion model, as is assumed in the MacArthur-Wilson theory—that is, uniform decrease in the density of propagules in all directions from a colonizing source. In marine environments, predictable seasonal current patterns during the peak reproductive periods will produce dispersal patterns that favor dispersal in some directions over others. Assuming higher rates of water transport offshore than inshore, it is likely that islands and peninsular points receive more diverse propagules than inshore protected areas. This factor may contribute significantly to the insular biogeographic nature of the cape region and its distant Indo-west Pacific faunal affinities. In this view, rocky reef and shoreline areas in protected embayments become isolates analogous to terrestrial islands. Marine islands may experience colonization rates that suggest a much closer proximity of each other than actually exists geographically. In fact, the gulf islands may maintain the species pool and may be the main source of propagules, whereas mainland rocky shores may be the isolated insular analogs.

Although the dynamics of propagule dispersal is poorly understood, it is clear that propagules must suffer enormous losses during passive transport as plankton. The fact that marine species in general exhibit an almost absolute predominance of reproductive strategies characterized by large numbers of propagules suggests that (1) habitats for juveniles are quite different from those for adults; (2) suitable habitats are so small and patchily distributed that they preclude the coexistence of parents and cohorts of their offspring; or (3) competition and predation are so intense in adult habitats that only in other (perhaps marginal) habitats do settling juveniles have a reasonable chance of establishment (Johannes 1978). The argument that only larvae and juveniles below some threshold size are at a disadvantage when attempting to establish in adult habitats does not seem to be an adequate explanation; if this were the case, then we might expect to see a far higher frequency of mouth-brooding and parental care, as with many African cichlids that occupy rocky habitats in large lakes. However, mouth-brooding is rare, and parental care of young is almost nonexistent in the sea. Clearly, the major selective forces on the reproductive fitness components in this community have been the transitional nature of the reef fishes' changing resource requirements throughout life and the high levels of reef predation. Given the limited ability of reef-species adults to traverse soft-bottom areas and open water and the presumed selective disadvantage of reproductive investment in large eggs and parental care in this kind of environment, it would seem that planktonic dispersal is the central mechanism ensuring that offspring will (1) be provided with the appropriate spectrum of age- and size-dependent resources outside the parent's habitat, (2) be dispersed to new reef patches, and (3) escape predation and competition in the adult's habitat. (See Barlow 1981 for a provocative discussion of patterns of dispersal and parental investments of coral-reef fishes.)

We suggest that lower extinction rates on reefs and rocky shorelines, even with high predation rates, may be facilitated by biological accommodations such as size- and age-dependent niche shifts, including ontogenetic and developmental changes in predator avoidance and behavior (Helfman 1978) and flexible growth rates (Smith 1978).

Another stochastic model, Sale's "lottery" hypothesis (1977, 1978, 1991b), predicts high variability in local community composition on account of random colonization by larvae, which is similar to island biogeography theory. The hypothesis also maintains that space is the only resource ever in short supply and that colonization success is decided by whomever gets there first rather than by competitive differences between species. Although Sale (1991) still maintains that his nonequilibrium view of reef fish community structure is supported by available data, others might disagree (see Ebeling and Hixon 1991). The Gulf of California is a warm temperate to subtropical sea, and, even though most studies have dealt with tropical reefs, no general, universal model of fish community regulation has emerged. For a thorough review of studies of reef fish community structure, see *The Ecology of Fishes on Coral Reefs* (Sale 1991a).

Disagreements will persist between those who see repeated patterns in nature and those who do not until logically complete theories are developed for both points of view (see Anderson et al. 1981 and Chesson and Warner 1981 for discussion of this problem). Ebeling and Hixon (1991) warn that any attempt to search for a single factor regulating reef fish community structure will be doomed to failure and that sweeping generalizations should be suspect.

Conclusions and Summation

The island–mainland distribution of resident rocky-shore fishes in the Gulf of California suggests a deterministic system where predictable events lead to stable, resilient communities. There is no evidence supporting either MacArthur-Wilson equilibrium or a nonequilibrium hypothesis of reef-fish community structure. Island rocky shores appear to provide a more optimal and benign environment for rocky-shore fishes than do mainland coasts, resulting in trends of greater diversity and density of resident fishes around islands. Geographic barriers to colonization are insignificant within the gulf, but ecological barriers such as sea temperatures play a dominant role in determining the biogeographic distributions of gulf shore fishes. The apparently more stable species equilibrium in island rocky-shore fish communities may be due to low local extinction rates coupled with high immigration rates. Mainland rocky shores seem to represent the insular habitats in an MacArthur-Wilson model owing to higher extinction rates caused by unpredictable physical disturbances that tend to increase species turnover and lower diversity. Hence, islands become "biogeographic mainlands" for reef fishes and may represent the chief source of propagules for recruitment. Finally, marine systems are fundamentally different from terrestrial systems, but the processes of immigration and extinction may resemble each other in short-term; small-scale recolonization events.

Acknowledgments We are grateful to the University of Arizona Foundation for use of the research vessel *La Sirena* and to its captain, Felipe Maldonado, for many eventful cruises in the Gulf of California to collect fishes. We thank all those who helped sort and identify fishes, especially Nancy M. (Moffatt) Kane, and we appreciate the fruitful discussions with Christine A. Flanagan, who also read parts of the manuscript. We thank Ted Case for his efforts in successfully producing a new edition of *Island Biogeography in the Sea of Cortés*.

References

Anderson, G.R.V., A.H. Ehrlich, P.R. Ehrlich, J.D. Roughgarden, B.C. Russell, and F.H. Talbot. 1981. The community structure of coral reef fishes. *Am. Nat.* **117**:476–495.

Bambach, R.K. 1978. Species richness in marine benthic habitats through the Phanerozoic. *Paleobiology* **3**:152–167.

Barlow, G.W. 1981. Patterns of parental investment, dispersal and size among coral-reef fishes. *Environ. Biol. Fish.* **6**:65–85.

Boden, B.P. 1953. Natural conservation of insular plankton. *Nature* **169**:697–699.

Bohnsack, J.A. 1979. The ecology of reef fishes on isolated coral heads: an experimental approach with emphasis on island biogeographic theory. Ph.D. dissertation, University of Miami, Coral Gables, FL.

Briggs, J.C. 1961. The East Pacific barrier and the distribution of marine shore fishes. *Evolution* **15**:545–554.

Brogan, M.W. 1994. Distribution and retention of larval fishes near reefs in the Gulf of California. *Mar. Ecol. Progr. Ser.* **115**:1–13.

Brown, J.H. and A. Kodric-Brown. 1977. Turnover rates in insular biogeography: effect of immigration on extinction. *Ecology* **58**:445–449.

Chesson, P.L., and R.R. Warner. 1981. Environmental variability promotes coexistence in lottery competitive systems. *Am. Nat.* **117**:923–943.

Clifford, H.T, and W. Stephenson. 1975. An introduction to numerical classification. New York: Academic Press.

Dawson, E.Y. 1960. A review of the ecology, distribution, and affinities of the benthic flora. Symposium: The biogeography of Baja California and adjacent seas. *Syst. Zool.* **9**:93–100.

Diamond, J.M. 1969. Avifaunal equilibria and species turnover rates on the Channel Islands of California. *Proc. Natl. Acad. Sci. USA* **64**:57–63.

Durham, J.W. and E.C. Allison. 1960. The geologic history of Baja California and its marine faunas. *Syst. Zool.* **9**:47–91.

Ebeling, A., and M.A. Hixon. 1991. Tropical and temperate reef fishes: comparison of community structures. In: P.F. Sale (ed.), *The Ecology of Fishes on Coral Reefs*. Academic Press, San Diego, CA; pp. 509–563.

Emery, A.R. 1972. Eddy formation from an oceanic island: ecological effects. *Caribb. J. Sci.* **12**:121–128.

Findley, L.T., P.A. Hastings, A.M. Van der Heiden, R. Guereca, J. Torre, and D.A. Thomson. 1999. Distribution de la ictiofauna endemica del mare de Cortes. VII Congreso de la Asociacion de investsigadores del Mar de Cortes. 25–28 Mayo, 1999, Hermosillo, Sonora.

Fisher, R.I., G.A. Rusnak, and F.P. Shepard. 1964. Submarine topography of Gulf of California. Chart I. In: T.H. van Andel and G.G. Shor, Jr. (eds.) *Marine Geology of the Gulf of California*. Amer. Assoc. Petrol. Geol. Mem. 3.

Garth, J.S. 1960. Distribution and affinities of the brachyuran Crustacea. Symposium: The biogeography of Baja California and adjacent seas. *Syst. Zool.* **9**:105–123.

Gilbert, C.R. 1972. Characteristics of the western Atlantic reef-fish fauna. *Q. J. Fl. Acad. Sci.* **35**:130–144.

Gilligan, M.R. 1976. Small marine animal collector for use by divers. *Prog. Fish-Cult.* **38**:40–41.

Gilligan, M.R. 1980a. Beta diversity of a Gulf of California rocky-shore fish community. *Environ. Biol. Fish.* **5**:109–116.

Gilligan, M.R. 1980b. Biogeography of rocky-shore fish communities in the Gulf of California. Ph.D. dissertation, University of Arizona, Tucson.

Glassell, S.A. 1934. Affinities of the brachyuran fauna of the Gulf of California. *J. Wash. Acad. Sci.* **24**:296–302.

Helfman, G.S. 1978. Patterns of community structure in fishes: summary and overview. *Environ. Biol. Fish.* **3**:129–148.

Horn, H.S. 1966. Measurement of "overlap" in ecological studies. *Am. Nat.* **100**:419–424.

Johannes, R.E. 1978. Reproductive strategies of coastal marine fishes in the tropics. *Environ. Biol. Fish.* **3**:65–84.

Lehner, C.E. 1979. A latitudinal gradient analysis of rocky-shore fishes of the eastern Pacific. Ph.D. dissertation, University of Arizona, Tucson.

Leis, J.M. 1991. The pelagic stages of reef fishes: the larval biology of coral reef fishes. In: P.F. Sale (ed), *The Ecology of Fishes on Coral Reefs*. Academic Press, San Diego, CA; pp. 183–230.

Leis, J.M., and J.M. Miller. 1976. Offshore distributional patterns of Hawaiian fish larvae. *Mar. Biol.* **36**:359–367.

MacArthur, R.H. and E.O. Wilson. 1967. *The Theory of Island Biogeography*. Princeton University Press, Princeton, NJ.

McCammon, R.B., and G. Wenninger. 1970. The dendrograph. Computer Contribution no. 48, State Geological Survey, University of Kansas, Lawrence.

Marliave, J.B. 1977. Substratum preferences of settling larvae of marine fishes reared in the laboratory. *J. Exp. Mar. Biol. Ecol.* **27**:47–60.

Molles, M.C., Jr. 1978. Fish species diversity on model and natural reef patches: experimental insular biogeography. *Ecol. Monogr.* **48**:289–305.

Pielou, E.C. 1975. *Ecological Diversity*. Wiley-Interscience; New York.

Randall, J.E. 1976. The endemic shore fishes of the Hawaiian Islands, Lord Howe Island and Easter Island. Colloque Commerson 1973, O.R.S.T.O.M. Traveaux et documents **47**: 49–73.

Randall, J.E. 1998. Zoogeography of shore fishes of the Indo-Pacific Region. *Zool. Stud.* **37**: 227–268.

Robins, C.R. 1971. Distributional patterns of fishes from coastal and shelf waters of the tropical western Atlantic. In: Symposium on Investigations and Resources of the Caribbean Sea and Adjacent Regions. FAO Fisheries Report 71–2; pp. 249–255.

Robinson, M.K. 1973. *Atlas of monthly mean sea surface and subsurface temperatures in the Gulf of California, Mexico*. San Diego Society for Natural History Memoir 5.

Rosenblatt, R.H. 1967. The zoogeographic relationships of the marine shore fishes of tropical America. *Stud. Trop. Oceanogr.* **5**:579–592.

Rosenblatt, R.H., J.E. McCosker, and I. Rubinoff. 1972. Indo-west Pacific fishes from the Gulf of Chiriqui, Panama. Los Ángeles County Natural History Museum Contributions in Science, no. 234, pp. 1–118.

Rosenblatt, R.H., and B.W. Walker. 1963. Marine shore fishes of the Galapagos Islands. *Calif. Acad. Sci. Occas. Pap.* **44**:97–106.

Ryther, J.H. 1969. Photosynthesis and fish production in the sea. *Science* **166**:72–76.

Sale, P.F. 1977. Maintenance of high diversity in coral reef fish communities. *Am. Nat.* **111**: 337–359.

Sale, P.F. 1978. Coexistence in coral reef fishes—a lottery for living space. *Environ. Biol. Fish.* **3**:85–102.

Sale, P.F. (ed). 1991a. *The Ecology of Fishes on Coral Reefs*. Academic Press, San Diego, CA.

Sale, P.F. 1991b. Reef fish communities: open nonequilibrial systems. In: P.F. Sale (ed), *The Ecology of Fishes on Coral Reefs*. Academic Press, San Diego, CA; pp. 564–598.

Sale, P.F., and R. Dybdahl. 1975. Determinants of community structure for coral reef fishes in an experimental habitat. *Ecology* **56**:1343–1355.

Schoener, A. 1974a. Experimental zoogeography: colonization of marine mini-islands. *Am. Nat.* **108**:715–738.

Schoener, A. 1974b. Colonization curves for planar marine islands. *Ecology* **55**:818–827.

Shannon, C.E., and W. Weaver. 1949. *The Mathematical Theory of Communication.* University of Illinois Press, Urbana.

Simberloff, D.S. 1969. Experimental zoogeography of islands. A model for insular colonization. *Ecology* **50**:296–314.

Smith, C.L. 1978. Coral reef fish communities: a compromise view. *Environ. Biol. Fish.* **3**: 109–128.

Smith, G.B. 1979. Relationship of eastern Gulf of Mexico reef fish communities to the species equilibrium theory of insular biogeography. *J. Biogeogr.* **1979**:49–61.

Snodgrass, R.E., and E. Heller. 1905. Shore fishes of the Revillagigedo, Clipperton, Cocos and Galapagos Islands. *Proc. Wash. Acad. Sci.* **6**:333–427.

Soulé, J.D. 1960. The distribution and affinities of the littoral marine Bryozoa (Ectoprocta). Symposium: The biogeography of Baja California and adjacent seas. *Syst. Zool.* **9**:100–104.

Talbot, F.H., B.C. Russell, and G.R.V. Anderson. 1978. Coral reef fish communities: unstable high diversity systems? *Ecol. Monogr.* **48**:425–440.

Thomson, D.A., and C.E. Lehner. 1976. Resilience of a rocky intertidal fish community in a physically unstable environment. *J. Exp. Mar. Biol. Ecol.* **22**:1–29.

Thomson, D.A., L.T. Findley, and A.N. Kerstitch. 2000. *Reef Fishes of the Sea of Cortez.* University of Texas Press, Austin.

Thomson, D.A. and S.L. Mesnick. 1994. Biodiversity of the marine fauna of the islands in the Gulf of California. Processed cruise report of University of Arizona expedition to the islands of the Gulf of California, 3–23 July 1993, Tucson, Arizona.

Thomson, D.A., S.L. Mesnick, and D.J. Schwindt. 1996. Human impact and biodiversity of islands in the Gulf of California. Processed cruise report of University of Arizona expedition to the islands of the Gulf of California, 9–30 July 1994, Tucson, Arizona.

Valentine, J.W., and E.M. Moores. 1974. Plate tectonics and the history of life in the oceans. *Sci. Am.* **230**:80–89.

Walker, B.W. 1960. The distribution and affinities of the marine fish fauna of the Gulf of California. Symposium: The biogeography of Baja California and adjacent seas. *Syst. Zool.* **9**:123–133.

Walker, B.W. 1966. The origins and affinities of the Galapagos shorefishes. In: R. I. Bowman (ed), *The Galapagos: Proceedings of the Symposia of the Galapagos Scientific Project.* University of California Press, Berkeley; pp. 172–174.

Wilhm, J.L. 1968. Use of biomass units in Shannon's formula. *Ecology* **49**:153–156.

Yensen, N.P. 1973. The limpets of the Gulf of California (Patellidae, Acmaeidae). M.S. thesis, University of Arizona, Tucson.

8

The Nonavian Reptiles

Origins and Evolution

ROBERT W. MURPHY
GUSTAVO AGUIRRE-LÉON

    Early in the history of systematic biology, scientists were interested in documenting the wonders of "the creation." Specimens procured on expeditions were placed in collections, and spectacular hand-colored plates graced giant monographs, showpieces of discovery and exploration. Darwin's work shifted interests to natural selection and the process of speciation. In addition to Darwin's volumes on evolution, whose main tenets were predicated on observations of island speciation patterns, Alfred Wallace's (1880) *Island Life* explained the great diversity of species on islands. Today, in an expanded concept islands remains the central focus for investigations of speciation and the mechanisms that drive change. Islands come in the form of Petri dish cultures of bacteria, bottles of *Drosophila*, mesic sky islands (mountaintop habitats isolated by intervening desert), and subaerial landmasses (surrounded by water). Evolution has remained the unifying principle of biology, and the concepts and methods associated with it have made their way into virtually all aspects of human culture.

    Several groups of islands have been instructional in the development of evolutionary theory. The Galapagos Islands clearly had the greatest impact. However, the islands in the Sea of Cortés have also significantly influenced our understanding of the speciation process. Within herpetology, studies have looked at the evolution of insular gigantism (Case 1978b; Petren and Case 1997), anatomical and genetic variability (Soulé et al. 1973), and species composition (Case 1975, 1983; Murphy 1983a). Others have addressed island biogeography in the manner of MacArthur and Wilson's (1963, 1967) model of colonization and extinction (Case 1975, 1983; Wilcox 1978, 1980) versus historical constraints imposed by plate tectonics (Murphy 1983). Various ecological attributes of insular populations have been compared to those of the founding source (reviewed in Case 1983). Controversies over peninsular effects have been evaluated (Taylor and Regal 1978, 1980; Seib 1980; Murphy 1991), as have the phylogenetic relationships of taxa (reviewed below). These are but a few examples of the herpetological investigations. The intensity of study in the Sea of Cortés has not waned in recent years. To the contrary, it has intensified.

Renewed interest in the evolution of the Baja California herpetofauna owes much to the nature of the peninsula and its associated islands. The peninsula affords the opportunity to study clinal variation or peninsular effects. The geological history provides opportunities to investigate the biological consequences of plate tectonics, especially as our understanding of the paleostratigraphy of the region advances. Clinal variation in climate and plant distributions facilitates investigations of ecological constraints on animal species distributions, and mounting paleoecological data fuel the fire. Isolated mountain regions and the associated islands of the peninsula provide laboratories for evaluating the genetic and anatomical consequences of isolation (i.e., speciation and divergence), both recent and old. Owing to these attributes, the region has recently received much attention from molecular systematists, in part because their data provide explicit hypotheses of genealogical relationships of individuals and the history of female dispersion.

Advances in Methods and Data

Concomitant with an increase in basic knowledge, sophisticated methods of data accumulation and analysis have been developed in the past 20 years. Molecular studies have advanced from gathering small isozyme databases to accumulating large amounts of sequence data (Hillis et al. 1996). More important, application of a rigorous philosophy of systematics (e.g., Popper 1934; Hennig 1966; Brooks and McLennan 1991), and recent refinements in methods of data analysis (e.g., Murphy 1993; Siddall and Kluge 1997; Murphy and Doyle 1998) and computer algorithms (Swofford 2000) have led to a revolution in our approach to evaluating the history of species. Phylogenies have been applied to a binomial taxonomy, a species name (Frost and Hillis 1990; Frost and Kluge 1994), thus allowing the instant retrieval of evolutionary history. Undoubtedly, the greatest impact of cladistic methodology has been the movement from storytelling to the scientific rigor of testing hypotheses using a refutationist philosophy (Popper 1934, 1959; Frost and Kluge 1994; Kluge 1997; Siddall and Kluge 1997). The congruence of phylogenies and models of geomorphic stratigraphic development of the peninsula and islands can now be tested using tools such as Brooks parsimony analysis (Wiley 1988; Brooks 1990).

Since the first publication of *Island Biogeography in the Sea of Cortez*, the number of significant databases has increased. Defendable phylogenetic hypotheses have been developed for a number of reptilian groups. DNA sequence studies have challenged our understanding of the evolution of the biota. Allozyme data evaluated within a phylogenetic framework have furthered our knowledge of the relationships of tree lizards (*Urosaurus*; Aguirre-León et al. 1999) and sand lizards (Murphy and Doyle 1998). They have also led to estimates of gene flow among populations of rock lizards (*Petrosaurus*; Aguilar-S. et al. 1988) and desert spiny lizards (*Sceloporus*; Grismer and McGuire 1996).

Paleoecology has developed as another line of investigation. Packrat midden records (Van Devender 1990; Bell and Whistler 1996) suggest that the distribution of xerophilic species remained constant at times of maximum glaciation in the Pleistocene. In contrast, the ranges of mesophilic species expanded and contracted concomitant with environmental fluctuations, including those associated with Pleistocene glacial events (Murphy 1983a,b, 1991; Grismer and McGuire 1993).

Baja California is not only an attractive area for biological investigations, it is of critical importance to geologists and geophysicists. Because of active plate interactions, the peninsula is the focus of much research on tectonic theory and change (chap. 2, this volume). The general parameters of the tectonic origins of the peninsula and the islands in the Sea of Cortés seem to be reasonably well established. Nevertheless, much controversy still exists, especially regarding the nuances of the stratigraphic history of the peninsula. Because of unresolved issues in the geological development of the region, an essential question emerges: Can phylogenies of organisms help geologists understand the tectonic and stratigraphic development of the area? We believe so.

Paleobiogeography of the Peninsular Archipelago Herpetofauna

Generalized Alternative Models

Because most island populations in the Sea of Cortés originated on the neighboring peninsula, an understanding of the peninsular fauna is critical. Two previous scenarios for the evolution of the peninsular herpetofauna differed significantly. In the first synthesis of data, Savage (1960) concluded that the composition of the fauna was due to waves of north-to-south invasion. Savage's brilliant argument centered on the geological concept of a permanent peninsula and a long existence of the Sea of Cortés. It built upon earlier studies by Matthews (1915), Nelson (1921), and Schmidt (1922, 1943). In contrast, Murphy (1983b, 1991) formed a vicariance scenario based on emerging pictures of plate tectonics: the peninsula broke away from mainland Mexico and it is still evolving tectonically. He correlated genetic-distance calculations with presumed tectonic events for a number of species assuming the dominant stratigraphic hypothesis to be true. However, Murphy merely fitted species to the most current stratigraphic model, and the study lacked rigorous phylogenetic methodology, which then did not exist for allozyme data. A number of subsequent studies have attempted to refine Murphy's scenario (e.g., Grismer 1994a,b). However, as with previous studies, these studies used a particular geological model for the peninsula, and the biological data were merely fitted to it. Their phylogenetic approach did not succeed because of the lack of methodological rigor and consistency. Frequently, one alternative, non-phylogenetic scenario for a species group was chosen, and all genetic data were ignored, including those demonstrating the absence of gene flow, at least within the framework of Mendelian principles (appendix 8.2).

New Databases

Before forming generalized biogeographic scenarios, several sound phylogenies are required (Brooks and McLennan 1991). Fortunately, we now know how to move beyond just-so storytelling, and we can do so at a very fine level of resolution by tracking female dispersion using mitochondrial DNA. Furthermore, allozyme data yield direct estimates of and evidence for gene flow.

Frequently, major discrepancies occur between molecular and anatomically based phylogenies and taxonomies. For example, dramatically incompatible sets of phylogenetic relationships have been hypothesized for chuckwalla lizards (*Sauromalus*) by Hollingsworth (1998) and Petren and Case (1997; app. 8.1). Equally divergent cladograms have been advanced for orangethroat whiptail lizards, *Cnemidophorus hyperythrus*, by Grismer (1999b) and Radtkey et al. (1997). Some of the problems relate to the use of nonheritable characters such as ecological preferences or, in anatomical studies, to the reliance on and distribution of primitive character states. Other problems are due to the incorporation of meristic and other "numerical" data and questionable methods of data coding (Murphy and Doyle 1998) and philosophically problematic analyses (Siddall and Kluge 1997). These problems exist in most cladistic anatomical analyses. Consequently, we have restricted our databases to molecular characters because of their unquestionable heritability and fine level of resolution, and to unweighted maximum parsimony analyses.

Below we briefly review the accumulated molecular cladograms and population genetics studies for species of reptiles from Baja California to discover common patterns of tree branching. We rely most heavily on DNA sequence data because the sequences yield particulate data amenable to cladistic evaluation and can potentially detail relationships among populations. We then use cladogenic patterns to propose a new hypothesis that best explains the repeating cladogenic patterns observed in the herpetofauna.

Even with greater methodological rigor, we emphasize that a firm synthesis of genetic cladogenic patterns with geological history remains aloof. This synthesis is only speculation due to the absence of data and analyses for many species groups and uncertainties about the stratigraphic and tectonic history of the region.

Interpreting the Cladograms

We can predict how DNA patterns will appear for peninsular populations through time (fig. 8.1). For nonterritorial species, recent linear dispersions appear as sequential bifurcations, with sister taxa spinning off as the species invade new areas (fig. 8.1A).

Figure 8.1 (Facing Page) Expectations of cladogenic patterns derived from mtDNA sequences for peninsular populations. (A) Initial cladogenic patterns following colonization at the southern end of the peninsula followed by rapid northward dispersion. A reverse pattern of cladogenesis is expected for southward dispersions. (B–D) Small regions accumulate mutations and eventually obfuscate the original dispersal pattern. Nearby populations become each other's nearest sister species, and (D) multiple haplotypes will eventually accumulate at single localities yielding homoplastic associations. (E) A vicariant division initially results in nonexclusive patterns of the two isolates. (F) In time, the isolates are recognized as exclusive lineages. If the populations reunite, the cladogenic patterns will eventually return to pattern D.

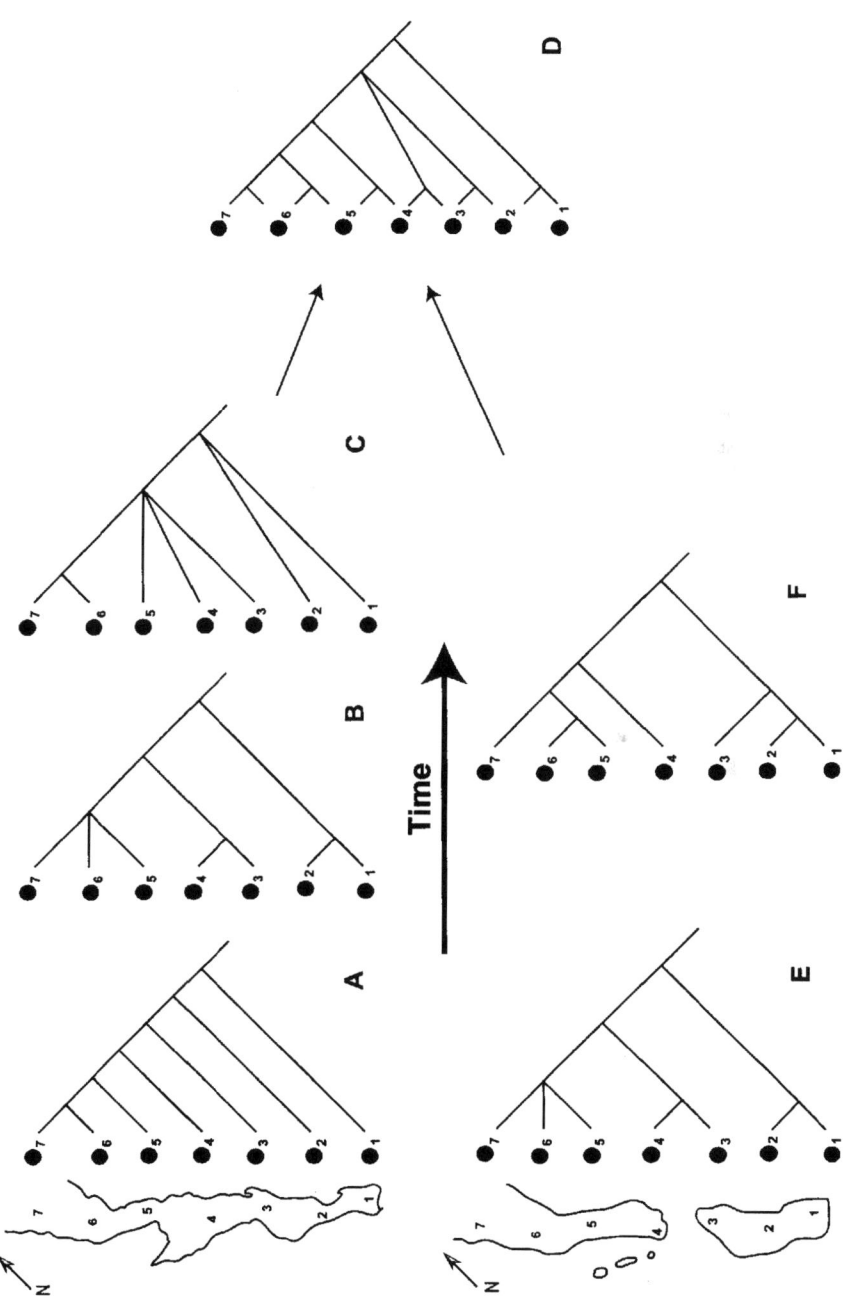

New populations are established in previously unoccupied sites, but the most recent bifurcations occur in areas of more recent cladogenic events. In time, local populations will accumulate mutations, and the genotypes will slowly spread. Dispersions and accumulated mutations will progressively obscure the original dispersal pattern (fig. 8.1B, C). Mutations accumulate over time in territorial species and small regions take on their own genetic identity. Derived genotypes will replace ancestral ones, eventually completely obscuring the original pattern. Nearby populations will be sister groups, and either homoplastic relationships or polytomies among individuals will appear on cladograms (fig. 8.1D). If the population becomes subdivided through a vicariance event, geographically proximal individuals from the two new groups will initially appear as sister taxa, to the exclusion of more geographically distant individuals in the same interbreeding group (fig. 8.1E). In other words, the populations will appear as a polyphyletic (nonexclusive) assemblage of individuals (Graybeal 1995). Time will erase this pattern (fig. 8.1F, D). We evaluate the accumulated data in light of these predictions below.

DNA Sequences

1. Side-blotched lizards, genus *Uta*. Upton and Murphy (1997) gathered 890 homologous nucleotide base pair (bp) sites of sequence data for these small saxicolous lizards from the mitochondrial (mt) DNA cytochrome *b* (cyt *b*) and ATPase 6 genes. They found that the peninsular populations of side-blotched lizards were a monophyletic group exclusive of Chihuahuan Desert populations (*Uta stejnegeri*). The peninsular populations formed two distinctive clades, one on the northern half of the peninsula, and the other on the southern half. The haplotype disconformity was narrowed to a 70-km wide region between San Ignacio and Santa Rosalia. Hollingsworth (1999) extended Upton and Murphy's study. He evaluated 1132 bp from cyt *b* and cytochrome oxidase III and narrowed the peninsular haplotype disconformity to a 10-km wide area near Santa Rosalia. Populations in Sonora arrived from the southward dispersion of more northern populations (fig. 8.2).

2. Chuckwallas, genus *Sauromalus*. Petren and Case (1997; updated in appendix 8.1) gathered 902 nucleotide sites mtDNA cyt *b* sequence data from a series of these large lizards. The peninsular samples of chuckwallas formed a clade separate from those north and east of the Peninsular Ranges. On the peninsula, two distinctive groupings occurred, with the division in the middle of the peninsula.

3. Western whiptail lizards, *Cnemidophorus tigris*. Radtkey et al. (1997) analyzed 887 nucleotide sites of mtDNA cyt *b* sequence data for western whiptails along the length of the peninsula. They observed three major clades for this group of highly vagile, nonterritorial lizards: an extreme northern clade, including Sonora, a mid- to northern Baja California (Norte) clade, and a Baja California Sur clade. The southern clade was further divided into Cape Region lizards and those farther north in the state of Baja California Sur (fig. 8.3).

4. Orangethroat whiptail lizards, *Cnemidophorus hyperythrus*. Radtkey et al. (1997) also evaluated mtDNA sequence data for this peninsula-restricted, highly vagile, nonterritorial species. The cladogram was pectinate in shape, a Hennigean comb, with the older cladogenic events occurring in the south, and the nearest sister populations farthest north. Reeder, Cole, and Dessauer (unpublished data) evaluated 12S and

The Nonavian Reptiles 187

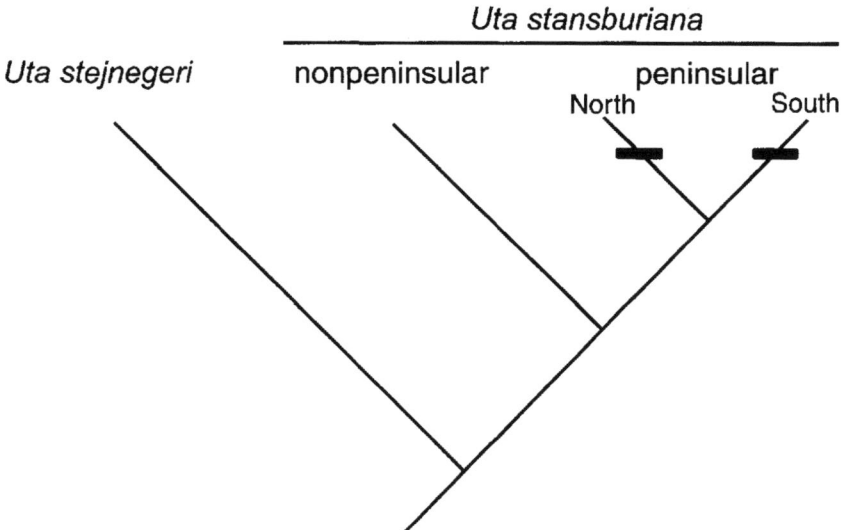

Figure 8.2 Reduced area cladogram for side-blotched lizards, genus *Uta*, on the peninsula of Baja California.

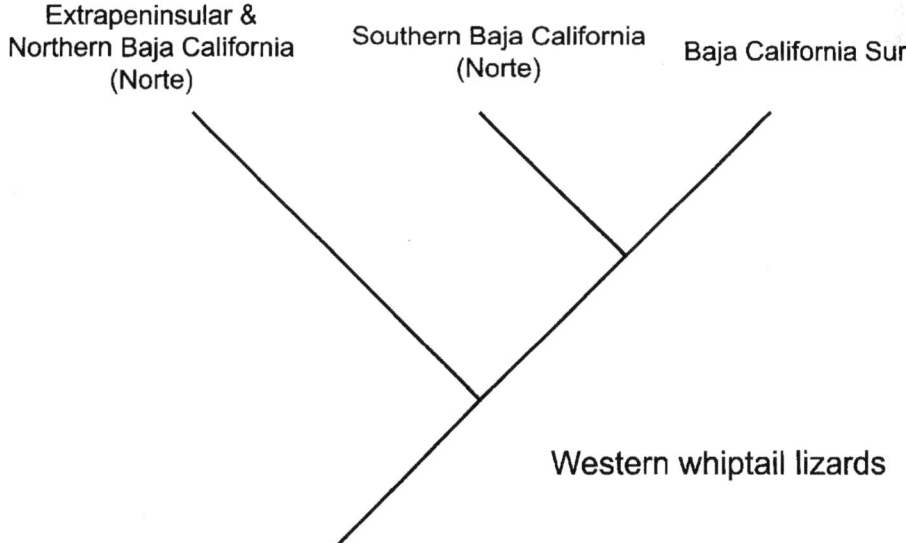

Figure 8.3 Reduced area cladogram for western whiptail lizards, *Cnemidophorus tigris*, on the peninsula of Baja California.

16S gene sequences from a number of whiptail lizards and found that this species is the sister taxon to a clade containing *C. deppei* and *C. guttatus* from mainland Mexico.

5. Pacific gopher snakes, *Pituophis catenifer*. Rodríguez-Robles and De Jesús-Escobar (2000) evaluated 893 nucleotide sites of mtDNA ND4 and 3 tRNA genes, of which 225 were potentially cladistically informative. All species of gopher snakes were examined, including the sister species *P. melanoleucus* and *P. ruthveni*. On the peninsula, *P. catenifer* formed northern and southern clades at the middle of the peninsula (fig. 8.4). Although Grismer (1994b) considers the peninsular snakes to be two species based on color pattern, the geographic distributions of the mtDNA haplotypes significantly conflict with his pattern-class alpha taxonomy.

6. California mountain kingsnakes, *Lampropeltis zonata*. This species occurs in the northern mountains of Baja California. Rodríguez-Robles et al. (1999) evaluated 105 potentially cladistically informative sites from the mtDNA ND4 and 3 tRNA genes. Individuals from the Peninsular Ranges of northern Baja California and southern California up to San Gorgonio Pass formed a distinctive clade. Populations north of San Gorgonio Pass formed a second clade, with two subclades: a coastal clade and an eastern Sierra Nevada clade (fig. 8.5).

7. Rattlesnakes, *Crotalus*. Murphy et al. (in press) gathered 2945 homologous sites of mtDNA sequence data from 5 genes for all but 2 of the 32 species of rattlesnakes. They found that the Baja California rattlesnake, *C. enyo*, was the sister species of the Neotropical rattlesnake, *C. durissus*. Excluding insular species, they confirmed that the western diamondback rattlesnake, *C. atrox*, was the sister species of the red diamond rattlesnake of Baja California, *C. ruber*. The speckled rattlesnake, *C. mitchellii*, which occurs on the peninsula and north, was likely the sister species of the western rattlesnake, *C. tigris*, of southern Arizona to Sinaloa.

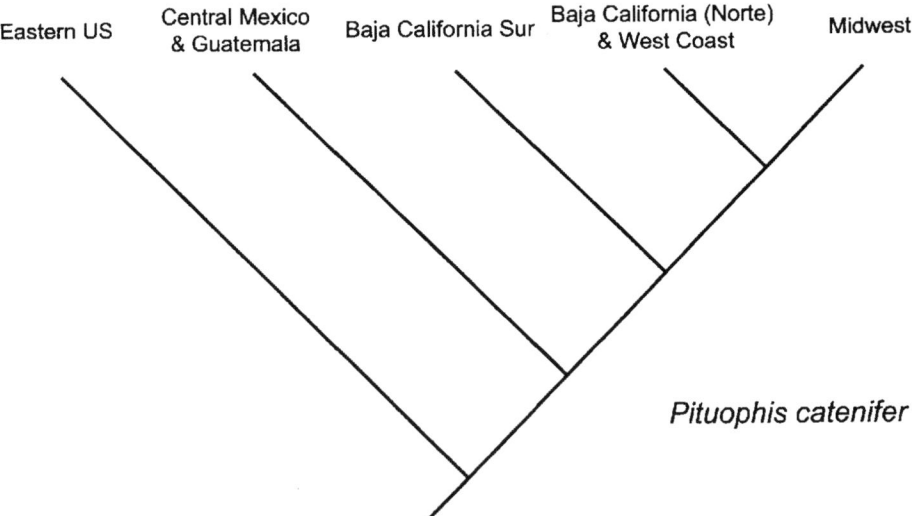

Figure 8.4 Reduced area cladogram for gopher snakes, *Pituophis catenifer*, on the peninsula of Baja California.

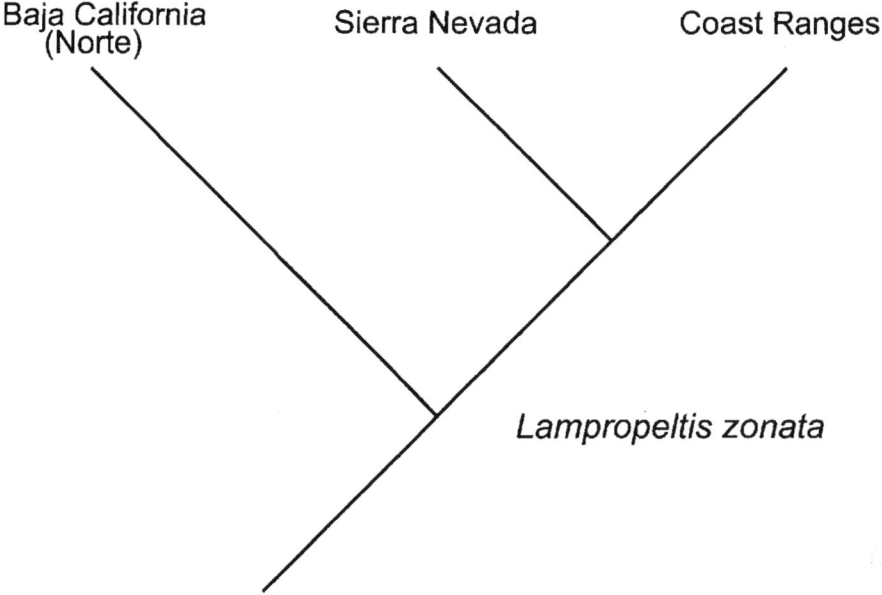

Figure 8.5 Reduced area cladogram for California mountain kingsnakes, *Lampropeltis zonata*, on the peninsula of Baja California.

8. Colorado Desert fringe-toed lizards, *Uma notata*. Trépanier and Murphy (in press) evaluated 1031 homologous sites of sequence data from cyt *b* and ATPase 6 mtDNA genes for 6 populations of these sand-dune–restricted lizards. Lizards around the head of the Sea of Cortés (the only place on the peninsula where they occur) west of the Colorado River are the sister group of *U. inornata* from near Palm Springs, California. In turn, individuals east of the Colorado are the sister group of *U. notata* and *U. inornata*, and lizards from the Mohawk Dunes in Arizona are the basal clade. Populations of Colorado Desert fringe-toed lizards in Sonora are southward dispersers from more northern populations.

9. San Lucan and western skinks, *Eumeces lagunensis* and *E. skiltonianus*. Richmond and Reeder (unpublished data) evaluated the phylogenetic relationships of these skinks using mtDNA ND4 gene sequences. *Eumeces lagunensis*, which occurs in Baja California Sur, is the sister taxon to populations of *E. skiltonianus* from Baja California (Norte) and southern California. In addition, both *E. skiltonianus* and Gilbert's skink, *E. gilberti*, are paraphyletic with respect to each other. *Eumeces lagunensis* is not closely related to *E. brevirostris*, as hypothesized by Grismer (1994b).

10. Iguanid genera. Reeder (1995), Reeder and Wiens (1996), and Schulte et al. (1998) evaluated relationships among the genera of phrynosomatines (Phrynosomatinae sensu Macey et al. 1997; Schulte et al. 1998). The most extensive data set is the total evidence evaluation of Schulte et al., and we use their generic phylogeny (fig. 8.6). The sequence data of both Reeder and Schulte et al. are congruent in finding that *Sator* is the sister group to *Petrosaurus* plus *Urosaurus* and most species of *Sceloporus*. Reeder's sequence data also supported a sister species relationship be-

190 The Biological Scene

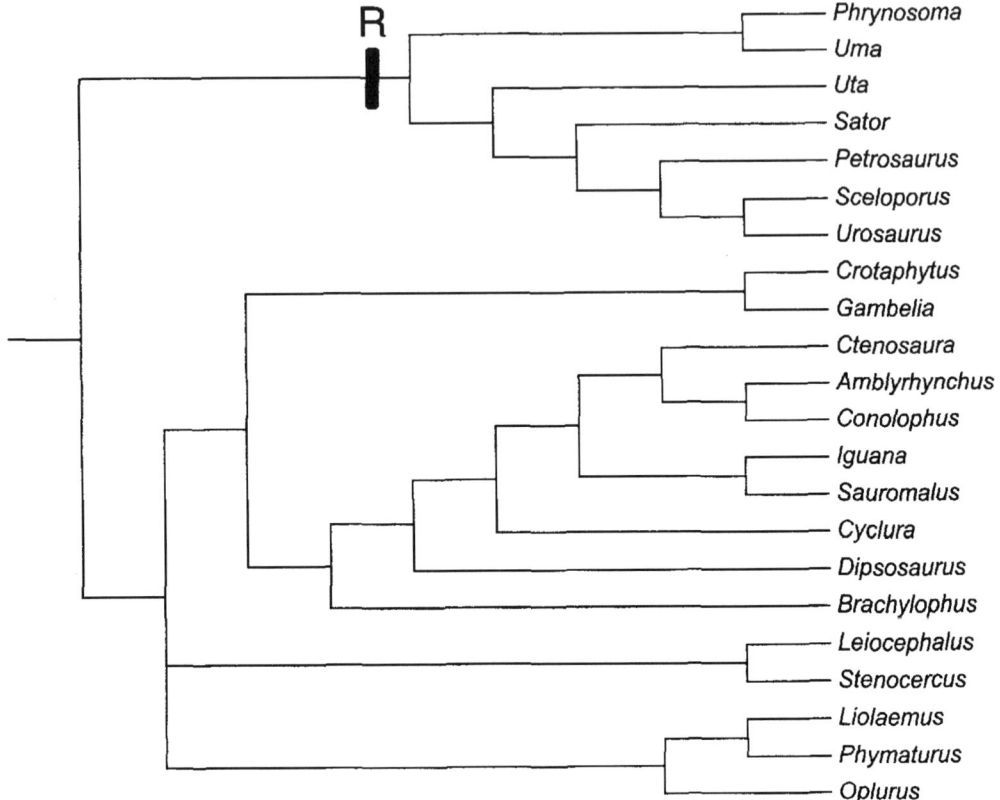

Figure 8.6 Composite phylogeny of iguanid genera based on the evaluations of Schulte et al. (1998) for the major lineages, and node "R" from Reeder (1996) and Reeder and Wiens (1996) for the phrynosomatines.

tween *Sator angustus* and *Sceloporus chrysostictus* + *Sceloporus variabilis*, species that Schulte et al. did not consider. Reeder and Wiens, based on total evidence including anatomical data, found that these two species of *Sceloporus* were members of a larger group that potentially also included *Sceloporus parvus*, *S. couchii*, *S. cozumelae*, *S. utiformis*, *S. siniferus, and S. squamosus*. Wyles and Gorman (1978) reported a sister relationship between *Sator* and *Sceloporus utiformis*. Given concordance among the sequence data sets, and problems in the evaluation of anatomical data (Murphy and Doyle 1998), the relationships among the critical group of *Sceloporus* remain uncertain. Nevertheless, the genus *Sceloporus* appears to be a paraphyletic assemblage of species. If tropical Mexican species of *Sceloporus* form the sister group of *Sator*, they will need to be placed in *Sator* to maintain monophyly of *Sceloporus*. Here we retain recognition of the genus *Sator* because of its apparent genealogical relationships and to preserve stability of the nomenclature until the required data are accumulated.

Relationships among the iguanine genera are based on the work of Sites et al. (1996) and Petren and Case (1997). The generic relationships have been placed on the phylogeny of Schulte et al. (1998), who did not evaluate all of the genera. Significantly, chuckwallas (*Sauromalus*), which are restricted to the Peninsular Ranges and areas farther north, but west of the Continental Divide, are the sister taxon of green iguanas (*Iguana*) from tropical America (fig. 8.6).

11. Zebra-tail lizards (*Callisaurus draconoides*). Lindell and Murphy (unpublished data) have sequenced multiple specimens of these lizards from throughout their range on the peninsula, the southwest United States, and northwest Mexico. Peninsular populations formed a monophyletic group exclusive of extra-peninsular populations. The mid-peninsular discontinuity in mtDNA sequences occurs.

12. Sand snakes (*Chilomeniscus*). These fossorial snakes occur throughout the arid regions of Baja California into the southwestern United States. Wong et al. (1998) sequenced a 380 bp segment of cytochrome *b*. They resolved a pectinate set of relationships, as observed in orangethroat whiptail lizards (see paragraph 4 above).

13. LeConte's thrasher (*Toxostoma lecontei*). Zink et al. (1997) evaluated 619 mtDNA sites from cyt *b* and ND6. They observed a large haplotype disconformity between birds from the Sierra Vizcaíno and those farther north.

14. Additional avian reptiles. mtDNA analyses of cactus wren (*Campylorhynchus brunneicapillus*) and verdin (*Auriparus flaviceps*) show a mid-peninsular discordance that is concordant with that reported by Zink et al. for LeConte's thrasher. However, Zink et al. (unpublished ms) found no phylogeographic divisions within the range of the California Gnatcatcher (*Polioptila californica*), although this species spans the same region.

15. See mammalian examples in chapter 12.

Allozymes

Previously, Murphy (1983b) reviewed allozyme studies. Below we note new data sets only.

1. Rock lizards, *Petrosaurus*. Aguilar-S. et al. (1988) evaluated 34 loci for samples of rock lizards on the peninsula, exclusive of *P. thalassinus*. The shortnose rock lizard, *P. repens*, and banded rock lizard, *P. mearnsi*, are genetically very similar in being separated by only two fixed and two nearly fixed loci. Within *P. repens*, gene flow appears to be common.

2. Desert spiny lizards, *Sceloporus magister* group. Grismer and McGuire (1996) presented allozyme data from the doctoral dissertation of Galen Hunsickler (1987), supplemented with anatomical data. They concluded that only a single species, *S. zosteromus*, occurred on the Peninsular Ranges, and that zones of anatomical intergradation occurred in the regions of the transpeninsular seaways. However, Murphy's (1983b) and Aguirre, Morafka, and Murphy's unpublished data do not support Grismer and McGuire's conclusion (app. 8.2).

3. Brush lizards, *Urosaurus*. Aguirre et al. (1999) used allozyme data to investigate nuclear DNA diversity among these lizards. They observed two areas where derived alleles (duplicated loci) appeared to be dispersing along the peninsula: from the Cape Region northward to near Isla San José, and from San Ignacio to south of Santa Rosalia. They further observed that the endemic Baja California brush lizard, *U. lahtelai* of the Cataviña region, central Baja California (Norte), shared derived allelic states with northern populations of blacktail brush lizard, *U. nigricaudus*, exclusive

192 The Biological Scene

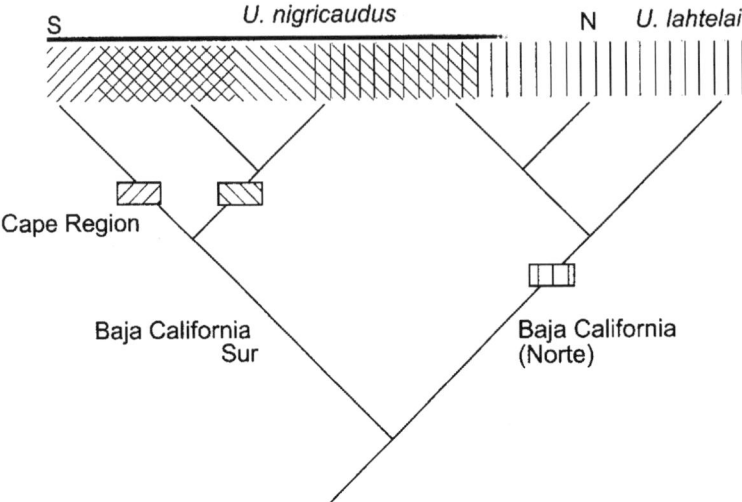

Figure 8.7 Reduced area cladogram for brush lizards, *Urosaurus nigricaudus*, on the peninsula of Baja California. The horizontal bar indicates the integradation of derived character states after reunification of ephemeral isolated populations.

of southern populations (fig. 8.7). (They also mistakenly reported the occurrence of *U. nigricaudus* on Isla Cedros.)

Emergent Patterns of Cladogenesis

From these molecular phylogenetic investigations, six patterns of relationships emerge:

1. Some endemic peninsular taxa have a sister group in tropical Mexico, particularly those species restricted to Baja California Sur (see also Papenfuss, 1982). Examples include chuckwallas and green iguanas, orangethroat whiptail lizards and mainland Mexico whiptail lizards, and Baja California and Neotropical rattlesnakes.

2. Some peninsular taxa share a sister relationship with more easterly taxa in northern Mexico and Arizona, and even more distant relationships to taxa in the Chihuahuan Desert of Texas and north-central Mexico. Examples include side-blotched lizards, fringe-toed lizards, gopher snakes, western diamondback and red diamond rattlesnakes, and speckled and western rattlesnakes.

3. A mid-peninsula genetic discontinuity is common. It occurs in side-blotched lizards, chuckwallas, western whiptail lizards, Pacific gopher snakes, San Lucan and western skinks, zebra-tail lizards, LeConte's thrasher, cactus wren, and verdin. Mammalian examples are reviewed in chapter 12.

4. At least two peninsular species have a comblike set of relationships with their nearest sister taxa occurring in the north, including orangethroat whiptail lizards and sand snakes.

5. A large genetic nonconformity occurs at San Gorgonio Pass. Examples include chuckwallas, Pacific gopher snakes, California mountain kingsnakes, and zebra-tail lizards. See also mammalian examples in chapter 12.

6. Xerophilic lizards at the head of the Gulf of California appear to have southward dispersal patterns with nearest sister relationships located in the south, such as fringe-toed lizards and side-blotched lizards.

A New Scenario for the Peninsular Herpetofauna

The origins of the native peninsular herpetofauna are intimately tied to the origin of the peninsula and to the changing environment. Much of our understanding of the development of the peninsula derives from plate tectonics as associated with stratigraphic information. Murphy (1983b) reviewed the development of the peninsula, and little has changed other than nuances of timing, tectonics, and additional documentation (chap. 2). Nevertheless, we briefly review the paleostratigraphy and paleoecology here.

In the early Miocene the Sea of Cortés had not formed and thus the peninsula of Baja California did not exist (fig. 8.8). A few islands occurred, and these are now part of the peninsula (Durham and Allison 1960; Gastil and Jensky 1973; Gastil et al. 1975; Minch et al. 1976). The continents were drifting northward. Concomitantly, from the Eocene onward, the Madro-Tertiary geoflora (Axelrod 1975) spread southward and replaced the tropical and subtropical flora. A mixture of dry tropical forest and tropical scrub likely characterized the flora of the proto-peninsular region (Axelrod 1979). Oak-pinyon woodlands likely dominated the northern flora because the area was quite moist, as evidenced by numerous, deep, Miocene streambeds (Gastil et al. 1975). Tropical floral elements dominated in the south.

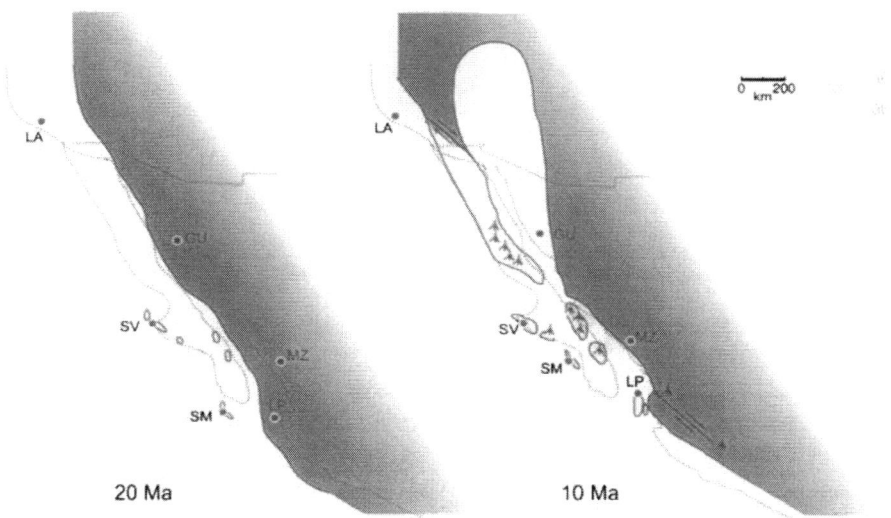

Figure 8.8 Paleographic reconstructions of western North America for the early Miocene (20 Ma) and mid-Miocene (10 Ma). The stippled areas symbolize the temporary formation of the Sea of Cortés. The Cape Region is 50 km farther south of its current position relative to the peninsula. Present locations of LA = Los Angeles; GU = Guaymas; SV = Sierra Vizcaíno; SM = Islas Santa Margarita and Magdalena; MZ = Mazatlán; LP = La Paz.

Around 12.5 million years ago, the southern peninsular region rotated westwardly away from the mainland (Mammerickx and Klitgord 1982; Hausback 1984). The displacement may have been as much as 100–150 km northwest from its original location. A temporary proto-Gulf of California formed (fig. 8.8; Atwater 1970; Atwater and Molner 1973; Gastil and Jensky 1973; Gastil et al. 1975). By the close of the Miocene, the Cape Region was located near the vicinity of the Islas Las Tres Marías (fig. 8.9; Atwater and Molner 1973; Gastil and Jensky 1973), and the proto-Gulf of California extended up to near Tiburón island (Jensky 1975; Gastil and Krummenacher 1977; Stock and Hodges 1989). The northern Peninsular Ranges had attained half the elevation that occurs today (Stock and Hodges 1989). The southern portions of the peninsula existed either as an island archipelago (Murphy 1983b), or as a solid peninsula (Grismer 1994b: fig. 12).

Around 4–5.5 million years ago, plate interactions accelerated, eventually moving the Peninsular Ranges some 300 km farther northwest relative to mainland Mexico. The Sea of Cortés permanently formed, and by 3 million years ago it extended northward at least to the San Gorgonio Pass of southern California and eastward into Arizona (figs. 8.9, 8.10). The southern peninsular regions initially formed as an island archipelago, but most of the landmasses were soon connected along the Peninsular Ranges (Hausback 1984). Although Murphy (1983b) suggested the possibility of a northern seaway connecting the Sea of Cortés and the inundated Los Angeles Basin (fig. 8.9), no geological evidence exists for this proposal (Boehm 1984, 1987). Around 3 million years ago, the Cape Region appeared as two islands (fig. 8.10). Upton and Murphy (1997), in the absence of geological evidence, proposed the occurrence of a

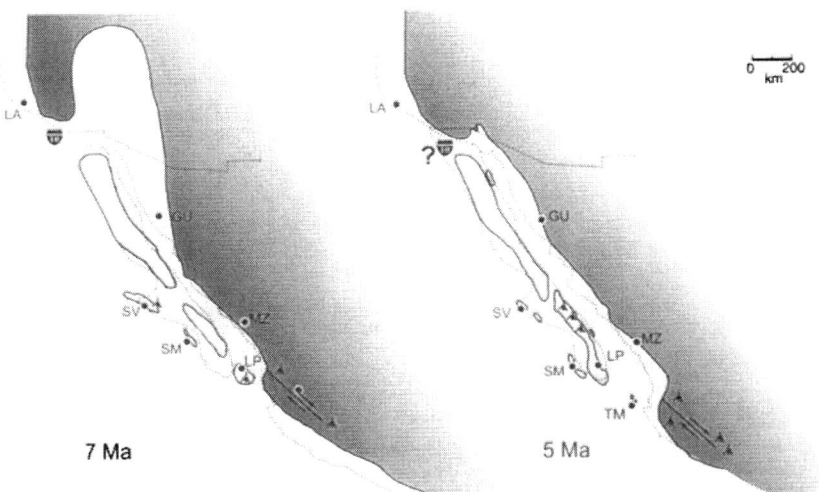

Figure 8.9 Paleographic reconstructions of western North America for the late Miocene (7 Ma) and the Miocene-Pliocene boundary (5 Ma). The stippled areas symbolize the temporary formation of the Sea of Cortés. A proposed seaway occurs at San Gorgonio Pass (Interstate Highway 10), and in the area of the mid-peninsula. Present locations of LA = Los Angeles; GU = Guaymas; SV = Sierra Vizcaíno; SM = Islas Santa Margarita and Magdalena; MZ = Mazatlán; LP = La Paz; TM = Islas Las Tres Marías.

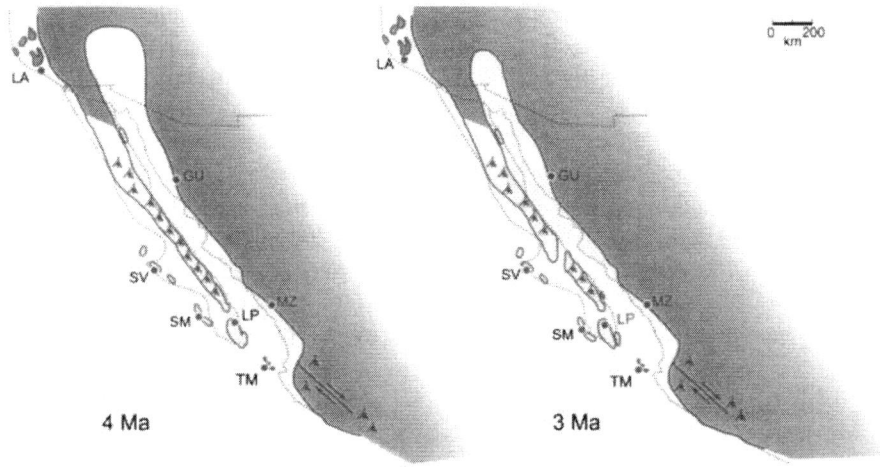

Figure 8.10 Paleographic reconstructions of western North America for the Early Pliocene (4 Ma) and the Late-Pliocene boundary (3 Ma). The mid-peninsula Vizcaíno Seaway temporarily reformed. Present locations of LA = Los Angeles; GU = Guaymas; SV = Sierra Vizcaíno; SM = Islas Santa Margarita and Magdalena; MZ = Maztalán; LP = La Paz; TM = Islas Las Tres Marías.

mid-peninsular seaway sometime before 1 million years ago, which we refer to as the Vizcaíno Seaway. Identification of volcanic intrusions in a marine environment near Santa Rosalia (Ochoa-Landin 1998) around 3 million years ago now support the prediction (fig. 8.10), but at an earlier date. Earlier deposits also suggest that the seaway was formed more than one time (figs. 8.9, 8.10).

The late Pliocene to Pleistocene saw the formation of the peninsula as it exists today, including environmental conditions (Morafka 1977). However, the deserts as we know them today did not develop until the end of the last interglacial (Axelrod 1979; Van Devender 1990). By the early Pleistocene, the Cape Island rejoined the peninsula, and the Vizcaíno Seaway dried up. The Peninsular Ranges attained their present great heights. During times of maximum glaciation, islands separated by a depth of less than 150 m were connected to the peninsula, and many embayments, such as Bahía de los Angeles and Bahía Concepción, disappeared.

The accumulated cladogenic patterns can be associated with tectonic events and the paleostratigraphy of the peninsula. Previously, both Murphy (1983b) and Grismer (1994b) associated species with various geomorphic events. Some taxa exhibited cladogenic patterns that fit the previous groups, but not others. The cladograms (e.g., figs. 8.2–8.7) depicted common geographic patterns and thus likely depict the sequence of vicariance events for taxa that share a common speciation history. By associating the cladograms with the tectonic history, three primary faunal associations can be hypothesized.

Transgulfian Vicariance

When the Cape Region and the Peninsular Ranges of Baja California separated from mainland Mexico about 5–12 million years ago, they likely isolated a fauna representative of mainland Mexico. Some cladogenic patterns derive peninsular species from tropical ancestors, and these are among the oldest representatives in the region. Extant representative taxa include the Baja California rattlesnake, *Crotalus enyo*, sister species to the Neotropical rattlesnake, *C. durissus*. The orangethroat whiptail lizard, *Cnemidophorus hyperythrus*, has the mainland sister species *C. deppei* and *C. guttatus*. The combined data for the genus *Sator* suggest that it was isolated from its sister taxa, *Sceloporus chrysostictus*, *S. variabilis*, *S. parvus*, *S. couchii*, *S. cozumelae*, *S. utiformis*, *S. siniferus*, and *S. squamosus*. The two-legged amphisbaenid, *Bipes biporus*, has two sister species on the mainland, *B. tridactylum* and *B. canaliculatus* (Kim et al. 1976; Papenfuss 1982). It is also possible that the lizards of the genera *Sauromalus* and *Petrosaurus* also had their origins at this time, as might have other species.

The East–West Split

A widespread northern fauna apparently occurred across subaerial regions of the southern United States, as exemplified today by species in the genera *Uta*, *Crotaphytus*, *Coleonyx*, and *Pituophis*. These east–west assemblages likely predated the formation of the permanent peninsula. Their subdivision formed two clades, one restricted to the region around the Sea of Cortés, and the other farther east in the Chihuahuan Desert and Mexican Plateau. The mesic, high mountain pass at the Continental Divide is partially responsible for the isolation of these faunal elements during the Pliocene (Morafka 1977). The separation of the east–west assemblage would be further reinforced, if not caused, by the barrier posed by San Gorgonio Pass sometime before 3 million years ago (Murphy 1983b).

The Peninsular Archipelago

Aguirre et al. (1999) summarized early evidence that the peninsula was an island archipelago at least 1 million years ago. Aguirre et al., who coined the term "peninsular archipelago," built on the evidence and hypotheses of Murphy (1983b) and Upton and Murphy (1997).

Unambiguous cladogenic patterns show that peninsular species became isolated from those farther north and east. Upton and Murphy (1997) and Hollingsworth (1999) showed that peninsular populations of side-blotched lizards form a distinct group. Rodriguez-Robles et al. (1999) found a significant mtDNA discontinuity in the California mountain kingsnake, *Lampropeltis zonata*. One group occurs on the northern Peninsular Ranges up to the San Gorgonio Pass, and another clade occurs farther north. The chuckwallas show an identical pattern of relationships: one clade is restricted to the Peninsular Ranges and another north and east (appendix 8.1).

Many species of reptiles have distributions on the Peninsular Ranges up to San Gorgonio, such as leaf-toed geckos (*Phyllodactylus*), rock lizards (*Petrosaurus*), and granite spiny lizards (*Sceloporus orcutti*). Consequently, Murphy (1983b) proposed the occurrence of the "San Gorgonio Filter Barrier" because these animals did not occur farther north. However, this proposal of a narrow mesic region fails to explain a generalized pattern for the isolation of both mesophilic and xerophilic species, as well as the restriction of many species to the Peninsular Ranges. If the San Gorgonio region were cool and moist (Axelrod 1937, 1950; Frick 1933), then a strong genetic discordance would not be expected in the California mountain kingsnakes (*Lampropeltis zonata*) and western skinks (*Eumeces skiltonianus*). And if the region were more xeric, then the same break would not be expected in side-blotched lizards, collared lizards (McGuire 1996; Murphy and Doyle 1998), and desert spiny lizards (*Sceloporus magister-zosteromus* complex). Many of these species on the Peninsular Ranges are not restricted to desert regions, and they coexist with other species associated with the head of the Sea of Cortés, such as fringe-toed lizards. Consequently, there must have been a universal barrier to dispersion and not just a filter. The most plausible explanation for the strong genetic patterns appears to be yet another temporary seaway at San Gorgonio Pass, appropriately termed the "San Gorgonio Seaway."

The peninsular archipelago developed. While the Vizcaíno Seaway was inundated, southern portions of the peninsula likely became more or less intact. Cladogenic patterns suggest that species dispersed along the developing Peninsular Ranges. Subsequently, the Vizcaíno Seaway and flooding at the Isthmus of La Paz again divided the peninsula. Both extensive genetic data and geological investigations now support the presence of at least these two seaways. The Vizcaíno Seaway formed the genetic discordance observed in both avian and nonavian reptiles. The genetic distinctiveness of the Cape Region whiptail lizard, *Cnemidophorus tigris maximus*, and the San Lucan rock lizard, *Petrosaurus thalassinus*, likely resulted from this more recent peninsular connection and subsequent disruption. It is expected that cladogenic patterns will be identical in other territorial species such as granite spiny lizards (*Sceloporus orcutti* complex), desert spiny lizards (*S. zosteromus* complex), and leaf-toed geckos (*Phyllodactylus xanti* complex) once data have been accumulated and analyzed.

After the formation of the Vizcaíno Seaway, the peninsula rejoined the mainland, forming San Gorgonio Pass. When the Sea of Cortés receded from its northern extension, newly developed habitats were invaded. Peninsular archipelagian species such as side-blotched lizards, *Uta stansburiana*, dispersed around the head of the Sea of Cortés and southward into Sonora, and western whiptail lizards, *Cnemidophorus tigris*, dispersed in the opposite direction. Species associated with the Colorado River, including species such as fringe-toed lizards, *Uma*, and Colorado River toads, *Bufo alvarius*, dispersed southward into the newly created delta.

During the Pleistocene, the peninsular islands reunified, forming the greatest amount of connected landmass in the history of the peninsula. Interestingly, the orangethroat whiptail, *Cnemidophorus hyperythrus*, appears to have originated on a peninsular island north of the Cape Region, likely in the vicinity of Islas Carmen and Danzante (Radtkey et al. 1997). These islands are separated by a channel depth of only 26 m (Gastil et al. 1983). After reunification of the peninsula, this species quickly dispersed northward. Additional evidence of a peninsular archipelago is provided by other taxa. Neither side-blotched lizards (Upton and Murphy 1997; Hollingsworth 1999), nor chuckwallas (appendix 8.1) from Carmen and Danzante cluster with nearby peninsular populations. Moreover, these insular populations cluster together, requiring either one additional peninsular island in the archipelago, or that these islands were joined only to each other. Given the reported ocean-channel depths (Gastil et al. 1983) and that only 10,000–15,000 years have elapsed since the last major continental glaciation, we believe the former explanation is more likely.

Evolution of the Insular Reptiles

The islands in the Sea of Cortés have divergent geological histories. Most are land-bridge islands, and thus the tectonic origin of these islands is likely inconsequential in terms of the composition of the herpetofauna. However, the tectonic history of the remaining deep-water islands is very important. Whereas deep-water islands of continental origin likely carried a complement of mainland faunal and floral elements, islands of oceanic origin must have been colonized by overwater dispersal, sometimes facilitated by humans.

Alternative Models

Few have tried to explain the origin and evolution of the insular herpetofauna. Schmidt (1922) considered the origins of the peninsular species and briefly discussed the problems of the insular populations. Savage (1960) noted that the insular herpeto-

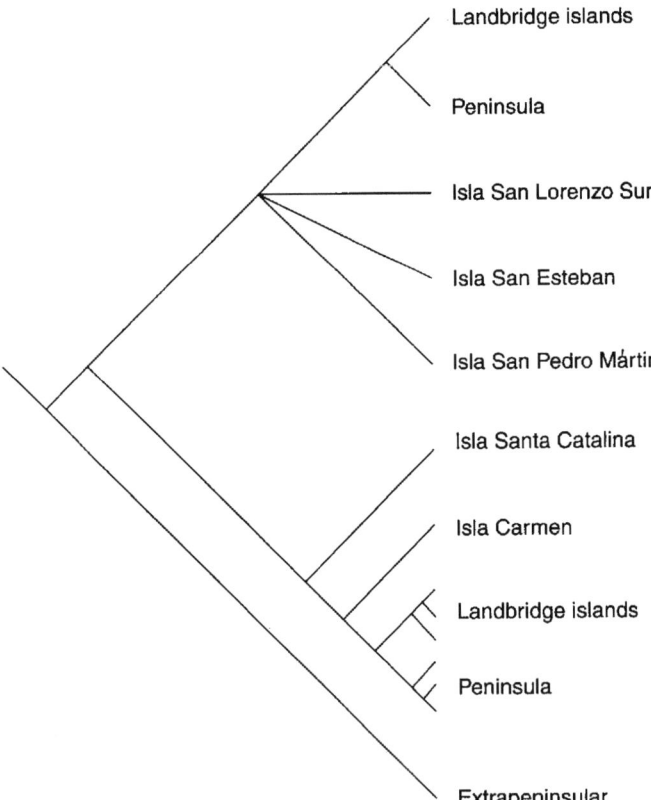

Figure 8.11 Reduced area cladogram for side-blotched lizards, *Uta*, on the islands in the Sea of Cortés based on Upton and Murphy (1997). Unsupported nodes have been collapsed. Although associations at unsupported nodes provide the best explanation of the data, they are considered tentative phylogenetic hypotheses that may change as data accumulate.

fauna formed interesting patterns. Soulé and Sloan (1966) first noted the biogeographic significance of landbridge islands and first considered the correlation between species numbers and island areas. Other treatments dealt with the evolution of particular groups of species (e.g., Robinson 1973; Smith and Tanner 1974). Murphy (1975, 1983a,b) provided the first synthetic reviews of the peninsular and insular herpetofauna. Grismer (1994c) updated Murphy's review of the insular fauna and speculated on alternative origins for a few insular species.

MacArthur and Wilson (1963, 1967) provided a monumental framework for interpreting the distributions of organisms on islands, particularly for oceanic islands. Their colonization and extinction model was applied to the fauna on the islands in the Sea of Cortés by Soulé and Sloan (1966), Case (1983, 1975), and Wilcox (1978, 1980). In contrast, Murphy (1983a) provided an alternative, deterministic argument for the evolution of the insular herpetofauna. Based on genetic-distance coefficients, patterns of species composition, and tectonic development, Murphy proposed that most populations on continental islands were there before isolation. The absence of particular species on these islands resulted from the absence of needed habitat. In contrast, the herpetofauna on oceanic islands (i.e., those islands that never had a terrestrial connection) resulted from somewhat random colonization. Grismer (1994c) haphazardly used phenetic anatomical studies, some of which predated phylogenetic theory and methodology, to conjecture origins of species on the islands. He concluded that all deepwater islands except for Isla San Pedro Mártir contained a mixture of peninsular and mainland species and therefore that these species assemblages were neither completely deterministic, nor generally reflective of colonization and extinction.

All of the above treatments are based on theoretical models and unsubstantiated assumptions; none is based on sound phylogenetic investigations or at the required fine-grained, population level. To demonstrate the insular origin of each species, precise data are required, and these must be evaluated within a phylogenetic framework.

New Databases

DNA Sequences

1. Side-blotched lizards, genus *Uta*. The phylogeny resolved by Upton and Murphy (1997) is shown in figure 8.11. Lizards on the northern deep-water islands in the Gulf of California are the sister group to peninsular and landbridge island populations, although this position is very tenuous. The side-blotched lizard from Isla Santa Catalina, *Uta squamata*, is the sister species to the southern peninsular clade, followed by the population from Isla Danzante. Hollingsworth (1999) observed nearly the same set of relationships as found or predicted by Upton and Murphy. Landbridge island populations generally had a nearest sister species relationship with a nearby mainland population, including *Uta nolascensis* from San Pedro Nolasco, and the three endemic subspecies on the northernmost islands in the Sea of Cortés (Upton and Murphy 1997). Hollingsworth also found that lizards from Carmen and Danzante formed a distinctive clade, and, like Upton and Murphy, observed that lizards on these landbridge islands did not share a sister relationship with nearby peninsular populations.

2. Chuckwallas, genus *Sauromalus*. Petren and Case (1997; app. 8.1) observed that the large chuckwallas from the northern Gulf of California formed the sister group to the northern peninsular populations. The Santa Catalina Island chuckwalla branched off from within the southern peninsular clade, and lizards from Danzante, Carmen, and Monserrat formed a separate clade (app. 8.1). The two giant chuckwallas were sister taxa, and their nearest sister group was the northern peninsular population. Populations of spiny chuckwallas from Isla San Lorenzo Sur branch off from within a clade from Isla Ángel de la Guarda (app. 8.1).

3. Western whiptail lizards, *Cnemidophorus tigris*. Radtkey et al. (1997) found five major clades: (1) northern Sonora plus northern Baja California and the deepwater islands of San Esteban, San Pedro Mártir, San Lorenzo Norte, and Salsipuedes; (2) Santa Catalina; (3) Baja California (Norte) plus associated landbridge islands, and Ángel de la Guarda and Partida Norte; (4) Baja California Sur and associated landbridge islands exclusive of the Cape Region; and (5) the Cape Region and associated landbridge islands Espíritu Santo and San Francisco. It is surprising that the population from Santa Catalina branched off as the sister taxon to the clade 1. However, reanalysis of these data including all potentially cladistically informative sites plus sites with missing data shows that the Santa Catalina population forms the sister group to all other western whiptail lizards (fig. 8.12). This reevaluation forms a far more plausible explanation of relationships.

4. Orangethroat whiptail lizards, *Cnemidophorus hyperythrus*. Radtkey et al. (1997) found that insular populations were derived in a south-to-north sequence of cladogenic events. The populations that rooted at the base of the tree occurred sequentially on Islas Carmen and Cerralvo. Populations on Islas San Francisco and San José formed the next clade, followed by Isla Monserrat, then a clade containing La Paz and nearby Isla Espíritu Santo, followed sequentially by Isla Coronados, Isla San Marcos, and finally northern peninsular populations. This pattern indicates a recent northern dispersion, one that occurred on the peninsula after the disappearance of the Vizcaíno Seaway, and possibly after the last glacial event (fig. 8.13).

5. Rattlesnakes, *Crotalus*. Murphy et al. (in press) gathered nucleotide base pair data from 2945 sites from 5 mtDNA genes for all but 2 of the 32 species of rattlesnakes. Within the diamondback group, they found that the Santa Catalina Island rattlesnake, *C. catalinensis*, was the sister taxon of *C. ruber*, as proposed by Murphy and Crabtree (1985). The mtDNA sequence cladogram placed the Tortuga rattlesnake, *C. tortugensis*, as a sister taxon to an undescribed species of *Crotalus* allied with *C. atrox*, from Santa Cruz. In turn, these two are the sister groups to the western diamondback rattlesnake from southern California and mainland Mexico (fig. 8.14), although the association is tenuous.

6. Mammalian cladograms are given in chapter 12.

Island Records

An updated checklist for the islands in the Sea of Cortés was recently published (Grismer 1999a). These records, plus two additional records, have been included in appendixes 8.2–8.4 for sake of completeness. However, much of the alpha taxonomy differs from that of Grismer (1999a) in order to reflect genealogical relationships.

The Nonavian Reptiles 201

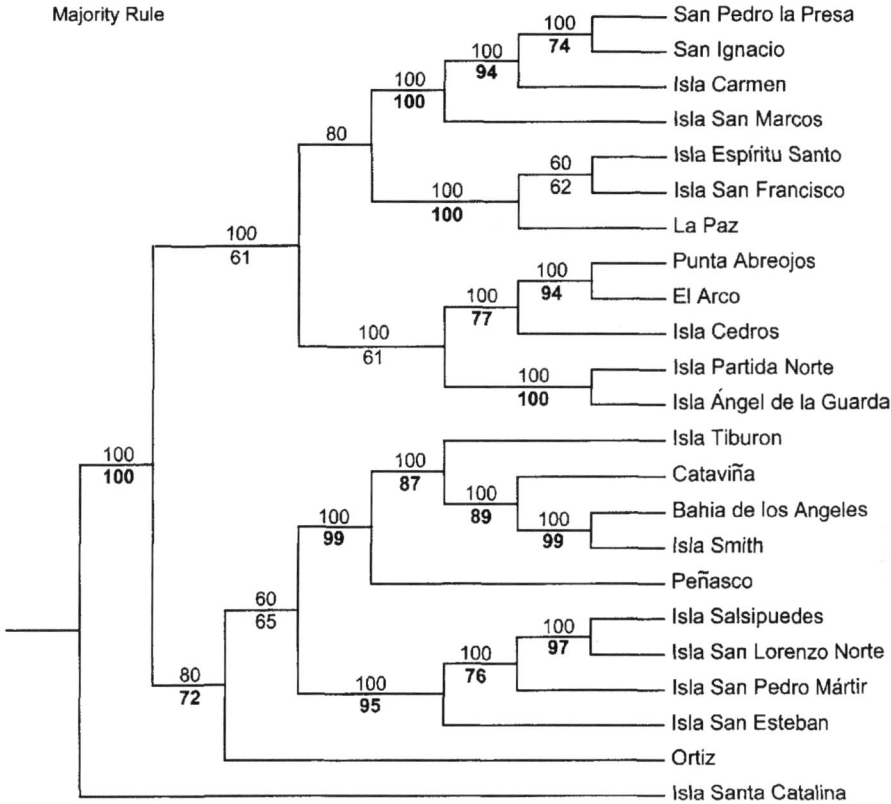

Figure 8.12 Revised cladogram for western whiptail lizards, *Cnemidophorus tigris*, on the islands in the Sea of Cortés based on an unweighted maximum parsimony evaluation of all available mtDNA sequence data, including those with missing information. Numbers above the lines denote frequency of the nodes among multiple most parsimonious trees. Bootstrap proportions are given below the lines; values >70% are in bold.

Interpreting the Cladograms

Most deep-water islands in the Sea of Cortés had a continental origin (chap. 2). The key to whether or not preference is given to a model of colonization and extinction, deterministic history, or a mixture of the two lies in both cladogenic sequences and patterns of species composition. If a deterministic history were responsible for the patterns, then we would expect to observe similar cladogenic patterns for multiple species and a faunal composition that reflects nested groups of species. For example, if the fauna of Isla Santa Catalina was already present when the island was formed, then the cladogenic patterns of relationships should be equivalent for all taxa so long as no recolonization followed extirpation. Using side-blotched lizards as a baseline reference, we would expect all other taxa to have a phylogeny similar to that in figure 8.11, if they had a common origin and equivalent territoriality. The determin-

202 The Biological Scene

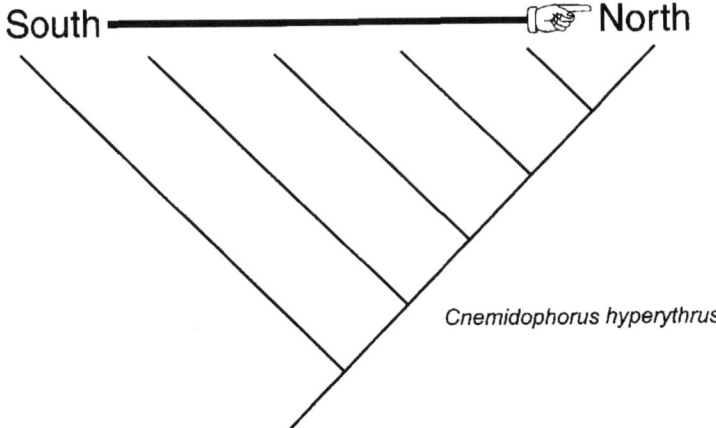

Figure 8.13 Reduced area cladogram for orangethroat whiptail lizards, *Cnemidophorus hyperythrus* group, on the peninsula of Baja California.

istic scenario is rejected by a cladogram depicting alternative relationships, such as a sister relationship with populations in mainland Mexico, with Baja California (Norte), or nested inside a group of landbridge islands (fig. 8.15).

Interpreting Patterns of Species Distributions

The theory of island biogeography (MacArthur and Wilson 1963, 1967) accounts for insular faunas by colonization and extinction, whereas a paleogeographic scenario explains the occurrences from the perspective of the history of the peninsula. Thus, whereas the island biogeography theory has a large stochastic component, the paleogeographic scenario predicts that the faunal elements arrive all at once and that extir-

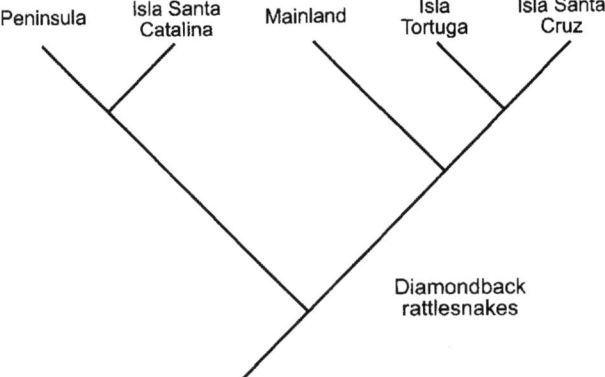

Figure 8.14 Reduced area cladogram for rattlesnakes, *Crotalus*, on the islands in the Sea of Cortés based on 2945 bp of mtDNA sequence data (Murphy et al. in press).

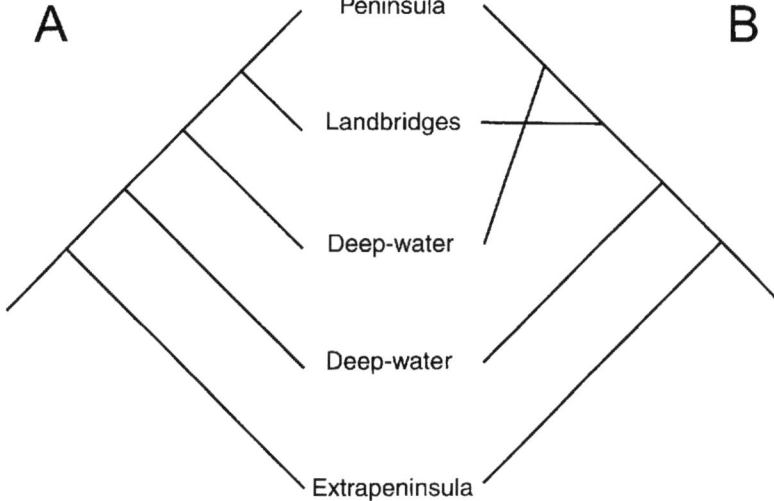

Figure 8.15 (A) Presumed tectonic history of some islands. (B) Hypothetical reduced area cladogram for insular organisms. Crossed lines show how noncorroboration of the tectonic data would look if the origins owed to dispersal or misclassification of islands based on channel depth (age).

pations are deterministic (Murphy 1983). If a mixture of the two extreme explanations is responsible for the origin of the faunas, then we still predict a stochastic composition for colonization, although extirpation might remain deterministic. Cladogenic patterns could serve to arbitrate among the scenarios, once additional data sets have been compiled.

A New Scenario?

The new scenario for the evolution of the insular herpetofauna might derive from our reexamination the evolution of the peninsular herpetofauna. The islands can be grouped into those having a continental origin and oceanic islands that were never connected to another landmass. The oceanic islands are Islas Tortuga, Rasa, and Monserrat. Three other islands may have had an oceanic origin: Islas Ángel de la Guarda, San Esteban, and San Pedro Mártir. Given the composition of the herpetofauna on Isla Ángel de la Guarda, a terrestrial connection is strongly indicated. The remaining islands are thought to have had a continental origin or to have been connected to a larger faunal source during the Pleistocene (chap. 2).

The Oceanic Islands and Overwater Colonization

Many species of reptiles colonize by overwater dispersion. Table 8.1 summarizes the distributions of taxa on oceanic islands in the region, and one island thought to be oceanic, Isla Monserrat. Upton and Murphy (1997) proposed that humans recently introduced the population of side-blotched lizards on Isla La Rasa, a Holocene volca-

Table 8.1 Successful, possible, and unsubstantiated overwater colonists on islands in the Sea of Cortés

Taxon	Oceanic Island Occurrence	No. of Gulf Islands on Which Taxa Occur
Successful Overwater Colonists		
1. *Phyllodactylus*	La Rasa	42
2. *Sceloporus orcutti* (*sensu lato*)	Tortuga	11
3. *Uta*	Tortuga	
	La Rasa[a]	49
	Revillagigedo	
4. *Hypsiglena*	Tortuga	18
5. *Lampropeltis getula*	Tortuga	12
6. *Crotalus tortugensis* (=*atrox*)	Tortuga	8
7. *Masticophis* (*flagellum* group)	Revillagigedo	11
Possible Overwater Colonists		
1. *Dipsosaurus dorsalis*	Monserrat	12
2. *Sauromalus*[a]	Monserrat	34
3. *Sceloporus* (*zosteromus* group)	Monserrat	6
4. *Cnemidophorus hyperythrus*	Monserrat	8
5. *Chilomeniscus cinctus*	Monserrat	9
6. *Phyllorhynchus decurtatus*	Monserrat	5
7. *Crotalus mitchellii*	Monserrat	11
8. *Crotalus ruber*	Monserrat	8
Unsubstantiated Overwater Colonists		
1. *Callisaurus draconoides*		15
2. *Crotaphytus insularis*		2
3. *Ctenosaura hemilopha*[a]		4
4. *Petrosaurus*		6
5. *Urosaurus microscutatus*		21
6. *Coleonyx variegatus*		10
7. *Cnemidophorus* (*tigris* group)		20
8. *Leptotyphlops humilis*		5
9. *Charina trivirgata*		4
10. *Bogertophis rosalea*		1
11. *Eridiphas slevini*		4
12. *Rhinocheilus ethridgei*[b]		1
13. *Salvadora hexalepis*		3
14. *Sonora mosaueri*		2
15. *Tantilla planiceps*		1
16. *Trimorphodon biscutatus*		6
17. *Crotalus enyo*		9
18. *Masticophis* (*lateralis* group)		3

[a]Possible human-facilitated dispersal.
[b]Not a potential colonizer of Tortuga and Monserrat.

nic island. There are no data on the Rasa Island leaf-toed gecko, *Phyllodactylus tinkeli*, to help determine whether its occurrence may be attributable to humans.

Isla Tortuga, an early Pleistocene volcanic island, contains a mix of species, all of peninsular origin except one from mainland Mexico, the Tortuga Island rattlesnake, *Crotalus tortugensis*. However, mtDNA sequence data associate the Tortuga Island rattlesnake with that from Isla Santa Cruz and very weakly with *C. atrox* from the mainland (fig. 8.14). Thus, most of the fauna of Isla Tortuga is associated with the nearby peninsula and not with distant mainland Mexico.

Among oceanic islands, Isla Monserrat is the largest and likely among the oldest (chap. 2). Its fauna has of a variety of nonavian reptiles (table 8.1), and apparently all of these have been successful overwater colonists. It is interesting that the chuckwallas from Monserrat were apparently derived from those on Isla Carmen, and not from either Danzante or Santa Catalina (app. 8.1), both of which are geographically closer to Monserrat. Ocean currents may have more influence than proximity in determining distributions.

About 20% of the nonavian reptilian groups have colonized by overwater dispersion, and an additional 20% likely did the same. We have no data on the colonizing ability of the remaining 60% of species groups; we cannot say if they did, could or will colonize by overwater dispersal. Although many species can colonize, one question remains. How frequently does this occur without human assistance?

Ángel de la Guarda Block

The island group with the best documented tectonic origin is the Ángel de la Guarda island chain in the northern part of the Sea of Cortés. The major islands include Islas Ángel de la Guarda, Partida Norte, Salsipuedes, San Lorenzo Norte, and San Lorenzo Sur. Satellite landbridge islands also occur in association with the large islands. Islas San Lorenzo Norte and San Lorenzo Sur are landbridge islands with respect to each other, and the herpetofaunal distributions suggest that Isla Salsipuedes also had a landbridge connection with them. All other large islands appear to have remained isolated after their formation, including during times of maximum Pleistocene glaciation. Figure 8.16 depicts the presumed tectonic history of the island group, which began its isolation about 1 million years ago (Moore 1973) from the peninsula (Phillips 1966). The side-blotched lizards from this group of islands form two distinctive clades, one associated with Ángel de la Guarda (an unnamed species; Upton and Murphy 1997) and the other, *Uta antiqua*, on Salsipuedes, San Lorenzo Norte, and San Lorenzo Sur. Western whiptail lizards also occur on these islands, and the cladograms show two distinctive groupings, although the southern group appears to have had its origin from Sonora. In contrast, the spiny chuckwallas, *Sauromalus hispidus*, on Isla San Lorenzo Sur branch off within the clade of chuckwallas from Isla Ángel de la Guarda indicating a recent colonization. Because Seri Indians moved populations of these large, edible lizards among islands in the Sea of Cortés (Felger and Moser 1985), it seems likely that the species was introduced from Ángel de la Guarda to the smaller Isla San Lorenzo Sur. Unfortunately, we do not know if chuckwallas originally occupied Isla San Lorenzo Sur and neighboring islands before this introduction. However, this seems likely, and, if true, then the original species likely became extinct or was genetically swamped out by introduced animals.

At least one species of rattlesnake may have dispersed over water to Isla Ángel de la Guarda. On the mainland, the red diamond rattlesnake, *Crotalus ruber*, is a much larger species than the speckled rattlesnake, *C. mitchellii*. However, on Isla Ángel de la Guarda, red diamond rattlesnakes are small, and speckled rattlesnakes are large. Insular populations of snakes generally become smaller on islands (Soulé and Sloan 1966). However, Cody (1974) and Case (1978b, 1983 chap. 9) proposed that the

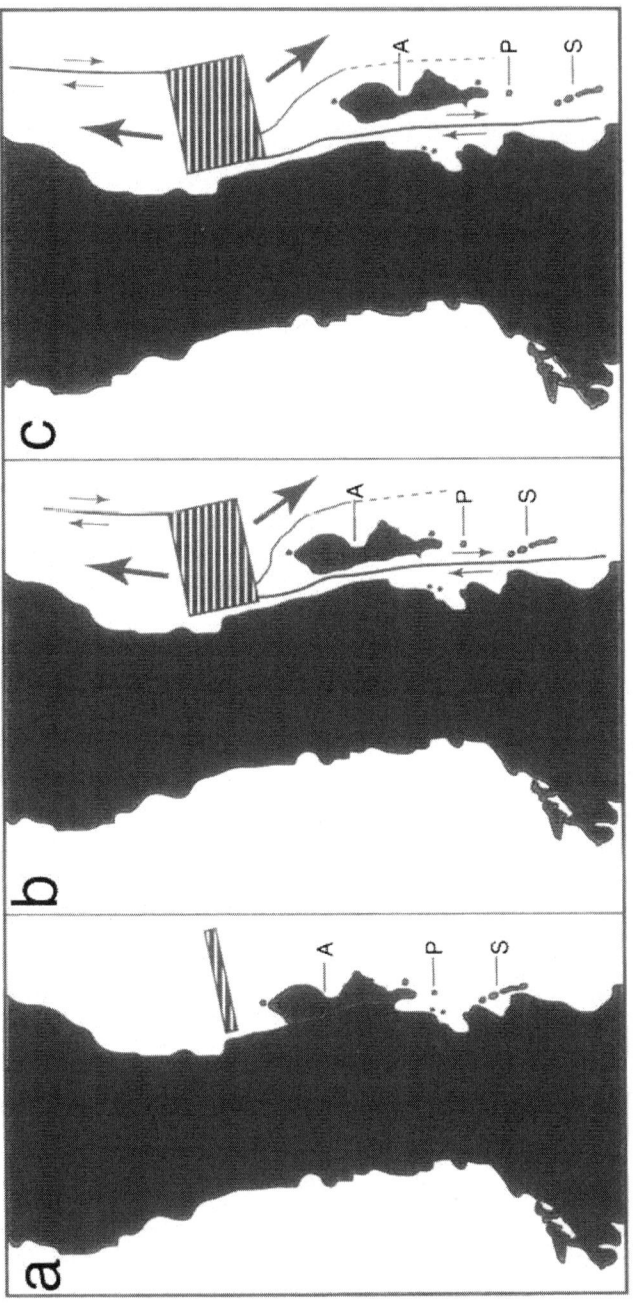

Figure 8.16 Paleogeographic reconstruction showing the formation of the Ángel de la Guarda island chain. (a) about 1 million years B.P. the island chain broke away from the peninsula. Diagonal lines indicate the formation of the Delfín Basin (= spreading center). (b) Reconstruction for about 0.5 million years ago. Arrows indicate relative plate movements. (c) Reconstruction of the present. Islands are indicated as follows: A = Ángel de la Guarda; P = Partida Norte; S = San Lorenzo Norte and San Lorenzo Sur.

speckled rattlesnakes became large on Ángel de la Guarda in order to feed on giant chuckwallas, and later the red diamond rattlesnake arrived, and to avoid competition, dwarfed. This explanation for the differences in body length is extremely interesting, particularly for studies of competition, but it remains to be tested from the perspective of cladograms.

The leaf-toed gecko on Isla Partida Norte, *Phyllodactylus partidus*, is an apparent sister species of the Sonoran gecko *P. homolepidurus*. Given that the species does not occur on other islands in the Sea of Cortés, it is possible that Seri Indians introduced the species. However, cladogenic data are required to confirm the species' genealogical relationships. In terms of Ángel de la Guarda and its associated islands, all available evidence points to a peninsular origin for the herpetofauna, including the origin of the giant chuckwallas. No cladogenic data suggest otherwise. In contrast, the more southern group of islands on the block appears to have a mixture of species. Whereas side-blotched lizards came from the peninsula and chuckwallas from Isla Ángel de la Guarda, the western whiptails appear to have their nearest sisters on other midriff islands, and then in Sonora (fig. 8.12; Radtkey et al. 1997). Additional data, both additional sequences from different genes and from more species, are required to clarify the history of the herpetofauna on this island group.

San Esteban

The tectonic history of San Esteban is not detailed, although it is thought to be of continental origin (Gastil et al. 1983; chap. 2, this volume). The island contains a mixture of peninsular and mainland species. The endemic piebald chuckwalla, *Sauromalus varius* (shown in fig. 16.1), had an origin from the peninsula (app. 8.1). Upton and Murphy (1997) and Hollingsworth (1999) found that San Esteban side-blotched lizards were the sister group of a large clade containing northern Baja Californian populations. The taxonomy of leaf-toed geckos (Dixon 1964; Murphy 1983b) suggests a peninsular origin. Spiny tail iguanas appear to have been introduced onto San Esteban by the Seri Indians who once lived on the island (Felger and Moser 1985); their presence is historically uninformative. The evolutionary relationships of western whiptail lizards are problematic. Apparently, they had an origin from mainland Sonora (fig. 8.12). They form the sister group to other insular populations, including those on deep-water Islas Salsipuedes, San Lorenzo Norte, and San Lorenzo Sur. Given the apparent introduction of the chuckwallas to Isla San Lorenzo Norte, it is also possible that western whiptails were also introduced, accidentally or intentionally. Finally, the blacktail rattlesnake, *Crotalus molossus*, appears to have arrived by overwater dispersal from Sonora. Nevertheless, given that the majority of the fauna of the island was derived from the peninsula, if the island had a continental origin it should have been with the peninsula.

San Pedro Mártir

This probably continental island had its origin from mainland Mexico (Henyey & Bischoff 1973; Bischoff & Henyey 1974). All nonavian reptilian species on this island are also thought to have their derivation from the mainland, and there is no apparent mixing of origins with the peninsula (Murphy 1983a; Grismer 1994a).

Santa Catalina

This continental island had an origin from the peninsula, but its age is unknown (Gastil et al. 1983). Based on tectonics, Murphy (1983a) proposed that the herpetofauna of this island was entirely representative of the peninsula. Grismer (1994c) challenged this assumption for both the side-blotched lizards and western whiptail lizards. The Santa Catalina Island side-blotched lizard, *Uta squamata*, belongs to the southern peninsular clade (Upton and Murphy 1997; Hollingsworth 1999). For the western whiptail lizards, the tree presented by Radtkey et al. (1997) conflicted with Murphy's proposal. However, our total evidence reevaluation of the sequence data does not conflict (fig. 8.12). Furthermore, the association of the Santa Catalina Island whiptail lizard at the base of the clade is not supported by significantly covaried character states (Fu and Murphy 1999). Some randomizations of the data yield a more parsimonious solution than the unrandomized data. We are confident that additional sequence data will clarify the conundrum in favor of a southern peninsular origin. The only other species thought to have had a mainland origin is the Santa Catalina Island kingsnake, *Lampropeltis catalinensis* (Blaney 1977). However, given the absence of a phylogenetic analysis Blaney's study does not provide evidence for a mixture of faunas.

Santa Cruz

This is one of the most biologically enigmatic islands in the Sea of Cortés, owing to the presence of the phrynosomatine lizard *Sator angustus*, and the western diamondback rattlesnake, *Crotalus atrox*. Boulder-dwelling *Sator* apparently exclude all other phrynosomatine lizards from occurring sympatrically through competitive exclusion and aggressive behavior (Case 1983), except for the sand-dwelling zebra tail lizard, *Callisaurus draconoides* (on Isla Cerralvo). This is also evident from the absence of other phrynosomatine lizards on landbridge Isla San Diego. The occurrence of *Sator* on deep-water Isla Santa Cruz and landbridge Isla San Diego has been attributed to both plate tectonics (Murphy 1975, 1983a,b) and extinction of a once widespread southern peninsular population (Grismer 1994c) leaving insular relicts. The peninsular extinction scenario seems unlikely given that other phrynosomatines apparently cannot coexist with *Sator* (chap. 9), and many of these species have had a long presence on the southern peninsula as evidenced by cladogenic patterns. The vicariant origin is also problematic. There are two insular species in the genus *Sator*, and they do not occur on nearby islands: one occurs on Islas San Diego and Santa Cruz, and the other on Isla Cerralvo, and yet they are sister species. They could have been derived from a single species of *Sator* on mainland Mexico, which has since become extinct. Any number of stories can be told but, at present, they are just stories even though we believe overwater colonization from one island to another remains the least problematic explanation. This conundrum may never have an irrefutable explanation.

The species of rattlesnake on Santa Cruz also has an enigmatic distribution. If its occurrence results from overwater dispersal from mainland Mexico, then more than 140 km of swimming or rafting is required. The distance is much greater if it dispersed from Isla Tortuga, or vice versa. Alternatively, if it was derived from a vicariant event

(Murphy 1983a), then its relationships should lay with the peninsula and not mainland Mexico. mtDNA sequence data (Murphy et al. in press) are equivocal as to whether the insular populations *C. atrox* are the sister clade to mainland *C. atrox* or the sister group of the peninsular *ruber* group. Thus, the origin of the herpetofauna remains enigmatic for Santa Cruz.

Cerralvo

The herpetofauna of Cerralvo appears to have a close relationship with the peninsula, an association that even includes two species of amphibians (app. 8.3). However, there are two exceptions. The lizard genus *Sator* is known from this island, and the problem of its occurrence was previously noted. The other enigmatic species is the Cerralvo longnose snake, *Rhinocheilus etheridgei*. Grismer (1994c) believes that it is a recent overwater colonist from mainland Mexico. Cerralvo may have formed 2–5 million years ago (Gastil et al. 1983), which coincides with the Cape Region breaking away from mainland Mexico (chap. 2). If Cerralvo is 5 million years old, then it could have carried nearby mainland representatives as it broke away from mainland Mexico. However, a more recent age of island formation seems likely given the extent of shared faunal elements, and the likelihood that the two amphibians did not arrive by overwater colonization, at least without human assistance. Another alternative explanation is that the southern peninsular population of longnose snake went extinct after the formation of the Vizcaíno Seaway was formed; this can be tested by cladogenic patterns.

Landbridge Islands

With only one exception, the landbridge islands in the Sea of Cortés have a subset of reptiles from the nearest larger landmass, be it the peninsula of Baja California, mainland Mexico, or a larger island. The exception is the occurrence of the leaf-toed gecko, *Phyllodactylus nocticolus*, on Isla Tiburón. If the current taxonomy reflects genealogical relationships, then the population must have arrived by island hopping. Given travel among the islands by Seri Indians, it seems likely that an accidental introduction occurred. This is being tested using a phylogenetic analysis of sequence data.

In terms of cladogenic patterns, all landbridge islands appear to have their closest genetic relationships with the nearby mainland populations, except for those on Carmen and Danzante. On these islands, the orangethroat whiptail lizard, Slevin's chuckwalla (*Sauromalus slevini*), and side-blotched lizard (*Uta*) do not have nearest sister relationships with adjacent peninsular populations. Furthermore, the highly vagile, nonterritorial orangethroat whiptail lizard, *Cnemidophorus hyperythrus*, appears to have undergone a recent, rapid northward dispersal that erased the nearby island–peninsula historical patterns (fig. 8.13). This pattern may owe to either a range shift or extinction of the former peninsular population, or to the peninsular genotype being swamped out by that of the insular-evolved lizards. In this case, as in several others noted above, we can only speculate about the possible explanations for the patterns owing to the absence of data and cladistic analyses.

Paleogeography or Equilibrium?

Colonization

If colonization were stochastic, we would expect to see divergent patterns of phylogenetic relationships expressed among cladograms. With few exceptions, the available cladograms are highly congruent, and we must assume that this congruence is due to identical origins. A few insular herpetofaunas are undoubtedly a mix of mainland and peninsular elements (apps. 8.2–8.4). However, many of the problematic occurrences may be due to human introductions and not to the animals arriving via overwater dispersal. Most insular occurrences appear to reflect a historical land connection. Consequently, colonization on the Sea of Cortés islands appears to be largely deterministic and not stochastic as proposed early in the development of equilibrium theory (Case 1983, 1975; Wilcox 1978, 1980). If colonization is not stochastic except for some occurrences on the three oceanic islands, then is extinction stochastic as predicted by island biogeographic theory (MacArthur and Wilson 1963, 1967)?

Patterns of Extinction

Murphy (1983a) reviewed patterns of extinction on islands in the Sea of Cortés and concluded that they were deterministic. His findings were based on a series of assumptions about the origins of the insular forms. Some of these assumptions have changed. Therefore, we briefly review the extinctions in light of new information.

Stochastic patterns of extinction should appear as random subsets of faunas on islands. In contrast, deterministic patterns of extinction should appear as nested subsets of co-occurring species. No information about extinction can be gained from evaluating oceanic islands because, in the absence of fossil data, it is impossible to determine which, if any, species have become extinct. Consequently, we can only examine landbridge and deep-water islands of continental origin to obtain data on extinction. Furthermore, even when connected to a larger faunal source, small islands may not have had suitable habitat for a full complement of mainland species. Thus, in explaining extinction we are limited to those taxa for which suitable habitat occurs on particular islands.

The species composition north of the Vizcaíno Seaway differs from that south of the area (Murphy 1991; Grismer 1994b). Given that the islands north and south of the seaway had different species pools, potential species compositions are not equivalent. The situation is further complicated by rainfall patterns. Islands in the northern Sea of Cortés are subject to extremes in temperatures and rainfall; the seasonal average of rainfall is <15 cm per year and highly unpredictable. In contrast, islands associated with the southern half of the peninsula have more predictable and equitable rainfall, averaging >15 cm per year (Soulé and Sloan 1966). The activities of the Seri Indians have also likely complicated historical analyses. Finally, although many of the islands are well surveyed for diurnal lizards, the occurrence of nocturnal species, especially snakes, is probably not well documented.

Considering these variables, we divided the islands into four more or less homogeneous groups: northern landbridge islands, northern deep-water continental islands, southern landbridge islands, and southern deep-water islands of continental origin (table 8.2). Among these groups, the southern islands form the most suitable group for examining potential patterns of extinction because, unlike for the northern islands, human-facilitated introductions have not been documented.

Table 8.3 shows distribution patterns for lizards on southern continental islands in the Sea of Cortés that have at least three species of diurnal lizards. The table also summarizes distribution data for lizards on oceanic islands. Taxa are ranked in order of decreasing number of insular occurrences. Islands are grouped by age and origin, and ranked in descending order of number of lizard species. Island areas are noted in parentheses. Table 8.3 shows a general trend for loss of species richness with decreasing island area, particularly for landbridge islands. Given a significant correlation in ranking ($p < .01$), most losses do not appear to be random, and thus occurrences are not stochastic. If extinction were simply stochastic, then we might expect some of these islands to have random subsets of the original species pool. However, this is not observed. The nested subsets of landbridge island species imply a regular, deterministic pattern of extinction.

Some exceptions to the pattern of extinctions occur. Western banded geckos, *Coleonyx variegatus*, have not been reported from Isla Carmen and San Francisco. However, these geckos are rare on some islands, such as Danzante, where there is only a sight record (Murphy and Ottley 1984). Thus, we believe the species likely occurs on the large island Isla Carmen, and possibly on smaller San Francisco, as appropriate habitat is available. The granite spiny lizard, *Sceloporus orcutti*, is not found on Isla

Table 8.2 Islands in the Sea of Cortés categorized into northern and southern distributions and into landbridge or deep-water status

Landbridge		Deep Water	
Island	Area (km^2)	Island	Area (km^2)
Northern Islands			
Tiburón	1225.53	Ángel de la Guarda	936.04
Mejia	2.26	Partida Norte	1.36
Estanque (Pond)	1.03	Salsipuedes	1.16
Miramar (El Muerto)	1.33	San Lorenzo (N & S)	37.29
Encantada Grande	6.85	San Esteban	40.72
		San Pedro Mártir	2.90
		San Pedro Nolasco	3.45
Southern Islands			
Espíritu Santo-Partida			
Sur	106.84	Santa Catalina	40.99
San José	187.16	Santa Cruz	13.06
Carmen	143.03	Cerralvo	140.46
San Marcos	30.07		
Los Coronados	7.59		
San Francisco	4.49		
Danzante	4.64		
San Ildefonso	1.33		
San Diego	0.60		

Islands with fewer than three species of diurnal lizards are not shown.

Table 8.3 Distribution of lizards on southern islands once connected to the peninsula of Baja California and southern oceanic islands in the Sea of Cortés

	Continental Islands												Oceanic Islands	
	Landbridge									Deep Water				
	Espíritu Santo-Partida Sur (107)	San José (187)	Carmen (143)	Coronados (7.6)	San Marcos (305)	San Francisco (4.5)	Danzante (4.6)	San Diego (0.6)	San Ildefonso (1.3)	Cerralvo (140)	Santa Catalina (40.9)	Santa Cruz (13.1)	Monserrat (19.9)	Tortuga (11.4)
Phyllodactylus	●	●	●	●	●	●	●	●	●	●	●	●	●	●
Uta	●	●	●	●	●	●	●	○	●	○	●	○	●	●
Sceloporus (orcutti group)	●	●	●	●	●	●	○	○	●	○		○		
Sauromalus	●	●	●	●	●	●	●	●	●	●		●	●	
Cnemidophorus (tigris group)	●	●	●	●	●	●	●			●	●			
Urosaurus	●	●	●	●	●	●	●			○				
Callisaurus	●	●	●	●	●	●	●			●				
Coleonyx variegatus	●	●	●		●									
Cnemidophorus hyperythrus	●	●	●		●	●				●	●		●	
Dipsosaurus	●	●	●	●	●	●				●			●	
Sceloporus (zosteromus group)	●	●	●	●				●					●	
Petrosaurus	●						●			●				
Sator										●		●		
Ctenosaura														

Islands with fewer than three species of lizards are not considered. Taxa are shown in order of decreasing number of insular occurrences. Islands are in order of decreasing number of lizard species. Island areas are in parentheses. Filled circles indicate species occurrences. Open circles indicate species likely eliminated by competitive exclusion.

Danzante. However, the Baja California rock lizard, *Petrosaurus repens*, occurs on Danzante, and given a 71.8 percent niche overlap (Case 1983), it is likely the ecological replacement of the granite spiny lizard. The island is likely too small to support both species (Case 1983; Murphy 1983a). Because no other missing records exist for the southern landbridge islands, except for San Diego (see below), the patterns of extinction do not appear to have been random, at least during the last 10,000–15,000 years.

Wilcox (1978) suggested that the landbridge islands were supersaturated. This was based in part on island area and distance estimates. Unfortunately, these estimates were not accurate (app. 1.1). We can empirically evaluate this possibility by examining the fauna on older continental islands. Cladogenic patterns revealed that Carmen and Danzante are older than the channel depths indicate, and yet they still display the fauna expected for landbridge islands (table 8.3). Thus, these data alone provide no indication of supersaturation. However, the lizard fauna of Isla Santa Catalina tells an alternative story. The presence of both the desert spiny lizard (*Sceloporus lineatulus*, a sister species of *S. monserratensis*) and the desert iguana (*Dipsosaurus catalinensis*) on Isla Santa Catalina indicates that the herpetofauna was once equivalent to that on Islas San José, Carmen, and Coronados (table 8.3). If true, then Santa Catalina has lost 5 of an original 11 species of lizard inhabitants, including a granite spiny lizard (*Sceloporus orcutti*), brush lizard (*Urosaurus*), zebra tail lizard (*Callisaurus*), western banded gecko (*Coleonyx variegatus*), and orangethroat whiptail (*Cnemidophorus hyperythrus*). Suitable habitat for all of these species appears to be present on Isla Santa Catalina. (If *Cnemidophorus hyperythrus* had an insular origin, as the cladogenic data suggest, then it is not a possible vicariant colonizer.) The nested subsets of landbridge islands do not predict this pattern of extinction. If Santa Catalina carried a representative fauna when it was formed, which seems likely given the presence of the fossorial western blind snake, *Leptotyphlops humilis*, then two important conclusions follow. First, if all species on the landbridge islands were available for colonization of Isla Santa Catalina, then extinction may have been a random event as predicted by island biogeography theory (MacArthur and Wilson 1963, 1967). Second, the landbridge islands are supersaturated (i.e., they have not reached equilibrium).

Islas Cerralvo, Santa Cruz, and San Diego appear to be special cases. Table 8.3 lists phrynosomatine lizards likely displaced by *Sator*, which occurs on these islands. San Diego is a landbridge island, and its complement of lizards is identical to nearby Isla Santa Cruz, which is a deep-water island. If *Sator* was absent, and the displaced phrynosomatines present, then both San Diego and Santa Cruz would have the expected complement of lizards (table 8.3). Therefore, extinction here appears to be deterministic via competitive exclusion or predation (Case and Cody 1983).

Cerralvo appears to be about 2–5 million years old. After considering the exclusion effects of *Sator* (table 8.3), two species appear to be randomly missing: the western banded gecko, *Coleonyx variegatus*, and the Cape Region whiptail lizard, *Cnemidophorus tigris maximus*. The absence of the western banded gecko could reflect insufficient collecting efforts, as noted above. However, on many islands, this species appears to be rare and could just be susceptible to extinction, particularly on older islands, because of low population densities. Nevertheless, these data also support the concept of supersaturated landbridge islands. The absence of *Cnemidophorus tigris maximus* is unexpected.

214   The Biological Scene

The oceanic islands are the final group to evaluate from southern Baja California. There is no evidence that humans introduced any of these insular species. Therefore, species must have arrived by overwater dispersal. Whether occurrences are randomized or not depends on each species' adaptation for colonizing. Amphibians are extremely poor oversea colonizers because of their permeable skin. Some ecological attributes of nonavian reptiles may predispose them to overwater travel. These might include an affinity for sleeping in trees and logs and ability to survive upon landing, such as relatively low metabolic rates, a high tolerance for salt, or a high tolerance for desiccation. If successful, overwater colonizers are exapted for insular survival, then they ought to be the best survivors. They should be the last species to become extinct on older islands, although limited empirical evidence suggests that long-term survival is also deterministic. Of six lizard species on Isla Santa Catalina, five have successfully colonized Isla Monserrate, presumably by overwater dispersal (table 8.3). Those absent from Santa Catalina are also absent from Monserrat. These data also suggest a predetermined ability to colonize (i.e., that colonization, like extinction, is deterministic).

Previously, Murphy (1983a) performed a similar evaluation of the northern midriff islands. However, as reviewed above, the available cladograms suggest that Seri Indians introduced some of the herpetofaunal species. In light of this new information, a similar evaluation would be futile.

Epilogue

We can now return to the original question. Is a new scenario required? Absolutely. Not only have sequence data significantly changed our perspective of the peninsula's evolution, they have also dramatically affected our understanding of the insular biota. Advances in tectonics have detailed the mechanics of the peninsula's origin, but they have not significantly impacted on its stratigraphic development. More than the geological data, the combined sequence data allow us to make predictions about how the peninsula must have evolved. Early predictions are being born out by corroborative paleostratigraphic data.

The concept of the peninsular archipelago, now better defined and more explicit, is most exciting for us. Although the peninsula might teach us about genetic clinal variation and linear transformations, this now appears to be rare. New data indicate that the peninsula will be more informative in terms of understanding the role of isolation in evolution and the evolutionary consequences of reunited populations and faunas. With additional data from key herpetological species and other comparative taxonomic groups, the peninsular archipelago and the associated islands provide us an invaluable vision of history, an understanding of the present, and a window to the future.

Acknowledgments   This study would not have been possible without the exemplary cooperation of Ted Case, Kenneth Petren, Javier Rodríguez-Robles, Robert Zink, Todd Reeder, Jonathan Richmond, Jay Cole, and Herbert Dessauer, who unselfishly contributed unpublished data and advanced copies of accepted papers. Amy Lathrop drew the paleogeographic reconstructions and cladograms. Bobby Fokidis proofread the distribution tables. Leslie Lowcock, David

Morafka, Ross MacCulloch, Andre Ngo, Hobart Smith, and Hans-Dieter Sues provided valuable editorial comments. This research was supported by grant A3148 from the National Sciences and Engineering Research Council of Canada, The Royal Ontario Museum (ROM) Foundation, the ROM Future Fund Today, the ROM Members Volunteer Committee to R.W.M. Funding was provided to G.A.L. by Consejo Nacional de Ciencia y Tecnologia (CONACYT) grant PCECBEN-002101, the American Museum of Natural History Center for the Biodiversity and Conservation, and Instituto de Ecologia, A. C. account 902-01 Addition funding came from the NTC Ft. Irwin through Mickey Quillman, W. Johnson, Steve Ahmann and David Morafka. Special thanks to the Direccion General de la fauna Silvestre for unfailing support of our research in Mexico and to G. Orzola of U.S. Fish and Wildlife Enforcement.

References

[Aguilar-S., (*sic*)] M.A., J.W. Sites Jr., and R.W. Murphy. 1988. Genetic variability and population structure in the lizard genus *Petrosaurus* (Iguanidae). *Journal of Herpetology* **22**: 135–145.

Aguirre-León, G., D.J. Morafka, and R.W. Murphy. 1999. The peninsular archipelago of Baja California: a thousand kilometers of tree lizard genetics. *Herpetologica* **55**:369–381.

Atwater, T. 1970. Implications of plate tectonics for the Cenozoic tectonic evolution of western North America. *Geological Society of America Bulletin* **81**:3513–3536.

Atwater, T., and P. Molner. 1973. Relative motion of the Pacific and American plated deduced from sea-floor spreading in the Atlantic, Indian, and South Pacific Oceans. In: R.L. Kovach and A. Nur (eds), *Proceedings of the Conference on Tectonic Problems of the San Andreas Fault System*. Stanford University Publications in the Geological Sciences, Stanford, CA; pp. 136–148.

Axelrod, D.I. 1937. A Pliocene flora from the Mount Eden Beds, southern California. *Carnegie Institution of Washington Publication* **476**:125–183.

Axelrod, D.I. 1950. Further studies on the Mount Eden Flora, Southern California. *Carnegie Institution of Washington Publication* **590**:73–117.

Axelrod, D.I. 1975. Evolution and biogeography of the Madrean-Tethyan sclerophyll vegetation. *Annals of the Missouri Botanical Garden* **62**:280–334.

Axelrod, D.I. 1979. Age and origin of the Sonoran Desert vegetation. *Occasional Papers of the California Academy of Sciences* **132**:1–74.

Bell, C.J., and D.P. Whistler. 1996. Fossil remains of the legless lizard, *Anniella* Gray, 1852 from Late Pleistocene deposits at Rancho La Brea, California. *Bulletin of the Southern California Academy of Sciences* **95**:99–102.

Bischoff, J.L., and T.L. Henyey. 1974. Tectonic elements of the central part of the Gulf of California. *Geological Society of America Bulletin* **85**:1893–1904.

Blaney, R.M. 1977. Systematics of the common kingsnake, *Lampropeltis getulus* (Linnaeus). *Tulane Studies in Zoology and Botany* **19**:47–103.

Boehm, M.C. 1984. An overview of the lithostratigraphy, biostratigraphy, and paleoenvironmenrts of the Late Neogene San Felipe marine sequence, Baja California, Mexico. In: V.A. Frizzell (ed), *Geology of the Baja California Peninsula*. Society of Economic Paleontologists and Mineralogists, Los Angeles, California; pp. 253–265.

Boehm, M.C. 1987. Evidence for a north-verging mid-to-late Miocene proto-Gulf of California. *Geol. Soc. Amer. Abst. Prog.*, 594.

Brooks, D.R. 1990. Parsimony analysis in historical biogeography and coevolution: methodological and theoretical update. *Systematic Zoology* **39**:14–30.

Brooks, D.R., and D.A. McLennan. 1991. *Phylogeny, Ecology, and Behavior: A Research Program in Comparative Biology*. University of Chicago Press, Chicago.

Case, T.J. 1975. Species numbers, density compensation, and colonizing ability of lizards on islands in the Gulf of California. *Ecology* **56**:3–18.

Case, T.J. 1978a. On the evolution and adaptive significance of postnatal growth rates in the terrestrial vertebrates. *Quarterly Review of Biology* **53**:243–282.

Case, T.J. 1978b. A general explanation for insular body size trends in terrestrial vertebrates. *Ecology* **59**:1–18.

Case, T. 1983. The reptiles: Ecology. In: T.J. Case and M.L. Cody (eds), *Island Biogeography in the Sea of Cortez*. Berkeley: University of California Press; pp. 159–209.

Case, T.J., and M.L. Cody. (eds). 1983. *Island Biogeography in the Sea of Cortéz*. University of California Press, Berkeley.

Cody, M.L. 1974. *Competition and the structure of bird communities*. Princeton University Press, Princeton, NJ.

Dixon, J.R. 1964. The systematics and distribution of lizards of the genus *Phyllodactylus* in North and Central America. *New Mexico State University Research Center Scientific Bulletin* **64**. New Mexico State University, Albuquerque.

Durham, J.W., and E.C. Allison. 1960. The geologic history of Baja California and its marine fauna. *Systematic Zoology* **9**:47–91.

Felger, R.S., and M.B. Moser. 1985. *People of the Desert and Sea*. University of Arizona Press, Tucson.

Frick, C. 1933. New remains of tripholodont-tetrabelodont mastidons. *Bulletin of the American Museum of Natural History* **59**:505–652.

Frost, D.R., and D.M. Hillis. 1990. Species in concept and practice: herpetological applications. *Herpetologica* **46**:87–104.

Frost, D.R., and A.G. Kluge. 1994. A consideration of epistemology in systematic biology, with special reference to species. *Cladistics* **10**:259–294.

Fu, J., and R.W. Murphy. 1999. Discriminating and locating character covariation: An application of permutation tail probability analyses. *Systematic Biology* **48**:380–395.

Gastil, R.G., and W. Jensky. 1973. Evidence for strike-slip displacement beneath the trans-Mexican volcanic belt. In: Kovach RL and Nur A (eds), *Proceedings of the Conference on Tectonic Problems of the San Andreas Fault System*. Stanford University Publications in the Geological Sciences, Stanford, CA; pp. 171–180.

Gastil, R.G., and D. Krummenacher. 1977. Reconnaissance geology of coastal Sonora between Puerto Lobos and Bahia Kino. *Geological Society of America Bulletin* **88**:189–198.

Gastil, R.G., J.C. Minch, and R.P. Philips. 1975. Reconnaissance geology of the state of Baja California. *Geological Society of America Memoir* **14**:1–170.

Gastil, R.G., J. Minch, and R.P. Phillips. 1983. The geology and ages of islands. In: T.J. Case and M.L. Cody (eds), *Island Biogeography in the Sea of Cortéz*. University of California Press, Berkeley; pp. 13–25.

Graybeal, A. 1995. Naming species. *Systematic Biology* **44**:237–250.

Grismer, L.L. 1994a. *The Evolutionary and Ecological Biogeography of the Herpetofauna of Baja California and the Sea of Cortés, Mexico*. Dissertation, Department of Biology, Loma Linda University, Loma Linda, CA.

Grismer, L.L. 1994b. The origin and evolution of the peninsular herpetofauna of Baja California, México. *Herpetological Natural History* **2**:51–106.

Grismer, LL. 1994c. Geographic origins for the reptiles on islands in the Gulf of California, México. *Herpetological Natural History* **2**:17–40.

Grismer, L.L. 1999a. Checklist of amphibians and reptiles on islands in the Gulf of California. *Bulletin of the Southern California Academy of Sciences* **98**:45–56.

Grismer, L.L. 1999b. Phylogeny, taxonomy, and biogeography of *Cnemidophorus hyperythrus* and *C. ceralbensis* (Squamata: Teiidae) in Baja California, México. *Herpetologica* **55**: 28–42.

Grismer, L.L., and J.A. McGuire. 1993. The oases of central Baja California, México. Part I. A preliminary account of the relict mesophilic herpetofauna and the status of the oases. *Bulletin of the Southern California Academy of Sciences* **92**:2–24.

Grismer, L.L., and J.A. McGuire. 1996. Taxonomy and biogeography of the *Sceloporus magister* complex (Squamata: Phrynosomatidae) in Baja California, Mexico. *Herpetologica* **52**: 416–427.

Hausback, B.P. 1984. Cenozoic volcanic and tectonic evolution of Baja California Sur, México. In: V.A. Frizzell (ed), *Geology of the Baja California Peninsula*. Society of Economic Paleontologists and Mineralogists, Los Angeles, California; pp. 219–236.

Henning, W. 1966. *Phylogenetic Systematics*. University of Illinois Press, Urbana.

Henyey, T.L., and J.L. Bischoff. 1973. Tectonic elements of the northern part of the Gulf of California. *Geological Society of America Bulletin* **84**:315–330.

Hillis, D.M., C. Moritz, and B.K. Mable (eds), 1996. *Molecular Systematics*. Sinauer Associates, Sunderland, MA.

Hollingsworth, B.D. 1998. The systematics of chuckwallas (*Sauromalus*) with a phylogenetic analysis of other Iguanid lizards. *Herpetological Monographs* **12**:38–191.

Hollingsworth, B.D. 1999. *The Molecular Systematics of the Side-Blotched Lizards* (Iguania: Phrynosomatidae: *Uta*). Ph.D. dissertation, Loma Linda University.

Jensky, W.A. 1975. *Reconnaissance Geology and Geochronology of the Bahia de Banderas Area, Nayarit and Jalisco, Mexico*. M.Sc. thesis, Department of Geology, University of California, Santa Barbara.

Kim, Y.J., G.C. Gorman, T. Papenfuss, and A.K. Roychoudhury. 1976. Genetic relationships and genetic variation in the amphisbaenian genus *Bipes*. *Copeia* **1976**:120–176.

Kluge, A.G. 1997. Testability and the refutation and corroboration of cladistic hypotheses. *Cladistics* **13**:81–96.

MacArthur, R.H., and Wilson, E.O. 1963. An equilibrium theory of insular zoogeography. *Evolution* **17**:373–387.

MacArthur, R.H., and Wilson, E.O. 1967. *The Theory of Island Biogeography*. Princeton University Press, Princeton, NJ.

Macey, J.R., A. Larson, N.B. Ananjeva, and T.J. Papenfuss. 1997. Evolutionary shifts in three major structural features of the mitochondrial genome among iguanian lizards. *Journal of Molecular Evolution* **44**:660–674.

Mammerickx, J., and K.D. Klitgord. 1982. Northern East Pacific Rise: evolution from 25 m.y. B.P. to present. *Journal of Geophysical Research* **87**:6751–6759.

Matthews, W.D. 1915. Climate and evolution. *Annals of the New York Academy of Sciences* **24**:171–318.

McGuire, J.A. 1996. Phylogenetic systematics of crotaphytid lizards (Reptilia: Iguania: Crotaphytidae). *Bulletin of the Carnegie Museum of Natural History* **143**.

Minch, J.C., R.G. Gastil, W. Fink, J. Robinson, and A.H. James. 1976. Geology of the Vizcaíno Peninsula. In: D.G. Howell (ed.), *Aspects of the Geologic History of the California Continental Borderland*. Pacific Section of the American Association of Petroleum Geologists, Miscellaneous Publication; 24: pp. 136–195.

Moore, D.G. 1973. Plate-edge deformation and crustal growth, Gulf of California structural province. *Geological Society of America Bulletin* **84**:1883–1906.

Morafka, D.J. 1977. A biogeographical analysis of the Chihuahuan Desert through its herpetofauna. *Biogeographica* **9**.

Murphy, R.W. 1975. Two new blind snakes (Serpentes: Leptotyphlopidae) from Baja California, Mexico with a contribution to the biogeography of peninsular and insular herpetofauna. *Proceedings of the California Academy of Sciences* **40**:93–107.

Murphy, R.W. 1983a. The reptiles: origins and evolution. In: T.J. Case and M.L. Cody (eds),

Island Biogeography in the Sea of Cortéz. University of California Press, Berkeley; pp. 130–158.

Murphy, R.W. 1983b. Paleobiogeography and genetic differentiation of the Baja California herpetofauna. *Occasional Papers of the California Academy of Sciences* **137**:1–48.

Murphy, R.W. 1991. Plate tectonics, peninsular effects, and the borderlands herpetofauna of western North America. In: P. Ganster and H. Walter (eds), *Environmental Hazards and Bioresource Issues of the United States-Mexico Borderlands*. Latin American Center, University of California, Los Angeles; pp. 433–457.

Murphy, R.W. 1993. The phylogenetic analysis of allozyme data: invalidity of coding alleles by presence/absence and recommended procedures. *Biochemical Systematics and Ecology* **21**:25–38.

Murphy, R.W., and C.B. Crabtree. 1985. Genetic relationships of the Santa Catalina Island rattleless rattlesnake, *Crotalus catalinensis* (Serpentes: Viperidae). *Acta Zoologica Mexicana* (n.s.) **9**:1–16.

Murphy, R.W., and K.D. Doyle. 1998. Phylophenetics: frequencies and polymorphic characters in genealogical estimation. *Systematic Biology* **47**:737–761.

Murphy, R.W., J. Fu, A. Lathrop, J.V. Feltham, and V. Kovac. In press. Phylogeny of the rattlesnakes (*Crotalus and Sistrurus*) inferred from sequences of five mitochondrial DNA genes. In: G.W. Schuett, M. Höggren, and H.W. Green (eds), *Biology of Vipers*. Eagle Mountain Publishing, Utah.

Murphy, R.W., and J.R. Ottley. 1984. Distribution of amphibians and reptiles on Islands in the Gulf of California. *Annals of Carnegie Museum* **53**:207–230.

Nelson, E.W. 1921. Lower California and its natural resources. *Memoirs of the National Academy of Sciences* **16**.

Ochoa-Landin, L. 1998. Geological, sedimentological and geochemical studies of the Boleo Cu-Co-Zn Deposit, Santa Rosalia Baja California, Mexico. Ph.D. dissertation, Department of Geosciences, The University of Arizona, Tucson.

Papenfuss, T.J. 1982. The ecology and systematics of the amphisbaenian genus *Bipes*. *Occasional Papers of the California Academy of Sciences* **136**:1–42.

Petren, K., and T.J. Case. 1997. A phylogenetic analysis of body size evolution and biogeography in chuckwallas (*Sauromalus*) and other iguanines. *Evolution* **51**:206–219.

Phillips, R.P. 1966. Reconnaissance geology of some of the northwestern islands in the Gulf of California (abstract). *Geological Society of America, Cordilleran Section Program*, **59**.

Popper, K.R. 1934 [1959]. *The Logic of Scientific Discovery*. Basic Books, New York.

Radtkey, R.R., S.K. Fallon, and T.J. Case. 1997. Character displacement in some *Cnemidophorus* lizards revisited: A phylogenetic analysis. *Proceedings of the National Academy of Science USA* **94**:9740–9745.

Reeder, T.W. 1995. Phylogenetic relationships among phrynosomatid lizards as inferred from mitochondrial ribosomal DNA sequences: substitutional bias and information content of transitions relative to transversions. *Molecular Phylogenetics and Evolution* **4**:203–222.

Reeder, T.W., and J.J. Wiens. 1996. Evolution of the lizard family Phrynosomatidae as inferred from diverse types of data. *Herpetological Monographs* **10**:43–84.

Robinson, M.D. 1973. Chromosomes and systematics of the Baja California whiptail lizards, *Cnemidophorus hyperythrus* and *C. cerralvensis* (Reptilia: Teiidae). *Systematic Zoology* **22**:30–35.

Rodríguez-Robles, J.A., and J.M. De Jesús-Escobar. 2000. Molecular systematics of New World gopher, bull, and pinesnakes (*Pituophis*: Colubridae), a transcontinental species complex. *Molecular Phylogenetics and Evolution* **14**:35–50.

Rodríguez-Robles, J.A., D.F. Denardo, and R.E. Staubs. 1999. Phylogeography of the California mountain kingsnake, *Lampropeltis zonata* (Colubridae). *Molecular Ecology* **8**:1923–1934.

Savage, J.M. 1960. Evolution of a peninsular herpetofauna. *Systematic Zoology* **9**:184–212.
Schmidt, K.P. 1922. The amphibians and reptiles of Lower California and the neighboring islands. *The American Museum of Natural History* **46**:607–707.
Schmidt, K.P. 1943. Corollary and commentary for "Climate and Evolution." *American Midland Naturalist* **30**:241–253.
Schulte, J.A.I., J.R. Macey, A. Larson, and T.J. Papenfuss. 1998. Molecular tests of phylogenetic taxonomies: a general procedure and example using four subfamilies of the lizard family Iguanidae. *Molecular Phylogenetics and Evolution* **10**:367–376.
Seib, R.L. 1980. Baja California: A peninsula for rodents but not for reptiles. *American Naturalist* **115**:613–620.
Siddall, M.E., and A.G. Kluge. 1997. Probabilism and phylogenetic Inference. *Cladistics* **13**: 313–336.
Sites, J.W., Jr., S.K. Davis, T. Guerra, J.B. Iverson, and H.L. Snell. 1996. Character congruence and phylogenetic signal in molecular and morphological data sets: a case study in the living Iguanas (Squamata, Iguanidae). *Molecular Biology and Evolution* **13**:1087–1105.
Smith, N.M., and W.W. Tanner. 1974. A taxonomic study of the western collared lizards, *Crotaphytus collaris* and *Crotaphytus inuslaris*. *Brigham Young University Science Bulletin* **19**:1–29.
Soulé M., and A.J. Sloan. 1966. Biogeography and distribution of the reptiles and amphibians on islands in the Gulf of California, Mexico. *Transactions of the San Diego Society of Natural History* **14**:137–156.
Soulé, M.E., S.Y. Yang, M.G.W. Weiler, and G.C. Gorman. 1973. Island lizards: the genetic-phenetic variation correlation. *Nature* **242**:191–193.
Stock, J.M., and K.V. Hodges. 1989. Pre-Pliocene extension around the Gulf of California and the transfer of Baja California to the Pacific Plate. *Tectonics* **8**:99–115.
Swofford, D.L. 2000. PAUP*: Phylogenetic Analysis Using Parsimony (*and Other Methods), version 4.0b4a. Sinauer Associates, Sunderland, MA.
Taylor, R.J., and P.J. Regal. 1978. The peninsular effect on species diversity and biogeography of Baja California. *American Naturalist* **112**:583–593.
Taylor, R.J., and P.J. Regal. 1980. Reply to Seib. *American Naturalist* **115**:621–622.
Trépanier, T.L., and R.W. Murphy. In press. The Coachella Valley fringe-toed lizard (*Uma inornata*): genetic diversity and phylogenetic relationships of an endangered species. *Molecular Phylogenetics and Evolution* **18**:327–334.
Upton, D.E., and R.W. Murphy. 1997. Phylogeny of the side-blotched lizards (Phrynosomatidae: *Uta*) based on mtDNA sequences: Support for a midpeninsular seaway in Baja California. *Molecular Phylogenetics and Evolution* **8**:104–113.
Van Devender, T.R., 1990. Late Quaternary vegetation and climate of the Sonoran Desert, United States and Mexico. In: J.L. Betancourt, T.R. Van Devender, and P.S. Martin (eds), *Packrat Middens. The last 40,000 years of Biotic Change*. The University of Arizona Press, Tucson; pp. 134–165.
Wallace, 1880.
Wilcox, B.A. 1978. Supersaturated island faunas: a species-age relationship for lizards on post-Pleistocene land-bridge islands. *Science* **199**:996–998.
Wilcox, B.A. 1980. Insular ecology and conservation. In: M. Soulé (ed), *Conservation Biology*. Sinauer Associates, Sunderland, MA; pp. 95–117.
Wiley, E.O. 1988. Parsimony analysis and vicariance biogeography. *Systematic Zoology* **37**: 271–290.
Wong, H., L.L. Grismer, B.D. Hollingsworth, and R.L. Carter. 1998. Mitochondrial DNA phylogeography of the sand snakes *Chilomeniscus* (Serpentes: Colubridae) from northwestern México and southern Arizona. In *Herpetology of the Californias*. San Diego Natural His-

tory Museum; San Diego, CA. Available at www.sdnhm.org/research/symposia/herpsym98ab.html

Wyles, J.S., and G.C. Gorman. 1978. Close relationship between the lizard genus *Sator* and *Sceloporus utiformis* (Reptilia, Lacertilia, Iguanidae): electrophoretic and immunological evidence. *Journal of Herpetology* **12**:343–350.

Zink, R.M., R. Blackwell, and O. Rojas-Soto. 1997. Species limits on the Le Contes Thrasher. *Condor* **99**:132–138.

9

Reptiles

Ecology

TED J. CASE

The reptiles of the islands of the Sea of Cortés have provided many opportunities to test ecological and biogeographical hypotheses because they support a diverse fauna with much insular endemism; are numerous and of varying ages and degrees of isolation; are relatively undisturbed by human activity and introduced species; and have a relatively well-understood geological history (see chap. 2). In particular, contrasts of mainland and island reptile populations in the region have resulted in significant progress in testing theories of island biogeography, principles of ecological character displacement, ecological release, density compensation, and vicariance biogeography (see chap. 8). The reptiles, being conspicuous in these arid habitats, have attracted relatively more research attention than other vertebrates, and today we have a reasonably complete picture of at least which species are on which islands. Since the first edition of this book, nearly 20 years ago, there have been only 15 new records for the major islands, of which all but one are of snakes.

In this chapter I review the basic elements of reptilian island biogeography in the Sea of Cortés with an emphasis on ecological factors shaping distributions and evolutionary trajectories. I first examine the patterns of species diversity and association across islands. I then take a closer look at some particular island forms, reviewing features of their life history that seem divergent from mainland relatives. In this regard I present some new data from a long-term study of two insular species of chuckwallas. Finally, I review patterns of population density across islands and their possible determinants.

Species Number

A recurrent debate in island biogeography centers on the relative importance of contemporary and ongoing ecological factors relative to historical circumstances in accounting for the number and the identities of species on islands. Historical biogeographers typically view the number of species on an island as being determined by the availability of appropriate habitats. They see changes in species composition chiefly as a consequence of alteration of the mix of habitats due to climatic change (e.g., Pregill and Olson 1980; Olson and Hilgartner 1982); extinctions are posited to occur in waves, as old habitats disappear and new ones become available. Colonizations, in this view, are greatly influenced by land connections, and if these disappear (because of sea level changes or the movement of islands by plate shifts), then colonization dries up as well. In contrast, the equilibrium theory assumes that immigration and extinction are on-going, random events which occur continually even in the absence of habitat change. An important result of this model is that, because colonization and extinction rates by definition will depend on the number of species already on an island, species number will approach a dynamic equilibrium where these two rates just balance one another (MacArthur and Wilson 1967). Whether such an equilibrium is reached will depend on the rates of colonization and extinction events on the one hand, relative to the time scale of the geological events that create, move, alter habitat, and destroy islands, on the other hand. When islands are young relative to the time required for an ecological equilibrium to be reached, then history must usually be consulted to understand species number and biotic composition. The utility of equilibrium theory suffers if history's fingerprint is so permanent that on a case-by-case basis a knowledge of the specific geological events affecting an island and its past connections is more valuable than present-day island size and isolation for a prediction of species number and composition.

An oceanic island formed by volcanic activity begins life far below its final species equilibrium, whereas an island recently connected to the mainland by a landbridge approaches its equilibrium from above (i.e., by the elimination of some species over time). Because of considerations like these, I (1975) analyzed landbridge and deep-water oceanic islands separately in a multiple regression of gulf island lizard species richness. Using the new presence–absence data for reptiles in appendixes 8.3 and 8.4, I performed a multiple regression analysis using log reptile species (LRS; = lizards plus snakes) number as a dependent variable. The following independent variables and their log transforms were considered: A = island area in square kilometers (log transformed); D = nearest distance to mainland in kilometers (log transformed); P = number of perennial plant species on island; L/O = a categorical variable to separate landbridge from oceanic islands (0 = landbridge; 1 = oceanic).

All islands in appendixes 8.3 and 8.4 that had at least one reptilian species were included. Monserrat is considered an oceanic island, and small satellite islands that are landbridge islands to larger oceanic islands are considered as landbridge islands.

The only two variables to enter a stepwise regression are log area ($p < .0001$), with a positive effect, and L/O ($p < .04$), with landbridge islands having more species than oceanic islands for their area. This model explained 67.1% of the variance. For landbridge islands alone, only the log of island area enters the model, and for deep-water islands alone, again, only the log of island area enters the model.

Landbridge islands contain more species on average than oceanic islands of similar geography, indicating that the species number of reptiles (for lizards, see Case 1975) and mammals (chap. 12) is still decreasing on landbridge islands. There is no similar landbridge effect among gulf land birds (chap. 10) or plants (chap. 4).

The relationship between log island area and the LRS is shown in figure 9.1. For LRS versus log area (LRS = $c + zA$), the coefficients z and c are:

Landbridge islands: $z = 0.222$ ($R^2 = 65.6\%$); $c = 0.727$
Oceanic islands: $z = 0.314$ ($R^2 = 79.2\%$); $c = 0.508$
All islands: $z = 0.223$ ($R^2 = 63.0\%$); $c = 0.681$.

For mammals these z-values are 0.310 (gulf landbridge), 0.119 (gulf oceanic), and 0.242 (all gulf islands), respectively. The most striking difference between these two taxa is in the z-value for oceanic islands. For reptiles, oceanic islands have a greater z-value than landbridge islands, whereas for mammals, the z-value of oceanic islands is less than one-half the comparable value for landbridge islands. This suggested to Lawlor and colleagues (1986, chap. 12, this volume) that mammalian faunas on oceanic islands are depauperate owing to the extreme rarity of colonization. In line with this conclusion, the degree of island isolation is a major factor accounting for differences in mammalian species number among oceanic islands and, to a lesser extent,

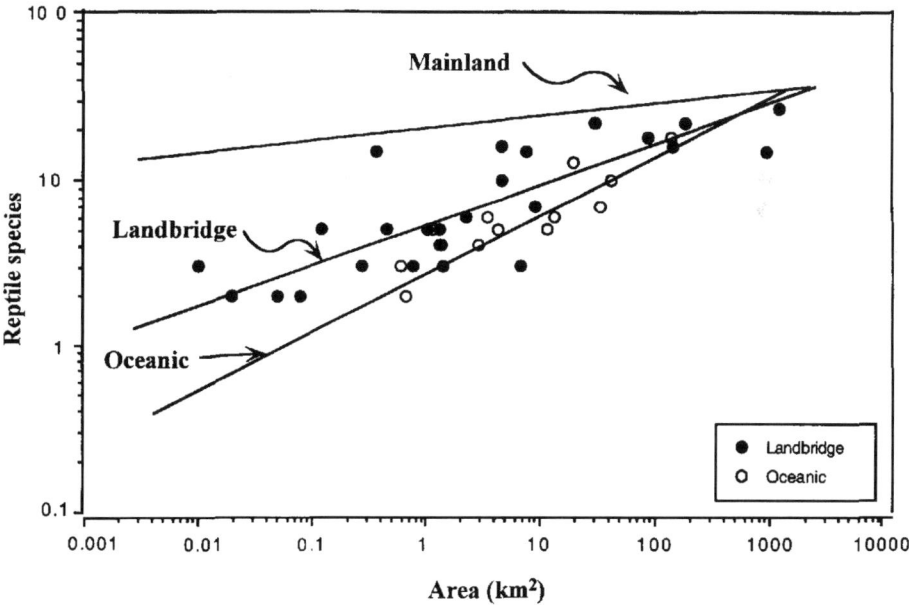

Figure 9.1 Reptile species number plotted against island area for oceanic and landbridge islands in the Sea of Cortés. Regression equations are in the text. The mainland species–area curve is from Richman et al. (1988): log mainland species number = 1.304 + 0.082 log area. The difference between the mainland line and the landbridge island line is used to infer the number of extinctions that may have taken place on the landbridge islands following their isolation (see fig. 9.3).

224 The Biological Scene

among landbridge islands. Island isolation plays no significant role in reptiles, which suggests that reptile colonization is more frequent and not as strongly hindered by distance as mammal colonization. Thus the discrepancy between reptilian diversity on large oceanic islands versus that on equally large landbridge islands is not nearly as great as the same discrepancy for mammals.

Occurrence Patterns on Oceanic Versus Landbridge Islands

Why are some species more broadly distributed on islands than others? For lizard species on Sea of Cortés islands, the number of oceanic islands each species occupies is highly correlated with the number of landbridge islands it occupies (fig 9.2; Pearson correlation, $p < .0001$), and only one form, *Sator*, occurs on more oceanic islands than on landbridge ones. Species that are frequent on oceanic islands evidence an ability for their populations to persist over long time periods and to be able to colonize over water (at least on oceanic islands without any ancient connections to either mainland). In contrast, species that are frequent on landbridge islands typically had a vicariant origin, mitigating the need for good overwater dispersal. A species that is relatively

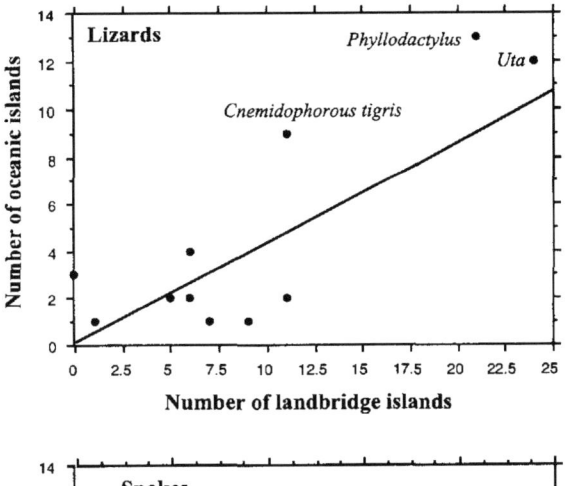

Figure 9.2 A plot of the number of oceanic islands occupied versus the number of landbridge islands occupied for lizards and snakes. *Sauromalus* is excluded because it has been redistributed by humans. Island endemic taxa are grouped with their ancestral species. These plots also exclude small satellite landbridge islands to oceanic islands.

frequent on landbridge islands presumably displays a lower extinction risk compared to other species. The apparent association of colonization ability (on oceanic islands) and extinction resistance (on all islands) across lizard species indicates that these two features may often go hand in hand. Both colonization and persistence require that a population be able to increase quickly when its numbers are small. The four lizard taxa that are most prevalent on islands (*Uta, Sauromalus, Phyllodactylus,* and *Cnemidophorus tigris*) have very different life histories, overall geographic ranges, and phylogenetic histories, but they share two attributes: they tend to be relatively abundant on the mainland and are often found in rocky habitats that abound on the islands. Pianka (1986) summarized the relative abundance of lizard species in five study sites in the Sonora Desert of the United States. Unfortunately, these sites do not include several species that occur in Baja California. Figure 9.3 is a plot of the summed counts from Pianka versus the number of Sea of Cortés islands occupied for the set of overlapping species in the two geographic regions. The correlation between relative abundance and island occurrence is highly significant ($p < .001$). This result matches that of Foufopoulos and Ives (1999), who examined inferred reptile extinction rates on landbridge islands in the Mediterranean Sea. Independent variables included body mass, longevity, habitat specialization, and population abundance. Only abundance and habitat specialization were useful predictors of extinction rates.

For snakes, the relationship between colonization success and persistence on oceanic islands is not so clear (fig. 9.2; $p = .18$). The kingsnakes have an exceptional distributional pattern on the islands. They have not been recorded from a single landbridge island, yet they are the single most ubiquitous snake on oceanic islands (on

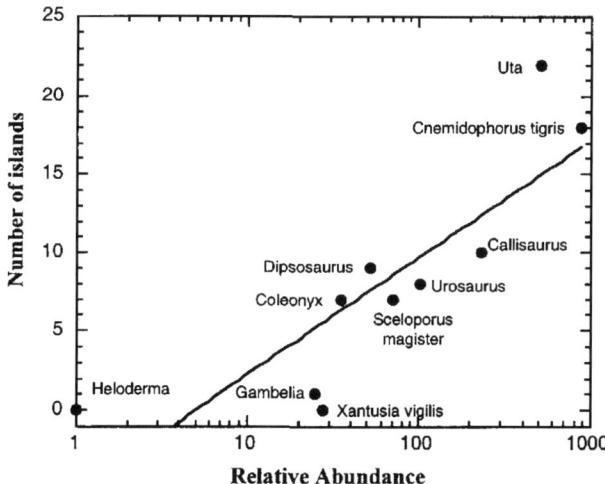

Figure 9.3 The number of island occurrences of lizards as a function of their relative abundance as determined at Pianka's (1986) five study sites in the Sonora Desert of the United States. Pianka's sites do not include the geographic ranges of several species that occur in Baja California, and so these species are excluded. *Sauromalus* is excluded because its insular distribution has been altered by people. The correlation between log relative abundance and island occurrence is highly significant ($p < .003$, $R^2 = 70.7\%$).

12). If they can persist on oceanic islands, why not on landbridge islands? Figure 9.2 excludes small satellite islands to oceanic islands, but their inclusion would not add records. Excluding the small islands, since kingsnakes are not found on any of these, there are 15 oceanic islands and 8 landbridge islands available; hence the fraction of landbridge islands is $8/23 = 0.3478$. The probability of a kingsnake not occurring on a single landbridge island given that it is on 12 islands total is $p = .0059$, by the binomial distribution. The probability of any of the 6 reptile taxa that occur on 12 or more islands having a distribution this extreme is $p = 6 (0.0059) = 0.0354$, which is still significant. There are four other snakes that occupy more oceanic islands than landbridge islands: *Phyllorhynchus, Rhinocheilus, Crotalus atrox* group, and *Crotalus ruber* group. I suspect that this pattern in snakes, but not in lizards, results because snakes more typically have a ancient vicariant origin even on islands that are now in deep water (see chap. 8). I also suspect that these snakes' persistence may be improved by an absence of competitors and predators, which is more typically found on oceanic islands, although even this explanation does not fully come to grips with the bizarre pattern in the kingsnakes, which remains a puzzle.

Inferred Extinctions

There have been no reptile extinctions on these islands in historical time, but various attempts have been made to estimate prehistorical Holocene extinctions using the relaxation method (Diamond 1972). Wilcox (1978, 1980b) made use of the fact that different landbridge islands in the Sea of Cortés became isolated from the Baja California mainland at different times after the last glacial maximum. Age estimates for the islands of Baja California were interpolated from a sea-level curve for the eastern U.S. seaboard (Milliman and Emery 1968). The rate of sea-level rise is roughly logistic, leveling off perhaps 6000 years ago. Crustal uplift or subsidence may significantly alter the channel depth, and consequent age estimates based on eustatic sea-level changes. However, vertical movements in areas not rebounding from the weight of continental ice sheets, such as the Baja peninsula or South Australia, are relatively insignificant. Kaula (1977) pointed out that sea-level changes due to glacial melting outweigh all other contributing factors by at least an order of magnitude, in time scales of 100–10,000 years. Further, although the Sea of Cortés is a tectonically active region, Ortleib (1977) found no evidence for rapid uplift (or subsidence) on either side of the peninsula in recent time. Based on existing channel depths separating the gulf islands from the mainland (i.e., similar to those in table 2.2), Wilcox (1978) reasoned that the degree of supersaturation of a landbridge island's lizard fauna (relative to an oceanic island) and the time since its last mainland connection should be negatively correlated. After partialing out the variance in species number due to island area and latitude, Wilcox found just such a result. Richman et al. (1988) determined a species–area curve for the Baja mainland, which allowed them to calculate an estimated number of extinctions for reptiles on landbridge islands in the gulf and off the Pacific coast as well as off Australia. They found that extinction rates in both regions were comparable and declined with increasing area.

Case and Cody (1987) showed that relaxation rates for mammals on Baja California islands were substantially higher than those for reptiles on the same islands, although both groups showed decreasing extinction rates with increasing island area. Schoener

(1983) also found that species turnover rates for reptiles are generally lower than those of birds and mammals and most arthropod systems, based on a wide review of the literature. Lizard populations should be expected to be more resistant to extinction because their lower metabolic rate should allow higher densities than either birds or mammals and thus larger population sizes.

Higher extinction rates for mammals are also seen in the historical record; there are four instances of presumed extinctions among insular forms in the Sea of Cortés within historical time (chap. 12), but none for reptiles. Case et al. (1998) reviewed the worldwide Holocene extinctions of reptiles and found that reptile extinctions were primarily on islands and that reptiles were probably less extinction prone during the Holocene than birds and mammals. Pregill (1986) correlated the human settlement time of islands with the extinction times of a number of insular reptiles. Worldwide, the arrival of humans on an island is closely associated with increased reptile extinction, especially of endemic species with large body size (also see Barahona et al. 2000).

Alteration of habitats on the islands by humans is minimal because the islands' extreme aridity has discouraged long-term settlement (Aschmann 1959 but see chaps. 14 and 15). The change in climate and vegetation on the Baja peninsula throughout the Holocene, drawn from studies of paleoclimate in deserts of the American Southwest, is summarized in Van Devender (1987) and Richman et al. (1988). It seems that the coastal deserts of Baja California were probably the areas least effected by pluvial climates in the arid Southwest. Consequently, Richman et al. (1988) favored the hypothesis that the inferred Holocene reptile extinctions on landbridge islands were more consistent with theories based on stochastic demography and environmental variability of small populations (MacArthur and Wilson 1967; Leigh 1981) rather than gross unidirectional climate change or human occupation.

Figure 9.4 updates these relaxation rates using the new island area estimates for and updated species records for the Sea of Cortés (apps. 1.1, 8.3, 8.4) and off the Pacific coast (Grismer 1993). We again see the familiar pattern of decreasing extinc-

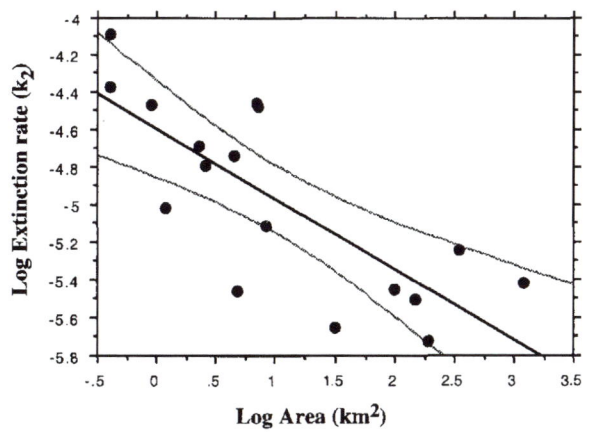

Figure 9.4 Inferred extinction rates (k_2) for landbridge islands based on a model of species decay since isolation: $dS/dt = -k_2 S^2$ and plotted as a function of present-day island area. The islands and their isolation times match those given in Richman et al. (1988), but areas and species lists have been updated. The regression line and 95% confidence intervals about the mean are shown. Log k_2 = −4.59 − 0.373 log area ($p < .0002$, $R^2 = 60.2\%$).

tion rates with increasing area. If we apply these landbridge island extinction rates to the more ancient oceanic islands of possible continental origin in the Sea of Cortés, one would predict that only a small proportion of the original species populations on an oceanic island (i.e., when it became isolated more than 1 million years ago) have remained to the present. This, in turn, would suggest that the present-day reptile populations, have, for the most part, arrived by colonization. There are difficulties, however, with such an extrapolation. First, we do not know the actual dynamics of extinction rate over time. There are two lines of evidence that suggests that extinction may start high and later fall to much lower levels. First, the study of Wilcox (1978) shows an inferred early species loss in lizards, settling in to a slower loss rate as time progresses. Second, when statistically fitting different models of extinction dynamics as a function of species number, the models that yield the best fit have extinction rate as an increasing function of species number:

$$\frac{dS}{dt} = -k\,S^n.$$

For $n = 1$, this model describes a process like radioactive decay, with a constant half-life. Instead, Richman et al. (1988) found that models with $n = 2$ or $n = 3$ produced better fits to the pattern of k versus log area. This implies an extinction rate that is initially high immediately after isolation, then as species number drops, it declines, as only a extinction-resistant set of species remains. Extrapolating the high initial rate to oceanic islands over the course of a million years would produce an exaggerated expectation of the number of species to go extinct over that time. By fitting a model with $n = 2$, we would predict that Ángel de la Guarda would go from about 35 species to 0.8 species in a million years, and Santa Catalina would go from 27 to 0.19 species in the same time. Today these islands have 15 and 10 species, respectively, so either colonization has made up the difference or extinction rates quickly fall to much lower levels as time goes on, or both. For example if $n = 4$, then the expected surviving species today are 6.8 for Ángel and 2.44 for Santa Catalina.

Elsewhere fossil evidence suggests that sometimes former vicariant occupations of islands become extinct and are replaced by new colonists. In Madagascar, late Cretaceous deposits of frogs and crocodilians share affinities with India and South American taxa (Buckley et al. 2000). At that time these land forms were close or connected to Madagascar. The present-day frogs and crocodiles on Madagascar are not descended from these forms, which are in different families. Instead, the extant fauna shares phylogenetic relationships with groups in Africa, suggesting that they arrived by subsequent colonization.

The high species turnover inferred from landbridge islands if extrapolated to oceanic islands seems contradictory to their high levels of endemism and relictual elements (such as *Sator*) and in some cases contradictory to the phylogenetic evidence for divergence times, where these can be clocked (see chap. 8). Unfortunately, we have little direct evidence for island colonizations, and clock-based divergence times from DNA have huge confidence intervals. Excluding the species clearly moved around by people, the only colonization that has been seen in historical times is that of *Callisaurus draconoides* on Danzante (Wong et al. 1995). The possibility that this species was simply overlooked for so many years cannot be entirely dismissed, but this seems remote given Danzante's small size and the amount of collecting attention that it has received.

Two totally volcanic oceanic islands in the Sea of Cortés are thought to be recent (Holocene): Rasa (0.68 km^2) and Tortuga (11.36 km^2; chap. 2). Rasa is inhabited by two reptilian species and Tortuga by five species, roughly the number of species expected for any oceanic islands of their size. Rasa has no mammalian species, but Tortuga has one endemic mouse, *Peromyscus dickeyi*. The typical oceanic island of their size would contain one and two mammalian species. Thus, these recent islands are depauperate in mammals but not in reptiles, again attesting to the superior colonization abilities of reptiles. The flora of Tortuga is not at all depauperate; in fact, it is surprisingly rich. Cody et al. (chap. 4) point out that Tortuga exceeds the expected number of plant species for an island of its size and isolation. The avifauna of Tortuga is not well studied, but again there are no indications that it is depauperate.

In summary, these statistical analyses reveal that reptiles are superior island colonists to mammals but less successful than land birds or plants. Extinctions of reptiles are evident and seem to occur at rates higher than those for mammals. For both oceanic and landbridge islands, island area is the best single predictor of reptile species number. The present-day reptile fauna of deep-water islands is a mixture of species of probable vicariant origin and others that colonized over water, but it is difficult to quantify the relative proportion of each.

Distributional Patterns

Faunal Mixing

The mainland herpetofauna on either side of the Sea of Cortés is similar at the species level but not identical (chap. 8), a fact that allows one to examine the relative contributions that Baja California and mainland Mexico make to the insular fauna. The islands may be divided into two groups for this analysis: Sonoran islands, which lie closer to mainland Mexico than to Baja California (Tiburón, San Pedro Mártir, San Esteban, and San Pedro Nolasco); and Baja California islands, which lie closer to the peninsula (all other islands). The reptiles may be divided into three groups: (1) those that range over both regions; (2) those restricted to Sonora and adjacent regions of Mexico and Arizona; and (3) those restricted to Baja California and the west U.S. coast. The latter two groups can be used as indicator species to judge the relative contributions that these two regions make to the island biotas. Table 9.1 summarizes this information. In addition, with the advent of molecular phylogenies, we may determine for some island forms whether they are descended from ancestors in mainland Mexico or Baja California even when the species has a circum-gulf distribution. Murphy and Léon provide estimates of these affinities in appendix 8.3. Figure 9.5 plots these estimates versus the distance of an island from the nearest point of the Baja peninsula. As the distance gets larger, the island lies increasingly closer to mainland Mexico. We can see that most islands are closer to the peninsula, and unsurprisingly have a fauna dominated by peninsular descendents. Baja California islands do not usually contain descendants of Sonoran endemics, and Sonoran islands do not usually contain descen-

Table 9.1 Reptilian species that are confined to only one side or the other of the Sea of Cortés and have broad distributions in those regions

Baja California Endemics		Sonora, Mexico Endemics	
Lizards			
Cnemidophorus hyperythrus	B-7	*Cnemidophorus burti*	N
Crotaphytus insularis	B-1	*Crotaphytus dickersonae*	S-1
Phrynosoma coronatum	N	*Phrynosoma solare*	S-1
Phyllodactylus nocticolus	B-17, S-5	*Phyllodactylus homolepidurus*	S-1
Sceloporus orcutti	B-7	*Sceloporus clarki*	S-2
Urosaurus nigricaudus	B-15	*Urosaurus ornatus*	S-1
Petrosaurus mearnsi	B-3	*Heloderma suspectum*	N
		Holbrookia maculata	N
Snakes			
Crotalus enyo	B-8	*Crotalus molossus*	S-2
Crotalus mitchelli	B-10	*Crotalus atrox*[b]	S-2, B-2
Crotalus ruber[a]	B-8	*Crotalus tigris*	S-1
Bogertophis rosaliae	B-1	*Elaphe triapsis*	N
Eridaphis slevini	B-4	*Masticophis bilineatus*	S-2
		Micruroides euryxanthus	S-1
		Thamnophis cyrtopsis	N
		Chionactis palarostris	N
		Arizona elegans (Sonoran subsp.)	N

The notation S-*x* or B-*x* after the species name indicate the number (*x*) of Sonoran islands (i.e., Tiburón, San Pedro Nolasco, San Esteban, Pelicano, Farallón, Lobos, Patos, or San Pedro Mártir) or Baja California islands (all other gulf islands), respectively. An "N" indicates no islands. This table excludes narrow endemics such as *Cnemidophorus labialis* (Baja Calif.), *Phyllodactylus unctus* (Baja Calif.), *Sceloporus hunsakeri* (Baja Calif.), and others.
[a]Includes the insular endemic derivative *C. catalinensis*.
[b]Includes the insular endemic derivative *C. tortugensis*.

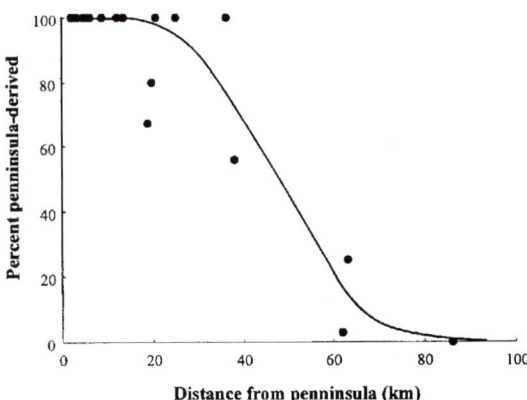

Figure 9.5 The percentage of an island's reptilian fauna derived from the Baja peninsula plotted as a function of the nearest distance from the island to the peninsula. There are only a few islands closer to mainland Mexico than to Baja California, so most islands are dominated by peninsular derivatives. The curve was drawn by eye.

dants of Baja California endemics. There are four interesting exceptions. The Baja California endemic gecko *Phyllodactylus noctiolus* is resident on the Sonoran island of Tiburón, and a derivative of the Sonoran rattlesnake *Crotalus atrox* is endemic to Isla Santa Cruz off Baja California. Despite its closest ancestors' locations on the other side of the gulf, populations of *Sator* exist on three islands in the Sea of Cortés that lie closer to the Baja peninsula. Previously *Sator* was considered an endemic genus to the Sea of Cortés, but due to taxonomic revisions, other noninsular species of *Sator* are now named on mainland Mexico (Wyles and Gorman 1978; Schulte et al. 1988; Murphy and Aguirre-Léon, chap. 8, this volume). Finally, based on mitochondrial DNA, it appears that the insular *Sauromalus varius* on San Esteban, which lies closer to Sonora, has its nearest relatives on Baja California islands (Petren and Case 1997; app. 8.1). The absence of more mainland Mexico derivatives on islands adjacent to Baja California suggests that many present-day island residents may have arrived after the Baja peninsula separated from the Mexican mainland from the nearest landmass.

Lawlor et al. (chap. 12) also find little evidence of faunal mixing for gulf island mammals, which reinforces the impression that mammalian colonization rates are generally quite low. Similarly, only one Sonoran bird, the curve-billed thrasher, gets as far as San Esteban, the most westerly Sonoran island.

Nestedness

Murphy and Aguirre-Léon have shown in striking fashion in table 8.3 (this volume) that presumed extinctions are not random across species in landbridge islands and on old, deep-water islands of presumed continental origin (see also chap. 10 for land birds). This was demonstrated by examining the rank-order incidence matrices. A pattern of nested subsets in species incidence occurs when smaller biotas contain a proper subset of the species in richer biotas. For a perfectly nested incidence matrix, all five-species islands, for example, would have the same five species, and all six-species islands would have these five species plus the same additional species, and so on. A perfectly nested incidence matrix would display an upper triangular form when islands, as columns, were ranked from most species-rich on the left to least species-rich on the right; and species, as rows, were ranked from most ubiquitous at the top to least at the bottom. The resulting rank-order incidence matrix forms a regular-progressing nested series. Murphy infers from the nested pattern that present-day species numbers are not maintained at a dynamic equilibrium because random species turnover would lead to more gaps in these totals. Such a result is explainable in terms of equilibrium theory if colonization and extinction curves are highly concave (MacArthur and Wilson 1967), which will result when species are different in their likelihoods of colonization and extinction (Diamond and Gilpin 1980). Whether species turnover is exactly zero or not is probably impossible to prove, but the high levels of reptile endemism on oceanic islands indicate turnover may be low. Reptiles on high-elevation habitat islands in western Arizona also show a regular ranked-order incidence matrix (Jones et al. 1985), which they believe results from selective extinctions and the inability of species to recolonize due to the inhospitable sea of arid habitats surrounding the mountain tops.

Departures from nestedness can be quantified in several ways based on the gaps in the rank-ordered incidence matrix. Atmar and Patterson (1993) and Wright et al.

Table 9.2 Nestedness for island subsets based on temperature

Island Set	Temperature	p Based on Monte Carlo
All	14.08	6.65e-28
Landbridge	5.97	1.37e-27
Landbridge without Tiburón	5.37	1.37e-30
Oceanic[a]	13.29	6.00e-10

[a]The oceanic island set here excludes satellite landbridge islands to larger oceanic islands.

(1998) reviewed these various indices and applied them to several different data sets. Many virtual and real island systems have been explored for patterns of nestedness because the pattern has conservation implications. Species that only occur in the most species-rich settings might serve as indicators of the presence of other species when the nestedness pattern is high. Also, a pattern of nestedness among islands means that several small islands would likely have fewer total species than a single large island of the same combined area. This is because the several small islands would likely have the same species composition, and new species would only be gained by going to still larger islands with more species. In a non-nested archipelago, the high species turnover from one small island to the next could produce a combined species count higher than a single large island.

Because the Sea of Cortés islands have two somewhat different mainland species pools, this archipelago, all else being equal, is expected to have a low nestedness index. However, because relatively few islands occur on the Sonoran side compared to the large number of islands closer to Baja California and dispersal distances are short, the archipelago could still be highly nested if the extinction order is relatively predictable. I calculated nestedness using the temperature (T) metric of Atmar and Patterson (1993) for various subsets of the islands; higher temperatures indicate lower nestedness (table 9.2).

We see significant nestedness at all levels, but landbridge islands are more nested than oceanic islands. This pattern was also found in general for a wide variety of other island systems where oceanic and landbridge islands could be compared (Wright et al. 1998). The T score of the Sea of Cortés landbridge islands is exceptional in being particularly low (i.e., high nestedness) relative even to other landbridge island groups. For the entire island set (which includes all the islands in apps. 8.2 and 8.3), the most idiosyncratic species, departing most from nestedness, is the kingsnake, followed by *Crotalus atrox* and *Sator*. The most deviant island is Tiburón, which contains several Sonoran species that do not make it to other islands. For mammals, oceanic islands alone do not show a significant T (chap. 12).

Incidence Curves

Given a set of species differing in their colonization and extinction probabilities and a heterogeneous collection of islands, several types of patterns may emerge. Certain species may be found only on the largest, most species-rich islands; other species will occur even on small, species-poor islands. For New Guinea birds, Diamond (1975) found a set of species (supertramps) that occur only on the smallest or most remote species-poor islands. He attributed their absence from larger islands to competitive exclusion by other species. Diamond presented these patterns in the form of incidence functions. Islands are grouped into categories based on their size or on the total number of species they possess. For each species, the percentage of islands occupied in each island-size category is plotted against these island categories (also see chaps. 4, 10). Incidence functions thus illustrate graphically how different species respond to differences in island size and species richness.

Incidence functions for the more common reptiles were prepared by dividing islands into the categories based on their number of reptilian species they contain. For this analysis I have lumped landbridge and oceanic islands. There is only one island with more than 21 reptile species, Tiburón, with a different source pool; hence this island is excluded. The incidence curves for seven of the most common lizards on gulf islands are shown in figure 9.6. This family of incidence functions forms a staggered sequence. *Uta* occurs even on species-poor islands, while at the other extreme the incidence of *Callisaurus draconoides*, *Urosaurus microscutatus*, and *Cnemidophorus hyperythrus* rapidly approaches zero on all but the largest, most species-rich islands. The final dip in the incidence function for *Uta* stems from its absence on Cerralvo, one of the five islands in the largest size category. This island is occupied by the endemic species *Sator grandaevus*. *Uta* occurs on every major island in the Sea of Cortés except those three islands occupied by *Sator*. The possible reasons for this negative association are discussed later. For now, notice that the absence of *Uta* on Cerralvo does not stem from diffuse negative interactions with a variety of other reptile species because it is present on the two most species-rich islands of Tiburón and San José, with 22 and 21 reptile species, respectively. The incidence functions for the most common snakes (fig. 9.7) are more erratic (nonmonotonic). This may reflect a greater stochastic element in snake colonizations and extinctions, but the possibility that island records are incomplete cannot be excluded.

Unlike the New Guinea birds studied by Diamond (1975), none of these reptiles exhibits a supertramp strategy characterized by high incidence on the most species-poor islands but low to zero incidence on species-rich islands. The kingsnakes, *Lampropeltis*, as mentioned earlier, are apparently absent from all of landbridge islands in the Sea of Cortés; because these islands include the largest and most species-rich, we see declines in the incidence of kingsnakes at higher species number. Those reptiles of Baja California ancestry occurring on only one or at most two islands in the entire gulf inevitably occupy islands that are large and species rich (table 9.3). All of these species typically occur at low population densities (see Case 1975).

Some species are not found on a single gulf island even though they are wide ranging on one or both of the adjacent mainlands. The desert night lizard (*Xantusia vigilis*) and the coast horned lizard (*Phrynosoma coronatum*) range all the way to the cape on the Baja peninsula. Both species are habitat specific, generally found in more mesic habitats (*P. coronatum*) on the west. Even in areas on the peninsula with apparently prime habitat, the desert night lizard can be rare or absent (Grismer et al. 1994). On the other side of the gulf, the gila monster (*Heloderma suspectum*), the giant spotted whiptail (*Cnemidophorus burti*), and the lesser earless lizard (*Holbrookia maculata*) also do not occupy even the large landbridge island of Tiburón. The Mexican

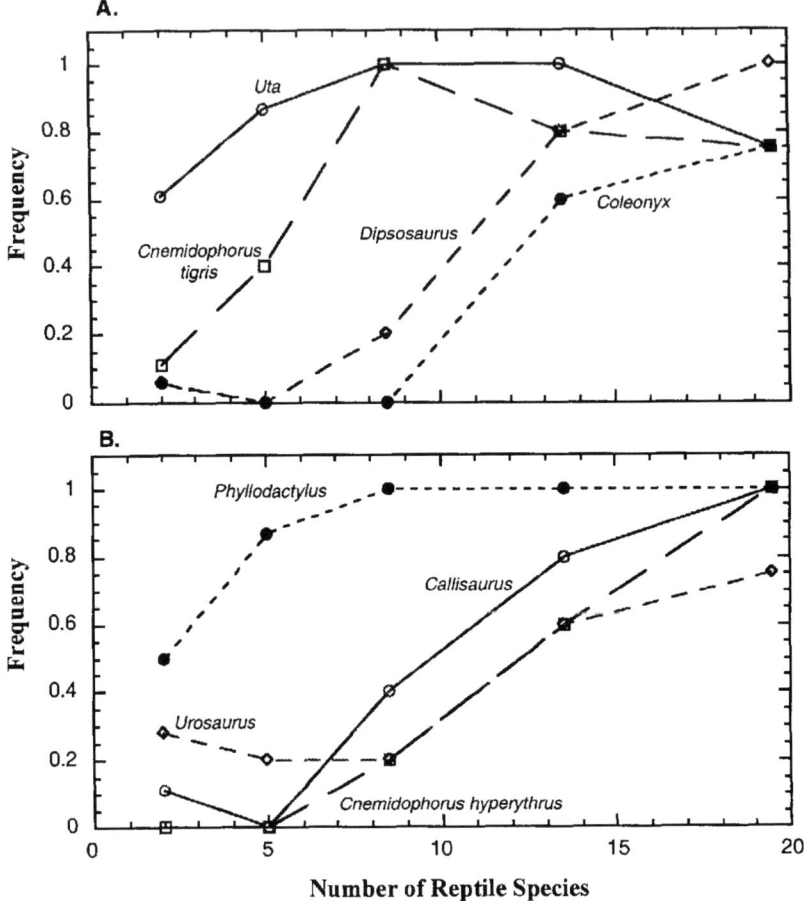

Figure 9.6 Incidence functions for common lizards on islands in the Sea of Cortés. Islands are divided into different categories based on the number of reptilian species they contain. The frequency of islands in each category that a lizard occupies is plotted on the ordinate. To avoid visual clutter, the 8 taxa are divided into two groups of 4 each in subfigures A and B. Islands were divided into the following categories based on their number of reptile species: 1–3; 4–6; 7–10; 11–15; and 16–21 reptilian species. There are 21, 11, 4, 4, and 5 islands, respectively, in each of these categories (apps. 8.1, 8.2).

lizards *Heloderma* and *C. burti* are also uncommon and are usually found in mesic desert arroyos, habitats noticeably lacking on most islands. Only *Holbrookia maculata* presents an anomaly. It is common in coastal regions of Sonora, Mexico (e.g., Kino Bay, Bahía San Carlos), so its absence from Tiburón is peculiar. There is nothing noticeably unique about its habits or life history that might account for its failure to occupy Sonoran islands. The similar but larger *Callisaurus draconoides* occurs on Tiburón and on 11 other gulf islands. Perhaps *Holbrookia* has only recently penetrated coastal mainland Mexico from a formerly more northern range, although I know of no fossil evidence to support this ad hoc conjecture.

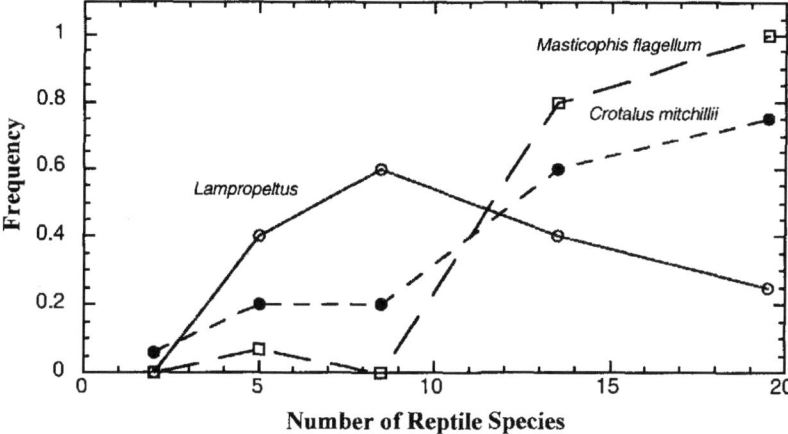

Figure 9.7 Incidence functions of common snakes on islands in the Sea of Cortés. Islands are divided into different categories based on the number of reptilian species they contain. The frequency of islands in each category that a lizard occupies is plotted on the ordinate. To avoid visual clutter, the 8 taxa are divided into two groups of 4 each in subfigures A and B. Islands were divided into the following categories based on their number of reptile species: 1–3; 4–6; 7–10; 11–15; and 16–21 reptilian species. There are 21, 11, 4, 4, and 5 islands, respectively, in each of these categories (apps. 8.1, 8.2).

Table 9.3 Island occurrence of reptilian species that inhabit Baja California and occur on only one or two islands in the Gulf of California

Species	Islands	No. of Reptilian Species on Island
Lizards		
Coleonyx switaki	San Marcos	22
Crotaphytus group[a]	Ángel de la Guarda	15
Gambelia wislizenii	Tiburón	27
Petrosaurus thalassinus	Danzante	16
	Espíritu Santo	19
Snakes		
Bogertophis rosaliae	Danzante	16
Masticophis aurigulus	Espíritu Santo	19
Pituophis catenifer	Tiburón	27
	San José	22
Rhinocheilus group[b]	Cerralvo	18
Sonora semiannulata	San Marcos	22
	San José	22
Tantilla planiceps	Carmen	16

[a]Includes *C. dickersonae* and *C. insularis*.
[b]Includes *R. etheridgei* and *R. lecontei*.

Among the snakes, the vine snake (*Oxybelis aeneus*), the Sonoran shovelnosed snake (*Chionactis palarostris*), the green rat snake (*Elaphe triapsis*), and the black-necked garter snake (*Thamnophis crytropsis*) are not found on a single Sonoran island. These snakes prefer more mesic habitats not generally found on any of the islands. *Chilomeniscus palarostris* is not a habitat specialist but has a restricted range, which only abuts the gulf coast in a small region of Sonora.

Species Associations and Distributional Anomalies

If species interact strongly, then we might be able to detect distributional associations (either positive or negative) between species pairs. Unfortunately, such associations can also arise for a number of other reasons, and statistical problems are severe (for reviews, see Gotelli and Graves 1996; Gotelli 2000). Case (1983b) compared the frequency of different species sets observed on gulf islands with respect to the average niche overlap, α, for that set of species. If island lizard communities are formed without reference to ecological overlaps between the species, then, on average, one-half of all observed island communities should have mean (within-community) niche overlaps falling above the median α and one-half should fall below. Instead, there were many more species sets of low mean overlap than expected. These results imply that lizard species are assorted into more niche-dissimilar sets than would be expected by chance alone. Moreover, there is a statistical preponderance of certain low niche-overlap species sets compared to other possible species sets with equally low overlaps. The chief reason behind this is that the most common sets of species involve *Uta*, *Sauromalus*, and *Phyllodactylus*. *Uta* is a small, diurnal, insectivorous habitat-generalist. *Phyllodactylus* is a small, insectivorous, nocturnal, rock-dwelling gecko. *Sauromalus* is a large, diurnal, rock-dwelling herbivore. Regardless of the niche categorizations used, any reasonable measure of niche overlap among these lizards is low. Moreover, these species have living requirements that tend to be satisfied even on very small islands. Hence the observed pattern may simply reflect an overrepresentation of the particular niche and habitat conditions on islands for these species, rather than negative interspecific interactions between these species and other species less ubiquitous on islands.

In some cases interspecific interactions seem important. One of the most curious distributional patterns among reptiles in the Sea of Cortés involves the lizard genus *Sator*. *Sator* only occupies three islands, but on these islands there are no *Uta*, *Urosaurus*, or *Sceloporus*. In fact, Santa Cruz and San Diego are occupied by no other diurnal, insectivorous lizard, even though the typical island of the same area would contain at least two or three such lizards. On the much larger island of Cerralvo, *Callisaurus* occurs along with *Sator* but is remarkably rare and seems restricted to sand dune areas at the southern end of the island (Soulé 1966; T. J. Case, unpublished observation). *Uta* occurs on each of the 25 or so major islands in the Sea of Cortés, except the 3 islands occupied by *Sator*. Using the hypergeometric distribution (see Wright and Biehl 1982), *Uta* and *Sator* are expected to share 2.6 islands. The χ^2 value for sharing zero islands is 2291, which is significant at $p < 10^{-5}$. Even after allowing for multiple comparisons given all the other species pairs, this is a significant negative association (see Case 1983a).

Many species pairs have a positive spatial association. The positive spatial associations between species pairs arise not because of pervasive mutualism, but simply because some species occur on many more islands than others, and some islands containing these lizards are much more species rich than others. Randomly formed incidence matrices with these constraints yield the same result. As the taxa under consideration become more restricted, these statistical effects are diluted and particular ecological interactions surface.

Few of the individual χ^2 values for the 210 lizard species pairs are significant (about 5%). The species pair *Uta–Sator* yields the highest χ^2 value with $p < 10^{-5}$. Because there are 210 lizard (and 990 reptile) species pairs, a deviation this extreme is not expected to occur even once by chance alone. Other highly significant χ^2 values indicating distributional segregation occur between *Sator* and *Urosaurus microscutatus*, *Sceloporus orcutti*, or the *Sceloporus magister* complex. Of the species pairs that share more islands than expected, none yield χ^2 values corresponding to $p < 10^{-2}$; thus their positive association may simply be due to chance.

Competition severe enough to result in distributional exclusions is often accompanied by overt interference between species members (Case and Gilpin 1974; Case et al. 1979). *Sator* are fiercely aggressive in the field. Stomach contents of large males taken from Santa Cruz contain, on occasion, smaller individuals of the same species. In a pilot study I paired *Sator* in large laboratory cages with smaller *Uta stansburiana* and *Urosaurus microscutatus*. In half of these experiments, the larger *Sator* ate the smaller lizard; in the remaining cases, the *Sator* chased and attacked the smaller lizard until the latter buried itself beneath the sand and remained there for the duration of the observation period (sometimes 2 days). When *Sator* are paired with *Uta palmeri* of roughly equivalent size, sustained attacks are not evident, nor are they in pairings of small *Uta* with *Urosaurus* (Case 1983a). Thus *Sator* aggressiveness may account for the absence of small phrynosomatids from *Sator* islands, but it cannot account for the absence of larger species such as *Sceloporus magister* or *S. orcutti*. The absence of *Sator* from the many other *Uta*-inhabited islands in the Sea of Cortés and the mainland may be due to the failure of *Sator* to disperse to these areas.

The herbivorous lizard guild cannot be interpreted as easily as that of the insectivorous diurnal lizards. The distributions of *Sauromalus* and perhaps *Ctenosaura* have been altered by humans. In the absence of native mammalian herbivores of most of the islands (e.g., deer, hares, rabbits), the reptilian herbivores often reach abundances far exceeding those commonly seen on the mainland. *Dipsosaurus* occupies 9 of the 25 major islands in appendix 8.3, *Sauromalus* occurs on 18, and *Ctenosaura* is found on only 3 islands. All three species are found on the mainland of both sides of the gulf. The diet of *Petrosaurus* on Danzante and Espíritu Santo contains nearly 50% plant material by volume. The insular endemics *Sator* and *Uta palmeri* also consume large portions of plant material. The distribution pattern of the herbivorous or semiherbivorous species does not suggest that interspecific competition has played an important within-guild role. Although no island contains all three specialized herbivores, eight islands have both *Sauromalus* and *Dipsosaurus* (one of these, Tiburón, is also inhabited by the desert tortoise). *Ctenosaura* and *Sauromalus* occur together on San Esteban, and their overlap in diet there is high, although ctenosaurs are more arboreal (Case 1982). The endemic *Sauromalus varius* (see fig. 16.1) is similar in maximum body size (≈ 2 kg) to that of the sympatric *Ctenosaura conspicuosa*.

The genus *Sauromalus* displays some distributional anomalies, suggesting humans may have played a recent role in dispersing individuals from island to island. *Sauro-*

malus hispidus occurs on Ángel de la Guarda and all its satellite islands, which were probably connected to the main island during the last glacial maximum 6,000–14,000 years ago. *Sauromalus hispidus* also occupies San Lorenzo Norte and San Lorenzo Sur, which although probably not connected to Ángel in the Pleistocene, were certainly much closer than they are today and were connected to each other less than 14,000 years ago. More odd, however, is that *Sauromalus hispidus* is found on a number of small landbridge islands in Bahía de los Ángeles, Baja California (e.g., Smith, La Ventana, Piojo, and Cabeza de Caballo). No detectable morphological or chromosomal divergence is evident between these populations and those on Ángel or the San Lorenzos (Robinson 1972), and mtDNA haplotypes are indistinguishable (Petren and Case 1997). Mainland *S. obesus* is absent from these landbridge islands and must have become extinct some time after the land connection severed. These landbridge islands did not have Pleistocene land connections with either Ángel or the San Lorenzos. The presence of the insular endemic *S. hispidus* and the absence of any introgression with *S. obesus* argues for a very recent introduction, perhaps by Seri Indians or native Mexicans who occasionally eat the lizards (Felger and Moser 1985, Nabhan, chap. 15, this volume).

The absence of a *Phyllodactylus* species from San Pedro Mártir and a number of minor landbridge islands in appendixes 8.2 and 8.3 may be due to insufficient collecting efforts. These islands possess adequate island rocky habitat for these ubiquitous geckos.

Body Size, Life Histories, and Social Behavior

About 25–35% of the reptiles on oceanic islands in the Sea of Cortés are endemic at the species level. Many of these endemics are strikingly different in morphology, life history, and behavior from their mainland relatives. As yet, only a few studies have adequately documented these differences, let alone explored their adaptive significance. In this section I review some of the most interesting and best studied examples. A syndrome of characteristics is often associated with an insular environment, including substantial increases (or sometimes decreases) in body size compared to mainland relatives, large population size, relaxed predator wariness, reduced reproductive effort, usually manifested as reduced litter or clutch sizes, and relaxed territoriality (Carlquist 1965; Gliwicz 1980; Stamps and Buechner 1985). Recurring patterns like these presumably result from the common environment shared by lizards on remote islands—a rugged terrain, loss of competitors and predators, and sometimes a more moderate climate.

Because body size is so intimately related to all aspects of an animal's ecology, selective explanations for body size variation are multifarious. For purposes of discussion I consider four broad and somewhat overlapping adaptive explanations for body size variation.

1. Models of optimal body size predict that body size should be directly related to food availability (Schoener 1969; Case 1978a; Brown et al. 1993). Factors that might increase food availability on islands and thus favor insular gigantism are the loss of competitors, a more benign climate, or more freedom in foraging activities because of a reduced threat from predators.

2. Reptiles as predators may respond not only to changes in overall food abundance but also to shifts in the size spectrum of this food. The loss of large-sized competitors (on an island) creates both more food and generally larger-sized food, thus favoring larger optimal body sizes. The loss of small-sized competitors does not necessarily favor a smaller body size; although the prey size distribution may now be shifted or skewed toward smaller-sized prey, which are more efficiently handled by smaller predators, the total abundance of food will also be greater, potentially overriding selection for small size.

3. Looking at the other side of the coin, reptiles are prey for raptors, mammalian carnivores, and other reptiles. The size distribution of these potential predators may shift from mainland to island situations. Usually the largest predators take the largest prey, yet these predators are often the poorest island colonizers. The predators of reptiles that do reach islands are often the smaller species that specialize on small prey. Hence, on islands there may be a selective premium for more rapid growth and large size to escape this window of high predator vulnerability. This premium is expected to be particularly intense in animals such as reptiles with indeterminate growth. Here clutch size increases with body size, yielding an extra payoff for large size in females once the risk of predation is removed.

4. Empirically, many insular lizard populations maintain densities greater than those of relatives on the mainland (Case 1975; Schoener and Schoener 1980; Stamps and Buechner 1985; Losos 1994). Regardless of whether the root of this phenomenon lies in reduced predation or in greater resource production, there are a number of potential social consequences that affect optimal body size. At least for polygamous/promiscuous species, intraspecific competition among males for females may become intensified. Because large males are usually socially dominant, hold larger or more productive territories, and mate with more females (Brattstrom 1974; Stamps 1977; Hews 1990), sexual selection may lead to large male body sizes under higher population densities.

These four hypotheses are neither all-inclusive nor mutually exclusive; parts of each may apply. These factors also probably interact. For example, relaxed predation may increase population sizes and thus the potential for interspecific competition. Fewer predators may allow more access to food resources otherwise unavailable. Sexual selection for male characters may be prohibited in a predator-rich environment.

Insular changes in reproductive effort have been studied more in birds than in reptiles. Among birds there are two widely observed geographic trends involving intra- and interspecific variation in clutch size (Cody 1966; Lack 1968): Clutch size is often greater in populations or species residing at higher latitudes, and clutch size of island populations at temperate latitudes is often lower than in closely related mainland forms. The causes of these trends are debated, but one favored set of explanations is based on the implicit trade-off between survival of adults and the juveniles under their care on the one hand, and reproductive effort on the other. Unlike birds, most reptiles lack postlaying parental care. Because the postlaying parental energy and time expenditure on homeothermic young are typically many times greater than the costs of simply producing and carrying eggs (Case 1978b), these trade-offs should be less severe in reptiles, and we might expect these clutch-size trends to be less pronounced.

In reptiles, clutch size is often positively correlated with body size. Because body size is much more labile in reptiles than in birds, interpopulational differences in clutch size may result from differences in body size rather than from differences in the allocation of energy for reproduction versus that for maintenance, growth, or survival of adults. Thus any comparison of clutch size differences among reptile populations must be adjusted for differences in body size.

The existence of an insular reduction in clutch size for reptiles has not been adequately reviewed. Certain endemic island species such as *Conolophus*, *Amblyrhynchus*, and *Tropidurus albemarlensis* on the Galápagos or *Galliotia* on the Canary Islands have much smaller clutch sizes than one would expect based on their body size (Tinkle et al. 1970; Arnold 1973). Other less dramatic examples include *Lacerta* on Mediterranean islands (Kramer 1946) and the cottonmouth snake (*Agkistrodon*) on the Florida Keys (Wharton 1966).

In the following section I examine body size, insular clutch size trends, and other life-history attributes for the three most thoroughly studied lizard genera in the Sea of Cortés: *Uta*, *Cnemidophorus*, and *Sauromalus*.

Uta

Differences in body size between populations of *Uta* in the Sea of Cortés can be dramatic. These size differences attracted the attention of Soulé (1966), who sought an adaptive explanation for their variation. He found that for 10 oceanic islands, *Uta* body size was inversely related to the square root of the number of other sympatric iguanid lizard species. Because *Uta* is one of the smallest diurnal lizards on the mainland, this result is consistent with both hypotheses 1 and 2 above.

Dunham et al. (1978) reanalyzed *Uta* body-size patterns. They performed a multiple regression analysis that attempted to relate *Uta* body size to the number of sympatric competing species, latitude, island area, number of plant species, and so on. Their methodology differed from that of Soulé (1966) in two important ways. First, they chose mean snout-vent length (SVL) as the dependent variable rather than a measure of maximum size. Mean body size is a poor choice because it is extremely sensitive to changes in population age structure. Second, Dunham et al. included very recent landbridge islands (with low levels of endemism, see section on levels of endemism), and old oceanic islands (with high levels of endemism) within the same analysis. In spite of this noise, there was still a highly significant negative correlation between *Uta* size and the number of competing species present on the island. This correlation persists even if *Uta palmeri*, the largest *Uta* on San Pedro Mártir, is removed. Because the independent variables were highly correlated, Dunham et al. cautioned against any definitive conclusions.

With *Uta* or with most other lizards, it is difficult to get an accurate measure of food availability so that one might test hypothesis 1. Over the years, I applied a combination of insect sampling techniques to census available arthropods because no one method seemed to yield a collection of prey matching that actually consumed by the lizards (see Case 1979, 1983a). More fundamentally, however, it is not possible to extrapolate from present differences in apparent arthropod abundance between islands to those in the evolutionary past. What we desire is a measure of prey abundance averaged over thousands or millions of years, and this is clearly beyond reach.

Another problem confounding interisland comparisons is the obvious change in feeding habits accompanying the large size of *Uta* on certain islands. Most of these populations display a definite shift toward eating more plant material, particularly flowers and fruits, but to a lesser extent foliage (Case 1983a). The endemic *Uta palmeri* (as well as the sympatric *Cnemidophorus martyris*) on San Pedro Mártir consume substantial amounts of fish scraps around seabird nests. Also associated with these nests is a unique arthropod fauna of bird mites, lice, ticks, and debris-breeding flies eaten by both lizards. Small islands with relatively high ocean edge compared to interior area, and particularly those used by breeding sea birds, have high ocean inputs to food webs, and some of these resources are utilized by lizards, contributing to their high densities (see chap. 13). *Uta* on the small islands off Bahía Gonzaga forage intertidally for isopods and develop large nasal salt glands that enable them to excrete excess salts (Grismer 1994; Hazard et al. 1998). These diet shifts make it ludicrous to apply standard insect sampling techniques, even though these techniques might yield adequate measures of food availability for lizards on larger islands.

On islands, the maximum SVL of male of *Uta* tends to be about 10–20% larger than that of females (Case 1983a). I found that the degree of size difference between the sexes, however, did not correlate with the maximum male size. From this I inferred that differences in the extent of sexual selection for male size probably were not driving the body-size differences between islands. However, since most secondary sexual characteristics lie on autosomal chromosomes, general phenotypic characters, such as body size, have high genetic correlations between the two sexes. Sexual selection for male size would lead to a correlated response in females. Thus it is not clear that the expectation would be met even if sexual selection was driving between-island body size differences.

Wilcox (1980b) found average densities of the giant endemic *Uta palmeri* to be 1300/ha, substantially higher than mainland densities of *Uta stansburiana*. Adult males maintain uniformly sized, nonoverlapping territories, whereas females and juveniles are more variable in space utilization and have extensively overlapping home ranges. Many females, juveniles, and nonterritorial males will share the territories of breeding males. Resident females spend most of their activity in small areas within a single male's territory. Wilcox's observations indicate that males appear to be defending courting arenas rather than feeding territories. The most intense male–male aggression occurs when an intruding male is courting a female, rather than simply passing through or feeding.

Hews (1990) later studied mate choice and sexual selection in *Uta palmeri*. She assessed male territory quality by counting the number of active booby nests and chicks in each territory and found that this index was also highly correlated with the number of arthropods caught in sticky traps and the mass of regurgitated fish near nests. Hews then related territorial quality and morphology to mating success to calculate selection differentials using the method of Lande and Arnold (1983). Territory quality and head depth (but not SVL, head width, mass, or jaw length) seemed to be under significant directional selection. If head size is heritable and genetically correlated with overall SVL, then direct selection on head size could still lead to a correlated evolutionary increase on SVL. Hews (1993) also found that *Uta palmeri* densities were correlated spatially with bird nest densities. The experimental addition of food (in the form of mashed fish) to some areas resulted in higher densities of females (but not males) and increased courtship rates for males.

Table 9.4 presents data on *Uta* clutch sizes taken from Wilcox (1980b), supplemented with additional data from Case (unpublished) and Wilcox (pers. commun.). In

Table 9.4 *Uta* clutch sizes

Island	N	Mean Clutch Size (2 SE)	Mean SVL of Largest Third of Population	
			Female	Male
San Pedro Mártir[a]	15	3.03 (0.18)	61.6	67.1
Partida Norte[a]	—	—	46.6	55.1
Salsipuedes[a]	39	2.76 (0.18)	49.5	57.0
Rasa[a]	6	2.80 (0.18)	54.0	56.5
San Esteban[a]	24	2.33 (0.15)	46.6	51.9
San Ildefonso	18	2.37 (0.12)	45.5	52.8
Tortuga[a]	11	1.72 (0.19)	48.0	55.6
Carmen	22	2.55 (0.13)	47.9	53.7
Santa Catalina[a]	10	1.62 (0.18)	47.0	52.4
Monserrat[a]	19	1.71 (0.19)	47.0	51.2
Danzante	24	2.51 (0.12)	45.2	55.0
Mainland				
Five southern locations, Arizona and New Mexico[b]	13–37	3.31–4.48	~44–51	~50–55
Baja California	17	—	45.7	49.4

Most of these data were gathered by Bruce Wilcox. SVL, snout vent length.
[a]Oceanic islands.
[b]Parker and Pianka (1975).

spite of the large body size of many island *Uta* populations, their clutch size is typically small compared to that of the mainland *Uta* populations studied by Parker and Pianka (1975). Eggs of *Uta palmeri* hatch in early September, and the initial growth rate of hatchlings averages about 0.5 mm snout-vent per day (Wilcox 1980b), extraordinarily fast compared to mainland *Uta*. After this initial rapid phase, however, growth greatly decelerates. Age at first reproduction is about 10 months in females, but males do not reach maturity until their second year. This contrasts with mainland *Uta*, where both sexes typically mature within their first year. *Uta palmeri* initiates mating and reproduction later in the season than populations (island or mainland) elsewhere in the gulf region, presumably so that reproductive activity coincides with the period of peak sea-bird nesting activity on the island beginning in May.

The decreased clutch size of *Uta palmeri* is accompanied by an increase in individual egg weight. The ratio of the total clutch weight to body weight (reproductive effort, RE) and egg weight to body weight (expenditure per progeny, EPP) are 0.23 and 0.07, respectively, in *Uta palmeri*. Both values, but particularly EPP, fall in the upper range of values calculated for mainland *Uta* by Parker and Pianka (1975). Thus, egg size has increased even faster than body size so that RE is relatively large in spite of a reduction in clutch size. By all indications, female *Uta palmeri* produce only one or two clutches per breeding season, which is fewer than the three or four clutches produced by the southern *Uta* populations studied by Parker and Pianka (1975). Hence the total RE per breeding season is probably lower in *Uta palmeri* females than in mainland *Uta*.

Large egg size may be favored under such conditions because of the competitive advantage that a larger hatchling has in high density populations where agonistic interactions are frequent and juvenile and egg mortality is potentially severe. Because adult survivorship is probably high, rapid growth would enable the juveniles to pass more quickly through periods of greater risk. A larger egg size should produce larger hatchlings with higher initial growth rates.

Cnemidophorus

While the *Uta* on San Pedro Mártir are relative giants, the *Cnemidophorus* on this same island are relative dwarfs. Body-size shifts in *Cnemidophorus* are complicated by what appears to be a classical case of character displacement involving *C. tigris* and *C. hyperythrus* (and their insular derivatives).

In the southernmost cape region of Baja California, a large *C. tigris maximus* (synonym, *C. maximus*) occurs sympatrically with the much smaller *C. hyperythrus* (fig. 9.8). (Because males and females differ in maximum size by a factor of only 1.06, which tends to be invariant between species and locations, the body-size figures and statistics lump both sexes together.) Proceeding north up Baja California, *C. tigris* becomes somewhat smaller in the form of *C. t. rubidus* and later *C.t. tigris*, both also sympatric with *C. hyperythrus*. In southern Sonora, *C. tigris aethiops* is also relatively small and is sympatric with a different subspecies of *C. burti*. Finally, proceeding south into northern Sinaloa, Mexico, *C. burti* drops out, and *C. costatus* overlaps the range of *C. tigris* along the Sonora–Sinoloa border. *C. costatus* is a large species and at the point of overlap; *C. tigris* is quite small—as small, in fact, as the *C. hyperythrus* on the Baja California cape directly across the Sea of Cortés.

Northern Sinaloa and the cape of Baja California are remarkably similar in vegetative structure; both consist of a short, semideciduous, subtropical thorn forest. Both places have similar weather patterns and are at similar latitudes (Shreve 1951; Hastings and Turner 1969; Hastings et al. 1972), yet in one place (the Baja cape), *C. tigris* has the largest body size and in the other (northern Sinaloa) has the smallest body size of any locality throughout its geographic range.

Although climatic and physical factors cannot readily account for the observed body size pattern in *C. tigris*, an explanation based on character displacement from congeneric competition can. That is, by hypothesis 2, the body size of *C. tigris* may be evolutionarily adjusted to competitive pressures from the sympatric congeners it meets.

This hypothesis may be examined further by comparing body sizes of *C. tigris* on islands in the Sea of Cortés or off the west coast of Baja California. On oceanic islands where the smaller *C. hyperythrus* is absent, *C. tigris* is consistently smaller in size relative to locations (island or mainland) where the two species occur sympatrically. On Ángel de la Guarda and Partida Norte this size difference is small, but on islands like San Pedro Mártir, Santa Catalina, the San Lorenzos, or San Pedro Nolasco it is great indeed. *C. hyperythrus* or insular endemic derivatives occur allopatrically on only two islands, Monserrat and Cerralvo. On Monserrat *C. hyperythrus* reaches a maximum size 5 mm larger than mainland relatives. On Cerralvo, the endemic *C. ceralbensis* is nearly 20 mm larger in snout vent, thus approximating the size of *C.*

244 The Biological Scene

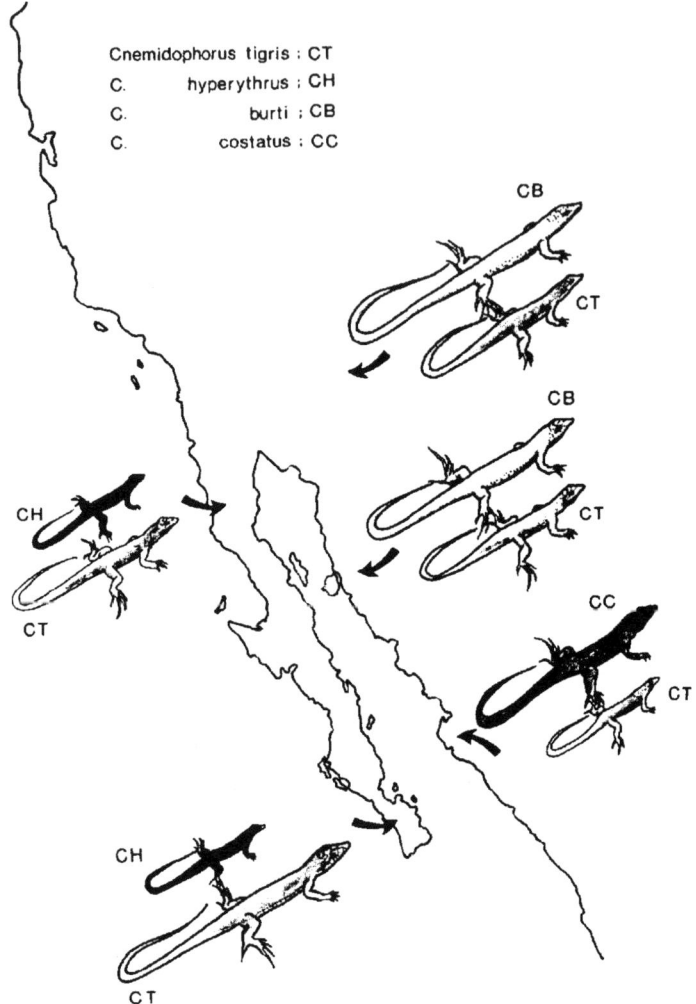

Figure 9.8 Relative body sizes of four species of lowland desert *Cnemidophorus* in the southwestern United States and Mexico. See text for details.

tigris on oceanic single-species islands. Thus, on oceanic islands, with only a few exceptions, we see a pattern similar to that of the anoles of the Lesser Antilles (Schoener 1970; Giannasi et al. 2000). Single-species islands have an intermediate-sized *Cnemidophorus*, while two-species islands have a small form syntopic with a large form. *Cnemidophorus tigris* also occurs alone on a few landbridge islands (Danzante, Smith, and islands in the Pacific). On these there is no marked reduction in body size (Case 1979).

Radtkey et al. (1997) inferred the evolutionary history of body size change and patterns of colonization in *Cnemidophorus* using a phylogeny based on nucleotide sequences of the cytochrome b gene. Their phylogenetic analysis suggested that (1) the oceanic islands have been colonized at least four times from mainland sources; (2) the monophyletic group of dwarf *C. tigris* on the oceanic islands San Esteban, San Pedro Martír, San Lorenzo Norte, and Salsipuedes shows one instance of character relaxation to a smaller size, followed by between-island colonization among these islands; (3) *C. tigris* shows an independent character relaxation on Isla Santa Catalina; (4) the large size, relative to *C. hyperythrus*, of *C. ceralbensis*, on Cerralvo, represents a retention of ancestral size; (5) *C. tigris* probably had two independent colonizations (one older and southern and one recent and northern) of the Baja peninsula, followed by two independent increases in body size; (6) the ancestors of *C. hyperythrus* colonized Baja California from the cape region and experienced a reduction in body size; and (7) *C. hyperythurs* has probably only recently expanded its range into northern Baja California and southern California. The latter observation is consistent with the absence of Pleistocene-Holocene fossils of this species in southern California at sites that abundantly contain most other sympatric species (Holman 1995) and its absence even from landbridge islands off Baja California Norte but not Baja California Sur.

Case (1979) measured niche overlap between *Cnemidophorus tigris* and *C. hyperythrus* along four niche axes, prey type, prey size, habitat, and along time and temperature of activity. Overlap is high along all niche dimensions, and overlap for prey sizes is influenced by the body size of the lizards, which in turn is geographically and ontogenetically variable. The overlap of a given large-sized lizard on a small lizard is usually greater than that of the small lizard on the larger. This follows because larger animals eat more prey per unit time and consume a greater diversity of prey sizes.

Using this information and an estimate of the relative abundance of various-sized prey, Case (1979) predicted the optimal body sizes of one- and two-species *Cnemidophorus* guilds. These predictions are closely in line with the observed sizes of *Cnemidophorus* lizards on the Sea of Cortés islands. Moreover, the quantitative changes in the size of prey consumed on different islands in the presence and absence of the other *Cnemidophorus* species closely matched the predictions, leading Case to suggest that competition for different-sized prey may account for the observed character displacement between these two species.

Prey-type overlaps between *Cnemidophorus tigris* and *C. hyperythrus* do not differ between areas of sympatry and allopatry. The prey-size and body-size distributions for these species, however, are much more divergent in sympatry than in allopatry (Case 1979).

As noted, *Cnemidophorus tigris* and its insular derivatives are usually dwarfed on oceanic islands in the Sea of Cortés. From that fact alone we expect their clutch size to be reduced. However, even beyond allometric constraints, *C. martyris* has a small mean clutch size (mean = 1.2 eggs; fig. 9.9) and relatively large egg size (12.0×19.5 mm). Based on the absence of hatchlings and subadults in his August 1967 sample, Walker (1980) believes *C. martyris* attains reproductive maturation in less than 1 calendar year. The latitudinal position of San Pedro Mártir and the presence of gravid females in August also indicate the production of multiple clutches per breeding season. Walker and Maslin (1969) reported a mean clutch size of 1.85 ($N = 48$) and a mean egg size of 9.2×16.4 mm for the dwarf *C. bacatus* on San Pedro Nolasco. Because these authors did not report the SVLs of the females involved, these clutches

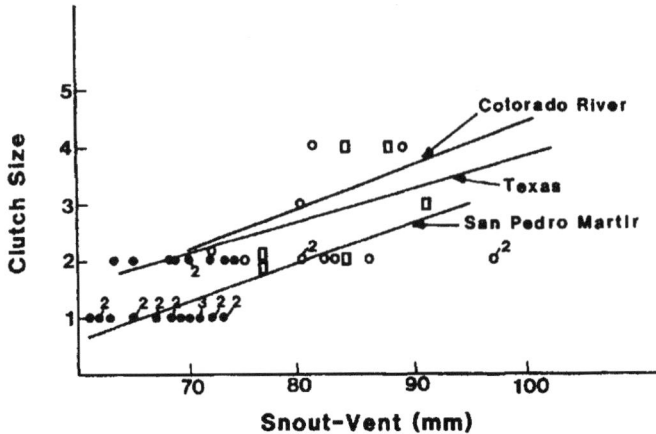

Figure 9.9 Clutch size of *Cnemidophorus tigris* or insular derivatives as a function of snout-vent length (mm). Points are the Baja California and Sea of Cortés individuals (the number of multiple points at the same position is indicated): (●) *C. martyris* (data from Walker 1980) on San Pedro Mártir, and other oceanic islands in the Sea of Cortés (present study); (○) landbridge island populations in the Sea of Cortés; (□) mainland population in Baja California. Superimposed are regression lines for *C. tigris* populations from the lower Colorado River (from Vitt and Ohmart 1977), western Texas (from Schall 1978), and for *C. martyris* (Walker 1980). Populations of *C. tigris* on San Pedro Mártir are dwarfs, and individuals from this island have lower than expected clutch sizes for their snout-vent length.

are not plotted in figure 9.9. Yet since the mean SVL of adult *C. bacatus* females is about 66 mm, an insular reduction in clutch size (beyond that due to smaller adult size) is not indicated.

Peninsular *Cnemidophorus hyperythrus* have a mean clutch of about 2.3 eggs (Bostic 1966). Unfortunately, there are few island clutches to compare to this figure. The scant data suggest that clutch size is not strongly reduced in island populations of *C. hyperythrus*. I have found only one gravid *C. ceralbensis* in museum specimens. Its clutch contained a single egg. This island endemic may be one of the oldest lizard species in the region (Savage 1960; Murphy and Aguirre-Léon, chap. 8, this volume), and it is morphologically very divergent; further studies of its ecology and life history would be revealing.

Sauromalus

Perhaps the best example of insular gigantism in the Sea of Cortés involves the lizard genus *Sauromalus* (the chuckwallas) (Shaw 1945). *Sauromalus hispidus* occupies islands in the Ángel de la Guarda block. *S. varius* occurs on San Esteban (see fig. 16.1) and has been introduced to Isla Pelicano (Alcatraz) off Kino Bay, Sonora, where it forms a hybrid swarm with *S. hispidus* and *S. obesus*. The typical upper-decile SVLs (UDSV) for these three species as well as for other island populations are given in Case (1982). In terms of body weight *S. obesus* on Baja California rarely surpasses

300 g, while *S. hispidus* reaches a maximum of about 1400 g and *S. varius* may exceed 1800 g.

Most islands farther south in the Sea of Cortés are also inhabited by endemic species or subspecies of *Sauromalus*, yet with the possible exception of *Sauromalus klauberi* on Monserrat, none of these island populations differ greatly in size from peninsular *S. obesus*.

Table 9.5 provides an overview of the life-history differences among insular *S. hispidus*, *S. varius*, and mainland *S. obesus*. This table incorporates data gathered from both museum collections and field studies. The hatchlings of all three species are similar in size (8–14 g), even though differences in adult body size are striking (fivefold: 380–1800 g). Clutch size is not reduced in the gigantic *Sauromalus* below allometric body-size predictions based on mainland *S. obesus*. In fact, if anything, clutch size seems somewhat high, reaching a maximum of 32 eggs. The average size of full-term oviducal eggs in *S. hispidus* (based on two clutches) is 25 × 24 mm, each weighing about 10 g. This is only slightly larger than eggs of *S. obesus* (8 g). *Sauromalus varius* has larger (18 g) and more oblong eggs, 40 × 28 mm. The reproductive effort (clutch weight/adult weight with eggs) is 37% for *S. varius* and 25% for *S. hispidus* and typically is between 35% and 40% in *S. obesus*.

Since the first edition of this book, much more has been learned about the genus *Sauromalus*, particularly on two fronts. First, the application of morphological and molecular phylogenetic techniques has produced a better picture of the relationship of the several island and mainland populations (see chap. 8; app. 8.1), and second, we have a more detailed understanding of the life history and population dynamics or the two island gigantic species, which will be presented here and is an extension of a

Table 9.5 Comparative life history and reproductive features of *Sauromalus* species

Character	*S. hispidus* Ángel de la Guarda	*S. varius* San Esteban	*S. obesus* (locations)	References[a]
Max. adult snout-vent length (mm)	M 298	M 338	M 180 (Amboy)	1
	F 304	F 330	F 178 (Amboy)	1
			M 223 (Little Lake)	1
			F 205 (Little Lake)	1
Max. adult body weight (g)	1400	1800	180 (Amboy)	1
			380 (Little Lake)	1
Avg. hatchling snout-vent (mm), weight (g)	70, 12	75, 14	54, 8 (China Lake)	2, 3
Avg. egg size (mm) and weight (g) near laying	25 × 24, 10	40 × 28, 18	20 × 15, 8 (Amboy)	1
Hatchling growth rates (g/day)	0.25	M 1.04[b] F 0.41[b]	0.10 (China Lake)	2, 3
Max. clutch size	29	32	13 (China Lake)	2
			15 (Chuckwalla Mtn)	3

M, males; F, females.
[a] References for *S. obesus* only: 1, Case (unpublished data); 2, Berry (1974); 3, Sylber (1985).
[b] Captive population.

study reported by Case (1982) based on mark–recapture techniques for *S. hispidus* on Ángel de la Guarda and *S. varius* on San Esteban. This study was begun in 1978 and continued through 1992 and 1991, respectively.

Ángel de la Guarda

The study site is in one of the most mesic parts of Ángel, in an extensive arroyo (Cañon de las Palmas) on the middle-east side. A total of 598 live individuals were captured, marked, and released from 1978 to 1992, representing 533 unique individuals (265 males and 228 females). The population experienced drastic fluctuations driven by drought and rescued by significantly higher rainfall, usually associated with El Niño events. Figure 9.10 shows the number of individuals found alive and dead over the years of the study and illustrates the large swings in abundance. As discussed in Case (1983), the skull length and jaw length of dead individuals were measured

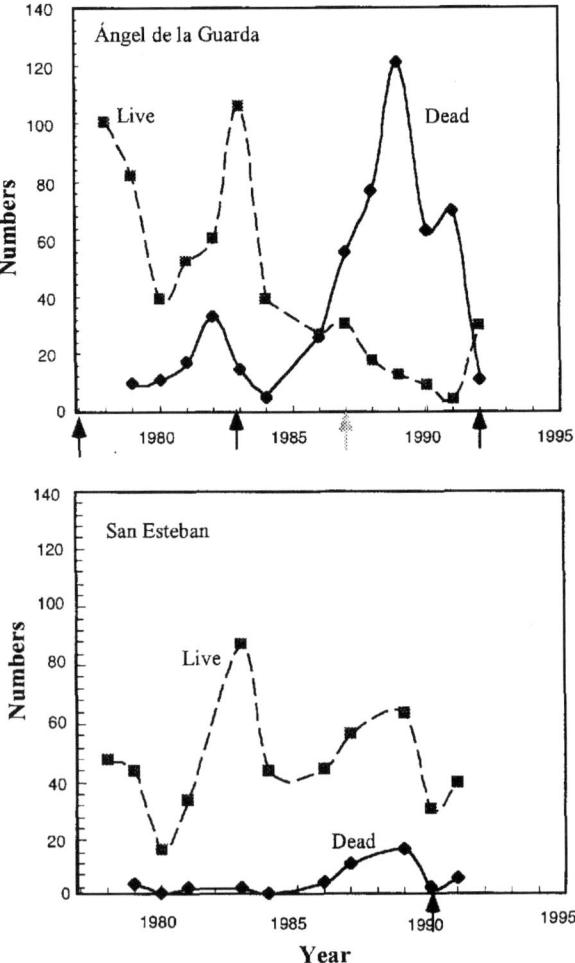

Figure 9.10 The number of live and dead chuckwallas observed on the two study sites each year. No surveys were conducted in 1985 on Ángel de la Guarda and 1982, 1985, and 1992 on San Esteban. Dead individuals were marked and not counted again in subsequent years; thus the number of dead represent only mortality during the intervening year. The El Niño years are shown by arrows at the bottom of each graph. Although 1987 was an El Niño year elsewhere, it did not result in significant new plant growth on either island. A tropical storm hit San Esteban in early 1990, producing flash floods that scoured large sections of the study site. Subsequent growth of annuals and vines was more extensive than in any other year of the study.

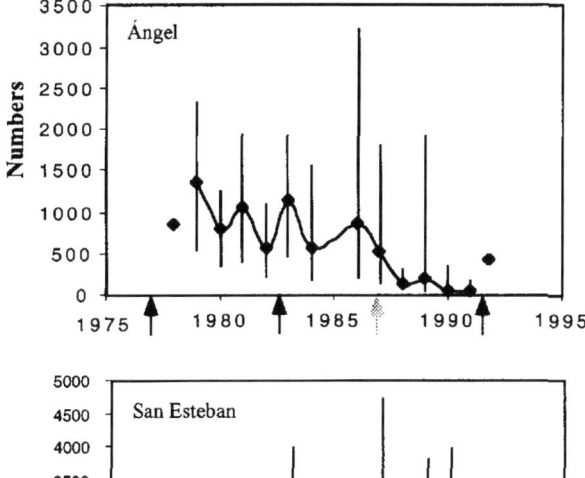
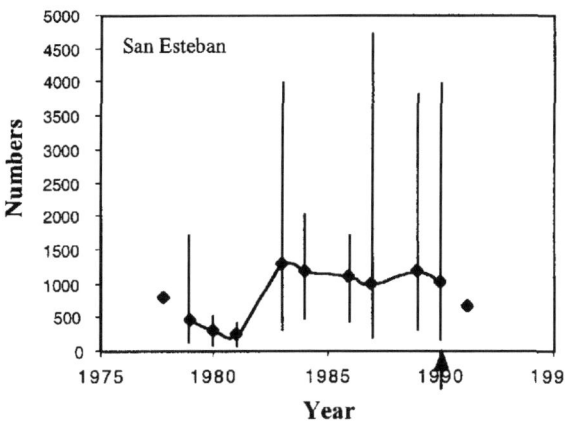

Figure 9.11 Estimated chuckwalla population sizes (and 95% confidence intervals) using the Jolly-Seber method as modified by Buckland (1980, 1982) for recaptures of dead individuals. This method does not produce estimates for the first and last sample. The estimates shown for these years were produced by multiplying the average probability of a live animal being captured (averaged over all intervening years) by the number of animals captured in the first and last years.

and the carcass marked so it would not be counted in subsequent years. The extended drought of 1985–1991 led to increasing numbers of dead animals per year in the first few years. Because the number of dead found in a year, however, represents the product of the death rate times the population size, as the population size declined, eventually the number of dead per year declined as well. I used the Jolly-Seber method, modified to allow for recaptures of dead individuals (using Buckland's [1980, 1982] computer program, Recapco). This procedure also modifies the maximum-likelihood function so that maximum survival rates can not exceed 1. Estimates were made of yearly survival rates and population sizes. Only 1 out of 51 marked hatchlings was subsequently recaptured, so they were excluded from these estimates. Recapture success for yearlings was also low, and consequently only reproductively mature individuals (> 200 mm SVL) were analyzed for population trends. These individuals are at least 3 years old. A total of 34 marked adults were later found as carcasses.

As is typical of mark–recapture studies, the confidence intervals around population size estimates are large, but we can still see a pattern of declining populations during extended droughts (fig. 9.11). This pattern is matched by decreasing body condition of individuals (fig. 9.12) and an absence of much reproduction (fig. 9.13A). Body condition rapidly improves after rainfall, as annual plants flourish and perennials leaf, bloom, and fruit in much greater abundance. Significant new cohorts of hatchlings

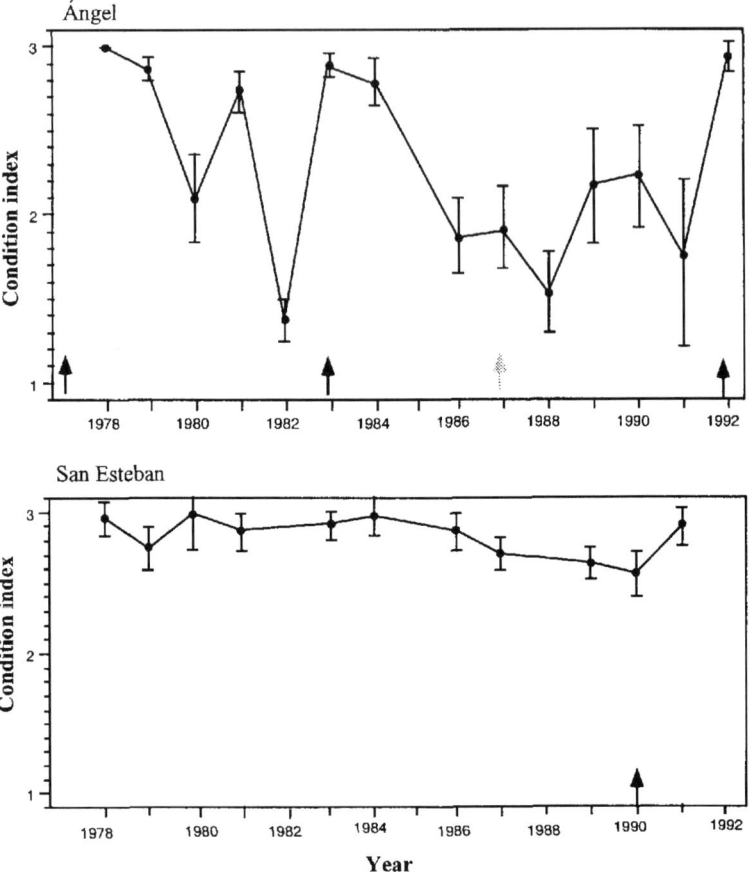

Figure 9.12 Chuckwalla condition across years. A condition index of 1 corresponds to individuals whose vertebrae are visible from the thorax through the first quarter of the tail and the tail is concave on the dorsal surface on each side of the vertebrae. Individuals with a score of 3 have plump convex tails and abundant stored fluid in their abdominal lymph sacs; individuals scored as 2 are intermediate. This visual metric was compared to a metric based on the residuals from a linear regression of the cube root of body weight against snout-vent length. The two condition measures were highly correlated ($p < .00001$).

were found only in 1978 and 1984 after exceptional rainfall (fig. 9.13A). Before this study I also visited several locations of the island in the drought years of spring/summer of 1972 and 1976 and found no evidence of recruitment. Despite the 1992 break in the previous extended drought, no hatchlings were observed, suggesting that the adults that did survive this exceptional drought still needed to accumulate additional energy reserves before reproducing. The body condition of animals responds quickly to rainfall (fig. 9.12). This visual metric of condition was compared to a metric based on the residuals from a linear regression of the cube root of body mass against SVL. The two condition measures were highly correlated ($p < .00001$). The weight-snout-vent residuals were used as the dependent variable in a multiple regres-

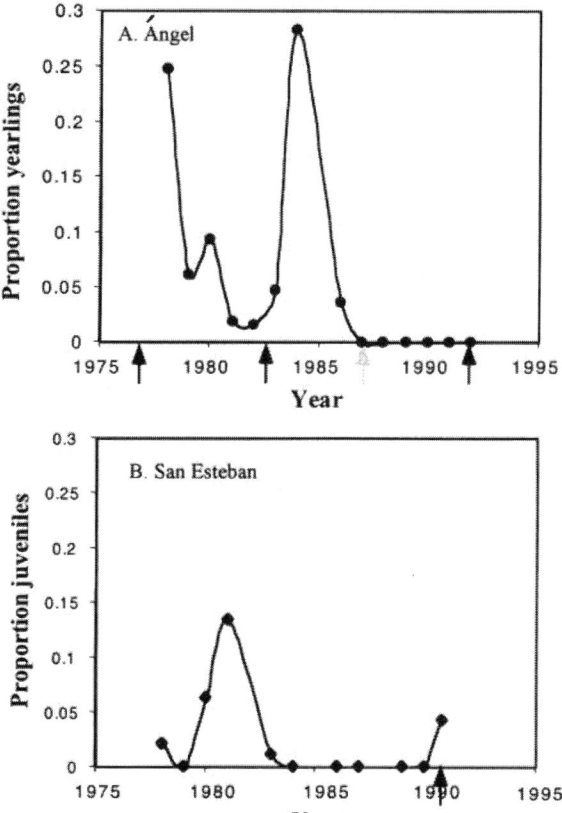

Figure 9.13 (A) The proportion of young chuckwallas (0 year olds) in each of the years in the sample captured on Ángel de la Guarda. These are counted as individuals with snout-vent length ≤123 mm. This cut-off can exclude some fast growing hatchlings and include some slow growing young that were produced the previous year (see Case 1982). (B) During the entire study only 3 first-year individuals were found on San Esteban. This plot shows the proportion of juveniles, scored as individuals <200 mm snout-vent length. We see a pulse in 1981 associated with a cohort of hatchlings that would have been produced in 1977–1979. Similarly, we see a pulse in juveniles in 1991, which most probably represents a cohort of hatchlings that were produced in 1989–90. The absence of smaller individuals before these pulses strongly suggests that hatchlings and yearlings live off the arroyos like the study site and then move onto them later as they age. Arrows indicate El Niño years in A and an additional heavy rainfall year at San Esteban in B.

sion with sex and recapture status (i.e., previously marked animals or new animals) as independent variables. Females had significantly higher weight residuals than males ($p < .01$), but animals that were previously marked did not differ significantly in condition from new animals, suggesting that the trauma of handling did not affect their subsequent weight gain.

Chuckwalla carcasses are particularly common on Ángel (Case 1982). They tend to be congregated under large cardon cacti or other high structural features in the landscape and appear more frequently during and after drought years than wet years (fig. 9.10). Observational evidence suggests that most carcasses of adult *S. hispidus* are scavenged by ravens and red-tailed hawks, where they are picked apart as the birds feed atop these cacti and other peaks. Small chuckwallas (< 130 mm SVL) are notably uncommon in these situations and thus may be eaten whole by these birds (one such attack on a hatchling by a raven was witnessed on Ángel) or torn apart so drastically that later recovery of their skeletons is unlikely. Dead males are as likely to be found as females (χ^2 test, $p < .398$), and because the sex ratio is equal in live individuals, survival rates do not differ between the sexes over these stage classes. Carcasses found under large cardons or trees are far more likely to be split open lengthwise on the ventral surface, presumably from avian predation or scavenging, compared to carcasses discovered in the open or within boulder crevices (χ^2 test, $p < .001$). This suggests that many lizards die from natural causes and are subsequently scavenged by raptors rather than dying directly from predation.

The sexes differ dramatically in the frequency of individuals with naturally missing toes. Figure 9.14 shows the frequency of individuals with one or more missing toes as a function of SVL for males and females. Below 200 mm the two sexes have similar frequencies of toe loss, but they diverge markedly at about 200 mm. Toe loss frequency in males reaches 1 for the largest individuals. I have not seen any male *S. hispidus* fighting in the field, although I have witnessed males *S. obesus* bite other males' toes. Such agonistic behavior may be occurring here as well, but the possible role of avian predators cannot be ruled out. Tail break frequencies are generally low, compared to mainland forms (Case 1982) and do not differ significantly different the two sexes (for SVL > 200 mm; males 23/202; females 9/145; $p = .1$).

The SVL of each dead individual was extrapolated from its skull length and dentary length (Case 1982). Comparisons of the size distribution of dead and live individuals were made for different periods of the study. Individuals <130 mm SVL were excluded for the reasons mentioned above. During wet years or in the first year of a drought cycle, the size distribution of live individuals does not differ significantly from that of individuals that died the same year (fig. 9.15). During years of prolonged drought, however, when recruitment is not occurring and thus not altering the size

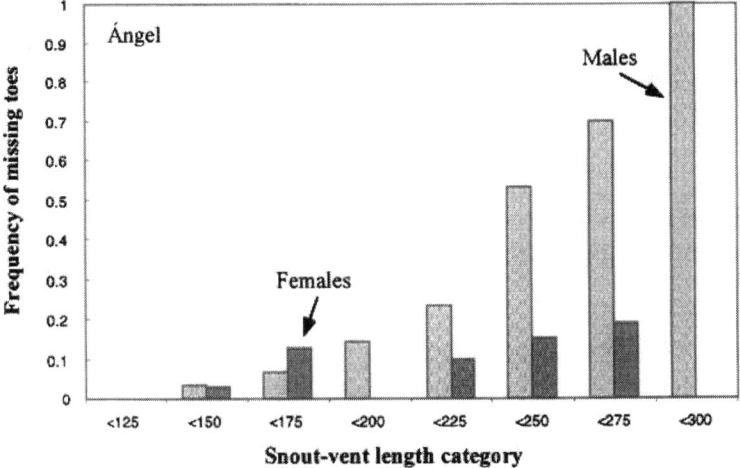

Figure 9.14 The frequency of chuckwallas on the Ángel de la Guarda study site with one or more missing toes as a function of their snout-vent length for males and females.

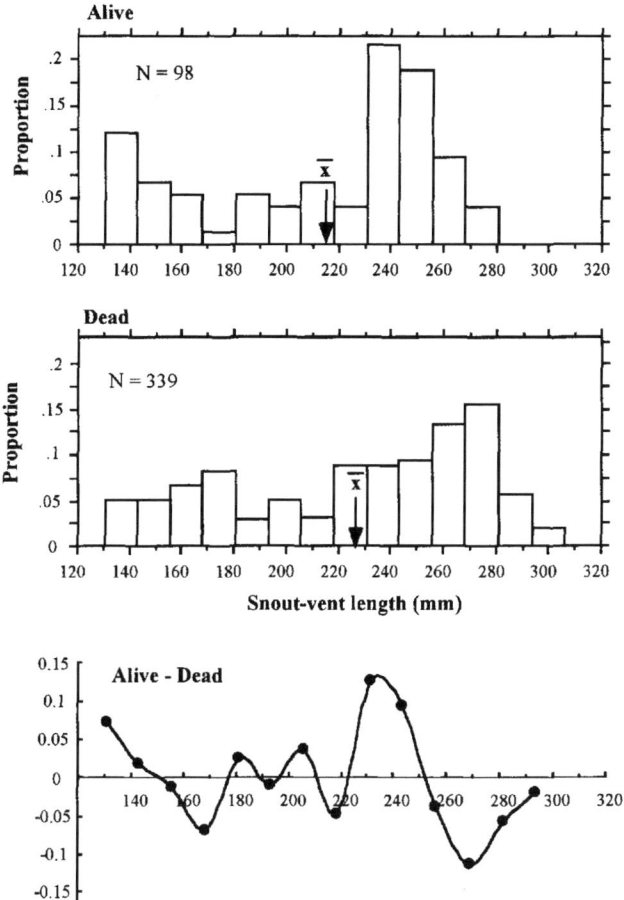

Figure 9.15 The frequency distribution of snout-vent length (SVL) (excluding individuals <130 mm SVL) for dead and live *S. hispidus* during the drought period of 1987–91. The means are indicated by arrows. The bottom panel shows the live proportion minus the dead proportion. Values above zero indicate size classes with relatively low mortality, and conversely values below zero have proportionately high mortality. The SVL of dead individuals was estimated from a nonlinear regression of based on individuals for which measurements of skull length, dentary length, and SVL were available. These ranged in size from 80 to 273 mm SVL. These relationships were fitted to a two-parameter sigmoidal curve, which explained more than 98% of the variance in bone length.

distribution of the live population (e.g., 1987–91), the mean SVL of the dead sample is significantly larger than that of contemporaneous live individuals ($p < .04$, ANOVA). As illustrated in figure 9.15, the shapes of the distributions for live and dead individuals are also significantly different (Kolmogorov-Smirnov test, $p < .011$; Wald-Wolfowitz runs test, $p < .0001$). Thus it appears that very large and presumably the oldest individuals suffer disproportionately during these drought periods. Because the age of these adult individuals is not known, we cannot say if this represents some

254 The Biological Scene

directional selection for smaller adult body size or simply senescence without any natural selection on size per se.

Yearly survival rates on Ángel tended to decline during the drought years (fig. 9.16A) and improve during or after El Niño years. Absolute ages of individuals are unknown, but all these recaptured individuals were first marked as adults. The survivorship curve in figure 9.16B is based on all marked adults, across all years of the

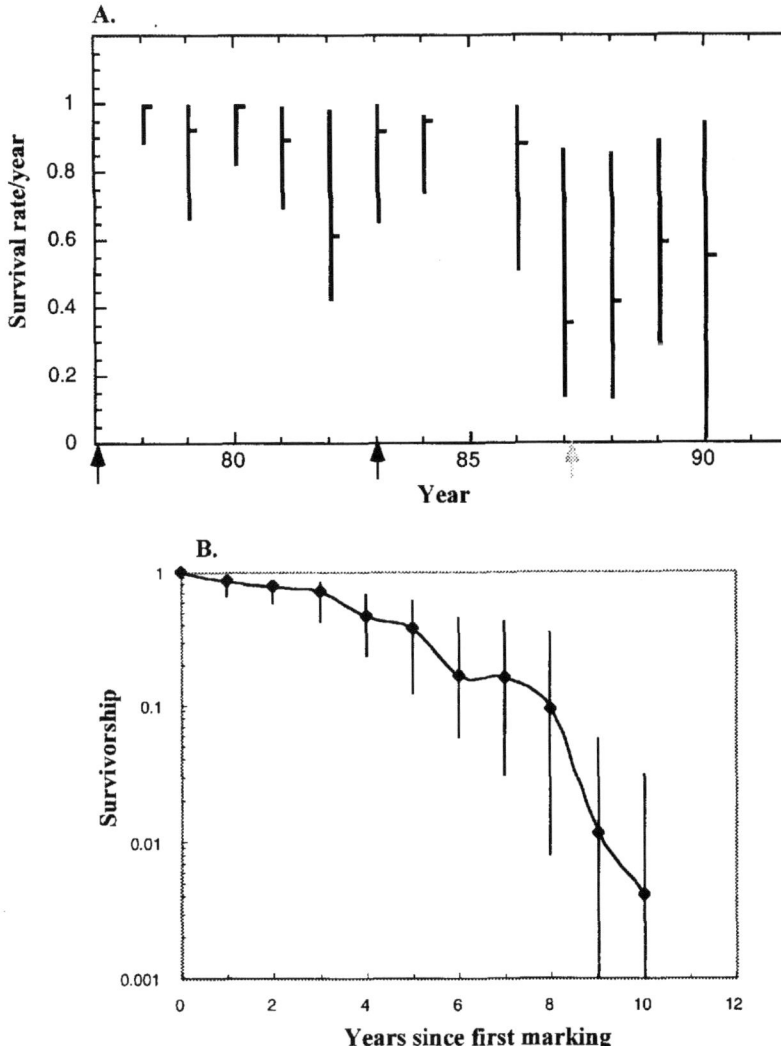

Figure 9.16 (A) Estimates (95% confidence intervals) for yearly survival rates of *S. hispidus* on Ángel. (B) The estimated survivorship curve (and 95% confidence intervals) for adults as a function of years since first capture. Constant mortality rates across ages on this semilog plot would produce a straight line. The observed downward-turning curve indicates senescence at older ages.

study, although most mortality occurs during drought years. We again see evidence of increased death rates in older individuals in the increasingly negative slope of the survivorship curve at older ages.

The median distance moved between recaptures was 57 m for males and 20 m for females. This is calculated only for recaptures that are separated by at least 6 days, since the locations of individuals are highly autocorrelated at shorter time intervals. The distribution of movement distances is highly skewed: 85% of the movements were <170 m, but occasionally both males and females moved 500 or even >1000 m and then set up a new home range characterized again by very small movement distances. This behavior was also observed in animals monitored by radio-telemetry during 1979–83.

San Esteban

The endemic *S. varius* is strikingly different from *S. hispidus* in appearance, yet the mtDNA phylogeny indicates a close sister-taxa relationship between the two species (Petren and Case 1997; app. 8.1). The study site was located in the most extensive arroyo on the island at the southeast corner of the island, just north of Ship Rock. A total of 543 live individuals were captured, marked, and released from 1978 to 1991, representing 469 unique individuals (249 males and 220 females). The yearly variation in population size (figs. 9.10, 9.11) does not appear as drastic as on Ángel, probably for two reasons. First, the eastern location of the island puts it in an area with more summer rainfall (Hastings and Turner 1969). Second, *S. varius* seems to be able to buffer environmental variation better. Dead individuals are found much less commonly than on Ángel (fig. 9.10), although mortalities are still concentrated in periods of drought. Like *S. hispidus* on Ángel, the population showed lower condition indices and lower survival rates (fig. 9.16) during droughts, but both measures were less variable than in *S. hispidus*. The visual condition index was highly correlated with the residuals from a cube root of weight–SVL regression ($p < .0001$) as they were for *S. hispidus*. Body condition was not affected by sex or recapture status.

During the entire study, only three hatchlings were found on and around the grid. Figure 9.13 shows the proportion of juveniles, scored as individuals <200 mm SVL for each year. There was a pulse in juveniles in 1981 associated with a cohort that would have been hatched in 1977–79. Yet during these years no yearlings were found. Similarly, another pulse in juveniles occurred in 1991, which most probably represents a cohort of hatchlings produced in 1989–90, although again no pulse in hatchlings was observed in those years, even though the year before (in 1989), I scored a large fraction of the adult females as gravid, based on palpating their abdominal cavities. Several gravid females were also observed that year excavating burrows in an area of loose, sandy soils in the arroyo at the end of a blind canyon. From all this I infer that hatchlings leave the relatively productive coastal arroyos after they hatch and move far up the canyon walls or perhaps into the interior of the island, only to reemerge to these richer bottom lands a few years later after they are substantially larger. Such migratory movements do not seem to occur in *S. hispidus* and have not been reported in other chuckwallas.

Tail break frequencies are higher than for *S. hispidus* on Ángel and not significantly different between the sexes (0.28 for males and 0.25 for females). Missing toes, however, are less common, and, as for *S. hispidus*, males have higher loss rates than females (males = 12.2%; females = 6.8% over all size categories; $p = .045$).

The median distance moved by individuals between recaptures (separated by more than 5 days) was 45.5 m for males and 46.1 m for females. As on Ángel, the distribution of movement distances is highly skewed with occasional long distance dispersal.

Sauromalus Gigantics: Conclusions

The two endemic chuckwallas share several life history features but differ in others. They both have similar population densities and exhibit mast years of reproduction which are initiated only when rainfall is abnormally high the previous winter, typically associated with an El Niño event. This is in striking contrast to *S. obesus*, which skips reproduction much less frequently. For example, in the Sonora Desert population studies by Apts (1987), at least 20% of the mature females bred in 4 out of 5 years, and more than 40% bred in 3 out of 5 years. In the intervening drought periods, body weight decreases, mortality rates increase, and ultimately population densities decline.

The two gigantic species also differ in several interesting ways. Although all three species are herbivorous and are not found too far from rock retreats, *S. hispidus* consumes more flowers and fruits than does *S. obesus*, whereas *S. varius* specializes almost exclusively on cactus fruits. The divergent yellow-orange and black picbald color pattern of *S. varius* makes this species especially cryptic beneath the prickly pear and *Stenocereus gummosus*, where it often forages. Most adults are found close to such cacti, and those that are farther away have small cactus spines (those that surround the fruit) in their mouth and tongue. Red and purple stains from the cacti fruits are on their chins. On the other hand, *S. hispidus* is more arboreal than the other forms, climbing cardons to reach their flowers and immature fruits. The arboreal niche on San Esteban is filled by *Ctenosaura*, which is about the same size as *S. varius* and occupies the same habitat types.

Both species can have relatively high population densities at the peak of the cycles. Converting the mean numbers shown in figure 9.11 to densities, I get ranges of 1.4–44.3/ha for *S. hispidus* and 8.3–45.4/ha for *S. varius*. This compares to values for *S. obesus* shown in table 9.6. *Sauromalus varius* does not undergo the extreme swings in population density seen in *S. hispidus*, nor does one typically find large numbers of carcasses. This relative temporal stability is also mirrored in the body condition indices of the lizards that do display the extreme swings seen in *S. hispidus* on Ángel.

Another notable difference between the two species is the apparent migratory behavior of hatchling *S. varius*. Movements up the canyon walls would take them away

Table 9.6 Population densities for *Sauromalus obesus*

Site	Mean Density (No./ha)	Reference
Chuckwalla Mountains, CA	23	Apts (1987)
China Lake, CA	13.8	Berry (1974)
Red Rock, CA	7.1	Johnson (1965)
Mojave Desert	10	Nagy (1973)

from most productive lowland habitats, but these also support the largest diversity of potential predators. Charles Sylber (pers. commun., 20 April 1985) described not being able to find any young in the arroyos in 1977 and again in 1978. However, when he searched higher up, between 400 and 1500 feet on the mountain slopes, in those same years, he did find hatchlings and juveniles. In 1978, he also saw a *Ctenosaura* catch a *S. varius* hatchling and swallow it whole. This apparent elevational migratory behavior, which at this point is only inferred, needs to be studied in more detail using radio-telemetry; but it could represent an adaptation to predation on young at low elevations.

It is possible that part of the observed changes in body size may be due to nongenetic factors such as phenotypic plasticity (e.g., Madsen and Shine 1993; Rhen and Lang 1995) or differing population age structures (e.g., King 1989). However, dentary ring comparisons suggest that age differences do not account for body size differences between *S. obesus* populations (Case 1976), and a common garden experiment (Tracy 1999) has shown that growth rate differences among southwestern U. S. populations of *S. obesus* have some genetic basis. Captive-bred individual *S. hispidus* and *S. varius* rapidly attain their gigantic size in captivity, whereas *S. obesus* do not become gigantic (Sylber 1985; Tracy 1999). This evidence, in the context of life-history studies of these species (Nagy 1973; Berry 1974; Case 1976, 1982; Apts 1987) supports the contention that body size differences between the island gigantics and smaller mainland forms are largely genetic.

Regardless of whether large body size evolved *in situ* on the islands or small size evolved on the mainland, we would like to identify the selective differentials that favor different adult sizes in these different circumstances. The gigantism of *S. hispidus* and *S. varius* is clearly not related to the absence of reptilian competitors because *Ctenosaura conspicuosa* occurs sympatrically with *S. varius* on San Esteban. These two species are nearly identical in body size, forage in the same habitats, and reach similar densities. There are no native mammalian herbivores on any of the gulf islands inhabited by the gigantic species, and the lack of such species might account in part for the niche and density release of these gigantic *Sauromalus*.

Similarly, all the islands inhabited by the gigantic *Sauromalus* (before transplants by humans) were oceanic islands lacking native mammalian carnivores. The potential predators that do exist are birds (kestrels, red-tailed hawks, and perhaps ravens) and snakes, which would take smaller-sized individuals preferentially. The single rattlesnake on San Esteban (*Crotalus molossus*) is a dwarf and is much too small to consume anything but hatchling or small juvenile *S. varius*. The same is true of the resident *Masticophis bilineatus*. This lack of large predators may confer a selective premium on large body size and rapid growth (hypothesis 3), and at the same time allow greater freedom in foraging activities, effectively increasing food availability (hypothesis 1). The gigantic *Sauromalus* are much more lax about regulating their body temperature and are active in a greater variety of habitats and under a wider range of temperatures than are mainland chuckwallas (Case 1982). Perhaps the best way to characterize this hypothesis about body size evolution in chuckwallas is to say that mainland forms are constrained to smaller body sizes by large predators, and these constraints are relaxed on islands where large predators are absent. Modes of escape used by mainland *Iguana* and *Ctenosaura* such as flight through vegetation and arboreality are not available to chuckwallas because of the sparsely vegetated desert habitat. *Amblyrhynchus*, *Conolophus*, and *Cyclura* are large, but they are also endemic to islands with historically no large predators (Petren and Case 1997).

Predation of *S. hispidus* on Ángel de la Guarda is probably greater than that on *S. varius*. Many fewer dead individuals are found on San Esteban than on Ángel (Case 1982). The endemic speckled rattlesnake *Crotalus mitchellii* on Ángel is capable of eating even the largest individuals. On a drizzly day in April 1981 at Palm Canyon on Ángel, the unusual cool weather brought out the rattlesnakes during the day. I observed two different *C. mitchellii* in the process of swallowing chuckwallas. One snake was 1410 cm long, roughly the maximum size of this race, and was having no trouble swallowing a large male *S. hispidus* (SVL 285 mm). Although predation pressure on Ángel is almost certainly less than that on the mainland, the large size of *S. hispidus* is matched by a compensatory size increase in the size of the speckled rattlesnake.

The reduction in predators on the large-sized individuals also may lead to an indirect increase in food availability by giving the lizards more access to food and more time to forage with less concern for predators. This, in turn, would have two effects important for the evolution of body size. Higher food availability selects directly for larger body size based on energetic models, and higher population densities associated with reduced predation may exaggerate the swings in food productivity associated with droughts and El Niño cycles. Large individuals in the population can better withstand short periods of starvation but seem particularly vulnerable during extended droughts. Because we cannot yet untangle the effect of age (and senescence) from body size per se, we cannot make any more quantitative predictions about selection's course.

Sex ratios in the insular gigantic *Sauromalus* are close to 1.0, and there is little difference between the sexes in maximum body size (see table 9.5). The maximum size of *S. hispidus* females is slightly larger than that of males, whereas in *S. varius* and *S. obesus* the reverse is true. This paucity of sexual dimorphism in the insular gigantics is also reflected in the total absence of coloration differences between the two sexes. Sexes are easily distinguished, however, on the basis of body shape and by the presence of femoral pores in males.

The lack of evolution of large size on the deep-water islands of Santa Catalina and Santa Cruz remains a mystery. The Isla Santa Catalina haplotype is 2.8% different from the southern peninsular haplotype (app. 8.1). This is comparable to differences among the gigantic and peninsular forms. If these populations have indeed been isolated for long times, other explanations would need to be invoked to account for small size, such as differences in selective pressures between the northern and southern deep-water islands or a genetic inability to evolve large size. Population densities on the southern islands do not appear to rival those of the gigantics on the northern islands.

Other Lizards and Snakes

Body-size differences of the magnitude of those among *Uta*, *Cnemidophorus*, and *Sauromalus* populations are not evident in most other gulf lizards. The possible exception to this rule is *Petrosaurus mearnsi*. In Baja California Norte, this lizard typically reaches a maximum SVL of about 85 mm and a mean SVL of 73.6 mm (Case 1983a).

The endemic derivative (*P. slevini*) on Ángel de la Guarda and its satellite island of Mejia reaches 100 mm and displays a mean SVL of 90.3 mm on Ángel and 91.3 on Mejia.

On the Baja California mainland, *Petrosaurus repens* in the middle and southern regions is substantially larger (maximum SVL, 132 mm), and *P. thalassinus* (maximum SVL, 152 mm) in the cape (Case 1983a) is larger still. Each of these forms occurs on the landbridge island. *P. repens* on Danzante and *P. thalassinus* on Espíritu Santo.

Sator angustus on the species-poor island of Santa Cruz and San Diego are no larger in size than *Sator grandaevus* on the relatively species-rich island of Cerralvo (Case 1983a). Sexual dimorphism in body size in *Sator* is more extreme than in other lizards in the gulf region and is accompanied by marked coloration differences between the sexes.

Populations of ctenosaurs are concentrated on extensive rock outcrops that provide crevices for predator escape. The *Ctenosaura* on San Esteban and Cerralvo are no larger than mainland races, probably because they are only recent colonists. Their densities are usually much higher than those commonly seen on either mainland. They also have shorter approach distances before they flee, consistent with a lack of large predators (Blázquez et al. 1997). Carothers (1981) studied a colony of 16 *Ctenosaura* on an isolated rock wall (area 40 m^2) on Cerralvo. The sex ratio of this colony was one male to four females, and individuals fell into four size classes: two juveniles, nine subadults, four adults, and one large adult male—the last being the colony "despot." Carothers observed a strict dominance hierarchy: the number of lizards dominated by a given individual is proportional to its body weight. Females were more aggressive than males except for the despot. Aggression was the main component of sociality, and agonism occurred more frequently between dominant individuals. Male–male agonism seems to be related to harem defense by the despot male. Intruding adult males and experimentally introduced males from other colonies are vigorously attacked by the despot.

Snakes provide striking examples of both insular gigantism and dwarfism. The tiger snakes, *Notechis*, have evolved both larger and smaller body sizes on different islands in directions that seem to parallel differences in the sizes of available prey (Schwaner 1985; Schwaner and Sarre 1990). The Crotalinae are often smaller on islands (Mertens 1934; Case 1978a) than on mainland locals. The deep-water islands in the Sea of Cortés and those off the west coast of Baja California provide examples (Klauber 1972; Case 1978a), although landbridge islands typically do not contain populations differing significantly in size from mainland relatives. Case (1978a) argued that most cases of snake dwarfism in the gulf were the result of lower prey availability and the generally smaller prey-sizes available. In the absence of typical mainland predators of snakes, particularly mammalian carnivores, snakes can reach high densities on islands. Examples include *Crotalus tortugensis* on Tortuga, *C. ruber* and *C. mitchellii* on Ángel de la Guarda, and *Mastiophis bilineatus* on San Esteban. Case speculated that food availability for snakes will be lower in such dense populations, thus favoring smaller size (hypothesis 1).

One problem with this explanation is that other dwarf snake populations in the Sea of Cortés are rather rare in their insular environment. *Crotalus catalinensis* on Santa Catalina and *C. molossus* on San Esteban are examples. On these islands snake prey in the form of rodents and lizards do not appear to be overexploited; on the contrary, rodents and lizards seem uncommonly abundant. For these populations, shifts in prey size (rather than overall prey abundance) may be more important determinants of snake size.

Many of the dwarf island-snakes have a greater proportion of lizards in their diet than do their mainland relatives, which concentrate on rodents. This could be due to the extreme abundance of lizards on some of these islands or to the cooler nighttime temperatures (particularly on the Pacific islands), which force otherwise nocturnal snakes to forage diurnally and thus encounter more lizards. Since the lizards on these islands are generally smaller than the average rodent, a smaller snake size is expected from hypothesis 2.

An interesting exception to the rule that snakes are generally small on islands is provided by the relative gigantism of the speckled rattlesnake, *Crotalus mitchellii*, on Ángel de la Guarda. On the peninsula, *C. ruber* is typically about twice as heavy as sympatric *C. mitchellii*, but on this island *C. ruber* has decreased in size while *C. mitchellii* has increased to the point that the insular *C. mitchellii* is now about twice as heavy as the insular *C. ruber*. Cody (1974) and Case (1978a) noted from meristic characters that *C. mitchellii* has diverged further from its mainland progenitors than has *C. ruber*, indicating that *C. mitchellii* arrived on the island first. This particular island is inhabited by lizards weighing nearly 1 kg (*Sauromalus hispidus*) and rodents weighing up to 150 g (*Neotoma lepida*). The early colonist *C. mitchellii* may have evolved a larger body size to use these large prey items efficiently. Later, when *C. ruber* arrived, prey were probably scarcer and of smaller size, possibly causing *C. ruber* to evolve a smaller size. It should be possible to test this hypothesis by comparing divergence times based on DNA-based phylogenies. If *C. mitchellii* has been on Ángel de la Guarda longer than *C. ruber*, then the former should possess more base-pair changes from its nearest relatives than *C. ruber* from its nearest relatives. Of course, this test would be confounded if the rate of evolution of DNA were different in the different species' lineages, but this can be determined and evaluated from the phylogeny (Sanderson 1997).

Two other island–mainland contrasts in *Crotalus* exhibit large body size differences. A solitary, dwarfed population ($\approx$630 mm total length) of *C. m. muertensis* is found on the landbridge island of El Muerto. Based on morphology, Klauber (1972) hypothesized that the El Muerto population is derived, not from mainland Baja, but from the giant *C. m. ángelensis* from Ángel de la Guarda. If this is indeed the case, and a molecular phylogeny could shed light on this hypothesis, this shift in body size is >700 mm in a few thousand years Finally, *C. ruber* (formerly *exsul*) of Isla de Cedros on the Pacific side is >200 mm smaller than on the Baja mainland (see Murphy et al. 1995). A more complete phylogenetic analysis could reveal the timing of isolation of both El Muerto and Isla de Cedros *Crotalus* populations and provide a qualitative estimate of the rate of morphological evolution.

Rattlesnakes, once freed from their mammalian predators, become the top-level predators on islands where they occur. On islands lacking rattlesnakes, one suspects that selection for defensive behaviors may be relaxed in the rodents and lizards. Generally, smaller rattlesnakes have more lizards in their diet than larger rattlesnakes, which rely more on rodents (Klauber 1972). Do rattlesnake body size shifts cause differential shifts in antipredator behavior? Do rattlesnakes lose their chemosensory acuity to prey absent from islands? It would be interesting to assay the chemosensory antipredator behavioral responses of rodents and lizards to rattlesnake odors on islands that differ in their rattlesnake fauna. Snake integumentary chemicals have been shown to be important to snake physiology and behavior (see Dial 1990). Chemical identification of snakes by rodents has been seen in ground squirrels (Hennessey and Owings 1978), kangaroo rats (Randall et al. 1995), and mice (*Mus musculus*; Dell'omo and Alleva 1994). Simon et al. (1981) found no evidence that the iguanid *Sceloporus jarrovi* could chemically detect a kingsnake predator; however, members of more chemically oriented lizard families, such as lacertids (Thoen et al. 1986; Van Damme et al. 1995) and geckos (Dial et al. 1989; Downes and Shine 1998), show definite behavioral antipredator responses to snake smells even in naïve individuals previously unexposed to these predators. The iguanid *Dipsosaurus dorsalis* and the Australian skink *Egernia stokesii* can recognize individual conspecifics (Alberts 1989, Dussault and Krekorian 1991; Bull et al. 2000). Studies of garter snakes (*Thamnophis*) clearly indicate that genetic variation for chemosensory discrimination of prey types exists and that geographic variation in prey type drive their evolution (Arnold 1981). The velvet gecko of Australia, *Oedura lesuerii*, responds defensively to odors of the broadheaded snake, which is a gecko predator but does not respond to the small-eyed snake, which eats mostly skinks (Downs and Shine 1998). Moreover, in areas lacking broadheaded snakes, the velvet geckos have much reduced defensive behaviors to the odors of these snakes.

A set of revealing contrasts for future study is shown in table 9.7. The design is to compare the same prey species between sister islands with and without rattlesnakes as top-level predators (none of these islands have mammalian carnivores). Sister islands in table 9.7 were connected to one another during the last glacial maxima, so any loss in chemosensory discrimination that may be evident, if shown to have a genetic basis, has occurred over roughly the last 10,000 years. Other comparisons could be made of prey between oceanic islands with rattlesnakes of different species or sizes (i.e., between rows). Finally, it would be interesting to measure the chemosensory abilities of the same rattlesnake species on different islands for specific prey (i.e., across-row comparisons) to determine if rattlesnakes lose the ability to discriminate odors of prey with which they no longer co-occur.

Two gulf island rattlesnakes have independently evolved a loss of rattles. Rattle loss is complete in the endemic Santa Catalina rattlesnake, *Crotalus catalinensis*. On San Lorenzo Sur, the *C. ruber lorenzoensis* population is variable. Some individuals have only a single rattle cell, while others have nearly complete rattles (Radcliffe and Maslin 1975).

Lizard Densities

Island habitats generally support fewer species than comparable mainland habitats. If the distribution and abundance of resources were comparable between islands and mainland, island species would each have a greater share of resources because of the loss of competing species. Concomitantly, some island species may expand their niches because of the absence of predators and/or species occupying overlapping niches on the mainland. All else being equal, the net result of these trends should be

Table 9.7 Comparative islands, their rattlesnakes, and target prey species

Islands *With* Rattlesnakes as Top Level Predators	Sister Islands *Without* Any Rattlesnakes[a]
Old, deep-water islands	
San Lorenzo Sur	**San Lorenzo Norte**
Rattlesnakes: *C. ruber*	**Salsipuedes**
Target prey:	(lacks *Chaetodipus spinatus*)
Peromyscus eremicus	
Chaetodipus spinatus	
Phyllodactylus nocticolus	
Cnemidophorus canus	
Uta antiqua	
Ángel de la Guarda	**Mejía**
Rattlesnakes: giant *C. mitchellii* and dwarf *C. ruber*	(lacks *Neotoma lepida* and *Cnemidophorus dickersonae*)
Target prey:	**Partida Norte**
Neotoma lepida	(lacks *Neotoma lepida*, *Petrosaurus slevini* and *Sauromalus hispidus*. Endemic *Phyllodactylus partidus* substitutes for *P. angelensis*)
Peromyscus eremicus	
Chaetodipus spinatus	
Phyllodactylus angelensis	
Cnemidophorus dickersonae	
Petrosaurus slevini	
Uta stansburiana	
Sauromalus hispidus	
Holocene landbridge islands to Baja California mainland	
Smith	**La Ventana**
Rattlesnakes: *C. mitchellii*	(lacks *Cnemidophorus tigris* and *Chaetodipus baileyi*)
Target prey:	
Peromyscus maniculatus	
Chaetodipus baileyi	
Phyllodactylus nocticolus	
Cnemidophorus tigris	
Uta stansburiana	
El Muerto	**San Luis** (= Encantada Grande)
Rattlesnakes: dwarf *C. mitchellii*	(rodent fauna not known, may lack *Phyllodactylus nocticolus*)
Target prey:	**Willard**
Peromyscus maniculatus	(may lack *Phyllodactylus nocticolus*)
Phyllodactylus nocticolus	
Uta stansburiana	

Comparisons of prey defensive behavior could be made across islands left to right or between islands in the left column with different species and/or sizes of rattlesnakes.

[a] In each case the sister islands were connected between 7000 and 13,000 years ago. Island pairs have the same set of target prey species except for the exceptions indicated. Islands have other rodents and lizards as well, but these do not occur on sister islands.

higher individual densities of species on islands, but lower summed densities across all species, as presumably the mainland species absent from the island were more efficient at utilizing the niche space into which the island residents have recently expanded (MacArthur et al. 1972). However, for several of the islands in the Sea of Cortés, summed lizard density indices are strikingly higher than on the mainland (excess density compensation) or on larger islands with more complete lizard faunas (Case 1975, 1983a; Polis et al., chap. 13, this volume).

Excess density compensation occurs in both insectivorous and herbivorous lizards (Case 1975, 1982). The reasons for this may be diverse. Certainly a portion of the apparent increase can be attributed to the fact that lizards on remote islands are less timid and therefore more visible than their mainland counterparts. Yet where mark-and-capture techniques have been applied to these populations, they have paralleled the results based on timed searches (Wilcox 1980b).

Exceedingly high lizard densities have also been noted for islands in the West Indies (Schoener and Schoener 1980; Wright 1981; Losos 1994), the South Pacific (reviewed in Case and Bolger 1991), small islands off New Zealand (Whitaker 1968, 1973; Crook 1973), the Seychelles (Brooke and Houston 1983; Gardner 1986), and small islands in the Panama Canal (Wright 1981). Wright (1979, 1981) observed that lizards were most dense on small islands with low bird densities. He suggested that the phenomenal lizard densities seen on certain small islands resulted from competitive release from the avifauna. This explanation does not seem to apply to the Sea of Cortés islands, however, because neither birds (Emlen 1979; George 1987; Cody and Velarde, chap. 10, this volume) nor rodents (Lawlor et al. chap. 12, this volume) are conspicuously abundant on small islands where lizards are rare.

The distributional data for avian, mammalian, and reptilian predators of lizards in the appendixes of this book yield no simple clues to island-by-island differences in lizard density. That is, one can find no single predator whose presence or absence accounts for major portions of the variance of lizard numbers. Rather, there seems to be a cumulative effect: the lower the number of potential predator species, the higher the observed lizard densities; the major portion of this density increase is usually composed of *Uta* (Case 1975, 1983a).

Even a casual visitor to San Pedro Mártir could not escape the immediate conclusion that this island supports an inordinate number of lizards (*Uta palmeri* and *Cnemidophorus martyris*) and an inordinate amount of lizard food, in the form of fish scraps around breeding booby (*Sula*) nests and flies. These high densities occur in spite of the presence of lizard predators in the form of kestrels, a rattlesnake (*Crotalus atrox*), and a kingsnake. On other small islands in the gulf, particularly those used by breeding sea birds, lizard densities are also high. These small islands receive relatively more energy input from the sea in the form of stranded plants and animals washed up on shore and food and excrement of sea birds. Lizards make use of these energy sources. At the same time, these islands lack most lizard predators on other larger islands. During El Niño years, the heavy rainfall increases plant growth on larger islands, and this leads to an increase in the herbivorous chuckwallas and probably other herbivores as well. This additional plant growth may find its way into higher numbers of arthropods (Polis and Hurd 1995; Polis et al. 1997; Stapp et al. 1999). The densities of insectivorous lizards and even carnivorous snakes might increase after a time lag of 1-2 years, although this has not been documented. In contrast, the higher ocean temperatures during El Niños can lead to breeding failures among sea birds and decreases in the amount of material that they bring onto the terrestrial systems of small islands. Polis et al. (chap. 13) make a case for the strong role of marine subsidies in increasing insular plant and animal abundance; however, their study islands are small (all but one is <1 km^2). For these islands the intertidal zone is a significant fraction of the entire island area. Because the importance of marine inputs declines with increasing island area (Polis and Hurd 1995), variation in lizard density on the major islands, such as San Esteban, Santa Cruz, Cerralvo, or Santa Catalina, is more likely due to differences in predation or terrestrial productivity differences.

My impression is that lizard predator densities are invariant or, if anything, vary directly, with lizard densities across islands, rather than inversely. Shrike and kestrel numbers vary little from island to island (chap. 10). The snake *Masticophis bilneatus* seems uncommonly dense on San Esteban, where small lizards are at least two or three times more numerous than on the mainland (Case 1975, 1983a). Isla Tortuga supports high densities of its endemic *Crotalus*, yet at the same time abounds in *Uta* and *Sceloporus*. Cerralvo supports relatively huge numbers of *Dipsosaurus*, *Sator*, and *Cnemidophorus*, yet potent lizard predators (kestrels, shrikes, 12 assorted snake species, and now feral cats) are also present.

Faced with a smaller subset of predator species on islands, lizards are able to exploit new resources and habitats that elsewhere are too risky for use. Evolutionary modifications will further enhance and solidify these niche expansions as evidenced by the endemic *Sauromalus* and intertidal foraging *Uta*. The predators reaching an island may benefit from the absence of their own competitors and reach higher densities. Over evolutionary time, they, too, might broaden their niches and become more potent lizard predators, as has *Crotalus mitchellii* on Ángel. Yet it seems doubtful that a few predators species could exert the same total predatory effect on their prey as a more diverse predator fauna. As more predator species are added, various niche options for the lizards are successively closed, and predator-free niche space is reduced.

High lizard densities coincident with high predator densities may also reflect the fact that even the large gulf islands (except Tiburón) lack carnivorous mammals (e.g., coyotes, ringtailed cats, foxes) which often act as top-level carnivores (i.e., as predators of lizard predators). Without this top trophic level, theory predicts an increase in overall biomass of all lower trophic levels, particularly when these lower trophic levels are partially self-limiting (Rosenzweig 1973; Abrams 1993).

Acknowledgments I thank Lee Grismer, Robert Murphy, Greg Pregill, Charles Sylber, and Tom Van Devender for sharing unpublished data and providing their insights. This research was supported by grants from the National Geographic Society and the National Science Foundation. I thank the Mexican government, the National Institute of Ecology, in Mexico, Angélica Narvaez of U.S. Embassy in Mexico City, and Emilio Bruna for assistance with the research, collecting, and export permit process. I am also grateful to Doug Bolger, Benita Epstein, Robert Fisher, Ken Petren, Steve Pruett-Jones, Ray Radtkey, and Rito Vale for valuable assistance in the field and to David Holway for valuable comments on the manuscript.

References

Abrams, P.A. 1993. Effect of increased productivity on abundance of trophic levels. *American Naturalist* **141**:351–371.

Apts, M.L. 1987. Environmental and variation in life history traits of the chuckwalla, *Sauromalus obesus*. *Ecological Monographs* **57**:215–232.

Arnold, E.N. 1973. Relationships of the palaearctic lizards assigned to the genera *Lacerta*, *Algyroides* and *Psammodromus*. *Bulletin of the British Museum of Natural History* **25**: 289–366.

Arnold, S.J. 1981. The microevolution of feeding behavior. In: A. Kamil and T. Sargent (eds), *Foraging Behavior*. Garland STPM Press, New York; pp. 409–453.

Aschmann, H. 1959. *The Central Desert of Baja California: Demography and Ecology*. Ibero-Americana no. 42. University of California Press; Berkeley.

Atmar, W., and B.D. Patterson. 1993. The measure of order and disorder in the distribution of species in fragmented habitat. *Oecologia* **96**:373–382.

Barahona, F., S.E. Evans, J.A. Mateo, M. Garcia-Marquez, and L.F. Lopez-Jurando. 2000. Endemism, gigantism, and extinction in island lizards: the genus *Gallotia* on the Canary Islands. *Journal of Zoology* **250**:373–388.

Berry, K.H. 1974. The ecology and social behavior of the chuckwalla, *Sauromalus obesus*. *University of California Publications in Zoology* **101**:1–60.

Blázquez, M.C., R. Rodriguez-Estrella, and M. Delibes. 1997. Escape behavior and predation risk of mainland and island spiny-tailed iguanas (*Ctenosaura hemilopha*). *Ethology* **103**: 990–998.

Bostic, D.L. 1966. A preliminary report of reproduction in the teiid lizard, *Cnemidophorus hyperthyrus beldingi*. *Herpetologica* **15**:81–90.

Brattstrom, B.H. 1974. The evolution of reptilian social behavior. *American Zoologist* **14**: 35–49.

Brooke, M.L., and D.C. Houston. 1983. The biology and biomass of the skinks *Mabuya sechellensis* and *Mabuya wrightii* on Cousin Island, Seychelles (Reptilia: Scincidae). *Journal of Zoology of London* **200**:179–195.

Brown, J.H., P.A. Marquet, and M.L. Taper. 1993. Evolution of body size: consequences of an energetic definition of fitness. *American Naturalist* **142**:573–584.

Buckland, S.T. 1980. A modified analysis of the Jolly-Seber capture-recapture model. *Biometrics* **36**:419–435.

Buckland, S.T. 1982. A mark-recapture survival analysis. *Journal of Animal Ecology* **51**:833–847.

Buckley, G.A., C.A. Brochu, D.W. Krause, and D. Pol. 2000. A pug-nosed crocodyliform from the late Cretaceous of Madagascar. *Nature* **405**:941–944.

Bull, C.M., C.L. Griffin, E.J. Lanham, and G.R. Johnston. 2000. Recognition of pheromones from group members in a gregarious lizard, *Egernia stokesii*. *Journal of Herpetology* **34**: 92–99.

Carlquist, S. 1965. *Island Life: A Natural History of the Islands of the World*. Natural History Press, Garden City, NJ.

Carothers, J.H. 1981. Dominance and competition in an herbivorous lizard. *Behavioral Ecology and Sociobiology* **8**:261–266.

Case, T.J. 1975. Species numbers, density compensation, and colonizing ability of lizards on islands in the Gulf of California. *Ecology* **56**:3–18.

Case, T.J. 1976. Body size differences between populations of the chuckwalla, *Sauromalus obesus*. *Ecology* **57**:313–323.

Case, T.J. 1978a. A general explanation for insular body size trends in terrestrial vertebrates. *Ecology* **59**:1–18.

Case, T.J. 1978b. Body size, endothermy and parental care in the terrestrial vertebrates. *American Naturalist* **112**:861–874.

Case, T.J. 1979. Character displacement and coevolution in some *Cnemidophorus* lizards. *Fortschritte Zoologie* **25**:235–282.

Case, T.J. 1982. Ecology and evolution of the insular gigantic *Sauromalus*. In: G. Burghardt and A.S. Rand (eds), *Iguanine Biology*. Noyes Publications, Park Ridge, NJ; pp. 184–211.

Case, T.J. 1983a. Reptile ecology. In: T.J. Case and M.L. Cody (eds), *Island Biogeography in the Sea of Cortéz*. University of California Press, Berkeley; pp. 159–209.

Case, T.J. 1983b. Assembly of lizard communities on islands in the Sea of Cortés. *Oikos* **41**: 427–433.
Case, T.J., and D.T. Bolger. 1991. The role of interspecific competition in the biogeography of island lizards. *Trends in Ecology and Evolution* **6**:135–139.
Case, T.J., D.T. Bolger, and A.D. Richman. 1998. Reptilian extinctions: The last ten thousand years. In: P.L. Fiedler and P. Kareiva (eds), *Conservation Biology for the Coming Decade*. Chapman and Hall, New York; pp. 157–186.
Case, T.J., and M.L. Cody. 1987. Biogeographic theories: tests on islands in the Sea of Cortés. *American Scientist* **75**:402–411.
Case, T.J., and M.E. Gilpin. 1974. Interference competition and niche theory. *Proceedings of the National Academy of Science, USA* **71**:3073–3077.
Case, T.J., M.E. Gilpin, and J.M. Diamond. 1979. Overexploitation, interference competition, and excess density compensation in insular faunas. *American Naturalist* **113**:843–854.
Cody, M.L. 1966. A general theory of clutch size. *Evolution* **20**:174–184.
Cody, M.L. 1974. *Competition and the Structure of Bird Communities*. Princeton University Press, Princeton, NJ.
Cole, K.L. 1986. The Lower Colorado River Valley: A Pleistocene desert. *Quaternary Research* **25**:392–400.
Crook, I.G. 1973. The tuatara, *Sphenodon punctatus* on islands with and without populations of the Polynesian rat, *Rattus exulans*. *Proceedings of the New Zealand Ecological Society* **20**:115–120.
Dell'omo, G., and E. Alleva. 1994. Snake odor alters behavior, but not pain sensitivity in mice. *Physiology and Behavior* **55**:125–128.
Dial, B.E., P.J. Weldon, and B. Curtis. 1989. Chemosensory identification of snake predators (*Phyllorhychus decuratus*) by banded geckos (*Coleonyx variegatus*). *Journal of Herpetology* **23**:224–229.
Diamond, J.M. 1972. Biogeographic kinetics: estimation of relaxation times for avifaunas of southwest Pacific islands. *Proceedings of the National Academy of Science, USA* **69**:3199–3203.
Diamond, J.E. 1975. Assembly of species communities. In: M.L. Cody and J.M. Diamond (eds), *Ecology and Evolution of Communities*, Harvard University Press, Cambridge, MA; pp. 342–444.
Diamond, J.M., and M.E. Gilpin. 1980. Turnover noise: contribution to variance in species number and prediction from immigration and extinction curves. *American Naturalist* **115**: 884–889.
Downes, S., and R. Shine. 1998. Sedentary snakes and gullible geckos: predator-prey coevolution in nocturnal rock-dwelling reptiles. *Animal Behaviour* **55**:1373–1385.
Dunham, A.E., D.W. Tinkle, and J.W. Gibbons. 1978. Body size in island lizards. A cautionary tale. *Ecology* **59**:1230–1238.
Dussault, M.H., and C.O. Krekorian. 1991. Conspecific discrimination by *Dipsosaurus dorsalis*. *Herpetologica* **47**:82–88.
Emlen, J.T. 1979. Land bird densities on Baja California islands. *Auk* **96**:152–167.
Felger, R.S., and M.B. Moser. 1985. *People of the Desert and Sea: Ethnobotany of the Seri Indians*. University of Arizona Press, Tucson.
Foufopoulos, J., and A.R. Ives. 1999. Reptile extinctions on land-bridge islands: Life history attributes and vulnerability to extinction. *American Naturalist* **153**:1–25.
Gardner, A.S. 1986. The biogeography of the lizards of the Seychelles Islands. *Journal of Biogeography* **13**:237–253.
George, T.L. 1987. Greater land bird densities on island vs. mainland: relation to nest predation level. *Ecology* **68**:1393–1400.

Giannasi, N., R.S. Thorpe, and A. Malhotra. 2000. A phylogenetic analysis of body size evolution in the *Anolis roquet* group (Sauria: Iguanidae): character displacement or size assortment. *Molecular Ecology* **9**:193–202.
Gliwicz, J. 1980. Island population of rodents: their organization and functioning. *Biological Reviews* **55**:109–138.
Grismer, L.L. 1993. The insular herpetofauna of the Pacific coast of Baja California, Mexico. *Herpetological Natural History* **1**:1–10.
Grismer, L.L. 1994. Three new species of intertidal side-blotched lizards (genus *Uta*) from the Gulf of California, Mexico. *Herpetologica* **50**:451–474.
Grismer, L.L., J.A. McGuire, and B.D. Hollingsworth. 1994. A report on the herpetofauna of the Vizcaino Peninsula, Baja California, Mexico, with a discussion of its biogeographic and taxonomic implications. *Bulletin of the Southern California Academy of Sciences* **93**: 45–80.
Gotelli, N.J. 2000. Null models of species co-occurrence patterns. *Ecology* **81**:2606–2621.
Gotelli, N.J., and G.R. Graves. 1996. *Null Models in Ecology*. Smithsonian Institution Press, Washington, DC.
Hastings, J.R., and R.M. Turner. 1969. *Climatological Data and Statistics for Baja California*, Technical Reports on the Meteorology and Climatology of Arid Regions 18, University of Arizona Institute of Atmospheric Physics, Tucson.
Hastings, J.R., R.M. Turner, and D.W. Warren. 1972. *An Atlas of Some Plant Distributions in the Sonoran Desert*. Technical Reports on the Meteorology and Climatology of Arid Regions 18, University of Arizona Institute of Atmospheric Physics, Tucson.
Hazard, L.C., V.H. Shoemaker, and L.L. Grismer. 1998. Salt gland secretion by an intertidal lizard, *Uta tumidarostra*. *Copeia* **1998**:231–234.
Hennesey, D.F., and D.H. Owings. 1978. Snake species discrimination and the role of olfactory cues in the snake-directed behavior of the California ground squirrel. *Behaviour* **65**:115–124.
Hews, D.K. 1990. Examining hypotheses generated by field measures of sexual selection on male lizards, *Uta palmeri*. *Evolution* **44**:1956–1966.
Hews, D.K. 1993. Food resources affect female distribution and male mating opportunities in the iguanian lizard *Uta palmeri*. *Animal Behaviour* **46**:279–291.
Holman, J.A. 1995. *Pleistocene Amphibians and Reptiles in North America*. Oxford University Press, New York.
Johnson, S.R. 1965. An ecological study of the chuckwalla, *Sauromalus obesus* Baird, in the western Mohave desert. *American Midland Naturalist* **73**:1–29.
Jones, K.B., L.P. Kepner, and T.E. Martin. 1985. Species of reptiles occupying habitat islands in western Arizona: a deterministic assemblage. *Oecologia* **66**:595–601.
Kaula, W.M. 1977. Problems in understanding vertical movements and earth rheology. In: N.-A. Morner (ed), *Earth Rheology, Isostasy and Eustasy*. John Wiley & Sons, New York; pp. 577–588.
King, R.B. 1989. Body size variation among island and mainland snake populations. *Herpetologica* **45**:84–88.
Klauber, L.M. 1972. *Rattlesnakes. Their Habits, Life Histories and Influence on Mankind*, 2nd ed, 2 vols. University of California Press, Berkeley.
Kramer, G. 1946. Veranderungen von Nachkommenziffer und Nachkommengrosse sowie der Altersverteilung von Inseleiderschsen. *Zeits. Naturforschung* **1**:77–710.
Lack, D. 1968. *Ecological Adaptations for Breeding in Birds*. London: Methuen.
Lande, R., and S.J. Arnold. 1983. The measurement of selection on correlated characters. *Evolution* **37**:1210–1226.
Leigh, E.G. 1981. Average lifetime of a population in a varying environment. *Journal of Theoretical Biology* **90**:213–239.

Losos, J.B. 1994. Integrative approaches to evolutionary ecology: *Anolis* lizards as model systems. *Annual Reviews of Ecology and Systematics* **25**:467–493.

MacArthur, R.H., J.M. Diamond, and J.R. Karr. 1972. Density compensation in island faunas. *Ecology* **53**:330–342.

MacArthur, R.H., and E.O. Wilson. 1967. *The Theory of Island Biogeography*. Princeton University Press, Princeton, NJ.

Madsen, T., and R. Shine. 1995. Phenotypic plasticity in body sizes and sexual size dimorphism in European grass snakes. *Evolution* **47**:321–325.

Mertens, R. 1934. Die Insel Reptilian. *Zoologica* **84**:1–205.

Milliman, J.D. and K.O. Emery. 1968. Sea levels during the past 35,000 years. *Science* **162**:1121–1123.

Murphy, R.W., V. Kovac, O. Haddrath, G.S. Allen, A. Fishbein, and N.E. Mandrak. 1995. mtDNA gene sequence, allozyme, and morphological uniformity among red diamondback rattlesnakes, *Crotalus ruber* and *Crotalus exsul*. *Canadian Journal of Zoology* **73**:270–281.

Nagy, K.A. 1973. Behavior, diet and reproduction in a desert lizard, *Sauromalus obesus*. *Copeia* **1973**:93–102.

Olson, S.L., and W.B. Hilgartner. 1982. Fossil and subfossil birds from the Bahamas. *Smithsonian Contributions to Paleobiology* **48**:22–60.

Ortlieb, L. 1977. Neotectonics from Marine terraces along the Gulf of California. In: Nils-Axel Morner (ed). *Earth rheology, isostasy and eustasy*. John Wiley & Sons, New York, pp. 497–504.

Parker, W.S., and E.R. Pianka. 1975. Comparative ecology of populations of the lizard *Uta stansburiana*. *Copeia* **1975**:615–632.

Petren, K., and T.J. Case. 1997. A phylogenetic analysis of body size evolution and biogeography in chuckwallas (*Sauromalus*) and other iguanines. *Evolution* **51**:206–219.

Pianka, E.R. 1986. *Ecology and Natural History of Desert Lizards*. Princeton University Press, Princeton, NJ.

Pianka, E.R., and W.S. Parker. 1972. Ecology of the iguanid lizard *Callisaurus draconoides*. *Copeia* **1972**:493–508.

Polis, G.A., and S.D. Hurd 1995. Extraordinarily high spider densities on islands – flow of energy from the marine to terrestrial food webs and the absence of predation. *Proceedings of the National Academy of Science, USA* **92**:4382–4386.

Polis, G.A., S.D. Hurd, C.T. Jackson, and F.Sanchez-Piñero. 1997. El Niño effects on the dynamics and control of an island ecosystem in the Gulf of California. *Ecology* **78**:1884–1897.

Pregill, G.K. 1986. Body size of insular lizards: A pattern of Holocene dwarfism. *Evolution* **40**:997–1008.

Pregill, G.K. and S.L. Olson. 1980. Zoogeography of West Indian vertebrates in relation to Pleistocene climate cycles. *Annual Reviews of Ecology and Systematics* **12**:75–98.

Radcliffe, C.W., and T.P. Maslin. 1975. A new subspecies of the red rattlesnake, *Crotalus ruber*, from San Lorenzo Sur Island, Baja California Norte, Mexico. *Copeia* **1975**:490–493.

Radtkey, R., S.M. Fallon, and T.J. Case. 1997. Character displacement in some Cnemidophorus lizards revisited: a phylogenetic analysis. *Proceedings of the National Academy of Sciences, USA* **94**:9740–9745.

Randall, J.A., S.M. Hatch, and E.R. Hekkala. 1995. Inter-specific variation in anti-predator behavior in sympatric species of kangaroo rat. *Behavioral Ecology and Sociobiology* **36**:243–250.

Rhen, T., and J.W. Lang. 1995. Phenotypic plasticity for growth in the common snapping turtle: effects of incubation temperature, clutch, and their interaction. *American Naturalist* **146**:726–747.

Richman, A., T.J. Case, and T. Schwaner. 1988. Natural and unnatural extinction rates for island reptiles. *American Naturalist* **31**:611–630.
Robinson, M.D. 1972. Chromosomes, protein polymorphism, and systematics of insular chuckwalla lizards (Genus *Sauromalus*) in the Gulf of California, Mexico. Ph.D. dissertation, University of Arizona.
Rosenzweig, M.L. 1973. Exploitation in three trophic levels. *American Naturalist* **107**:275–294.
Sanderson, M.J. 1997. A nonparametric approach to estimating divergence times in the absence of rate constancy. *Molecular Biology and Evolution* **14**:1218–1231.
Savage, J.M. 1960. The evolution of a peninsular herpetofauna. *Systematic Zoology* **9**:184–212.
Schall, J.J. 1978. Reproductive strategies in sympatric whiptail lizards (*Cnemidophorus*): Two parthenogenetic and three bisexual species. *Copeia* **1978**:108–116.
Schoener, T.W. 1969. Models of optimal size for solitary predators. *American Naturalist* **103**:277–313.
Schoener, T.W. 1970. Size patterns in West Indian Anolis lizards. II. Correlations with the sizes of particular sympatric species – displacement and convergence. *American Naturalist* **104**:155–174.
Schoener, T.W. 1983. Rate of species turnover decreases from lower to higher organisms: a review of the data. *Oikos* **41**:372–377.
Schoener, T.W., and A. Schoener. 1980. Densities, sex ratios and population structure in four species of Bahamian *Anolis* lizards. *Journal of Animal Ecology* **49**:19–53.
Schulte J.A.I., J.R. Macey, A. Larson, and T.J. Papenfuss. 1998. Molecular tests of phylogenetic taxonomies: a general procedure and example using four subfamilies of the lizard family Iguanidae. *Molecular Phylogenetics and Evolution* **10**:367–376.
Schwaner, T.D. 1985. Population structure of black tiger snakes, *Notechis ater niger*, on offshore islands of South Australia. In: G. Grigg, R. Shine, and H. Ehrmann (eds), *Biology of Australian Frogs and Reptiles*. Surrey Beatty and Sons, Sydney; pp. 35–46.
Schwaner, T.D., and Sarre, S.D. 1990. Body size and sexual dimorphism in mainland and island tiger snakes. *Journal of Herpetology* **24**:320–323.
Shaw, C.E. 1945. The chuckwallas, genus *Sauromalus*. *Transactions of the San Diego Society of Natural History* **10**:269–306.
Shreve, F. 1951. *Vegetation and flora of the Sonoran Desert*. Vol. I. Vegetation. *Carnegie Institute of Washington Publication* **591**:1–192.
Simon, C.A., K. Gravelle, B.E. Bissinger, I. Eiss, and R. Ruibal. 1981. The role of chemoreception in the iguanid lizard *Sceloporus jarrovi*. *Animal Behavior* **29**:46–54.
Soulé, M. 1966. Trends in the insular radiation of a lizard. *American Midlands Naturalist* **100**:47–64.
Stamps, J.A. 1977. Social behavior and spacing patterns in lizards. In: C. Gans and D.W. Tinkle (eds), *Biology of the Reptilia*, Vol. 7. *Ecology and Behavior*. Academic Press, New York; pp. 265–321.
Stamps, J.A., and M. Buechner. 1985. The territorial defense hypothesis and the ecology of insular vertebrates. *Quarterly Review of Biology* **60**:155–181.
Stapp, P, G.A. Polis, and Francisco Sanchez-Piñero. 1999. Stable isotopes reveal strong marine and El Niño effects on island food webs. *Nature* **401**:467–469.
Sylber, C.K. 1985. Eggs and hatchlings of the yellow giant chuckwalla and the black giant chuckwalla. *Herpetology Review* **16**:18–21.
Thoen, C., Bauwens, D., and R.F. Nerheyen. 1986. Chemoreceptive and behavioral responses of the common lizard *Lacerta vivipara* to snake chemical deposits. *Animal Behavior* **34**:1805–1813.
Tinkle, D.W., H.M. Wilbur, and S.G. Tilley. 1970. Evolutionary strategies in lizard reproduction. *Evolution* **24**:55–74.

Tracy, C.R. 1999. Differences in body size among chuckwalla (*Sauromalus obesus*) populations. *Ecology* **80**:259–271.
Van Damme, R., D. Bauwens, C. Thoen, D. Vanderstighelen, and R.F. Veheyen. 1995. Responses of naïve lizards to predator chemical cues. *Journal of Herpetology* **29**:38–43.
Van Devender, T.R. 1977. Holocene woodlands in the Southwestern Deserts. *Science* **198**: 189–192.
Van Devender, T.R. 1987. Holocene vegetation and climate in the Puerto Blanco Mountains, southwestern Arizona. *Quaternary Research* **27**:51–72.
Vitt, L.J., and R.D. Ohmart. 1977. Ecology and reproduction of lower Colorado River lizards: II. *Cnemidophorus tigris* (Teiidae), with comparisons. *Herpetologica* **33**:223–234.
Walker, J.M. 1980. Reproductive characteristics of the San Pedro Mártir whiptail, *Cnemidophorus martyris*. *Journal of Herpetology* **14**:431–432.
Walker, J.M., and T.P. Maslin. 1969. A review of the San Pedro Nolasco whiptail lizard *Cnemidophorus bacatus* (Van Denburgh and Slevin). *American Midland Naturalist* **82**:127–139.
Wharton, C.H. 1966. Reproduction and growth in the cottonmouths, *Agkistrodon piscivorus* Lacepede, of Cedar Keys, Florida. *Copeia* **1966**:149–161.
Whitaker, A.H. 1968. The lizards of the Poor Knights Islands, New Zealand. *New Zealand Journal of Science* **11**:623–651.
Whitaker, A.H. 1973. Lizard populations on islands with and without Polynesian rats, *Rattus exulans*. *Proceedings of the New Zealand Ecological Society* **20**:121–130.
Wilcox, B.A. 1978. Supersaturated island faunas: A species-age relationship for lizards on post-Pleistocene land-bridge islands. *Science* **199**:996–998.
Wilcox, B.A. 1980a. Insular ecology and conservation. In: M.E. Soulé and B.A. Wilcox (eds), *Conservation Biology*. Sinauer Associates, Sunderland, MA; pp. 95–117.
Wilcox, B.A. 1980b. Aspects of the biogeography and evolutionary ecology of some island vertebrates. Ph.D. dissertation, University of California, San Diego.
Wong, H., E. Mellink, and B.D. Hollingsworth. 1995. Proposed recent overwater dispersal by *Callisaurus draconoides* to Isla Danzante, Gulf of California, Mexico. *Herpetological Natural History* **3**:179–182.
Wright, D.H., B.D. Patterson, G. Mikkelson, A.H. Cutler, and W. Atmar. 1998. A comparative analysis of nested subset patterns of species composition. *Oecologia* **113**:1–20.
Wright, S.J. 1979. Competition between insectivorous lizards and birds in Central Panama. *American Zoologist* **19**:1145–1156.
Wright, S.J. 1981. Extinction-mediated competition: The *Anolis* lizards and insectivorous birds of the West Indies. *American Naturalist* **117**:181–192.
Wright, S.J., and C.C. Biehl. 1982. Island biogeographic distributions: testing for random, regular and aggregated patterns of species occurrences. *American Naturalist* **119**:345–347.
Wyles, J.S., and G.C. Gorman. 1978. Close relationship between the lizard genus *Sator* and *Sceloporus utiformis* (Reptilia, Lacertilia, Iguanidae): electrophoretic and immunological evidence. *Journal of Herpetology* **12**:343–350.

10

Land Birds

MARTIN L. CODY
ENRIQUETA VELARDE

Early Exploration in the Gulf Region

Very few of the early scientific explorers in the Gulf of California had much to say about the land birds. There might be two reasons for this: first, the land birds in arid, desert regions are sparse and in general unbecoming, and second, the species encountered are by and large those seen in the much more accessible regions of southwestern North America. Chapter 1 introduced János Xántus, who is recognized as the pioneer ornithologist (or at least bird collector) in the cape area of Lower California, whose contributions (e.g., 1859, in which the first description of the Gray Thrasher, *Toxostoma cinereum*, was published) are appropriately commemorated in the Xantus Hummingbird, the most spectacular endemic on the peninsula. Lawrence (1860) first described the species as *Amazilia xantusi* (thence *Hylocharis xantusii*, and now *Basilinna xantusii*), and P. L. Sclater announced the discovery to *Ibis* readers in the same year. By the end of the nineteenth century, several ornithologists had collected in the southern peninsula and reported their findings (e.g., Baird 1870; Belding 1883; Bryant 1889; Ridgway 1896), but very little of this work referred to the islands in the gulf. Brewster's (1902) report on the cape region avifauna was the most comprehensive of the earlier studies.

Serious attention was first paid to the gulf island birds by Maillard (1923) and Townsend (1923), and the latter's 1911 island-hopping trip in the *Albatross* served as a model for many similar expeditions later. The first distributional synthesis of their work, and especially that of Nelson (1921), Lamb (e.g., 1924), and Thayer (e.g., 1907), was published by Joseph Grinnell in 1928 in a monograph that is still the standard reference for the peninsula and gulf area. The last 50 years have seen little progress beyond the accumulation of further distributional records and the description of new subspecies (e.g., van Rossem 1929, 1932; Banks 1963a,b,c, 1964, 1969). The island birds remain rather poorly known; even species lists are likely to be incomplete, and ecological studies of the island populations have scarcely begun.

In this chapter we report on the results largely of our own field work. This book's earlier edition depended heavily on Cody's fieldwork on the gulf islands and vicinity, conducted on island trips in 1970, 1973, 1975, 1976, 1977, 1981, and 1990. Since that time, Velarde and colleagues have conducted extensive surveys throughout the gulf, and thus the data set is now much more comprehensive. New island surveys not only expand and sharpen the distributional record, but also serve as a comparison to earlier data where they exist. Many of the islands not covered by Cody earlier have been visited recently, and the present data set includes a much more complete island roster.

Following a section on the faunistics of the gulf-region birds, we discuss diversity and its controls in shoreline birds, raptors, and especially in the smaller land birds of the islands. Next we present what information we have on the structure and organization of the island communities, and we end with a synthesis of what can be learned of island biogeographical processes from the gulf island land birds.

Avifauna of the Gulf Region

Distribution and Affinities

The Sonoran Desert avifauna is at best an indistinct entity, for most of the birds that breed within the geographic area that supports Sonoran Desert vegetation either breed beyond this area, or breed also in other types of vegetation, or both. There are very few species with distributions that coincide with the map of this vegetation in figure 4.4; a few candidates such as Gila Woodpecker (*Melanerpes uropygialis*) and Curve-billed Thrasher (*Toxostoma curvirostre*) occur also in thorn scrub much farther south, leaving the "gilded" race of the Northern Flicker (*Colaptes cafer chrysoides*) as perhaps closest to endemic in this habitat. A majority of the common Sonoran Desert birds are either extremely wide in geographic distribution (e.g., Northern Mockingbird, *Mimus polyglottos*; Mourning Dove, *Zenaida macroura*; Loggerhead Shrike, *Lanius ludovicianus*; Red Cardinal, *Cardinalis cardinalis*) or else, if more restricted geographically, they occur in a wide range of habitats (e.g., House Finch, *Carpodacus mexicanus*; Violet-green Swallow, *Tachycineta thalassina*; Black-chinned Hummingbird, *Archilochus alexandri*; Ash-throated Flycatcher, *Myiarchus cinerascens*). Many Sonoran Desert birds are shared with other low, arid, and open habitats of the Southwest; the 14 common breeding species in the Mohave Desert (Cody 1974) breed also throughout the Sonoran Desert, with the single exception of the Mohave Desert near-endemic, LeConte's Thrasher (*Toxostoma lecontei*), which occurs only locally in the Sonoran Desert in the northeastern peninsular lowlands and in the Desierto de Vizcaino region of central western Baja California (Grinnell 1928), where the endemic isolate *T. l. arenicola* is a candidate for full species status (Sheppard 1996).

As an introduction to the Sonoran Desert avifauna, the results of censuses at 12 mainland and peninsular sites are given in appendix 10.1, all taken in similar habitat during the breeding season in 5- to 8-ha plots. These sites range from Organpipe Cactus National Monument, Arizona, near the head of the gulf, south down the peninsula to Puertecitos, Rancho Arenoso, and Cataviñá, at El Arco and Casas Blancas in the central peninsula, and in the south at San Juan Road and near La Paz. The series is continued to the east across the gulf at Huatabampo, thence north to Hermosillo and Tucson. A site much farther south, at Ciudad Guzmán, Jal, is included because, not only is the vegetation in this dry, interior valley essentially Sonoran in character (though it is well beyond the range of vegetation so classified), but it serves to illustrate the broad geographic ranges of many typically Sonoran Desert bird species.

The striking aspect of these censuses, from widely spaced locations, is their similarity. Leaving aside the Jalisco site, 10 species occur at all 11 Sonoran Desert sites, and a similar number of additional species occurs at three-quarters of the sites. Thus, the same 20 or so bird species can be seen and censused throughout the Sonoran Desert, whether one is in Baja California, Arizona, or Sonora. Notably, 10 of these Sonoran "core" species also occur in the desert scrub near Ciudad Guzmán around 1000 km farther south, and all but the larger two woodpecker species are regularly constituents of the Mojave Desert also.

The site totals, measures of α-diversity, are relatively modest at 21–31 species, with the difference between the smaller and the larger counts attributable to rarer or more local species (19/21 species at Huatabampo, the least diverse site, occur in at least 4 other sites). Not only is local diversity modest, but censuses change with changing habitat only by additions or subtractions from the same basic species list, with few species turnovers; thus β-diversity is also low.

Some of the differences between the censuses in appendix 10.1 are due to the presence of ecological counterparts in different regions of the desert (the γ-diversity component). Thus, each site has a single quail species, but three are listed (app. 10.1; hereafter see appendixes for scientific names). California Quail occurs in desert habitats on the peninsula, Gambel's Quail in similar habitats elsewhere (Arizona, Sonora), and the Banded Quail in the Jalisco desert. One site, Puertecitos, lists two species, but California and Gambel's Quail form hybrids there. Similarly, all sites have just one species of thrasher, even though, again, three species (Gray, Curve-billed, and Bendire's) are listed, and most sites have two of the three hummingbird species censused. The ubiquitous Cactus Wren is recorded at every Sonoran Desert site, but it is replaced by a congener, Rufous-naped Wren, in Jalisco, where there is also a different oriole (Streak-backed) from those found typically in the north (Hooded, Scott's). Subtle differences in habitat might account for the presence of generally one or the other of Poorwill or Lesser Nighthawk, and the very widely distributed Black-throated Sparrow is replaced in the brushier desert of southern Sonora by the Rufous-winged Sparrow and by other *Aimophila* sparrows in Jalisco. Thus, some of the species turnovers between sites follow from the loss of species with restricted geographic ranges (e.g, Xantus Hummingbird, Lucy's Warbler), and others because some species have only a local occurrence in this desert habitat (e.g., Western Scrub-jay, Black-chinned Hummingbird, Lesser Goldfinch).

In appendix 10.2, the γ-diversity differences, in terms of the average number of species changed among the 12 mainland and peninsular desert sites, are summarized. These turnover figures are computed from the formula $100\{1 - C[(S_1 + S_2)/2S_1S_2]\}$, where two censuses of S_1 and S_2 species hold C species in common. The percentage turnover ranges from 7% to 67%, with the lowest turnovers manifest among the sites closer together at similar latitude (Tucson, Organpipe, Puertecitos, Hermosillo). A plot of γ-diversity against the distances between the census sites is shown in figure 10.1. The largest turnovers in bird species occur generally between sites farthest apart; for pairs of sites where intersite distance is entirely over land, 59% of the variance in γ-diversity is explained by intersite distance. However, many sites are separated by the gulf and have intersite distance partly over the sea. Water gaps enhance species turnover between sites; the regression line for sites with water gaps >200 km (species turnover, SPTURN = 18.38 + 0.019 (DIST); $p = .017$, $n = 22$) has an intercept significantly higher than that for sites with water gaps <200 km (SPTURN = 12.84 + 0.019 (DIST); $p < .001$, $n = 44$), although their slopes are the same. The difference in intercept, 5.5%, amounts to a 7% reduction in the number of species held in common between sites with extensive water gaps between them. The proportion of separation distance over water is significantly and positively correlated with the residuals in figure 10.1 ($r = .40$, $p < .01$), the absolute size of water gap less strongly correlated ($r = .27$, $p \approx .01$). Further, deviations from the regression line are weakly correlated with habitat differences, as estimated by differences between sites in the proportion of annual precipitation that occurs as summer rainfall ($r = .23$, $p \approx .07$; see appendix 10.3 for statistical details).

Generally, the different components of diversity have different controlling factors (see, e.g., Cody 1983b). Among those factors hypothesized to account for the overall low diversity (α-, β- and γ-diversity components included) are the low productivity and the open nature of the vegetation, both of which limit the number of ecological opportunities and therefore limit α-diversity. Another factor is that the area of the Sonoran Desert and of other similar arid habitats in the southwest is not great now, and it was much more restricted during the Pleistocene pluvials; the constriction of the areal extent would limit both α-diversity and especially β-diversity (Cody 1975, 1983b, 1993, 2000), although fragmentation of the deserts into different and isolated refugia during wetter and cooler periods provides speciation opportunities and promotes γ-diversity (see Hubbard's 1974 treatise on speciation patterns in arid-zone birds). Finally, this desert is of relatively recent evolution in North America (see chap.

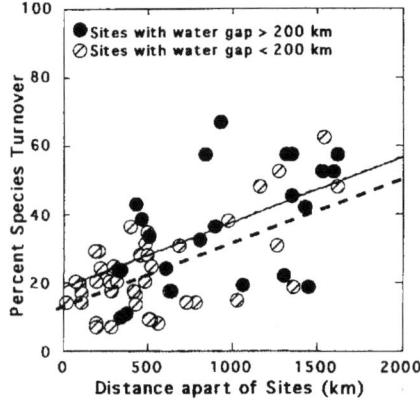

Figure 10.1 Bird species turnover among Sonoran Desert sites around the Sea of Cortés as a function of the distance apart of the sites. Appendixes 10.1–10.3 give details of the data and analysis. Note that species turnover is greater (upper regression line, solid symbols) among sites separated by an extensive water than among sites with no or a smaller water gap (lower dashed regression line, hatched symbols).

4), and thus bird diversity and its components are likely to be incipient and increasing rather than representing equilibrial values.

The Thorn Scrub

Around the southern and eastern periphery of the gulf, where summer rainfall exceeds 300 mm (see fig. 4.4), the Sonoran Desert merges into the denser, generally taller and brushy habitat known as thorn scrub. This vegetation is usually dominated by *Acacia*, along with other leguminous trees and shrubs (chap. 4), among which *Prosopis* (mesquites), *Lysiloma, Erythrina* (coral tree), and *Bursera* are common. Nine mainland sites at which this habitat was censused are listed in appendix 10.4, from the Chiricahua Mountains (Arizona) in the north, and south through Sonora and Sinaloa to Jalisco. Two sites in the cape region of Lower California are also included, at San Bartolo and San Matías. The bird lists are given in appendix 10.4, showing that species richness is somewhat higher in general in the thorn scrub relative to the desert habitats, and it reaches higher values at mainland sites than on the peninsula. The inland sites in Sonora (Nácori Chico to the Rio Cuchujaqui) support the highest α-diversities, and this is where the thorn-scrub vegetation is most extensive. There are fewer species at the coastal site in Sinaloa, the southern sites in Jalisco, and in the cape region (i.e., in locations where the thorn scrub is more restricted in areal extent). Many of the thorn-scrub species of Sonora are restricted to this habitat or else occur here and also in tropical deciduous woodland (typical species would be Elegant Quail, White-tipped Dove, Nutting's Flycatcher, Sinaloa Wren, Black-capped Gnatcatcher, etc.). But these and other typical thorn-scrub species are largely absent from the cape region, where only Yellow-billed Cuckoo and Varied Bunting can be so classified. Two other species common in cape thorn scrub are fairly cosmopolitan (Common Ground Dove and Blue-gray Gnatcatcher, in more open and taller habitat, respectively), and as many as 25 of the 29 species listed are those also recorded in the Sonoran Desert sites. This proportion (86%) is nearly as high in the Arizona mesquite scrub (81%), but it is much lower farther south on the mainland (Sonora and Sinaloa: 45%; Jalisco: 55%), where desert habitats are shrinking and thorn scrub is expanding. In comparing species turnover between desert and the nearest thorn-scrub censuses, there is no significant distance effect (see above). However, the difference in habitat corresponds to a combined gain (of "new" thorn-scrub species) and loss (of "old" desert species) of 21.7 ± 8.7 (SD) species on the mainland, versus 11.0 ± 2.2 species in the cape. This means that there are many fewer typical thorn-scrub species available as colonists adjacent to those gulf islands, especially Cerralvo and San José, which do have some representation of this habitat.

The Cape Region Endemics

Like the lowland thorn-scrub habitats, the high-elevation pine-oak woodlands in the cape region of Lower California are dramatically different in avifaunal composition from similar habitats in mainland Mexico. Because this vegetation type does not occur on the gulf islands, we do not discuss its bird species at length here. However, whereas in mainland Mexico the highland pine-oak woodlands have an exceedingly diverse bird fauna with many species endemic to the habitat (see Cody 2001), the same vegetation in the cape region (despite the similarity in pine and oak species to those in the highlands of Colima, with which the cape region once abutted; see chap. 2) is populated largely with bird species typical of the oak woodlands of southern California and the pine-oak woodlands and pine forests of the Sierra San Pedro Mártir in northern Baja California. It is not possible to say for how long the original Mexican highlands avifauna survived in the cape region as it rifted from the mainland, but presumably both the direct access provided to the northern bird species down the northward-drifting peninsula, and the southward expansion down the peninsula of the chaparral and pine-oak woodland vegetation and their bird avifaunas during Pleistocene sealed the fate of the original inhabitants of the cape pine-oak forest.

Remnants of the mainland avifauna still survive in the cape region. One example is the endemic junco, *Junco bairdi* (which some authorities, e.g., Sullivan [1999], consider conspecific with *J. phaeonotus*), restricted to the higher cape mountains largely above 1000 m, where it is common. This junco is considered more closely related to Central American than to northern relatives in California (Miller 1941; Davis 1959; Stager 1960); it is found neither in thorn scrub nor the Sonoran Desert. A second cape region endemic is Xantus Hummingbird *Basilinna* (ex. *Hylocharis*) *xantusii*, which again has its closest affinities with a Mexican highland species, the White-eared Hummingbird *Basilinna* (*Hylocharis*) *leucotis*. Unlike the first species, Xantus Hummingbird has an expanded range and breeds in both the thorn scrub and desert. It is found north throughout the peninsula in desert habitat as far as the Mediterranean-climate zone at 31°N in the northwest and as far as Cataviñá in the center of the peninsula, but it avoids the more arid gulf coast north of Santa Rosalia. A third cape endemic is the robin, which previously enjoyed full species status as *Turdus confinis*; its relationships apparently are with the northern *T. migratorius* (Davis 1959; Sallabanks and James 1999) rather than with Central American robins, and it is now classified as a subspecies of the latter, *T. m. confinis*.

The remaining noteworthy peninsular endemic is the Gray Thrasher *Toxostoma cinereum*, a bird of the desert and scrub rather than the mountains, and it is discussed below.

Peculiarities of the Peninsular Deserts

Four bird species in particular distinguish the avifauna of the peninsular deserts from similar habitats to the north and east: Western Scrub-jay, California Quail, Xantus Hummingbird, and Gray Thrasher. The last two species are endemic, to the southern two-thirds and three-quarters of the peninsula, respectively. The hummingbird extends into the desert from presumed highland origins in the cape region and complements Costa's Hummingbird in a role comparable to that of the Black-chinned Hummingbird elsewhere. The Gray Thrasher is a close relative of the Bendire's Thrasher (Engels 1940; England and Laudenslayer 1993), which serves as its counterpart in Sonora and Arizona, as does the much more widely distributed Curve-billed Thrasher. The Gray Thrasher presumably differentiated in the isolation of a Sonoran Desert refuge on the peninsula during the last (Wisconsin) glaciation (Hubbard 1974) or perhaps earlier according to accumulating data from molecular genetics (e.g., Zink and Slowinski 1995).

The other two species, Western Scrub-jay and California Quail, are both species typical of the California chaparral and oak woodland. Presumably they gained entry into desert habitats when these northern habitats covered most of the peninsula during the Pleistocene and thereby provided these birds with a broad base from which to move into the lower and drier desert refugia. California Quail provides an obvious counterpart to its congener, Gambel's Quail, which is widespread in the Sonoran Desert elsewhere, but the scrub-jay has no ecological equivalent in the deserts of Sonora or Arizona, and its role in the peninsular desert is a unique one.

Range maps of three representative species groups, thrashers, hummingbirds, and quail, are shown in figures 10.2–10.4. The maps illustrate the ranges of the peninsular endemics and the distributions of ecological counterparts across the Sea of Cortés.

Patterns of Differentiation

Before evaluating patterns of differentiation in the birds of the gulf islands, we first summarize these patterns in peninsular birds so that island variants might be put into better perspective. Most of the widely ranging peninsular species are divided into subspecies on the basis of slight morphological differences in such characteristics as coloration, size, and proportions. Grinnell (1928) compiled the ranges and some attributes of these subspecies, and it is apparent that each corresponds to a particular climate or vegetational zone. The contiguous areas of Baja California to which distinct subspecies have been attributed are the same or similar in many different species: (1) the northwest Pacific slope, with its Mediterranean-climatic influence; (2) the northeast ("San Felipe") desert along the gulf coast south to about 30° latitude; (3) the peninsula south of about 28–29°; and (4) the cape region proper. For example, the Cactus Wren *Campylorhynchus brunneinucha* is divided into subspecies in regions (1) and (2) (*C. b. bryanti, C. b. couesi*) and a third that occupies the rest of the peninsula. Grinnell's Black-tailed Gnatcatcher (*Polioptila melanura*) is a complex taxon that occupies practically the whole peninsula, with the exception only of the higher mountains. His four subspecies included a cape region subspecies (*Polioptila melanura abbreviata*), a second (*P. m. margaritae*) in the desert north of the cape to around latitude 29°30′, a third in the Mediterranean climatic zone of the northwest (*P. m. californica*), and a fourth in the San Felipe desert (*P. m. melanura*). The ranges of at least the first three taxa are contiguous, each intergrades into the adjacent subspecies over limited areas, and within the range of each subspecies the characteristics of the type are apparently conserved. These northwestern and peninsular races are now given full-species status, as California Gnatcatcher *P. californica*, with nominal Black-tailed Gnatcatchers (*P. m. melanura*) retained in the northeast and also across the gulf on the mainland (Arizona, Sonora).

This pattern of contiguous ranges of subspecific populations is repeated in most peninsular species, with the following generalizations. Just six of the species that breed in the Sonoran Desert on the peninsula are not divided into subspecies (Common Raven, Costa's Hummingbird, Purple Martin, Scott's Oriole, Canyon and Rock Wrens); all six have wide geographic ranges, and at least the first three are quite cosmopolitan in habitat preference. A further 12 desert species are represented by a single subspecies throughout the southern part of the peninsula, north to between

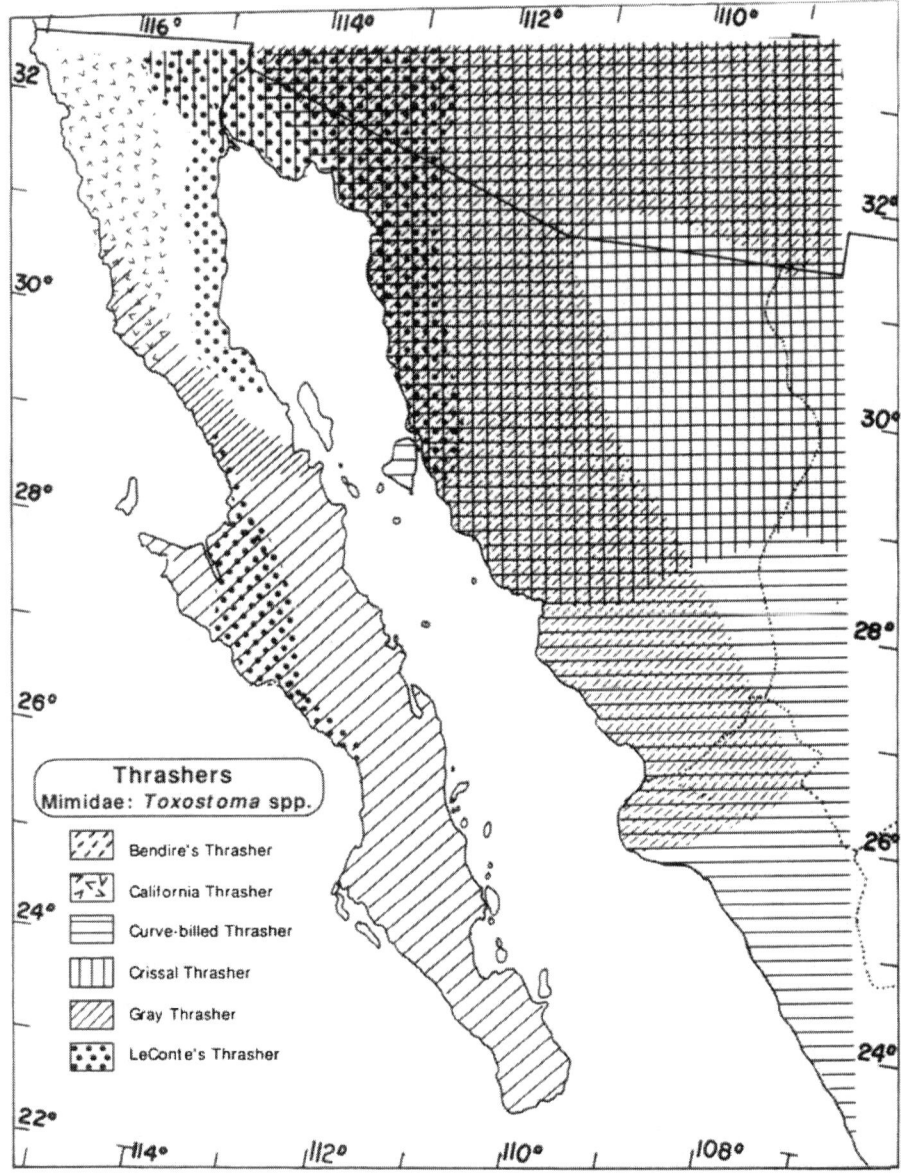

Figure 10.2 Breeding ranges of thrashers (*Toxostoma* spp.) around the Gulf of California.

28°30′ and 29°30′N, whereupon their subspecific designations change. This is the latitude at which the desert changes to coastal sage and chaparral to the northwest, to the drier desert of the San Felipe area to the northeast, or to mountain woodland and forest in the center of the peninsula. Some of these desert species extend their ranges only a limited distance into the Mediterranean vegetation of the northwest with differ-

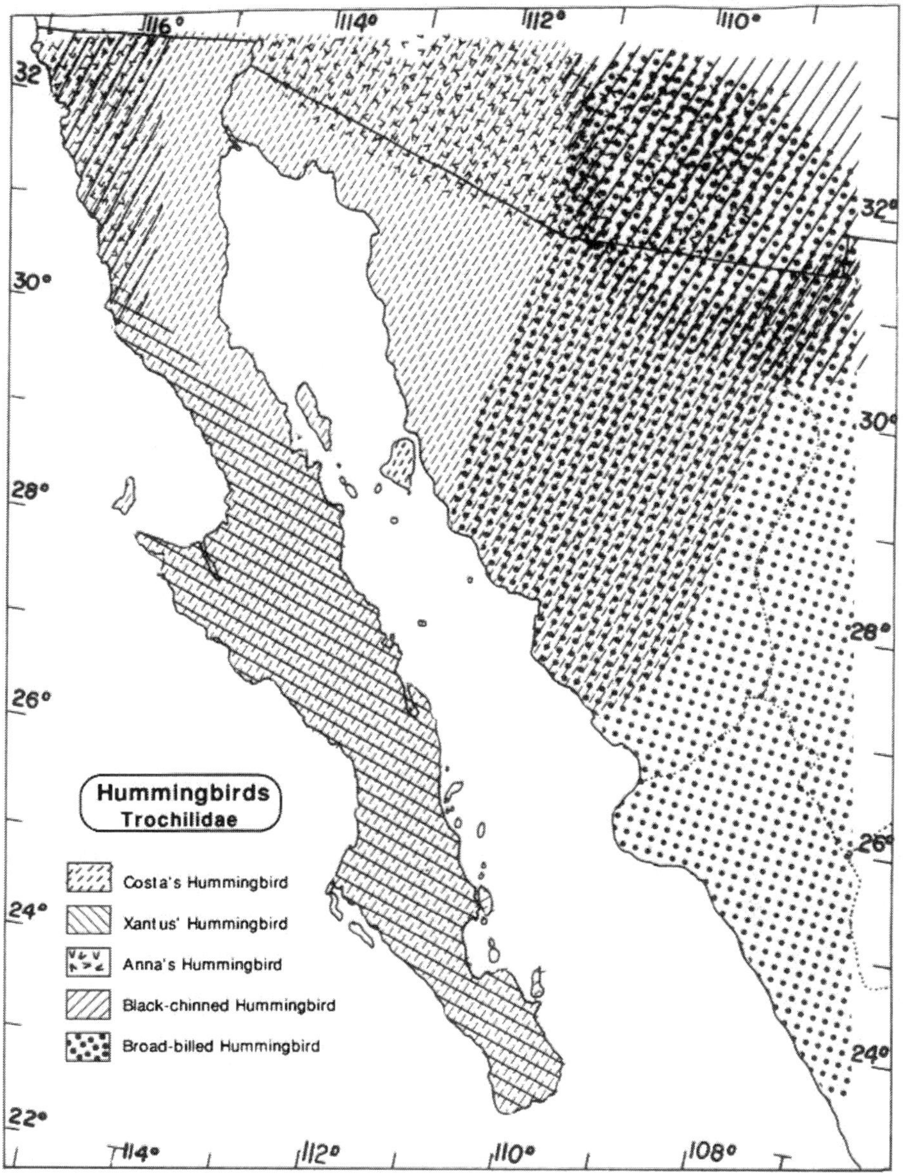

Figure 10.3 Breeding ranges of hummingbirds (family Trochilidae) around the Gulf of California.

ent subspecies (e.g., Gray Thrasher, Cactus wren, Northern Flicker); others with different subspecies in this vegetation extend into California (California Gnatcatcher, Hooded Oriole, Violet-green Swallow, Brown Towhee, Western Scrub-jay). Yet other species are represented by a distinct subspecies throughout the northern part of the peninsula (House Finch, Ash-throated Flycatcher, Loggerhead Shrike), and another

280   The Biological Scene

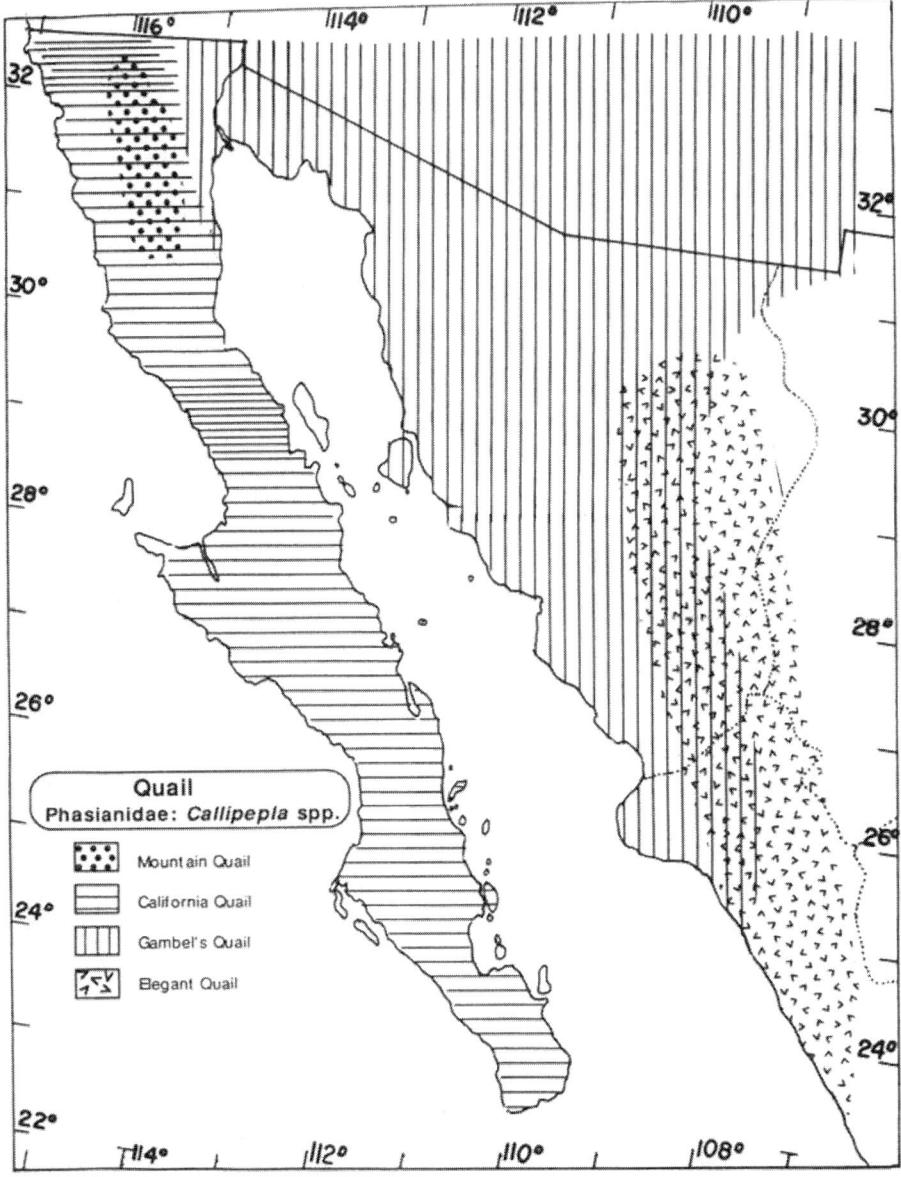

Figure 10.4 Breeding ranges of quail species (*Callipepla* spp.) around the Gulf of California.

species set avoids the Mediterranean climatic zone and extends east of the northern mountains toward the Río Colorado, in the basin of which their subspecific identities again change (Gila Woodpecker, Ladderbacked Woodpecker, "Gilded" Flicker, Verdin).

Two characteristic species of the Sonoran Desert do not extend into the drier deserts of the northern peninsula: Red Cardinal and Pyrrhuloxia. The former is common only south of 27°N; the latter extends only sporadically north of the cape region, to around Bahía Concepción; both are represented by endemic peninsular subspecies. Two other species, Black-throated Sparrow and Brown Towhee, are represented by different subspecies south of latitude 27°, a range corresponding to the limits of a brushier desert with more summer rain in the south. The desert species endemic to the peninsula (Gray Thrasher), and subspecies of desert birds endemic to the southern part of the peninsula (Cactus Wren, Black-throated Sparrow, California Gnatcatcher, Brown Towhee, House Finch, the three desert woodpeckers, etc.), and the presence of desert species restricted to and differentiated in the southern part of the peninsula (Red Cardinal, Pyrrhuloxia) indicate a possible refugium for Sonoran Desert habitat in the southwestern peninsula during Pleistocene pluvial periods, perhaps in the Magdalena Plains area (see chap. 4 for corroboration from plant distributions).

Island Endemics

There are no endemic bird species on the islands in the Gulf of California, and taxonomic divergence in the island populations is limited to a few weakly differentiated subspecies. Generally, the islands support species whose populations are indistinct from those on the closest adjacent peninsula or mainland. Thus, the Black-throated Sparrows on Monserrat, Coronado, Santa Catalina, Danzante, San Marcos, and Espíritu Santo have been identified as the southern peninsular *Amphispiza bilineata bangsi* (van Rossem 1945a), but with distinct races on Cerralvo (*belvederei*), Carmen (*carmenea*) and San Esteban (*cana*). The Red Cardinals of Cerralvo have also been described as different (*Cardinalis cardinalis clintoni* Banks), but those on Coronado, Carmen, and Santa Catalina are the same as the southern peninsular *C. c. ignea*. A third species with a comparable degree of subspeciation is the Ladderback Woodpecker, with five races around the gulf and one of these, *Dendrocopus scalaris soulei*, endemic to Cerralvo.

In general, bird species are somewhat more differentiated (at the subspecies level) on two particular islands within the gulf: on the large southern island of Cerralvo (Banks 1963b,c) and on the central midriff island of San Esteban (Banks 1969). Cerralvo is an island with a good representation of thorn-scrub vegetation, yet it is populated with exclusively Sonoran Desert bird species, and it is not a landbridge island. Thus its isolation might permit the morphological divergence of its birds in response to the somewhat different conditions of climate and vegetation, a divergence that would be precluded on the peninsula where these same habitats and bird species interbreed with populations of the same species from their more typical desert habitats. San Estéban however, with its central gulf location, might be, and has been, colonized by birds either from the peninsula or from Sonora, and thus several species there are intermediate between otherwise allopatric subspecies to the west and east. Blue-gray Gnatcatcher, House Finch, Ash-throated Flycatcher, and Verdin are in this category. As mentioned, the Black-throated Sparrow on San Esteban is described as distinct and (subspecifically) endemic, and certainly the island is more physically isolated than most others in the gulf. Several subspecies are described from Tiburón (van Rossem 1932), and one of these, the Curve-billed Thrasher *Toxostoma curvirostre insularum*, is endemic to both Tiburón and San Esteban.

The paucity of well-differentiated island birds is attributable to the fact that the physical conditions on the gulf islands, in terms of both climate and vegetation structure, are generally similar to those of peninsular or mainland areas from which ancestral populations might have been derived. A second factor that would preclude the formation of distinct island forms is a combination of short residence times (before extinction) coupled with a continuous influx of mainland or peninsular relatives that could swamp incipient divergence on the islands. The importance of such an effect cannot be evaluated, but as some endemic races do exist on the islands, including *Amphispiza bilineata carmenae* on the landbridge island of Carmen, it appears that birds in at least some populations are relatively sedentary. A lack of divergence on the islands, then, may be owing more to the similarity of abiotic environments between islands and mainland. A further aspect of this similarity is in the biotic conditions on the islands, with similar plant species (chap. 4) and similar sets of potential competitors. The numbers and identities of bird species on the gulf islands are comparable to those of mainland areas, especially when one considers the bird diversity within a given vegetation type such as riparian scrub or desert scrub; thus, on the islands selection operates on the same species in the same settings.

Endemism among the gulf island bird populations contrasts with that among reptiles and mammals on the same islands. Whereas there are several bird subspecies on Tiburón, a recent landbridge island, there are virtually no reptiles endemic there. Conversely, several gulf islands that are relatively old or well isolated (e.g., Cerralvo, San Pedro Mártir) have high endemism rates in mammals and reptiles, whereas their bird endemism levels are very low. Clearly, the selective forces that produce local morphotypes differ among these taxonomic groups, at least in their outcome. The low-vagility reptiles and mammals adapt and respond to local environmental conditions with a far greater frequency than do the high-vagility birds, in which differentiation occurs over broader climatic gradients (e.g., east to west across Tiburón into the midriff islands, from less to more mesic regions) and does not occur as a function of local isolation.

A second factor that precludes more differentiation among the island birds is the fact that there are many fewer differences between mainland and island bird communities than there are between mainland and island lizard or mammal communities. Thus, the biotic component of selection for local differentiation is weaker in birds, contributing to their lower level of endemism.

Bird Diversity on the Islands

The Shoreline Birds

One group of land birds uses not the interior habitats of the gulf islands but rather the shorelines, which provide them with suitable nesting habitat and appropriate foraging opportunities. Most of these species are large (herons, egrets), widely ranging, and not at all unaccustomed to dispersing over distances such as those that isolate the gulf islands.

The breeding shoreline birds are listed in appendix 10.6, in which records from a variety of sources, including our own observations, are combined. These species are resident on the islands in general, but their commuting abilities are such that nesting birds have been recorded on fewer islands than have foraging birds. Thirteen species, mostly herons and egrets (Ardeidae) breed or feed regularly on the islands; the most ubiquitous shoreline species are the Great Blue Heron *Ardea herodias* (30 islands), American Oystercatcher *Haematopus palliatus* (28 islands), Snowy Egret (24 islands), and Belted Kingfisher (16 islands), four species that can be seen on virtually every larger island. Other species are more patchily distributed. The larger southern land-bridge islands, San José and Espíritu Santo in particular, with their variety of shoreline habitats beyond ubiquitous rocky shores, including sandy beaches and mangrove-fringed lagoons, are attractive to the largest numbers of species (12 and 10 respectively), and attract especially a wide variety of herons and egrets. Numbers of breeding shorebirds are related to island area [$\text{LogSPP} = 0.521 + 0.119*\text{Log}(\text{AREA})$; $R^2 = .34$, $p < .001$], unrelated to the degree of isolation per se ($p > .05$), and apparently enhanced by the shoreline characteristics around the large, southern islands.

Many other shorebirds may be seen in season on the islands, and their shores provide suitable habitat for plovers and sandpipers (Charadriidae, Scolopacidae) that overwinter in the gulf or are passage migrants. Many of the three dozen species of this group listed by Grinnell (1928) for the Lower California peninsula have been recorded on gulf islands, but in contrast, relatively few of the two dozen ducks, geese, and swans (Anatidae) Grinnell noted have island records. Presumably most species of this group require the larger peninsular estuaries and lagoons to support wintering populations or, as in others, wintering populations make use only of the outer (Pacific) coast of the peninsula.

The Raptors

A second group of land birds on the gulf islands is segregated ecologically and taxonomically from the smaller species of desert habitats, and we treat it independently for this reason. The raptors are, in general, species of wide and cosmopolitan distribution, catholic as to habitat preference, and are widely dispersing and capable of traveling long distances. The species list for the island raptors is given in appendix 10.6 and includes vultures, eagles, hawks, and owls (and purely for convenience the goatsuckers). Most of these predatory or scavenging birds are resident (16 listed species) and breed on the islands on which they have been recorded, but some undoubtedly do not breed on all islands of record, especially the smaller ones, to which they may commute to forage. The distinction is not made in our analyses of distribution and diversity. In addition, a few migrants have been recorded, notably the Northern Harrier *Circus cyaneus* and two *Accipiter* species on Tiburón. Bald Eagles, which persisted in the southern gulf up to the 1920s, are locally extinct there now. However, the Peregrine Falcon *Falco peregrinus* is very widely distributed throughout the gulf region and is regularly sighted on the islands.

The raptors censused in the mainland Sonoran Desert sites of appendix 10.1 are included in appendix 10.7 for comparative purposes. All of the mainland species have been recorded on the islands; indeed, the island raptors are those common over most of western North America and even beyond. Given the propensity of these species for long-distance movements and the requirements of most species for large breeding areas, it is understandable that more species tend to occur on larger and topographically more diverse islands. The species of most regular occurrence are the Turkey Vulture, Red-tailed Hawk, Peregrine Falcon, and American Kestrel. The owls, of which five species are listed, are generally far less predictable and likely more sendentary on a given island.

The species–area relation for raptors is shown in figure 10.5 and demonstrates strong dependence of diversity on area and an insignificant effect of landbridge status and of isolation in general. While area is a dominant influence, different species show different minimum area requirements. Incidence functions for 10 species of raptors are shown in figure 10.6, including the most widely distributed species. All show an orderly relation between distribution (incidence on islands ranked from smallest to largest in eight categories) and island size class. The two "shoreline" raptors, the Osprey and Peregrine Falcon, extend farthest onto the smallest islands, and occupy even some rather tiny islets. Red-tailed Hawks also show broad distribution and occur on virtually every island >2 km^2 in size. The incidence of Turkey Vulture decreases sooner, falling to 50% on class 3 islands (2–8 km^2); they are still unreported from Ángel de la Guarda, but this is most probably an oversight (though a rather perplexing one). American Kestrel falls out sooner still; the species occurs on all islands >20 km^2 and few smaller (e.g., Partida Norte, San Pedro Mártir).

The two most widely distributed owls are the Great Horned Owl and Burrowing Owl (fig. 10.6). Whereas the latter is recorded patchily over a wide range of islands, the former shows a tighter restriction to the largest islands (>100 km^2), although it is not (yet) recorded on non-landbridge Cerralvo (141 km^2). The Western Screech Owl reflects historical aspects of the islands and occurs only on the largest two landbridge islands (Tiburón and San José), but it is absent from non-landbridge islands of larger or similar size (Ángel de la Guarda, Cerralvo). The Elf Owl is found on the largest island, Tiburón (landbridge), the small and purportedly landbridge island Monserrat, and small, isolated Tortuga. Given the ubiquity of cardons on even the tiniest islands, and therefore the ready source of nest sites for this hole-nesting species, we can exclude the possibility that nest sites might constrain the distributions of hole-nesting birds on islands.

Finally, Barn Owls are recorded, very spottily, on seven islands of a very wide range of areas, down to Rasa (0.68 km^2). This is a similar number of islands to Great

Figure 10.5 Species–area curve for breeding raptor species on the islands in the Gulf of California. Distributional data are given in appendix 10.7. Species numbers are significantly related to island area, and the landbridge status of the island does not affect species richness.

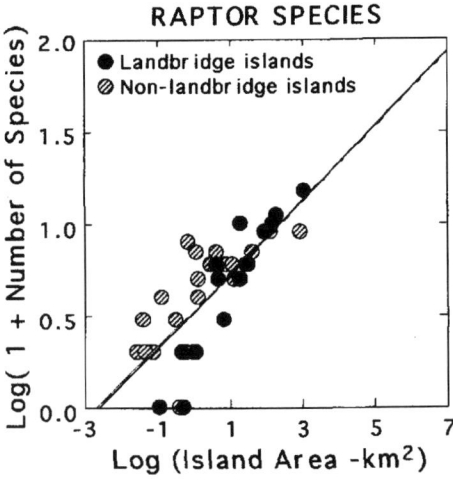

Land Birds 285

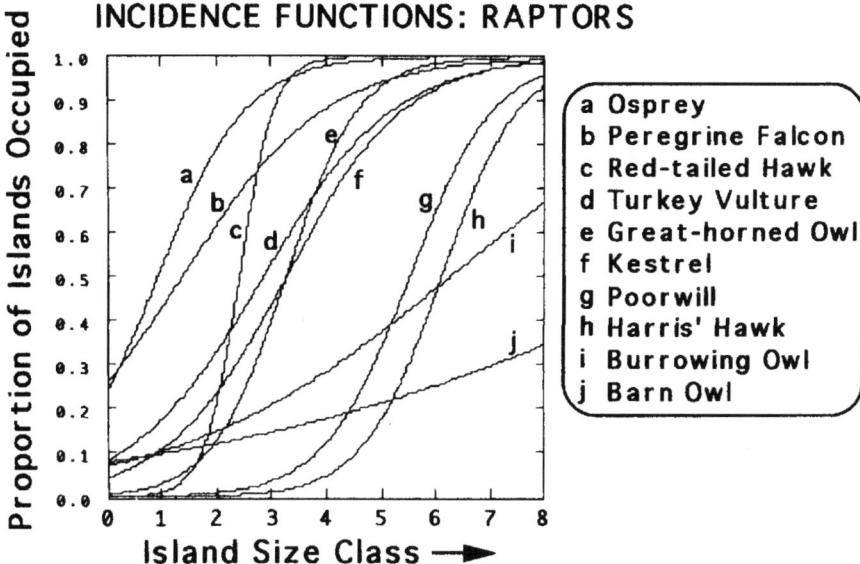

Figure 10.6 Incidence functions for breeding raptors on gulf islands. The most widespread species are Osprey and Peregrine Falcon, both of which breed and hunt along shorelines. Some species such as the Red-tailed Hawk show strong area dependence with steeply falling incidence at critical island area, whereas others, especially owls, have patchy island occurrence with weak area effects. Island size classes: 7, >500 km² $(n = 2)$; 6, 100–500 km² $(n = 3)$; 5, 32–100 km² $(n = 4)$; 4, 10–32 km² $(n = 5)$; 3, 2–10 km² $(n = 6)$; 2, 0.5–2 km² $(n = 8)$; 1, <0.5 km² $(n = 10)$.

Horned Owls (eight) but Great Horned and Barn Owl coexist only on Tiburón, a situation that is thought-provoking but too likely to be occasioned by chance alone to be further explored. It is possible that the somewhat spotty distribution records of the owls, including the several large-island absences, are due to oversight in these nocturnally active birds.

The Small Land Birds

The Northern Islands

The small land birds on the gulf islands are listed in appendix 10.8 for northern islands and appendix 10.9 for southern islands. This seems a natural subdivision for the bird species because the islands south of San Marcos, the adopted dividing line, are somewhat brushier with denser vegetation than the northern islands, and bird species are affected by this difference. Most species are distributed differently among the northern as compared to the southern islands, as is seen in the following tables and discussed below.

Our own observations are supplemented in the appendixes by museum and collection records assembled by Navarro et al. (*Atlas de las Aves de México*, in prep.). In addition, reference is made to records of earlier surveyors, but only when species were not seen on an island by both of us, nor by our colleagues. These observations are segregated into three time periods: pre-1970s, 1970s–early 1980s, and 1985–1990s. Note that not all islands were censused in all three time periods. Differences in bird species between survey periods seems more safely attributed to chance and the vagaries of the year, season, location, and intensity of census effort, rather than to species turnover with time. On the three northern largest islands where the two of us have surveyed the land birds, Ángel de la Guarda, San Esteban, and San Lorenzo, Cody (pers. obs.) has recorded 0.81, 0.80, and 0.80 of the listed species, respectively, and Velarde and associates (pers. obs.) have recorded 0.76, 0.85, and 0.73 of the species. We expect by chance that the two series would discover 12.9, 13.6, and 8.8 species in common, and, in fact, they record 13, 14, and 8 species, respectively, numbers clearly indistinguishable from those expected by chance. Further, we would expect independent surveys to miss a proportion $(1-p)(1-q)$ of the species, which amounts to 1.0, 0.6, and 0.8 expected species missed, and, in fact, just two species, the Blue-gray Gnatcatcher on San Esteban and the Black-tailed Gnatcatcher on Ángel, were recorded on the islands earlier but not by us. Thus the surveys seem relatively complete.

Some 39 bird species are recorded from the group of northern islands, and the confirmed or probable breeding species are ranked in table 10.1 from those species that occur on a wide range of different-sized islands, such as the Common Raven (17 islands) and Rock Wren (14 islands), to those 9 species that occur only on the largest island in the group, Tiburón. Islands likewise are ranked from those with most species (Tiburón, 30 spp.) to those with fewest (several islands with a single species). Tiburón, with its large area and minimal as well as recent isolation, is not very islandlike. Separated from the Sonoran mainland by the Infiernillo channel just 2 km wide, it has been an island only for the last 10,000 years. The size of its bird fauna reflects this fact. Table 10.1 incorporates all records from this island, and its 30 species are essentially those that would be censused in a mainland area of comparable size and topographic diversity, or even in the much smaller mainland censuses of appendix 10.1. Note that several species that are resident on Tiburón breed on no other islands in the northern gulf, although their ranges extend much farther south in most cases; the "Gilded" Flicker, Brown Towhee, Pyrrhuloxia, Purple Martin, Red Cardinal, and Common Ground Dove are examples, though the last three do occur on some southern islands. In additional, Gambel's Quail occurs on no other island, being replaced on the peninsula by California Quail, which is recorded on the two largest southern landbridge islands, and a single record of Pyrrhuloxia from a single southern island is discussed below.

Table 10.1 ranks bird species by number of islands occupied and islands according to numbers of species present. In ordering the matrix of species occurrences in this way, the islands are sorted close to their size order (Spearman $r = .884$, $p < .001$), and a record of species by island is obtained that is concentrated to the upper left and displays rather few gaps or holes. This means that not only is species number predominantly related to island size, but species identities are similarly predictable. That is, each species tends to occurs on a range of islands from the largest down to a certain size, and each island tends to support a set of species down to a certain level in the table. This characteristic is called "nestedness," and species in table 10.1 are significantly nested relative to null models (Wilcoxon rank-sum tests; $p < .05$ that the table as a whole is nested; $p < .05$ for individual species except Common Raven and Rock Wren). This feature indicates that there are very few species turnovers with increasing island size, but rather the species lists on the larger islands are built up simply by adding to the species that occur on smaller islands. Data of this sort were seen also in chapter 8 with the herpetofauna; they indicate quite specifically that (1) there are few ecological substitutes among the island birds (sp. A ↔ sp. B), and so an island of a given size supports a given species set with very few alternatives; and (2) there must be few species turnovers in time, for if an extinction occurs on an island, it is filled presumably by colonists of the same species (sp. A → sp. A) rather than of another species or perhaps by density compensation by other residents (i.e., an increase in density of the remaining species on the island).

In table 10.1, one of the rows (species rankings) is accorded to two species of gnatcatchers, Black-tailed and California, which until recently were considered conspecific. Now different species in this taxon, rather than subspecies, occur allospecifically on Tiburón (Black-tailed Gnatcatcher) versus on islands toward the western peninsula (California Gnatcatcher). Note that, within a broader taxon such as wrens (Troglodytidae), species show very different patterns of island occurrences. Some of the largest islands, as well as many mainland sites, have three species—Cactus, Canyon, and Rock Wrens; however, relatively few larger islands support more than two of these species, but islands down to the size of a few hundred square meters accommodate Rock Wrens, which are exceptionally good island colonists and inhabit virtually every island in the gulf. The Cactus Wren was cited previously as a good example of a poor island colonist, and until recently it was recorded on no gulf island except Tiburón (despite the abundance of cacti on the islands; see chap. 4). However, within the last decade the Cactus Wren has been recorded on several smaller and non-landbridge islands, both north and south. One wonders if this apparent expansion is attributable perhaps to some genetic novelty favoring good dispersal, or perhaps to changes in distribution or density in the source populations, as it seems unlikely that altered conditions on the islands over the last decade favor the persistence of the Cactus Wren for the first time. It is unlikely also that it has been present but overlooked before now.

Although, in general, isolation does not significantly contribute to variation in species numbers over islands, it does have relevance to the ragged edge between occupied and unoccupied islands in table 10.1. San Luis, Estanque, Turners, and Cardonosa are all landbridge islands, recently connected to the mainland, or Ángel de la Guarda, or Tiburón, respectively. The islands of Rasa and San Pedro Mártir have received intensive work from sea bird researchers, and some of the land bird records there may be of vagrants (app. 10.8) or casual breeders (table 10.1) revealed by the disproportionate amount of time we have spent there. Both the very precipitous San Pedro Mártir and the very level Rasa have extensive sea bird colonies. Previous censuses on San Marcos were judged likely incomplete (Banks 1963a, Case 1978), but with recent work (Velarde et al.; this chap.) the roster has filled out.

Notably, recent field data have substantiated nearly all of the older island records; only 17, or <10%, of the entries in appendix 10.8 (and 13 in app. 10.9) are exclusively from the older literature or collections. Among the northern islands, these include no recent verification of the White-winged Dove and Rock Wren on San Marcos, of the Gilded Flicker, Scott's Oriole, and Pyrrhuloxia on Tiburón, the Blue-gray Gnatcatcher on San Esteban, the House Finch on Animas Norte, and the Ash-throated Flycatcher

Table 10.1 Northern islands: nested matrix of species by islands

Species	Tiburon	Ángel	San Esteban	San Lorenzo	Animas N	Partida N.	San Marcos	Tortuga	S. F Mar
Corvus corax	●	●	●	●	●	●	●	●	●
Salpinctes obsoletus	●	●	●	●	●	●	●	●	●
Tachycineta thalassina	●	●	●	●	●	●	●	●	
Carpodacus mexicanus	●	●	●	●	●	●	●	●	●
Calypte costae	●	●	●	●	●		●	●	
Amphispiza bilineata	●	●	●	●	●	●	●	●	
Zenaida macroura	●	●	●	●	●	●	●		●
Mimus polyglottos	●	●	●	●	●	●	●	(●)	●
Auriparus flaviceps	●	●	●	●	●		●	●	
Lanius ludovicianus	●	●	●	●	●	●			●
Myarchus cinerascens	●	●	●	●	●	●	●	●	
Picoides scalaris	●	●	●	●	●	●		●	
Polioptila mel/calif	●	●	●	●			●	●	
Catherpes mexicanus	●	●			(●)	(●)			
Zenaida asiatica	●	●	●				●		
Phainopepla nitens	●	●	●	(●)					
Campylorhynch. brunn.	●		●		●				
Toxostoma curvirostre	●		●						
Aeronautes saxatalis		●		●		(●)			
Icterus parisorum	●	●	●						
Polioptila caerulea		(●)	●						
Melanerpes uropygialis	●		●						
Molothrus ater	●	●							
Pipilo fuscus	●								
Icterus cucculatus	●								
Callipepla gambeli	●								
Columbina passerina	●								
Colaptes auratus	●								
Progne subis	●								
Cardinalis cardinalis	●								
Cardinalis sinuatus	●								
Dendroica petechia	●								
No. of species	30	20	20	15	14	13	12	10	

Tabulated entries include all those of likely breeding species, judged by observations, breeding range, and habitat requirements more tentative records are those within parentheses.

Rasa	Salsi-puedes	San Luis	Estanque	Turners	Carde-nosa	Peli-cano	Granito	Encan-tada	R. Partida	R. Rasa	No. of Islands Occupied
●	●	(●)	(●)		(●)		(●)		(●)	(●)	17
●	●	●				●		●			14
●	●										10
●											10
		●		●							9
				●							9
	●										9
											9
			●								8
			●								8
											8
											7
											6
											4
											4
											4
											4
											3
											3
											3
											2
											2
											2
											1
											1
											1
											1
											1
											1
											1
											1
											1
4	4	3	3	2	1	1	1	1	1	1	

Table 10.2 Southern islands: nested matrix of species by islands

Species	San José	Espíritu Santo	Carmen	Cerralvo	Monserrat	S. Catalina	Partida Sur	Coronado
Amphispiza bilineata	●	●	●	●	●	●	●	●
Corvus corax	●	●	●	●	●	●	●	●
Salpinctes obsoletus	●	●	●	●	●	●	●	●
Carpodacus mexicanus	●	●	●	●	●	●	●	●
Picoides scalaris	●	●	●	●	●	●	●	●
Auriparus flaviceps	●	●	●	●	●	●	●	●
Myarchus cinerascens	●	●	●	●	●	●	●	●
Calypte costae	●	●	●	●	●	●	●	●
Polioptila mel./californ.	●	●	●	●	●	●	●	●
Cardinalis cardinalis	●	●	●	●	●	●	●	●
Mimus polyglottos	●	●	●	●	●	●	●	●
Zenaida asiatica	●	●	●	●	●	●	●	●
Polioptila caerulea	●	●	●	●	●	●	●	●
Lanius ludovicianus	●	●	●	●	●	●		●
Aeronautes saxatalis	●	●	●		●	●		
Zenaida macroura	●	●	●	●		●		●
Tachycineta thalassina	●	●	●	●	●		●	●
Melanerpes uropygialis	●	●	●	●	●	●	●	
Icterus cucculatus	●	●	●	●		(●)		
Progne subis	●				●			
Catherpes mexicanus	●	●		●			●	
Columbina passerina			●	●	●			
Dendroica petechia	●	●					●	
Toxostoma cinereum	●			●		(●)		
Hylocharis xantusi	●	●		●				
Sayornis nigricans	(●)	(●)	(●)					
Campylorhynchus brunn.	(●)							
Icterus parisorum			●	●				
Carduelis psaltria		●			●			
Callipepla californica	●		●					
Pyrocephalus rubinus		(●)						
Geothlypis beldingi	(●)							
No. of species:	28	24	24	23	20	19	17	16

Tabulated entries include all those of likely breeding species, judged by observations, breeding range, and habitat requirement. The more tentative records are those within parentheses.

Danzante	Santa Cruz	S. Francisco	San Ildefonso	San Diego	Gallo	Animas Sur	Los Islotes	Gallina	No. of Islands Occupied
●	●	●	●	●	●			●	15
●	●	●	●	●	●	(●)			15
●	●	●	●	●		●			14
●	●	●	●	●	●				14
●	●	●	●		●				13
●	●	●	●						12
●	●	●							11
●	●	●							11
●	●	●							11
●	●	●							11
●	●								10
	●								9
●									9
	●								8
●		●							7
		●							7
									7
									7
									5
●							(●)		4
									4
									3
									3
									3
									3
									3
				(●)					2
									2
									2
									2
									1
									1
14	13	12	6	5	4	2	1	1	

on Tortuga. This concordance supports the notion that species turnovers and extinctions on the gulf islands are low.

The Southern Islands

Records of the breeding land birds of the southern gulf islands are assembled in appendix 10.9. Cody's records from the four largest islands averaged just 70% of their listed avifaunas, whereas Velarde et al.'s records (with 87% of records verified) were more thorough. Only one older record of species expected to breed, the Mourning Dove on San José, has not been recently corroborated; indeed, this species seems to have been noticeably scarce of gulf islands recently. On smaller southern islands, Danzante up to Partida Sur (and excluding Coronado), Cody's records cover 86% of the listed species, and an intensive study on Carmen, Monserrat, Coronado, and Danzante (George 1987a) contributed two new island records and verified several others. Nevertheless, a number of older reports lack recent verification, especially those of Banks (1963a) of White-winged Dove on Santa Cruz, Mourning Dove and Red Cardinal on San Francisco, and Ladderbacked Woodpecker on San Ildefonso. Alternative explanations for these and other older records not recently confirmed, that these species have been overlooked in recent surveys or that the island populations have become locally extinct, cannot be distinguished.

In table 10.2, derived from appendix 10.9, the census results from the gulf islands south of San Marcos are compiled, with species and islands ranked as in table 10.1. Thirty-two species likely breed on southern islands, the same as in the north. However, table 10.1 includes Gambel's Quail and Curve-billed Thrasher, replaced in table 10.2 by California Quail and Gray Thrasher, with Phainopepla, Brown-headed Cowbird, Brown Towhee, Gilded Flicker, and Pyrrhuloxia on the northern list only, and Xantus Hummingbird, Black Phoebe, Lesser Goldfinch, Vermillion Flycatcher, and Belding's Yellowthroat on the southern list only. Of these species, the only ones with any appreciable island representation are Phainopepla in the north, and Yellow (Mangrove) Warbler in the mangrove thickets of the south, plus Tiburón in the north.

The landbridge island of San José is the most diverse of the southern islands, lacking only five species in the list (table 10.2). Nevertheless, there are several common mainland species, censused, for example, at San Juan Road and La Paz (app. 10.1), that do not reach any of the southern islands: the Greater Roadrunner, Western Scrub-jay, Brown Towhee, Phainopepla, Gilded Flicker, and perhaps Pyrrhuloxia. Of these, only the Greater Roadrunner may lack the ability to fly across water gaps, and that between San José and the peninsula is <5 km. The Pyrrhuloxia was reportedly seen by Emlen (1979) on Monserrat, but there have been no other reports from this oft-visited island, and the species has been omitted from the table.

Once again, there is a striking correspondence between islands ordered by species richness and the areas of the islands (Spearman $r = .97$; $p < .05$), and the species distributions are significantly nested (for the table as a whole, and for all species on 11 or fewer islands except Purple Martin; Wilcoxon rank-sum tests).

Differences Between Northern and Southern Islands

In comparing tables 10.1 and 10.2, some obvious differences are apparent in species by island distribution. For example, the Red Cardinal is widely distributed on southern islands as small as 2.5 km², but it occurs on no northern islands except Tiburón. The Gila Woodpecker and Hooded Oriole occur on Tiburón and the former also on San Esteban in the north, but both are found on a wider range of island sizes, including landbridge and non-landbridge, in the south. These three species all prefer thicker and brushier habitats of the sort more prevalent on the southern islands.

The distributional differences between the northern and southern islands are shown by comparison of their rankings in figure 10.7, in which northern and southern ranks are plotted against each other. The Black-throated Sparrow ranks first on the southern islands but is ranked sixth in the north; this species appears to be more successful on the southern islands with their thicker vegetation, a difference that is accentuated on the smallest islands. Other species that move up in ranking on the southern islands

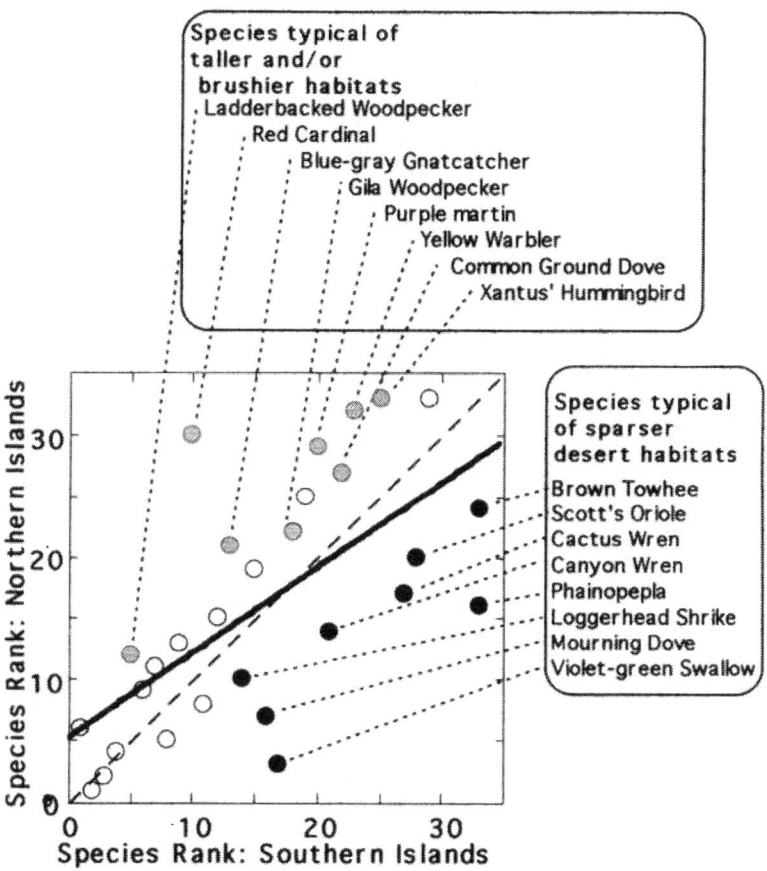

Figure 10.7 Many breeding land birds have different ranks on northern versus southern islands (see tables 10.1, 10.2), with many reverses between the two island groups. The overall regression line (bold, continuous) is not significantly different from 45° line (dashed). Species that are more widespread on northern islands (lower rank; inset box at right) are those typical of more open, drier habitats, and many widely distributed species on brushier southern islands (inset, upper box) rank much higher on the drier and sparsely vegetated northern islands.

include woodpeckers, cardinals, and gnatcatchers, species that do well in the thornscrub habitat on the peninsula and therefore can make use of the habitat differences between northern and southern islands. The species that move down in rankings between northern and southern islands, conversely, are those species with preferences for more open and drier habitats that are more sparsely vegetated. Examples of species with more restricted occurrences in the south are the Mourning Dove, Loggerhead Shrike, and Phainopepla (see fig. 10.7), all of which are more widely distributed in the northern islands. Indeed, Phainopepla occurs on no southern islands, but ranks 16th in the northern group.

A pictoral view of the differences in species distributions over island size is given by incidence functions for the northern and southern islands, plots of the proportion of islands within a size group that is occupied by a given taxon. Figure 10.8 gives a broad picture of species incidences over island size. Note the striking feature that on southern islands, but not on northern islands, the lower limit of island area on which a wide range of species reaches distributional limits varies among species and is quite precisely defined. The figure indicates that island area has a dominant influence on species incidence, even though area limits different species at different absolute values. A possible reason why this pattern is obscured on northern islands is discussed below.

Incidence functions for species that are similarly distributed over island size among northern and southern islands are shown in figure 10.9, which includes species of wide and rather cosmopolitan distribution in general. Species more broadly distributed on southern islands are shown in figure 10.10, and those of wider occurrence on northern islands in figure 10.11.

Species–Area Relations

Northern and southern islands are combined in the plots of bird species numbers versus island size shown in figure 10.12. A linear relation between island area and bird species number on logarithmic scales is highly significant ($R^2 = .75$; $p < .001$); 75% percent of the variability in bird species is attributable to the area of the island. The addition of island isolation as an independent variable does not improve the regression, adding <1% to the explained variance in the dependent variable. To test for the possibility of historical effects, we compare the regression lines for landbridge islands and non-landbridge islands (fig. 10.12). The regression lines are nearly identical, showing that there remain, unlike the case with reptiles and mammals, no effects of island history on the land bird diversity. However, there is a significant difference in species numbers between northern and southern islands, which are plotted and fitted with separate regression lines in figure 10.12. While the slopes of the two lines are indistinguishable, the southern intercept (0.813 vs. 0.646 in the north) is significantly higher ($p = .015$). This means that a southern islands of 1, 10, and 100 km^2 support, on average, 6.5, 13, and 25 species, and northern islands of these sizes support 4.5, 8.7, and 17 species.

Previous analysis of a less complete breeding bird data set (Cody 1985) showed a somewhat higher percentage (79%) of the variance in bird species numbers accounted for by island areas. We might construe this to mean that additional sampling effort turns up additional but rarer species, as the commoner species are already known and

Land Birds 295

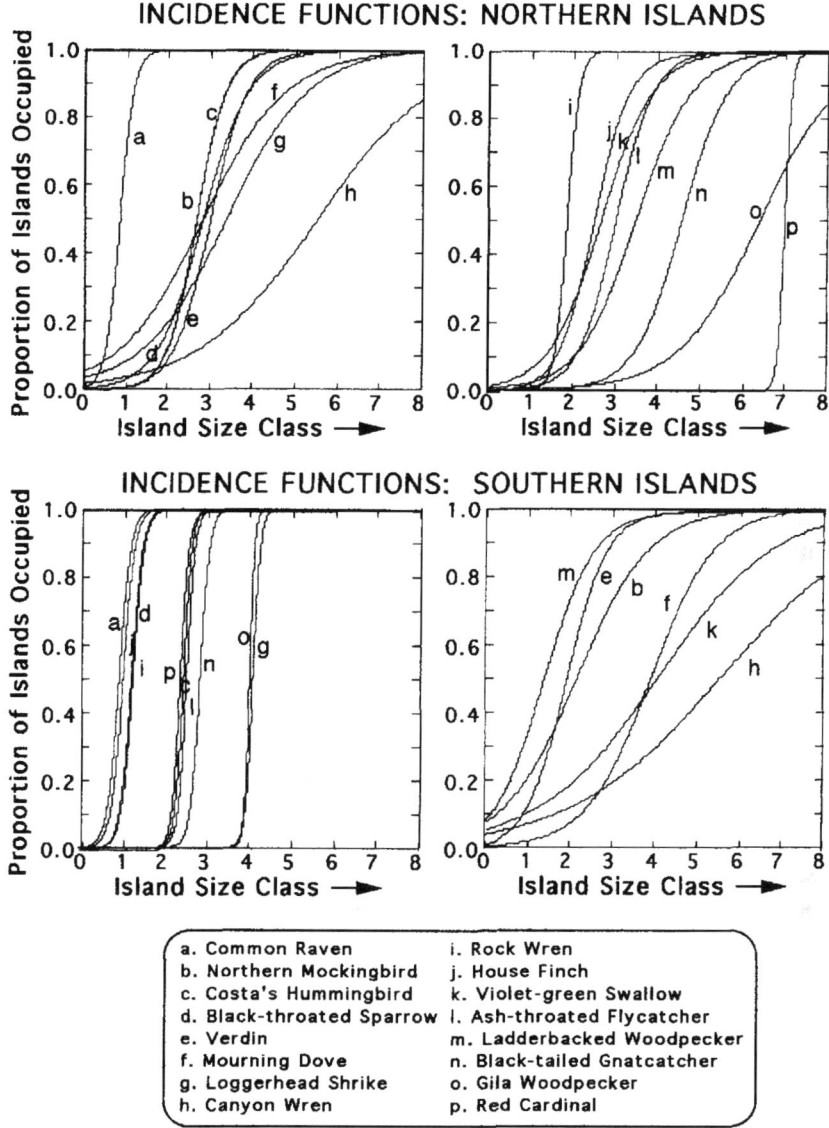

Figure 10.8 Incidence functions for breeding land birds on northern (upper) and southern (lower) islands. Note the shifts in incidence between the two island groups, the strong island area dependence of many southern species (lower left), and the general lack of such effects on northern islands (upper left, right). Island size classes as in Fig. 10.6.

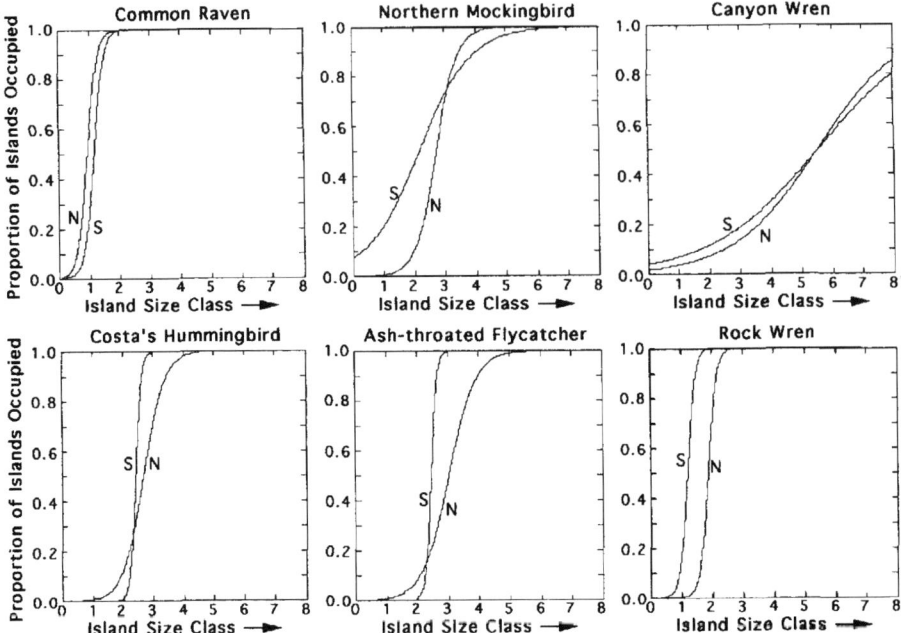

Figure 10.9 Incidence functions of species that have relatively similar distributions over island size in northern and southern islands.

recorded. It may be that inclusion of rare species muddies rather than sharpens the species–area relationship; area better predicts the common island species rather than the size of an exhaustive list that includes peripheral species of little account, ecologically speaking. Statistical details of species–area curves are given in table 10.3.

Other Factors and Other Diversity Models

Island area alone can be used to make reasonably accurate predictions of bird species numbers on gulf islands, but presumably area is a surrogate for habitat types, their availability, and their bird resources which, if measured directly, would presumably be just as effective as predictors or more so. Among other potentially relevant factors, the need to drink free water might affect the island distributions of some species. Two conspicuously successful island birds, the Black-throated Sparrow and the Rock Wren, do not need to drink, especially when they have access to green vegetation and insects (Smith and Bartholomew 1966). House Finches, in contrast, need to drink water to avoid weight loss when fed a dry seed diet (Bartholomew and Cade 1956). This may apply only to finches in more mesic areas (the tested birds were caught on the University of California-Los Angeles campus), but if it holds generally, the House Finches on the dry gulf islands must either use the islands seasonally when there is abundant green vegetation or commute to water sources at some distance, perhaps off-island.

Inspection of figure 10.12 shows that the log-log model, although it provides a reasonable fit to the data, does not simulate some of the obvious features of the bird

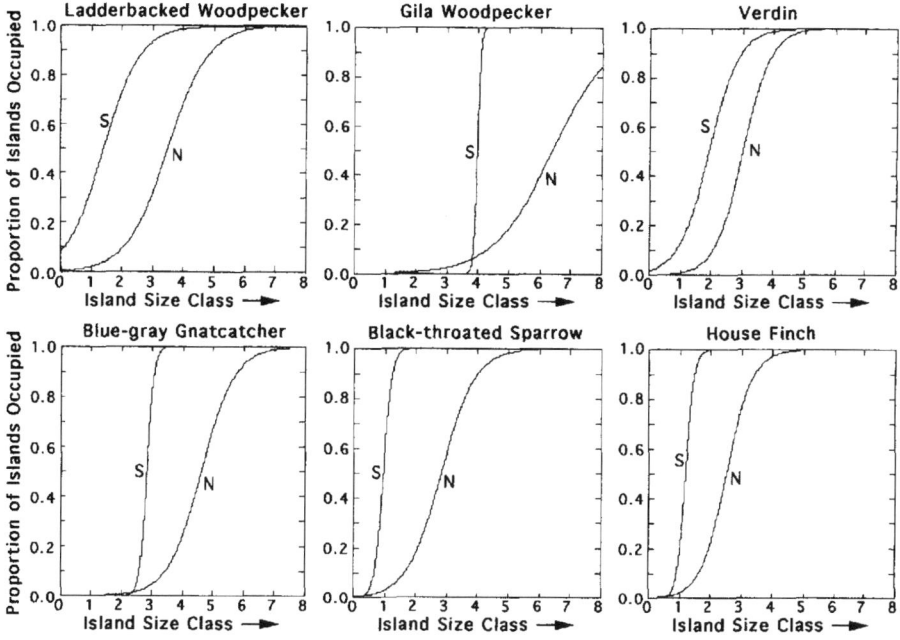

Figure 10.10 Incidence functions for species that are more broadly distributed over smaller islands in the southern gulf relative the northern gulf.

diversity data. Bird species numbers seem to increase toward larger islands in fits and starts, even though the distribution of island sizes is relatively smooth and Gaussian (fig. 10.13). Bird species numbers appear to cluster around certain modes, vary little within quite wide island size ranges, and make transitions to higher modes only with substantial increase in island area. For example, the two richest islands (30, 28 species) are Tiburón and San José, with widely disparate areas of 1224 and 187 km^2. The four islands with 19–20 species (Ángel de la Guarda, San Esteban, Santa Catalina, and Monserrat) vary in area (20–936 km^2) by a factor of nearly 50. Species counts fall dramatically on islands in the size range of 1–5 km^2, and below this point bird species counts are extremely variable. This parallels the abrupt decrease in plant species numbers in the same range of island sizes and indicates that islands larger than about 5 km^2 are fundamentally different from islands smaller than about 2 km^2 both for plants (chap. 4) and for birds, which presumably track differences in vegetation structure brought about by the differences in plant species.

The plant species added with increasing island size are those that require the products of the larger drainage basins that characterize only the larger islands: deeper alluvial soils and larger channels that support phreatophytes. It seems that the explanation for the step in both the plant and the bird curves is owed largely to the hydrology of these arid islands, to the formation of drainage basins, and to the way this process changes with increasing island size. Given a certain area of watershed A, the discharge from the drainage basin is proportional to A^m, where $0.5 < m < 1.0$ (Strahler 1964);

298   The Biological Scene

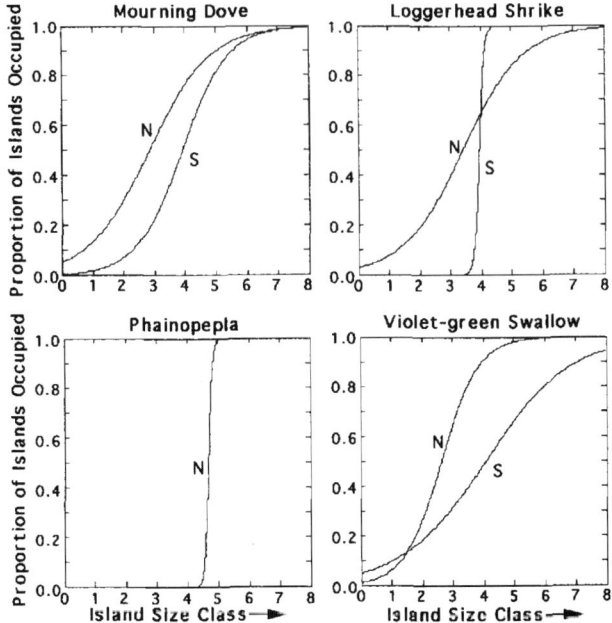

Figure 10.11 Incidence functions for species that are more broadly distributed over smaller islands in the northern gulf relative the southern gulf. Phainopepla does not occur on southern islands but is predictable on large northern islands.

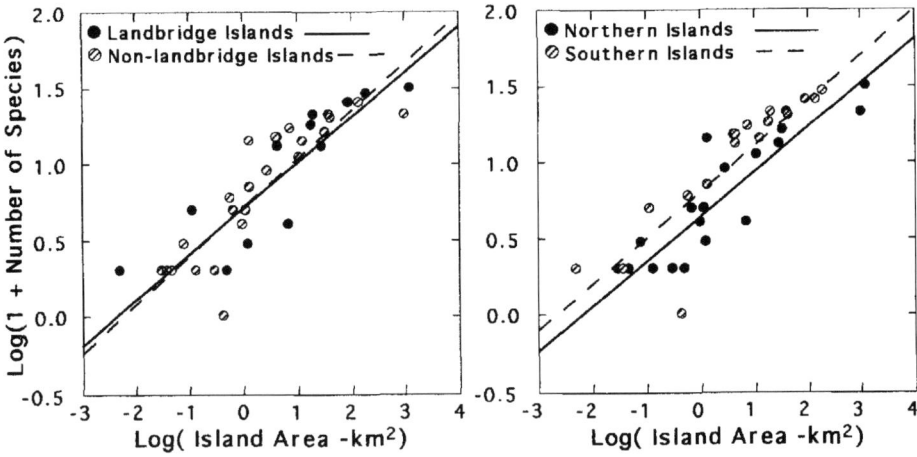

Figure 10.12 Species–area curves for breeding land birds. Effects of island history are unimportant, as species richness is the same on landbridge and non-landbridge islands (left). However, southern islands support significantly more species than northern islands of the same size (right), as well as showing a stronger area relationship.

Land Birds 299

Table 10.3 Statistics of species–area relations for birds on islands in the Sea of Cortés

Taxon	n	Slope ± SE	Intercept ± SE	R^2	p	Isolated	Landbridge
Breeding raptors	38	0.234 ± 0.029[a]	0.405 ± 0.040	.66	<.001	NS	NS
Breeding shoreline species (herons, etc.)	38	0.119 ± 0.028	0.521 ± 0.037	.34	<.001	NS	NS
Breeding landbirds							
Northern islands	22	0.293 ± 0.039[a]	0.646 ± 0.050[b]	.74	<.001	NS	NS
Southern islands	17	0.300 ± 0.040[a]	0.813 ± 0.057[b]	.79	<.001	NS	NS
Nonbreeding landbirds, Migrants + wintering spp.	38	0.354 ± 0.036	0.406 ± 0.048	.72	<.001	NS	NS
Passage migrants	38	0.217 ± 0.038	0.190 ± 0.050	.50	<.001	NS	NS
Wintering species	38	0.305 ± 0.032	0.361 ± 0.042	.71	<.001	NS	NS

Regression parameters labeled with the same letter are not significantly different (NS).

channel width increases as the square root of the discharge from the basin (Slatyer and Mabbutt 1964) and thus approximately with the cube root of the basin's area.

While discharge and channel characteristics vary smoothly with area of the drainage basin, other factors provide threshold effects. Thus, channels are classified from first order, the smallest and uppermost in the basin, to higher orders formed by the joining of channels of lower order. Most of the empirical rules that appear to regulate

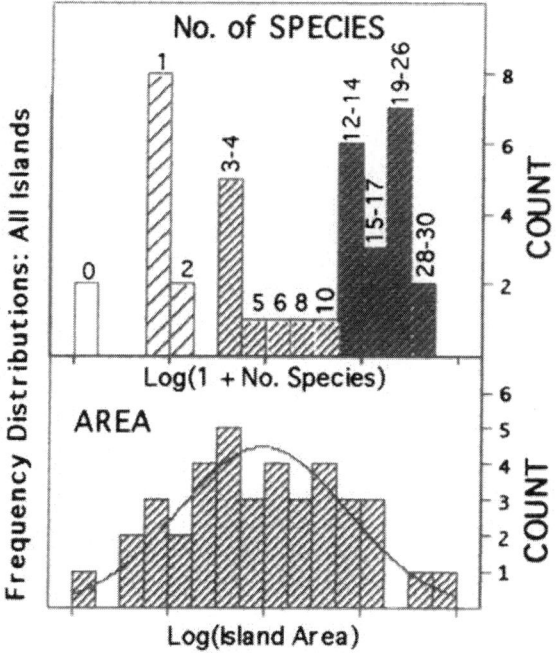

Figure 10.13 While island areas are relatively continuously distributed (below), island bird species numbers are much less so (above). Many islands of dissimilar size have similar species counts, whereas relatively small differences in island size are related to shifts between one species mode and another (see text).

the morphometry of drainage basins are called Horton's laws (Horton 1932), and one of these relates the length of streams (channels) of adjacent order: channels of order n exceed the length of channels of order $n - 1$ (smaller channels) by a constant factor, R_L (>1.0). This means that, as island size increases, the island will support (assuming a comparable topography, etc.) larger channels of higher orders in stepwise fashion, with the steps at equal intervals on a log (area) scale. As each size threshold is passed, the larger channel of the higher-order stream-bed might well accommodate a new suite of plant species appropriate to the additional range of channel environments, and the new vegetation will in turn support a new suite of bird species. This reasoning may help explain the somewhat stadial nature of the size of island bird communities.

Interpretation

The distinguishing features of landbird distributions on the gulf islands are (1) the predictability of species numbers from island area, (2) the predictability of species identities from the size of the island bird community, and (3) the somewhat stadial structure of the bird communities, with poor predictability of species richness on smaller islands. We interpret these facts to mean that an island of a given area provides resources (habitats, food, nesting sites, foraging opportunities) for a certain set of bird species, in much the same way that Lack (1976) attributed variation in species numbers over size in the Caribbean islands. Larger islands provide a more diverse range of resources and support bird species in addition to those present on the smaller islands; only species present on smaller islands find appropriate habitats and foraging niches throughout a wider range of island sizes.

The nestedness of species subsets over increasing island sizes, with relatively few gaps in the pattern, suggests three inferences. First, problems in reaching the islands are not a major factor in bird distribution, but the main issue is whether the island provides appropriate resources for the birds once they have reached it. Second, extinctions, if they occur (and we usually assume that extinctions are a way of life [or death] on small islands), must be followed by the same species' recolonization if the pattern is to hold. Third, there are few if any ecological substitutes that preclude invasion of suitable islands, nor apparently is invasion tempered by niche expansion of the resident species into part or all the niche space vacated by the extinct population. Note that there are few taxonomically related species that potentially might act as ecological substitutes, and thus dissimilarity in resource use might act in conjunction with a shortage of time during a temporary local extinction to readily permit reinvasion. If this scenario is correct, we would expect minimal density compensation among the birds of relatively species-poor islands.

From considerations of the hydrology of channel formation, bird resources are not expected to increase smoothly with island area but are likely enhanced in a stepwise fashion as thresholds of island size are passed. Thence, adding a new component to an island's vegetation is likely to add not just a single bird species but a set of additional species, each of which can use a different aspect of this new feature. Thus, species number approximately doubles on islands larger than San Ildefonso (1.3 km^2) in the south and San Pedro Mártir (2.9 km^2) in the north. The size increase especially adds species of more specialized insectivores (flycatchers, gnatcatchers, larger woodpeckers) typical of denser, shrubbier vegetation that is taller and thicker than the low and open desert scrub used by Black-throated Sparrows, House Finches, wrens, and Ladderbacked Woodpeckers on the smaller islands.

An important point to remember with respect to insular distributions is that the resources available to bird species on a given circumscribed area (e.g., 1 km^2) are very different between a mainland site and an island of the same size. This is not because of the isolation per se of the island, for, as we have seen, isolation contributes little toward an understanding of island bird diversity. It is because the mainland site is part of a far larger drainage system and catchment basin by virtue of being contiguous with similar land areas, whereas the island is a self-contained system with a potential for supporting vegetation (and birds) completely dependent on its own area and topography. Thus, there are island effects in terms of reduced species diversities, but these are specifically attributable to the reduced range of environments supportable by isolated land areas rather than to the isolation per se. Part of this same explanation, of course, can be applied to the slope of the species–area curve in general.

The Migrants

The Gulf of California lies on the migration pathway of many bird species that breed in western North America and winter in western Mexico or farther south, and not surprisingly a substantial number of migrant species has been seen on the gulf islands. Although desert environments in general are not particularly popular as overwintering habitats, a few bird species, especially emberizid sparrows, spend their nonbreeding seasons in the Sonoran Desert of the peninsula and are abundant on the gulf islands as well.

The records of nonbreeding birds on the gulf islands are assembled in appendix 10.10, which includes 69 species (exlusive of shorebirds and raptors). Of these, half have been recorded on just one or two islands and are passage migrants of irregular occurrence en route to or from wintering grounds in central or southern Mexico. A number of hummingbirds, flycatchers, and warblers are seen more regularly on the islands, though infrequently or sporadically, in line with brief pauses in passage. Passage migrants that winter in mainland Mexico include Rufous and Allen's Hummingbirds, Hermit Thrush, and Wilson's Warbler, and all make cursory use of the islands. Both Gray and Pacific Flycatchers have been reported wintering in the Cape region, where the former is especially abundant in the leafless thorn scrub throughout the winter and into spring. A number of warbler species, including Yellow-rumped, MacGillivray's, and especially the Orange-crowned Warbler, are quite common on the islands as passage migrants and winter from southern Baja California and northern Mexico south.

In a small number of species the gulf islands are latitudinally within or very close to both breeding and wintering ranges. These species may breed on some islands, winter on these or on others, or visit yet others in transit between breeding and wintering sites. Often we lack critical information, as inevitably our visits are too short or too prescribed, to resolve each species' status on an island, and our records in the appendices contain an element of judgment. Species included in this somewhat equivocal category are Poorwill, Black-chinned Hummingbird, several flycatchers, Bewick Wren, Gray Vireo, Yellow Warbler, and perhaps Varied Bunting and Lesser Goldfinch.

Some wintering birds on the gulf islands favor open habitats but are not easily seen or identified and therefore may be more common than the records show (e.g., Water

Pipit, Horned Lark); others in similar habitats are conspicuous and social (e.g., Lark Bunting) and are unlikely to be overlooked. Some have quite specific habitat requirements but are crepuscular in habit (e.g., Sedge Wren) or are clearly limited in range by their preferred overwintering habitat (e.g., Northern Waterthrush). The Northern Waterthrush winters in southern Baja California and is especially common in the mangrove fringe around lagoons in the southern gulf. Its presence on the two islands where this habitat is common, Partida Sur and Espíritu Santo, has been noted since Townsend (1923).

For a smaller number of species, particularly finches and sparrows in the family Emberizidae, the islands together with the Sonoran Desert on either side of the gulf apparently are important wintering grounds, as these birds can be seen there in substantial numbers in the nonbreeding season. Lark Buntings winter mainly in Sonora and Sinaloa and have shown up on four gulf islands as well as in southern Baja California. Their occurrence on islands is probably irregular except on Tiburón, where they seem predictable all winter. The commonest wintering emberizines are Green-tailed Towhee, Brewer's Sparrow, White-crowned Sparrow, Savannah Sparrow, Lincoln's Sparrow, and Chipping Sparrow, in that order; their incidence functions are shown in figure 10.14. Brewer's Sparrows and to a lesser extent Chipping Sparrows are widespread and common throughout the peninsular and mainland deserts, where they form the large winter flocks that are a conspicuous component of the habitat's winter avifauna. The former appears more common on the islands in general, and the latter is quite scarce on the southern islands.

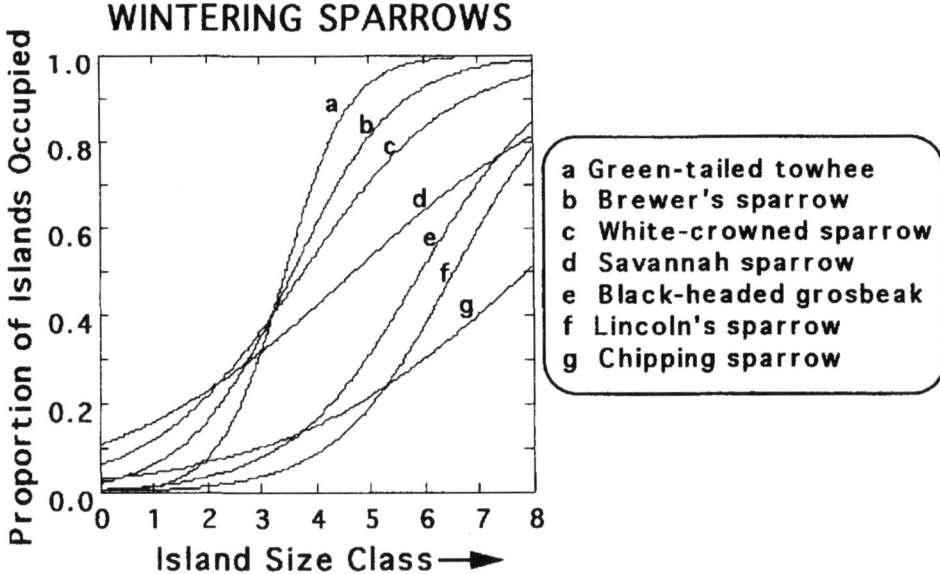

Figure 10.14 Incidence function of wintering sparrows (subfamily Emberizinae) on gulf islands. The islands appear to be important wintering grounds for several species, some of which (Green-tailed Towhee, Savannah Sparrow) seem commoner on the islands than the mainland. Black-headed Grosbeaks may be more frequently observed as a passage migrant than a wintering species.

The White-crowned Sparrow is abundant on virtually all gulf islands north of Carmen, and it also seems scarcer on the southern islands. Equally widespread is the Green-tailed Towhee, recorded by most island-visiting ornithologists from a total of 15 islands; its characteristic call-note is one of the first bird sounds one hears during late winter and spring island visits. The species occurs from Ángel de la Guarda in the north to Cerralvo in the south. It is found on all large islands with brush-choked drainage channels but is absent from the smaller and more openly vegetated islands. This species seems particularly common on the islands relative to either peninsular or mainland sites, and we speculate that Green-tailed Towhees winter preferentially on the islands because, with the exception of Tiburón, the usual desert Brown Towhees are absent there.

The Savannah Sparrow is a common wintering bird on gulf islands. There are strictly resident subspecies in the coastal marshes on northern Pacific coast of the peninsula (*Passerculus sandwichensis beldingi*), and on the Sonoran coast (*P. s. atratus*), and a scattering of winter records, in Sonora and on the peninsula, of several subspecies that breed to the north as far as Alaska (*P. s. anthinus, P. s. alaudinus, P. s. crassus, P. s. nevadensis*). It is the large-billed subspecies, *P. s. rostratus*, that apparently constitutes the great majority of the island records (Grinnell 1928, van Rossem 1945b). Birds of this race breed in marshes on the outer Pacific coast of the lower peninsula, on San Benito Island (northwest of Cedros) and also in the Colorado River delta region and repair to gulf islands in winter. This involves a shift south from the head of the gulf to Tiburón, but a more remarkable easterly migration from the Pacific across the peninsula to the wide range of gulf islands, on which this race can be found in season. Although the wintering status of other Savannah Sparrow races remains debatable (since Grinnell 1905), the species is regularly seen in open sites on most gulf islands, where birds feed mainly along the shoreline, often even within the intertidal zone.

The diversity of wintering bird species and passage migrants shows the usual relationship with island area, both combined and individually. However, wintering species produce a significantly higher slope to the species–area relation than do passage migrants (fig. 10.15), and the relation diverges from that of breeding birds in that relatively fewer wintering species occur on smaller islands. This might conceivably be a sampling effect, given the shorter residence times of the migrants and wintering birds. The species–area curves for the various bird groups discussed here are compared in table 10.3. The steepest species–area relationships are seen in nonbreeding birds, the shallowest in shoreline birds.

Bird Densities

Determinants of Density

Density Compensation

The density that a given species maintains (number of pairs or individuals per unit area) is a function of several variables, one of which may be the number and density of competing species. Island communities, in which species numbers are low relative

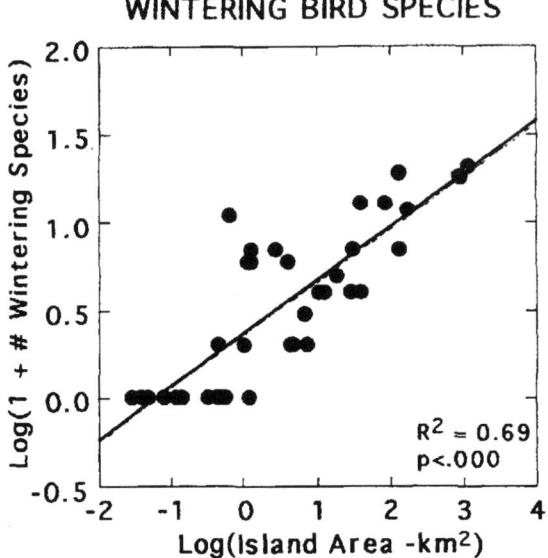

Figure 10.15 The numbers of wintering bird species show a similar dependence on island area as breeding birds. Island isolation does not contribute to variation in species numbers.

to comparable mainland sites, might be expected to support more individuals per species in response to less competition for food resources than would be found on the mainland among more species. Thus on islands in the northwestern United States, Song Sparrows *Melospiza melodia* live on increasingly smaller territories where increasingly fewer competitors share the habitat (Yeaton and Cody 1974). In habitat fragments of Afro-montane woodlands in South Africa, total bird density remains nearly constant as bird species numbers vary by a factor of three (Cody 1983a). Most studies of density compensation fail to control either for the structural or productivity aspects of the habitat in island–mainland comparisons; therefore, well-documented cases are scarce. We recognize theoretically that islands with fewer species may reach total bird densities equal to mainland densities (complete density compensation), they may fall short of the overall mainland density but exceed the mainland density summed over the island species subset (partial density compensation), or they may even in some circumstances exceed the total density on the mainland (overcompensation; see Case et al. 1979). Wright (1980) reviewed this literature.

Sonoran Desert bird communities are poor candidates for demonstrating density compensation. The habitat is open, with a great deal of horizontal variability; the life forms of the plant species are diverse in the extreme (Cody 1989), and invariably the vegetation cover is highly discontinuous. This set of circumstances decrees that each of the land bird species pursues a species-specific role and occupies a distinct niche, with little overlap with other species in the community. Some species have morphological and behavioral attributes adaptive to digging under bushes, others to foraging in the canopy of riparian trees, and still others to sallying from the tops of bushes or to probing in cracks and crevices. The variety of bird phenotypes thus covaries with both foraging behavior and with microhabitat, and in turn with the use of discrete food resources. Where foraging sites are dissimilar and are spaced apart, as are the bushes in Sonoran Desert, each species becomes relatively specialized, and the payoffs from encroaching on another's speciality are slight at best.

The opposite is the case in communities of more generalized species, where habitats are more structurally uniform with continuous vegetation cover that is homogeneous in both horizontal and vertical axes. The chaparral in southern California and its islands, investigated by Yeaton (1974), along with the Song Sparrow case and Afromontane habitats mentioned earlier, qualify on these grounds, and in these examples there is a good deal of density compensation in depauperate islands or islandlike communities.

Effects of Habitat

Sonoran Desert birds are sensitive to habitat structure, and breeding bird populations vary in density as the habitat changes. For example, consider the bird species seen along a transect that extends from a flat limestone bench with short, open vegetation on the southern tip of Carmen some 5 km northwest to the foothills into taller and denser habitat (Cody, unpublished data). Initially the vegetation is dominated by *Larrea*, *Lycium*, *Bursera microphylla*, and *Abutilon*, but eventually paloverdes, mesquites, and the diverse cacti of the riparian desert scrub are reached. More and more bird species are recorded as one proceeds along this transect, and all except Rock Wrens are still present in the tallest vegetation. Different species, however, have peak densities in points along the transect. Black-throated Sparrows decline in density in vegetation above roughly 3 m and Ash-throated Flycatchers decline in somewhat taller vegetation, but most of the remaining species increase in abundance with increasing vegetation height. Only when the vegetation reaches 3–4 m are Gila Woodpeckers and Red Cardinals encountered. Clearly, unless similar habitats are compared, one does not expect to record similar densities of particular bird species.

Effects of Productivity

In the arid regions of southwestern North America, productivity is determined largely by precipitation. Thus, in wet years one finds more annual plants, greater growth and production in perennial plants, better growth and survival of the insects that feed on these plants, and so on. In the pine-oak woodland of southeastern Arizona, the number of insects caught on Tanglefoot boards is directly proportional to the deviation of the annual rainfall from the long-term average (Cody 1981). In response to the year-to-year fluctuations in food, the numbers, identities, and densities of the insectivorous bird community vary correspondingly. Rainfall in the Sonoran Desert is both lower in amount and less predictable from year to year, and thus there is a potentially great variation in yearly food supplies for the birds within the same habitat.

As already discussed, the composition of desert bird communities is relatively constant over large geographic areas. Therefore, if the bird communities respond at all to climatic variations from year to year, they do so by changes in densities rather than by changes in species identities. In these deserts there appear to be no bird species like the opportunistic fugitives found in the Kalahari and especially in the central Australian deserts, birds which disperse far and wide in search of suitable breeding conditions that develop wherever rain happens to fall, and which reproduce fast and furiously once good breeding situations are found.

Most Sonoran Desert birds, which are habitat-specific residents, must weather the ups and downs of productivity more or less *in situ* because birds like thrashers, verdins, woodpeckers, and wrens are not known for their long-distance dispersal prowess. Others, such as Ash-throated Flycatchers and hummingbirds, depend on aspects of productivity (plenty of large, flying insects and a concentration of nectar-producing plants, respectively) that are manifest only in the better seasons and sites in the desert, and such species do undertake extensive seasonal movements and migrations.

Productivity differences between gulf islands and the mainland or peninsula might be estimated indirectly using precipitation data, but given the absence of island weather stations and the local variability in precipitation, this is not a realistic approach. Mean annual precipitation interacts with topography to produce the drainage channels and riparian vegetation already discussed, but the degree to which a shrub dependent on recent rainfall rather than on deeper water will support more or less growth and more or fewer insects will affect its utility to the insectivorous birds. Again, this will be an extremely local phenomenon.

Direct measures of food abundance for birds in desert habitats are aggravated by habitat heterogeneity and discontinuity and by the wide variation in species-specific foraging sites and modes. Further, such estimates are relevant only to the time and place they are assessed. Tanglefoot catches made by T. J. Case (pers. commun., 1975) and insect counts made by Emlen (1979) suffice to show that there are no gross dissimilarities between the gulf islands and the peninsula, but little more can be said about this.

Bird Densities on Islands and Mainland

Data on breeding bird densities from seven mainland desert sites and eight gulf islands are given in appendix 10.10. Three of the mainland sites are on the Baja California peninsula, two in southern Arizona, and two more in Sonora. The islands for which we have good bird density data are a variety of southern islands from Carmen south. At the mainland sites, detailed studies of the breeding bird communities on 5-ha mapped sites were undertaken (by Cody), and densities of breeding birds are known in pairs per hectare. In desert habitats of low bird density and large territory sizes, densities can be estimated more efficiently by line-transect methods, on which there is now a considerable literature (e.g., Bibby et al. 1992). Transect methods yield counts of individuals seen per hour (I/h), made as the observer walks slowly (~ 1 km/h) through the habitat, attempting not to overlook any of the birds present.

The translation of line-transect counts to estimates of breeding pairs per unit area is not straightforward because different species vary in conspicuousness. The five data points in appendix 10.10 where both density measures are available indicate a more or less linear relation ($I/h = 7.49 \pm 0.52$ SE*Pr/ha) between the two. The line-transect counts, however, are strictly comparable among sites and between islands and mainland, and reliably reflect the way a given species varies in abundance between sites.

As shown in figure 10.16a, bird density (I/h) bears a close relation to the number of bird species (SPP) known to be present at the site or on the island ($I/h = -0.97 + 2.57$ (SPP); $R^2 = .73$, $p < .001$). In the second part of the figure, the number of bird

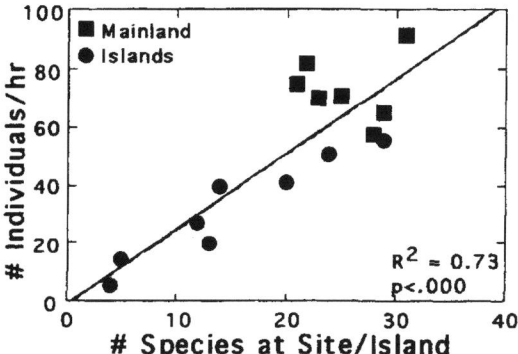

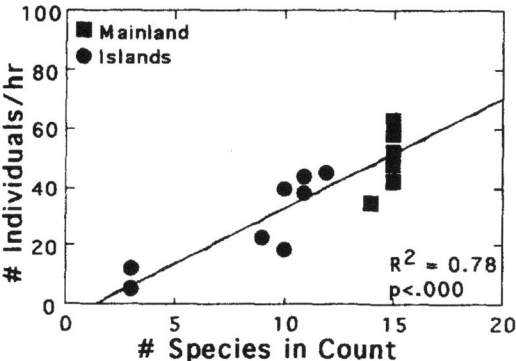

Figure 10.16 Breeding bird density on gulf islands is a linear function of the number of species there, either total species numbers recorded (upper) or the subset of species found on the density surveys (lower). The regression lines depart insignificantly from a pass through the origin; as species fall out on smaller islands, density falls proportionately.

species actually encountered during the density survey is used as the independent variable, with similar results. That is, as species are added among sites, additional individuals are accumulated in fixed proportion, at a rate of about eight more individuals seen per hour for each three additional species. The regression lines pass through the origin (intercepts are insignificantly different from zero) and are essentially linear. These two features indicate that, in the overall picture, there is little or nothing in the way of density compensation. If fewer species were able to use some of the resources of absent competitors, then the data in figure 10.16 would show slopes that leveled off with increasing species numbers, and linear models would intersect above the origin on the ordinate. There is no significant difference between the mainland and island sites in terms of the residuals, and islands are just as likely as mainland sites (but no more so) to be rather high in bird density and lie above the regression line. The density data conform to the hypothesis that sites (island and mainland) differ in their ranges of habitats and bird food resources and that larger islands come with additional resources that support more species and a higher total bird density.

More insight is gained into density variations when the densities of individual species are examined. In figure 10.17 the densities of nine broadly distributed species are shown as functions of the number of species in the density counts. The island mean density is bracketed by a vertical line representing 2 SDs around it. A reasonable null hypothesis of island densities in line with mainland densities and decreasing with

308   The Biological Scene

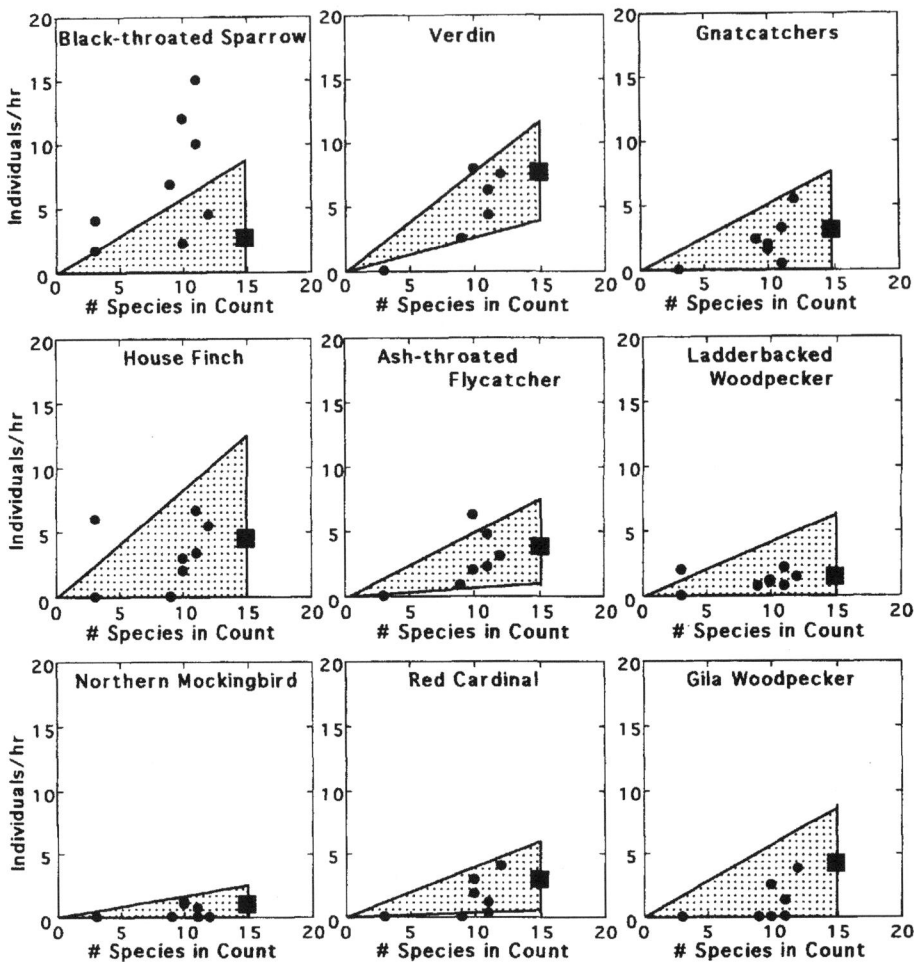

Figure 10.17 Island densities of nine common species on seven mainland sites (mean represented by square symbol) and eight southern gulf islands (dots). The vertical line through the mainland mean represents 2 SDs about the mean; lines joining its extremities to the origin demark a stippled zone in which island bird densities are expected to lie (with no density change or compensation). Notably, the Black-throated Sparrow occurs on several islands at densities significantly higher than expected; see text for discussion.

fewer island species predicts that the island densities will lie within the zone defined by the 2 SD line around the mainland mean and its connections to the origin. For six of the nine species, island densities are in line with those predicted (Verdin, gnatcatchers, Ash-throated Flycatcher, Northern Mockingbird, Red Cardinal, and Gila Woodpecker). There is evidence, however, that on the tiny island of Gallo densities of House Finches and perhaps Ladderbacked Woodpeckers are higher than expected.

One species, the Black-throated Sparrow, has conspicuously high densities on several islands. Of eight islands surveyed, five show exceptionally high densities of this species; listed in decreasing density values, these are Monserrat, Danzante, Carmen, San Francisco, and Gallo, with counts 2- to 2.5-fold larger than predicted. A priori, the Black-throated Sparrow seems one of the better candidates for density compensation; like the Song Sparrow, it is a generalized insectivore and capable of eating seeds when insects are unavailable. It ranks first among species reaching the southern islands and is high also in the north. It utilizes a broad range of habitats and forages both on the ground and in shrubs.

It was among Black-throated Sparrows particularly that Emlen (1979) believed competitive release occurred on the southern gulf islands, where he recorded densities some three times those of the mainland. Our own data confirm that this species is particularly common on islands he surveyed (Danzante, Monserrat, and Carmen). Subsequently, George (1987a,b) studied these islands together with Coronado and substantiated the higher island Black-throated Sparrow densities despite generally similar total bird densities on islands and similar or reduced island densities in other taxa. He confirmed Emlen's finding of higher island Loggerhead Shrike densities, and in addition found that island Blue-gray Gnatcatchers maintained higher densities than on the mainland where California ("Black-tailed") Gnatcatchers shared their habitat. George supported the hypothesis that enhanced island densities are owing to reduced interspecific competition in the gnatcatcher (lacking its usual congener), but this seems less likely in the sparrow. Perhaps the closest species ecologically to the sparrow is the House Finch, but deviations from predicted densities of the finch on the islands do not correlate with those of the sparrow. The explanation favored by George is reduced nest predation on the islands; he found that predation on sparrow nests and on artificial nests was higher on the mainland than on Isla Coronado. His work also illustrates the opportunistic responses of Black-throated Sparrows in tracking resources that follow rainfall (George 1987c); although the response is not island-specific, these birds are capable of January and February breeding after heavy winter rainfall in the preceeding months.

A final point is made with the 3 islands that have 12–14 land birds each: Danzante, San Francisco, and Santa Cruz. Although all are similar in the numbers and identities of birds they support, the bird densities differ considerably among them. Danzante is a landbridge island close to the shore, San Francisco a landbridge island somewhat more isolated, and Santa Cruz is a more isolated and non-landbridge island. Bird densities are far higher on Danzante (39 *I*/h) and lowest on Santa Cruz (18 *I*/h). Since Santa Cruz is the largest of the three islands, there seems little reason for suggesting that productivity might differ among them. It appears that the positions of the islands, with Danzante close to the peninsula and San Francisco close to neighboring San José, might influence bird density, though apparently not bird diversity. The movements of bird species among these relatively accessible islands warrants further study, as do many other aspects of avian ecology and biogeography in the Gulf of California.

Acknowledgments We thank a number of colleagues who have contributed to recent field work in the gulf: H. Benitez, F. Eccardi, A. Navarro, X. Noemí, and J. Sarukhán. A range of funding agencies has supported the field research: IBUNAM, TNC-Int., C I México, CONACYT, and NSF.

References

Baird, S.F. (ed) 1870. Ornithology. v. 1 Land Birds. In: *Ornithology of California*. Publ. by the Legislature [of California], Cambridge, Mass.

Banks, R.C. 1963a. Birds of the Belvedere expedition to the Gulf of California. *Trans. San Diego Soc. Nat. Hist.* **13**:49–60.

Banks, R.C. 1963b. The birds of Cerralvo Island, Baja California, Mexico. *Condor* **65**:300–312.

Banks, R.C. 1963c. New birds from Cerralvo Island, Baja California, Mexico. *Calif. Acad. Sci. Occ. Pap.* **37**.

Banks, R.C. 1964. Birds and mammals of the voyage of the "Gringa." *Trans. San Diego Soc. Nat. Hist.* **14**:179–184.

Banks, R.C. 1969. Relationships of the avifauna of San Esteban Island, Sonora. *Condor* **71**: 88–93.

Bartholomew, G.A., and T.J. Cade. 1956. Water consumption of house finches. *Condor* **58**: 406–412.

Belding, L. 1883. Catalogue of a collection of birds made near the southern extremity of the peninsula of Lower California. *Proc. U.S. Nat. Mus.* **5**:532–550.

Bibby, C.J., N.D. Burgess, and D.A. Hill. 1992. *Bird Census Techniques*. Academic Press, London.

Boswall, J., and M. Barrett. 1978. Notes on the breeding birds of Isla Raza, Baja California. *Western Birds* **9**:93–108.

Brewster, W. 1902. Birds of the Cape Region of Lower California. Bull. *Mus. Comp. Zool. Harvard Coll.* **41**(1):1–241.

Bryant, W.E. 1889. A catalogue of the birds of Lower California, Mexico. *Proc. Calif. Acad. Sci. Ser.* **1889**:237–320.

Case T.J., M.E. Gilpin, and J.M. Diamond. 1979. Overexploitation, interference competition, and excess density compensation in insular faunas. *Am. Nat.* **113**:843–854.

Cody, M.L. 1974. *Competition and the Structure of Bird Communities*. Monographs on Population Biology. Princeton University Press, Princeton, NJ.

Cody, M.L. 1975. Towards a theory of continental diversity: Bird distributions on Mediterranean habitat gradients on three continents. In: M.L. Cody and J. M. Diamond (eds), *Ecology and Evolution of Communities*. Harvard University Press, Cambridge, MA: pp. 214–257.

Cody, M.L. 1981. Habitat selection in birds: the roles of vegetation structure, competition, and productivity. *Bioscience* **31**:107–113.

Cody, M.L. 1983a. Bird species diversity and density in Afromontane woodlands. *Oecologia* **59**:210–215.

Cody, M.L. 1983b. Continental diversity patterns and convergent evolution. In: F. Kruger and B. Hartley (eds), *Nutrients and Mediterranean-type Ecosystems*. Springer Verlag, Berlin; pp. 357–402.

Cody, M. L. 1985. The land birds. In: T.J. Case and M.L. Cody (eds), *Island Biogeography in the Sea of Cortéz*. University of Calif. Press, Berkeley; pp. 210–245.

Cody, M.L. 1989. Growth form diversity in desert plants. *J. Arid Environ.* **17**:199–209.

Cody, M.L. 1993. Bird diversity patterns and components across Australia. In: R.E. Ricklefs and D. Schluter (eds), *Species Diversity in Ecological Communities: Historical and Geographical Perspectives*. University of Chicago Press, Chicago; pp. 147–158.

Cody, M.L. 2001. Bird diversity components in oak and eucalytus woodlands. *Auk* **118**:443–456.

Davis, J. 1959. The Sierra Madrean element of the avifauna of the Baja California Region. *Condor* **61**:75–84.

Emlen, J.T. 1979. Land bird densities on Baja California islands. *Auk* **96**:152–167.

Engels, W.L. 1940. Structural adaptations in thrashers (Mimidae: genus *Toxostoma*), with comments on interspecific relationships. *Univ. Calif. Publ. Zool.* **42**:341–400.

England, A.S., and W.F. Laudenslayer Jr. 1993. Bendire's thrasher *(Toxostoma bendirei)*. In: A. Poole and F. Gill (eds), *Birds of North America*, #71. Academy of Natural Sciences, Philadelphia, PA.

George, T.L. 1987a. Factors influencing the abundance of land birds on Baja California Islands: tests of alternative hypotheses. Ph.D. thesis, University of New Mexico.

George, T.L. 1987b. Greater land bird densities of island vs. mainland: relation to nest predation level. *Ecology* **68**:1393–1400.

George, T.L. 1987c. Nesting phenology of landbirds in Baja California, Mexico. *Condor* **89**: 920–923.

Grinnell, J. 1905. Where does the large billed sparrow spend the summer? *Auk* **22**:16.

Grinnell, J. 1928. A distributional summary of the ornithology of Lower California. *Univ. Calif. Publ. Zool.* **32**:1–300.

Henny, C.J., and D.W. Anderson. 1979. Osprey distribution, abundance and status in western North America. III. The Baja California and Gulf of California population. *S. Calif. Acad. Sci.* **78**:89–106.

Horton, R.E. 1932. Drainage basin characteristics. *Trans. Am. Geophys. Union* **13**:350–361.

Hubbard, J.P. 1974. Avian evolution in the arid lands of North America. *Living Bird* **1973**: 155–196.

Lack, D.L. 1976. *Island Biology, Illustrated by the Birds of Jamaica*. Blackwell, Oxford.

Lamb, C.C. 1924. Lower California notes. *Oologist* **61**:63.

Lawrence, G.N. 1860. Description of three new species of hummingbirds of the genera *Heliomaster*, *Amazilia* and *Mellisuga*. *Am. Lyc. Nat. Hist. NY* **7**:107–111.

Maillard, J. 1923. Expedition of the California Academy of Sciences to the Gulf of California in 1921. The birds. *Proc. Calif. Acad. Sci.*, ser. **12**:443–456.

Miller, A.H. 1941. Speciation in the avian genus *Junco*. *Univ. Calif. Publ. Zool.* **44**:173–434.

Nelson, E.W. 1921. Lower California and its natural resources. *Nat. Acad. Sci. Mem.* **16**:1–194.

Ridgway, R. 1896. *A Manual of North American Birds*, 2nd ed. Lippincott, Philadelphia, PA.

Sallabanks, R., and F.C. James. 1999. American Robin *(Turdus migratorius)*. In: A. Poole and F. Gill (eds), *Birds of North America*. Academy of Natural Sciences, Philadelphia, PA.

Sclater, P.L. 1860. Notice of discovery of *Hylocharis xantusi*. *Ibis* **2**:309.

Sheppard, J.M. 1996. LeConte's thrasher *(Toxostoma lecontei)*. In: A. Poole and F. Gill (eds), *Birds of North America*. #230 Academy of Natural Sciences, Philadelphia, PA.

Slatyer, R.O., and J.A. Mabbutt. 1964. Hydrology of arid and semiarid Regions. In: V.T. Chow (ed), *Handbook of Applied Hydrology*. McGraw-Hill, San Francisco, CA. pp. 24-1–24-46.

Smith, M., and G.A. Bartholomew. 1966. The water economy of the black-throated sparrow and the rock wren. *Condor* **68**:447–458.

Stager, K.E. 1960. The composition and origin of the avifauna. *Syst. Zool.* **9**:179–183.

Strahler, A.N. 1964. Quantitative geomorphology of drainage basins and channel networks. In: V.T. Chow (ed), *Handbook of Applied Hydrology*, McGraw-Hill, San Francisco, CA; pp. 4-39–4-76.

Sullivan, K.A. 1999. Yellow-eyed junco *(Junco phaeonotus)*. In: A. Poole and F. Gill (eds), *Birds of North America*. no. 464. Academy of Natural Sciences, Philadelphia, PA.

Thayer, J.E. 1907. Catalog of birds collected by W.W. Brown Jr. in middle Lower California. *Condor* **9**:135–140.

Thayer, J.E., and O. Bangs. 1909. Description of new subspecies of the snowy heron. *Proc. N. Engl. Zool. Club* **4**:39–41.

Townsend, C.H. 1923. Birds collected in Lower California. *Am. Mus. Nat. Hist. Bull.* **48**:1–26.

van Rossem, A.J. 1929. The status of some Pacific Coast clapper rails. *Condor* **31**:213–215.

van Rossem, A.J. 1932. The avifauna of Tiburon Island, Sonora, Mexico, with descriptions of four new races. Trans. *San Diego Soc. Nat. Hist.* **7**:119–150.

van Rossem, A.J. 1945a. Preliminary studies in the black-throated sparrows of Baja California, Mexico. *Trans. San Diego Soc. Nat. Hist.* **10**:237–244.

van Rossem, A.J. 1945b. A distributional survey of the birds of Sonora, Mexico. *Occ. Papers Mus. Zool. La. State Univ.* **21**:1–379.

Vaughan, T.A., and S.T. Schwartz. 1980. Behavioral ecology of an insular woodrat. *J. Mammal.* **61**(2):205–218.

Wauer, R.M. 1978. *The breeding avifauna of Isla Tiburon, Sonora, Mexico.* U.S. Fish and Wildlife Service.

Wright, S.J. 1980. Density compensation in island avifaunas. *Oecologia* **45**:385–389.

Xantus, J. 1859. Descriptions of supposed new species of birds for Cape St. Lucas, Lower California. *Proc. Phil. Acad. Nat. Sci.* **1859**:297–299.

Yeaton, R. 1974. An ecological analysis of chaparral and pine forest bird communities on Santa Cruz Island and mainland California. *Ecology* **55**:959–973.

Yeaton, R., and M.L. Cody. 1974. Competitive release in island song sparrow populations. *Theor. Popul. Biol.* **5**:42–58.

Zink, R.M., and J.B. Slowinski. 1995. Evidence from molecular systematics for decreased avian diversity in the Pleistocene Epoch. *Proc. Nat. Acad. Sci. USA* **92**:5832–5835.

11

Breeding Dynamics of Heermann's Gulls

ENRIQUETA VELARDE
EXEQUIEL EZCURRA

Islands are landmarks for sea birds, whether for orientation, as resting points during foraging and migration trips, or most importantly as nesting sites. This is due to the isolation that islands offer, rendering them free of many of the continental predators. If, additionally, islands are located in the midst of highly productive waters, they provide sea birds with abundant food, which is particularly valuable during the nesting season. This is the case in the northern Sea of Cortés. Not surprisingly, we find that the islands of this region are nesting sites for more than 90% of the world's populations of Heermann's Gulls (*Larus heermanni*) and Elegant Terns (*Sterna elegans*), and for about 90% of the global populations of the Least Storm-petrel (*Oceanodroma microsoma*), the Craveri's Murrelet (*Synthliboramphus craveri*), and the Yellow-footed Gull (*Larus livens*). The midriff island area of the Gulf of California also shelters approximately 70% of the world's Black Storm-petrel (*O. melania*) and, at the subspecific level, provides breeding grounds for about 50% of the California Brown Pelicans (*Pelecanus occidentalis californicus*), 50% of the Blue-footed Boobies (*Sula nebouxii nebouxii*), and 40% of the Brown Boobies (*S. leucogaster brewsteri*).

A combination of characteristics in one particular island, Rasa, has made it a natural breeding sanctuary for Heermann's Gulls. Besides the two traits mentioned above (lack of land predators and high marine productivity), these characteristics include (1) its characteristic flat topography from which it derives its name (*rasa* means "flat" in Spanish), and (2) its sparse vegetation cover, resulting from the extensive coverage of the island with guano that hinders vegetation growth.

The Heermann's Gull is the only North American representative of the group of White-hooded Gulls (Anderson 1983; Moynihan 1959; Storer 1971). The only other member of this group of gulls and hence its closest relative is the Grey Gull (*Larus modestus*), which inhabits the Pacific coast of South America along Chile and Peru and breeds inland in the Atacama Desert some 50–100 km away from the coast (Howell et al. 1974; Howell 1978). These two species nest in very arid environments and present several morphological and behavioral adaptations that allow them to survive in these extreme environments (Howell et al. 1974; Howell 1978; Bartholomew and Dawson 1979; Bennett and Dawson 1979; Rahn and Dawson 1979; Velarde González 1989). In this chapter, we review the factors that affect the nesting of the Heermann's Gull population in Isla Rasa and analyze the annual fluctuations of breeding success and the condition of the nesting birds (as estimated by body mass) for a 10-year period.

Isla Rasa, Its Surrounding Waters, and the Heermann's Gull

Rasa is situated in the midriff island region of the Gulf of California (28°49'24"N, 112°59'03"W), approximately 60 km southeast from Bahía de los Ángeles, Baja California. The island dates from the Holocene (Gastil et al. 1983). It was formed by an eruption flow probably originating from Partida Island, a larger volcanic island located some 8 km northwest of Rasa, and is composed of basaltic conglomerates and agglomerates with a few interspersed sedimentary deposits. The whole island is being slowly uplifted by tectonic movements.

The island is flat with a maximum elevation of 35 m, whereas the mean elevation of nearby islands is approximately 500 m. It has an approximate area of 56 ha and consists mainly of low hills of volcanic rock and large valleys with deep guano deposits. The island shores are rocky, with cliffs mainly in the eastern side and large rock boulders in the rest of the periphery. There are three tidal lagoons at the western side, one of them directly connected with the ocean, which empties at low tide and fills up at high tide.

The flat sedimentary valleys of Isla Rasa provide prime habitat for nesting Heermann's Gulls. Soil core analyses suggest that the higher valleys originated directly from the accumulation of bird guano on the basaltic substrate (Vidal 1967). The lower valleys might have originated from ancient tidal lagoons similar to the ones present there now. The process of tectonic uplifting caused these former lagoons to emerge and fill up with sediments and guano, where vegetation started to establish later. Vidal (1967) described the origin of the sedimentary valleys of Isla Rasa based in the sequence found in the soil profiles. These layers consisted, in descending order, of soft guano, compact guano, and a mixture of sand and sediments from the Upper Pleistocene with mollusks in the process of fossilization. In contrast, the higher valleys had soft guano, hard guano, and clay and brackish water. Evidently, as the area between the hills filled up with guano through the centuries, the valleys enlarged.

The climate is hot and very dry, with the scanty rain distributed throughout the year but mainly between August and December (type BWx' in Köeppen's classification; see García 1964). The mean ocean surface temperature in this area varies between 14°C in February and 30°C in August (Robinson 1973). Botanically, the island lies within the sarcocaulescent deserts of the Gulf of California (Wiggins 1980). The most abundant plants are two species of cholla (*Opuntia cholla* and *O. alcahes* var. *alcahes*) which, together with the saltbush (*Atriplex barclayana*), cover large areas of the island. There are also a few dozen cardons (*Pachycereus pringlei*) and a few individuals of sour pitahaya (*Stenocereus gummosus*) and senita (*Lophocereus schottii*), as well as some shrubs (*Lycium brevipes* and *Cressa truxillensis*). Along the coastal areas of the tidal lagoons, there are three halophytes (*Salicornia pacifica*, *Sesuvium verrucosum*, and *Abronia maritima*).

Only two aerial predators, the Yellow-footed Gull and the Peregrine Falcon, occasionally prey upon the Elegant Terns and the Heermann's Gulls nests of Rasa. The island has no native mammals. Until early 1995, there were two species of rodents, the black rat (*Rattus rattus*) and the house mouse (*Mus musculus*), which are believed to have been introduced in the late nineteenth century (Bahre 1983). These two rodents were eliminated in February 1995 through a successful eradication program (Ramírez et al., unpublished data). There are only two reptiles: the side-blotched lizard (*Uta stansburiana*) and the leaf-toed gecko (*Phyllodactylus tinklei*). More than 80 bird species have been registered on or in the waters surrounding the island (Velarde Gonzaléz 1989; Cody and Velarde, chap. 10, this volume). However, only a few land and coastal water birds have been confirmed to nest in Rasa at one time or another: the Reddish Egrett (*Egretta rufescens*), the Pacific Oystercatcher (*Haematopus palliatus*), the Osprey (*Pandion haliaetus*), the Peregrine Falcon (*Falco peregrinus*), the Barn Owl (*Tyto alba*), the Raven (*Corvus corax*), the Rock Wren (*Salpinctes obsoletus*), and the House Finch (*Carpodacus mexicanus*).

The most conspicuous birds nesting in Rasa are the sea birds, but no reliable historical population records of these sea birds exist. In 1999 there were approximately 260,000 Heermann's Gulls, 200,000 Elegant Terns (representing >95% of the world populations of each of these two species), and 10,000 Royal Terns (*Sterna maxima*) nesting in the island. The number of these three species has varied greatly over 21 years of systematic censuses. Royal Terns have fluctuated between 10,000 and 17,000 individuals. Elegant Terns have increased dramatically, from approximately 30,000 individuals in the early 1980s to 200,000 in 1999. In contrast, the Heermann's Gull populations have had small fluctuations. Even during their worst years (e.g., 1998), most adult individuals were present in their nesting territories even though reproductive success was low in those periods.

The Influence of Oceanic Anomalies in the Midriff Region

Isla Rasa is situated in one of the most productive oceanic regions in the world (Álvarez-Borrego and Lara-Lara 1991). This high marine productivity is the result of a combination of coastal and tidal upwellings (due, respectively, to the wind along the coastal areas and to tidal action against the coastal underwater walls). Marine productivity varies between years, being particularly low during El Niño years (also known as El Niño Southern Oscillation or ENSO events), but these variations are not as strong in the midriff as in the Pacific Ocean or the southern Gulf of California (Álvarez-Borrego and Lara-Lara 1991).

The El Niño cycle is generated by an alternating variation in surface pressure in the tropical Pacific. The cycle is commonly measured by comparing simultaneous readings of sea level pressure (SLP) anomalies at Tahiti, in the east-central tropical Pacific, and Darwin, on the northwest coast of Australia. The phenomenon shows oscillations with a quasi-period of 2–5 years. When SLP is low at Tahiti and high at Darwin, the El Niño, or warm phase of the cycle, is taking place. When El Niño conditions prevail, trade winds slacken and warm water accumulates in the eastern Pacific Ocean. This in turn decreases the coastal upwellings along the American continental coast and results in a collapse of the primary productivity of the sea. The cold phase of the cycle, called "La Niña," is characterized by high pressure in the eastern equatorial Pacific, low pressure in the west, and a resultant strong westward flow in the tropical belt of the Pacific Ocean. When La Niña conditions occur, the westward deflection of tropical waters along the coasts of the American continent generates the upwelling of deep, nutrient-rich and cool waters, and the productivity of the eastern Pacific Ocean is enhanced. It is widely recognized that these oscillating conditions have striking consequences for the productivity of the oceans.

Pacific sardines (*Sardinops caeruleus*) and northern anchovies (*Engraulis mordax*) migrate from the Guaymas Basin into the midriff island region in spring and summer (Sokolov 1974; Hammann et al. 1988; Hammann 1991) in response to the high productivity conditions (i.e., strong upwellings, low surface-water temperatures, and high nutrient contents) that prevail in this area (Badan-Dangon et al. 1985). These two small pelagic fish constitute 75–100% of the diet of Heermann's Gulls and Elegant terns (Velarde et al. 1994) and are important in the diet of many other seabird species (Anderson et al. 1980). Therefore, many of the islands of the region have important sea bird nesting colonies (Velarde and Anderson 1994).

The Heermann's Gull Cycle of Migration and Breeding

Heermann's Gulls start arriving in the midriff island region in mid-February. By late March about 200,000 have reached the area. Every day at dusk they gather around Rasa to occupy the island in a spectacular communal flight. The gulls stay all night on the island, courting and defending their nesting territories, which range over some 1.5 m^2, only to leave with the first light of dawn. This goes on until the last days of March or the first days of April, when the gulls remain on the island for longer periods and finally stay on Rasa permanently. Nesting density in the valley areas is approximately 70 nests/100 m^2 and on the rocky hills is 9.5 nests/100 m^2 (Velarde 1992, 1999). Some egg laying begins even before the gulls permanently occupy the island and peaks about 10 days afterward. Nesting is synchronous, a behavior that has been shown to reduce predator efficiency (Velarde 1992, 1993). Synchronous nesting also coincides with the period when food is most abundant in the area (Sokolov 1974; Hammann et al. 1988; Hammann 1991).

The general breeding biology of the Heermann's Gull in Isla Rasa, and the relationship between breeding success and factors such as food availability have been described by Velarde (1999). However, no long-term analysis of the annual fluctuations of the breeding success and of related factors, such as the condition of the breeding birds, has been published.

Between 1984 and 1993 (with the exception of 1988 and 1992, when breeding success was almost zero), we banded 4000 fledgeling Heermann's Gulls every year, using aluminum bands provided by the U.S. Fish and Wildlife Service. These individuals started to return to Isla Rasa as breeders for the first time in 1988. Between 1989 and 1992 and between 1995 and 1999, we monitored the breeding success of banded, known age Heermann's Gulls in Rasa. Because of logistic difficulties, we did not monitor the populations during 1993 and 1994.

Nests where banded individuals were found were marked with numbered wooden stakes at the beginning of each nesting season in the valley areas. Banded individuals and, when possible, their nesting mates, were captured at the end of the incubation period. Aluminum bands were replaced by Incaloy bands, which last for several decades. Banded individuals and their mates (when captured) were weighed.

Marked nests were monitored daily, or at least every other day, from the day the first egg was laid until all chicks were at least 20 days old, an age at which they are past peak mortality (Gonzalez-Peralta et al. 1988). Data on clutch size, egg mass, egg survival, and chick survival were obtained. From these data we calculated breeding success (number of fledging chicks/number of eggs laid) and nesting success (number of fledgeling chicks in each nest) for each nest and averaged these values for the whole group of nests in each year.

Measuring the El Niño Cycles

The Southern Oscillation Index (SOI; see NCEP 2000) is a measure of air pressure anomalies in the Pacific Ocean (i.e., El Niño or ENSO conditions) and is calculated from SLP readings in Tahiti ($\sim 18°$S and $150°$W) and Darwin (northern Australia, around $12°$S, $132°$E), as follows:

$$\text{SOI}(\text{month},\text{year}) = [T(\text{month},\text{year}) - D(\text{month},\text{year})]/S$$

where T and D are values of SLP in Tahiti and Darwin, respectively, and S is the standard deviation of the numerator for all months combined for the years 1951–1980. In short, the SOI is the normalized difference in atmospheric pressure between Tahiti and Darwin and is used as an indicator of the warm phase. Normally pressure at Tahiti is the higher of the two, and equatorial trade winds blow toward Australia; during the warm El Niño phase this pressure difference reverses, and westerly wind bursts arise in the western equatorial Pacific (Chelliah 1990). For our study, we obtained the monthly SOI values for the 1988–1999 period. We changed the sign of the SOI index, so that high positive values would indicate the warm phase (El Niño events) and high negative values the cold phase.

Taking into consideration that Heermann's Gulls nest during April–June each year, we calculated the average of the monthly SOI indices during the winter and early spring (December–April; i.e., the months before the arrival and during the early nesting of the birds). We expected these winter-spring SOI values to give an indication of the conditions for marine productivity prevailing in the eastern Pacific and the Gulf of California in the months before nesting and therefore to predict the food availability for the breeding gulls. Because the SOI index is normalized (i.e., it is measured in standard deviations), we arbitrarily defined a breeding season as having strong El Niño conditions when the mean winter-spring anomaly values were greater than 2.

The Relationship Between El Niño Cycles and the Reproductive Success of Heermann's Gulls

A linear model ANOVA was used to test for differences in reproductive success. As predicted, or dependent, variables, we used both the breeding and nesting successes measured in 1298 nest follow-ups accumulated during the 9 years of our study (on average, we followed some 144 nests per year). As predictors we tested (1) the fixed effect of the year, (2) the SOI index, (3) the body mass of the female parent, (4) the body mass of the male parent, (5) the mass of the eggs, (6) the valley site within the island (to test for fixed site effects), and (7) the age of the parents. The graphical relationship between simple pairs of variables was described and tested by means of linear regression. When both the independent and the dependent variables were subject to experimental error, we used major axis regression instead of traditional least squares (Sokal and Rohlf 1995).

Statistical Relationships

Two El Niño events occurred during our study period (fig. 11.1). One occurred in the winter-spring of 1991–92 and the other during 1997–98. The body mass of the birds fluctuated from year to year and decreased significantly during El Niño years (fig. 11.2). The maximum weight was achieved in the 1990 nesting season, and the minimum during the 1992. Males and females increased and decreased their body mass in similar amounts along the years (fig. 11.2a). Consequently, male and female mean body mass showed a high positive relationship ($r = .96$, $p \leqslant .001$). The body mass of female Heermann's Gulls was 81–84% that of the males. However, the relationship varies from year to year, and females seem to suffer more than males during critical

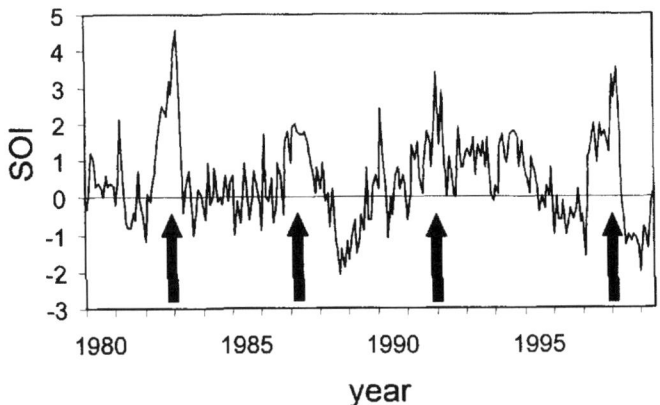

Figure 11.1 Monthly values of the Southern Oscillation Index (SOI) from January 1980 to June 1999. The arrows indicate the four El Niño events during the period. Note that El Niño events, as indicated by high SOI values, tend to occur in late fall, winter, and early spring, and that a typical El Niño year shows a succession of months with SOI around or greater than 2. Our study period included the two events of the 1990s decade: 1991–92 and 1997–98.

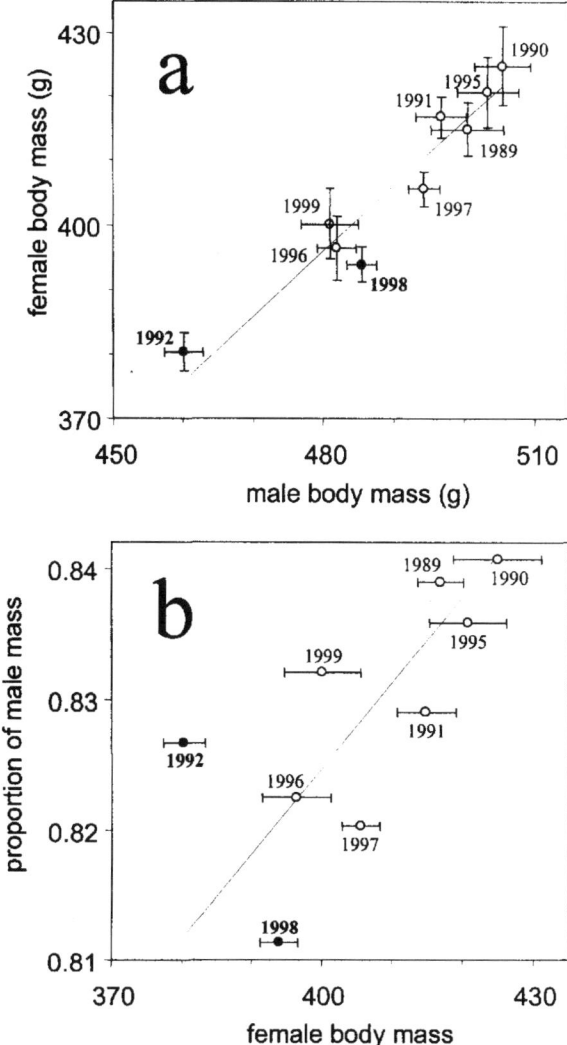

Figure 11.2 (a) Relationship between Heermann's Gull male and female body masses ($r = .96$; $p < .001$). The slope of the relationship is 1.01, indicating that both males and females lose similar amounts of weight in critical years. However, because male body mass is larger than female body mass, the relative variation of female body mass is proportionally larger. (b) Female-to-male body mass ratio as a function of female body mass ($r = .68$; $p = .04$). With decreasing body masses, the relative weight of females also decreases. In both graphs, solid points and bold labels indicate El Niño years (1992 and 1998). The lines were fitted by principal axis regression. The sample sizes were 69, 63, 133, 203, 96, 237, 325, 70, and 93 nests for years 1988, 1989, 1990, 1991, 1992, 1995, 1996, 1997, and 1998, respectively.

periods. During years of low food availability, females have about 81% of the mass of the males, whereas in good years the proportion increases to 84% (fig. 11.2b).

Clutch size and egg mass were not significantly related to El Niño conditions; the number of eggs laid at the beginning of the reproductive season and the mean mass of the eggs seemed to be largely independent of the physical conditions of the ocean. Reproductive success, in contrast, varied greatly between El Niño and normal years. Both nesting success and breeding success showed their lowest values during 1992 and 1998, the two El Niño years of our study period (fig. 11.3). These remarkable drops in effective reproduction coincide with positive anomalies in the Pacific Ocean. Consequently, a significant ($p < .05$) negative linear relationship was found between SOI values and both breeding and nesting success (fig. 11.4). On a closer inspection, however,

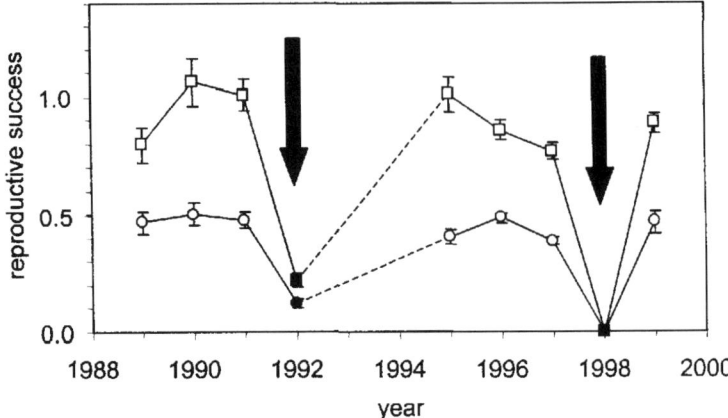

Figure 11.3 Reproductive success during the study period. Circles indicate breeding success, squares indicate nesting success. The dotted lines between 1992 and 1995 indicate the lack of data for 1993 and 1994. The black arrows show the El Niño events of 1991–92 and 1997–98.

we found that these plots consist of two distinct, separate data clusters, one containing the data points from normal years, and the other containing the data points from El Niño years. That is, the system seems to have two distinct outcomes of the reproductive season: When the winter-spring SOI index is >2 (strong El Niño conditions), the reproductive success collapses to near-zero values. When the index is <1 (cool-phase conditions), breeding success is around 0.4–0.5 and nesting success is around 0.8–1.0.

A linear model ANOVA on the whole data set (1298 nests) confirmed this result. In all cases, the best predictor of both breeding success ($F_{1,1287} = 289.0$; $p \ll .000001$) and nesting success ($F_{1,1287} = 72.9$; $p \ll .000001$) was a qualitative, binary variable describing the breeding seasons as "El Niño" (SOI > 2; years 1992 and 1998) or "normal" (SOI < 1; all other years). Once this predictor was included into the stepwise linear model, all other variables (i.e., the fixed effect of the year, the SOI index, the body mass of the female parent, the body mass of the male parent, the mass of the eggs, the fixed effect of valley site, and the age of the parents) showed only marginally significant or mostly nonsignificant effects. A full 100% of the variation explained by the best stepwise linear model for breeding success and almost 90% of the explained variation for nesting success were attributable to the simple qualitative effect of El Niño years. The SOI index was in all cases a poorer predictor than its binary simplification, suggesting once again that breeding success is a threshold phenomenon. A similar two-cluster distribution was found between the body mass of females and males and both nesting and breeding success. This is exemplified in figure 11.5 for the case of female body mass and breeding success.

Oceanic Conditions, Body Mass, and Breeding Success of Heermann's Gulls

During years of low food availability, females show decreased body mass to a greater degree than males. This nonlinear relationship in body mass indicates that females

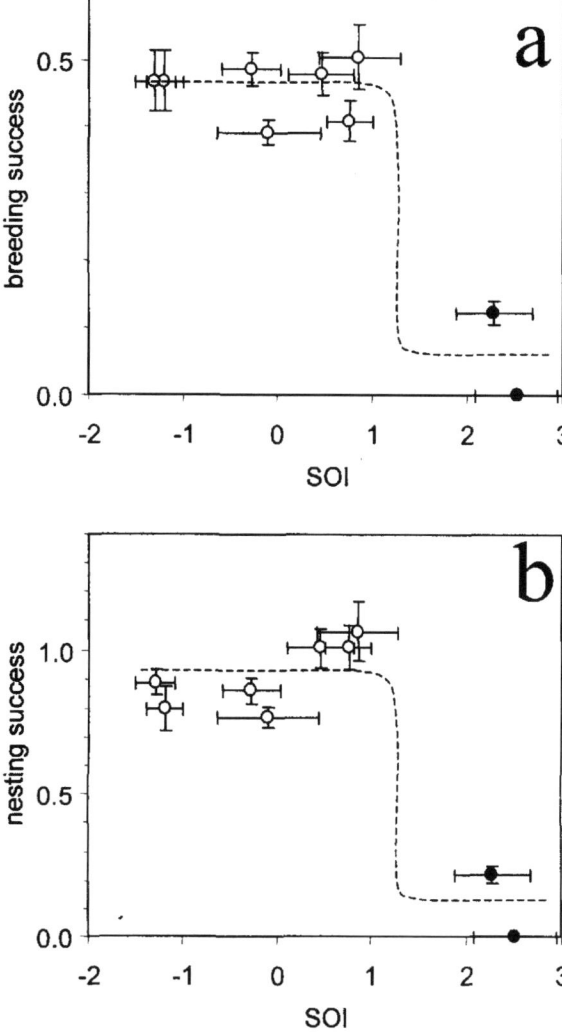

Figure 11.4 (a) Relationship between the winter-spring Southern Oscillation Index (SOI) and breeding success. A significant negative linear correlation was found for the whole data set ($r = -.81; p = .01$). However, if the data set is divided into two treatments ("normal" years and "El Niño" years), the data set forms two disjunct clusters with no within-cluster trend. The SOI index has no capacity to predict breeding success within normal years, nor within El Niño years. The resulting threshold-type relationship is sketched in the broken line, showing the abrupt collapse in breeding success that occurs when the winter-spring SOI anomaly reaches values >2. (b) Relationship between the winter-spring SOI values and nesting success. The pattern is similar to that of breeding success in panel a. Although the data set is correlated ($r = -.68; p = .04$), it can also be described by a threshold-type relationship separating normal years from El Niño periods. In both graphs, solid dots indicate EL Niño years (1992 and 1998).

suffer more from the stress of El Niño years than males. During courtship, it is usual for males to provide food for the females. However, the amounts of food regurgitated by males to feed females tend to decrease during El Niño years, and it is likely that, for the females, this factor may further aggravate the famine conditions during anomalous years.

The fact that neither clutch size nor egg mass were found to be related to oceanic conditions suggests that (1) clutch success is possibly more related to conditions in the gull's summer grounds than to local conditions in the Gulf of California, and (2) the El Niño cycle affects the Heermann's gulls once they have settled in the midriff of the Gulf of California.

Perhaps the most important result of this analysis is the tight relationship detected between reproductive success, measured both as nesting success and breeding success,

Figure 11.5 Relationship between female mass and breeding success, showing a similar pattern to that described in figure 11.2. A significant positive linear correlation exists for the whole data set ($r = .68$; $p = .04$), but the data points are distributed in two distinct clusters. During normal years, females show on average a larger body mass (ca. 415 g), and their breeding success is around 0.5; during El Niño years, the mean body mass of females decreased to around 380–390 g, and their breeding success collapsed to 0–0.12. Solid dots indicate El Niño years.

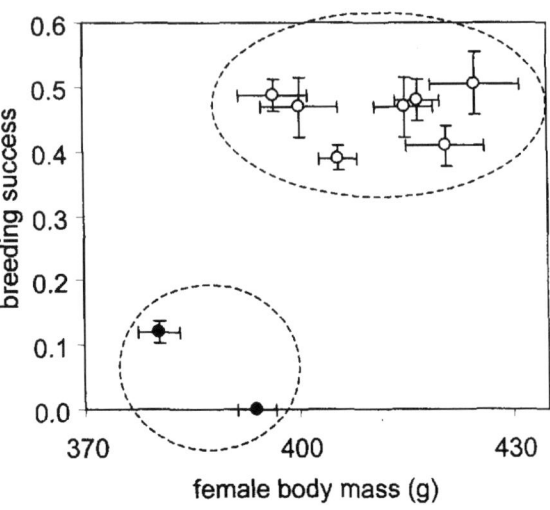

and the ENSO phenomenon. Intense El Niño anomalies, such as those observed in 1992 and 1998, seem to produce a breeding collapse in the colony of Heermann's Gulls at Rasa. Furthermore, the phenomenon seems to follow a threshold, or trigger, behavior. In nonanomalous, or normal, years, when the winter-spring SOI values are low (<1), the breeding success estimates seem to be independent of the SOI index, in spite of the large variation observed in its values, ranging from -1.2 to 0.86. However, an increase in the SOI index to values ≥ 2 is sufficient to collapse the population's reproductive success. During our study period we did not observe SOI values between 1 and 2, and it is difficult to hypothesize about the possible reproductive outcome for the gull population if such conditions occurred. We can conclude, however, that the collapse of the gull's reproduction occurs along a narrow range of SOI values. Thus, the anomaly seems to have a nonlinear, threshold-type effect on this biological system. A gradual increase in the anomaly will force an abrupt collapse in reproductive success once the threshold is reached. This may be related to the enclosed oceanographic conditions of the Sea of Cortés that seem to be able to dampen down the effects of the Pacific warming during El Niño seasons (see chap. 3).

Effects of the 1997 ENSO event on other sea bird species that nest in the midriff island region were also evident (Anderson et al. 1999). Although several hundred Elegant Terns were observed in the vicinity of Rasa, only 180 were actually nesting in Rasa, as compared to an average of about 40,000 of the 9 previous years. In the case of California Brown Pelicans, there were only 1200 nesting attempts as compared to a long-term average of some 25,000 nests, representing less than 5% of the normal productivity. Only one nest was reported to have produced young. In the case of Brandt Cormorants, Double Crested Cormorants, and Yellow-footed Gulls, the same trends were observed, with only a few young produced, but always below 5% of the normal productivity. Ainley et al. (1988) reported similar effects for the 1982–83 ENSO. It has also been reported that the number of Blue-footed Boobies, Brown

Boobies, and nesting Brown Pelicans on the nearby island of San Pedro Mártir declined dramatically between 1990 (a normal year) and 1992 (after an el Niño season; Tershy 2000). Analogous demographic collapses during strong El Niño years have been reported for the Galapagos Islands (e.g., Anderson 1989; Valle et al. 1987) and also for the central Pacific Ocean (e.g., Schreiber and Schreiber 1984).

Conclusions: Global Oceanic Phenomena and Their Impact on the Sea of Cortés

The dynamics of the breeding demography of Heermann's Gulls is tightly connected to the global cycle of SLP/ocean-temperature anomalies. During the warm phase of the El Niño cycle, female gulls lose weight and the survival of chicks drops almost to zero. The alternating variation in surface pressures in the tropical Pacific Ocean that drives the El Niño cycle is echoed by the reproductive output of the gull population in an extraordinary manner. These results highlight the importance of teleconnections in global ecology: a gradient of air pressures developing over the tropical Pacific Ocean may critically affect the reproductive output of a population that breeds in one of the tiniest islands of the Gulf of California.

References

Ainley, D.G., H.R. Carter, D.W. Anderson, K.T. Briggs, M.L. Coulter, F. Cruz, J.B. Cruz, C.A. Valle, S.I. Fefer, S.A. Hatch, E.A. Schreiber, R.W. Schreiber, and N.G. Smith. 1988. Effects of the 1982–83 El Niño-Southern Oscillation on Pacific Ocean bird populations. *Proceedings of the International Ornithological Congress* **29**:1747–1758.

Álvarez-Borrego, S., and R. Lara-Lara. 1991. The physical environment and primary productivity of the Gulf of California. In: J.P. Dauphin and B. Simoneit (eds), *The Gulf and Peninsular Province of the Californias*. American Association of Petroleum Geology Memoir No. 47. pp. 555–567.

Anderson, D.J. 1989. Differential responses of boobies and other seabirds in the Galapagos to the 1986–87 El Niño-Southern Oscillation event. *Marine Ecology Progress Series* **52**(3): 209.

Anderson, D.W. 1983. The Seabirds. In: T.J. Case and M.L. Cody (eds.), *Island Biogeography in the Sea of Cortéz*. University of California Press, Berkeley; pp. 246–264.

Anderson, D.W., F. Gress, K.F. Mais, and P.R. Kelly. 1980. Brown Pelicans as anchovy stock indicators and their relationships to commercial fishing. *CalCOFI Reports* **21**:54–61.

Anderson, D., J. Keith, E. Palacios, E. Velarde, F. Gress, and K.A. King. 1999. El Niño 1997–1998: Seabird responses from the Southern California Current and Gulf of California. Pacific Seabird Group 26th. Annual Meeting. *Pacific Seabirds* **26**:22.

Badán-Dangón, A., C.J. Koblinsky, and T. Baumgartner. 1985. Spring and Summer in the Gulf of California: observations of surface thermal patterns. *Oceanologica Acta* **8**:13–22.

Bahre, C.J. 1983. Human Impact: The Midriff Islands. In: T. Case and M. Cody (eds), *Island Biogeography in the Sea of Cortez*. University of California Press, Berkeley; pp. 290–306.

Bartholomew, G.A., and W.R. Dawson. 1979. Thermoregulatory behavior during incubation in Heermann's Gulls. *Physiological Zoology* **52**:422–437.

Bennett, A.F., and W.R. Dawson. 1979. Physiological responses of embrionic Heermann's gulls to temperature. *Physiological Zoology* **52**:413–421.

Chelliah, M. 1990. The global climate for June-August 1989: A season of near normal conditions in the tropical Pacific. *Journal of Climate* **3**:138–160.

García, A.E. 1964. *Modificaciones al Sistema de Köeppen para adaptarlo a las condiciones de la República Mexicana*. Instituto de Geografía, UNAM, México.

Gastil, G., J. Minch, and R.P. Phillips. 1983. The geology and ages of islands. In: T.J. Case and M.L. Cody (eds), *Island Biogeography in the Sea of Cortéz*. University of California Press, Berkeley; pp. 13–25.

Gonzalez-Peralta, L., G. Lozano, and E. Velarde. 1988. Mortality factors of Heermann's Gull (*Larus heermanni*) chicks during the prefledging period in Rasa Island, Baja California. *Pacific Seabird Group Bulletin* **15**:29.

Hammann, M.G. 1991. Spawning habitat and egg and larval transport, and their importance to recruitment of Pacific sardine, *Sardinops sagax caeruleus*, in the Gulf of California. In: T. Kawasaki, S. Tanaka, Y. Toba, and A. Taniguchi (eds), *Long-term variability of pelagic fish populations and their environment*. Pergamon Press, Oxford; pp. 271–278.

Hammann, M.G., T.R. Baumgartner, and A. Badán-Dangón. 1988. Coupling of the Pacific sardine (*Sardionps sagax caeruleus*) life cycle with the Gulf of California pelagic environment. *CalCOFI Reports* **29**:102–109.

Howell, T.R. 1978. *Ecology and reproductive behavior of the gray gull of Chile and of the red-tailed tropicbird and white tern of Midway Islands*. National Geographic Society Research Report, 1969 Project V, National Geographic Society, Washington, D.C.; pp. 251–284.

Howell. T.R., B. Araya, and W.R. Millie. 1974. Breeding biology of the gray gull *Larus modestus*. *University of California Publications in Zoology 104*, Berkeley.

Moynihan, M. 1959. A revision of the family Laridae (Aves). *American Museum Novitates* **1928**:1–42.

NCEP. 2000. Southern Oscillation Index (SOI), January 1866–June 1999 (online database). Climate Prediction Center, National Center for Environmental Prediction. National Oceanic and Atmospheric Administration, Washington, DC.

Rahn, H., and W.R. Dawson. 1979. Incubation water loss in eggs of Heermann's and Western gulls. *Physiological Zoology* **52**:451–460.

Robinson, M.K. 1973. *Atlas of monthly mean sea surface and subsurface temperatures in the Gulf of California, Mexico*. Scripps Institution of Oceanography, San Diego.

Schreiber, R.W., and E. A. Schreiber. 1984. Central Pacific seabirds and the El Niño Southern Oscillation: 1982 to 1983 perspectives. *Science* **225**:713–716.

Sokal, R.R., and F.J. Rohlf. 1995. *Biometry*, 3rd ed. W.H. Freeman and Co., New York.

Sokolov, V.A. 1974. Investigaciones biológico pesqueras de los peces pelágicos del Golfo de California. *CalCOFI Reports* **17**:92–96.

Storer, R.W. 1971. Classification of birds. In: D.S. Rarner and J.R. King (eds), *Avian Biology*, vol. I. Academic Press, New York; pp. 1–18.

Tershy, B.R. 2000. A natural history of Isla San Pedro Mártir. Available online at http://herb.bio.nau.edu/~cortez/IslaSPM.htm. Island Conservation & Ecology Group, Institute of Marine Sciences, University of California, Santa Cruz.

Valle, C.A., F. Cruz, J.B. Cruz, G. Merlen, and M.C. Coulter. 1987. The impact of the 1982–1983 El Niño-southern oscillation on seabirds in the Galapagos Islands, Ecuador. *Journal of Geophysical Research* **92**(C13):14,437–14,444.

Velarde González, M.E. 1989. Conducta y ecología de la reproducción de la Gaviota Parda *Larus heermanni* en Isla Rasa, Baja California. Doctoral dissertation, Facultad de Ciencias, UNAM. México, D.F.

Velarde, E. 1992. Predation of Heermann's Gull (*Larus heermanni*) chicks by Yellow-footed Gulls (*L. livens*) in dense and scattered nesting sites. *Colonial Waterbirds* **15**:7–13.

Velarde, E. 1993. Predation of nesting larids by Peregrine Falcons at Isla Rasa, Baja California, Mexico. *Condor* **95**:706–708.

Velarde, E. 1999. Breeding biology of Heermann's Gulls on Isla Rasa, Gulf of California, Mexico. *Auk* **116**:513–519.

Velarde, E., and D.W. Anderson. 1994. Conservation and management of seabird islands in the Gulf of California: Setbacks and successes. In: D.N. Nettleship, J. Burger, and M. Gochfeld (eds), *Seabirds on Islands: Threats, Case Studies and Action Plans*. BirdLife Conservation Series No.1, BirdLife International, Cambridge, UK; pp. 721–765.

Velarde, E., M.S. Tordesillas, L. Vieyra, and R. Esquivel. 1994. Seabirds as indicators of important fish populations in the Gulf of California. *CalCoFi Reports* **35**:137–143.

Vidal, N. 1967. Aportación al conocimiento de la Isla Rasa, Baja California. B.Sc. thesis, Facultad de Ciencias, UNAM. México, D.F.

Wiggins, I.L. 1980. *Flora of Baja California*. Stanford University Press, Palo Alto, CA.

12

The Mammals

TIMOTHY E. LAWLOR
DAVID J. HAFNER
PAUL STAPP
BRETT R. RIDDLE
SERGIO TICUL ALVAREZ-CASTAÑEDA

In *The Log of the Sea of Cortez*, that memorable treatise of science, adventure, and philosophy, John Steinbeck (1951) made bare mention of mammals. Of course, the main purpose of that effort was to chronicle a trip to the Gulf of California to collect invertebrates in the company of Steinbeck's friend and scientist, Ed Ricketts. The party visited four islands—Tiburón, Coronados, San José, and Espíritu Santo. At anchor off Isla Tiburón, Steinbeck reported a swarm of bats that approached their boat. One bat was collected but, to the best of our knowledge, it was never identified or preserved.

Aside from some descriptions of taxa (e.g., Burt 1932), relatively little was known at the time about mammals from islands in the Sea of Cortés. There is now a reasonably rich history of systematic and biogeographic studies of mammals in and adjacent to the Sea of Cortés (for general reviews, see Orr 1960; Huey 1964; Lawlor 1983; and Hafner and Riddle 1997). Here we summarize much of that information and explore biogeographic patterns that emerge from it, add important recent records of bats, and evaluate new evidence about the origins of insular faunas and the ecological processes and human impacts that affect colonization and persistence of mammals on gulf islands.

Composition of the Mammalian Fauna

The terrestrial mammalian fauna of islands in the Sea of Cortés (including islands off the Pacific coast of Baja California) comprises 45 species, of which 18 currently are recognized as endemics (but see below), representing 5 orders, 9 families, and 14 genera (app. 12.1). Collectively they share relationships with mainland representatives on both sides of the gulf and are divisible into 28 clades of species or species groups (app. 12.2). Rodents are disproportionately represented, constituting a total of 35 species and 76 of 97 total insular occurrences, and they are the only nonvolant mammals to become established on distant oceanic islands (table 12.1). In addition, except for the few species of lagomorphs, which occur only on landbridge islands, a greater proportion of mainland species of rodents occurs on islands than is the case for other groups of mammals (table 12.1). Overall, large mammals (jackrabbits, carnivores, artiodactyls) are poorly represented. Only larger landbridge islands contain them, and no species is recorded from more than four islands.

Peromyscus is the most widely represented genus. Seven mainland species of these seemingly ubiquitous omnivores contribute to island faunas (see app. 12.1). Only *Peromyscus truei*, a species restricted to woodlands and chaparral, does not occur on islands. Pocket mice (*Chaetodipus*), which are largely granivorous, also are well represented. Six mainland species are found on gulf islands (a seventh, *C. fallax*, inhabits Cedros, a Pacific island).

Aside from large landbridge islands (e.g., Tiburón, and the Pacific islands Magdalena and Santa Margarita), it is rare to find more than one species of a genus occupying a given island. Surprisingly, exceptions are on small islands. Willard Island supports *Peromyscus maniculatus* and *P. crinitus*, and San Pedro Nolasco is inhabited by *Peromyscus boylii* and *P. pembertoni* (however, the latter, likely a derivative of *P. merriami*, is probably extinct; see below).

Bats are poorly known from the Sea of Cortés. Twelve species, representing 3 families and 10 genera, are recorded from 29 islands (app. 12.3; Reeder and Norris 1954; Banks 1964a,b,c; Huey 1964; Orr and Banks 1964; Villa-R. 1967; Sanchez-Hernandez 1986; Alvarez-Castañeda and Patton 1999; M. A. Bogan, pers. commun., 1999; M. L. Cody, pers. commun., 1980; T. A. Vaughan, pers. commun., 1980; Alvarez-Castañeda et al., pers. obs.). The most commonly observed species, the fish-eating bat (*Myotis vivesi*), has been reported from 24 islands (and probably occurs on

Table 12.1 Representation on gulf islands of orders and species of mammals

		Mainland Species Representation on Islands					
		All Islands		Landbridge Islands		Oceanic Islands	
Order of Mammals	No. of Mainland Species[a]	No.	%	No.	%	No.	%
Nonvolant							
Insectivora	1	0	0.0	0	0.0	0	0.0
Lagomorpha	4	3	75.0	3	75.0	0	0.0
Rodentia	26	19	73.1	18	69.2	9	34.6
Carnivora	10	3	30.0	3	30.0	0	0.0
Artiodactyla	4	2	50.0	2	50.0	0	0.0
Volant							
Chiroptera	47	12	25.5	12	25.5	7	12.8

Pacific islands are not included. For nonvolant mammals, only desert-adapted species are considered.
[a]Sonora + Baja California.

virtually all islands); it is the only bat on 17 of 29 islands from which bats are recorded. There are no recognized endemic forms, consistent with the high mobility of bats and proximity of islands to mainland areas.

With few exceptions, insectivorous bats predominate. In addition to the fish-eating bat, two frugivorous species are present: *Leptonycteris curasoae*, which consumes nectar, pollen, and fruit, and derives large quantities of water from nectar (Carpenter 1969); and *Macrotus californicus*, which has a mixed diet of fruit and insects. In general, more species of bats occur on larger, more densely vegetated islands (including eight species on Isla Tiburón, the largest gulf island); no bats are known from Isla Ángel de la Guarda, a large but sparsely vegetated island.

The relatively modest contribution of bat species to insular faunas (table 12.1; app. 12.3) is due mostly to incomplete sampling for these mammals. For example, one night's netting on an easily accessible island (Espíritu Santo) in 1999 (Riddle, Hafner, and Alvarez-Castañeda pers. obs.) yielded four bats of three families and four genera, two of which (*Nyctinomops femorosaccus* and *Lasiurus xanthinus*) were new records for gulf islands. There also has been little effort to collect bats on Pacific islands (only one is recorded; app. 12.3). Increased trapping effort is likely to yield many additional records of bats.

Origins and Evolution of Insular Faunas

The mammalian faunas of the gulf islands and surrounding mainland areas have been influenced primarily by the evolution of North American regional deserts and geological events associated with the formation and changing shorelines of the Sea of Cortés. In view of the aridity of islands in the gulf and their recent origin, it is not surprising that their mammalian inhabitants are virtually all desert species, each apparently with closely related sister taxa on the closest mainland. However, this straightforward interpretation of the origin of insular mammalian faunas has been complicated somewhat by the recent discoveries of trenchant evolutionary divergences within supposedly circum-gulf species that are concordant with Pliocene and early Pleistocene geological events (Riddle et al. 2000a,b). Thus, as with the herpetofauna of the region (Murphy 1983a,b; Grismer 1994), source populations surrounding gulf islands are more varied than previously recognized. Further, comparison of sequence data from mtDNA of insular populations has provided evidence that the specific source for each island population has not always been from the adjacent mainland (Hafner et al. 2001). Modern mammalian faunas represent a sampling of ancient mainland events in the evolution of North American regional deserts. Earlier views of the gulf and the Baja California peninsula portrayed them largely as peripheral elements in the diversification of regional deserts (e.g., Shreve 1942; Orr 1960; Savage 1960; Wiggins 1960, 1980; Axelrod 1979). However, emerging geological and biogeographic information suggests a greater role for faunas of these areas in the Neogene and Pleistocene assembly of arid-adapted floras and faunas (Murphy 1983a,b; Grismer 1994; Hafner and Riddle 1997; Riddle et al. 2000a,b).

As discussed in chapters 2, 8, and 9, islands in the gulf fall into two categories: old, deep-water oceanic islands and young, shallow-water landbridge islands. This division has important implications for biogeographic analyses of the mammals on the islands. First, it means that well-isolated oceanic islands were colonized by waif dispersal, whereas extant populations on landbridge islands likely originated from fragmentation of widespread mainland populations as the postglacial rises in sea level isolated the islands. The picture is one of two types of faunas: one on oceanic islands made up only of forms that can colonize across water and the other on landbridge islands, consisting chiefly of relictual populations. Expressed in terms of insular biogeographic theory, landbridge islands reach an equilibrium number of species from above, via extinction (they are "oversaturated"), and oceanic islands from below, via dispersal (they are "undersaturated"). The difference in manner of formation of the two types of islands should have important effects on species composition, species diversity, degree of endemism, and amount of morphological variation. Also, the nearly simultaneous (in geological time) separation of landbridge islands from the mainland following the close of the last glacial interval offers an opportunity to reconstruct the full-glacial distribution and species composition of desert refugia surrounding the Sea of Cortés. In effect, landbridge islands serve as genetic and morphological museums of their ancestral founders.

The geologic history of the gulf coincides with the evolution of the North American regional deserts, and insular mammalian faunas consist primarily of desert-adapted species. Among peninsular mammals, and in strong contrast to reptiles, there may be no relicts of more tropical forms in the cape region resulting from initial formation of the gulf approximately 5.5 Ma (Grismer 1994). Instead, relatively mesic-adapted species of mammals occurring today in the cape region probably arrived by dispersal down the higher-elevation spine of the peninsula during pluvial intervals of the Pleistocene, as recently as 10,000 years ago (e.g., *Sorex ornatus*, *Peromyscus truei*) or may have resulted from recent introduction by humans (*Oryzomys couesi*; Nelson 1921; see also Alvarez-Castañeda 1994). Thus, there was little opportunity for dispersal to either landbridge or oceanic islands by other than arid-adapted mammals, particularly from Baja California.

Present-day mammalian faunas of the gulf islands are relatively depauperate subsets of the arid-adapted faunas of the adjacent peninsular and Sonoran regional deserts. Their evolution can reasonably be attributed to events and consequences in geological versus ecological time. Geological events influenced the assembly of the mainland sources for island colonization and resulted in initial isolation of populations on landbridge islands. Subsequent extinction, waif dispersal, and perhaps competition among insular populations then modified the composition of insular faunas.

Historical Biogeography

Pertinent Geological Events

Mainland mammalian faunas (of both the peninsular and Sonoran regional deserts) that served as sources for the gulf islands contained species that resulted from ancient, as well as recent, geologic events in the gulf region (Hafner and Riddle 1997; Riddle et al. 2000a,b). These events are of four discrete time periods: Miocene, late Pliocene, early Pleistocene, and late Pleistocene. During the late Miocene to early Pliocene (5.5–4 Ma), the gulf began to form, either due to the separation of Baja California

(Lonsdale 1989; Stock and Hodges 1989) or to subsidence as a result of basin and range extension in North America (Gastil et al. 1983; summarized in Grismer 1994). During the late Pliocene (ca. 3 Ma), northern extensions of the gulf up the Salton Trough formed the San Gorgonio constriction (Boehm 1984; Ingle 1987), and along the course of the Colorado River formed the Boues embayment (Blair 1978; Eberly and Stanley 1978; Buising 1990), effectively isolating the peninsula from continental regions. At the same time, the cape region was isolated from the rest of the peninsula by the Isthmus of La Paz (McCloy 1984). There is circumstantial evidence for a mid-peninsular seaway, across the present-day Viscaíno Desert, in the early Pleistocene, about 1 Ma (Riddle et al. 2000a,b; summarized in Upton and Murphy 1997). Finally, sea levels fluctuated and ecological zones shifted in response to repeated waxing and waning of the late Pleistocene glacial–interglacial climatic cycles, which became markedly longer and more extreme about 700 Ka (Webb and Bartlein 1992). During glacial intervals, sea levels fell, connecting landbridge islands to the adjacent mainland and exposing continental shelves; regional deserts were compressed to the south into isolated refugia (Betancourt et al. 1990), which probably included newly exposed shelves, emergent landbridges, and landbridge islands. Arid-adapted mammals were either restricted to these desert refugia or persisted in pockets of sclerophyllous woodland or marginal grassland habitats. Arid-adapted taxa spread from these refugia during warmer interglacials, providing the opportunity for dispersal into neighboring arid and sclerophyllous regions, and populations on landbridge islands were isolated as ocean levels rose.

Phylogeographic Variation in Mitochondrial DNA

Recent advances in molecular techniques, particularly the increased facility to sequence large sections of the mitochondrial genome, have provided powerful tools to examine both evolutionary relationships and detailed distributional changes (historical biogeography) of plants and animals. Molecular techniques provide sensitive indicators of relationships among taxa and relative estimates of the timing of divergence events, especially for species with restricted ecological distributions and low vagility ("biogeographic indicator species"; Harris 1985; Sullivan 1988; Hafner 1993). Initial studies of arid-adapted rodents in the gulf region revealed old divisions within and between related taxa of peninsular and Sonoran deserts (Riddle et al. 2000a,b) and supported most of the insular–mainland associations previously recognized while clarifying the relationships of certain problematic insular taxa (Hafner et al. 2001; fig. 12.1).

Old historical (Neogene) divisions between peninsular and Sonoran taxa and among peninsular subgroups thus far revealed by data from mtDNA sequences seem associated with at least three events of the late Neogene: temporary existence of a mid-peninsular seaway across the Viscaíno Desert approximately 1 Ma; northern extensions of the gulf approximately 3 Ma; and simultaneous formation of the Isthmus of La Paz (Riddle et al. 2000a,b,c). A mid-peninsular seaway has long been hypothesized (see Nelson 1921). However, as summarized by Upton and Murphy (1997), evidence to date for the seaway appears largely circumstantial, based on low mid-peninsular elevations, limited stratigraphic data from adjacent regions, and the distribution of marine organisms on either side of the peninsula. According to J. Minch

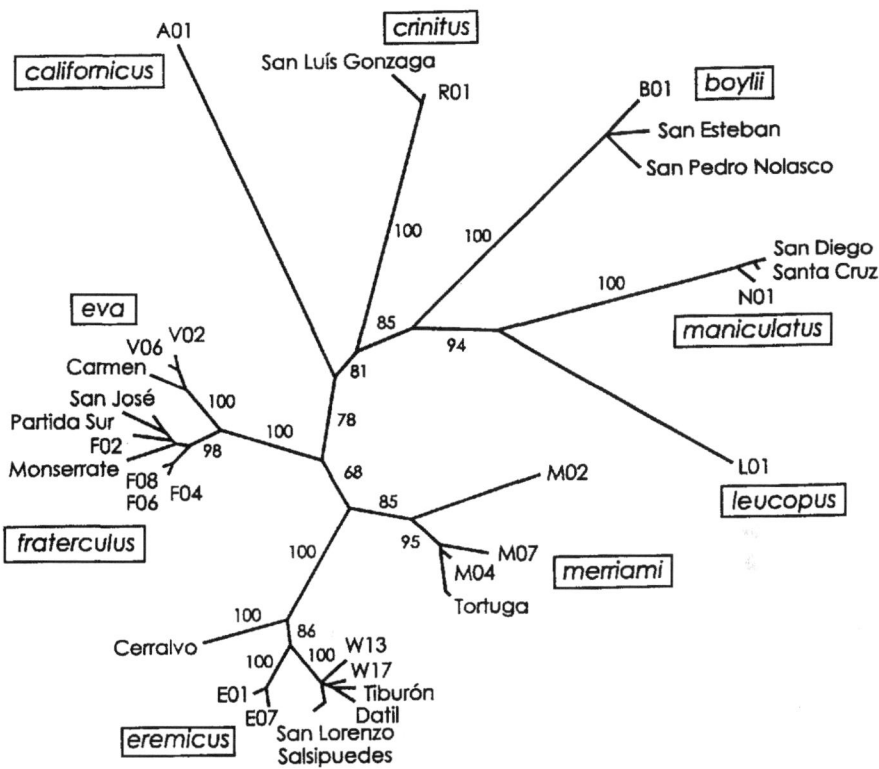

Figure 12.1 Unrooted neighbor-joining tree (Hafner et al. 2001) comparing 15 insular populations of *Peromyscus* (indicated by island name) to 16 populations of 9 mainland species of *Peromyscus*, based on estimates of sequence divergence among variable haplotypes from a 699 base-pair fragment of the mtDNA COIII gene. Nonparametric bootstrapping values (percent of 500 replications) summarize relative levels of branch support on the neighbor-joining tree. Representative reference sequences are indicated by haplotype number from Riddle et al. (2000a). The clade labeled "*fraterculus*" identifies a new taxon proposed for certain populations of peninsular *Peromyscus* (Riddle et al. 2000a; see text).

(pers. commun., 1997), sedimentary evidence for such a seaway is not likely to be forthcoming because the limestone deposits in the Viscaíno Desert are virtually fossil-free (lacking even foraminiferans) and directly overlay Paleocene deposits.

Upton and Murphy (1997) reported a genetic discontinuity between northern and southern populations of *Uta stansburiana* that is concordant with the existence of such a past seaway, which they estimated (based on a combination of mtDNA phylogeography and dated geologic events) as having been formed about 1 Ma. Five rodent species or species groups (*Peromyscus eremicus, Chaetodipus baileyi, C. arenarius, Dipodomys merriami,* and *Ammospermophilus leucurus*) exhibit signatures of a mid-peninsular seaway (Riddle et al. 2000c). Levels of sequence divergence in these forms are intermediate (1.8–4.1% sequence divergence) to those separating related taxa on opposite sides of the Gulf (8.7–10.5% sequence divergence) and to values of insular versus mainland taxa (~1% sequence divergence; Hafner et al. 2001). Concordant patterns may be expected in another rodent (*Neotoma lepida*; Plantz 1992) and have been reported for a bird (*Toxostoma lecontei*; Zink et al. 1997) and three additional reptiles: *Pituophis melanoleucus* (Rodriquez-Robles and De Jesús-Escobar 2000), *Urosaurus nigricaudus* (Aguirre et al. 1999), and the rock lizards *Petrosaurus mearnsi* and *P. thalassinus* (Aquilars-S. et al. 1988; Grismer 1999).

Most likely, rising sea levels that led to northern extensions of the Sea of Cortés and formation of the Isthmus of La Paz simultaneously separated peninsular from Sonoran desert populations and cape region from other peninsular populations. Subdivision of peninsular and Sonoran populations into separate phylogroups consistent with the northern gulf extensions is present in three rodent groups (*P. eremicus* and *C. baileyi* [Riddle et al. 2000c] and *N. lepida* [Plantz 1992]) and the red-spotted toad (*Bufo punctatus*; Riddle et al. 2000c). Peninsular subdivisions of the former two rodents appear to be cryptic species (*P. fraterculus* [Riddle et al. 2000a] and *C. rudinoris* [Riddle et al. 2000b]). Separation of the cape region from remaining peninsular populations coincident with the Isthmus of La Paz is present in at least *C. arenarius* (Riddle et al. 2000c) and *Urosaurus* (Aguirre et al. 1999), and may be present in *Petrosaurus* (Aquilars-S. et al. 1988; Grismer 1999).

Thus, as a result of northern extensions of the gulf (3 Ma) and a mid-peninsular seaway (1 Ma), there were at least three mammalian faunal sources available to colonize landbridge islands that were last separated from the mainland as recently as several thousand years ago: coastal Sonoran, northern peninsular, and southern peninsular. A potential fourth faunal source (cape region) may have resulted from formation of the Isthmus of La Paz (3 Ma) or initial formation of the gulf (5.5–4 Ma).

Faunal Relationships

Distributions of mainland species of terrestrial mammals and their island derivatives are shown in appendixes 12.1 and 12.2. For completeness, Pacific island forms are also included. (These do not include two unverified occurrences reported in Dickey [1983]: "two small rodent species" on San Ildefonso and "the white-footed mouse" on Isla Estanque [= Vibora or Pond].) An obvious pattern emerges from these distributions. In general, insular faunas resemble continental faunas of the nearest mainland area. With three suggested exceptions, ancestral forms of eastern gulf islands are represented by extant Sonoran species, and those of western islands by peninsular species (Hafner et al. 2001; app. 12.2). Evidence of cross-gulf colonization is limited to proposed derivatives of Sonoran *Peromyscus* (subgenus *Haplomylomys*) on deepwater oceanic islands that are closer to the peninsular mainland: (1) *Peromyscus dickeyi*, a derivative of *Peromyscus merriami*, a Sonoran species, on Isla Tortuga, and (2) derivatives of Sonoran *Peromyscus eremicus* on the midriff islands at least as far west as Isla Salsipuedes and the San Lorenzos (populations currently labeled *P. interparietalis*), and on Isla Cerralvo (*P. eremicus avius*).

Identification of *Peromyscus eremicus avius* of Isla Cerralvo as a possible derivative of Sonoran *P. eremicus* rather than peninsular *P. eva* or *P. eremicus* (putatively *P. fraterculus*; see above) appears to parallel the distribution of the lizard genus *Sator* in the southern gulf islands (Murphy 1975, 1983b). Murphy (1975) interpreted the distribution of *Sator* as resulting from entrapment of the genus on islands during the initial Miocene formation of the gulf. Indeed, *P. eremicus avius* appears to be a somewhat distant relative of other *P. eremicus*, with a divergence predating separation of Sonoran and Chihuahuan clades of *P. eremicus* (Riddle et al. 2000a). However, mtDNA divergence values indicate a separation of *P. e. avius* subsequent to separation of peninsular (*P. eva* + *P. fraterculus*) and Sonoran (*P. merriami* + *P. eremicus*) species in the subgenus. If correct, a more likely explanation for the origin of *P. eremicus* on Cerralvo is overwater dispersal from mainland Sonora during the Pliocene.

The fact that most populations of mammals on landbridge islands are clearly allied with mainland populations in closest proximity indicates that isolation from mainland populations followed the most recent rise in sea level at the end of the Wisconsin glaciation, in some populations as recently as several thousand years ago (Case 1975; Wilcox 1978). Thus, divergence attained during previous cycles of colonization and isolation may have been obliterated by confluence of landbridge island and mainland populations during the Wisconsin glacial interval. Available data on degree of divergence in mtDNA sequences of landbridge island versus mainland populations of *Peromyscus* (Hafner et al. 2001) are consistent with this interpretation.

Shifts in the distribution of mainland species subsequent to isolation of populations on landbridge islands at the close of the Wisconsin glaciation are indicated by comparing the current distribution of mainland populations with derivatives on landbridge islands. The brush rabbit (*Sylvilagus bachmani*), a chaparral species, may have enjoyed a broader distribution along the Baja California peninsula during the more mesic Wisconsin glaciation, because the nearest mainland relatives of the population on Isla San José (*S. mansuetus*) are located nearly 400 km to the north (Santana) and 200 km to the south (Santa Anita). *Peromyscus eremicus* (putatively *P. fraterculus*; see above) evidently has retreated northward, as indicated by populations on Islas Espíritu Santo and San José and an apparently relict population on the peninsula east of La Paz, at Las Cruces (Lawlor 1971b). Similarly, populations of *Chaetodipus baileyi* and *Peromyscus eremicus* on oceanic islands close to the Baja California peninsula, and of *P. boylii* on oceanic islands on the Sonoran side of the gulf may indicate that source populations formerly occurred along the respective adjacent coasts during the Wisconsin maximum. Reconstruction of former mainland distributions based on oceanic islands is far more tentative, as indicated by the possible long-distance dispersal of *P. eremicus* to Isla Cerralvo and of *P. merriami* to Isla Tortuga (Hafner et al. 2001).

All species of bats thus far recorded from Sea of Cortés islands occur on both the peninsular and the Sonoran-Sinaloan mainlands, but the extent of movement between islands and the mainland, or between Sonoran and peninsular mainlands, is unclear. Genetic (Wilkinson and Fleming 1996) and reproductive (Ceballos et al. 1997) data indicate that large, able fliers such as *Leptonycteris curasoae* regularly migrate across the gulf between the southern Baja California peninsula and Jalisco (see also Rojas-Martinez et al. 1999). Sahley et al. (1993) documented nightly flights of 20–30 km by *L. curasoae* from day roosts on Tiburón to the adjacent Sonoran mainland. Certainly, the presence of *Lasiurus* on oceanic islands, including Hawaii and the Galapagos Islands, attests to the occasional, accidental dispersal of strong, rapidly flying bats, particularly those that regularly undertake long-distance migrations. It is likely that some species of bats dispersed across the gulf (deliberately or accidentally) and became established on the peninsula. For example, four species (*Natalus stramineus, Balantiopteryx plicata, Pteronotus davyi,* and *Myotis peninsularis*) have disjunct distributions in the cape region (Woloszyn and Woloszyn 1982); their nearest relatives are across the gulf. However, the large degree of variation in flight speeds and endurance found in bats (e.g., Hayward and Davis 1964) would probably lead to varying degrees of isolation among species of insular bats. T. A. Vaughan (pers. commun., 1980) reported that *Antrozous pallidus* from Isla Carmen is smaller than its peninsular counterpart, despite a relatively narrow (6 km) separating seaway. Some species of bats may be reluctant to set out over open water, even if they are capable of flying the distance, and genetic interchange between island and mainland populations may result only from rare, chance dispersal in these species, as with nonvolant small mammals.

Pleistocene Refugia

Hafner and Riddle (1997) speculated that there may have been three distinct refugia for arid-adapted mammals on the Baja California peninsula during glacial maxima: a California refugium along the northern Pacific coast; a Magdalena refugium along the southern Pacific coast and extending out onto newly exposed continental shelf; and a gulf coast refugium along the southern gulf coast. These remained separate from the continuous Mohavia refugium at the head of the gulf and the Sinaloa refugium along the southern Sonoran and Sinaloan coasts. Initial results from mtDNA analysis of 11 island populations of one group (*Peromyscus* subgenus *Haplomylomys*) provide tentative support for the existence of a separate gulf coast refugium and for northward expansion of the gulf coast, Mohavia, and Magdalena refugia after the end of the Wisconsin glacial interval. Specifically, two populations from islands opposite the southern gulf coast of the peninsula evidently are derivatives of putative *P. fraterculus* (which now occurs to the north) rather than *P. eva*, which occupies the adjacent coast; and all of the midriff-island populations thus far analyzed appear to be derivatives of the Sonoran species, *P. eremicus*.

Analysis of additional island populations of *Haplomylomys* (recorded from 10 other islands surrounding the peninsula), as well as island populations of *P. maniculatus* (12 islands), *Chaetodipus spinatus* (14 islands), and the *Neotoma lepida* species-group (13 islands) should provide informative tests of the hypothesis of separate full-glacial refugia. We expect that relationships among populations of the latter two species, which frequent arid habitats, should conform to the refugium hypothesis, whereas the former, which is widespread in many habitats within its geographic range, should not.

Ecological Biogeography

Ecological Characteristics of the Mammalian Fauna

Many ecological factors affect distributions of mammals, including dispersal abilities, habitat quality and diversity, productivity and availability of food sources, natural-history traits, and extent of competition and predation. That colonizing abilities differ among mammals is demonstrated by examination of worldwide patterns of continental and insular distributions of major groups of mammals (Lawlor 1986, 1996). The best colonizers are small, relatively abundant terrestrial species with catholic diets and high population growth rates, and bats. Landbridge islands typically have attenuated continental faunas, whereas faunas on oceanic islands are ecologically incomplete (Simpson 1956; Baker and Genoways 1978; Lawlor 1996). The latter conclusion holds among archipelagoes regardless of differences in the magnitude of isolation by overwater distances or size ranges of islands and relates directly to the manner in which the islands were colonized—either by waif dispersal (oceanic islands) or by fragmentation from mainland faunas (landbridge islands) (Lawlor 1996). These generalizations are supported in gulf island faunas by the fact that only small rodents and bats inhabit oceanic islands, whereas ecologically diverse faunas that include insectivores, lagomorphs, carnivores, and artiodactyls, in addition to rodents and bats, characterize landbridge islands.

The variety of habitats suitable for mammals on gulf islands is limited because of the arid climate and virtual absence of fresh water. Relatively unproductive rocky habitats predominate (see chap. 4). This overall lack of habitat diversity, among other things, suggests that resource coupling is the important determinant of lizard and bird species numbers on gulf islands (Case 1975, 1983, chap. 9, this volume; Case and Cody 1987). For the above reasons, terrestrial mammals preferring situations other than rocky habitats are rare or nonexistent in the Sea of Cortés except on the largest and most diverse islands, which almost invariably are also landbridge islands (e.g., Tiburón, Carmen, San José, Magdalena, Santa Margarita, Espíritu Santo).

The overwhelming majority of mammalian species found on gulf islands are associated with rocky habitats in both mainland and island settings. The species that do not prefer such habitats are broadly adapted (e.g., *Peromyscus maniculatus*), found only on relatively large islands rich in habitat or prey-species diversity (e.g., *Lepus* spp., *Perognathus penicillatus*, *Thomomys bottae*, *Dipodomys merriami*, *Canis latrans*, *Urocyon cinereoargenteus*, *Bassariscus astutus* [see fig. 16.4], *Odocoileus hemionus*), or may have become extinct recently (e.g., *Peromyscus pembertoni* on San Pedro Nolasco).

Diversity, density, and productivity of plant species (see chap. 4) vary considerably on the islands. As discussed below for islands in Bahía de los Ángeles, seasonal and El Niño patterns of rainfall in the Sea of Cortés strongly influence the annual seed production on which many species depend. Consequently, the ability of many species to persist on an island is related directly to seasonal changes in availability of food. Lawlor (1983) argued that the ephemeral nature of seed availability should be much more limiting to granivores such as pocket mice (*Chaetodipus*) than to deer mice (*Peromyscus*), which have a varied diet that also includes terrestrial and marine invertebrates. Two pieces of evidence support this contention. First, one-species islands (those inhabited by only a single species of mammal) are occupied only by *Peromyscus*, never by *Chaetodipus* or, for that matter, by any other mammals. Overall, species of *Peromyscus* and *Chaetodipus* frequent 42 and 25 major islands, respectively (these values become 51 and 28 if occurrences on minor islands [table 12.2] are added). The absence of *Chaetodipus* on such small islands is a significant departure from that expected from the proportion of total mouse populations (consisting of *Peromyscus* and *Chaetodipus*) on gulf islands made up by pocket mice (25/67, or 37.3%; $p <$ .0001, using a test of equality of percentages based upon the arcsine transformation). In other words, the absence of pocket mice from one-species islands is not attributable to an overall paucity of these rodents on gulf islands. Nor is their absence due simply to a lack of sandy habitats on small islands, because the most common species of

Table 12.2 Characteristics of Bahía de los Ángeles and Bahía de las Ánimas islands inhabited by rodents

Island	Area (km^2)	Distance from Peninsula (km)	*Peromyscus Maniculatus*	*Chaetodipus Baileyi*
Blanca	0.03	0.90	X	
Bota	0.09	2.85	X	
Coronadito	0.10	3.09	X	
Flecha	0.16	2.79	X	
Pata	0.18	2.55	X	X
Mitlan	0.19	2.00	X	X
Piojo	0.57	6.30	X	
Cabeza de Caballo[a]	0.70	2.02	X	
Ventana	1.41	3.15	X	
Smith	9.13	2.18	X	X

Islands surveyed that lack mice: Bahía de las Ánimas: Las Ánimas Norte, Las Ánimas Sur, Pescador; Bahía de los Ángeles: Gemelos East, Gemelos West, Cerraja, Llave, Jorobado. Island area and isolation from Murphy and Aguirre-Léon (chap. 8, this volume) and Polis et al. (1998).
[a]Rodents were not detected on Cabeza de Caballo until late 2000 and therefore were not included in analyses.

pocket mice (*C. spinatus* and *C. baileyi*) occur in rocky habitats on both islands and the mainland.

Second, congeneric pairs of *Peromyscus* and *Chaetodipus* species occur on five and three islands, respectively. These associations are mostly on large islands (e.g., Tiburón), but *Peromyscus* pairings are known from two small islands (*P. crinitus* and *P. maniculatus* on Willard, and *P. boylii* and *P. pembertoni* on San Pedro Nolasco). Though not significant, this finding is consistent with the especially strong competition for limited seed resources expected among the more highly specialized *Chaetodipus*, especially on small islands.

The structure of insular rodent communities comprising *Peromyscus* and *Chaetodipus* is reminiscent of species assembly patterns described for desert rodent communities in the southwestern United States (Patterson and Brown 1991; Fox and Brown 1993). There, pocket mice and deer mice form hierarchical arrangements in which the more specialized heteromyids frequent fewer locations than their generalized murid counterparts. However, the extent to which this hierarchy of occurrences has been shaped by competition or simply by coexistence of species with distinct dietary adaptations remains an open question.

Rainfall patterns also determine availability of fresh water. Among rodents, woodrats (*Neotoma*) are the least capable of tolerating dehydration (MacMillen 1964). Woodrats depend heavily on plant tissues for free water. The large seasonal and annual fluctuations in densities of *Neotoma lepida* and related endemics on gulf islands (Lawlor, pers. obs.; T. A. Vaughan, pers. commun., 1980), which match those often seen in mainland settings (Lee 1963; MacMillen 1964; Lieberman and Lieberman 1970), may be due in part to availability of preformed water from plants.

Published natural history information for mammalian populations on gulf islands is confined largely to a few studies of the feeding ecology and behavior of rabbits and woodrats. Jackrabbits (*Lepus californicus*) are common on several islands, where they inhabit a variety of habitats and exploit an expanded diet of cacti and shrubs compared to their mainland counterparts (Hoagland 1992; Cervantes et al. 1996). Woodrats (*Neotoma lepida*) on Danzante are scansorial herbivores with diets restricted largely to ironwood (*Olneya tesota*) leaves (Vaughan and Schwartz 1980), but periodic use is also made of other shrubs and cholla cacti. They occur at low densities on the island, although their biomass exceeds that of the only other species there (a pocket mouse, *Chaetodipus spinatus*), and they occupy unusually large territories for the species. Vaughan and Schwartz (1980) argued that reduced numbers of predators on Danzante produced a mating system based on resource-defense polygyny in these woodrats. In contrast, species of *Peromyscus*, which are not strongly territorial, seemingly exhibit excess density compensation rather than increased territory size on some depauperate islands (Lawlor 1983).

Case Study: Insular Rodents on Small Islands near Bahía de los Angeles

Stapp has been studying rodent populations in the northern midriff islands, in Bahía de los Ángeles (BLA), with the objective of determining how the availability of foods of terrestrial and marine origin influence insular rodent populations. Compared to other islands in the Sea of Cortés, the landbridge islands in BLA are mostly small (<1 km^2 in area; table 12.2) and relatively close to the peninsula. Nevertheless, because rodents are poor overwater dispersers, ecological processes shaping rodent assemblages on BLA islands should largely parallel those affecting rodents throughout the region.

Resource Availability and Dietary Breadth

Like most small islands in the Sea of Cortés, the terrain of BLA islands is very steep and rocky, with little sand except on small, isolated beaches. Island substrate determines the availability of refuges for rodents and also has an important indirect effect on food availability by restricting plant establishment and growth. Perennial shrubs and cacti are present on most islands, but plant cover is relatively low (5–17%) in dry years (Sánchez Piñero and Polis 2000). Terrestrial resources available to rodents on these islands include seeds and leaves of perennial plants (primarily *Opuntia*, *Pachycereus*, and *Atriplex*), as well as land arthropods (mostly tenebrionid beetles and ground dwelling spiders) and small vertebrates such as lizards (*Uta stansburiana*) and the eggs and nestlings of songbirds.

Foods derived from the ocean also provide important nutrients and energy for island consumers, supporting higher population densities than would be possible based on terrestrial resources alone (Polis and Hurd 1996). Marine resources include invertebrate communities of the rocky intertidal and supralittoral zones, algae and carcasses of marine animals that wash ashore, plus the scavengers and detritivores that are attracted to them. Sea birds are another important conduit of marine-based energy. On nesting-colony islands, sea birds contribute eggs, chicks, and food scraps to island food webs. Guano deposited by sea birds on both roosting and nesting islands fertilizes plant growth, providing nutrients for insular terrestrial communities (Stapp et al. 1999). Marine-based resources may be critical to island populations because terrestrial productivity is seasonally ephemeral and, in most years, relatively low overall, and because inputs from the ocean are less easily depleted than terrestrial resources. In other words, consumers cannot control resource delivery and renewal rates (Polis and Hurd 1996). Consequently, mammals that use marine-based foods may be buffered from extreme year-to-year variability in the availability of terrestrial resources, which in turn may allow them to become more widespread and persistent on islands than species that depend solely on terrestrial foods.

Interspecific differences in the ability of rodents to take advantage of marine-based resources may in part explain patterns of distribution on the Sea of Cortés islands, and perhaps worldwide. Rodents inhabit 10 of the 19 islands in the BLA region, which range in size from 3 ha to 9.1 km^2 in area. Like most other small islands in the Sea of Cortés, BLA islands are inhabited chiefly by two types of rodents: omnivores (*Peromyscus*) and granivores (*Chaetodipus*). Deer mice (*Peromyscus maniculatus*) occur on all 10 islands with rodents; pocket mice (*Chaetodipus baileyi*) are restricted to three islands (table 12.2). As in other regions of the gulf, *Chaetodipus* never occurs alone on BLA islands, which Lawlor (1983) attributed to the scarcity of suitable seed resources on most islands, especially small ones. The diversity and availability of seeds may indeed explain the presence of *Chaetodipus* on two small BLA islands (Pata, Mitlan); both islands have extremely high plant species diversity for their size (West, chap. 4, this volume). Although the presence of *Chaetodipus* on Mitlan may also be explained by its close proximity to Smith, the largest island in the area, the presence of *Chaetodipus* on Pata, which is small and relatively isolated, was unexpected; Pata is the smallest gulf island occupied by a granivorous rodent.

Earlier studies of coastal populations have shown that *Peromyscus* feeds on marine invertebrates in intertidal and sandy-beach habitats (McCabe and Cowan 1945; Osborne and Sheppe 1971; Thomas 1971; Herman 1979; Lawlor, pers. obs.). Whereas these past studies have combined behavioral observations with traditional methods for assessing food habits, Stapp et al. (1999) used stable isotope analysis (SIA) to document use of marine resources by *Peromyscus* and other island consumers (e.g., lizards, arthropods). By comparing carbon and nitrogen isotope ratios of consumer tissues to those of potential prey items, SIA provides an estimate of an animal's dietary composition and of its relative trophic position in food webs (Schoeninger and DeNiro 1984; Peterson and Fry 1987). The advantage is that SIA quantifies diets of organisms that are difficult to study with conventional techniques. Furthermore, depending on the tissue analyzed, SIA provides a cumulative record of the relative importance of different food sources through time. Analyses of tail (bone and skin) tissue from *Peromyscus* and *Chaetodipus* provide an integrated index of the contributions of marine and terrestrial-based resources throughout an individual's life, which typically is 1 year or less (Stapp, unpublished data).

SIA results confirm that *Peromyscus* is a dietary generalist and that individuals living close to shore eat marine invertebrates (fig. 12.2). Further inland, the elevated nitrogen isotopic signatures suggest that deer mice apparently consume terrestrial arthropods, other small invertebrates, and in wet years on some islands, seeds (Stapp et al. 1999). In contrast, pocket mice are strictly granivorous both near the shore and inland, consuming a mixture of seeds from plants with C3 and C4/CAM photosynthesis (fig. 12.2).

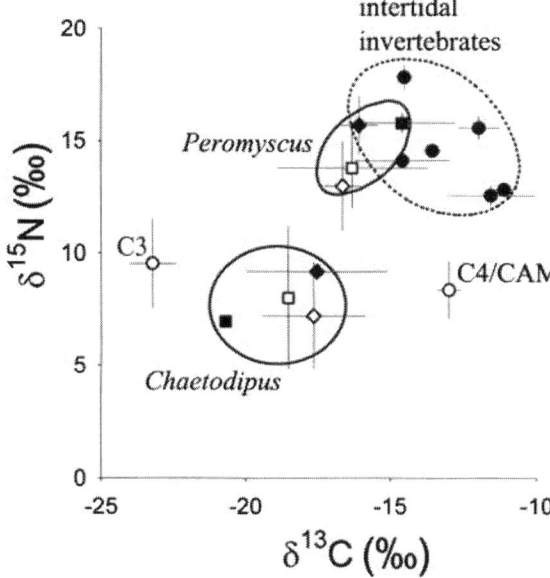

Figure 12.2 Stable carbon (^{13}C) and nitrogen (^{15}N) isotopic signatures (mean + SE) of *Peromyscus* and *Chaetodipus* near shore (filled symbols) and in inland areas (>100 m) from shore; open symbols) on Bahía de los Ángeles islands ($n = 1-5$ islands). Squares indicate samples collected in a typical dry year (1997); diamonds are from 1998, an El Niño-Southern Oscillation year. Isotopic signatures of terrestrial C3 and C4/CAM plants and intertidal invertebrates are provided for reference (Stapp et al. 1999).

Subsidies provided by the ocean to island populations of *Peromyscus* influence patterns of abundance of mice on and among islands. During typical dry years, *Peromyscus* tends to be most abundant near shore, especially on islands occupied by both mouse species (fig. 12.3). *Chaetodipus* tends to be more numerous inland, presumably because coastal areas are barren of most perennial, seed-bearing plants. Although competition probably contributes to this pattern (see below), the high-protein foods available near shore may permit relatively high densities of *Peromyscus* in these areas. Moreover, the effects of marine resources on *Peromyscus* populations are evident on an island-wide basis (fig. 12.4). Small islands, which tend to have large shoreline perimeters relative to their areas and, hence, relatively larger fractions of marine inputs, tend to support higher densities of mice than larger islands, at least during normal, dry years (1997, 1999, 2000). High population densities are a common feature of many insular rodent populations (Gliwicz 1980), which has been attributed to lower predator pressure, fewer competitors, or frustrated dispersal (fence effects) on islands. Although all these may contribute to high densities of *Peromyscus* on BLA islands, the disproportionately large input of marine resources to small islands, many of which support roosting and nesting sea-bird colonies, provides another explanation. The success of *P. maniculatus* on small rocky islands in BLA probably reflects both competitive and predatory release, as well as the abundance of marine resources on islands.

Periods of unusually high rainfall can have dramatic and long-lasting effects on terrestrial primary and secondary productivity (Noy-Meir 1973; Polis 1991). At Bahía de los Ángeles, El Niño-Southern Oscillation (ENSO) events that occur every 5–8 years bring large amounts of winter precipitation, switching the system from one dominated by marine resources to one driven more by terrestrial productivity (Polis et al. 1997). Average plant cover on BLA islands increases 30–49% (Sánchez Piñero

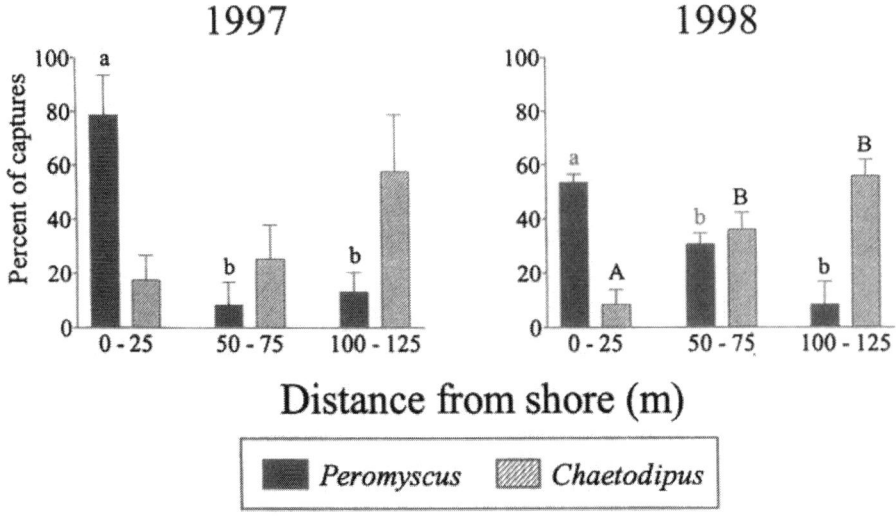

Figure 12.3 Distribution of captures of *Peromyscus* and *Chaetodipus* at different distances from the ocean during a typical dry year (1997) and a wet El Niño Southern Oscillation year (1998). Within a species and year, means with same letter are not statistically different (nested ANOVA, Tukey test, $p < .05$). Values are means (+1 SE; $n = 2-3$ islands).

and Polis 2000) and vegetation is dominated by annual plants. This pulse in annual plant growth also supports higher densities of herbivorous insects and their predators (Polis et al. 1997), which serve as prey for deer mice on many islands (fig. 12.2). Unfortunately, there are no accurate estimates of changes in seed production for most plants, but seed output of the dominant perennial on islands (*Atriplex barclayana*) was three times higher in 1998, an intense ENSO year, than in the previous dry year (Stapp et al. 1999). The increase in seed production associated with ENSO events may have wide-reaching effects on island granivores such as pocket mice because some seeds enter soil seed banks that may persist during intervening dry years. Further, *Chaetodipus baileyi*, like other heteromyid rodents, caches large numbers of seeds in extensive burrow systems (Stapp, pers. obs.).

ENSO-related changes in resources have a significant impact on population densities of both species, but the numerical response of *Peromyscus* is much more pronounced. On islands surveyed in both dry (1997) and wet ENSO (1998) years, mean abundance of *Peromyscus* increased by 372% (SE = 91%) immediately following the 1998 ENSO. In contrast, *Chaetodipus* numbers increased only 81% (SE = 58%) from 1997 to 1998 and declined by 33% on Pata, where *Peromyscus* abundance increased sixfold. *Peromyscus* not only increased significantly in abundance but also invaded inland areas of islands during the 1998 ENSO (fig. 12.4). The pulse of terrestrial resources in response to the 1998 ENSO also reduced the relative importance of marine inputs for whole-island abundance, so that population densities were high on all islands, regardless of island perimeter-to-area ratio (fig. 12.4).

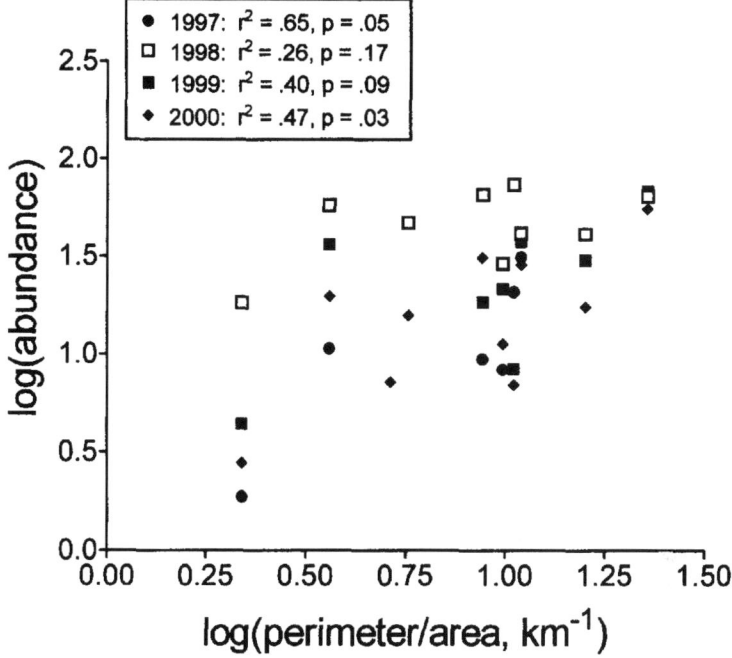

Figure 12.4 Relative abundance of *Peromyscus* during three dry years (1997, 1999, 2000) and an El Niño Southern Oscillation year (1998) on islands as a function of island perimeter-to-area ratio, an index of amount of marine input. Small islands have relatively more shoreline for their size, creating a greater target area for drift of algal wrack and carcasses, and more area for intertidal invertebrate communities. Statistics are regression coefficients.

Competition

Patterns of spatial distribution of *Peromyscus* and *Chaetodipus* on islands where they co-occur suggest distinct habitat preferences that reflect the availability of food resources used by each species. Deer mice tend to be restricted to near shore, whereas pocket mice are more evenly distributed but most common farther inland (fig. 12.3). However, on islands occupied only by *Peromyscus*, there were no significant differences in either 1997 or 1998 in the relative abundance at different distances from shore (Stapp, unpublished data), which suggests that the presence of *Chaetodipus* affects habitat use and local abundance of *Peromyscus*. This hypothesis is consistent with our knowledge of the ecology of these species in other systems: pocket mice (*Chaetodipus*, *Perognathus*) tend to be aggressive and territorial (Jones 1993), whereas *Peromyscus* is behaviorally subordinate to other species in many rodent communities (e.g., Grant 1972; Bowers et al. 1987; Kaufman and Kaufman 1989; Heske et al. 1994; Stapp 1997). We also speculate that, in addition to their seed-caching abilities, the desert-adapted *Chaetodipus* is more efficient than *Peromyscus* at locating and harvesting seeds that are scarce and widely scattered on islands in most years.

Although *Chaetodipus* may have an advantage during arid years typical of the region, *Peromyscus* seems to be more effective at converting the ENSO-related pulse of terrestrial resources into offspring, which subsequently spill over into inland areas inhabited by *Chaetodipus* (fig. 12.3). During these periods of unusually high abundance of *Peromyscus*, we captured individuals of both species simultaneously in the same trap on six occasions. In all cases, deer mice killed and partially consumed pocket mice (by comparison, no mortalities were observed in 32 instances that year of multiple captures involving only a single species). The nature and outcome of competitive interactions between *Chaetodipus* and *Peromyscus* thus appears to switch between wet and dry years. Experimental manipulations of seed resources or population densities are needed to assess the degree of and relative symmetry of competition between island populations of *Peromyscus* and *Chaetodipus*.

Dispersal Ability

The species composition of rodent assemblages on small landbridge islands near BLA is more likely a reflection of differential extinction of species present before island formation than the consequence of regular recolonization over water. To the extent that dispersal may account for compositional differences, characteristics of deer mice appear to make them more effective overwater dispersers than pocket mice. First, the close proximity to and high densities of *Peromyscus* near shore mean that dispersing *Peromyscus* are more likely to head out over water than *Chaetodipus*, which are more closely associated with inland habitats. Second, the generalized diet of *Peromyscus*, which may involve foraging in large algal mats and in marine animal carcasses, may predispose it to dispersal should these rafts be set adrift. Third, human occupancy and visitation of islands may have increased exchange of *Peromyscus* among islands during historic times. During periods of high densities, *Peromyscus* overrun fishing, tourist, and research camps on beaches (Lawlor and Stapp, pers. obs.).

Two Different Strategies

Taken together, these ecological factors offer a mechanistic explanation for the biogeographic distribution of rodents on islands in the Sea of Cortés. *Peromyscus* and *Chaetodipus* appear to have different strategies for coping with the extreme and unpredictable insular environments characteristic of the region, and these strategies reflect differences in diet, life history, competitive ability, and dispersal (table 12.3).

Peromyscus is widely distributed among islands because of its ability to persist on islands once established and its capacity to disperse short distances across water. Physical differences among islands, including the availability of suitable refuges and the amount of habitable area, ultimately determine whether an island can support a rodent population. The ability of *Peromyscus* to use a wide variety of food resources, notably the largely unlimited resources available from the ocean, buffers populations against periods of low terrestrial productivity and, ultimately, extinction. *Peromyscus* responds numerically to pulses of terrestrial resources via dramatic increases in reproduction, leading to large temporal variations in population density, but the degree to which these individuals are recruited and contribute to long-term future survival of the population is not clear. On islands with competitors such as *Chaetodipus*, however,

Table 12.3 Ecological influences on the distribution of *Peromyscus* and *Chaetodipus* among islands in the Sea of Cortés

Trait	*Peromyscus*	*Chaetodipus*
Diet	Terrestrial and marine foods	Seeds
Distribution of resources among islands	Widespread	Restricted; dependent on plant diversity and abundance
Competitive ability	Inferior	Superior
Response to ENSO[a] pulse in resources	High reproductive output	Seed caching; modest reproduction
Temporal variation in population size	High (CV = 84%)[b]	Low (CV = 48%)
Explanation for persistence	Buffered during low terrestrial productivity by ability to use marine-based resources	Seed storage, foraging efficiency, and conservative life history; susceptible to extended drought
Dispersal ability and opportunities	Good	Poor
Distribution among islands	Widespread	Less common

[a] El Niño Southern Oscillation.
[b] mean coefficient of variation in relative abundance between 1997 and 1998.
From Stapp (in prep.).

Peromyscus may be restricted to areas near the shore by interspecific territoriality and/or the scarcity of preferred terrestrial foods. Similarly, *Peromyscus maniculatus* seems to be unsuccessful in competitor-rich communities on the adjacent mainland (Stapp, unpublished data), underscoring the importance of these islands in maintaining regional species diversity.

In contrast, *Chaetodipus* occurs on a small subset of islands, usually larger islands with a greater variety of plant associations or sandy substrates, or on small islands with unusually high plant diversity. Critical to persistence of *Chaetodipus* is its ability to exploit efficiently the scarce seed resources available during dry years and to monopolize and store large amounts of seeds produced by periodic ENSO rainfall events. Compared to *Peromyscus*, the life-history traits of *Chaetodipus* and other heteromyids are relatively conservative (e.g., long life spans, low reproductive effort; Brown and Harney 1993), which makes them well-adapted for survival under harsh and unpredictable conditions typical of desert islands. Although population densities are probably more stable over time than those of *Peromyscus*, we speculate that insular populations of *Chaetodipus* would be more susceptible to extinction in an extended drought period in which the soil seed bank was not replenished by ENSO winter rainfall. Islands undergoing extinctions would be less likely to be recolonized by *Chaetodipus* because of their limited ability and opportunities for overwater dispersal and because fewer islands have sufficient plant cover or diversity to sustain granivore populations (Lawlor 1983).

Diversity Patterns

Mammals are the least conspicuous and diverse vertebrates frequenting Sea of Cortés islands. Assemblages on oceanic islands are most impoverished, containing no more than three species of nonvolant mammals. Although one landbridge island supports 14 species (Tiburón), others have 7 or fewer (although Magdalena, a Pacific island, contains 9; apps. 12.1, 12.2).

Species–Area Relationships

The power model of the relationship between species number and island size takes the typical form $\log(S) = c + z(\log A)$, where $\log(S)$ is log species number, A is island area, c is a constant, and z is the slope of the line described by the equation. In the equilibrium model of island biogeography (MacArthur and Wilson 1963, 1967), high z-values (e.g., >0.40) are associated with well-isolated archipelagos in which rates of colonization are low and rates of extinction are high and vary inversely with island size. Low z-values (e.g., <0.20) are characteristic of interconnected faunas in which immigration rates more than offset effects of extinction, as in the case of nonisolated samples on continents. Intermediate, or equilibrial, z-values (0.20–0.35) typify insular situations in which rates of colonization and extinction are equal or nearly so.

For a variety of reasons, interpretations of the species–area relation must be made with caution (see, e.g., Rosenzweig 1995). Both c and z depend on the taxa and types of islands being examined (Preston 1962; MacArthur and Wilson 1967; Brown 1978), and in a power model c and z are not wholly independent (Lomolino 1989). Moreover, a variety of factors can interact to produce identical z-values, so by itself a single species–area curve cannot discriminate among biogeographic hypotheses.

The z-values for mammals from all islands, including those along the Pacific coast off Baja California, and for islands confined to the Sea of Cortés, are 0.265 ($n = 34$) and 0.242 ($n = 26$), respectively; both values resemble those typical of equilibrial faunas. In contrast, dividing major islands into shallow-water landbridge and deep-water oceanic islands reveals markedly different values. The z-value for landbridge islands (0.365; $n = 24$; 0.310 if only gulf landbridge islands are plotted) conforms to established values for such fragmented insular faunas, but the value for oceanic islands (0.119; $n = 14$) departs from theoretical prediction (fig. 12.5).

The distinction between z-values for the two types of islands is a departure from that predicted by equilibrium theory. However, it is consistent with qualitative differences among species–area curves for mammalian faunas on oceanic and landbridge islands worldwide that Lawlor (1986, 1996) attributed to the legacy of events shaping the assembly of mammalian faunas on islands (waif dispersal vs. extinction, respectively). In all cases in which intra-archipelago comparisons can been made, z-values are steeper for landbridge islands than for oceanic islands.

Previous analyses using stepwise multiple regression (Lawlor 1983) revealed that island area was the best predictor of species richness in the Sea of Cortés; isolation was not significantly correlated with diversity. In contrast, no single variable contributed overwhelmingly to explanations of variation in species numbers on oceanic islands, although measures of both island area and isolation were significantly correlated with richness.

Incorporation of species numbers from the small BLA islands (table 12.2) into species–area analyses of landbridge-island faunas reveals a prominent small-area effect (fig. 12.6). Species richness is relatively stable (one to three species) and independent of area for roughly four orders of magnitude of the smallest islands. MacArthur

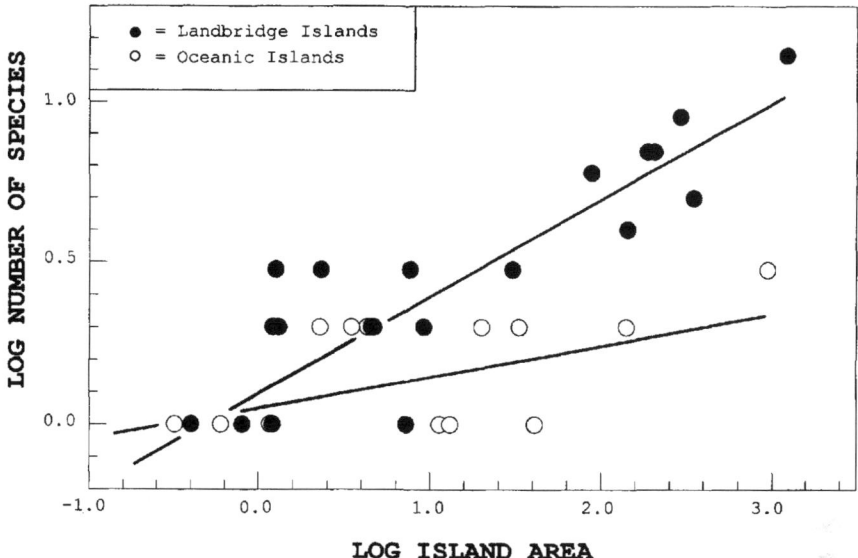

Figure 12.5 Log-log plot of species–area relations for terrestrial mammals occurring on major landbridge and oceanic islands ($n = 20$ and 14, respectively) along the peninsular Pacific coast and in the Sea of Cortés. Species numbers are provided in appendix 12.2.

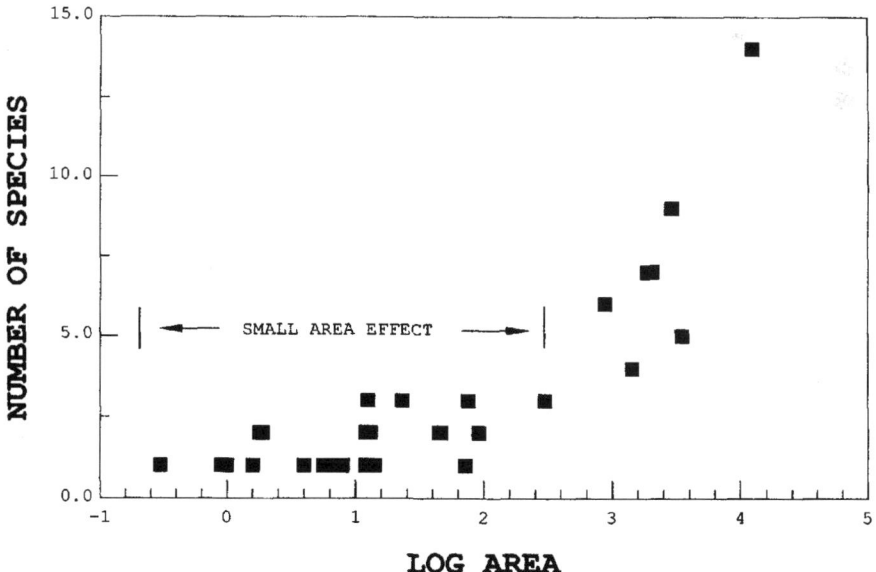

Figure 12.6 Semi-log plot of species–area relations for terrestrial mammals occurring on major and minor landbridge islands along the peninsular Pacific coast and in the Sea of Cortés ($n = 29$). Note the prominent small-island effect. Species numbers are provided in appendix 12.2.

and Wilson (1967) attributed this pattern to population instability on very small islands. In his species-based, heirarchical model of island biogeography, Lomolino (2000) argued that the effect may be characteristic of a more general sigmoidal relationship between species and area in archipelagos containing a wide range of island sizes. The only mammals frequenting resource-limited small islands in the gulf are rodents of the genera *Peromyscus*, *Chaetodipus*, and *Neotoma*.

Species Assemblages

For the historical-legacy model to be sustained, compositional differences between assemblages on oceanic and landbridge islands also must be consistent with it. Three predictions of the model are relevant here. First, terrestrial mammals on oceanic islands should be confined to a much narrower range of taxonomic and ecological groupings of mainland fauna than for landbridge islands (Lawlor 1996). This prediction is confirmed from the facts that oceanic islands contain only rodents, whereas landbridge islands are occupied by a much wider range of taxa with a greater diversity of ecological roles.

Second, occurrence frequencies (the number of available islands occupied by each species) and average incidence per species of such occurrences should be lower on oceanic than on landbridge islands. This prediction follows because mammals have poor dispersal capabilities over water, so species on deep-water islands should have relatively low probabilities of invading all islands.

These expectations are confirmed. Oceanic-island faunas are characterized by a highly disproportionate number of species that occupy only one island (10 of 14 total species, or 71%); the maximum number of islands inhabited by any given species is 4 (1 species). In contrast, on landbridge islands, incidences range from 10 species occupying only a single island (42% of 24 species) to one species present on 11 islands. Average incidences (the average fraction of total landbridge and oceanic islands inhabited by each species) are 15.2 and 11.1, respectively (Lawlor 1996).

Third, faunas on different-sized landbridge islands should constitute a graded series of nested subsets, whereas those on oceanic islands should be characterized by unordered assemblages of species. This, too, is the case, although the narrow range of variation of species numbers on oceanic islands (one to three) does not provide a robust test. As reported by other authors using the matrix of species occurrences provided by Lawlor (1983), only terrestrial mammalian faunas on landbridge islands are significantly nested (Patterson and Atmar 1986; Patterson 1987, 1990; Wright and Reeves 1992; Wright et al. 1998). This distinction is sustained, regardless of which landbridge islands are targeted, using the updated species matrix (app. 12.2) and Atmar and Patterson's (1993, 1995) nestedness temperature calculator (fig. 12.7).

Equilibrium or Historical Legacy?

Compared to lizards and birds (reviewed in Case 1983; Case and Cody 1983, 1987), diversity patterns in mammals are much less dynamic, reflecting what appears to be a much greater impact of historical influences on species numbers and assemblages. Insular isolation has marked effects both on invasion likelihood and extinction rates of terrestrial mammals, especially for those on small islands. Contemporary ecological

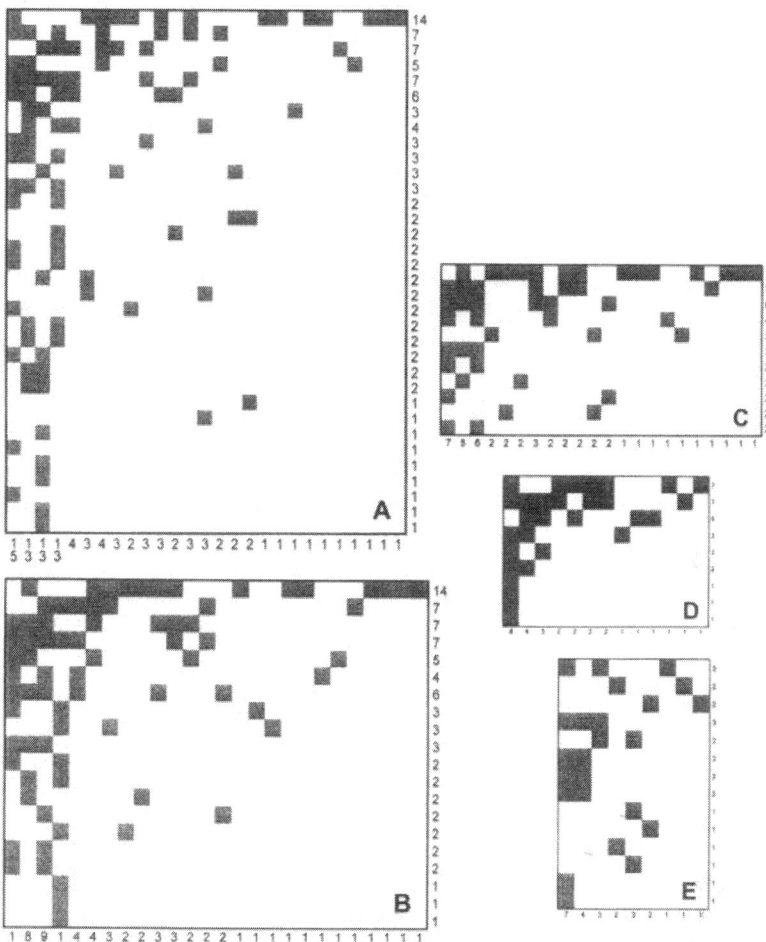

Figure 12.7 Nested species arrangements of clades (species or species groups; app. 12.2) from major islands in the Sea of Cortés determined by use of Atmar and Patterson's (1993, 1995) nestedness temperature calculator. Data are ordered by number of species per island (from most to fewest; y-axis) and number of insular occurrences per species (from most to fewest; x-axis). Individual plots are (A) all islands (including both Sea of Cortés and Pacific islands); (B) all landbridge islands; (C) landbridge islands in Sea of Cortés only; (D) Pacific landbridge islands only; (E) oceanic (deep-water) islands only. Except for oceanic islands (E), all sets are significantly ordered.

effects, while not insignificant, have evidently shaped biogeographic patterns of insular faunas to a lesser extent in terrestrial mammals than in other vertebrates.

This overall interpretation is supported by combined evidence derived from both species-richness patterns and analysis of species assemblages in mammals from islands in the Sea of Cortés. It is further supported by worldwide patterns. For instance, low z-values characterize oceanic faunas regardless of archipelago age and relative isolation (Lawlor 1996), suggesting that subsaturation is pervasive.

Extinctions and Introductions

Although colonizing abilities may explain the initial appearance of a species on an island, they do not account fully for the persistence of a species after its arrival. For example, landbridge islands in the gulf probably had a more or less complete continental fauna of desert-adapted species when formed. Extinctions have since pared down each fauna to its present composition. As previously discussed, faunas on these islands have collapsed since separation from the mainland and, as expected, relaxation has been most pronounced for faunas on small islands. Using estimated extinction rates, Case and Cody (1987) confirmed this effect and demonstrated that mammals became extinct faster than reptiles on landbridge islands in the Sea of Cortés.

It is also clear that extinctions have not been random among taxa. Relatively small omnivorous, granivorous, and herbivorous species predominate on all gulf islands. Hence, in addition to being good colonists, these species are also more resistant to extinction once established, as exemplified by rodents on BLA islands. Of course, the ultimate reason for this is their ability to sustain relatively large population sizes, as predicted for such species by MacArthur and Wilson (1967).

Extinctions of mammalian populations on islands in the Sea of Cortés have not been limited to prehistoric or ecological influences, however. Surveys conducted by Alvarez-Castañeda, together with Centro de Investigaciones Biologicas del Noroeste (CIBNOR) and others, have identified eight taxa of mammals that evidently have disappeared recently (app. 12.1) and several others that are vulnerable (e.g., *Neotoma lepida* on Danzante and San Francisco) or in imminent danger of extinction (*Dipodomys merriami insularis* and *Neotoma albigula varia* on San José and Turner, respectively) (Schultz et al. 1970; Lawlor 1983; Mellink 1992a,b; Smith et al. 1993; Alvarez-Castañeda 1994, 1997a; Alvarez-Castañeda and Cortés-Calva 1996, 1999, in press; Bogan 1997; Cortés-Calva et al. 2001a,b). Of course, many taxa on islands in the Sea of Cortés are vulnerable to extinction because they have distributions restricted to one or a few islands, limited population sizes on islands of relatively small size, or both. Yet most modern extinctions are uniquely attributable to human activities involving habitat disturbances or introductions of non-native species. The latter include predators (chiefly domestic cats, *Felis cattus*) and non-native rodent competitors (rats, *Rattus rattus*; and house mice, *Mus domesticus*), in addition to other domestic species (table 12.4).

Four extinctions involve species endemic to single islands; the remaining ones are local extirpations of otherwise more widespread species (app. 12.1). Domestic cats, which have well-known negative impacts in island settings (Owen 1977; Konecny 1987a,b; Van Rensburg and Bester 1988), and introduced rodents are implicated in four and three disappearances, respectively. Human activities such as guano mining (e.g., Granito, San Roque), poison-control efforts to eradicate introduced rats (e.g., San Roque), establishment of fishing camps (e.g., Granito), and tourist visitations (e.g., Danzante, Espíritu Santo) doubtless have had impacts as well. The remaining island, San Pedro Nolasco, is one of five islands (and one of only two small ones; see above) containing two species of *Peromyscus* (*P. pembertoni* and *P. boylii*). Although it will never be known for certain, the apparent disappearance of *P. pembertoni* from San Pedro Nolasco may be the result of competitive interactions with the other *Peromyscus* species.

Table 12.4 Introduced domestic mammals recorded from islands in the Sea of Cortés

	Introduced Mammals							Native	Extinct
Island	Cats	Rodents	Goats	Rabbits	Cattle	Horses	Dogs	Mammals	Mammals
Sea of Cortés									
Willard							Coyote?	2	
Mejía		X						2	1
Granito		X						1	1
Angel de la Guarda	X							3	
San Pedro Nolasco	X							2	1
Tiburón							Coyote	14	
San Marcos	X	X	X				X	2	
Coronado	X							3	
Carmen	X		X					4	
Monserrat	X							2	1
Catalina	X							1	
Santa Cruz	X							1	
San José	X		X				X	7	
San Francisco			X					2	
Espíritu Santo	X		X		X	X		6	
Cerralvo	X		X					2	
Pacific Ocean									
San Martín	X							3	1
San Gerónimo	X							1	1
Cedros	X						X	5	
Natividad	X							1	
Asunción	X							1	
Magdalena							Coyote	9	
Santa Margarita	X	X					X	7	

For sources of data, see text. Islands are arranged from north to south in each category. Note that domestic species are absent on those islands inhabited by coyotes.

Another replacement may have occurred on Santa Catalina. In 1995, Alvarez-Castañeda failed to locate the native species, *Peromyscus slevini*, a *maniculatus*-like form (Burt 1932; see app. 12.2). Instead, he found an *eremicus*-like mouse previously unknown from the island (Hafner et al. 2001). The status of these species is undergoing further exploration (Carleton and Lawlor in prep.).

Predictably, small islands demonstrate disproportionate numbers of extinctions compared to large ones. In fact, in all cases in which potentially competing species (*Mus, Rattus*) have been introduced to very small islands (those <3.0 km^2), the single native species has been extirpated. Because only *Peromyscus* occupies single-species islands, members of this genus are most heavily affected (islas Granito, Mejía, San Roque). Introduced species are reported from no small single-species islands on which *Peromyscus* persists (San Diego, Santa Catalina).

On slightly larger islands (3–10 km^2) containing woodrats (*Neotoma*) and one or two other rodent species, woodrats characteristically disappear first (Todos Santos, San Martín, Turner, Coronado). This is consistent with their greater susceptibility to predation from domestic cats or to competition from introduced *Rattus* with more generalized diets. By contrast, woodrats are common on Espíritu Santo, despite the presence of introduced cats and native ring-tails (*Bassariscus astutus*) (Alvarez-Castañeda et al. submitted). Could interactions with established predators preadapt insular woodrats for coexistence with domestic cats?

It is impossible to establish with certainty that extinctions have occurred, but in each of the above instances intensive effort has been expended without success to establish the presence of populations. In some instances, many surveys involving thousands of trap-nights have been undertaken by different investigators over many seasons and years (decades in a few cases). For example, the last known collection of *Peromyscus pembertoni* (San Pedro Nolasco) was made by W. H. Burt in 1931, and attempts to locate this species, involving more than 7000 trap-nights and several investigators, have failed to find it (Lawlor 1983). Moreover, most disappearances are reported for small islands, on which thorough sampling has been conducted in every habitat and at all locations.

Intentional or inadvertent introductions by humans are common on gulf islands (table 12.4). Domestic cats are especially widespread. It is interesting that cats are absent on three islands on which coyotes are present (table 12.4), an observation consistent with the absence of these mesopredators in the face of large carnivores elsewhere (e.g., Crooks and Soulé 1999).

Introductions are implicated in two occurrences of antelope ground squirrels (*Ammospermophilus* spp.) on Sea of Cortés islands (Espíritu Santo and San Marcos). These highly visible squirrels were first reported from San Marcos in 1987 (J. Reichman, pers. commun., 1989), despite several earlier visits by biologists. They are kept as pets by people in Santa Rosalía and Loreto, on the peninsular coast (Hafner, pers. obs.; Wright 1965). Isla San Marcos is only 5 km from the mainland and, with fresh water provided by brackish wells, was once the site of a small tannery (Slevin 1923). It is currently the site of a gypsum mine that supports mine workers and their families on the island, and squirrels are especially numerous in the canyon immediately north of the mine. If ground squirrels were introduced as escaped pets on San Marcos, the same might be true of the population on Espíritu Santo. Only 6 km from La Paz, this island supported several hundred Indians during aboriginal times, was the site of experimental pearl farming operations in the early 1900s, and is routinely visited by fishermen from the peninsular mainland (Dickey 1983).

Patterns of Variation

Population Variation

In general, Sea of Cortés mammals exhibit only modest degrees of differentiation from related mainland species. The numbers of endemic species and subspecies (see above and app. 12.1) are relatively low but are high compared to those of lizard faunas on the same islands (Case and Cody 1983; Murphy 1983b, Grismer 1994; chaps. 8, 9, this volume). Predictably, the most divergence among terrestrial mammals has taken place on oceanic islands, presumably because of their greater isolation.

Among gulf mammals, variation in members of the genus *Peromyscus* has been studied most intensively (Banks, 1967; Lawlor 1971a,b, 1983; Avise et al. 1974; Gill 1981; Riddle et al. 2000a; Hafner et al. 2001) and to a lesser degree in *Chaetodipus* (Gill 1981; Riddle et al. 2000b). In contrast to landbridge island forms, populations on oceanic islands are characterized by relatively large deviations in body size (see below), peculiar characteristics of the skull and phallus, and unique chromosomal and biochemical (mtDNA and allozyme) attributes.

Hafner (1981) quantified the occurrence of upper premolars (P^3) in antelope ground squirrels (*Ammospermophilus*), the size and occurrence of which are extremely variable in the rodent family Sciuridae. The tooth is reduced and occasionally absent in all *Ammospermophilus*, its absence varying in frequency from 2% to 15%. In *A. insularis* from Espíritu Santo, fully 55% of specimens lacked one or both of the premolars. Howell (1938) considered this to be one of the major distinctive features of this insular endemic species.

Interisland morphological variation among terrestrial mammals is demonstrated in both body size and pelage coloration. As in mammals elsewhere, pelage color typically matches the color of the surrounding background. Dark substrates on volcanic islands promote melanism in several island populations. Examples include the black jackrabbit (*Lepus insularis*) on Espíritu Santo, occasional melanistic individuals of rock squirrels (*Spermophilus variegatus*) on Tiburón, a gray morph in pocket mice (*Chaetodipus spinatus*) on Mejía, and relatively dark populations of *Peromyscus guardia* and *Peromyscus pseudocrinitus* on Mejía and Coronado, respectively (Burt 1932; Banks 1967; Lawlor 1971a; Gill 1981).

Although important for assessing the impact of genetic bottlenecks and amounts of gene flow among insular populations, intra-island variation in gulf mammals has been little explored. Patterns of allozymic, morphometric, and discrete character variation in *Peromyscus* seem most consistent with the hypothesis that genetic bottlenecks, either recently or historically, have led to reduced intrapopulation variability (Lawlor 1983). Using two measures of morphological variation (coefficients of variation and epigenetic character variation), Lawlor (1983) found that populations on oceanic islands and small landbridge islands were less variable than those on large landbridge islands and the mainland, and that intrapopulation variation was least in morphologically more divergent oceanic-island populations. Additionally, unique chromosomal, allozymic, and morphological traits are found only in populations on oceanic and small landbridge islands (Lawlor 1971a; Avise et al. 1974). Overall, Lawlor (1983) found no support for hypotheses based on indices of habitat complexity, competition, or resource availability. That marked reductions in population size occur periodically on small islands is verified by measured density changes of rodents on islands in Bahía de los Ángeles and the disproportionate number of disappearances of small-island populations recorded in the Sea of Cortés (see above).

Variation in Body Size

Body size varies in regular ways among insular mammals. When assessed across taxa on a worldwide basis, gigantism prevails in rodents, whereas dwarfism predominates in carnivores and in the herbivorous lagomorphs, artiodactyls, and elephants, although there are exceptions. Generally, small species exhibit gigantism and large ones exhibit dwarfism. Van Valen (1973a) coined the "island rule" to recognize the common occurrence of this phenomenon. Hypotheses to explain insular size variation center on an interplay among competition, predation, and resource limitation (e.g., Foster 1963; Van Valen 1973a,b; Case 1978, 1979; Heaney 1978; Lawlor 1982; Angerbjorn 1985, 1986; Lomolino 1985; Smith 1992), although other explanations have been posited (Foster 1965; Wassersug et al. 1979; Melton 1982; Lomolino 1985). These relationships vary among species and locations (summarized in Dayan and Simberloff 1997).

Body-size (as indexed by body length) shifts in mammals on islands in the Sea of Cortés are not wholly consistent with other studies (fig. 12.8). Lagomorphs tend to be larger, not smaller, on gulf islands (*contra* Hoagland 1992). Within rodents, departures from mainland relatives tend toward larger sizes and are striking in some populations of *Peromyscus* and *Neotoma*; a maximum body length increase of 27.5% is achieved in woodrats (*Neotoma bunkeri*) on Coronado (Smith 1992). In contrast, smaller sizes obtain in insular *Chaetodipus*, with a maximum decrease of 12.7% in *Chaetodipus spinatus* on San Marcos (Lawlor 1982). Data for multiple congeneric species occurring on the same island suggest that, whatever the selective factors for determining body size, they may operate consistently in closely related species that share similar diets. For example, all three species of seed-eating pocket mice occurring on Tiburón (*Chaetodipus baileyi*, *C. intermedius*, and *C. penicillatus*) show similar decreases in body length (Lawlor 1982).

Lawlor (1982) proposed that body-size differences in small mammals from gulf islands evolved in response to differing levels of competition and available resources in low-productivity environments. Assuming that predators and overall food sources are reduced on islands, there should be different consequences for species competing for particulate foods (granivores) than for those with generalist diets or that feed on vegetative matter (omnivores, folivores). Insular conditions should induce dwarfism in specialists because they exploit a resource that is distributed discontinuously in space and time, but such conditions should favor gigantism in generalist species, whose food supply is more homogeneously distributed and perhaps enhanced in competitor-poor environments.

Among genera of rodents, the order of specialist to generalist was expected to be *Chaetodipus-Peromyscus-Neotoma*. Supporting data for this pattern stem mostly from

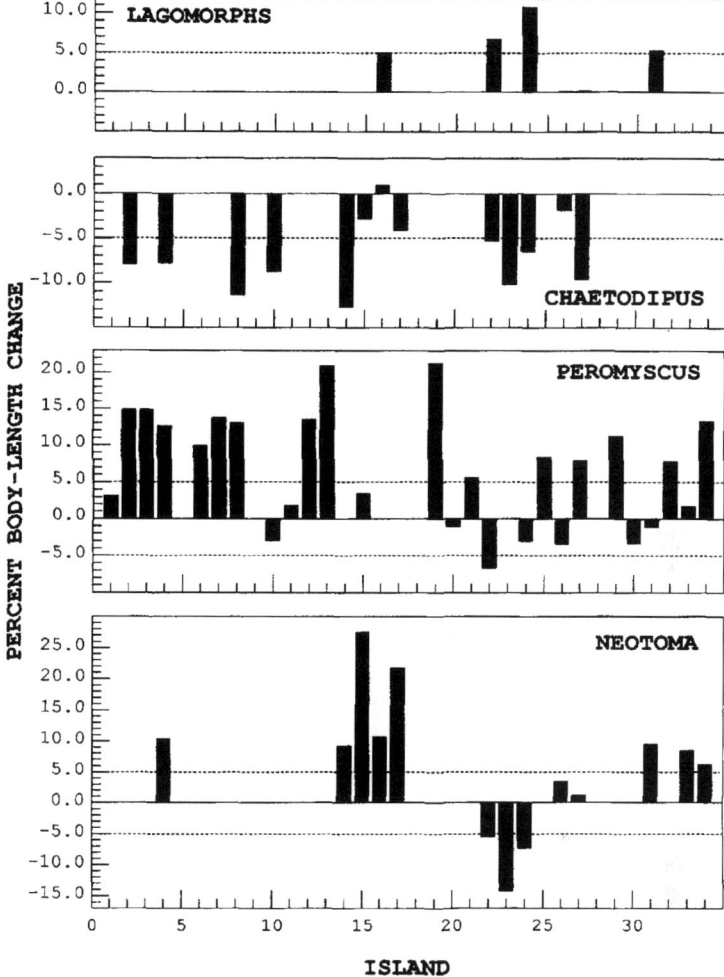

Figure 12.8 Changes in body length for rodents and lagomorphs occurring on Pacific and Sea of Cortés islands. Islands (number designations on x-axis) are identified in appendix 12.1. Data are summarized from Lawlor (1982) and Smith (1992).

three predictions (Lawlor 1982). First, among widespread species of similar body sizes and diets, seed specialists (*Chaetodipus*) should inhabit fewer islands than omnivores (*Peromyscus*), and this effect should be most pronounced on very small islands (islands on which seed depletion should be most severe). As reported above, pocket mice frequent many fewer islands than deer mice, a difference due entirely to the absence of pocket mice on small, one-species islands, on which only *Peromyscus* occur. Contrary to the suggestion of Angerbjorn (1985), this pattern is not explicable simply by the presence of fewer potential mainland species of *Chaetodipus* available to colonize islands compared to *Peromyscus* (six and seven, respectively).

Second, there should be fewer congeneric associations of pocket mice than deer mice. Congeneric pairings of *Peromyscus* and *Chaetodipus* occur on five and three islands, respectively; those of pocket mice are limited to only large landbridge islands.

Third, effects of increasing intensity of competition, as indexed by the extent of overlapping food habits and number of potential competitors that they encounter, should produce more marked effects in granivores (*Chaetodipus*) than in generalists (*Peromyscus, Neotoma*). Simply put, dwarfism should be induced in *Chaetodipus* in the presence of fewer potential competitors than in *Peromyscus*, which in turn should exhibit reduced body size in the face of fewer competitors than *Neotoma*. Alternatively, gigantism should be rare or nonexistent in pocket mice and relatively more common in deer mice and woodrats in the presence of competing species. In fact, dwarfism in pocket mice occurs in the presence of as few as one species, and there are no demonstrable examples of gigantism. In contrast, aside from one exception in each, gigantism prevails in *Peromyscus* and *Neotoma* in all significant departures from mainland body sizes (Lawlor 1982; Smith 1992).

Although these data generally support the central hypothesis, the hypothesis requires modification. Smith (1992) confirmed that body size does increase in insular *Neotoma*. She also demonstrated a significant relationship between body size and reduction of predation pressure but no relation with presence of other rodent species, and, citing data from Vaughan and Schwartz (1980), argued persuasively that woodrats on gulf islands are probably specialists (on leaves and twigs) rather than generalists. Despite this correction, the size patterns in *Neotoma* remain in agreement with the hypothesis because their strictly herbivorous diet still predicts large body sizes for insular populations.

Patterns of morphologic variation in the Sea of Cortés agree with Lomolino's (1985) general explanation of insular body-size changes. In a review of worldwide trends in mammals, he proposed that factors promoting gigantism (e.g., competitive release, physiological advantages) should decrease with larger body sizes, whereas those influencing dwarfism (e.g., resource limitation) should increase with larger body sizes. If true, the combined effects of selective forces should produce a tendency for insular body size to decrease with increasing body sizes of species on the mainland. In the Sea of Cortés, pocket mice are an apparent exception to the suggested patterns (i.e., despite their small size overall, they become smaller, not larger, on islands), but their reduced body size is explained by the limited quantities and ephemeral qualities of seed resources on gulf islands.

Conclusions

The biogeographical distribution of rodents on islands in the Sea of Cortés reflects a combination of evolutionary and ecological processes acting over both short and long time scales. In combination, the patterns of distributions, species richness, and variation and differentiation represent a historical legacy of colonization differences among populations on landbridge and oceanic islands. Populations on landbridge islands are largely relict derivatives of previously cosmopolitan populations, whereas those on oceanic islands are evidently descendants of the chance characteristics of a small number of overwater dispersers (the "sweepstakes" winners of Simpson [1940]).

References

Aguirre, G., D.J. Morafka, and R.W. Murphy. 1999. The peninsular archipelago of Baja California: a thousand kilometers of tree lizard genetics. *Herpetologica* **55**:369–381.
Alvarez-Castañeda, S.T. 1994. Current status of the rice rat, *Oryzomys couesi peninsularis*. *Southwestern Naturalist* **39**:99–100.
Alvarez-Castañeda, S.T. 1997a. Diversidad y conservación de pequeños mamíferos terrestre de B.C.S. Ph.D. thesis, Universidad Nacional Autónoma de México.
Alvarez-Castañeda, S.T. 1997b. Peromyscus pseudocrinitus. *Mammalian Species* **601**:1–3.
Alvarez-Castañeda, S.T., and P. Cortés-Calva. 1996. Anthropogenic extinction of the endemic deer mouse, *Peromyscus maniculatus cineritius*, on San Roque island, Baja California Sur, Mexico. *Southwestern Naturalist* **41**:99–100.
Alvarez-Castañeda, S.T., and P. Cortés-Calva. 1999. Familia Muridae. In: S.T. Alvarez-Castañeda and J. Patton (eds), *Mamíferos del Noroeste Mexicano*. Centro de Investigaciones Biológicas del Noroeste, S.C., La Paz, Mexico; pp. 445–566.
Alvarez-Castañeda, S.T., and P. Cortés-Calva. In press. Extirpation of Bailey's pocket mouse, *Chaetodipus baileyi fornicatus* (Heteromyidae: Mammalia), from Isla Montserrat, Baja California Sur, Mexico. *Western American Naturalist*.
Alvarez-Castañeda, S.T., and J.L. Patton. 1999. Mamíferos del Noroeste de México. Centro de Investigaciones Biológicas del Noroeste, La Paz, S.C., México.
Angerbjorn, A. 1985. The evolution of body size in mammals on islands: some comments. *American Naturalist* **125**:304–309.
Angerbjorn, A. 1986. Gigantism in island populations of wood mice (Apodemus) in Europe. *Oikos* **47**:47–56.
Aquilars-S, M.A., J.W. Sites, and R.W. Murphy. 1988. Genetic variability and population structure in the lizard genus *Petrosaurus* (Iguanidae). *Journal of Herpetology* **22**:135–145.
Atmar, W., and B.D. Patterson. 1993. The measure of order and disorder in the distribution of species in fragmented habitat. *Oecologia* **96**:373–382.
Atmar, W., and B.D. Patterson. 1995. The nestedness temperature calculator: a visual basic program, including 294 presence-absence matrices. AICS Research, Inc., University Park, New Mexico, and The Field Museum, Chicago.
Avise, J.C., M.H. Smith, R.K. Selander, T.E. Lawlor, and P.R. Ramsay. 1974. Biochemical polymorphism and systematics in the genus *Peromysus*. Insular and mainland species of the subgenus *Haplomylomys*. *Systematic Zoology* **23**:226–238.
Axelrod, D.I. 1979. Age and origin of Sonoran Desert vegetation. *Occasional Papers, California Academy of Sciences* **132**:1–74.
Baker, R.J., and H.H. Genoways. 1978. Zoogeography of Antillean bats. *Special Publication of the Philadelphia Academy of Sciences* **13**:53–97.
Banks, R.C. 1964a. Birds and mammals of the voyage of the "Gringa." *Transactions of the San Diego Society of Natural History* **13**:177–184.
Banks, R.C. 1964b. The mammals of Cerralvo Island, Baja California. *Transactions of the San Diego Society of Natural History* **13**:397–404.
Banks, R.C. 1964c. Range extensions for three bats in Baja California, Mexico. *Journal of Mammalogy* **45**:489.
Banks, R.C. 1967. The *Peromyscus guardia-interparietalis* complex. *Journal of Mammalogy* **50**:501–513.
Betancourt, J.L., T.R. Van Devender, and P.S. Martin. 1990. *Packrat Middens: The Last 40,000 Years of Biotic Change*. University of Arizona Press, Tucson.
Blair, W.N. 1978. Gulf of California in Lake Mead area in Arizona and Nevada during the late Miocene time. *Bulletin of the American Association of Petroleum Geologists* **62**:1159–1170.

Boehm, M.C. 1984. An overview of the lithostratigraphy, biostratigraphy, and paleoenvironments of the Late Neogene San Felipe marine sequence, Baja California, Mexico. In: V.A. Frizzell (ed), *Geology of the Baja California Peninsula*. Field Trip Guidebook—Pacific Section, Society of Economic Paleontology and Mineralogy **39**:253–265.

Bogan, M.A. 1997. On the status on *Neotoma varia* from Isla Datil, Sonora. In: T.L. Yates, W.L. Gannon, and D.E. Wilson (eds), *Life among the Muses: Papers in Honor of James S. Findley*. Special Publication, The Museum of Southwestern Biology, University of New Mexico, Albuquerque; **3**:81–87.

Bowers, M.A., D.B. Thompson, and J.H. Brown. 1987. Spatial organization of a desert rodent community: food addition and species removal. *Oecologia* **72**:77–82.

Brown, J.H. 1978. The theory of insular biogeography and the distribution of boreal birds and mammals. *Great Basin Naturalist Memoirs* **2**:209–227.

Brown, J.H., and B.A. Harney. 1993. Population and community ecology of heteromyid rodents in temperate habitats. In: H.H. Genoways and J.H. Brown (eds), *Biology of the Heteromyidae*. Special Publication no. 10, American Society of Mammalogists, Provo, Utah; pp. 618–651.

Buising, A.V. 1990. The Bouse Formation and bracketing units, southeastern California and western Arizona: implications for the evolution of the proto-Gulf of California and the lower Colorado River. *Journal of Geophysical Research* **95**:20,111–20,132.

Burt, W.H. 1932. Descriptions of heretofore unknown mammals from islands in the Gulf of California, Mexico. *Transactions of the San Diego Society of Natural History* **16**:161–182.

Carpenter, R.E. 1969. Structure and function of the kidney and water balance of desert bats. *Physiological Zoology* **42**:288–302.

Case, T.J. 1975. Species numbers, density compensation, and colonizing ability of lizards on islands in the Gulf of California. *Ecology* **56**:3–18.

Case, T.J. 1978. A general explanation for insular body size trends in terrestrial vertebrates. *Ecology* **59**:1–18.

Case, T.J. 1979. Optimal body size and an animal's diet. *Acta Biotheoretica* **28**:54–69.

Case, T.J. 1983. The reptiles: ecology. In: T.J. Case and M.L. Cody (eds), *Island Biogeography in the Sea of Cortéz*. University of California Press, Berkeley; pp. 159–209, 438–454.

Case, T.J., and M.L. Cody. 1983. Synthesis: pattern and process in island biogeography. In: T.J. Case and M.L. Cody (eds), *Island Biogeography in the Sea of Cortéz*. University of California Press, Berkeley; pp. 307–341.

Case, T.J., and M.L. Cody. 1987. Testing theories of island biogeography. *American Scientist* **75**:402–411.

Ceballos, G., T.H. Fleming, C. Chávez, and J. Nassar. 1997. Population dynamics of *Leptonycteris curasoae* (Chiroptera: Phyllostomidae) in Jalisco, Mexico. *Journal of Mammalogy* **78**:1220–1230.

Cervantes, F.A., S.T. Alvarez-Castañeda, B. Villa-R., C. Lorenzo, and J. Vargas. 1996. Natural history of the black jackrabbit (*Lepus insularis*) from Espíritu Santo Island, Baja California Sur, México. *Southwestern Naturalist* **41**:186–189.

Cortés-Calva, P., Eric Yensen, and S.T. Alvarez-Castañeda. 2001a. Neotoma martinensis. *Mammalian Species* **657**:1–3.

Cortés-Calva, P., S.T. Alvarez-Castañeda, and E. Yensen. 2001b. Neotoma anthonyi. *Mammalian Species* **663**:1–3.

Crooks, K.R., and M.E. Soulé. 1999. Mesopredator release and avifaunal extinctions in a fragmented system. *Nature* **400**:563–566.

Dayan, T., and D. Simberloff. 1997. Size patterns among competitors: ecological character displacement and character release in mammals, with special reference to island populations. *Mammal Review* **28**:99–124.

Dickey, K.J. 1983. *A Natural History Guide to Baja California*. Published by the author, Chula Vista, California.
Eberly, L.D., and T.B. Stanley, Jr. 1978. Cenozoic stratigraphy and geologic history of southeastern Arizona. *Bulletin of the Geological Society of America* **89**:921–940.
Foster, J.B. 1963. The evolution of the native land mammals of the Queen Charlotte Islands and the problem of insularity. Ph.D. dissertation, University of British Columbia, Vancouver.
Foster, J.B. 1965. The evolution of the mammals of the Queen Charlotte Islands, British Columbia. *Occasional Papers, British Columbia Provincial Museum* **14**:1–130.
Fox, B.J., and J.H. Brown. 1993. Assembly rules for functional groups in North American desert rodent communities. *Oikos* **67**:358–370.
Gastil, G., J. Minch, and R.P. Phillips. 1983. The geology and ages of the islands. In: T.J. Case and M.L. Cody (eds), *Island Biogeography in the Sea of Cortéz*. University of California Press, Berkeley; pp. 13–25.
Gill, A.E. 1981. Morphological features and reproduction of *Perognathus* and *Peromyscus* on northern islands in the Gulf of California. *American Midland Naturalist* **106**:192–196.
Gliwicz, J. 1980. Island populations of rodents: their organization and functioning. *Biological Reviews of the Cambridge Philosophical Society* **55**:109–138.
Grant, P.R. 1972. Interspecific competition among rodents. *Annual Review of Ecology and Systematics* **3**:79–106.
Grismer, L.L. 1994. The origin and evolution of the peninsular herpetofauna of Baja California, Mexico. *Herpetological Natural History* **2**:51–106.
Grismer, L.L. 1999. An evolutionary classification of reptiles on islands in the Gulf of California, México. *Herpetologica* **55**:446–469.
Hafner, D.J. 1981. Evolutionary relationships and historical zoogeography of antelope ground squirrels, genus *Ammospermophilus* (Rodentia: Sciuridae). Ph.D. dissertation, University of New Mexico, Albuquerque.
Hafner, D.J. 1993. North American pika (*Ochotona princeps*) as a late Quaternary biogeographic indicator species. *Quaternary Research* **39**:373–380.
Hafner, D.J., and B.R. Riddle. 1997. Biogeography of Baja California peninsular desert mammals. In: T.L. Yates, W.L. Gannon, and D.E. Wilson (eds), *Life among the Muses: Papers in Honor of James S. Findley*. Special Publication, The Museum of Southwestern Biology, University of New Mexico, Albuquerque; **3**:39–68.
Hafner, D.J., B.R. Riddle, and S.T. Alvarez-Castañeda. 2001. Evolutionary relationships of white-footed mice (*Peromyscus*) on islands in the Sea of Cortéz, Mexico. *Journal of Mammalogy* **82**:775–790.
Hall, E.R. 1981. *The Mammals of North America*, 2nd ed. John Wiley and Sons, New York.
Harris, A.H. 1985. *Late Pleistocene Vertebrate Paleoecology of the West*. University of Texas Press, Austin.
Hayward, B., and R. Davis. 1964. Flight speeds in western bats. *Journal of Mammalogy* **45**:236–242.
Heaney, L.R. 1978. Island area and body size of insular mammals; evidence from the tricolored squirrel (*Callosciurus prevosti*) of southwest Asia. *Evolution* **32**:29–44.
Herman, T.B. 1979. Population ecology of insular *Peromyscus maniculatus*. Ph.D. dissertation, University of Alberta, Edmonton.
Heske, E.J., J.H. Brown, and S. Mistry. 1994. Long-term experimental study of a Chihuahuan desert rodent community: 13 years of competition. *Ecology* **75**:438–445.
Hoagland, D.B. 1992. Feeding ecology of an insular population of the black-tailed jackrabbit (*Lepus californicus*) in the Gulf of California. *Southwestern Naturalist* **37**:280–286.
Howell, A.H. 1938. Revision of the North American ground squirrels, with a classification of the North American Sciuridae. *North American Fauna* **56**:1–256.

Huey, L.M. 1964. The mammals of Baja California. *Transactions of the San Diego Society of Natural History* **13**:85–168.

Ingle, J.C., Jr. 1987. Paleooceanographic evolution of the Gulf of California: foraminiferal and lithofacies evidence. *Abstracts with Programs, Geological Society of America* **19**:721.

Kaufman, D.W., and G.A. Kaufman. 1989. Population biology. In: G.L. Kirkland, Jr., and J.N. Layne (eds), *Advances in the Study of Peromyscus (Rodentia)*. Texas Tech University Press, Lubbock; pp. 233–270.

Konecny, M.J. 1987a. Home range and activity patterns of feral house cats in the Galapagos Islands. *Oikos* **50**:17–23.

Konecny, M.J. 1987b. Food habits and energetics of feral house cats in the Galapagos Islands. *Oikos* **50**:24–32.

Lawlor, T.E. 1971a. Evolution of *Peromyscus* on northern islands in the Gulf of California, Mexico. *Transactions of the San Diego Society of Natural History* **16**:91–124.

Lawlor, T.E. 1971b. Distribution and relationships of six species of *Peromyscus* in Baja California and Sonora, Mexico. *Occasional Papers, Museum of Zoology, University of Michigan* **661**:1–22.

Lawlor, T.E. 1982. The evolution of body size in mammals: evidence from insular populations in Mexico. *American Naturalist* **119**:54–72.

Lawlor, T.E. 1983. The mammals. In: T.J. Case and M.L. Cody (eds), *Island Biogeography in the Sea of Cortéz*. University of California Press, Berkeley; pp. 265–289, 480–500.

Lawlor, T.E. 1986. Comparative biogeography of mammals on islands. *Biological Journal of the Linnean Society* **28**:99–125.

Lawlor, T.E. 1996. Species numbers, distributions, and compositions of insular mammals of mammals. In: H.H. Genoways and R.J. Baker (eds), *Contributions in Mammalogy. A Memorial Volume Honoring Dr. J. Knox Jones, Jr.* Museum of Texas Tech University, Lubbock; pp. 285–295.

Lee, A.K. 1963. Adaptations to arid environments in woodrats. *University of California Publications in Zoology* **64**:57–96.

Lieberman, M., and D. Lieberman. 1970. The evolutionary dynamics of the desert woodrat, *Neotoma lepida*. *Evolution* **24**:560–570.

Lomolino, M.V. 1985. Body size of mammals on islands: the island rule reexamined. *American Naturalist* **125**:310–316.

Lomolino, M.V. 1989. Interpretations and comparisons of constants in the species-area relationship: an additional caution. *American Naturalist* **133**:277–280.

Lomolino, M.V. 2000. A species-based theory of insular biogeography. *Global Ecology and Biogeography* **9**:39–58.

Lonsdale, P. 1989. Geology and tectonic history of the Gulf of California. In: E. L. Winterer, D.M. Hussong, and R.W. Decker (eds), *The Eastern Pacific Ocean and Hawaii*. The Geological Society of America; pp. 499–521.

MacArthur, R.H., and E.E. Wilson. 1963. An equilibrium theory of insular biogeography. *Evolution* **17**:373–387.

MacArthur, R.H., and E.E. Wilson. 1967. *The Theory of Island Biogeography*. Princeton University Press, Princeton, NJ.

MacMillen, R.E. 1964. Population ecology, water relations, and social behavior of a southern California semidesert rodent fauna. *University of California Publications in Zoology* **71**:1–66.

McCabe, T.T., and I.M. Cowan. 1945. *Peromyscus maniculatus macrorhinus* and the problems of insularity. *Transactions of the Royal Canadian Institute* **1**:172–215.

McCloy, C. 1984. Stratigraphy and depositional history of the San Jose del Cabo trough, Baja California Sur, Mexico. In: V.A. Frizzell (ed), *Geology of the Baja California Peninsula*.

Field Trip Guidebook—Pacific Section, Society of Economic Paleontology and Mineralogy; 39:267–273.
Mellink, E. 1992a. The status of *Neotoma anthonyi* (Rodentia, Muridae, Cricetinae) of Todos Santos Islands, Baja California, Mexico. *Bulletin of the Southern California Academy of Science* **91**:137–140.
Mellink, E. 1992b. Status de los heterómyidos y cricétidos endémicos del estado de Baja California. Comunicaciones académicas, serie ecología, Centro de Investigaciones Científicas y de Educación Superior de Ensenada. Ensenada, Mexico.
Melton, R.H. 1982. Body size and island *Peromyscus*: a pattern and a hypothesis. *Evolutionary Theory* **6**:113–126.
Murphy, R.W. 1975. Two new blind snakes (Serpentes: Leptotyphlopidae) from Baja California, Mexico with a contribution to the biogeography of peninsular and insular herpetofauna. *Proceedings of the California Academy of Sciences, Series 4*, **40**:93–107.
Murphy, R.W. 1983a. Paleobiogeography and patterns of genetic differentiation of the Baja California herpetofauna. *Occasional Papers of the California Academy of Sciences* **137**:1–48.
Murphy, R.W. 1983b. The reptiles: origins and evolution. In: T.J. Case and M.L. Cody (eds), *Island Biogeography in the Sea of Cortéz*. University of California Press, Berkeley; pp. 130–158.
Nelson, E.W. 1921. Lower California and its natural resources. *Memoirs of the National Academy of Science* **16**:1–194.
Noy-Meir, I. 1973. Desert ecosystems: environments and producers. *Annual Review of Ecology and Systematics* **4**:25–41.
Orr, R.T. 1960. An analysis of the recent land mammals. Symposium: The biogeography of Baja California and adjacent seas. *Systematic Zoology* **9**:171–179.
Orr, R.T., and R.C. Banks. 1964. Bats from islands of the Gulf of California. *Proceedings of the California Academy of Sciences* **30**:207–210.
Osborne, T.O., and W.A. Sheppe. 1971. Food habits of *Peromyscus maniculatus* on a California beach. *Journal of Mammalogy* **52**:844–845.
Owen, O.S. 1977. *Conservación de Recursos Naturales*. Editorial Pax-México, Distrito Federal, México.
Patterson, B.D. 1987. The principle of nested subsets and its implications for biological conservation. *Conservation Biology* **1**:323–334.
Patterson, B.D. 1990. On the temporal development of nested subset patterns of species composition. *Oikos* **59**:330–342.
Patterson, B D., and W. Atmar. 1986. Nested subsets and the structure of insular mammalian faunas and archipelagos. *Biological Journal of the Linnean Society* **28**:65–82.
Patterson, B.D., and J.H. Brown. 1991. Regionally nested patterns of species composition in granivorous rodent assemblages. *Journal of Biogeography* **18**:395–402.
Peterson, B.J., and B. Fry. 1987. Stable isotopes in ecosystem studies. *Annual Review of Ecology and Systematics* **18**:293–320.
Plantz, J.V. 1992. Molecular phylogeny and evolution of the American woodrats, genus *Neotoma* (Muridae). Ph.D. dissertation, University of North Texas, Denton.
Polis, G.A. 1991. Desert communities: an overview of patterns and processes. In: G.A. Polis (ed), *The Ecology of Desert Communities*. University of Arizona Press, Tucson; pp. 1–26.
Polis, G.A., and S.D. Hurd. 1996. Linking marine and terrestrial food webs: Allochthonous input from the ocean supports high secondary productivity on small islands and coastal land communities. *American Naturalist* **147**:396–423.
Polis, G.A., S.D. Hurd, C.T. Jackson, and F. Sanchez Piñero. 1997. El Niño effect on the dynamics and control of an island ecosystem in the Gulf of California. *Ecology* **78**:1884–1897.

Preston, F.W. 1962. The canonical distribution of commoness and rarity. *Ecology* **43**:185–215, 410–432.

Reeder, W.G., and K.S. Norris. 1954. Distribution, type locality, and habits of the fish-eating bat, *Pizonyx vivesi*. *Journal of Mammalogy* **35**:81–87.

Riddle, B.R., D.J. Hafner, and L.F. Alexander. 2000a. Phylogeography and systematics of the *Peromyscus eremicus* species group and the historical biogeography of North American warm regional deserts. *Molecular Phylogenetics and Evolution* **17**:145–160.

Riddle, B.R., D.J. Hafner, and L.F. Alexander. 2000b. Comparative phylogeography of Bailey's pocket mouse (*Chaetodipus baileyi*) and the *Peromyscus eremicus* species group: historical vicariance of the Baja California Peninsular Desert. *Molecular Phylogenetics and Evolution* **17**:161–172.

Riddle, B.R., D.J. Hafner, L.F. Alexander, and J.R. Jaeger. 2000c. Cryptic vicariance in the historical assembly of a Baja California peninsular desert biota. *Proceedings of the National Academy of Sciences, USA*, **97**:14438–14443.

Rodriquez-Robles, J.A., and J.M. De Jesús-Escobar. 2000. Molecular systematics of New World gopher, bull, and pinesnakes (Pituophis: Colubridae), a transcontinental species complex. *Molecular Phylogenetics and Evolution* **14**:35–50.

Rojas-Martinez, A., A. Valiente-Banuet, M. del Coro Arizmendi, A. Alcántara-Eguren, and H.T. Arita. 1999. Seasonal distribution of the long-nosed bat (*Leptonycteris curasoae*) in North America: does a generalized migration pattern really exist? *Journal of Biogeography* **26**:1065–1077.

Rosenzweig, M.L. 1995. *Species Diversity in Space and Time*. Cambridge University Press, Cambridge.

Sahley, C.T., M.A. Horner, and T.H. Fleming. 1993. Flight speeds and mechanical power outputs of the nectar-feeding bat, *Leptonycteris curasoae* (Phyllostomidae: Glossophaginae). *Journal of Mammalogy* **74**:594–600.

Sanchez-Hernandez, C. 1986. Noteworthy records of bats from islands in the Gulf of California. *Journal of Mammalogy* **67**:212–213.

Sánchez Piñero, F., and G.A. Polis. 2000. Bottom-up dynamics of allochthonous input: direct and indrect effects of seabirds on islands. *Ecology* **81**:3117–3132.

Savage, J.M. 1960. Evolution of a Peninsular herpetofauna. Symposium: The biogeography of Baja California and adjacent seas. *Systematic Zoology* **9**:184–212.

Schoeninger, M.J., and M.J. DeNiro. 1984. Nitrogen and carbon isotopic composition of bone collagen from marine and terrestrial animals. *Geochimica et Cosmochimica Acta* **48**:625–639.

Schultz, T.A., F.R. Radovsky, and P.D. Budwiser. 1970. First insular record of *Notiosorex crawfordi*, with notes on other mammals of San Martin Island, Baja California, Mexico. *Journal of Mammalogy* **51**:148–150.

Shreve, F. 1942. The desert vegetation of North America. *Botanical Review* **8**:195–246.

Simpson, G.G. 1940. Mammals and land bridges. *Journal of the Washington Academy of Science* **30**:137–163.

Simpson, G.G. 1956. Zoogeography of West Indian land mammals. *American Museum Novitates* **1759**:1–28.

Slevin, J.R. 1923. Expedition of the California Academy of Sciences to the Gulf of California in 1921. General account. *Proceedings of the Southern California Academy of Sciences* **12**:55–72.

Smith, F.A. 1992. Evolution of body size among woodrats from Baja California, Mexico. *Functional Ecology* **6**:1–9.

Smith, F.A., B.T. Bestlelmeyer, J. Biardi, and M. Strong. 1993. Anthropogenic extinction of the endemic woodrat, *Neotoma bunkeri* Burt. *Biodiversity Letters* **1**:149–155.

Stapp, P. 1997. Community structure of shortgrass-prairie rodents: competition or risk of intraguild predation? *Ecology* **78**:1519–1530.
Stapp, P., G.A. Polis, and F. Sánchez Piñero. 1999. Stable isotopes reveal strong marine and El Niño effects on island food webs. *Nature* **401**:467–469.
Steinbeck, J. 1951. *The Log from the Sea of Cortéz*. Viking Press, New York.
Stock, J.M., and K.V. Hodges. 1989. Pre-Pliocene extension around the Gulf of California and the transfer of Baja California to the Pacific Plate. *Tectonics* **8**:99–115.
Sullivan, R.M. 1988. Biogeography of Southwestern montane mammals: an assessment of the historical and environmental predictions. Ph.D. dissertation, University of New Mexico, Albuquerque.
Thomas, B. 1971. Evolutionary relationships among *Peromyscus* from the Georgia Strait, Gordon, Goletas, and Scott Islands of British Columbia, Canada. Ph.D. dissertation, University of British Columbia, Vancouver.
Upton, D.E., and R.W. Murphy. 1997. Phylogeny of the side-blotched lizards (Phrynosomatidae: *Uta*) based on mtDNA sequences: support for a midpeninsular seaway in Baja California. *Molecular Phylogenetics and Evolution* **8**:104–113.
Van Rensburg, P.J.J., and M.N. Bester. 1988. The effect of cats *Felis catus* predation on three breeding Procellariidae species on Marion Island. *South Africa Journal of Zoology* **15**: 261–278.
Van Valen, L. 1973a. Body size and the numbers of plants and animals. *Evolution* **27**:27–35.
Van Valen, L. 1973b. Pattern and the balance of nature. *Evolutionary Theory* **1**:31–49.
Vaughan, T.A., and S.T. Schwartz. 1980. Behavioral ecology of an insular woodrat. *Journal of Mammalogy* **61**:205–218.
Wassersug, R.J., H. Yang, J.J. Sepkoski, Jr., and D.M. Raup. 1979. The evolution of body size on islands: a computer simulation. *American Naturalist* **114**:287–295.
Webb, T., III, and P.J. Bartlein. 1992. Global changes during the last 3 million years: climatic controls and biotic responses. *Annual Review of Ecology and Systematics* **23**:141–173.
Wiggins, I.L. 1960. The origin and relationships of the land flora. Symposium: The biogeography of Baja California and adjacent seas. *Systematic Zoology* **9**:148–165.
Wiggins, I.L. 1980. *Flora of Baja California*. Stanford University Press, Palo Alto, CA.
Wilcox, B.A. 1978. Supersaturated island faunas, a species-age relationship for lizards on post-Pleistocene land-bridge islands. *Science* **199**:996–998.
Wilkinson, G.S., and T.H. Fleming. 1996. Migration and evolution of lesser long-nosed bats *Leptonycteris curasoae*, inferred from mitochondrial DNA. *Molecular Ecology* **5**:329–339.
Wilson, D.E., and D.M. Reeder (eds). 1993. *Mammal Species of the World. A Taxonomic and Geographic Reference*, 2nd ed. Smithsonian Institute Press, Washington, DC.
Woloszyn, D., and B.W. Woloszyn. 1982. Los mamíferos de la Sierra de la Laguna, Baja California Sur. Consejo Nacional de Ciencia y Technologia, México, D.F.
Wright, D.H., and J.H. Reeves. 1992. On the meaning and measurement of nestedness of species assemblages. *Oecologia* **92**:416–428.
Wright, D.H., B.D. Patterson, G.M. Mikkelson, A. Cutler, and W. Atmar. 1998. A comparative analysis of nested subset patterns of species composition. *Oecologia* **113**:1–20.
Wright, N.P. 1965. *A Guide to Mexican Mammals*. Minutiae Mexicana, México, D. F.
Zink, R.M., R.C. Blackwell, and O. Rojas-Soto. 1997. Species limits in the Le Conte's thrasher. *Condor* **99**:132–138.

13

Island Food Webs

GARY A. POLIS
MICHAEL D. ROSE
FRANCISCO SÁNCHEZ-PIÑERO
PAUL T. STAPP
WENDY B. ANDERSON

    Most of this book focuses on the biogeography and ecology of plants and various animal taxa on islands in the Sea of Cortés. These chapters highlight the historical and biogeographical factors that contributed to the patterns of species distribution and co-occurrence among islands. However, these patterns also reflect the action of ecological processes because the species present interact, directly or indirectly, within the food web that occurs on any given island. Island food webs may also be unique from other communities in the degree to which their structure and dynamics are also strongly influenced by the surrounding ocean. We believe that a deeper appreciation of the trophic connections between the sea and the land, and the resulting effects on the structure and dynamics of island food webs, is key to understanding the biogeography of species on islands.

    Many factors that operate through the food web can enhance or depress populations in a way that affects their local distribution and persistence, and, as a consequence, affects patterns of diversity on a biogeographical scale. Of these, we recognize three as being particularly important: the availability and quality of resources, competition, and consumption (i.e., by herbivores, predators, parasites). Bottom-up factors (nutrients, primary productivity, and food availability to consumers) set limits on island productivity and hence on the potential abundance of a particular group. Within a given community, secondary productivity and population density are subsequently constrained by top-down (i.e., consumption) and competitive effects. One of our goals in this chapter is to show how processes that influence productivity of gulf islands determine patterns of abundance of organisms on islands and affect interactions among species and trophic levels in these systems.

Our second goal is to demonstrate the importance of spatial and temporal variability in productivity in determining the structure and dynamics of island food webs. Using our long-term studies of plants and consumers on islands in the northern gulf, we show that productivity varies greatly, both among years and islands, as a result of both local conditions and global climatic factors. Such variable productivity markedly affects food web dynamics and ultimately the abundance of species on the islands in the Sea of Cortés.

Overview of the Study System

Upon entering the gulf for the first time, one is struck by the extreme and highly variable nature of the environment of the islands and the adjacent peninsula. The region comprises some of the driest places in North America (Cody et al. 1983). Mean annual precipitation ranges from only 59 mm at Bahía de los Ángeles to 75–125 mm in the central Midriff and Bahía de Loreto (Polis et al. 1997b; Sánchez-Piñero and Polis 2000). As a consequence, plant productivity is among the lowest on the continent. Most islands are extremely rocky, which limits soil formation and opportunities for establishment of plants. The barren terrestrial landscape is in strong contrast to the adjacent marine system. Some regions of the gulf are characterized by upwelling during the summer (e.g., Bahía de Loreto) or year-round (e.g., Midriff), resulting in highly productive waters which support large sea-bird colonies on many islands (e.g., Rasa, San Lorenzo Norte).

Our studies in the Sea of Cortés began in 1990 and have focused primarily on midriff islands and, to a lesser extent, on the southern islands in the vicinity of Bahía de Loreto. Our principal study area includes 14 islands in Bahía de los Ángeles, Isla Ángel de la Guarda, Isla Mejia, four islands near Bahía Animas, and the adjacent Baja California peninsula (fig. 13.1). In addition, we also have studied five islands in the central midriff region and six islands near Bahía de Loreto (fig. 13.1).

The results we present here are based on ongoing research, representing 7–15 weeks of field work each year between 1990 and 1999. Many of the important elements influencing island food webs were measured. This includes the phenology and quantity of precipitation; nutrient availability to plants; plant cover and diversity; and annual net primary productivity (ANPP). We also have conducted regular surveys of abundance and diversity of a broad range of island consumers, including insects of various trophic groups (herbivores, detritivores, predators, parasitoids), spiders, lizards, rodents, and songbirds. Some of these groups have been studied for all or most of the duration of our research in the gulf (e.g., insects, spiders), whereas others have been sampled less intensively or are more recent studies (birds, rodents). Details of our standardized methods and results described in this chapter are presented elsewhere (e.g., Polis and Hurd 1995, 1996a,b; Polis et al. 1997b, 1998, in press; Anderson and Polis 1998, 1999; Rose and Polis 1998, Stapp et al. 1999, Sánchez-Piñero and Polis 2000).

Based on our research to date, several generalizations can be made about the ecology of our study system. First, our observational and comparative studies have revealed high levels of spatial variability, among as well as within islands. Major find-

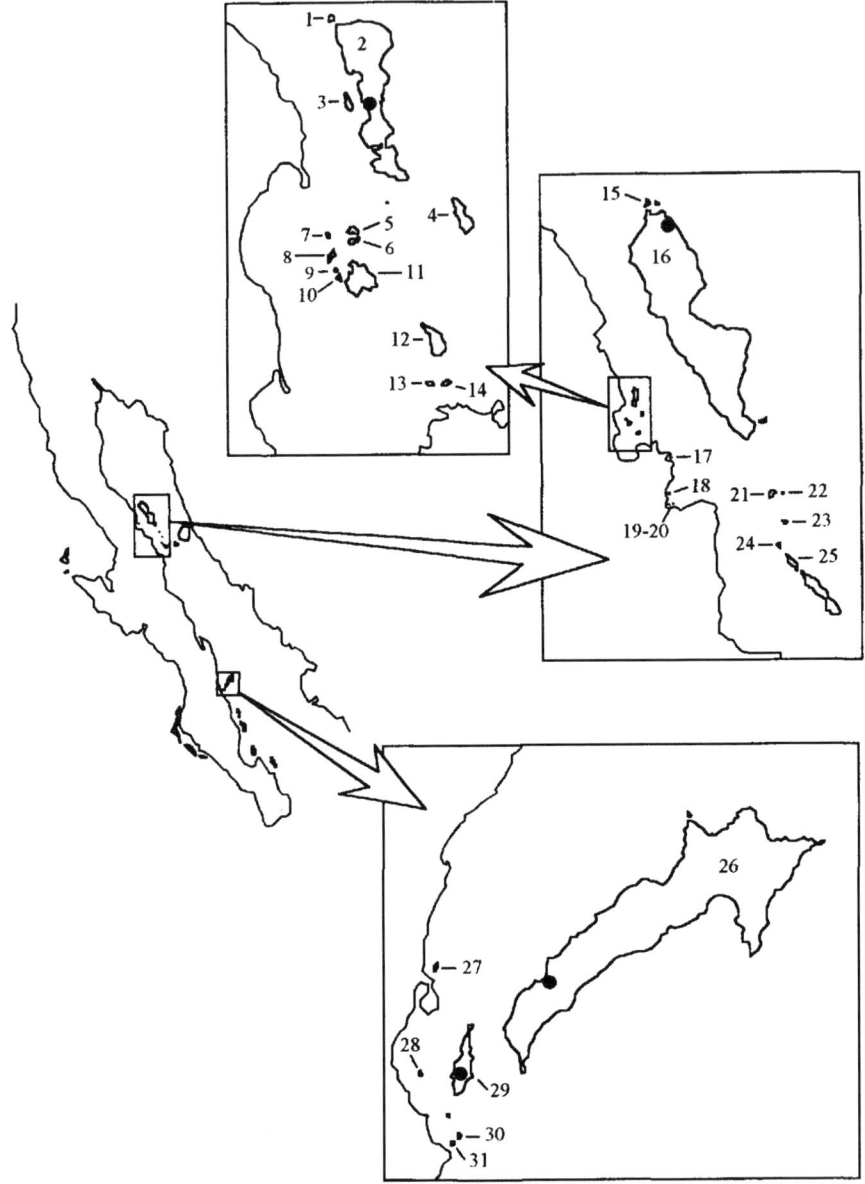

Figure 13.1 Map of Baja California showing the location of the islands in the Midriff region (including the Bahía de los Ángeles area) and Bahía de Loreto considered in this study. Bahía de los Ángeles: (1) Coronadito, (2) Smith, (3) Mitlan, (4) Piojo, (5) Pata, (6) Bota, (7) Jorobado, (8) Flecha, (9) Llave, (10) Cerraja, (11) Ventana, (12) Cabeza de Caballo, (13) Gemelitos West, (14) Gemelitos East; Midriff: (15) Mejia, (16) Ángel de la Guarda, (17) Pescador, (18) Blanca, (19–20) Islotes de Bahía de Las Animas, (21) Partida, (22) islote east of Partida, (23) Rasa, (24) Salsipuedes, (25) San Lorenzo Norte; Bahía de Loreto: (26) Carmen, (27) Mestiza, (28) Blanco, (29) Danzante, (30) Tijeras, (31) Pardo. Solid circles indicate study site locations on the larger islands.

ings include: (1) Islands vary greatly in species composition and diversity. As expected, small islands have fewest species (see chaps. 5, 6). (2) ANPP differs enormously among islands. Usually, most islands are relatively barren, except for perennial shrubs and succulents. However, when water is available, islands show striking differences in ANPP. These differences are related to sea-bird guano, which is rich in nitrogen, fertilizes plants, and stimulates ANPP. (3) Plant and consumer abundance also varies considerably, even among adjacent (<1 km apart) islands that, in principle, should receive the same levels of precipitation. Plant productivity and consumer abundance may vary by one to three orders of magnitude among islands. (4) Abundance of most consumers and potential prey availability are often greatest on small islands, on islands where sea birds roost and nest, and along the coast. They tend to be lowest on large islands, the mainland, away from the coast, and on islands not used by sea birds. We speculated that spatial variability in plant productivity and consumer abundance might be due to differences in marine input among islands, and this hypothesis has driven much of our research (see below).

Second, as in most desert environments, rainfall and its resulting effects on plant growth are highly variable in our system from year to year. Little or no measurable precipitation occurs in most years, but episodic rainfall, often associated with El Niño-Southern Oscillation (ENSO) events, makes temporal variability in rainfall among the highest in the world. Winter rains during 4 recent ENSO years (1992, 1993, 1995, 1998) averaged 109 ± 26 mm. In contrast, dry years average 12 ± 13 mm (1990–91, 1994, 1996–97, 1999). Variation in availability of moisture causes high variability in ANPP and herbivore abundance from year to year. In the wet year 1998, for example, ANPP was 9.3 times greater than in 1999, a dry year.

Overall, we hypothesize that there are two main drivers in this system. First, spatial differences in marine inputs contribute significantly to the great spatial variability in terrestrial ANPP and consumer abundance among and within islands. Second, variable plant productivity associated with alternating dry and ENSO periods is an important determinant of consumer abundance and food web structure on islands.

For this chapter, we first describe three sources of marine input in coastal and insular systems that allows marine resources to influence primary and secondary productivity on gulf islands. Next, we discuss the major factors that regulate marine input onto islands and the significance of marine resource subsidies to both primary and secondary productivity, including the distribution and abundance of both plants and consumer populations. We then show how ENSO-driven changes in precipitation interact with marine inputs to determine population and community dynamics, specifically within the contexts of abundance, stability, and food web interactions. We conclude by speculating how trophic connections between marine and terrestrial systems affect the biogeography of island populations.

Trophic Flow at the Coastal Ecotone

Marine nutrients and carbon enter terrestrial food webs via three major conduits. First, coastal areas receive organic matter via shore drift of carrion and detrital algae (hereafter referred to as "shoreline effects"). Polis and Hurd (1996a) conservatively estimated that about 28 kg/m/year macroalgae entered the shores of islands near Bahía de los Ángeles. Carrion and stranded algae are converted into large populations of specialized intertidal and supralittoral detritivores, scavengers and predators that inhabit sandy beaches and rocky shores. These include isopods, Orchestoidea amphipods, beetles, flies, spiders, and scorpions. Populations can reach astounding densities—for example, *Vaejovius littoralis* Williams, a littoral specialist on many islands, occurs at densities of 8–12 scorpions/m^2, the highest density of any scorpion (Due and Polis 1985).

Second, many animals feed in the sea but use land. On gulf islands, sea birds and pinnipeds (figs. 16.5 and 16.3 show photographs of common species) import organics and nutrients into the terrestrial web through their carcasses, reproductive by-products (placentae, egg shells), food scraps, dead eggs, feathers, and waste products such as guano (Polis and Hurd 1996a). Worldwide, sea birds annually transfer 10^4–10^5 tons of phosphorus in the form of guano to land (Hutchinson 1950). Inputs by marine vertebrates, principally sea birds, increase the abundance of many terrestrial plant and animal species and form the base of a productive food web on coasts and islands (see below).

Third, a surprising amount of marine carbon, nutrients, and inorganic elements is transported inland via windblown sea foam and spray (Noller in press). These marine materials especially affect areas near the shore but can travel great distances as dust and aerosol. The potential impact of sea foam and marine aerosols is not known, but the continual input may substantially alter the soils and primary productivity of systems from the coast to hundreds of kilometers inland. Because we have not quantified the effects of these inputs on our islands, the remainder of our discussion will deal with shoreline drift and sea bird effects.

What Determines the Degree and Amount of Marine Input?

The amount of materials that drift ashore and that are deposited by sea birds is a function of several factors. Shore drift is related to the amount of off-shore productivity, its form, and how much material washes ashore (e.g., currents, winds). Terrestrial factors regulate how much and how far marine productivity penetrates inland: coastal topography (e.g., beach slope, cliffs), location (e.g., windward vs. leeward), consumer mobility (restricted to coast vs. widely roaming), and efficiency of the shore biota (plants, predators, scavengers, detritivores) to convert input to terrestrial tissue.

The ratio of edge to interior (i.e., perimeter-to-area; P/A) is a major determinant of allochthonous input to any habitat (Polis et al. 1997a). P/A relationships are paramount to understanding the effects of marine inputs on terrestrial systems (Polis and Hurd 1995, 1996a). On islands, the amount of drift is a function of shore perimeter; its relative impact is a function of island P/A. Thus, although both small and large islands should receive, on average, the same mass of drift per unit area of shoreline, small islands receive much more marine biomass per total unit area versus large islands and continents.

Sea bird contributions on islands depend on their use by the birds. Only some islands are used for nesting (with an important input from sea bird remains and guano), whereas other islands are used only as roosting places (receiving mainly guano), and a number of islands are negligibly used by sea birds (with minimal inputs). Nesting sea bird presence and abundance on the islands is a combination of predator-free nesting habitat with accessible food supplies over the breeding season, although other variables (e.g., behavioral differences among species and island topography) are also important (Anderson 1983). Sea birds in the Sea of Cortés prefer smaller, isolated islands for nesting, traits related to reduced predation pressure and disturbance. Smaller and isolated islands have fewer predators and lower human disturbance (a main factor affecting nesting sea birds in the gulf) (Anderson 1983; Everett and Anderson 1991; Velarde and Anderson 1994). The traits of islands related to their use as roosting places is less obvious and is not clearly related to local food availability (O'Driscoll 1998; Parrish et al. 1998).

Effects of Marine Inputs on the Terrestrial Ecosystem

For the terrestrial ecosystem of the islands, marine inputs represent resources that contribute to higher primary productivity and population densities than can be supported by *in situ* terrestrial resources alone. Thus, consumers in areas with high marine inputs (small islands, coastal areas off kelp beds, near marine vertebrate colonies) reach higher abundance than similar areas without inputs. Below we review the effects of marine inputs on terrestrial consumers and plant productivity, with special reference to our ongoing studies on desert islands in the Sea of Cortés (Polis and Hurd 1996a; Polis et al. 1997a,b, 1998; Rose and Polis 1998; Anderson and Polis 1999; Sánchez-Piñero and Polis 2000).

Shoreline Effects

Diverse taxa are subsidized by marine input via shore wrack (detrital algae and carrion) (fig. 13.2). Our Gulf of California studies show that supralittoral arthropods are one to two orders of magnitude more abundant than terrestrial arthropods of adjacent inland regions (Due and Polis 1985; Polis and Hurd 1996a). These high numbers

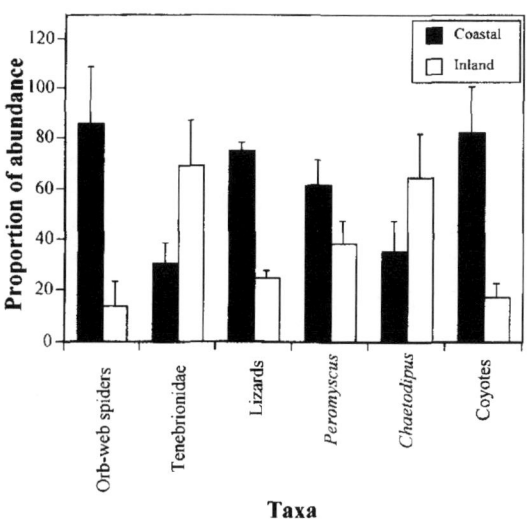

Figure 13.2 Proportion of abundance of different taxa in shore versus inland areas of islands and coastal sites in the Gulf of California. Orb-web spiders, arthropodivorous lizards, omnivorous *Peromyscus* rodents, and coyotes show higher abundances on shore areas, whereas the opposite pattern is shown by detritivorous tenebrionid beetles and granivorous *Chaetodipus* rodents. Sources: Polis and Hurd (1996b); Sánchez-Piñero and Polis (unpublished data); Polis et al. (unpublished manuscript); Stapp and Polis (unpublished data); Rose and Polis (1998).

of intertidal and supralittoral arthropods subsidize populations of entirely terrestrial arthropods in and above the supralittoral. For example, orb-web spider reach densities more than six times higher in supralittoral zones (feeding on kelp flies) than farther inland (Polis and Hurd 1995, 1996a,b; fig. 13.2).

The high densities of coastal invertebrates, combined with abundance of marine carrion and algal wrack, attract a variety of vertebrates that prey on coastal invertebrates or scavenge marine carcasses. Our studies of coastal and insular populations of vertebrates in Baja California demonstrate how a wide range of terrestrial species takes advantage of rich shoreline resources. We recorded more than 20 species of resident and migratory land birds foraging for coastal arthropods (e.g., hummingbirds, sparrows, flycatchers, wrens; Polis and Hurd 1996a). Abundant *Uta* and *Callisaurus* lizards feed heavily on intertidal isopods and/or sea-bird parasites (Wilcox 1980; Grismer 1994). Some lizards establish feeding territories within carcasses of beached pinnipeds and feed on abundant carrion flies and beetles. Densities of arthropodivorous lizard tend to be higher (96.2 ± 16.2 lizards/h) on gulf islands with high shore and sea-bird input than other islands and the mainland (19.1 ± 3.0 and 5.0 ± 1.6 lizards/h, respectively), and 2–37 times higher along the shore versus inland regions of the same island (Polis et al., unpublished manuscript; fig. 13.2).

Omnivorous rodents (*Peromyscus*) also inhabit many islands in the Gulf of California and eat marine-based resources (Lawlor et al., chap. 12, this volume; P. Stapp, unpublished data). Small islands with sea bird colonies support high densities of these mice (see below). On islands without sea birds and on the mainland, *Peromyscus* are typically two to eight times more abundant along the shore than farther inland (fig. 13.2). In contrast, granivorous rodents (*Chaetodipus*), which occupy fewer islands as well as the adjacent mainland, show no differences in abundance between shoreline and inland areas or are two to four times more numerous inland (fig. 13.2). Stable isotope analyses of *Peromyscus* tissues (Stapp et al. 1999) suggest that coastal mouse populations are subsidized by marine-based resources.

Marine foods also support dense coastal populations of mammalian carnivores. On the coast of the Baja California peninsula, coyotes (*Canis latrans*) are subsidized by diverse marine resources (Rose and Polis 1998). Consequently, coyotes were 3–13 times more frequent on the coast than inland. More than 69% of all coastal scats contain marine items (compared to <1% inland), resulting in more even and diverse diet composition when compared to inland regions. Marine birds (in 16.5% of scats), fish (12%), and crustaceans (17.7%) composed a significant portion of the coastal diet, whereas rodents (48%) and rabbits (15.9%) were the major foods inland. Higher foraging activity of predators such as coyotes may result in higher predation risk for terrestrial prey near coastal shores. Preliminary seed-tray foraging experiments suggest this may be the case: granivorous pocket mice (*Chaetodipus*) near shore spent relatively less time foraging in exposed seed trays than those under protective cover and harvested fewer seeds overall than individuals farther inland (P. Stapp, unpublished data).

Sea Bird Effects

"Wherever a large bird colony exists, the birds may be regarded as dominants in the sense that . . . they determine the nature of the community of the area occupied by the colony" (Hutchinson 1950, p. 369). The overwhelming impact of nesting and roosting sea birds on the nutrient budgets of islands is well documented (see Anderson and Polis 1999). Marine inputs via sea bird guano deposited on nesting and roosting islands significantly affect nutrient concentrations, plant biomass, and species assemblies. On naturally nutrient-poor islands in the Gulf of California, more than 50 islands benefit from guano-derived increased nitrogen and phosphorus concentrations (Hutchinson 1950; Anderson 1983). Soil on sea bird islands (those where sea birds nest or roost) averaged seven times more nitrogen and five times more phosphorus than non–sea-bird islands (Anderson and Polis 1999). These nutrients then increase plant biomass and enrich plant nutrient levels. However, two important caveats must be considered. First, nutrients only stimulate primary productivity during wet periods, when such resources become available to plants. For example, in wet years (1992, 1993, and 1995) sea bird islands supported 1–2 times greater plant cover (annuals and perennials) and 10–12 times greater plant biomass (annuals) than non–sea-bird islands (Polis et al. 1997b; Anderson and Polis 1999). Second, plant productivity is not linearly related to guano concentrations, as shown by the decrease of plant cover on islands with high guano cover (see Sánchez-Piñero and Polis 2000).

Tissue nutrient concentrations of plants are positively correlated with guano cover and soil nutrients. Plants store or assimilate nitrogen and phosphorus as a function of soil nutrient availability. Thus, in 1995, annuals had two to six times higher nitrogen and four times higher phosphorus concentrations on seabird versus non–sea-bird islands (Anderson and Polis 1999). These results correlated with the high nitrogen concentrations of soils on sea bird islands covered by guano and the depleted nitrogen content of soils of non–sea-bird islands.

In addition, guano deposition affects plant diversity and community composition on the islands. Assemblages of annual plants on sea bird islands are depauperate (1–2 spp./0.25 m^2 plot) compared to islands without birds and are dominated by nitrophilous species (*Amaranthus, Chenopodium*). Non–sea-bird islands support a higher diversity of plants (6–>10 spp./0.25 m^2 plot), with greater evenness (Anderson and Polis, unpublished data).

Marine vertebrates subsidize diverse trophic groups of arthropods in three ways. First, parasites feed on vertebrate hosts (e.g., ticks, and parasitic flies; Heatwole 1971; Duffy 1983, 1991; Duffy and Campos de Duffy 1986; Boulinier and Danchin 1996). Second, by providing animal tissue (carcasses, food scraps, etc.), sea birds enhance scavenger populations (e.g., Tenebrionidae, Dermestidae; Hutchinson 1950; Heatwole 1971; Heatwole et al. 1981; Polis and Hurd 1996a). Third, guano, through its fertilizing effects on plant productivity and quality, enhances populations of herbivores and detritivores (e.g., oribatid mites, lepidopterans, and scolytid beetles; Onuf et al. 1977; Lindeboom 1984; Mizutani and Wada 1988; Ryan and Watkins 1989). In turn, abundant arthropod communities around marine vertebrate colonies support a diversity of secondary consumers (fig. 13.3).

Our trapping on Gulf of California islands shows that arthropods are more abundant on sea bird islands versus those not used by sea birds. Primary consumers, such as detritivorous tenebrionid beetles, are more abundant on islands where sea birds breed or roost and inside than outside colonies (fig. 13.3). Parasites and commensals (e.g., *Paraleucopsis mexicana* flies and *Ornithodoros* ticks) are also abundant on some islands due to the presence of nesting sea-bird colonies. These bottom-up effects reticu-

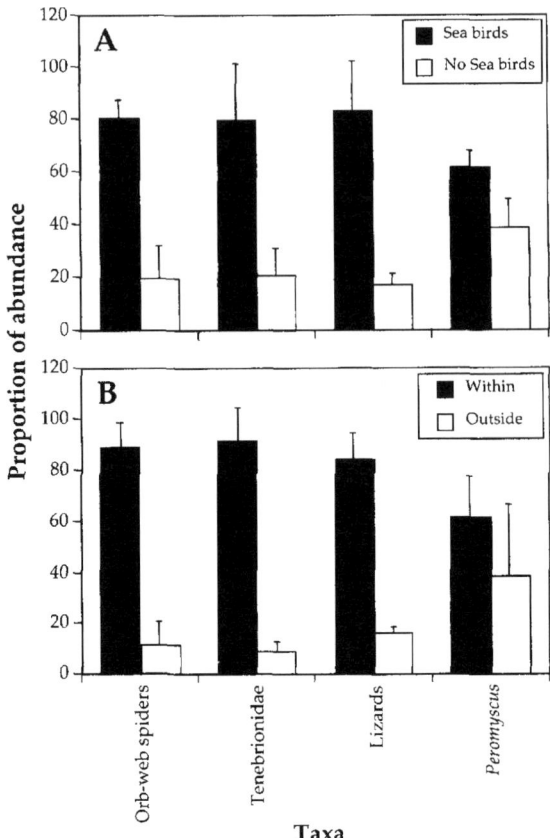

Figure 13.3 Proportion of abundance of different taxa: (A) on sea bird versus non–sea-bird islands and (B) within versus outside sea bird colonies. Sources: Polis and Hurd (1996b); Sánchez-Piñero and Polis (2000); Polis et al. (unpublished manuscript); Stapp and Polis (unpublished data).

late through the food web to produce similar patterns of high abundance in predatory arthropods (e.g., spiders; fig. 13.3). Similarly, arthropodivorous and scavenging vertebrates respond to both sea bird carrion and the increased abundance of arthropods. Arthropodivorous lizards are more abundant on islands with sea birds (Polis et al., unpublished manuscript). On the subset of these islands large enough to support rodent populations, *Peromyscus* densities are 1.5–4 times higher than on islands without seabird colonies and 18–21 times higher than on the adjacent peninsula (fig. 13.3A; P. Stapp, unpublished data). Moreover, unlike non-colony islands and the mainland, *Peromyscus* tend to be uniformly dense on sea bird islands, regardless of distance from shore.

Sea bird effects are evident even at the microhabitat scale. *Peromyscus* are captured frequently near stick-and-debris nests constructed by pelicans. Lizard, tenebrionid beetle, and spider densities are 5–20 times greater within versus outside of bird colonies (Polis and Hurd 1996a; Sánchez-Piñero and Polis 2000; Polis et al., unpublished manuscript; fig. 13.3B).

Our work on detritivorous tenebrionid beetles (Sánchez-Piñero and Polis 2000) illustrates the complex but critical role of sea birds on Gulf of California islands. Tenebrionids are about five times more abundant on islands where sea birds roost and

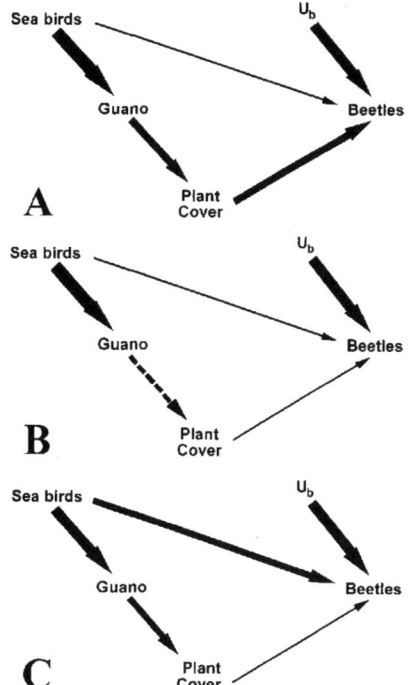

Figure 13.4 Direct and indirect effects of sea birds on tenebrionid beetles in (A, B) roosting and (C) nesting islands of the Gulf of California after wet 1995 and dry 1996. (A) On roosting islands during wet years, bird effects occur indirectly: guano enhances plant productivity, which increases beetle abundance. (B) During dry years, sea birds did not show any significant effect on beetle populations on roosting islands, either directly or indirectly. Beetle abundance remained high on these islands compared to islands without sea birds because plant detritus accumulates from previous pulses of productivity during wet years. (C) On nesting islands, direct effects from seabird carrion most strongly affected beetles; this effect was consistent in both years.

nest than on non–sea-bird islands. Large beetle populations are related to direct and indirect effects of seabirds. First, on islands where birds roost but do not nest, bird carrion is rare, and beetle abundance is correlated with plant cover, an indirect function of seabirds (fig. 13.4A). On these islands, guano increases nutrient concentration in the soil (Anderson and Polis 1999), and, in wet years, these nutrients enhance plant growth and quality and productivity is very high. In dry years, sea bird guano produces no significant effect on either plants or beetles (fig. 13.4B); however, populations are still high, presumably because beetles eat plant detritus from previous wet years (Polis et al. 1997b; Stapp et al. 1999; Sánchez-Piñero and Polis 2000).

Second, on nesting islands, sea birds may primarily affect beetle populations directly through carrion availability and, secondarily, via the guano–plant–detritus pathway. The carrion pathway becomes relevant because the large amount of sea bird tissue provides food on breeding islands and because, on islands with large nesting colonies, sea birds trample vegetation and excess guano may scorch plants, reducing plant growth even in wet years. Thus, beetle abundance likely is directly related to the number of nesting sea birds and carrion availability (fig. 13.4C). The interesting point is that the same factor (here, sea birds) may produce similar consequences (e.g., high beetle populations) but via two distinct pathways (via guano and plants or via carrion).

El Niño Effects on Island Dynamics

Polis and Hurd (1995, 1996a,b) showed that, in dry years, allochthonous productivity arising from the surrounding resource-rich marine system heavily influences secondary productivity and consumer dynamics (fig. 13.5), whereas *in situ* primary productivity was relatively unimportant. However, this situation changes in wet years that accompany ENSO conditions every 3–7 years, during which winter precipitation is 260–790% above normal (Polis et al. 1997b). At these times, plant productivity increases dramatically. For example, ANPP decreased 9.3-fold from wet 1998 to dry 1999. Wet ENSO periods seem to switch the system from one whose energy flow is derived primarily from allochthonous marine-based resources to one driven to a greater extent by *in situ* terrestrial productivity. However, on islands used by sea birds, increased nutrient levels in soils by guano deposition are responsible for enhanced plant productivity and emphasize the importance of marine inputs on islands even during wet El Niño years (fig. 13.5).

Wet ENSO pulses also change the abundance and species composition of local arthropod communities. For instance, the wet year 1992–93 saw two times greater numbers of insects than dry 1991 and 1994; insect densities grew 70% from 1991 to 1992 (Polis et al. 1997b). Herbivorous insect abundance increased 40–190 times in the 1992–93 ENSO compared to 1990 and 1991 levels. Consumers such as spiders

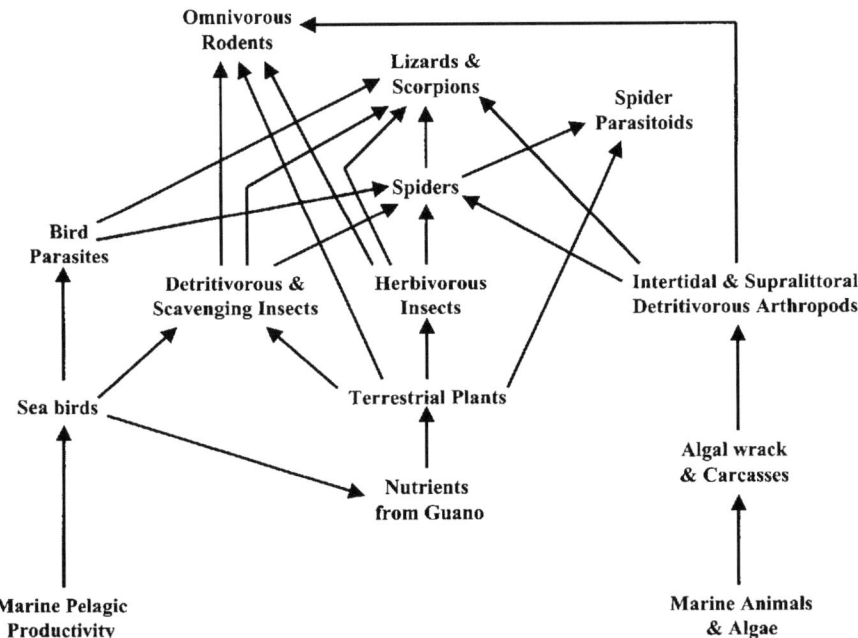

Figure 13.5 Simplified terrestrial food web on Gulf of California islands. Marine input enters via shore drift (algal wrack and carcasses of marine animals) and sea birds at different levels in the terrestrial food web: producers (via nutrient inputs from guano deposition by seabirds), primary consumers (e.g., scavenging insects from sea-bird remains, and detritivorous and herbivorous insects from guano-enhanced plant productivity) and secondary consumers (spiders, arthropodivorous lizards and omnivorous rodents, via increased prey availability from both shore drift and seabird effects). Notice that nutrient inputs from sea bird guano are available for plant growth during wet (El Niño) years.

and rodents benefit from greater prey abundance: spider density increased 99% from 1991 to 1992, *Peromyscus* rodents increased 372% from dry 1997 to wet 1998 (Polis et al. 1998; Lawlor et al., chap. 12, this volume). These dynamics suggest that ENSO events are a large and important component of the dynamics and energetics of these desert island communities.

Although the ENSO effects we describe are dramatic and immediate, these events may have even greater consequences in the long term. ENSO periods supply large reserves of living plant tissue and detritus that are stored and slowly released over many years. These pulse-reserve dynamics are central to desert communities (Noy-Meir 1973, 1974; Polis 1991) because they allow primary producers and consumers to persist during periods of little or no terrestrial productivity. Persistence is sustained in plants via living tissue amassed during wet periods (perennials) and dormant seeds in soil (annuals); in primary consumers through stored fat, standing living plant tissue, dormant seeds and/or detritus; in predators via metabolized fat reserves and detritivores and granivores. Reserve effects are suggested by detritivorous tenebrionid beetles that maintain high abundances on sea bird islands (ca. five times greater than non–sea-bird islands) because of the augmented amount of plant detritus left over from productive wet years (Sánchez-Piñero and Polis 2000). Such pulses of productivity have potential long-term influences on the structure and dynamics of populations, food webs, and insular communities (see Polis 1991; Polis et al. 1996).

Sometimes community interactions are not intuitive. Polis et al. (1998) argued the presence of a hidden trophic interaction that emerges only after wet years. Enhanced plant productivity associated with the 1992–93 ENSO event reportedly led to increased production of nectar and pollen, the principal food of adult pompilid wasps, parasitoids that prey on spiders. After 1992 rains, spider densities doubled in association with an abundance of herbivorous prey. But spider populations crashed on 19 islands (mean of 75% reduction) in 1993, despite continued high plant productivity and, presumably, high prey availability. This decline correlated temporally and spatially with increases in pompilid abundance (Polis et al. 1998). Polis et al. speculated that, in normal dry years, larval food for these parasitoids (i.e., spiders) were abundant, but a shortage of adult foods (pollen and nectar) could limit pompilid populations. During the El Niño in 1992–1993, the pulse of flowering plants may have provided ample adult food to complement the high levels of larval food. This could explain the observed increase of pompilid densities on most islands between 1992 and 1993. The dramatic drop in spider densities (Polis et al. 1998) may reflect sporadic, nonequilibrial impacts that contribute to population and community variations as a direct result of ENSO events.

Marine Inputs: Population and Community Effects

The results described above suggest that, on our island study sites, marine inputs affect the food web of the adjacent terrestrial system on islands and coastal areas (fig. 13.5). Although we recognize that every island or archipelago has a unique ecological and biogeographic history, we propose that our studies of gulf islands provide several important lessons for understanding the ecology of island populations and communities in other systems.

First, marine subsidies may increase the population density of species at different trophic positions throughout the food web (fig. 13.5). These effects are most pronounced near the coast, on small islands, and on islands where sea birds nest or roost in large numbers. Because abundant and protein-rich marine resources may allow high reproductive output, shoreline populations may potentially act as source populations that send dispersers to interior or sink habitats, increasing abundance throughout the island. This phenomenon could explain high densities on islands (e.g., MacArthur et al. 1972; Case et al. 1979), but to date this has not been investigated.

Second, in some cases, subsidized species may decrease the abundance of other species, either via consumption or competition (Polis et al. 1997). Polis and Hurd (1996b) showed that coastal spiders, subsidized to high numbers by shore flies, greatly decreased the abundance and diversity of herbivorous insects on coastal plants.

Third, allochthonous input may affect competitive interactions. Subsidized populations of omnivorous species also eat terrestrial items, reducing the amount of resources available to species or individuals that use only terrestrial foods. Thus, input may favor generalists at the expense of specialists, especially in areas with high marine inputs, such as coastlines and small islands. For example, on our gulf islands, omnivorous rodents (*Peromyscus*) not only eat terrestrial and marine arthropods but also seeds, the main food of granivores (*Chaetodipus*). The availability of marine foods to *Peromyscus* thus may intensify competitive interactions between it and *Chaetodipus*. On one of the few gulf islands to support both species, *Chaetodipus* numbers decreased by 50% during an El Niño-related increase in *Peromyscus* abundance. This result was surprising because heavy rains presumably increased seed abundance, the preferred food of *Chaetodipus*. *Peromyscus* may either monopolize this pulse of seeds more effectively or interfere with *Chaetodipus* foraging. In either case, the ability of *Peromyscus* to use marine resources likely permits it to persist during periods of seed scarcity and likely contributes to its success on islands compared to *Chaetodipus* (chap. 12).

Fourth, marine inputs greatly affect the numerical stability of many island populations (Anderson and Polis in press). There are two possible mechanisms for this. First, nutrient input from guano tends to destabilize populations of fertilized plants. ANPP is more variable (i.e., shows a higher coefficient of variation) on sea bird islands; productivity is high in wet years, but low in dry ones. On non–sea-bird islands, ANPP increases in wet years, but less than on sea bird islands; in dry years, ANPP decreases, but is higher than on sea bird islands because more perennials and biennials occur on non–sea-bird islands. Although we have less reliable data on insect herbivores, they seem to follow the same trends, with populations being less stable where sea birds are present.

The second mechanism for the effect of marine inputs of population stability is that these inputs may stabilize populations of higher level consumers. Allochthonous resources may stabilize recipient populations by serving as food refuges, which act as a floor below which populations do not drop (Polis et al. 1997a). Thus, marine input may allow populations to be relatively stable, despite great fluctuations in *in situ* primary productivity. We have some support for these predictions: populations of spiders and lizards tend to be more stable in areas of marine input (small islands, at the coast) than those in areas less affected by marine input (i.e., large islands and inland areas). Additional studies are needed to test the importance of these mechanisms.

Marine Resources and Island Biogeography

Through its direct and indirect effects on the abundance and stability of populations of individual species, marine input potentially greatly influences island ecology and biogeography, on gulf islands and worldwide. The most obvious mechanism for these effects is that the abundance and stability of many species' populations are altered by marine inputs. In most cases, abundance and stability are enhanced by marine resources. There are two implications. First, changes in abundance and stability change patterns of persistence and thus distribution of species on islands. Large, stable populations are less likely to go extinct from either genetic causes or demographic stochasticity. Second, we speculate that marine input is an important mechanism underlying high insular densities of many consumers on islands compared to mainland areas (Polis and Hurd 1995; Polis et al., unpublished manuscript), although a variety of other, interrelated factors also are certainly involved (e.g., competitive and predatory release, fence effects; MacArthur et al. 1972; Case 1975; Case et al. 1979).

In some cases, marine input allows many species to exist on islands with little or no *in situ* production by land plants. Many consumer species live on barren desert islands and coral cays (Heatwole 1971; Polis and Hurd 1996a; Polis et al. 1997a). For example, there are no plants on Islote de Rasa, a barren rock north of Isla Rasa that harbors sea lions and marine birds. Nevertheless, at least two species of spiders and a species of *Scolopendra* centipede occur on this island. The first successful colonists to Krakatoa were predators (spiders and lizards) that persisted on prey from the supralittoral web before any plants or herbivores were present (Thornton and New 1988). The occurrence of highly subsidized consumers on islands is predicted to alter patterns of classic assembly rules (Holt et al. 1999).

Marine resources may affect species diversity on islands. Although much theory concerns how various factors, including primary productivity, influence diversity (Huston 1994; Rosenzweig 1995), little or no theory addresses how allochthonous input affects diversity. In principle, at least three factors associated with marine input may affect diversity. First, the number of species on an island reflects the probabilities of persistence or extinction, which may be altered by marine inputs. For example, some species of plants and herbivores may not persist on small, productive bird islands because their populations are too unstable. Second, theoretical models suggest that marine input may relax assembly rules for generalist consumer species to colonize islands (Holt et al. 1999); predators, scavengers, and detritivores can live on islands with few or no land plants and herbivores. Third, allochthonous nutrients from sea bird guano appear to depress plant diversity. Gulf of California islands that receive the most sea bird guano are dominated by a few species of highly productive plants (e.g., *Amaranthus*, *Chenopodium*), whereas islands without such fertilizing input are more diverse but less productive. Small islands may support fewer species of plants (and herbivores) than expected because of the physiological constraints imposed by high soil nitrogen concentrations or levels of disturbance. Specialist, nitrophilous plants that can cope with these stressors may persist on these islands and outcompete other plants. (Cody et al. [1983] note that smaller gulf islands are the only ones with fewer plant species than expected from mainland species–area curves. Most small islands in our core study area are active or historical sea bird roosting areas.)

Marine subsidies may influence the probability that harmful species become established on islands. Many studies show that introduced feral populations of rats, cats, sheep, and cattle use marine resources (Power et al. in press). In fact, rats and cats are a problem on several islands on both the Pacific and gulf side of Baja California (D. Croll and B. Tershy, pers. commun. 1999) and have been implicated as contributing to the extinction of native rodents in the gulf (Smith et al. 1993; Alvarez-Castañeda and Cortes-Calva 1996). The extent to which these exotic species use marine resources has not been determined, but the subsidization of these pests by ocean foods would have important conservation implications.

Conclusions

In this chapter we have shown that physical and biotic connections between the sea and land affect the ecology of Gulf of California islands in myriad ways. Marine nutrients from birds fertilize plants and increase primary productivity. Marine prey and detritus support dense populations of diverse consumers on the coast and many islands. Marine vertebrates, particularly sea birds, provide a rich source of food that alters the community of entire islands. Allochthonous marine inputs not only affect recipient species, but also may percolate through the food web to affect the dynamics of interior areas of islands and coastal areas.

We emphasize that the processes revealed by our studies in the Gulf of California may operate in other island systems as well. For example, many coastal and island populations of lizards and tuatara are heavily subsidized by marine foods (Polis et al., unpublished manuscript). Omnivorous small mammals such as mice, rats, and shrews eat the abundant arthropod prey in both rocky intertidal and sandy beach habitats (Koepcke and Koepcke 1952; Osborne and Sheppe 1971; Thomas 1971; Gleeson and van Rensburg 1982; Navarrete and Castilla 1993; Teferi and Herman 1995). Terrestrial songbirds that feed on arthropods also commonly forage along the coast (Sealy 1974; Egger 1979; Wolinski 1980; Burger 1985; Zann et al. 1990; McCann et al. 1993). Larger mammalian carnivores and birds (vultures, condors, corvids, eagles) eat carcasses of fish, sea birds, and marine mammals or prey on intertidal invertebrates (Koepcke and Koepcke 1952; Zabel and Taggart 1989; Brown and McLachlan 1990; Jimenez et al. 1991; see Rose and Polis 1998). Even large herbivores such as red deer (Clutton-Brock and Albon 1989), cattle (Hall and Moore 1986), and sheep (Paterson and Coleman 1982) feed on seaweed in the intertidal zone when terrestrial plants are scarce. These fully terrestrial species then interact, directly and indirectly, with other species in the food web. Thus, such marine subsidies may work their way through the food web to affect diverse aspects of the ecology and biogeography of island and coastal communities.

We conclude by stressing that it is not possible to understand the demography of coastal and island species, nor the structure and dynamics of coastal communities and ecosystems, nor the diversity and biogeography of islands without considering the potentially substantial energetic and nutrient contributions of the ocean. Land and water, although existing separately and readily recognized as distinct biological communities, are very real extensions of each other. This insight must govern research, conservation, and management efforts in regard to islands and coastal ecotones.

References

Alvarez-Castañeda, S.T., and P. Cortes-Calva. 1996. Anthropogenic extinction of the endemic deer mouse, *Peromyscus maniculatus cineritius*, on San Roque Island, Baja California Sur, Mexico. *Southwestern Naturalist* **41**:459–461.

Anderson, D.W. 1983. The seabirds. In: T.J. Case and M.L. Cody (eds), *Island Biogeography in the Sea of Cortéz*. University of California Press, Berkeley; pp. 246–264.

Anderson, W.B., and G.A. Polis. 1998. Marine subsidies of island communities in the Gulf of California: evidence from stable carbon and nitrogen isotopes. *Oikos* **81**:75–80.

Anderson, W.B., and G.A. Polis. 1999. Nutrient fluxes from water to land: seabirds effects on plant quality on Gulf of California islands. *Oecologia* **118**:324–332.

Anderson, W.B., and G.A. Polis. In press. Allochthonous nutrient and food inputs: consequences for ecosystem stability. In: G.A. Polis, M.E. Power, and G.R. Huxel (eds), *Food Webs at the Landscape Scale*. University of Chicago Press, Chicago.

Boulinier, T., and E. Danchin. 1996. Population trends in kittiwake *Rissa tridactila* colonies in relation to tick infestation. *Ibis* **138**:326–334.

Brown, A.C., and A. McLachlan. 1990. *Ecology of Sandy Shores*. Elsevier, Amsterdam.

Burger, A.E. 1985. Terrestrial food webs in the Sub-Antarctic: Island effects. In: W.R. Siegfried, P.R. Condy, and R.M. Laws (eds), *Antarctic Nutrient Cycles and Food Webs*. Springer-Verlag, Berlin.

Case, T.J. 1975. Species numbers, density compensation, and colonizing ability of lizards on islands in the Gulf of California. *Ecology* **56**: 3–18.

Case, T.J., M.E. Gilpin, and J.M. Diamond. 1979. Overexploitation, interference competition, and excess density compensation in insular faunas. *American Naturalist* **113**:843–854.

Clutton-Brock, T.H., and S.D. Albon. 1989. *Red Deer in the Highlands*. BSP Professional Books, Oxford.

Cody, M.L., R. Moran, and H. Thompson. 1983. The plants. In: T.J. Case and M.L. Cody (eds), *Island Biogeography in the Sea of Cortez*. University of California Press, Berkeley, pp. 49–97.

Due, A.D., and G.A. Polis. 1985. The biology of *Vaejovis littoralis*, an intertidal scorpion from Baja California, Mexico. *Journal of Zoology* **207**:563–580.

Duffy, D.C. 1983. The ecology of tick parasitism on densely nesting Peruvian seabirds. *Ecology* **64**:110–119.

Duffy, D.C. 1991. Ants, ticks, and seabirds: dynamic interactions? In: J.E. Loye and M. Zuk (eds), *Bird-Parasite Interactions: Ecology, Evolution and Behaviour*. Oxford University Press, Oxford; pp. 242–257.

Duffy, D.C., and M.J. Campos de Duffy. 1986. Tick parasitism at nesting colonies of blue-footed boobies in Peru and Galapagos. *Condor* **88**:242–244.

Egger, M. 1979. Varied thrushes feeding on talitrid amphipods. *Auk* **96**:805–806.

Everett, W.T., and D.W. Anderson. 1991. Status and conservation of the breeding seabirds on offshore Pacific islands of Baja California and the Gulf of California. In: J.P. Croxall (ed), *Seabird Status and Conservation: A Supplement*. International Council for Bird Preservation Technical Publication no. 11; pp. 115–139. Cambridge, U.K.

Gleeson, J.P., and P.J.J. van Rensburg. 1982. Feeding ecology of the house mouse *Mus musculus* on Marion Island. *South African Journal of Antarctic Research* **12**:34–39.

Grismer, L.L. 1994. Three new species of intertidal side-blotched lizards (genus *Uta*) from the Gulf of California, Mexico. *Herpetologica* **50**:451–474.

Hall, S.J.G., and G.F. Moore. 1986. Feral cattle of Swona, Orkney Islands. *Mammal Review* **16**:89–96.

Heatwole, H. 1971. Marine-dependent terrestrial biotic communities on some cays in the Coral Sea. *Ecology* **52**:363–366.

Heatwole, H., T. Done, and E. Cameron. 1981. *Community Ecology of a Coral Cay*. Junk, The Hague.

Holt, R.D., J. Lawton, G.A. Polis, and Neo D. Martinez. 1999. Trophic rank and the species-area relation. *Ecology* **80**:1495–1504.

Huston, M.A. 1994. *Biological Diversity*. Cambridge University Press, Cambridge.

Hutchinson, G.E. 1950. Survey of existing knowledge of biogeochemistry. III. The biogeochemistry of vertebrate excretion. *Bulletin of the American Museum of Natural History* 96.

Jimenez, J.E., P.A. Marquet, R.G. Medel, and F.M. Jaksic. 1991. Comparative ecology of Darwin's fox (*Pseudalopex fulvipes*) in mainland and island settings of southern Chile. *Revista Chilena de Historia Natural* **63**:177–186.

Koepcke, H.W., and M. Koepcke. 1952. Sobre el proceso de transformacion de la materia organica en las playas arenosas marinas del Peru. *Publicaciones del Museo de Historia Natural "Javier Prado"* **8**:49–25.

Lindeboom, H.J. 1984. The nitrogen pathway in a penguin rookery. *Ecology* **65**:269–277.

MacArthur, R.H., J.M. Diamond, and J.R. Karr. 1972. Density compensation in island faunas. *Ecology* **53**:330–342.

McCann, J.M., S.E. Mabey, L.J. Niles, C. Bartlett, and P. Kerlinger. 1993. A regional study of coastal migratory stopover habitat for neotropical migrant songbirds: Land management implications. *Transactions of North American Wildlife and Natural Resources Conference* **58**:398–407.

Mizutani, H., and E. Wada. 1988. Nitrogen and carbon isotope rations in seabird rookies and their ecological implications. *Ecology* **69**:340–349.

Navarrete, S.A., and J.C. Castilla. 1993. Predation by Norway rats in the intertidal zone of central Chile. *Marine Ecology Progress Series* **92**:187–199.

Noller, J.S. In press. The importance of nutrient transports via wind. In: G.A. Polis, M.E. Power and G.R. Huxel (eds), *Food Webs at the Landscape Scale*. University of Chicago Press, Chicago.

Noy-Meir, I. 1973. Desert ecosystems: environment and producers. *Annual Review of Ecology and Systematics* **4**:25–41.

Noy-Meir, I. 1974. Desert ecosystems: higher trophic levels. *Annual Review of Ecology and Systematics* **5**:195–214.

O'Driscoll, R.L. 1998. Description of spatial pattern in seabird distributions along line transects using neighbour K statistics. *Marine Ecology Progress Series* **165**:81–94.

Onuf, C.P., J.M. Teal, and I. Valiela. 1977. Interactions of nutrients, plant growth and herbivory in a mangrove ecosystem. *Ecology* **58**:514–526.

Osborne, T. O., and W. A. Sheppe. 1971. Food habits of *Peromyscus maniculatus* on a California Beach. *Journal of Mammalogy* **52**:844–845.

Parrish J.K., N. Lemberg, and L. South-Oryshchyn. 1998. Effects of colony location and nekton abundance on the at-sea distribution of four seabird species. *Fisheries Oceanography* **7**: 126–135.

Paterson, I.W., and C.D. Coleman. 1982. Activity patterns of seaweed-eating sheep of North Ronaldsay, Orkney. *Applied Animal Ethology* **8**:137–146.

Polis, G.A. 1991. Desert communities: an overview of patterns and processes. In: G.A. Polis (ed), *The Ecology of Desert Communities*. University of Arizona Press, Tucson; pp. 1–26.

Polis, G.A., W.B. Anderson, and R.D. Holt. 1997a. Towards an integration of landscape and food web ecology: the dynamics of spatially subsidized food webs. *Annual Review of Ecology and Systematics* **29**:289–316.

Polis, G.A., R.D. Holt, B.A. Menge, and K.O. Winemiller. 1996. Time, space, and life history: influences on food webs. In: G.A. Polis and K.O. Winemiller (eds), *Food Webs: Integration of Patterns and Dynamics*. Chapman and Hall, New York; pp. 435–460.

Polis, G.A., and S.D. Hurd. 1995. Extraordinarily high spider densities on islands: flow of energy from the marine to terrestrial food webs and the absence of predation. *Proceedings of the National Academy of Sciences, USA* **92**:4382–4386.

Polis, G.A., and S.D. Hurd. 1996a. Linking marine and terrestrial food webs: allochthonous input from the ocean supports high secondary productivity on small islands and coastal land communities. *American Naturalist* **147**:396–423.

Polis, G.A., and S.D. Hurd. 1996b. Allochthonous input across habitats, subsidized consumers and apparent trophic cascades: examples from the ocean-land interface. In: G.A. Polis and K.O. Winemiller (eds), *Food Webs: Integration of Patterns and Dynamics*. Chapman and Hall, New York; pp. 275–285.

Polis, G.A., S.D Hurd, C.T. Jackson, and F. Sánchez-Piñero. 1997b. El Niño effects on the dynamics and control of a terrestrial island ecosystem in the Gulf of California. *Ecology* **78**:1884–1897.

Polis, G.A., S.D. Hurd, C.T. Jackson, and F. Sánchez-Piñero. 1998. Multifactor population limitation: variable spatial and temporal control of spiders on Gulf of California islands. *Ecology* **79**:490–502.

Power, M.E., G.A. Polis, M. Vanni, and P. Stapp. In press. Subsidy effects on populations in human dominated ecosystems: implications for conservation, sustainable harvest, and control. In: G.R. Huxel and M.E. Power (eds), *Food Webs at the Landscape Scale*. University of Chicago Press, Chicago.

Rose, M.D., and G.A. Polis. 1998. The distribution and abundance of coyotes: The effect of allochthonous food subsidies from the sea. *Ecology* **79**:998–1007.

Rose, M.D., and G.A. Polis. 2000. On the insularity of islands. *Ecography* **23**:693–701.

Rosenzweig, M.L. 1995. Species diversity in space and time. Cambridge University Press, Cambridge.

Ryan, P.G., and B.P. Watkins. 1989. The influence of physical factors and ornithogenic products on plant and arthropod abundance at an inland nunatak group in Antarctica. *Polar Biology* **10**:151–160.

Sánchez-Piñero, F., and G.A. Polis. 2000. Bottom-up dynamics of allochthonous input: direct and indirect effects of seabirds on islands. *Ecology* **81**:3117–3132.

Sealy, S.G. 1974. Ecological segregation of Swainson's and hermit thrushes on Langara Island, British Columbia. *Condor* **76**:350–351.

Smith, F.A., B.T. Bestelmeyer, J. Biardi., and M. Strong. 1993. Anthropogenic extinction of the endemic woodrat, *Neotoma bunkeri* Burt. *Biodiversity Letters* **1**:149–155.

Stapp, P., G.A. Polis, and F. Sánchez-Piñero. 1999. Stable isotopes reveal strong marine and El Niño effects on island food webs. *Nature* **401**:467–469.

Teferi, T., and T.B. Herman. 1995. Epigeal movements by *Sorex cinereus* on Bon Portage Island, Nova Scotia. *Journal of Mammalogy* **76**:137–140.

Thomas, B. 1971. Evolutionary relationships among *Peromyscus* from the Georgia Strait, Gordon, Goletas, and Scott Islands of British Columbia, Canada. Ph.D. thesis, University of British Columbia, Vancouver.

Thornton I.W., and T.R. New. 1988. Krakatau invertebrates: the 1980s fauna in the context of a century of recolonization. *Philosophical Transactions of the Royal Society of London B* **328**:131–165.

Velarde, E., and D.W. Anderson. 1994. Conservation and management of seabird islands in the Gulf of California: setbacks and successes. In: D.N. Nettleship, J. Burguer, and M. Gochfeld (eds), *Seabirds on Islands. Threats, Case Studies and Action Plans*. BirdLife Conservation Series 1, BirdLife International, Cambridge, U.K.; pp. 229–243.

Wilcox, B.A. 1980. Species number, stability, and equilibrium status of reptile faunas on the California Islands. In: D.M. Power (ed), *The California Islands: Proceedings of a Multidisciplinary Symposium*. Santa Barbara Museum of Natural History, Santa Barbara, CA; pp. 551–564.

Wolinski, R.A. 1980. Rough-winged swallow feeding on fly larvae. *Wilson Bulletin* **92**:121–122.

Zabel, C.J., and S.J. Taggart. 1989. Shift in red fox, *Vulpes vulpes*, mating system associated with El Niño in the Bering Sea. *Animal Behavior* **38**:830–838.

Zann, R.A., M.V. Walker, A.S. Adhikerana, G.W. Davison, E.B. Male, and Darjono. 1990. The birds of the Krakatau Islands (Indonesia) 1984–86. *Philosophical Transactions of the Royal Society of London* **328**:29–54.

PART III
THE HUMAN SCENE

14

Human Impact in the Midriff Islands

CONRAD J. BAHRE
LUIS BOURILLÓN

Although many students of island biogeography consider the midriff islands one of the world's last major refuges of pristine desert-island biota, humans have been a part of that ecosystem for possibly 10,000 years or more. Humans have long affected the terrestrial and inshore marine biota, but the most serious injuries they have caused apparently began in the mid- to late nineteenth century with the start of guano mining on Patos, Rasa, and San Pedro Mártir islands. Since then, most of the major human impacts affecting the Midriff are related to rapid population growth in northwestern Mexico and increasing demands for the Midriff's fishery and tourist resources. This chapter offers both historical and ecological perspectives on the human occupancy of the Midriff, so that a cultural–historical foundation will be available for ecological studies in the region as well as for land-use planning and conservation.

The Midriff, located between 28° and 29°45′N and 112° and 114°W, includes the coasts of Lower California and Sonora and 39 islands and islets. Tiburón, with an area of 1223.53 km² and a maximum elevation of 1219 m, and Ángel de la Guarda, with an area of 936.04 km² and a maximum elevation of 1315 m, are among the largest and most mountainous islands of Mexico. The other major islands of the Midriff, in order of decreasing size, are San Esteban (40.72 km²), San Lorenzo (33.03 km²), Smith (Coronado) (9.13 km²), San Lorenzo Norte (Las Ánimas) (4.26 km²), San Pedro Mártir (2.9 km²), Mejía (2.26 km²), Partida Norte (1.36 km²), Dátil (Turner) (1.25 km²), Alcatraz (Tassne or Pelícano) (1.2 km²), Salsipuedes (1.16 km²), Estanque (Pond) (1.03 km²), Rasa (0.68 km²), and Patos (0.45 km²) (Murphy, unpublished data).

The entire region is extremely arid, and Tiburón is the only island that has permanent potable water, found in a few springs or in *tinajas*, although several tinajas on

Ángel de la Guarda may contain water for long periods. The only island permanently inhabited since initial European contact is Tiburón, the historic stronghold of the Seri Indians or Comcáac, once a seminomadic, nonagricultural, seafaring, hunting, fishing, and gathering people.

The flora and fauna reflect both desert and marine environments, and the islands have unique and fragile ecosystems. Many islands are important sea bird rookeries, especially Rasa, the breeding site for more than 90% of the world's Heermann's gulls (*Larus heermanni*) and elegant terns (*Sterna elegans*), and San Pedro Mártir, possibly the world's largest colony of brown boobies (*Sula leucogaster*) and blue-footed boobies (*Sula nebouxii*) (B. Tershy, pers. commun., November 1999). Other important sea bird rookeries are Partida Norte, San Lorenzo, San Lorenzo Norte, and Alcatraz. Small mammals, land birds, and reptiles, discussed in the preceding chapters, make up the vertebrate fauna of the islands; only Tiburón has large nonmarine mammals, such as desert mule deer (*Odocoileus hemionus sheldoni*), coyote (*Canis latrans*), ring-tailed cat (*Bassariscus astutus*), and white-sided jackrabbit (*Lepus alleni*) (Burt 1932; Figueroa and Castrezana 1996) and, since their introduction in 1975, bighorn sheep (*Ovis canadensis mexicana*) (Montoya and Gates 1975; see chap. 12). The gulf also has a rich diversity of fishes, sharks, sea turtles, crustaceans, and mollusks.

To the casual observer, the flora and fauna of the midriff islands seem little affected by humans. That is not true, however, especially for the marine animals and the islands mined for guano. The sea bird rookeries have been subjected to egg collecting and sea bird hunting since the aboriginal period; certain plants and animals have been moved between the islands and the mainland by the Seri and probably the Cochimí Indians; guano mining has been conducted sporadically on Rasa, San Pedro Mártir, and Patos; introduced predators have negatively affected sea bird and reptilian populations on certain islands and may have resulted in the extinction of endemic terrestrial mammals; sea turtles, especially the black sea turtle (*Chelonia mydas agassizi*), and totoaba or totuava (*Totoaba macdonaldi*) have been harvested to the point of commercial extinction; and commercially valuable shellfish and fish, other than totoaba, have been severely depleted.

Spanish and Mexican settlement in the region was impeded by the deficiency of permanent potable water, the lack of precious metals, and the hostility of the Seri. Indeed, the aboriginal population of the Midriff at the time of the first European contact exceeded the area's modern population until the 1930s. Today's settlements of Bahía Kino and Bahía de los Ángeles were *rancherías* and/or landings in the eighteenth and nineteenth centuries (Bahía de los Ángeles was a supply station for the mission of San Borja) and were not established as towns until the 1930s. In fact, Bahía de los Ángeles was little more than a fish camp with an airstrip and lodge in the early 1950s. At present, Bahía Kino (population 4038) is the largest settlement on the mainland adjacent to the Midriff, and Bahía de los Ángeles (population 462) is the largest settlement on the peninsula nearest to the Midriff. Both are growing rapidly, as are the other villages, towns, and cities in or near the region such as Puerto Libertad (population 3009), Punta Chueca (population 411), Desemboque de los Seris (population 309), Santa Rosalía (population 10,451), Guaymas (population 134,625), and Hermosillo (population 559,154) (INEGI 1995a,b). Until the 1950s the Midriff was isolated; now the Sonoran and Baja California coasts are easily accessible by paved highways, and thousands of commercial and sport fishermen and tourists visit the islands annually.

Physical Setting

The geology of the midriff islands has been studied seriously only since the 1970s (Gastil et al. 1975; Gastil and Krummenacher 1977; see chap. 2). The islands consist largely of late Mesozoic granitic plutons veneered by Tertiary and Quaternary extrusive volcanics (basalts and andesites). Repeated volcanism and faulting have left a landscape of high-angle fault planes bounding irregular horst and graben structures. Also present are a number of minor depositional basins of several Cenozoic ages. The topography, especially of Tiburón, San Lorenzo, San Esteban, and Ángel de la Guarda, is rugged. The bedrock uplands are scarred by the drainage ways of ephemeral streams, while the lower depositional slopes consist of coalescing alluvial fans and broad valley fills (Helgren and Bahre 1981).

Weather stations are few and weather records are brief. According to U.S. Geological Survey climatic data, average annual precipitation in the Midriff ranges from 61.6 mm at Bahía de los Ángeles to 120.4 mm at Bahía Kino. Average annual precipitation for the period (1971–1976) at Desemboque de los Seris was 54.7 mm.

Precipitation is biseasonal in pattern and fluctuates greatly from year to year. Infrequently, precipitation from west-coast tropical hurricanes, or *chubascos*, may amount to 150 mm or more in 24 h with no more precipitation for several years. The rainshadow and orographic effects of the mountain ranges on Tiburón and Ángel de la Guarda appear to affect local vegetation; for example, patches of thorn scrub, usually associated with more mesic environments, occur at high elevations in the Sierra Kunkaak on Tiburón (Felger and Moser 1985).

Rainfall is heaviest from August to October (peaking at Baja California stations in September and at Bahía Kino in August), and there is precipitation between late October and early March (peaking at all stations in December). Late summer precipitation usually results from concentrated, brief convectional thunderstorms (largely on the Sonoran side of the Midriff) or from rare chubascos; winter precipitation results from passing cold fronts. Hurricanes, the most significant hydrological events, strike between late May and early November, but are most common in September and early October (Ives 1952; Roden 1964; R. Minnich, pers. commun. April 1999). For example, 215.9 mm of precipitation was recorded at El Barríl in September 1966 in the wake of two hurricanes that struck the Midriff just weeks apart. Since 1975 four major hurricanes—Kathleen (1976), Doreen (1977), Lester (1992), and Nora (1997)—have struck the Midriff, causing considerable damage in local areas. Lester had wind speeds of about 65 knots when it struck Tiburón; it destroyed the lighthouse at Punta Willard on Tiburón, knocked down cardón cactus (*Pachycereus pringlei*) and cleared arroyos of brush on San Esteban (T. Bowen, pers. commun. October 1999), and left 5000 people homeless in a large area just west of Hermosillo.

The Midriff has high summer heat and cool winters. Diurnal temperature ranges are usually large in spite of marine influences. Maximum daytime temperatures from June to September regularly exceed 38°C and sometimes rise above 43°C. The highest and lowest mean monthly temperatures for Bahía de los Ángeles are 30.5°C in July and 15.9°C in January and for Bahía Kino 29.2°C in August and 13.3°C in January. The lowest temperatures, rarely below freezing, occur in January and February (R. Minnich, pers. commun. July 1999); two unusual periods of freezing temperatures occurred along the Canal del Infiernillo in 1972 and 1978, killing large numbers of mangroves east of Punta Sargento.

Floristically, the Midriff is a part of the Central Gulf Coast subdivision of the Sonoran Desert as defined by Shreve (1951). Vegetation and flora are discussed in chapter 4. The dominant vegetation type is desert scrub, with three communities being particularly significant: creosote bush (*Larrea tridentata*) scrub, mixed desert scrub (small-leaved and drought-deciduous shrubs and small desert trees), and cactus scrub (dominated primarily or in part by columnar cacti). The most abundant life forms are medium-sized shrubs, small perennials, and ephemerals.

History of Human Occupance

Prehistory

The Midriff is largely an archaeological *terra incognita*. Evidence of prehistoric habitation has been found on the Midriff coasts of Sonora and Baja California and on the four largest islands: Tiburón, Ángel de la Guarda, San Lorenzo, and San Esteban. Tom Bowen's books, *Seri Prehistory—The Archaeology of the Central Coast of Sonora, Mexico* (1976) and *Unknown Island: Seri Indians, Europeans, and San Esteban Island in the Gulf of California* (2000), María Elisa Villalpando's survey of San Esteban (1989), Eric Ritter's survey of the Bahía de los Ángeles area (1998), and Ritter et al.'s (1994) survey of the coast at Bahía de las Ánimas are the most current and best summaries of the prehistoric archaeology of the Midriff. Other archeological investigations are either site reports or studies of artifact collections. Rock clusters, cairns, summit enclosures, habitation caves, pit ovens, stone circles, rock enclosures, and other features of probable prehistoric origin are abundant on Tiburón and San Esteban and along the Midriff's Sonoran and Lower California coasts (Bowen 1976, 2000; Ritter 1998).

The archaeology of the western midriff islands remains little known, although casual observations suggest that Ángel de la Guarda may have numerous prehistoric sites. Arnold (1957) has described a lithic site of unknown age on Ángel de la Guarda, and Ives (1963) claimed that middens exist in rock shelters along the eastern coast of Ángel de la Guarda. During a trip to Ángel de la Guarda in 1978, Bahre recorded numerous rock cairns at Punta Víboras, near its southern tip, as well as several middens having scattered obsidian flakes at Punta Refugio at its northern end. The age of the cairns and the middens is unknown. San Lorenzo appears to have little more than a few shoreline camps, lithic scatters, and outline figures (Bowen 2000).

The pre-ceramic period is particularly muddled, and only an imperfect chronology has been advanced. Bowen (1976) proposed a Clovis Big Game Hunter occupation for the central coast of Sonora, on the basis of three Clovis points found at different sites along the coast. No other Clovis points are reported in the Midriff, although a fluted point from the Central Desert of Baja California was identified by Aschmann

(1952: 262) "as closely resembling a Clovis-type projectile." Moriarty (1968) and Bendímez Patterson et al. (1993) indicate that marine-oriented peoples appear to have been well established along the coast of Bahía de los Ángeles about 6000 years ago. Evidence of possible pre-ceramic occupation, including outline figures, has been found on San Esteban and Tiburón (Bowen 1976, 2000). If the figures are contemporaneous to those made by the Malpais and San Dieguito I cultures in the deserts of southern California and northwestern Sonora, they could place humans on Tiburón and San Esteban 8000 years ago or conceivably much earlier (Rogers 1966; Bowen 1976; Hayden 1976). Those finds are especially interesting because the islands could only have been reached by watercraft. The earliest radiocarbon date for human bone on Tiburón is 1400 ± 300 years ago (Haynes et al. 1966), but no detailed excavations have been completed. Also, a shell midden at Tecomate appears to be pre-ceramic at its lowest levels (Bowen 1976).

Intensive occupation of the central Sonoran coast and Tiburón followed the development of Tiburón Plain Ware (commonly called "egg-shell pottery") perhaps 500–1000 years ago (Bowen 1976). Bowen believes the makers of Tiburón Plain Ware were ancestors of the modern Seri.

The late archaeological culture of the Midriff coast of Baja California is called Comondú. This culture includes material remains of the historic Cochimí (Massey 1949, 1966), and both share material traits with the makers of Tiburón Plain Ware and the historic Seri.

When the first Europeans landed in the sixteenth century, the Seri inhabited the central coast of Sonora, Tiburón, and possibly San Esteban; the Cochimí lived on the Baja California coast. No Indians were recorded on the remaining midriff islands, where the lack of permanent potable water probably precluded permanent settlement. The tule *balsa*, the indigenous watercraft found throughout Baja California and the Midriff, must have been widely established in the region prehistorically (Heizer and Massey 1953), especially as balsas were discontinuously distributed prehistorically from southern Chile to northern California. The Seri used reeds (*Phragmites australis* and later the non-native *Arundo donax*) for raft construction until the turn of the twentieth century, and traveled by balsa between the Sonoran coast, Tiburón, San Esteban, San Lorenzo, and the Lower California coast (McGee 1898; Bowen 2000). Bowen describes several types of Seri balsas and reviews Seri oral traditions concerning prehistoric trips by balsa to the Midriff islands and the Baja California coast. The Cochimí also used balsas, but the extent of their travels is unrecorded.

It is not known whether the modern Seri are a remnant of widespread earlier hunting and gathering Hokan-speaking peoples, who once occupied much of Sonora from the Colorado River to the Río Mátape and were pushed into the central Sonoran coast by Uto-Aztecan-speaking farmers, or whether they arrived on the Sonoran coast from Baja California across the Midriff. Seri oral traditions and historic records indicate that Seri in the early nineteenth century traveled west across the Midriff to raid missions and Cochimí *rancherías* along the Baja California coast from Bahía de los Ángeles to as far south as Mulegé and Loreto (Sheridan 1996, 1999; Bowen 2000). The modern Seri seemingly know a great deal about the western Midriff and infrequently visit the Lower California coast and some western midriff islands. Nevertheless, there are no historic records or archaeological investigations that indicate the Seri lived on any western midriff islands.

Little is known about Seri–Cochimí contacts across the Gulf of California, but both prehistoric and historic contacts seem highly likely (Bowen 2000). Open-water distances from Punta San Francisquito on the peninsula to coastal Sonora do not exceed 25 km by way of San Lorenzo, San Esteban, and Tiburón. Sheridan (1996) points out that the Seri bear little resemblance, linguistically or culturally, to the other Indians of Sonora, and that their way of life most closely resembles that of the Baja California Indians, particularly the Cochimí.

Early European Contacts

The first Europeans to visit the Midriff were Francisco de Ulloa in 1539 (Hakluyt 1810; Wagner 1929) and Hernando de Alarcón in 1540 (Clavigero 1937). Both sailed through the channel between San Esteban and Tiburón on their way to the head of the gulf. After Alarcón's expedition, no major explorations of the Midriff were recorded for more than 100 years, although some claim that Sebastian Vizcaino may have reached the Midriff in 1596, and, according to Bowen (2000), several unlicensed pearl-hunting expeditions visited the Midriff between 1596 and 1648. Between 1648 and 1700, however, several important European explorers and missionaries traveled to the Midriff, mostly to the Sonoran side: Admiral Pedro Porter y Casanate, Father Eusebio Kino, Captain Blas de Guzmán, and Father Adamo Gilg. For a review of early European exploration of the Midriff, see Bowen (2000).

The possibility of discovering rich pearl-oyster (*Pteria sterna*) beds was a major incentive for early Spanish exploration of the Midriff. Although Sebastian Vizcaino was supposedly given pearls from the Midriff in 1596, there are no reliable records of pearling there until the eighteenth century (Bowen 2000). Around 1720, according to Mosk (1931), rich pearl-oyster beds were worked along the Sonoran coast from Cabo Tepopa to the mouth of the Río Altar, in an area then known as Tepoca. Pearling at Tepoca proved to be difficult and was for the most part abandoned in 1733 because of Spanish–Seri hostilities (Mosk 1931; Sheridan 1999; Bowen 2000). During the eighteenth century, pearl-oyster beds were also recorded along the Canal del Infiernillo (Mosk 1931). Manuel de Ocio, a soldier at San Ignacio in the early 1740s, is reputed to have gathered 400 pounds of pearls in the central part of the gulf from 1743 to 1744 (Bowen 2000). The pearl-bearing oyster (*Pteria sterna*) of the upper gulf, which was an inferior species for pearls compared to the one in the lower gulf, was, according to Mosk (1931), exhausted by the mid-eighteenth century, after which all commercial pearling was restricted to areas south of Santa Rosalía. The harvesting of oyster shells for the mother-of-pearl industry, however, was a major industry in the Midriff in the 1830s and probably wreaked havoc on the oyster beds (Bowen 2000).

Numerous references were made before 1735 to the Seri and their supposed subdivisions (the Salineros, Tepocas, Tiburones, Tastioteños, etc.) living along the Sonoran coast and on Tiburón (Bahre 1980; Sheridan 1982, 1999). Little mention was made of the Cochimí on the Baja California coast. In 1746, Father Fernando Consag described a large *ranchería* of Cochimí collecting shellfish at Bahía de los Ángeles, and in 1765 Indians at that same encampment were described by Father Wenceslaus Linck, founder of the mission of San Borja in 1762. Linck, after hearing reports of fires on Ángel de la Guarda, visited the island but found no inhabitants, animals, or water (Clavigero 1937).

Between 1752 and 1767, the Jesuits established three missions among the Cochimí in the Central Desert, whose territories bordered or included the Midriff: Santa Gertrudis in 1751; San Borja in 1762; and Santa María in 1767. According to Aschmann (1959), the Cochimí population of those missions and their territories totaled about 8000, of whom only a few hundred were found on the Midriff coast. The Cochimí were introduced to mission life with disastrous effects. Epidemic diseases soon nearly annihilated them; of the 3193 Indians at these missions in 1770, only 329 survived in 1794 (a little more than 3% of their numbers when the Europeans first arrived) (Aschmann 1953, 1959). In 1865 the last Cochimí died. After their collapse, the Lower California coast of the Midriff remained virtually a *despoblado* until Bahía de los Ángeles was reestablished in the 1930s.

At the time of initial Spanish contact, the Seri were living along the Sonoran coast from the mouth of the Río Mátape, near present-day Guaymas, north to Cabo Tepoca, and on Tiburón (Bahre 1967; Bowen 1983). The Seri were less tractable than the Cochimí, and Seri–Spanish hostilities soon occurred. The Jesuit mission system bypassed the Seri, avoiding the desert coast and taking an inland route up the fertile Sonoran river valleys. Nevertheless, the Jesuits and their Franciscan successors continued to entertain the idea of introducing the Seri to mission life in central Sonora, and of constructing a mission in Seri territory. Only the Franciscans managed to build a mission (in 1772, near Pozo Carrizal), but the Seri, who killed the resident priest (Arricivita 1792), immediately destroyed it. Estimates of Seri numbers in the seventeenth and eighteenth centuries range from 1000 to 5000 (Sauer 1935; Di Peso and Matson 1965; Sheridan 1979). The Seri must have suffered from the same epidemics that struck the Cochimí, but there are few records of their losses. Sporadic warfare also took a heavy toll on Seri life. After the last large-scale Mexican military expedition against the Seri in 1904 (Davis 1933; Bowen 2000), the Seri numbered fewer than 200 persons.

Major Economic Activities: Nineteenth Century to Present

The most significant non–Indian economic activities on the midriff islands in the nineteenth and early twentieth centuries were guano mining, sea-bird egg collecting, and sea lion hunting. In addition, during this period the Mexican government awarded concessions to private groups to make mezcal liquor from the agaves (*Agave cerulata dentiens*) on San Esteban and to colonize and develop mining, agriculture, fishing, and ranching on Tiburón, Ángel de la Guarda, and San Esteban (Bowen 2000). The mezcal concessions were apparently unsuccessful, and the concessions to colonize Tiburón and the other midriff islands never got off the ground. All of the colonization concessions (e.g., the Gulf of Cortez Land and Fish Company) were the brain child of General Guillermo Andrade, who, with the help of American capitalists, schemed unsuccessfully for nearly 20 years to colonize Tiburón, Ángel de la Guarda, and San Esteban. One speculator made the absurd proclamation that Tiburón had sufficient water and carrying capacity for 25,000 cattle (Bowen 2000).

Guano mining, which began in the Midriff in the 1850s (about 10 years after the use of guano as fertilizer began booming in the United States and Europe), peaked in the 1880s and continued sporadically on Patos until the early 1990s. Although Patos was the first and last island in the Midriff to be mined for guano, the largest amounts were mined from Rasa and San Pedro Mártir islands during the late nineteenth century.

In 1884, the French-controlled Boleo Mining Company began large-scale copper mining at Santa Rosalía. The population boomed, and residents began to collect sea bird eggs and to hunt sea turtles. Before the mines played out, in 1923, Santa Rosalía had more than 10,000 inhabitants and was the largest market in northwestern Mexico for sea turtles and sea bird eggs from the Midriff.

California sea lions (*Zalophus californianus*), the only resident pinnipeds in the Midriff, were hunted commercially in the gulf, largely for oil, from the early nineteenth century until the 1960s (Bowen 2000; Zavala González and Mellink 2000). Today, they are particularly abundant on San Esteban, San Pedro Mártir, and Ángel de la Guarda, but can be found on nearly all of the Midriff islands. Edward Nelson (1921), who visited Lower California in 1905 while employed by the U.S. Biological Survey, noted that sea lions were hunted around San Pedro Mártir, Ángel de la Guarda, and San Lorenzo. During the 1950s and 1960s, the oil from the sea lions killed in the Midriff was sold to buyers in Guaymas and Bahía de los Ángeles (Bowen 2000). Both Antero Diáz (pers. commun., August 1977) and Brig Arnold (pers. commun., March 1977) reported seeing the fires of oil-rendering operations on Ángel de la Guarda in the 1950s. Commercial sea lion hunting was always unorganized and sporadic in the Midriff. Until the late 1960s, when commercial sea lion hunting largely ceased, the areas most frequented by sea lion hunters were San Esteban, San Pedro Mártir, and the north end of Ángel de la Guarda. Sea lions continue to thrive in the Midriff (Le Boeuf et al. 1983; Bourillón et al. 1988; Zavala González 1999), but little is known about their abundance in the nineteenth century. Although sea lions were protected from hunting by foreigners in 1976 and legally protected from all hunting in 1982 (Zavala González 1990), a few are still killed by fishermen for shark bait (Zavala González and Mellink 1997, 2000).

Major Mexican settlement in the Midriff began with the growth of the totoaba and shark-fishing industries in Bahía Kino in the early 1930s (Bahre et al. 2000). The totoaba is endemic to the Gulf of California and reportedly can reach 2 m in length and weigh up to 100 kg (Barrera Guevara 1990). The totoaba is particularly vulnerable to fishermen in the Colorado delta, where it migrates each spring to spawn. In about 1924, Mexicans began to export large amounts of totoaba flesh to the United States (principally by train from Guaymas and by truck from El Golfo de Santa Clara and San Felipe to San Diego and San Pedro, California, and Tucson, Arizona [Chute 1928; Scofield 1930]). Before that, totoaba was caught almost solely for the gas bladder (*buche*) of the female, which was sold in dried form to Chinese and other Asians for use in glutinous soups (Craig 1926; Chute 1930). Commercial totoaba fishing for the buche originally started about 1910 in the area between Guaymas and the mouth of the Río Fuerte. Shortly after World War I, commercial operations moved first to San Felipe and then around the head of the gulf to El Golfo de Santa Clara, Puerto Peñasco, and then south along the Sonoran coast to Puerto Libertad, and finally to Bahía Kino (Chute 1928; Huey 1953).

In 1942 the totoaba catch peaked at 2261 metric tons. In 1975, however, after the catch had fallen to about 58 metric tons, the Mexican government placed an indefinite closed season on totoaba fishing (Flanagan and Hendrickson 1976). In 1979 the species was placed on the U.S. Endangered Species list, and in 1994 it was placed on the *Norma Oficial Mexicana* (NOM-059 ECOL Mexican endangered species list) (Bahre et al. 2000). Nevertheless, totoaba is still caught illegally in gillnets and often sold as a type of corvina known locally as *cabicucho* (*Altractoscion othonopterus*), and juvenile totoaba are taken as incidental catch by shrimp trawlers. For example, during the mid-1980s it was estimated that 6200 adult totoaba were lost annually to poaching and that 120,300 juvenile totoaba died annually as shrimp fishery by-catch (Cisneros-Mata et al. 1995).

In the early 1930s, Mexicans and Seri began fishing commercially for totoaba along the Sonoran coast of the Midriff between Bahía Kino and Cabo Tepopa. In fact, commercial fishing, largely for totoaba, and the establishment of the Kino Bay Club (a sportsmen's club) brought large numbers of Mexicans and Americans to Bahía Kino and initiated permanent Mexican settlement on the Sonoran coast of the Midriff (Browne 1931; Spicer 1962). Even as late as 1970 Mexicans and Seri were still fishing commercially for totoaba there, and Bahre remembers Mexicans catching tons of totoaba in gillnets just north of Kino Viejo at Punta San Ignacio in 1963.

In about 1937, commercial fishing for shark developed. At that time, shark liver (a major source of vitamin A), shark skin, and shark fins were much in demand (Byers 1940; Walford 1945). After the vitamins derived from shark liver were synthesized in 1941, the world markets for shark liver declined. Even so, the Seri continued commercial fishing for shark liver until at least 1947 (Shipp 1947).

Although the interest in shark fishing for shark liver subsided, demand for shark fins (a delicacy in Asian soups) continued. Shark fins obtain high prices; in fact, when the fins of large sharks were the major products derived from shark fishing in the 1950s and early 1960s, the fishermen usually cut the fins off the sharks and discarded the rest of the fish in the water. Shark fishing is still important in the Midriff, stimulated by the demand not only for shark fins, but also for shark cartilage (used in medicines to treat cancer), flesh, and skin in both Mexico and the United States. Groups of Mexican shark fishermen from gulf ports and ports as far south as Tapachula, Chiapas, follow the shark migration from southern Mexico to the northern gulf and are frequently encountered in the Midriff during the shark-fishing season (May–September).

The areas most heavily fished for shark are the deep waters near San Pedro Mártir, Tiburón, San Lorenzo, Ángel de la Guarda, and San Esteban. According to Thomson et al. (1996), sharks are caught in gillnets or on long-lines (see also Brown and Jaffe 1995) by both large- and small-scale commercial fishing operations. The small-scale, independent shark fishermen operate their *pangas* (a locally made fiberglass skiff) out of temporary fishing camps, such as San Francisquito. They usually salt and dry the shark flesh before selling it in Guaymas and other ports, mostly for Mexican markets. The major commercial fishing operators, working mainly out of Guaymas and Puerto Peñasco, usually pack the finned and gutted shark bodies in ice for delivery to U.S. and Mexican markets (see Thomson et al. 1996 for a description of current shark fishing operations in the gulf).

After the collapse of the totoaba fishery and the shark liver industry, both Seri and Mexicans emphasized commercial sea turtle hunting. Of the five different species of sea turtles in the Midriff, the most common and commercially favored was the black sea turtle, an endemic subspecies of the green sea turtle (Alvarado and Figueroa 1992; Nichols and Seminoff 1994). Sea turtles have always been featured in Seri diet and mythology (Smith 1974; Felger and Moser 1985). With the introduction of outboard motors and fiberglass boats (pangas), commercial turtle hunting became a big business in the Midriff, and large numbers of turtles were sold throughout Sonora. Although Mexicans also hunted turtles in the Midriff before the early 1970s, the Seri accounted for most of the take in the Sonoran Midriff. Because of overharvesting and the destruction of nesting areas in southern Mexico, mainly by turtle egg collectors, sea turtles are now listed as endangered in the gulf. Until the late 1970s, when sea turtles became protected, Mexicans used scuba equipment to hunt sea turtles wintering on the ocean floors surrounding some of the major Midriff islands (Felger et al. 1976).

Since 1970, the Seri have derived most of their income from commercial fishing and selling handicrafts. The Seri now live along the Sonoran coast between Punta Chueca and Desemboque de los Seris on an *ejido* (peasant corporate community chartered by the Mexican government) that includes Tiburón. Currently, however, no Seri live on the island; the only inhabitants are a dozen Mexican navy personnel who live at Punta Tormenta.

Mexican fishermen continue their efforts in the Midriff to meet expanding markets for seafood in Mexico and the United States. Since the early 1970s, commercial fishing, especially for shrimp, shark, Pacific sardines (*Sardinops sagax caerulea*), northern anchovies (*Engraulis mordax*), lobster (*Panulirus* sp.), scallops (*Lyropecten subnudosus* and *Agropecten circularis*), giant sea cucumbers (*Parastichopus fuscus*), groupers (*Mycteroperca* spp.), snappers (*Lutjanus* spp.), and cabrillas (*Paralabrax* spp.), has increased dramatically (Hammann and Cisneros-Mata 1989; Thomson et al. 1996). The bulk of the Mexican sardine fleet, once located in Ensenada, moved to Guaymas in the early 1970s, and since 1974 most of Mexico's sardine production has been obtained from the Gulf of California. According to Velarde and Anderson (1994), commercial sardine fishing has caused a drastic decrease in the percentage of Pacific sardines in the diet of the breeding seabirds on Rasa. Shrimp yields per manhour of fishing continue to decline, and the shrimp catch has tumbled to 40 or 50% percent of what it was in 1986. For a (grim) perspective on the growth and effects of commercial fishing in the Gulf of California, see the series "A Dying Sea" by Tom Knudson and José Luis Villegas which appeared in the *Sacramento Bee* (1995).

Human Impacts

Subsistence Hunting and Gathering Activities of the Aborigines

About 700 Seri are now concentrated at two small villages, Punta Chueca and Desemboque de los Seris, on the Sonoran coast north of Bahía Kino. Before the 1950s, the Seri moved with the seasonal plant harvests and game and rarely went inland except to trade, raid, or otherwise exploit European settlements and missions. They have now largely abandoned their traditional hunting and gathering pursuits and buy their foodstuffs from Mexican stores with money earned from commercial fishing and the sale of handicrafts (shell necklaces, baskets, and ironwood and stone carvings). Ironwood carving was their major source of income from the early 1960s to the mid-1990s. They now buy ironwood (*Olneya tesota*) from other parts of Sonora because they and the Mexican woodcarvers in Bahía Kino have exhausted most of the accessible ironwood along the central Sonoran coast (St. Antoine 1994).

Even though the Seri have largely abandoned their traditional subsistence patterns, Seri elders still know intimately their natural environment, especially its plant and animal resources (Felger and Moser 1985). Their extinct neighbors, the Cochimí, most likely had a subsistence pattern similar to that of the historic Seri.

For assessing the impact of these hunting and gathering peoples on the environment, two quantitative estimates of their diets are presented (tables 14.1, 14.2). Neither estimate, however, is based on data systematically collected from Seri or Cochimí. Both diets offer some idea of the quantity and quality of the foodstuffs they exploited. McGee's (1898) diet estimate for the Seri appears to place too much emphasis on sea turtles and other animal foods and too little emphasis on plant foods. Nevertheless, Felger and Moser (1985), in their studies on Seri use of wild plants, concluded that animal foods constituted a larger portion of the historic Seri diet than plant foods. Felger and Moser's data are based on information gathered from modern Seri; Aschmann's (1959) data are necessarily taken from early accounts and missionary records; and McGee (1898), who spent very little time with the Seri, based his estimates mostly on cursory evidence.

According to Felger and Moser (1973, 1974, 1976, 1985), the Seri once used a large number of animal species and 94 species of wild plants for food. Their pharmacopoeia encompassed 17 species of animals and at least 106 species of wild plants. The major wild plant foods of the Seri were columnar cacti fruit, sea grass seeds (*Zostera marina*), agave hearts, and the seeds of various ephemerals (*Chenopodium murale*, a non-native, and *Amaranthus* spp.) and desert trees and shrubs (*Prosopis* spp., *Cercidium* spp., and *Lycium* spp.). The major wild plant foods of the Cochimí were identical where the same plant species occur in both Lower California and Sonora. The use of *Zostera* seeds for food is not recorded in Lower California; however, the plant occurs there, and may have been utilized by prehistoric peoples.

No wild plant species appears to have been entirely exterminated by the collecting activities of the Seri and Cochimí. In fact, some species of *Chenopodium* and *Ama-*

Table 14.1 Quantitative estimate of the diet of the Seri Indians

Type of Food	Percentage of Diet	Pounds Per Person Per Year	Number Per Person
Turtles	25	250	2½
Pelicans	5	50	4
Other waterfowl and eggs	8	80	
Fish	15	150	
Shellfish	10	100	
Large land game	7	70	⅔
Other land game	8	80	
Stolen domestic livestock	6	60	2 to 15
Tunas (edible cactus fruit)	9	90	720
Other vegetable foods	5	50	
Miscellaneous, including second harvest	2	20	
Total	100	1000	

After McGee (1898).

ranthus were spread through human use, and *Agave subsimplex, Opuntia phaeacantha*, and *O. violacea* were intentionally transplanted by the Seri (Felger and Moser 1985; see chap. 15, this volume). Aschmann (1959) noted that in some areas of the Central Desert, agaves were eliminated by Indian collectors, especially near the old missions, but that agaves survived elsewhere because the distance from drinking water to areas of agave concentration were too great. Gentry (1972) observed that *Agave subsimplex* is rare in some parts of coastal Sonora, presumably because of Seri collecting. *Agave cerulata dentiens* (Gentry 1982), which is endemic to San Esteban, is esteemed by the Seri, who made trips to the island to collect it for food along with the piebald (blotched) chuckwalla (*Sauromalus varius*) and the spiny-tailed iguana (*Ctenosaura hemilopha*). Richard Felger (pers. commun., August 1977) believes the Seri had little impact on agave reproduction because they harvested only poling agaves, considered the sweetest and best for eating (see also Bahre and Bradbury 1980). Grismer (1994) postulates that the Seri or their ancestors introduced the spiny-tailed iguana to San Esteban (see chaps. 8, 9, and 15).

Leaves of certain agaves were also cooked and chewed as an emergency source of water by the Cochimí (Aschmann 1959) and Seri (Felger and Moser 1976, 1985). In addition, the pulp of the barrel cactus (*Ferocactus wislizenii*) was an emergency source of potable water for the Seri, and McGee (1898) speculated that its scantiness along the central Sonoran coast was due to Seri collecting. Felger and Moser (1985: 85) note that in the distant past during the long dry season from late spring to early summer "an entire [Seri] camp sometimes depended on barrel cactus juice for days or weeks at a time."

Table 14.2 Typical annual food intake for the Indians of the Central Desert (percent)

Food	Percentage
Vegetable stuffs	57
Agave	28
Cactus fruit	12
Leguminous seeds and fruits	6
Other small seeds	6
Roots	3
Other items	2
Land animals	18
Rodents and reptiles	8
Insects	5
Deer and other large animals	4
Birds	1
Marine animals	25
Shellfish	11
Fish	5
Sea mammals	5
Turtles	2
Sea birds	1
Bird and turtle eggs	1

After Aschmann (1959).

The only comprehensive studies on the ethnozoology of the Seri are by Malkin (1962), the section on animals as food in Felger and Moser (1985), and chapter 15 of this volume. The ethnozoology of the Cochimí is found only in documentary records. The historic Seri and Cochimí ate almost any animal they could catch, including mammals, birds, reptiles, insects, mollusks, and fish. The Cochimí reportedly had taboos against eating mountain lion (*Felis concolor*), badger (*Taxidea taxus*), horned lizard (*Phrynosoma* spp.), and some grubs (Aschmann 1959), and the Seri, according to Felger and Moser (1985), will not kill or eat common ravens (*Corvus corax*), coyotes, or leatherback turtles (*Dermochelys coriacea*). In addition, the Seri avoid the meat of bull's-eye pufferfish (*Spheroides annulatus*) because it is toxic (Felger and Moser 1985). It is noteworthy, however, that the remains of leatherback turtle and bull's-eye pufferfish are found in prehistoric burial sites around Bahía de los Ángeles (Ritter 1998).

Marine animals, especially sea turtles, shellfish, and mollusks, were important in the aboriginal diet. At the time of initial European contact, the Seri fished with spears and poison and did not use fishhooks or nets; the Cochimí used both (Aschmann 1959). Mangrove estuaries, particularly numerous along the shores of the Canal de Infiernillo, were frequented most by aborigines for collecting inshore marine resources: such areas abound in crabs, mollusks, birds, and fish. Mollusks, in particular, contributed to the aboriginal diet, as the extensive distribution of shell middens in the Midriff attests.

The Cochimí and Seri hunted brown pelicans (*Pelecanus occidentalis*) for food and skins (Coolidge and Coolidge 1939; Watkins 1939; Aschmann 1959, 1966; Quinn and Quinn 1965; Felger and Moser 1985). The skins were used to make shelters, kilts, blankets, and war shields. Other sea birds were hunted for their skins and meat, and sea bird eggs were collected for food. The Seri once frequently visited both Alcatraz and Patos to kill pelicans and to collect sea bird eggs. Before the advent of rifles, the Seri and Cochimí probably killed few large land and sea mammals (Sheldon 1979).

Modern Mexican and Seri rifle-hunters apparently killed off a large proportion of the mule deer and bighorn sheep on the central Sonoran coast, and Mexican poachers wiped out the pronghorn antelope (*Antilocapra americana mexicana*) on the Sonoran coast decades ago. In 1963 Tiburón was taken from the Seri and declared a Natural Reserve Zone and National Wildlife Refuge (Diario Oficial de la Federación 1963; Lindsay 1966). In 1975 twenty bighorn sheep were taken from the Sonoran coast just north of Punta Chueca and put on Tiburón (Montoya and Gates 1975). Shortly thereafter, the island was returned to the Seri, although the sheep continued to be protected. The bighorn have done well, and permits are now auctioned to the highest bidder(s) to hunt them. A portion of the money obtained from the permits is used to support wildlife research and conservation on Tiburón (see chap. 16).

The extent to which the hunting and gathering activities of the Seri and Cochimí influenced the ecological balance of the region cannot be stated precisely, and any changes in the distribution and density of the vegetation and fauna as a result of aboriginal hunting and gathering are largely unrecorded. It appears, however, that the Seri and other prehistoric peoples spread plants and moved large iguanid lizards between the islands (Grismer 1994; see chap. 15, this volume), but as Sheridan (1996: 196) aptly points out: "They [the Seri] survived by knowing their environment, not transforming it." The same can be said of the Cochimí.

Guano Mining

Mexican and American companies began to mine guano in the central gulf in about 1850. James Hobbs (1875), who visited Guaymas in that year, noted that large amounts of guano from an island about 40 km from Guaymas (San Pedro Nolasco?) were being shipped to England for use as fertilizer. The first record of guano mining in the Midriff is contained in a description of a ship that caught fire and sank while taking on guano at Patos in 1858 (Bowen 2000). According to Seri oral history, ships sporadically visited Patos for guano between the 1850s and early 1900s (Bowen 2000). See Bowen (2000) for a history of guano mining in the Midriff.

The most extensive, large-scale guano mining operations in the Midriff took place on Rasa and San Pedro Mártir in the late nineteenth century. The guano concession to Rasa, which until 1873 had been held by two individuals from Mulegé who apparently had removed little guano from the island, was transferred to the Gulf of California Phosphate Company, an English firm that had contracted to deliver 5000–10,000 tons of guano annually for a 5-year period beginning in 1874 to a firm in Germany (Bowen 2000). In March 1875, Commander George Dewey (Belden 1880), while surveying the Gulf of California, reported that more than 10,000 tons of guano had been removed from Rasa in the first 2 years of mining. Thomas Streets (1877), on the same expedition with Dewey, noted that an estimated 60,000 tons of guano remained to be removed. In 1878, apparently after completing their contract with the German firm, the Gulf of California Phosphate Company was dissolved (Bowen 2000), and mining continued on Rasa under the auspices of the Mexican Phosphate and Sulphur Company of San Francisco (Goss 1888; Bowen 2000).

In about 1885 the Mexican Phosphate and Sulphur Company began extensive guano mining operations on San Pedro Mártir. According to Nordhoff (1888), the Mexican Phosphate and Sulphur Company shipped about 1000 tons of guano each month to Europe and San Francisco from both Rasa and San Pedro Mártir, using two steamers, several sailing vessels, and 350 men in its operations. In 1888 the company had 135 Yaqui Indians mining guano on San Pedro Mártir (Goss 1888). These Yaqui apparently lived on the island year-round with their families. Slevin (1923) and Maillard (1923) attributed the paucity of sea birds on San Pedro Mártir during a 1921 visit to the island to past guano mining. Tershy et al. (1992:26) claim "that almost every rock [on San Pedro Mártir] has been moved, either to scrape off the guano or in an apparent attempt to facilitate the deposition and accumulation of new guano. The entire microtopography of the island was thus rearranged and the erosion and loss of plant cover that resulted must have been tremendous."

Large-scale guano mining seems to have ended on Rasa and San Pedro Mártir in the 1890s (Bowen 2000), but small-scale guano mining on both islands continued well past the 1900s. For example, Villa Ramírez (1976) mentions guano mining on Rasa in 1911. The amounts of guano obtained after 1900 were nowhere as great as those secured between the mid-1880s and the early 1890s.

In June 1943, the Mexican government, in an attempt to revive the guano industry, divided the area between Acapulco and the Midriff into a number of "guano zones" and established a Mexican guano development corporation, Guanos y Fertilizantes de México (Schaben 1943). Shortly thereafter the government commissioned William Vogt and B.F. Osorio Tafall (1944) to appraise the guano-producing potential of the gulf. Vogt (1946) made a 3-day airplane inspection trip to the region in February 1944. He noted that the gulf lacked good guano-producing birds and that the guano was generally low in nitrogen and full of dirt and rocks because the territoriality of the sea birds caused the guano cover to be thin. He also noted that the industry might be more productive if the islands were reconditioned by filling in gorges, blasting out obstructions, putting down concrete caps, and initiating a protection campaign for the sea birds. From April to July of the same year, Osorio Tafall (1944) visited the islands and completed an excellent study of the sea birds of the gulf. His observations paralleled those of Vogt. Shortly thereafter Patos was cleared of vegetation to facilitate mining, and, according to Gentry (1949), the Peruvian guanay bird was introduced. Mining operations on Patos were short-lived, however, and the guanay bird died off, if indeed it was ever introduced. The flora of Patos, largely because of the destruction of the vegetation by guano miners, is the most depauperate of the major Sonoran islands (Felger 1966; Felger and Lowe 1976). According to local fishermen interviewed by Bourillón, small-scale guano mining last occurred on Patos in the early 1990s.

Guano miners not only cleared large areas of vegetation but left masonry structures, rock piles, and terraces on San Pedro Mártir, Rasa, and Patos. For example, Murray and Poole (1965) estimate the number of rock piles left by guano miners on Rasa to be in the thousands. The activities of the guano miners undoubtedly affected the nesting patterns of the sea birds, especially when they mined the islands year-round. In addition, guano miners were probably responsible for introducing Old World rats (*Rattus rattus* and *Rattus norvegicus*), mice (*Mus musculus*), cats, and other nonnatives on the islands. In the fall and winter of 1994–95, a program was initiated to eliminate *Rattus norvegicus* and *Mus musculus* on Rasa (Tershy 1995) and, more recently, a similar eradication program has been proposed for San Pedro Mártir. Fortunately, Old World rats have not been introduced on Partida Norte, which would endanger the island's large population of ground-nesting petrels (*Oceanodroma melanis* and *O. microsoma*) and fishing bats (*Myotis vivesi*). Tershy believes that the introduced rats on San Pedro Mártir may explain the paucity of black or least storm petrels breeding on the island. He has also documented extensive predation by rats on booby eggs and chicks on the island. Feral cats and dogs seemingly survive only on Tiburón, although feral cats have been seen on Ángel de la Guarda. Goats were apparently never established on the midriff islands. However, the Seri report that a lone goat was shot on San Esteban around 1915 by a Seri who concluded that the goat was left there by Mexicans who had gone to the island to make mezcal (Bowen 2000).

Sea-Bird Egg Collecting

Sea bird eggs have probably been collected for food in the nesting rookeries in the Midriff since the advent of humans there. Particularly prized by Mexicans are the eggs of the gulls (*Larus*) and terns (*Sterna*) because of their supposed nutritive, curative, and aphrodisiac properties (Villa Ramírez 1976). Mexicans have two easy ways of separating out partially hatched eggs: (1) Eggs are placed in a bucket of water. Those that float are partially incubated and hence thrown away; eggs that sink are kept (Walker 1951). (2) Every egg they can find in a small area of the rookery is smashed early in the morning or evening; after 10 or 12 h, the fresh eggs are picked up on the return trip.

The greatest threat to the Midriff sea bird population from egg collecting developed around 1900, when Mexicans began large-scale harvesting of sea bird eggs for markets in Santa Rosalía. Rasa was the rookery visited most frequently for this activity. When Maillard (1923) and Slevin (1923) visited Rasa in 1921, they noted that Mexican egg collectors were completely sweeping the island twice a day for eggs. In the late 1940s eggs from Rasa were sold in Hermosillo, San Felipe, Guaymas, and Bahía Kino (Walker 1965). Walker (1951) mentioned meeting a Mexican egg collector who claimed to have sold more than 27,000 eggs from Rasa in the markets of Santa Rosalía in 1947. Sea-bird egg collecting is still conducted on some major rookeries, and Dan Anderson (pers. commun., May 1999) thinks that the paucity of nesting sea birds on Alcatraz and Patos is because of past extensive egg collecting.

Commercial egg collecting on Rasa eventually caused a serious decline in certain sea bird populations in the gulf, and in 1964 the Mexican government established Rasa as a Natural Reserve Zone and Seabird Refuge (Diario Oficial de la Federación 1964; Velarde and Anderson 1994). In spite of Rasa's reserve status, tern and gull eggs were still collected there in the late 1970s. In the fall of 1978, Bahre learned from a Mexican fisherman that in the previous spring more than 5000 eggs were collected on Rasa by a group of fishermen from Bahía de los Ángeles. According to Tershy et al. (1997), commercial fishermen still collect eggs of the Yellow-Footed Gull (*Larus livens*) and Heermann's Gull on San Pedro Mártir. Probably no two historic human activities have had a greater impact on the Midriff sea bird rookeries than guano mining and large-scale egg collecting.

Temporary Fishing Camps

An impact of increasing significance in the Midriff, especially on the islands, is the proliferation of temporary fishing camps established by small-scale commercial or panga fishermen (Bourillón 1996). Because these panga fishermen are usually too far from home to make it economically feasible to return every day, they often establish temporary fishing camps on the islands and mainland coasts of the Midriff near productive fishing areas. From 1993 to 1994, Bourillón (1996) counted 73 temporary fishing camps in the Midriff, 15 of which were then currently being used on the islands (mostly on Salsipuedes and Tiburón) by about 95 fishermen.

Few of the fishermen in these camps, however, venture more than 100 m inland from the shoreline unless they are looking for fuel wood or poaching wildlife. Nevertheless, these fishermen have had a major impact in some areas and have in several cases negatively affected island ecology by introducing exotics, disturbing sea-bird nesting areas, poaching, dumping trash, harvesting fuel wood, and trampling vegetation. For example, exotic plants such as sandbur (*Cenchrus brownii*) and watermelon (*Citrullus lanatus*) were collected at the camping site of El Perro at the southern end of Tiburón. Although fishermen occasionally bring dogs with them, these dogs, if left behind, could only survive on Tiburón, the only island with permanent fresh water. Besides, until recently the Seri always left dogs on Tiburón when they moved between the island and the mainland. Some biologists claim that there are coyote–dog hybrids on Tiburón (J. Campoy, pers. commun., October 1994; see also Kennelly 1978).

The placement of temporary fishing camps in or near sea-bird nesting areas appears to have caused sea birds to change their nesting patterns or even abandon major nesting areas. For example, Brown Pelicans and Brandt's Cormorants (*Phalacrocorax penicillatus*) have abandoned traditional nesting areas on Salsipuedes because of disturbances associated with fishing camps. Similar changes in nesting caused by fishermen and eco-tourists have also been noted for ospreys (*Pandion haliaethus*) on islands in Bahía de los Ángeles (D. Anderson, pers. commun., June 1997) and for Yellow-Footed Gulls on San Esteban, San Pedro Mártir, and Partida Norte.

Poaching of mule deer and bighorn sheep on Tiburón, largely by fishermen, continues. Before bighorn sheep were introduced, illegal hunting of mule deer on the island was commonplace and, when Bahre visited Tecomate in the spring of 1965, two mule deer had been killed about 100 m from the well by "gun-runners." At that time, non-Seri fishermen and others frequently visited Tiburón to kill deer. At present, the Seri are entrusted by the government to patrol the island to curtail poaching. It is reported, however, that wealthy Mexican and American trophy hunters continue to hire fishermen from Bahía Kino to assist them in poaching trophy bighorn on Tiburón. Sport fishermen from Bahía Kino (Rowell 1995) and Bahía de los Ángeles also frequently visit the waters adjacent to the Midriff, but rarely land on the islands.

Eco-tourism, Scientific Field Research, and Illegal Collecting

Eco-tourism appears to be a promising economic alternative to the depressed fishing economy of the gulf, and eco-tourism in the southern part of the gulf has become a major business (Tershy et al. 1999). For the most part, however, eco-tourism in the Midriff is still in its infancy, although other types of tourism have played a major role in the rapid development of Bahía Kino and Bahía de los Ángeles since 1970. Antero Diáz, who offered cruises of the Midriff out of Bahía de los Ángeles in the late 1950s or early 1960s, was probably the first to start eco-tourism in the Midriff, although most of his cruises were for fishing and diving. Until recently four companies offered cruises to the islands in the Midriff with stops to observe and photograph wildlife at San Pedro Mártir, San Esteban, San Lorenzo Norte, Ángel de la Guarda, Tiburón, and Rasa. Today, one company offers cruises with visits to the islands, and about nine companies organize sea kayaking trips, mainly to the islands in Bahía de los Ángeles (A. Reséndiz, pers. commun., June 1999) and around Tiburón (T. Pfister, pers. commun. March 1999). Except for Rasa, which has resident biologists for short periods, no island has wardens or guides to monitor visits.

Since 1970, numerous colleges and universities have used the midriff islands as field laboratories, and several colleges from the United States now have permanent field stations in Bahía Kino and Bahía de los Ángeles. Unfortunately, some scientific field research on the islands has negatively affected the animals and plants. For example, on San Pedro Mártir in the 1980s geologists cleared an area of columnar cacti and landed a helicopter during the sea-bird nesting season, and in 1991, herpetologists, while collecting reptiles, damaged sea-bird nesting areas (Tershy et al. 1997). On San Esteban in the 1970s an airstrip was cleared for scientists who were collecting chuckwallas (D. Cornejo, pers. commun., January 1998); and on Piojo island in 1997 a group of entomologists put insect traps in Brown Pelican nesting areas (A. Reséndiz, pers. commun., June 1997). Other impacts have been incurred by commercial filming crews who have disturbed nesting areas and by illegal collectors of reptiles (Mellink 1995) and plants (C. Espinoza, pers. commun., June 1997).

Conclusions

The Midriff is faced with a number of existing and potential ecological disturbances resulting from the tremendous human population growth in northwestern Mexico in the last 40 years, along with subsequent demands on the economic and aesthetic resources of the region and the potential for marine pollution. Paved highways make Bahía de los Ángeles less than a day's drive from Los Angeles or Tijuana; Bahía Kino is a little more than an hour from Hermosillo and only 6 h from Tucson; Puerto Libertad is easily accessible by paved road from both Caborca and Hermosillo; and a gravel road runs from Bahía Kino north along the coast to Punta Chueca, Desemboque de los Seris, and Puerto Libertad. Bahía Kino and Bahía de los Ángeles have become major tourist centers, and trips by commercial fishermen, sport fishermen, and tourists to the islands are commonplace. Travel to the Midriff is easier than it has ever been, and the natural beauty of the region promises even greater tourism in the future.

In anticipation of the environmental dangers resulting from tourism, pollution, and unregulated economic exploitation, on 2 August 1978, the Mexican government declared 52 islands in the Gulf of California, including all of the midriff islands, Reserve Zones and Refuges for Migratory Birds and Wildlife (Diario Oficial de la Federación 1978). The management plan for the reserve is near completion (see chap. 16, this volume).

According to Thomson et al. (1996), demand for seafood from the Gulf of California, with the consequent rapid development of industrial fleets with advanced fishing technologies, is leading to the rapid depletion of several important commercial species in the Midriff and will eventually affect the health of the entire ecosystem. In addition, the threat of water pollution and oil spills increases daily, especially now that an oil tanker regularly travels through the Midriff carrying fuel oil to the power plant in Puerto Libertad. The major challenge for Mexico is to conserve this valuable ecosystem while sustaining local development dependent on the natural resources of the Midriff.

Acknowledgments We thank Dan Anderson, Tom Bowen, Tom Sheridan, Bernie Tershy, Donald Thomson, and Grady Webster for reviewing the manuscript; Eric Ritter for providing copies of his latest publications on the Midriff; and Elizabeth Dutzi, Tadeo Pfister, the late Jesús Ramírez, Antonio Reséndiz, William Neil Smith, Marisol Tordesillas, Alfredo Zavala, and the fishermen of Bahía Kino and Bahía de los Ángeles for contributing to our research. A Faculty Research Grant from the University of California, Davis, to C.J.B. and grants from the World Wildlife Fund—Mexico, Special Expeditions, Inc. and Conservation International–Gulf of California Program to L.B. facilitated the study.

References

Alvarado, J., and A. Figueroa. 1992. Recapturas post-anidatorias de hembras de tortuga marina negra (*Chelonia agassizi*) marcadas en Michoacán, México. *Biotropica* **24**:560–566.

Arnold, B.A. 1957. *Late Pleistocene and Recent Changes in Land Forms, Climate, and Archaeology in Central Baja California*. University of California Publications in Geography, vol. 10, no. 4. University of California Press, Berkeley.
Arricivita, J.D. 1792. *Crónica seráfica y apostólica del Colegio de Propaganda Fide de la Santa Cruz de Querétaro en la Nueva España*. Part II. Don Felipe de Zúñiga y Ontivéros, México, D.F.
Aschmann, H.H. 1952. A fluted point from central Baja California. *American Antiquity* **27**: 262–263.
Aschmann, H.H. 1953. Desert genocide. *El Museo*, n.s. **1**:3–15.
Aschmann, H.H. 1959. *The Central Desert of Baja California: Demography and Ecology*. Ibero-Americana. vol. 42. University of California Press, Berkeley
Aschmann, H.H. (trans, ed). 1966. *The Natural and Human History of Baja California*. Dawson's Book Shop, Los Angeles, CA.
Bahre, C.J. 1967. The reduction of Seri Indian range and residence in the state of Sonora, Mexico (1536–present). Master's thesis, University of Arizona, Tucson.
Bahre, C.J. 1980. Historic Seri residence, range, and sociopolitical structure. *Kiva* **45**:197–209.
Bahre, C.J., L. Bourillón, and J. Torre. 2000. The Seri and commercial totoaba fishing (1930–1965). *Journal of the Southwest* **42**:559–575.
Bahre, C.J., and D.E. Bradbury. 1980. Manufacture of mescal in Sonora, Mexico. *Economic Botany* **34**:391–400.
Barrera Guevara, J.C. 1990. The conservation of *Totoaba macdonaldi* (Gilbert), (Pisces: Scianidae), in the Gulf of California, Mexico. *Journal of Fish Biology* **37** (suppl. A):201–202.
Belden, S. (comp). 1880. *The West Coast of Mexico from the Boundary Line between the United States and Mexico to Cape Corrientes, Including the Gulf of California*. U.S. Hydrographic Office, Bureau of Navigation, Publication 56. Government Printing Office, Washington, DC.
Bendímez Patterson, J., M.A. Téllez, and J. Serrano. 1993. Excavaciones arqueológicas en el Poblado de Bahía de los Ángeles. *Estudios Fronterizos* **31–32**:175–216.
Bourillón, L. 1996. Actividad humana en la región de las grandes islas del Golfo de California, Mexico. Master's thesis, Instituto Tecnológico y de Estudios Superiores de Monterrey, Campus Guaymas, Sonora, México.
Bourillón, L., A. Cantú, F. Ecardi, E. Lira, J. Ramírez, E. Velarde, and A. Zavala. 1988. *Islas del Golfo de California*. Dirección General del Gobierno de la Secretaría del Gobernación e Instituto de Biología de la Universidad Nacional Autónoma Nacional de México, México, D.F.
Bowen, T. 1976. *Seri Prehistory: The Archaeology of the Central Coast of Sonora, Mexico*. Anthropology Papers, University of Arizona, no. 27. University of Arizona Press, Tucson.
Bowen, T. 1983. Seri. In: A. Ortiz (ed). *Handbook of North American Indians*, vol. 10. Smithsonian Institution, Washington, DC; pp. 230–249.
Bowen, T. 2000. *Unknown Island: Seri Indians, Europeans, and San Esteban Island in the Gulf of California*. University of New Mexico Press, Albuquerque.
Brown, L., and C. Jaffe. 1995. Tragedy in the gulf: a two-year look at the shark fishing industry in the Sea of Cortez. *Ocean Realm* (April):34–40.
Browne, P.E. ca 1931. *Kino Bay*. Imprenta B. Valencia, Hermosillo, Sonora, México.
Burt, W.H. 1932. Descriptions of heretofore unknown mammals from islands in the Gulf of California, Mexico. *Transactions of the San Diego Society of Natural History* **7**:161–182.
Byers, R.D. 1940. The California shark industry. *California Fish and Game Bulletin* **26**:23–38.
Chute, G.R. 1928. The totuava fishery of the California Gulf. *California Fish and Game Bulletin* **14**:275–281.
Chute, G.R. 1930. Seen Kow, a regal soup-stock. *California Fish and Game Bulletin* **16**:23–35.

Cisneros-Mata, M.A., G. Montemayor-López, and M.J. Román-Rodríguez. 1995. Life history and conservation of *Totoaba macdonaldi*. *Conservation Biology* **9**:806–814.

Clavigero, F.J. 1937. *The History of (Lower) California*. Trans. by S.E. Lake and A.A. Gray. Stanford University Press, Palo Alto, CA.

Coolidge, D., and M.R. Coolidge. 1939. *The Last of the Seris*. E.P. Dutton, New York.

Craig, J.A. 1926. A new fishery in Mexico. *California Fish and Game Bulletin* **12**:166–169.

Davis, E.H. 1933. Juan Tomás, warchief of the Seris. *Touring Topics* **25**:20–21.

Diario Oficial de la Federación. 1963. Decreto del 13 de marzo. Creación de la Zona de Reserva Natural y Refugio para la Fauna Silvestre de Isla Tiburón, Sonora. Gobierno Federal de los Estados Unidos Mexicanos.

Diario Oficial de la Federación. 1964. Decreto del 30 de mayo. Creación de la Zona de Reserva Natural y Refugio de Aves de Isla Rasa, Baja California. Gobierno Federal de los Estados Unidos Mexicanos.

Diario Oficial de la Federación. 1978. Decreto del 2 de agosto. Creación de la Zona de Reserva y Refugio de Aves Migratorias y Fauna Silvestre Islas del Golfo de California, Baja California y Baja California Sur. Gobierno de los Estados Unidos Mexicanos.

Di Peso, C.C., and D.S. Matson (trans). 1965. The Seri Indians in 1692 as described by Adamo Gilg, S.J. *Arizona and the West* **7**:35–56.

Felger, R.S. 1966. Ecology of the Gulf Coast and islands of Sonora, Mexico. Dissertation, University of Arizona, Tucson.

Felger, R.S., K. Cliffton, and P.J. Regal. 1976. Winter dormancy in sea turtles: independent discovery and exploitation in the Gulf of California by two local cultures. *Science* **191**: 283–285.

Felger, R.S., and C.H. Lowe. 1976. The island and coastal vegetation and flora of the northern part of Gulf of California. *Natural History Museum of Los Angeles County, Contributions to Science no. 285*, Los Angeles, CA.

Felger, R.S., and M.B. Moser. 1973. Eelgrass (*Zostera marina* L.) in the Gulf of California: discovery of its nutritional value by the Seri Indians. *Science* **181**:355–356.

Felger, R.S., and M.B. Moser. 1974. Seri Indian pharmacopoeia. *Economic Botany* **28**:414–436.

Felger, R.S., and M.B. Moser. 1976. Seri Indian food plants: desert subsistence without agriculture. *Ecology of Food and Nutrition* **5**:13–27.

Felger, R.S., and M.B. Moser. 1985. *People of the Desert and Sea: Ethnobotany of the Seri Indians*. University of Arizona Press, Tucson.

Figueroa, A., and B. Castrezana. 1996. Recommendations for conducting tours in the Gulf of California islands. Conservation International-Gulf of California Program, Guaymas, Sonora, México.

Flanagan, C.A., and J.R. Hendrickson. 1976. Observations on the commercial fishery and reproductive biology of the totoaba, *Cynoscion macdonaldi*, in the northern Gulf of California. *Fishery Bulletin* **74**:531–554.

Gastil, R.G., and D. Krummenacher. 1977. Reconnaissance geology of coastal Sonora between Puerto Lobos and Bahía Kino. *Geological Society of America Bulletin* **88**:189–198.

Gastil, R.G., R.P. Phillips, and E. C. Allison. 1975. *Reconnaissance Geology of the State of Baja California*. Geological Society of America. Memoir 140. Geological Society of America, Boulder, CO.

Gentry, H.S. 1949. *Lands Plants Collected by the Valero III, Allan Hancock Pacific Expeditions 1937–1941*. Allan Hancock Pacific Expeditions, vol. 13, no. 2. University of Southern California, Los Angeles.

Gentry, H.S. 1972. *The Agave Family in Sonora*. Agriculture Handbook. no. 399. Agricultural Research Service, United States Department of Agriculture. Government Printing Office, Washington, DC.

Gentry, H.S. 1982. *Agaves of Continental North America.* University of Arizona Press, Tucson.
Goss, N.S. 1888. New and rare birds found breeding on the San Pedro Mártir Isle. *Auk* **5**: 240–244.
Grismer, L.L. 1994. Geographic origins for the reptiles on islands in the Gulf of California, Mexico. *Herpetological Natural History* **2**:17–40.
Hakluyt, R. 1810. *Hakluyt's Collection of the Early Voyages, Travels, and Discoveries of the English Nation*, vol. 3. Evans Mackinlay, and Priestley, London.
Hammann, M.G., and M.A. Cisneros-Mata. 1989. Range extension and commercial capture of the northern anchovy, *Engraulis mordax* Girard, in the Gulf of California, Mexico. *California Fish and Game* **75**:49–53.
Hayden, J.D. 1976. Pre- altithermal archaeology in the Sierra Pinacate, Sonora, Mexico. *American Antiquity* **41**:274–289.
Haynes, C.V., Jr., P.E. Damon, and D.C. Grey. 1966. Arizona radiocarbon dates VI. *Radiocarbon* **8**:1–21.
Heizer, R.F., and W.C. Massey. 1953. Aboriginal navigation off the coasts of Upper and Baja California. *Bureau of American Ethnology Bulletin* **39**:285–311.
Helgren, D.M., and C.J. Bahre. 1981. Reconnaissance geomorphology of the central coast of Sonora, Mexico. *Zeitschrift für Geomorphologie* **25**:166–179.
Hobbs, J. 1875 [1872]. *Wild Life in the Far West, Personal Adventures of a Border Mountain Man.* Wiley, Waterman, and Eaton, Hartford, CT.
Huey, L.M. 1953. Fisher folk of the Sea of Cortez. *Pacific Discovery* **6**:8–13.
INEGI. 1995a. Baja California. Resultados definitivos: tabulados básicos. Instituto Nacional de Estadística Geografía e Informática, México, D.F.
INEGI. 1995b. Sonora. Resultados definitivos: tabulados básicos. Instituto Nacional de Estadística Geografía e Informática, México, D.F.
Ives, R.L. 1952. Hurricanes of the west coast of Mexico. *Proceedings of the 7th Pacific Science Congress* **3**:21–31.
Ives, R.L. 1963. The problem of the Sonoran littoral cultures. *Kiva* **28**:28–32.
Kennelly, J. J. 1978. Coyote reproduction. In: M. Bekoff (ed), *Coyotes: Biology, Behavior and Management.* Academic Press, New York; pp. 79–93.
Knudson, T., and J.L. Villegas. 1995. A dying sea. *Sacramento Bee.* December 10–13.
Le Boeuf, B.J., D. Aurioles, R. Condit, C. Fox, R. Grisiner, R. Romero, and F. Sinsel. 1983. Size and distribution of the California sea lion population in Mexico. *Proceedings of the California Academy of Sciences* **43**:77–85.
Lindsay, G.E. 1966. The Gulf Island Expedition of 1966. *Pacific Discovery* **19**:2–11.
Maillard, J. 1923. Expedition of the California Academy of Sciences to the Gulf of California in 1921: the birds. *Proceedings of the California Academy of Sciences*, Ser. 4. **12**:443–456.
Malkin, B. 1962. *Seri Ethnozoology.* Occasional Papers of the Idaho State College Museum, No. 7, Pocatello, ID.
Massey, W.C. 1949. Tribes and languages of Baja California. *Southwestern Journal of Anthropology* **5**:272–307.
Massey, W.C. 1966. Archaeology and ethnohistory of Lower California. In: R. Wauchope (ed), *Handbook of Middle American Indians*, vol. 4. University of Texas Press, Austin; pp. 38–58.
McGee, W.J. 1898. *The Seri Indians.* 17th Annual Report of the Bureau of American Ethnology, Part I. Government Printing Office, Washington, DC.
Mellink, E. 1995. The potential effect of commercialization of reptiles from Mexico's Baja California Peninsula and its associated islands. *Herpetological Natural History* **3**:95–99.
Montoya, B., and G. Gates. 1975. Bighorn capture and transplant in Mexico. *Desert Bighorn Council Transactions* 28–32.

Moriarty, J.R. 1968. Climatologic, ecologic, and temporal inferences from radiocarbon dates on archaeological sites, Baja California, Mexico. *Pacific Coast Archaeological Society Quarterly* **4**:11–38.
Mosk, S.A. 1931. Spanish voyages and pearl fisheries in the Gulf of California: a study in economic history. Dissertation, University of California, Berkeley.
Murray, S., and R. Poole. 1965. *PowerBoating the West Coast of Mexico*. Desert Southwest, Palm Desert, CA.
Nelson, E.W. 1921. *Lower California and Its Natural Resources*. National Academy of Sciences Memoirs, vol. 16. Government Printing Office, Washington, DC.
Nichols, W.J., and J.A. Seminoff. 1994. Turtles in the Sea of Cortez: where are they? *CEDO News* **6**:20–21, 28–29, 33.
Nordhoff, C. 1888. *Peninsular California, Some Account of the Climate, Soil Production, and Present Condition Chiefly of the Northern Half of Lower California*. Harper, New York.
Osorio Tafall, B.F. 1944. La expedición del M. N. Gracioso por aguas del extremo Noroeste Mexicano. *Anales de la Escuela Nacional de Ciencias Biológicias* **3**:331–360.
Quinn, C.R, and E. Quinn (eds). 1965. *Edward H. Davis and the Indians of the Southwest United States and Northwest Mexico*. E. Quinn, Downey, CA.
Ritter, E.W. 1998. Investigations of prehistoric behavioral ecology and culture change within the Bahía de los Angeles region, Baja California. *Pacific Coast Archaeological Society Quarterly* **34**:9–43.
Ritter, E.W., J.W. Foster, R.I. Orlins, L.A. Payen, and P.D. Bouey. 1994. Archaeological insights within a marine cornucopia: Baja California's Bahía de las Animas. *Pacific Coast Archaeological Society Quarterly* **30**:1–24.
Roden, G.I. 1964. Oceanographic aspects of the Gulf of California. In: T. van Ardel and G. Shor (eds), *Marine Geology of the Gulf of California*. American Association of Petroleum Geologists Memoirs, vol. 3. American Association of Petroleum Geologists, Tulsa, OK; pp. 30–58.
Rogers, M.J. 1966. The ancient hunters—who were they? In: R.F. Pourade (ed), *Ancient Hunters of the Far West*. Union-Tribune, San Diego, CA; pp. 23–108.
Rowell, K. 1995. Characteristic activities and catch composition of the sport fishing fleet of Bahía Kino, Sonora, México. Senior Project, Prescott College, Prescott, AZ.
Sauer, C.O. 1935. *Aboriginal Population of Northwestern Mexico*. Ibero-Americana, vol. 10. University of California Press, Berkeley.
Schaben, L.J. 1943. Mexico to develop guano industry. *Agriculture in the Americas* (United States Department of Agriculture, Office of Foreign Agricultural Relations) **3**:214–216.
Scofield, N.B. (ed). 1930. Commercial fishery notes (transporting totuava by trains). *California Fish and Game Bulletin* **16**:186.
Sheldon, C. 1979 [1912–1925]. *The Wilderness of Desert Bighorns and Seri Indians*. Edited by D.E. Brown, P.M. Webb, and N.B. Carmony. Arizona Desert Bighorn Sheep Society, Phoenix, AZ.
Sheridan, T.E. 1979. Cross or arrow? The breakdown in Spanish-Seri relations 1729–1750. *Arizona and the West* **21**:317–334.
Sheridan, T.E. 1982. Seri bands in cross-cultural perspective. *Kiva* **47**:185–213.
Sheridan, T.E. 1996. The Comcáac (Seris): people of the desert and sea. In: T.E. Sheridan and N.J. Parezo (eds), *Paths of Life*. University of Arizona Press, Tucson; pp. 187–211.
Sheridan, T.E. 1999. *Empire of Sand: The Seri Indians and the Struggle for Spanish Sonora, 1645–1803*. University of Arizona Press, Tucson.
Shipp, C. 1947. The stone-age men of 1947. *Coronet* **21**:42–46.
Shreve, F. 1951. *Vegetation of the Sonoran Desert*. Carnegie Institution of Washington Publication 591. Government Printing Office, Washington, DC.

Slevin, J.R. 1923. Expedition of the California Academy of Sciences to the Gulf of California in 1921. *Proceedings of the California Academy of Sciences*, Ser. 4 **12**:55–72.
Smith, W.N. 1974. The Seri Indians and the sea turtles. *Journal of Arizona History* **15**:139–158.
Spicer, E.H. 1962. *Cycles of Conquest: the Impact of Spain, Mexico, and the United States on the Indians of the Southwest, 1533–1960*. University of Arizona Press, Tucson.
St. Antoine, S. 1994. Ironwood and art: lessons in cultural ecology. In: G.P. Nabhan and J.L. Carr (eds), Ironwood: an ecological and cultural keystone of the Sonoran Desert. *Occasional Papers in Conservation Biology*, no. 1. Conservation International, Washington, DC; pp. 69–85.
Streets, T.H. 1877. Contributions to the natural history of the Hawaiian and Fanning Islands and Lower California. *Bulletin of the United States National Museum*, no. 7. Government Printing Office, Washington, DC.
Tershy, B.R. 1995. Island conservation and introduced vertebrates in northwestern Mexico. *IUCN Invasive Species Specialist Group Newsletter* **2**:20–21.
Tershy, B.R., L. Bourillón, L. Metzler, and J. Barnes. 1999. A survey of ecotourism on islands in northwestern Mexico. *Environmental Conservation* **26**:212–217.
Tershy, B.R., D. Breese, A. Angeles-P, M. Cervantes-A., M. Mandujano-H., E. Hernandez-N., and A. Cordoba-A. 1992. Natural history and management of Isla San Pedro Mártir. Conservation International-Mexico, 59-A Col. Miramar, Guaymas, Sonora, México.
Tershy, B.R., D. Breese, and D.A. Croll. 1997. Human perturbations and conservation strategies for San Pedro Mártir Island, Islas del Golfo de California Reserve, Mexico. *Environmental Conservation* **24**:261–270.
Thomson, D.A., S.L. Mesnick, and D.J. Schwindt. 1996. Human impact and biodiversity of islands in the Gulf of California (University of Arizona Marine Ecology Course, 9–30 July 1994). Department of Ecology and Evolutionary Biology, University of Arizona, Tucson.
Velarde, E., and D.W. Anderson. 1994. Conservation and management of seabird islands in the Gulf of California: setbacks and successes. In: D.N. Nettleship, J. Burger, and M. Gochfeld (eds), *Seabirds on Islands*. BirdLife Conservation Series no. 1. Birdlife International, Cambridge; pp. 229–243.
Villa Ramírez, B. 1976. Isla Raza, Baja California, enigma y paradigma. *Supervivencia* **2**: 17–29.
Villapando C., M.E. 1989. Los que viven en las montañas: correlación arqueológico-etnográfica en Isla San Esteban, Sonora, México. *Noroeste de México*. Numero 8. Centro Regional de Sonora, Instituto Nacional de Antropología e Historia, Hermosillo, Sonora, México.
Vogt, W. 1946. Report of an airplane inspection trip made over the islands in the Gulf of California and off the Pacific Coast of Lower California. In: *Report on Activities of the Conversation Section, Division of Agricultural Cooperation, Pan American Union (1943–1946)*. Pan America Union, Washington, DC; pp. 110–116.
Wagner, H.R. 1929. *Spanish Voyages to the Northwest Coast of America in the Sixteenth Century*. California Historical Society, San Francisco.
Walford, L.A. 1945. Observations on the shark fishery in the central part of the Gulf of California. *Fishery Leaflet* no. 121. U.S. Department of Interior, Fish and Wildlife Service, Washington, DC.
Walker, L.W. 1951. The sea birds of Isla Raza. *National Geographic* **94**:239–248.
Walker, L.W. 1965. Baja's island of birds. *Pacific Discovery* **18**:27–31.
Watkins, F.E. 1939. Seri Indian pelican-skin robes. *Masterkey* **13**:210–213.
Zavala González, A. 1990. La población de lobo marino común, *Zalophus californianus* [Lesson 1828], en las islas del Golfo de California, México. Bachelor of Science thesis, Universidad Nacional Autónoma de México, México, D.F.
Zavala González, A. 1999. El lobo marino de California (*Zalophus californianus*) y su relación con la pesca en la región de las Grandes Islas, Golfo de California, México. Dissertation, CICESE, Departamento de Ecología, Ensenada, Baja California, México.

Zavala González, A., and E. Mellink. 1997. Entanglement of California sea lions, *Zalophus californianus californianus*, in fishing gear in the central-northern part of the Gulf of California, Mexico. *Fisheries Bulletin* **95**:180–184.

Zavala González, A., and E. Mellink. 2000. Historical exploitation of the California sea lion, *Zalophus californianus*, in México. *Marine Fisheries Review* **62**:35–40.

15

Cultural Dispersal of Plants and Reptiles

GARY PAUL NABHAN

The equilibrium theory of island biogeography (MacArthur and Wilson 1967) gives little attention to the human forces that have contributed to shape the biota of archipelagos. Most of the studies that have been done to test the theory, however, concentrated mostly on natural forces and less on the ancient influences of sea-faring cultures on island biodiversity. Although many biologists have followed MacArthur and Wilson's lead by charting the natural processes shaping the island biogeography of the midriff islands in the Sea of Cortés (Soulé and Sloan 1966; Case and Cody 1983), the cultural dispersal of native plants and animals across the gulf has hardly been taken into account in these pattern analyses of the region's biota.

Nevertheless, new opportunities have emerged. Analyses made possible by novel genetic tools can now be combined with recent revelations of oral history from Seri Indian seafarers who have frequented the midriff islands and who know of their ancestors' activities on the islands. Archaeologists have found indigenous remains on San Esteban, Ángel de la Guarda, San Lorenzo Norte and Sur, and Tiburón, with dateable occupation sequences on San Esteban for a minimum of 350 years (Bowen 2000). We can now begin to reconcile data from cultural geography, genetics, and biogeography to track cultural dispersal with new precision. A cohesive but curious story has begun to emerge from this unlikely partnership of genetic analyses performed in laboratories and oral history documentation in the field: historic seafarers of this arid region have carried with them flora and fauna that became established on islands other than those accessible by natural routes of dispersal (Grismer 1994; Petren and Case 1996, 1997; Nabhan in press).

This should come as no surprise to scientists who read beyond their own area of interest: similar cultural dispersal dynamics have been documented in Polynesia and

Melanesia (McKeown 1978; Fisher 1997; Austin 1999) and in Central America and the Caribbean (Bennett 1992; Case 1996). Factoring indigenous cultural dispersal into island biogeography has led to very different views of biotic origins and migrations than those offered by a purely biological perspective. Sophisticated genetic analyses of culturally dispersed biota on various islands now open the way for a quantitative ethno-science, through which hypotheses regarding the timing and routes of intentional and accidental cultural dispersal of biota, as well as the probability of establishment of small populations, can be rigorously tested (Austin 1999). I will call this new subdiscipline "island ethno-biogeography."

Most biogeographers can intellectually admit that no island in the world is pristine in the sense that it is truly isolated from human influences. However, with some exceptions (e.g., Steadman 1995), researchers have often assumed that most of these influences are recent, or that they obscure the tests of their theory within any archipelago. Although it is obvious that the midriff islands are not yet overrun by plant and animal introductions, there are at least 12 introduced vascular plant species and 8 vertebrate species that have become established on one or more of the islands over the last few centuries (Mellink, in press; West and Nabhan, in press). It has been abundantly documented that Tiburón was not the only island aboriginally inhabited for some time (Bowen 1976, 2000; Villalpando 1989), that Europeans as well as the Seri had been visiting and foraging on the many midriff islands for centuries (Bahre 1983; Sheridan 1999), and that the Seri had probably moved reptiles and mammals between the islands (Grismer 1994; Lowe and Norris 1955; Soulé and Sloan 1966; Petren and Case 1997; Lowe et al., unpublished ms.). Bahre (1983, p. 291) commented that "to the casual observer, the flora and fauna of the Midriff islands seem little affected by man. Although that is largely true for the terrestrial flora and fauna, it is not for seabirds and marine mammals." The idea that human influences may have largely affected sea birds and marine vertebrates but not the terrestrial flora and fauna has found little support in more recent research, which has shown that that the Seri efficiently hunted mule deer, jackrabbits, chuckwallas, tortoises, iguanas, and game birds for centuries, possibly affecting the interaction ecology of many floral and faunal inhabitants of the islands (Felger and Moser 1985; Bowen 2000).

I will focus most of this discussion on just two culturally dispersed groups of organisms, iguanids and cacti. Other than the Seri, or *Comcáac*, an indigenous population whose numbers may have varied from 180 to 3500 individuals through time, the identity of other prehistoric cultures that visited the islands is limited (Aschmann 1967; Bowen 2000). I therefore limit my discussion to historic Seri practices and oral histories consistent with what is known of their archaeological record (Villalpando 1989; Bowen 2000). In the Conclusion, I return to the issues of what other organisms may have been affected by historic cultural management and use of the islands, and where the Seri themselves may have come from.

Cultural Dispersal of Cacti

To understand the potential magnitude of cultural dispersal of cacti to various midriff islands, I will first review translocation practices affecting mainland as well as island populations of certain cacti and elephant trees (*Bursera* spp.). The most frequent reason offered to me by contemporary Seri for transplanting and translocating these plants is to mark the place where a newborn's placenta is buried, using a live specimen which has the probability of persisting at least as long as a human lifespan. The Seri use the term *hant haxp m-ihiip* "place/time (commemorating) birth" to describe where a child's placenta is buried by his or her parents, then covered with ashes and herbs, and marked with a long-growing succulent plant brought to that spot by godparents or grandparents. While earlier ethnographers did not recognize that some of these living birth-markers were translocations, Felger and Moser (1985, p. 254) did document in detail how cardon (*Pachycereus pringlei*) and saguaro (*Carnegiea gigantea*) could be used to mark the place of placental burials: "The placenta of a newborn was buried at the base of a cardón or saguaro. Five small plants of any species were buried with it. Ashes were put on top of the burial to keep coyotes from locating it. The cactus served to mark the spot. In later years, one might visit the spot of his placenta burial to put green branches on it for good luck."

In addition, my own interviews with Seri elders have confirmed that (1) they often mark the growth tip of these plants with a wooden or metal rod, so that the growth of the cactus can later be compared with the growth of the child; and (2) the cacti and elephant trees to be transplanted are often carried as 0.3–0.6 m tall dry-rooted plantlets unearthed from the *ihizitim* or "ancestral grounds" of the newborn's grandparents, in a metaphorical gesture to connect the child with that place and his or her family legacy within it.

This cultural ritual would at first seem to have little to do with island biogeography, until one begins to relate it to the many disjunct populations of cacti found among the midriff islands and adjacent mainland areas. Because there were only about 180 Seri surviving at the beginning of the twentieth century, and 650 living at the end of the twentieth century, perhaps 1000–2000 Seri births were marked with living succulents (not all of them transplanted) over the last hundred years. But because columnar cacti such as cardons might persist as much as 300 years (Turner et al. 1995; Yetman and Búrquez 1996), there are potentially 3000–5000 living birth-markers still persisting around Seri camps in central gulf coast habitats, including the midriff islands. We still do not know whether these translocations established cactus species on islands where they had not been previously disseminated via natural modes of dispersal. Nevertheless, this cultural dispersal process may help explain the many disjunct populations of cacti found on the Sonoran mainland and the midriff islands. Even if the cultural dispersals only increased the number of cactus populations found on certain islands, they may have altered the availability of cactus flowers and fruits for pollinators and dispersers migrating along stepping-stone corridors, thereby altering the interaction diversity evident in the midriff islands (Thompson 1999).

The distributions of cacti on mainland Sonora and adjacent islands have only recently been elaborated in a general manner, and few species have had detailed maps of their distributions drawn (Turner et al. 1995; Paredes-Aguilar et al. 2000). However, Felger and Moser (1985) noted the peculiar distributions of several columnar cacti, prickly pears, and chollas. The organ pipe, *Stenocereus thurberi*, is common on the mainland and on Tiburón, but extremely rare on San Esteban. The sina, *Stenocereus alamosensis*, seldom reaches into the coastal plains beyond Hermosillo and San Carlos, but the Seri know of disjunct populations near Tastiota on the coast and at Siete Cerros, 50–125 km away from other (natural) populations. The Siete Cerros population appears to be a single clone, intermixed with a single clone of a hybrid pencil cholla, *Opuntia arbuscula* × *thurberi*, with extremely large, tart, edible fruit. The Seri call this cholla *heem icös cmasl* "yellow-spined cholla cactus," and use the same name for the chollas on San Esteban (*Opuntia alcahes* var. *alcahes*; Paredes-Aguilar et al. 2000).

The chollalike cactus, *O. marenae*, is considered relatively rare in Seri territory, occurring at widely scattered localities from Caborca south to Kino. A related species, *O. reflexispina*, is known only at coastal fishing camps near Los Arrieros, El Sahuaral, and Tastiota, where its thick roots "were cooked in ashes and eaten as a cure for diarrhea" (Felger and Moser 1985, p. 271). The disc-shaped prickly pear, *O. engelmanni* var. *engelmanni* "is uncommon and highly localized, e.g., on Alcatraz Island and in sandy places near the base of Punta Sargento." Felger and Moser added that this latter prickly pear was one of the species "planted at Punta Sargento by the people of that region," although elsewhere they repeatedly refer to the Seri as nonagricultural and nonhorticultural.

In speculating on the origins of disjunct cactus populations in the Sierra Libre of central Sonora, Yetman and Búrquez (1996) have added the cardon or sahueso (*Pachycereus pringlei*) and the pithaya agria (*Stenocereus gummosus*) to the list of plants possibly translocated by the Seri. They suggest that the cardon in the Sierra Libre are more than 40 km inland from any other known population, under climatic conditions aberrant for them, but in an area where the historic Seri spent several decades as refugees more than two centuries ago. The population is uniform in its age, which they take as evidence that the entire cohort germinated at the same time, perhaps during a period when the Seri were still actively occupying this isolated mountain range. Yetman and Búrquez (1996, p. 29) did not talk to Seri about their speculations, but concluded that "Whether the plants sprouted from randomly scattered (lost or spilled) seeds, germinated from seeds imbedded in fecal matter, or were deliberately planted by Seris who anticipated an extended stay in the Sierra Libre and wanted some of the conveniences of home, we can only imagine."

Yetman and Búrquez (1986) argue that the reason the Seri seeded or transplanted the cacti was because the direct consumption of the fruits was important to their diet (Felger and Moser 1974). However, it would have taken hundreds of cacti as well as many decades to obtain a sizeable crop of fruit. Instead, I find it more plausible that they translocated a single cactus whenever a newborn needed a birth-marker, and that these living birth-markers gradually accumulated around their camps.

How does this relate directly to island biogeography? At least nine species of cacti and one species of elephant tree were live-transported by the Seri for distances of 50–100 km and translocated at temporary camps on the mainland and the midriff islands; and at least six of these species have peculiar distributions on Tiburón and San Esteban, often being found out of typical habitat in even-aged stands. For instance, two small patches of organ pipe cacti were found around two prehistoric or protohistoric Seri camps on San Esteban in the 1960s and 1970s by archaeologist Tom Bowen, the largest in the northwestern corner of the island known as *Cofteocl lifa*, "Chuckwalla Peninsula." They were reported to Richard Felger, who confirmed that they were the only ones on the entire 4.5 km-square island (Felger and Moser 1985). In December 1999, I relocated one cluster of ancient organpipes persisting in the middle of San Esteban in the area separate from the small population in the northwestern corner of the island described by Bowen (2000). Despite the remarkable thoroughness of his work in general, Moran (1983) did not list organpipes on San Esteban in his checklist of the vascular plants of the midriff islands, although he did include other cacti which the Seri translocated (e.g., cardon and pitahaya agria).

Such culturally influenced anomalies were not at all taken into account by the first set of biogeographic analyses in the midriff islands. The probability that nine species were intentionally dispersed, and the possibility that additional species were accidentally dispersed by prehistoric and protohistoric seafarers, is not a moot point when one realizes that the majority of the midriff islands have floras of fewer than 100 species. Even distributional anomalies should be reevaluated. For example, consider that the boojum or cirio (*Fouquieria columnaris*) population on Ángel de la Guarda occurs only above 700 m, whereas the peninsular populations less than 20 km away are found all the way down to sea level. Could they have been culturally dispersed to the summit of the island by the presumed *cachanilla* relatives of the Seri living in the Bahía de los Ángeles area (Aschmann 1967)? Because the Seri believe boojums to be "a kind of people" (along with chain-fruit chollas and leatherback turtles), they attribute to them supernatural powers which relate to their origins among mythic giants of Baja California. Embedded in such legends may be some recognition of cultural dispersal which could explain the disjunct populations on both Ángel de la Guarda and the Sonoran mainland (the latter at Punta Cirio in the Sierra Bacha).

Cultural Dispersal of Iguanids

In a survey of the herpetofauna of Seri Indian homelands on the Sonoran mainland and the midriff islands, I documented 49 terrestrial reptile species, at least 5 of which now dwell in locations where they do not necessarily appear to be naturally occurring (Nabhan in press). Perhaps the most notable examples of these distribution anomalies are (1) the hybrid swarm of chuckwallas (*Sauromalus* spp.) on Alcatraz in Bahía Kino (Lowe and Norris 1955; Lowe et al., unpublished ms.); (2) the co-occurrence of two large iguanids (the piebald chuckwalla *Sauromalus varius* and the spiny-tailed iguana *Ctenosaura hemilopha*) on San Esteban Island (a photograph of *S. varius* appears in fig. 16.1); and the peculiar distribution of leaf-toed geckoes (*Phyllodactylus* spp.) throughout the midriff islands (Grismer 1994). I will put aside the issue of the island biogeography of geckoes for the moment, but it is worth introducing the issue of cultural dispersal of iguanids (fig. 15.1) with Lowe and Norris's (1955, p. 93) early speculation that both chuckwallas and spiny-tailed iguanas "probably reached these islands [Alcatraz and Cholludo] by transport by man or by birds. The Seri Indians and some Mexicans eat both of these lizards."

Lowe et al. (1995) later documented that the Alcatraz Island populations of chuckwallas involved genetic exchange between three species of *Sauromalus* (*S. varius, S. obesus,* and *S. hispidus*), presuming that all three species were introduced as an emergency food reserve by either Seri or Mexican fishermen. This hypothesis has not only gained favor with Grismer (1994) and others, but Petren and Case (1997) argue that cultural dispersal is the most plausible explanation for *S. hispidus* populations found on small islets up-current from source populations on the Baja California side of the gulf.

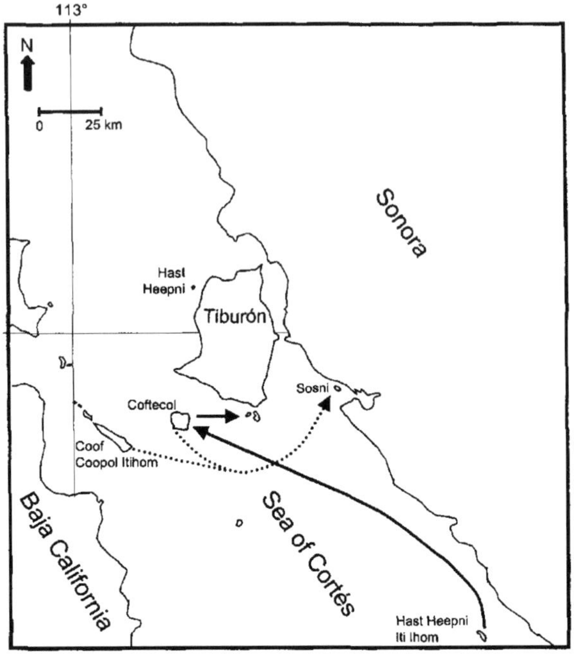

412   The Human Scene

Figure 15.1 Hypothesized cultural dispersal of iguanids, based on Seri oral histories and place-name etymologies recorded by Nabhan (in press). Continuous lines represent the routes for the spiny-tailed iguana (*Ctenosaura hemilopha*) and dotted lines for chuckwallas (*Sauromalus*).

To help resolve this issue, I asked several Seri elders in the winter of 1997–98 and fall of 1998 to explain their oral history of these introductions. I recorded these oral histories while we were working together on a captive breeding project for piebald chuckwallas (*S. varius*) in Punta Chueca, Sonora. These Seri elders claimed that at least some of the chuckwallas on Alcatraz were brought there during the *totoaba* fishing boom in the late 1920s and 1930s. During that time, fish buyers from Hermosillo would buy the Seri catch if it were brought to Bahía Kino, so the Seri established a permanent camp there. But because they would sometimes get stranded on Alcatraz during windy or stormy weather, they needed an emergency food supply during their stayovers. While fishing near San Lorenzo, which they call *Coof Coopol Itihom* "Black Chuckwalla's Home Ground," they went to that island, live-captured black chuckwallas (*S. hispidus*), and brought them to Alcatraz for release.

My Seri consultants did not know whether the same translocation of piebald chuckwallas from San Esteban was done much earlier or around the same time, but they knew it had been done. They did not know for sure whether the mainland common chuckwalla (*S. obesus*) was culturally or naturally dispersed to the island. They understood that different chuckwallas on Alcatraz expressed a mix of traits derived from these three parent populations, but that overall, the body mass of most individuals was much larger than that of the nearby mainland populations of the common chuckwalla. Body masses of the black chuckwalla and piebald chuckwalla are about five times that of the common chuckwalla (Petren and Case 1997), so their translocation of these species to Alcatraz resulted in a much greater meat harvest per animal. The

Arizona-Sonora Desert Museum is continuing to monitor this hybrid swarm, one that might rightly be called a culturally selected breed (or *raza criolla*) of chuckwallas, one for which the Seri should retain cultural use (or intellectual property) rights.

The Seri also claim that for similar reasons, their ancestors intentionally translocated spiny-tailed iguanas from San Esteban to Cholludo on the southeast side of Tiburón. Although no one I have spoken with retains oral history of the translocation of spiny-tailed iguanas from San Pedro Nolasco to San Esteban, they find this hypothesis entirely plausible: their name for San Pedro Nolasco is *Hast Heepni Itihom* "Rocky (Island) Home Ground of Spiny-tailed Iguanas." Grismer (1994) has independently arrived at this hypothesis and suggests that genetic evidence is consistent with it in the sense that the San Esteban population is more similar to San Pedro Nolasco's than to other populations sampled to date.

When one factors in Petren and Case's (1997) conclusion that there were also historic translocations of black chuckwallas to several small islands closer to the Baja California peninsula, it opens up the possibility that seafarers in addition to the Seri were involved in such a cultural dispersal process. The contemporary Seri claim that translocated chuckwallas could not survive on all islets where they could be potentially dispersed, and they also dismiss the possibility that desert tortoises could have survived on other islands if they had been dispersed there. This suggests a much more rational process than the "casual cultural dispersal" of live-captured chuckwallas which Seri fishermen would toss to the women and children as they returned to camp from San Esteban (Bahre 1983). Because most piebald chuckwallas brought from this island to Tiburón or the mainland for eating had already had their legs broken so they could not escape during transport, they would have provided poor founder populations. In short, it appears that the Seri had articulated a rationale for their translocation efforts and that the satellite populations they established could have been derived from multiple translocation events rather than from accidental escapes.

Conclusions

In this view of human influences on the island biogeography of the Sea of Cortés, the indigenous seafaring hunter-gatherers have not been passive recipients of local wild resources, but active agents of dispersal for certain culturally significant resources within their regional domain. However, there also exists the possibility that "camp-followers," such as *Amaranthus watsonii* and *Chenopodium murale* were unintentionally dispersed to several islands, where they were then used as wild greens by the Seri (Felger and Moser 1985). In addition, Seri or other aboriginal seafarers may have transported nocturnal stowaways such as *Phyllodactylus* geckoes, just as other seafarers have dispersed gekkonid species elsewhere in the world (Dye and Steadman 1990; Case et al. 1993; Fisher 1997). These are but three genera of organisms that could insinuate their way into fishermen's camping gear and boats.

Such stowaway organisms as these may be potential indicators of cultural dispersal routes followed by the Seri and/or their presumed Hokan-speaking relatives in Baja California. The mtDNA sequences of anthropophilic species can be sampled and their distribution patterns can be analyzed in much the same way as Austin (1999) did to test the "express train to Polynesia" hypothesis of cultural migration using descendants of *Lipinia noctua* stowaways as his indicators. In addition, the interpretation of distributional patterns of allozymes undertaken by Fischer (1997) for the anthropophilic gecko *Gehyra mutilata* could serve as a model for the degree to which leaf-toed geckoes' distributions were modified by human dispersal. Through randomly amplified polymorphic DNA analysis of wild chile genomic DNA, we have demonstrated that even after an anthropophilic species undergoes genetic bottlenecks due to long-distance dispersal by humans or other vertebrates, new genetic variation can arise fairly rapidly in disjunct populations where unique selection pressures work on novel mutations (Votava et al. in press).

Using mtDNA to determine monophyletic and polyphyletic populations of culturally dispersed species on the islands, it might be possible to refute or confirm whether the Seri migrated by island hopping from Baja California, as many anthropologists suspect, or alternatively, whether they migrated by land from much farther away, as recent genetic analyses suggest. One recent molecular analysis of HLA alleles suggests that the Seri are more closely related to the Warao Indians in Venezuela and other indigenous groups in Argentina than extant tribes in Mexico (Infante et al. 1999). The Seri have long been recognized to be a relictual language isolate group in a sea of Uto-Aztecan speaking tribes with different cultural origins. Because of the early extinction of many of the peoples of Baja California that may have shared their peculiar origins, direct linguistic comparison of the Seri with nearby groups has not been possible as a means to understand their phylogeny and history.

Finally, biogeographers should not rule out the possibility of finding now-extinct land bird or reptile remains in caves and in archaeological sites on the midriff islands, as is the case in the Polynesian islands (Steadman 1995). It is plausible that faunal extinctions occurred on these islands before 1900, but sufficiently intense search efforts have not been made to rule out such extirpations (J. Mead and P. Martin, pers. commun.). A more intensified search for zooarchaeological materials on the midriff islands could help determine if such extirpations occurred, whether humans were a factor in them, and if so, when. Given the extraordinary attention to human-induced extinctions in the Pacific islands (Olson and James 1984; Steadman 1995) and elsewhere around North America (Steadman and Martin 1984), it is remarkable that no one has addressed this issue in the Sea of Cortés. In Bennett's (1992) view, the biogeography of certain regions in the Americas

> can be properly understood only if the anthropogenic elements are included in our investigations. One of the tasks requiring early attention is a typology of ecosystems that jettisons the long-prevailing myth of ecosystem pristinenness. In other words, an effort should be made to integrate all information obtainable with reference to current and past human activities in the area along with the [biotic and abiotic] data that are usually collected, [so that the former are] investigated with the same objectivity. (p. 2)

The possible roles humans have played in the introduction and extirpation of native biota in the midriff islands needs to be better integrated into a quantitative theory for predicting species richness and composition of islands. A better understanding of the role of cultural dispersal in the introduction and extirpation of the native biota will allow island biogeography to continue to advance in the Sea of Cortés.

Acknowledgments I am grateful to E. Ezcurra for inviting me to prepare this chapter, for badgering me to complete it, and for giving me the faith that human dimensions might ultimately be integrated into quantitative island biogeography theory. Patty West assisted me in innumerable ways in the preparation of this chapter. I am also grateful to T. Bowen, M.B. Moser, E. Mellink, L. Grismer, C. Ivanyi, H. Lawler, D. Yetman, A. Búrquez, J. Hills, J. Mead, L. Monti, and C. Bahre for scholarly discussions and to A. López Blanco, A. Robles, E. Molina, J.J. Moreno, J.R. Torres, A. Burgos, and A. Astorga for sharing their traditional ecological knowledge with me. This work was greatly facilitated by support from the Kelton Foundation, with Richard Kelton in particular providing access to the islands, from Agnese Haury and the Packard Foundation to S. Lanham of Environmental Flying Services, and to the Arizona-Sonora Desert Museum.

References

Aschmann, H. 1967. *The Central Desert of Baja California: Demography and Ecology*. Manessier Publishing Co., Riverside, CA.

Austin, C.C. 1999. Lizards took express train to Polynesia. *Nature* **397**:113–115.

Bahre, C.J. 1983. Human impact: the midriff islands. In: T.J. Case and M.L. Cody (eds), *Island Biogeography in the Sea of Cortéz*. University of California Press, Berkeley; pp. 290–305.

Bennett, C.F. 1992. Human activities on the Central American land bridge and their relevance to the region's biogeography. In: S.P. Darwin and A.L. Welden (eds), *Biogeography of Mesoamerica*. Tulane Studies in Zoology and Botany, Supplemental Publication 1, Tulane University, New Orleans; pp. 1–9.

Bowen, T. 1976. *Seri Prehistory: The Archaeology of the Central Coast of Sonora*. University of Arizona Anthropological Paper 27, Tucson.

Bowen, T. 2000. *Unknown Island: The Seri Indians, Europeans, and San Esteban Island in the Gulf of California*. University of New Mexico Press, Albuquerque.

Case, T.J. 1996. Global patterns in the establishment and distribution of exotic birds. *Biological Conservation* **78**:69–96.

Case, T.J., D.T. Bolger, and K. Petren. 1993. Invasions and competitive displacement among house geckos in the Tropical Pacific. *Ecology* **75**:464–477.

Case, T.J., and M.L. Cody. 1987. Island biogeographic theories: tests on islands in the Sea of Cortez. *American Scientist* **75**:402–411.

Dye, T., and D.W. Steadman. 1990. Polynesian ancestors and their animal world. *American Scientist* **78**:207–215.

Felger R.S., and M.B. Moser. 1974. Columnar cacti in Seri Indian culture. *The Kiva* **39**:257–256.

Felger, R.S., and M.B. Moser. 1985. *People of the Desert and Sea: Ethnobotany of the Seri Indians*. University of Arizona Press, Tucson.

Fisher, R.N. 1997. Dispersal and evolution of the Pacific Basin gekkonid lizards *Gehyra oceanica* and *Gehyra mutilata*. *Evolution* **51**:906–921.

Grismer, L.L. 1994. Geographic origins for the reptiles on islands in the Gulf of California, Mexico. *Herpetological Natural History* **2**:17–41.

Infante, E., A. Olivo, C. Alaez, F. Williams, D. Middleton, G. de la Rosa, M.J. Pujol, C. Durán, J.L. Navarro, and C. Gorodezky. 1999. Molecular analysis of HLA class I alleles in the Mexican Seri Indians: implications for their origin. *Tissue Antigens* **54**:35–42.

Lowe, C.H. and K.S. Norris. 1955. Analysis of the herpetofauna of Baja California, Mexico. III. New and revised subspecies of Isla San Esteban, Gulf of California, Sonora, Mexico, with notes on other satellite islands of Isla Tiburón. *Herpetologia* **11**:89–96.

MacArthur, R.H., and E.O. Wilson. 1967. *The Theory of Island Biogeography*. Princeton University Press, Princeton, NJ.

McKeown, S. 1978. *Hawaiian Reptiles and Amphibians*. Oriental Publishing Company, Honolulu.

Mellink, E. In press. Invasive vertebrates on the islands in the Sea of Cortez. In: B. Tellman (ed), *Biological Invasions in the Sonoran Desert Region*. University of Arizona Press, Tucson.

Moran, R. 1983. Vascular plants of the Gulf islands. In: Case, T.J., and M.L. Cody (eds) *Island Biogeography in the Sea of Cortez*. University of California Press, Berkeley; pp. 348–381.

Nabhan, G.P. In press. *Singing the Turtles to Sea: Comcáac Art and Science of Reptiles*. University of California Press, Berkeley.

Olson, S.L., and H.F. James. 1984. The role of Polynesians in the extinction of the avifauna of the Hawaiian Islands. In: P.S. Martin and R. Klein (eds), *Quaternary Extinctions: A Prehistoric Revolution*. University of Arizona Press, Tucson; pp. 768–783.

Paredes-Aguilar, R., T.R. Van Devender, R.S. Felger, G.P. Nabhan, and A.L. Reina-Guerrero. 2000. *Cactáceas de Sonora, México: Su Diversidad, Uso y Conservación*. Arizona-Sonora Desert Museum Press/IMADES, Tucson.

Petren, K. and T.J. Case. 1996. An experimental demonstration of exploitation competition in an ongoing invasion. *Ecology* **77**:118–132.

Petren, K. and T.J Case. 1997. A phylogenetic analyses of body size evolution and biogeography of chuckwallas (*Sauromalus*) and other iguanids. *Evolution* **51**:206–219.

Sheridan, T. E. 1999. *Empires in the Sand: Seri Ethnohistory*. University of Arizona Press, Tucson.

Soulé, M., and A.J. Sloan. 1966. Biogeography and distribution of the reptiles and amphibians on islands in the Gulf of California, Mexico. *Transactions of the San Diego Society of Natural History* **14**:137–156.

Steadman, D.W. 1995. Prehistoric extinctions of Pacific Island birds: biodiversity meets zooarchaeology. *Science* **267**:1123–1131.

Steadman, D.W., and P.S. Martin. 1984. Extinction of birds in the Late Pleistocene of North America. In: P.S. Martin and R. Klein (eds), *Quaternary Extinctions: A Prehistoric Revolution*. University of Arizona Press, Tucson; pp. 466–481.

Thompson, J.N. 1999. The evolution of species interactions. *Science* **284**:2116–2118.

Turner, R.M., J.E. Bowers, and T.L. Burgess. 1995. *Sonoran Desert Plants: An Ecological Atlas*. University of Arizona Press, Tucson.

Villalpando, C.M.E. 1989. *Los Que Viven en Las Montañas*. Instituto Nacional de Antropología e Historia, Noroeste de México, Hermosillo, Sonora.

Votava, E., G.P. Nabhan, and P.W. Bosland. In Press. High levels of genetic diversity and similarity revealed with molecular analysis among and within wild populations of chiltepín. *Proceedings of the National Academy of Sciences USA*.

West, P. and G.P. Nabhan. In press. Invasive weeds: their occurrence and possible impact on the Central Gulf Coast of Sonora and the Midriff Islands in the Sea of Cortez. In: B. Tellman (ed), *Biological Invasions in the Sonoran Desert Region*. University of Arizona Press, Tucson.

Yetman, D.A., and A. Búrquez. 1996. A tale of two species: speculation on the introduction of *Pachycereus pringlei* in the Sierra Libre, Sonora Mexico by *Homo sapiens*. *Desert Plants* **12**(1):23–31.

16

Ecological Conservation

EXEQUIEL EZCURRA
LUIS BOURILLÓN
ANTONIO CANTÚ
MARÍA ELENA MARTÍNEZ
ALEJANDRO ROBLES

History of Conservation Efforts in the Gulf of California

In 1973, George Lindsay, one of Baja California's most eminent botanists, visited the islands of the Sea of Cortés together with Charles Lindbergh, Joseph Wood Krutch, and Kenneth Bechtel. Lindbergh, one of the most celebrated popular heroes of the twentieth century, had become by that time a committed conservationist, interested in the preservation of whales and in the conservation of nature at large. Joseph Wood Krutch, a naturalist, had written *The Forgotten Peninsula*, one of the first natural history descriptions of Baja California. George Lindsay had helped organize a series of scientific explorations into the Sea of Cortés and the peninsula of Baja California, first from the San Diego Natural History Museum, and later from the California Academy of Sciences (Banks 1962a,b; Lindsay 1962, 1964, 1966, 1970; Wiggins 1962).

Kenneth Bechtel, a philanthropist from San Francisco, had given financial support to the Audubon Society in the 1950s and 1960s to study the sea bird rookery at Isla Rasa, which had been decreed a protected area by the Mexican government in 1962. Bechtel was interested in showing the Sea of Cortés to people who might be aroused by its astounding natural beauty (see figs. 16.1–16.9) and who might help to protect it. For this purpose, he organized the trip and invited Lindbergh to visit the region.

The group flew a chartered Catalina flying-boat that allowed them to get to small and remote islands. They landed in the water and then piloted up to the beach so they could have shade under the wing. They visited many of the islands, starting from Consag north of Bahía de los Ángeles, and ending up in Espíritu Santo, east of the Bay of La Paz. It was a wonderful and memorable trip.

Figure 16.1 The endemic chuckwalla (*Sauromalus varius*) of San Esteban. (Photo by Fulvio Eccardi.)

Figure 16.2 The gecko (*Coleonyx switaki gypsicolus*) on San Marcos. (Photo by Fulvio Eccardi.)

Figure 16.3 A sea lion (*Zalophus californianus*) colony on Granito. (Photo by Fulvio Eccardi.)

Figure 16.4 Ring-tailed cat (*Bassariscus astutus*) on Tiburón. (Photo by Fulvio Eccardi.)

Figure 16.5 Heermann's Gulls (*Larus heermanni*) mating on Rasa Island. (Photo by Fulvio Eccardi.)

Figure 16.6 Colonies of Royal Terns (*Sterna maxima*), Elegant Terns (*Sterna elegans*), and Heermann's Gulls (*Larus heermanni*) on Rasa. (Photo by Fulvio Eccardi.)

422   The Human Scene

Figure 16.7 Los Angeles Bay with Coronado (Smith) in the foreground. (Photo by Fulvio Eccardi.)

Figure 16.8 Cardon cacti on San Esteban. (Photo by Fulvio Eccardi.)

Figure 16.9 Giant endemic barrel cacti on Santa Catalina. (Photo by George Lindsay.)

Two or three months later, both Lindbergh and Lindsay traveled to Mexico City to watch the Mexican premiere of a documentary film on the Sea of Cortés by the California Academy of Sciences that Kenneth Bechtel had sponsored (see chap. 1). Taking advantage of the opportunity, and also of his immense popularity, Charles Lindbergh requested to see the president of Mexico, Luis Echeverría. Unfortunately, the president was abroad, on a foreign tour to Asia. The president's secretary met them, possibly expecting to hear some ideas from Lindbergh with respect to aviation in Mexico. Much to his amazement, he heard Lindbergh ranting passionately about Baja California, and especially about the islands of the Sea of Cortés.

Later, Lindbergh called the American ambassador and asked him to organize a press conference for representatives from the Mexican media. Shortly after, the startled George Lindsay saw the editors of about five major Mexican newspapers come into their suite at the Hotel del Prado, in front of the *Alameda* in downtown Mexico City. The media leaders wanted to meet Lindbergh, expecting an interview on aviation and perhaps on Lindbergh's heroic solo flight across the Atlantic. With astonishment, they heard him preach about the immense natural wealth and the beauty of the Sea of Cortés.

A few months after that trip, on 26 August 1974, Charles Lindbergh died. He never saw the gulf islands under any type of legal protection. Four years after his surprising appearance in Mexico City, however, a decree was issued protecting all of the islands of the Sea of Cortés. George Lindsay firmly believes that Lindbergh's intervention helped to promote the necessary governmental awareness for the decision to take place and conservation measures to ensue (SDNHM 1996). He is probably right: The slow buildup of the individual efforts of devoted conservationists and scientists has brought the islands of the Sea of Cortés under increasing levels of conservation. Many naturalists have devoted their best efforts to the protection of the Sea of Cortés, also known as the Gulf of California. In this chapter we analyze the results of these actions.

The Beginnings

Possibly the first efforts to protect the islands of the Sea of Cortés started in 1951 with the publication of Lewis Wayne Walker's popular paper on the seabirds of Rasa in the *National Geographic* magazine (see timeline of conservation events in table 16.1). Walker was at that time a researcher at the San Diego Natural History Museum, and later became associate director at the Arizona-Sonora Desert Museum. He was very knowledgeable about the natural history of the region and possessed first-hand field experience of Baja California and the islands of the Sea of Cortés, especially of Rasa. He wrote many popular articles on the natural history of the region, and through these publications he popularized the plight of Rasa (Walker 1951, 1965).

Kenneth Bechtel (the same philanthropist who later organized the flying-boat expedition described earlier) was at the time a trustee of the Audubon Society. In the early 1950s he donated $5000 for the preservation of Rasa. This started Walker's research on Rasa, which was later also supported with a grant from the Belvedere Scientific Fund (also related to the Bechtel family). This financial support also reached Bernardo Villa's laboratory at the Institute of Biology in the National University of Mexico (Universidad Nacional Autónoma de México, or UNAM; for a list of acronyms, see table 16.2). The funds were used to maintain a biologist and a field station on Rasa.

The results of these investigations soon reached the Direction of Forestry and Wildlife in the Mexican federal government, which in the late 1950s was headed by Enrique Beltrán, an eminent Mexican conservationist. Beltrán's own interest on the issue (and the public notoriety that Rasa had achieved through popular publications and through the field trips of many biologists) helped prepare the way for the first federal decree protecting the insular ecosystems of the Sea of Cortés: In 1964 the official governmental federal register (*Diario Oficial de la Federación*) published a decree declaring Rasa a nature reserve and a refuge of migratory birds (DOF 1964).

The work at Rasa was later supported with donations from the Roy Chapman Andrews Fund at the Arizona-Sonora Desert Museum. This and other funds contributed to maintain the presence on Rasa of researchers and students from Bernardo Villa's laboratory. Many of these students later became leading conservationists in the Sea of Cortés. Villa's work in the early 1980s effectively combined research with conservation. One of Villa's students at that time, Enriqueta Velarde, decided to extend the idea to other islands of the Sea of Cortés. With the scientific support of Lindsay and Daniel Anderson from the University of California, Davis, and the financial and conservationist support of Spencer Beebe from The Nature Conservancy, Velarde, at that time at the Institute of Biology of UNAM, launched the first conservation project for the islands. The project produced, among many other applied results, the book *Islas del Golfo de California* (Bourillón et al. 1988), printed by UNAM and

Table 16.1 Time line of significant conservation events in the Gulf of California, 1951–2000

Year	Event
1951	Lewis Wayne Walker publishes "Sea birds of Isla Raza" in the *National Geographic* magazine.
1952	Kenneth Bechtel donates 5,000 dollars to the Audubon Society for the preservation of Rasa, effectively setting in motion L.W. Walker's and Bernardo Villa's long-term work on the island.
1963	Tiburón Island is decreed a Wildlife Refuge and Nature Reserve by President Adolfo López Mateos.
1964	The Mexican Federal Government decrees Rasa a protected "Nature Reserve and Refuge of Migratory Birds" (Reserva Natural y Refugio de Aves Migratorias).
1965	A small, two-room stone field station is built on Rasa as part of Villa's ongoing research and conservation work.
1967	To recover Sonoran wildlife, white-collared peccari and pronghorn antelopes are introduced to Tiburón Island by Mexico's federal government. The species do not prosper.
1973	In spring, Kenneth Bechtel organizes a flying-boat expedition to the islands of the Sea of Cortés with Charles Lindbergh, George Lindsay, and Joseph Wood Krutch.
1973	Inspired by the flying-boat expedition, Lindbergh and Lindsay travel to Mexico City to see President Luis Echeverría. They meet with the president's cabinet and the editors of the major Mexican newspapers and urge them to preserve the islands of the Sea of Cortés.
1975	A decree is issued by President Luis Echeverría, restituting Tiburón Island to the Seri as part of their communal property and declaring the coastal waters of Tiburón Island for the exclusive use of the Seri and off-limits for other fishermen.
1975	Bighorn sheep are introduced to Tiburón Island as a part of a federal program to study and protect the Sonoran subspecies.
1978	The Mexican government issues a decree protecting all the islands of the Sea of Cortés under the category of "Wildlife Refuge" (Refugio de Vida Silvestre).
1979	Enriqueta Velarde takes over the research and conservation tasks in Rasa, inspired by Villa's pioneer work. Since then, biologists have been present in Rasa during each seabird breeding season.
1982	Mexico's first Environment Ministry (Secretaría de Desarrollo Urbano y Ecología) is created. All natural protected areas are put under its jurisdiction.
1983	Publication of Case and Cody (1983)'s book *Island Biogeography in the Sea of Cortez.*
1988	As a result of the work of Velarde's team, the National University of Mexico and the federal government publish the book *Islas del Golfo de California*, bringing national attention to the islands of the Sea of Cortés and their conservation problems.
1988	Mexico's first environmental law ("Ley General del Equilibrio Ecológico y la Protección al Ambiente") is passed.
1992	The government of Mexico obtains the approval of a $25 million donation from the global Environment Fund for the management and the conservation of ten protected areas, including Islas del Golfo de California.
1993	On June 10, at Cerro Prieto near Puerto Peñasco, the president of Mexico decrees the establishment of the first marine reserve in Mexico, the Biosphere Reserve of the upper Gulf of California and delta of the Colorado River. This opens discussions on protecting the waters surrounding the islands in the Sea of Cortés.
1993	A successful program is initiated by Jesús Ramírez to eradicate introduced rats and mice from Isla Rasa using modern rodenticides.
1995	Complete eradication of introduced rodents in Isla Rasa is achieved.
1995	UNESCO's Man and the Biosphere (MAB) Program dedicates the Protected Area Islas del Golfo de California as an international Biosphere Reserve.
1995	The environmental impact statement for a hotel in Coronado Island is strongly challenged by local and national conservation and citizen groups, leading to the abandonment of the project.

(*continued*)

Table 16.1 (*Continued*)

Year	Event
1995	The Mexican company Salinas del Pacífico introduces the Baja California subspecies of the desert bighorn to Isla del Carmen. In contrast with the support that the Sonoran bighorn program in Tiburón had enjoyed, the del Carmen plan is received with criticism by conservationists in the peninsula.
1996	The Parque Nacional Bahía de Loreto is created by a federal decree to protect the Bay of Loreto, including its five islands, from large fishing fleets.
1996	Mexico's environmental law, the "Ley General del Equilibrio Ecológico y la Protección al Ambiente" is amended. The new law recognizes eight categories of natural protected areas, demanding recategorization of all previously decreed natural protected areas into the new system.
1997	The Fund for Natural Protected Areas is created with $16.48 million that still remain from GEF's original $25 million grant for 10 protected areas in Mexico. The protected area Islas del Golfo de California Natural starts receiving part of the financial revenues generated by the fund, a process that ensures long-term financing for conservation.
1998	The management plan of Espíritu Santo, the first island-specific conservation plan in the Sea of Cortés, is completed with the joint participation of local research centers, conservation organizations, and Ejido Bonfil. The ejido landowners accept constraints on the development in their own insular lands.
1998	The Seri start auctioning permits for sport hunting of bighorn sheep in Tiburón. Half of the proceeds go to support research, conservation and management actions, the other half goes to the Seri tribe.
1998	With the active participation and involvement of the local community, a binational group of researchers, governmental resource managers, and the Ejido Tierra y Libertad draft a management and conservation plan for the islands surrounding Bahía de los Ángeles
2000	The Mexican federal government recategorizes the islands of the Sea of Cortés into an "Area for the Protection of Wildlife."

the Mexican federal government, which was extremely influential in bringing attention to the islands and their conservation problems.

Many of the biologists that participated in this early team are now crucial players in the conservation of the Sea of Cortés. The team included, among others, Alfredo Zavala, now Baja California regional director for the Protected Area of the Islands of the Gulf of California; the late Jesús Ramírez Ruiz, who in the early 1990s eradicated introduced rodents from Rasa, and María Elena Martínez, Luis Bourillón, and Antonio Cantú, who head two conservation organizations in the region (see Bourillón et al. 1988). In many ways, it can be said that the conservation work at Rasa was the catalyst that started most of the other conservation work in the Sea of Cortés.

Chronologically, however, Tiburón was the first island of the Sea of Cortés to receive official status as a protected area, through a decree published a year before that concerning Rasa. The largest island of the Sea of Cortés, Tiburón occupies 120,756 ha. In pre-Hispanic times Tiburón was an important part of the territory of the Seri Indians (or Comcaác, in their own language; Felger and Moser 1985). Because of this, the island is not only an important natural site but also harbors important historic, archaelogic, and cultural elements. Although in the twentieth century the Seri have not permanently lived on the island, they have always used it as their main fishing

Table 16.2 List of acronyms used in the text

CEMDA	*Centro Mexicano de Derecho Ambiental*; Mexican Center of Environmental Law
CES	*Centro Ecológico de Sonora*; a research center in Sonora
CICIMAR	*Centro de Investigaciones de Ciencias del Mar*; a local research center at La Paz, Baja California Sur
CIDESON	*Centro de Investigación y Desarrollo de los Recursos Naturales de Sonora*; a research center in Sonora
DOF	*Diario Oficial de la Federación*; Mexico's governmental federal register
FANP	*Fondo de Áreas Naturales Protegidas*; Mexican Fund for Natural Protected Areas
FMCN	*Fondo Mexicano para la Conservación de la Naturaleza*; Mexican Fund for the Conservation of Nature
GEA	*Grupo Ecologista Antares*; a conservation group based in Loreto, Baja California Sur
GEF	Global Environmental Facility, an international funding organization created as a result of the Rio Summit (UNCED)
INE	*Instituto Nacional de Ecología*; Mexico's National Institute of Ecology
ISLA	*Conservación del Territorio Insular Mexicano*; a conservation group based in La Paz, Baja California Sur
MAB	UNESCO's Man and the Biosphere Program
RAMSAR	United Nations' International Convention of Wetlands
SARH	*Secretaría de Agricultura y Recursos Hidráulicos*; Mexico's Ministry of Agriculture and Water Resources
SDNHM	San Diego Natural History Museum
SEDUE	*Secretaría de Desarrollo Urbano y Ecología*; Mexico's first Environmental Ministry, 1983–1992
SEMARNAP	*Secretaría del Medio Ambiente, Recursos Naturales y Pesca*; Mexico's Ministry of the Environment, 1994–2000
SRA	*Secretaría de la Reforma Agraria*; Mexico's Ministry of Agrarian Reform
UABCS	*Universidad Autónoma de Baja California Sur*; State University of Baja California Sur
UC	University of California
UC-MEXUS	UC's Mexico-U.S. Program
UMA	*Unidad de Manejo de la Vida Silvestre*; locally operated Units for the Management and Sustainable Use of Wildlife
UNAM	*Universidad Nacional Autónoma de México*; Mexico's National University
UNCED	United Nations Conference on Environment and Development (also known as the Rio Summit)
UNESCO	United Nations Educational, Scientific and Cultural Organization

camp, hunting ground, and plant-collecting territory and have always considered it part of their tribal land.

On 15 March 1963, Tiburón was decreed a Wildlife Refuge and Nature Reserve by President Adolfo López Mateos (DOF 1963). This first decree was issued as a result of an initiative by Beltrán. The ruling, however, was based on biological and ecological grounds and failed to take into consideration the needs and demands of the Seri. Twelve years later, in 1975, the secretary of the Agrarian Reform gave the Seri formal possession of Tiburón as part of an *ejido* (i.e., communal land) allotment for the tribe. This was the first recognition by the federal government of the Seri's right to their ancestral land. On 11 February 1975, a decree was issued by President Luis Echeverría, restituting Tiburón to the Seri as part of their communal property. Although this decree was basically issued as part of a series of governmental actions to empower native peoples within their traditional lands, it also had conservationist implications for the island as well as the mainland coast. The decree established that the coastal waters of the island could be only used by the Seri, and by their fishing cooperative, the Sociedad Cooperativa de la Producción Pesquera Seri (INE 1994), and declared Tiburón off-limits for other fishermen.

Biosphere Reserves and Biological Diversity

In the early 1970s, roughly at the time of Lindbergh's trip to Mexico City, many changes were occurring within the Mexican scientific and conservation groups. These scientific transformations also helped to protect the islands. In 1974 the Instituto de Ecología, A.C. (Institute of Ecology, a nonprofit research organization) was created in Mexico City, and soon after it started to promote the concept of biosphere reserves in the country. Although widely accepted at present, the idea of biosphere reserves, which had been developed by a group of ecologists in the United Nations' Man and the Biosphere Program (MAB), was radically new in 1975. Biosphere reserves were conceived as natural protected areas where the indigenous populations living inside the area or in the surrounding buffer zones were encouraged to use their natural resources in a sustainable manner. The new approach departed radically from the natural park concept, which basically advocates for pristine areas free of human influence. Rather, biosphere reserves promoted sustainable use as an effective tool for conservation.

Many of the concepts discussed at the United Nations Conference on Environment and Development (UNCED; also known as the Rio Summit) in 1992 were already operational in MAB's concept of biosphere reserves almost 20 years before, including (1) the global approach to conserve biodiversity through a planetary network of protected areas; (2) the preservation of cultural diversity together with natural diversity; (3) the involvement of local populations in the protection of natural resources; and (4) the promotion of the sustainable use of nature.

Three Mexican biosphere reserves were among the first in the world to become part of the MAB network of biosphere reserves. One of these reserves was in the Chihuahuan desert, in the arid north of the country. This simple fact created consciousness among decision makers in the federal government that protected areas could be created in what was previously perceived as barren wastelands (previously, National Parks in Mexico had been located almost unfailingly in montane areas with temperate forests). Additionally, the international success of the Mexican biosphere reserves caught the attention of the environmental authorities, who realized that large, natural areas could be protected under the new scheme as it did not preclude resource use but rather pursued a judicious utilization. Although the islands of the Sea of Cortés were initially not conceived as biosphere reserve but rather as a wildlife refuge (Refugio de la Vida Silvestre), it was in the wake of these changes that the decree protecting them was issued in 1978 (DOF 1978).

Present Status of the Islands as Protected Areas

Mexico's environmental legislation, the *Ley General del Equilibrio Ecológico y la Protección al Ambiente* (DOF 1988, 1996b) recognizes eight categories of natural protected areas that can be established by the federal authority: (1) biosphere reserves (*reservas de la biosfera*), (2) national parks (*parques nacionales*, including both terrestrial and marine parks), (3) natural monuments (*monumentos naturales*), (4) areas for the protection of natural resources (*áreas de protección de recursos naturales*), (5) areas for the protection of wildlife (*áreas de protección de flora y fauna*), and (6) natural sanctuaries (*santuarios*).

Because of the use of the same denomination of biosphere reserves for protected areas as defined by the Mexican law and also for areas integrated within MAB's network of protected areas, some reserves in Mexico are considered biosphere reserves under Mexican law but have not yet fulfilled the conditions to be incorporated under MAB's international system, while other Mexican protected areas are formally recognized by MAB as biosphere reserves but do not have formal biosphere status under the Mexican legislation (see SEDUE 1989; Gómez-Pompa and Dirzo 1995).

At present, the MAB Program has integrated 12 Mexican protected areas into its international network of biosphere reserves: (1) Mapimí, located in the core of the Chihuahuan Desert (dedicated in 1976), (2) La Michilía, in the temperate forests of the Western Sierra Madre in Durango (1976), (3) Montes Azules, a large expanse of tropical rainforest in Chiapas (1978), (4) El Cielo, on the slopes of the eastern Sierra Madre in Tamaulipas (1986), (5) Sian Ka'an, on the coast of Quintana Roo (1986), (6) Sierra de Manatlán, in the mountains of Jalisco and Colima (1988), (7) Calakmul, in the Mayan forests of Campeche (1993), (8) El Pinacate y Gran Desierto de Altar, in the core of the Sonora Desert (1993), (9) El Triunfo, a cloud forest in the Soconusco region of Chiapas (1993), (10) El Vizcaíno, in the central part of the Baja California (1993), (11) Alto Golfo de California y Delta del Río Colorado, in the upper Sea of Cortés (1995, also dedicated as a site of global significance within the International Convention of Wetlands or RAMSAR Convention), and (12) Islas del Golfo de California (1995), the islands of the Sea of Cortés. Of these 12 areas recognized internationally as biosphere reserves, 2 do not have a formal recognition as biosphere reserves under Mexican federal law: El Cielo is a state-decreed protected area, and the islands of the Sea of Cortés, originally decreed as a reserve zone and refuge for wildlife and migratory birds, were recategorized in June 2000 into an area for the protection of wildlife (Área de Protección de Flora y Fauna Islas del Golfo de California; see DOF 2000).

On the other hand, the Mexican legislation also recognizes 15 additional biosphere reserves, which are currently waiting for submission to the MAB program for formal recognition within the international network of MAB's biosphere reserves. These areas are (1) Archipiélago de Revillagigedo (Mexican Pacific Ocean), (2) Arrecifes de Sian Ka'an (coast of Quintana Roo), (3) Banco Chinchorro (coast of Quintana Roo), (4) Chamela-Cuixmala (Jalisco), (5) La Encrucijada (Chiapas), (6) La Sepultura (Chiapas), (7) Lacan-Tun (Quintana Roo), (8) Los Tuxtlas (Veracruz), (9) Mariposa Monarca (Michoacán and state of Mexico), (10) Pantanos de Centla (Tabasco), (11) Ría Lagartos (Yucatán), (12) Sierra Gorda (Querétaro), (13) Sierra de Abra Tanchipa (San Luis Potosí), (14) Sierra de la Laguna (Baja California Sur), and (15) Tehuacán-Cuicatlán (Puebla and Oaxaca).

As a wildlife protection area, under Mexican law the islands of the Sea of Cortés do not enjoy the same strict restrictions that are imposed on biosphere reserves. The reasons to designate the islands within Mexican legislation with a different category to the one they hold internationally is possibly related to the large size and the spatial complexity of the whole archipelago, and the difficulties involved in strict law enforcement within the whole protected area. In spite of their less restrictive status under Mexican law, the islands of the Sea of Cortés are in practice managed as a large reserve, and substantial efforts are devoted to their protection (INE 1994; Breceda et al. 1995). The relevance given by federal authorities to the islands of the Sea of Cortés is possibly the result of an effort to fulfil the Mexican government's commitment with the MAB network and with the Global Environmental Facility (GEF), which has funded part of the conservation work on the islands. In 1996 the administration of the natural protected area was divided into three regional headquarters: the southern islands are managed from an administrative office at La Paz; Tiburón and San Esteban are managed from an office in Guaymas; and the western midriff islands area is managed from headquarters in Bahía de los Ángeles.

Case Studies: Main Conservation Efforts in the Sea of Cortés

The Northern Gulf

The Preservation of the Alto Golfo

From the mid-1950s it has become well known that the upper Sea of Cortés (known in Spanish as the *Alto Golfo*) and the delta of the Colorado River are important sites for the reproduction and breeding of many species of birds and fish. This very productive region, however, has been under heavy fishing pressure. In 1975, the totoaba fish (*Totoaba macdonaldi*) was facing extinction through overfishing. This problem forced the federal government to decree a moratorium for totoaba harvest in the Sea of Cortés.

Other problems kept mounting. In the mid-1980s marine mammalogists began to be concerned about the population status of the vaquita porpoise (*Phocoena sinus*), which is endemic to the upper Gulf of California. The vaquita is indeed a rare marine mammal. Described in 1958, only a few specimens have been studied. The occurrence of vaquita specimens as incidental take in gill nets started to signal an alert to Mexican and international conservation groups.

In the early 1990s, the population of vaquita was estimated as less than 500. The vaquita was classified as endangered, and the International Whaling Commission labeled it as one of the highest priority marine mammals in the world. It was then that the Mexican federal government created, through the Secretary of Fisheries, the Technical Committee for the Protection of the Totoaba and the Vaquita (Comité Técnico para la Preservación de la Totoaba y la Vaquita), with the purpose of evaluating and studying the issue and recommending adequate measures for the conservation of both endangered species. Bernardo Villa, one of the Mexican biologists who had dedicated much time to studying the fauna of the Sea of Cortés, was named president of the committee, and it enjoyed the participation of several leading Mexican biologists and conservationists. Samuel Ocaña, formerly governor of Sonora and a devoted conservationist, was appointed technical secretary of the group. After a few sessions, it became evident that serious discrepancies existed between various constituents of the committee. Although some members favored immediate action to protect the upper Gulf of California from the devastating effects of overfishing, others were of the opinion that regulating fisheries in any way would harm the local economy.

In June 1992, an international meeting was organized in San Diego by the University of California Mexico-U.S. Program (UC-MEXUS) to discuss two conservation issues of great relevance for marine mammals: the problem of dolphin incidental take in Mexican tuna fisheries, and the totoaba–vaquita extinction threat. The meeting was called by Arturo Gómez Pompa, a professor at UC Riverside, and also at that time special advisor on environmental matters to the president of Mexico. Thus, the problem of overfishing in the Sea of Cortés started to appear in the international arena, harming Mexico's reputation on conservation and natural resource management.

In 1992, a severe crisis struck the fishermen of El Golfo de Santa Clara and Puerto Peñasco, in Sonora, and San Felipe, in Baja California, all located in the upper Gulf of California. Their shrimp catches had fallen precipitously (Arvizu 1987), and the fishermen blamed the federal authorities in general, and the Secretary of Fisheries in particular, for failing to enforce fishing bans to allow the recovery of the resource. The idea started to grow among fishermen that the sea had to rest and its fisheries had to recover.

In the summer of 1992, the technical committee met in Hermosillo, Sonora. At this meeting, both the director general of Natural Resources (Dirección General de Aprovechamiento Ecológico de los Recursos Naturales) of Mexico's National Institute of Ecology, Exequiel Ezcurra, and Gómez Pompa expressed their support for the idea of establishing a natural protected area in the upper gulf. Most members of the committee showed sympathy for the proposal, but the representatives of both the Secretary of Fisheries (Secretaría de Pesca) and the National Institute of Fisheries (Instituto Nacional de la Pesca) expressed their complete opposition. As a result, it was decided to request two of the most recognized centers in Sonora, the Centro Ecológico de Sonora (CES) and the Centro de Investigación y Desarrollo de los Recursos Naturales de Sonora (CIDESON), to develop and elaborate a feasibility study for a Biosphere Reserve.

Toward the end of 1992, the study was completed. The next step was to gain the approval and the consensus of the fishing communities in El Golfo de Santa Clara, Puerto Peñasco, and San Felipe, as well as the *ejido* communities in the delta of the Colorado River. In the first months of 1993 the costs and benefits of a protected area were discussed with these communities. Slowly, the people in the area started first to accept and later to support the idea. In March 1993, Sven Olof Lindblad, owner of Lindblad Expeditions, donated a week of usage time in his boat *Sea Bird* for conservation projects. In collaboration with the World Wildlife Fund, the National Institute of Ecology (Instituto Nacional de Ecología, or INE) from Mexico's federal government used the opportunity to assemble businessmen, scientists, conservationists, social leaders from the small-scale fisheries, and traditional authorities from the indigenous peoples around the Sea of Cortés, and bring all these sectors together to discuss the issues around the sustainable management of the region. As a result of the *Sea Bird* cruise, a joint declaration was issued, signed by all the invited participants, urging the federal government to protect the habitat of the vaquita by declaring a marine reserve in the upper gulf. Finally, the project was presented to the Secretary of Social Development in the federal government, Luis Donaldo Colosio, a native of northern Sonora who was much interested in the idea. With the support of Colosio, the project moved forward.

On 10 June 1993, at a memorable occasion at Cerro Prieto, a volcanic mountain in the Gran Desierto near Puerto Peñasco, the president of Mexico, Carlos Salinas de Gortari, decreed the establishment of the biosphere reserve of the upper Gulf of California and delta of the Colorado River (Reserva de la Biosfera del Alto Golfo de California y Delta del Río Colorado; see DOF 1993). The project had strong support from both the local population and conservation groups. Important decision makers attended the ceremony, including many cabinet members from the Mexican federal government, the governors of Sonora, Baja California, and Arizona, the U.S. Secretary of the Interior Bruce Babbitt, and the traditional governor of the Tohono O'Odham (Papago) people, whose lands extend on both sides of the Mexico–U.S. border.

The objectives of the establishment of this reserve were the conservation of endangered species both from the Sea of Cortés and the Colorado River estuary, including the vaquita, the totoaba, the desert pupfish (*Cyprinodon macularius*), and the Yuma clapper rail (*Rallus yumanensis longirostris*) The establishment of the reserve also intended to protect the reproduction and breeding of many other species in the zone. Perhaps more important, this was the first marine reserve established in Mexico. In spite of the opposition of the fisheries authorities, it opened the way for new marine protected areas in the Sea of Cortés, in the Mexican Pacific Ocean, and on the other coasts of Mexico. Specifically, it opened the door for the discussion of the possibility of protecting the waters surrounding each of the islands in the Sea of Cortés. Although this latter measure has not yet been achieved, the upper Gulf of California debate established these discussions as an ongoing process. Specifically, it facilitated current efforts by various organizations to extend the decreed protection into the waters adjacent to some important islands such as Espíritu Santo, San Pedro Mártir, San Lorenzo, Las Ánimas, Salsipuedes, Rasa, and Partida.

The Midriff Islands

Isla Tiburón

Tiburón was the first island in the Sea of Cortés protected by presidential decree, closely followed by Rasa. The main purpose for protecting Tiburón in 1963 was to create a mule deer (*Odocoileus hemionus sheldoni*) refuge, protecting the species from the extensive poaching that prevailed in the Sonoran mainland (Quiñonez and Rodríguez 1979). The 1963 sanctuary was put under the management of the Secretaría de Agricultura y Recursos Hidráulicos (SARH; Secretariat of Agriculture and Water Resources) an agency of the federal government. SARH built basic facilities for poaching control, 130 km of dirt roads, two airstrips, a small wildlife research station, and some water reservoirs to improve habitat quality for game species. At that time, the hunting habits of Seri Indians were considered a threat to the game species and an essential part of the problem of game conservation on the island. As a result, no hunting permits were granted initially to the tribe, despite the fact that Tiburón had always been part of their traditional hunting territory.

Bighorn sheep (*Ovis canadensis mexicana*, the Sonoran desert subspecies; see

Monson 1980) were introduced in 1975 as a part of a federal program to study and protect bighorn in Sonora (Becerril-Nieva et al. 1988; cited in Hernández-Alvídrez and Campoy-Favela 1989). Twenty sheep (four males) were introduced to the island, captured by staff from the New Mexico Department of Game and Fish in the mainland mountain ranges in front of the island (Montoya and Gates 1975). Before the sheep introduction, in 1967 two other species of mammals were introduced from the Sonoran mainland; 20 white-collared peccari (*Tayassu tajacu*) and 17 pronghorn antelopes (*Antilocapra americana*; Quiñónez and Rodríguez 1979).

In 1975, however, the government approach to Indian issues changed, and the interests of the Seri people on the wildlife of Tiburón were taken into account for the first time. Under the administration of Mexican President Luis Escheverría, the island was returned to Seri ownership, although it still remained for a time under federal control. For 2 years (1975–77) marines (the Mexican Navy had permanent presence on the island on small outposts since the 1970s) and game wardens prohibited Seri from landing on the island (Olivera and López 1988).

The bighorn sheep transplant was the only successful introduction. In fact, the population grew to a number estimated in 1993 to range between 480 and 967, as evaluated through an aerial census (Lee and López-Saavedra 1994). More recent aerial surveys have estimated a population of 650 animals (Pallares 1999).

In the late 1980s biologists from the Centro Ecológico de Sonora (CES; Ecological Center of Sonora, a state research center) continued the wildlife studies that SARH biologists initiated in the mid-1970s. Wildlife research on the island restarted with great vigor in 1995 when an ambitious project to study and manage bighorn sheep was launched. Scientists from the National University of Mexico, the Arizona Department of Game and Fish, and staff from two conservation organizations, Unidos para la Conservación and Agrupación Sierra Madre, surveyed and studied the bighorn population. Under an innovative scheme for research and conservation funding, half of what is earned during the international auctioning of the sport hunting permits goes to support research by UNAM's scientists and for conservation and management actions for the bighorn sheep population on Tiburón. The other half goes to the Seri tribe. In 1998 the prices paid for the permits during an auction in Reno, Nevada, rose to unprecedented levels: American hunters bid up to $395,000 for two permits. The 1999 auction resulted in $150,000 for two more permits (Navarro 1999).

The Seri community has been actively involved in this project, hiring a professional wildlife biologist for local field coordination and training a team of young Seri men as field technicians. Seris are also bringing to the project their traditional ecological knowledge about the wildlife of the island. The entire Seri community is expected to benefit from this sheep-hunting program. Money raised from hunting permits is deposited in a trust fund administered by a Seri technical committee. These monies are used for health, educational, and cultural projects, as well as for supporting the operational cost of the Seri traditional government.

The bighorn sheep project in Tiburón also has the objective of providing animals to repopulate former bighorn sheep distribution ranges in Sonora, Chihuahua, and Coahuila. Thus, with all this funding and support, Tiburón is now contributing in an enormous way to the conservation of bighorn sheep in mainland Mexico, and at the same time it is generating another source of income for the Seri people. This innovative management project also plans to fund studies of other important species on the island, such as the endemic mule deer, apparently suffering from habitat competition by the sheep, and the large coyote (*Canis latrans*) population. Under a new federal Program for Wildlife Conservation and Diversified Productive Use of Land, Tiburón is now managed as a Unit for the Management and Sustainable Use of Wildlife (UMA is its acronym in Spanish).

Isla Rasa

Rasa, the major site for breeding sea birds in the Sea of Cortés (Bancroft 1927; Case and Cody 1983), has served as a role model for successful ecological conservation of islands in the gulf. Immediately after its legal protection in 1964 (DOF 1964), the sea birds on this island were studied intensively and protected thereafter thanks to an enormous commitment of effort by almost three generations of Mexican biologists and the support from numerous research and conservation organizations (Tobías 1968; Velázquez-Noguerón 1969; Velarde 1988, 1993; Velarde et al. 1994).

In the early 1960s a concern for the protection of Rasa grew following drastic reductions in the population numbers of nesting sea birds caused by egg collecting (see chap. 14). In 1940, the population of sea birds (all species combined) was estimated to be 1 million (Walker 1965), in the late 1960s it had reduced to 25,000 (Barreto 1973), and possibly reached a historic low with an estimated number of 5000 in 1973 (Villa 1983). The pioneering conservation efforts and lobbying of Louis Wayne Walker of the Arizona Sonora Desert Museum and the National Audubon Society, George Lindsay and Robert Orr of the California Academy of Sciences, Bernardo Villa from the National University of México, and some leading residents of Bahía de los Ángeles, notably Antero Díaz (Velarde et al. 1985), led the federal government to declare the island a sea bird sanctuary.

Once the island was declared a sanctuary, biologists working for the Mexican Direction of Wildlife spent time on the island during the sea birds' breeding season to provide on-the-ground protection and to collect baseline data on population numbers. In 1965 a two-room stone house was build on Rasa (Vidal 1967). The lengths of stay of researchers on the island became more prolonged every year, as UNAM students working under Villa's guidance became involved in the conservation efforts. Villa spent many seasons on the island between 1975 and 1985, and in 1979 one of his students, Velarde, took over the research and conservation tasks inspired by Villa's remarkable pioneering work. Under Velarde's direction, biologists have been present during the sea bird breeding season (mid-March to early July) from 1979 to the present time. These scientists and students have researched in detail the sea birds' breeding ecology and behavior and the island's natural history. They also have helped prevent possible disturbance by the 300 or more ecotourists that visit the island every year (Villa et al. 1979, 1980; Velarde and Anderson 1994), and deter fishermen from collecting eggs and from landing or hiking in the nesting areas. The success of this island research and protection is evident. Sea bird populations have rebounded dramatically: Heermann's Gulls maintained an estimated average of 260,000–350,000 birds throughout the 1990s (Vermeer et al. 1993; Velarde and Ezcurra, chap. 11, this volume), while the Elegant Terns increased from 45,000 individuals in the early 1990s to around 200,000 birds in 1999 (Velarde and Anderson 1994; Velarde and Ezcurra, chap. 11, this volume). Despite numerous obstacles, since 1979, the sea bird population numbers and annual breeding success have been monitored systematically by a dedicated band of researchers and graduate students.

In 1993–1994, a program to eradicate introduced rats (*Rattus rattus*) and mice (*Mus musculus*) was initiated by the late Jesús Ramírez, who also did the pioneering research to start the bighorn sheep program on Tiburón. The complete eradication of introduced rodents using modern rodenticides was achieved by 1995. Researchers have found no sign of rodent activity while doing monitoring working during recent field seasons (1998, 1999, and 2000; E. Velarde, pers. commun., November 2000). Now that the island apparently is free of introduced mammals, scientists are monitoring ecological changes in the nesting colonies, the vegetation, and insect populations, and they have plans to restore the populations of burrow-nesting sea birds that may have once used this island (such as the Craveri's Murrelet *Synthliboramphus craveri*, the Black Storm-Petrel *Oceanodroma melania*, and Least Storm-Petrel *O. microsoma*, and perhaps even the Black-vented Shearwater *Puffinus opisthomelas* that was present in the island in the early 1920s) (Bancroft 1927; Boswall and Barrett 1978).

Small Islands in Bahía de los Ángeles

The concern for the protection of the islands inside the Bahía de los Ángeles was fueled by members of the local Ejido Tierra y Libertad, especially the late Antero Díaz and his family, and in collaboration with several American researchers working in this area since the late 1950s. In the early 1970s the oceanographer Antonio Reséndiz arrived at Bahía de los Ángeles to start a small program devoted to sea turtle research, with the help of Grant Bartlett of the Laboratory of Comparative Biochemistry in San Diego. Reséndiz's work was not limited to sea turtles: he collaborated extensively with almost all researchers that arrived at Bahía de los Ángeles and founded Campo Archelon, a research center. With the support of his wife Betty since 1985, the center developed into a focal point for local conservation efforts in the bay area. He is the current president of the Ejido Tierra y Libertad.

In 1988, efforts by residents led by Carolina Espinoza culminated in the construction of the local Museum of Natural History and Culture. This exceptional museum has functioned as an information center for visitors describing the natural and cultural history of the area, highlighting the ecological importance of the islands, and providing environmental education opportunities for the local people.

The proximity of Bahía de los Ángeles to southern California and its small islands inside protected waters make this area an ideal place for natural history trips and ecotourism. The growth in human use and the impacts on the bay and its islands prompted a binational group of researchers, governmental resource managers, and key local people to draft a management plan for the islands (Bourillón and Tershy 1997). A prioritized set of actions was proposed, which involved placing information signs on the common landing and camping sites in the islands, defining and delimiting hiking trails, and starting an information/orientation/registration system for island visitors. With financial support from a U.S. Fish and Wildlife Service grant under a joint program with Mexico's National Institute of Ecology, and with the support of Dan Anderson and the late Gary Polis, a group of members of the ejido established a committee in 1998 to fully implement the plan (Jiménez et al. 1999). This management plan has shown how the active participation and involvement of a local community can effectively increase local conservation through simple and low-cost management actions, usually not adequately considered in large-scale programs for island conservation.

The Lower Gulf

Conservation in the Bay of Loreto

Loreto is a small town on the central gulf coast of Baja California Sur; its bay is protected by five islands (Coronado, del Carmen, Danzante, Monserrat, and Santa Catalina). The growing pressures of large, industrial fishing boats (mainly shrimp bottom-trawlers) on the fisheries of the bay led to a collapse of the local fisheries in the 1970s and 1980s, and in doing so generated a concern among the local small-scale, or *panga*, fishermen and the sport fishing operators as to how to protect the bay from overfishing.

Many measures were proposed. Some extreme ones including the sinking of boats into the bay to destroy the shrimp boat dragnets, but little was done at first. In 1995 the municipal president of Loreto, Alfredo García Green, started meeting with conservationists and concerned citizens to discuss this issue. Under the leadership of Grupo Ecologista Antares (GEA; a local conservation group), a proposal was prepared and submitted to the federal government to protect the Bay of Loreto, including its five islands. On 19 July 1996, the *Parque Marino Nacional Bahía de Loreto* was created by a federal decree (DOF 1996a). In December 1996, when Mexico's environmental law was changed, the marine park was recategorized as a national park (DOF 1996b).

Conceptually, this 296,580-ha national park was a major breakthrough. Although the islands of the Sea of Cortés had legal protection since 1978, the sea surrounding them did not. Many conservationists have argued that the decree protecting the gulf islands should be extended to the marine ecosystems around them, but fierce opposition from the large-scale fisheries lobby has prevented this from happening. In the Bay of Loreto, the concerns of the local community proved to be stronger than the short-term economic interests of the fishing industry. In 1999, the park had modest financing for a director, vehicles, basic equipment, and an office. Some additional support came from The Nature Conservancy to develop basic operational activities. It also had a working management plan and the full support of the local people, two facts that promote the park as a leading coastal protected area in Mexico.

Isla Coronado

In 1995, two Italian investors proposed building a hotel on Isla Coronado, a small, 850-ha island situated 11 km north of the town of Loreto. Because this island is part of the protected area of the islands of the Gulf of California, the project required an environmental impact statement (EIS). The hotel of 40 rooms was designed to have all the technological advances to be environmentally friendly: solar panels for water heating, sewage treatment, water desalinating plant, and careful management of solid waste. The project was presented by the proponents as a viable alternative for the protection of the island. It planned to concentrate on low-impact tourism and to enable a research station to be built next to the hotel that would monitor the island's wildlife and provide guidelines for a conservationist management of the site (Inmobiliaria Isla Coronados 1995). The project was strongly challenged by local and national conservation and citizen groups and was finally abandoned.

Nevertheless, the development of the proposal, the discussion that arose around the project, and the arguments for its final cancellation offer another example of the strong attraction islands exert for tourist resorts and of the rationale used to justify their development. The Coronado project is also a prime example of the problems, conflicts of interest, and dangers for island conservation that can be caused by lack of clear governmental policies.

The main argument used by the developers was the limited conservation and management actions that were in place on Coronado despite its legal protection since 1978. The presence of introduced animals (cats and sheep) and the high levels of human use by fishermen and tourists were used as strong arguments to propose that a research station, to be sponsored by the hotel and managed by the University of Baja California Sur (Universidad Autónoma de Baja California Sur, or UABCS), was the only and best option to protect the island and ameliorate its deteriorated ecological status. The probable, but not yet certain, extinction of the endemic pack rat of Coronado (*Neotoma bunkeri*) due to 30 years of cat predation was another argument used to support the claim that the island was disturbed and could be developed.

However, most of the opponents to the project argued that once the place is connected to the coast by means of daily transport of food, personnel, tourists, baggage, people, machinery, trash, and so on, it would be almost impossible to ensure that no new exotic animals or weedy plants will reach the island. Coronado is now part of the Parque Nacional Bahía de Loreto and it is better protected than it was before. Additionally, the introduced species are being successfully eradicated (Arnaud 1998).

However, there are ongoing development plans and growing pressures to develop other gulf islands for tourism, mainly those in which private individuals can claim ownership, such as San José, Cerralvo, Carmen, and Espíritu Santo. The same arguments brandished for Coronado, that protection of the island was not being enforced and that an environmentally minded development could protect the island better, are being used to pursue other development proposals in the Sea of Cortés.

Isla Carmen

Possibly the most beautiful and scenic of the islands of the Bay of Loreto, Carmen, an island with private ownership claims, has been for many years a favorite destination for ecological tourists and plays a major role both in the protected area of the islands of the Sea of Cortés in general, and in the Loreto Bay National Park in particular.

Based on the apparent success of the breeding program of the Sonoran bighorn sheep on Isla Tiburón, the National Institute of Ecology of Mexico decided to promote a similar program in Isla Carmen, this time aimed at the Baja California subspecies (*Ovis canadensis weemsi*), an endemic taxon that is found only in the Sierras El Mechudo, La Giganta, and Las Tres Vírgenes within the peninsula of Baja California. In 1995, the Mexican company Salinas del Pacífico, S.A., presented a project to recover the peninsular populations of bighorn by breeding them in the protected environment of Isla Carmen (Instituto Nacional de Ecología 1995). The National Institute of Ecology supported the plan and gave the company a permit to capture 15 adult bighorn, 12 females and 3 males, in the Sierra de El Mechudo. The master plan establishes that when the herd reaches 175 individuals in an estimated time of 10 years, adults will be captured and used to repopulate the peninsular mainland. The plan, however, was received with criticism by conservationists in the peninsula. First, the peninsular population was evaluated in the early 1980s, and the count gave some 5000–7000 individuals, a large number by any estimation. Second, there is no evidence that the mainland population is under any important threat. Third, an island population size of less than 175 does not seem sufficiently large to ensure the recovery of a peninsular population of many thousand. And, last, although the Baja California bighorn is indeed a rare and valuable wildlife species, trying to ensure its long-term survival through a program based on the introduction of these game animals into fragile island environments is not justifiable. In short, although the bighorn program in Isla Carmen was done with a conservationist justification, there is reason to doubt the value of the enterprise from a truly conservationist perspective.

Isla Espíritu Santo

Espíritu Santo is a large, 10,200-ha island surrounded by a set of smaller islets known as Isla Partida, Los Islotes, La Ballena, El Gallo, and La Gallina. It lies some 20 km north of the city of La Paz. Because of its proximity to La Paz, the island has been intensely used in the past and is still the most intensely visited island of the Sea of Cortés. It is filled with evidence of pre-Hispanic occupancy by the Pericú people. In the nineteenth century Don Gastón Vivés established here the first pearl-oyster farm in the world.

In the 1960s the ports in the state of Baja California Sur were granted legal status as duty-free areas, and the economy made a rapid transition from ranching and agriculture into international commerce. As a result, communications developed rapidly: ferry ports, airports, and the transpeninsular highway were built, and tourism started to take off. Some pioneering entrepreneurs who loved the natural beauty of the region started to develop at that time a new brand of nature tourism that involved low-impact activities such as whale watching, kayaking, camping, and visits to the islands. Ecotourism was then a new and revolutionary concept in Mexico, and its development in the Sea of Cortés has since set an example for other regions of the country. Espíritu Santo became a major destination for local nature tours. Coastal fishing in open panga boats also developed in the 1960–70s, and the fishermen started to use the island for temporary camps and were still using it for that purpose in the 1990s. Twenty-two camps were in operation in Espíritu Santo in 1999. Finally, the academic and researcher sector also developed with the new economy in Baja California Sur. The University of Baja California Sur and two research centers were established in La Paz in the 1970s, and researchers started to visit the island and use it as a convenient research area and field station. In spite of these uses, the island is still in an extremely good state of conservation, an empirical fact that shows that low-impact nature tourism can indeed meet its declared goals.

In spite of the common use by many stakeholders, the island, however, has individual legal owners. In 1976, 2 years before the decree protecting the islands of the Sea of Cortés was issued, the Secretaría de la Reforma Agraria (SRA, or Secretariat of

Agrarian Reform) gave legal tenure of the island to the ejido Alfredo Bonfil from La Paz. In the 1990s, facing a crisis of underground water depletion in the mainland, the ejido started to look for alternative, nonagricultural uses for their land and turned to Espíritu Santo. In 1992 the Mexican constitution was amended to allow communal ejido lands to be privatized, and the ejido Bonfil obtained authorization from SRA to parcel out 90 ha of island acreage for development. Thus, an obvious conflict arose between the presidential decree declaring the islands of the Sea of Cortés a natural protected area and the development authorization for Espíritu Santo. This led in turn to a sort of legal stalemate, in which the ejido people have the right to develop the island, but have not been able to get the permits from the Secretariat of the Environment (Mexico's Secretaría de Medio Ambiente, Recursos Naturales y Pesca, or SEMARNAP).

To resolve this situation, the ejido decided to cooperate in the preparation of a management plan for the island. They rapidly realized that their biggest chance of obtaining some income from the island's land surface lay in being able to use it in a manner compatible with its status as a natural protected area. The management plan was drawn up by CICIMAR, a local research center, with the participation of the management authorities of the protected area Islas del Golfo de California (ISLA, a conservation organization), The Nature Conservancy, the University of Baja California Sur, the Mexican Center of Environmental Law (Centro Mexicano de Derecho Ambiental, or CEMDA), and, most important, the ejido Bonfil. A number of workshops were held, and in 1998 the final document was ready. The management plan for Espíritu Santo was the first island-specific plan to be finished in the Sea of Cortés, and it has become a landmark for regional conservation. The fact that the ejido landowners agreed to participate (and in doing so accepted the potential consequence of there being restrictions to development in their own insular lands) was indeed a turning point and a lesson for the growing pressures for traditional tourism development that the islands of the Sea of Cortés are facing.

Financing Conservation

As a response to increasing loss of biodiversity, at the beginning of the 1990s the government of Mexico started to explore new strategies to preserve natural protected areas. In 1992, in the wake of the Rio Summit, the government of Mexico obtained a $25 million donation from the Global Environment Fund (GEF) to help in the management and the conservation of a group of outstanding protected areas within the system of natural protected areas in Mexico. The reserves approved for GEF funding were Calakmul, El Triunfo, El Vizcaíno, Isla Contoy, a small island in the coast of Quintana Roo, Islas del Golfo de California, Mariposa Monarca, Montes Azules, Ría Lagartos, Sian Ka'an, and Sierra de Manantlán.

The extremely bureaucratic nature of the administrative system imposed by the GEF donation hindered the arrival of the GEF funds at the field sites, and little change was initially observed in the administration of the islands of the gulf as a natural protected area. Most of the GEF's donation moneys were initially spent in a very cumbersome system of administration in Mexico City. In 1996, the Mexican federal government established the National Council for Natural Protected Areas, composed by representatives from conservationist, academic, business, social, and indigenous communities, which functions as an adviser to the Secretariat of the Environment. The limited capacity to mobilize GEF's financial resources into the 10 selected reserves drove the council to study the issue and recommend the creation of a permanent conservation fund within an already existing private organization, independent from the government. The Mexican Fund for the Conservation of Nature (Fondo Mexicano para la Conservación de la Naturaleza, or FMCN) was chosen for this purpose. Mexico's Secretariat of the Environment negotiated and finally restructured the agreement with GEF, agreeing to deposit the remaining capital of the donation in a special trust fund that would allow the long-term operation of the nature reserves.

On 21 July 1997, the FMCN received $16.48 million that still remained from GEF's original grant (the other $8.52 million had already been used by the federal administration between 1993 and 1997 for the management of the biosphere reserves included in the program) and invested these moneys into a newly created Fund for Natural Protected Areas (the Fondo de Áreas Naturales Protegidas or FANP). The financial revenues generated are administered by the FANP's staff, which has its own board within the umbrella of the FMCN. The administrators and managers of the natural protected area Islas del Golfo de California (like all other reserves included in the GEF donation) now present an annual operation program to obtain financial support from the fund. The financed activities have to correspond to the management program, which identifies the major threats to the conservation of biodiversity and specific actions to mitigate them.

In short, a radical change has occurred during the 1990s in the management of some protected areas in Mexico, including the islands of the Sea of Cortés. From "paper parks" with no fixed budget, the islands now have been endowed by a modest financial fund that permanently supports their operations. Additionally, as part of the negotiations with GEF, the Mexican federal government has agreed to permanently support the salary of the basic managerial staff. In the case of the gulf islands, and because of the sheer size of the protected area, three administrative offices were established in 1996, operating out of La Paz, Baja California Sur, Guaymas, Sonora, and Bahía de los Ángeles/Ensenada, Baja California. A fourth office was established in Loreto, Baja California Sur, to supervise the Loreto Bay National Park. Within the protected area, the GEF funds support programs for protection of biodiversity, actions for the strengthening of management, actions to promote the involvement and the participation of fishermen, *ejidatarios*, the Seri people, and eco-tourism companies in the management and operation of the reserve, and additional fund-raising.

Thanks to these actions, the local inhabitants of all the gulf islands (temporary residents totaling some 700 or less at any given time) and of those of the surrounding towns are slowly changing their economic activity from traditional fishing to tourism services. The general management plan of the protected area Islas del Golfo de California, finally approved in October 2000, aims to promote carefully channeled eco-tourism, with emphasis on environmental education.

Concluding Remarks

Conservation in the islands of the Sea of Cortés has progressed through the support of researchers, nongovernmental organizations, local communities, and local, state, and federal governments. The involvement of local groups as allies in conservation has possibly been the single most important element in island conservation. Local commitment has been the driving force of environmental protection in the islands and the key to the success of conservation programs.

Most of the attempts to develop the islands have been based on the argument that some of the islands show a certain level of environmental degradation, mostly as a result of introduced species and of exploitation of natural resources. However, development of the islands would only generate more degradation. It is entirely false that by developing one or a few islands the rest may be better conserved. Empirical experience in other regions of Baja California such as Los Cabos and Nopoló has shown that once development programs are established, others soon follow. No island is dispensable, and all must be protected effectively. If some of the islands are degraded, then the conclusion must be that the island should be restored, not that it should be developed. The experience of eradication of introduced fauna in Isla Rasa—and the subsequent recovery of the marine bird populations—shows that restoration ecology can be very effective in these environments. No island in the Sea of Cortés is unimportant for conservation.

Finally, the incorporation of all islands of the Sea of Cortés into MAB's international network of Biosphere Reserves generates an immense responsibility for the Mexican government and the local communities. The islands are now a recognized part of the global heritage of the world's biological diversity, and efforts must be made to protect them effectively as such.

References

Arnaud, G. 1998. Erradicación de especies exóticas de Isla Coronados, Parque Nacional Bahía de Loreto. Notas Generales. *Insulario* (Gaceta Informativa de la Reserva Islas del Golfo de California) **6**:16–17.

Arvizu, M.J. 1987. Fisheries activities in the Gulf of California, Mexico. *CALCOFI Report* **28**: 26–32.

Bancroft, G. 1927. Notes on the breeding coastal and insular birds of Central Lower California. *Condor* **29**:188–195.

Banks, R.C. 1962a. A history of exploration for vertebrates in Cerralvo Island, Baja California. *Proceedings of the California Academy of Sciences* **30**(6):117–125.

Banks, R.C. 1962b. Birds of the Belvedere expedition to the Gulf of California. *Transactions, San Diego Society of Natural History* **13**:49–60.

Barreto, J.A. 1973. Isla Rasa, B.C., refugio de gaviotas y gallitos de mar. *Bosques y Fauna* **10**(4):3–8.

Becerril-Nieva, N., R. López-Estudillo, and A. Arzola. 1988. La caza del borrego cimarrón *Ovis canadensis mexicana* en Sonora, temporada 1983–1988. Unpublished report, Secretaría de Desarrollo Urbano y Ecologia, México, D.F.

Boswall, J., and M. Barrett. 1978. Notes on the breeding birds of Isla Rasa, Baja California. *Western Birds* **9**(3):93–108.

Bourillón, L., and B. Tershy. 1997. A model management planning process for protected Gulf of California Islands: The Bahía de los Angeles area. U.S. Fish and Wildlife Service, Washington, DC.

Bourillón, L., A. Cantú, F. Eccardi, E. Lira, J. Ramírez, E. Velarde, and A. Zavala. 1988. *Islas del Golfo de California*. SG-UNAM, México, D.F.

Breceda A., A. Castellanos, L. Arriaga, and A. Ortega. 1995. Nature conservation in Baja California Sur Mexico. *Protected Natural Areas Journal* **15**(3):267–273.

Case, T.J., and M.L. Cody. 1983. *Island Biogeography in the Sea of Cortés*. University of California Press, Berkeley.

DOF. 1963. Decreto por el que se declara zona de reserva natural y refugio para la fauna silvestre, la Isla de Tiburón, situada en el Golfo de California. *Diario Oficial de la Federación*, 15 March 1963.

DOF. 1964. Decreto que declara zona de reserva natural y refugio de aves a Isla Rasa, estado de Baja California. *Diario Oficial de la Federación*, 30 May 1964.

DOF. 1978. Decreto por el que se establece una zona de reserva y refugio de aves migratorias y de la fauna silvestre en las islas que se relacionan, situadas en el Golfo de California. *Diario Oficial de la Federación*, 2 August 1978.

DOF. 1988. Ley General del Equilibrio Ecológico y la Protección al Ambiente. *Diario Oficial de la Federación*, 28 January 1988.

DOF. 1993. Decreto por el que se declara area natural protegida con el carácter de Reserva de la Biosfera, la Región conocida como Alto Golfo de California y Delta del Río Colorado, ubicada en aguas del Golfo de California y los Municipios de Méxicali, B.C., de Puerto Peñasco y San Luis Río Colorado, Sonora. *Diario Oficial de la Federación*, 10 June 1993. pp. 24–28.

DOF. 1996a. Decreto por el que se declara área natural protegida con el carácter de parque marino nacional, la zona conocida como Bahía de Loreto, ubicada frente a las costas del Municipio de Loreto, Estado de Baja California Sur, con una superficie de 11,987-87-50 hectáreas. *Diario Oficial de la Federación*, 19 July 1996.

DOF. 1996b. Decreto que reforma, adiciona y deroga diversas disposiciones de la Ley General del Equilibrio Ecológico y la Protección al Ambiente. *Diario Oficial de la Federación*, 13 de diciembre de 1996.

DOF. 2000. Acuerdo que tiene por objeto dotar con una categoría acorde con la legislación vigente a las superficies que fueron objeto de diversas declaratorias de áreas naturales protegidas emitidas por el Ejecutiveo Federal. *Diario Oficial de la Federación*, 7 June 2000.

Felger, R.S., and M.B. Moser. 1985. *People of the Desert and Sea. Ethno-botany of the Seri Indians*. University of Arizona Press, Tucson.

Gómez-Pompa, A., and R. Dirzo. 1995. *Reservas de la bisofera y otras áreas naturales protegidas de México*. INE/CONABIO, México, D.F.

Hernández-Alvidrez, R., and J. Campoy-Favela. 1989. Observaciones recientes de la población de borrego cimarrón en Isla Tiburón, Sonora, México. *Ecológica* **1**(1):1–29.

Inmobiliaria Isla Coronados. 1995. Desarrollo Turístico Hotel Posada Caracol y Estación de Investigación y Monitoreo, Isla Coronados, Loreto. B.C.S. Environmental Impact Statement prepared by Universidad Autónoma de Baja California Sur for Immobiliaria Isla Coronados. La Paz, BCS.

INE. 1994. *Programa de manejo de la Reserva Especial de la Biosfera islas del Golfo de California*. SEDESOL, México, D.F.

INE. 1995. Proyecto de recuperación del borrego cimarrón (*Ovis canadensis weemsi*) en la Península de Baja California Sur. Introducción del borrego (*O. c. weemsi*) a la Isla del Carmen, B.C.S. Project presented by Salinas del Pacífico, S.A., to the Instituto Nacional de Ecología. México, D.F. October 1995.

Jiménez, R.B., F. Verdugo L., G. Smith V., R. Verdugo L., and Ceseña. P. 1999. Informe final de las acciones de señalización, difusión y comunicación para las Islas de Bahía de los Angeles, Baja California, México. Report for Pronatura B.C., prepared by the Island Committee of the Ejido Tierra y Libertad, Bahía de los Angeles, B.C.

Krutch, J.W. 1961. *The Forgotten Peninsula: A Naturalist in Baja California.* W. Sloane Associates, San Francisco.
Lee, R., and E.E. López-Saavedra. 1994. A second helicopter survey of desert bighorn sheep in Sonora, Mexico. *Desert Bighorn Council Transactions* **38**:12–13.
Lindsay, G.E. 1962. The Belvedere expedition to the Gulf of California. *Transactions of the San Diego Society of Natural History* **13**(1):144.
Lindsay, G.E. 1964. Sea of Cortez expedition of the California Academy of Academy of Sciences, June 20–July 4, 1964. *Proceedings of the California Academy of Sciences* **30**(11): 211–242.
Lindsay, G.E. 1966. The Gulf Islands expedition of 1966. *Proceedings of the California Academy of Sciences* **30**(16):309–355.
Lindsay, G.E. 1970. Some natural values of Baja California. *Pacific Discovery* **23**(2):1–10.
Monson, G. 1980. Distribution and abundance. In: G. Monson and L. Sumner (eds), *The Desert Bighorn.* University of Arizona Press, Tucson; pp. 40–51.
Montoya, B., and G. Gates. 1975. Bighorn capture and transplant in Mexico. *Desert Bighorn Council Transactions* **19**:28–32.
Navarro, F. 1999. Unidad de Manejo y Aprovechamiento Sustentable de la Vida Silvestre (UMA) in Isla Tiburón: Un ejemplo de Desarrollo Sustentable. *Insulario* (Gaceta Informativa de la Reserva islas del Golfo de California) **8**:13–15.
Olivera, J.L.D., and A. López. 1988. Isla Tiburón para los "Tiburones". *Etnias* **1**(2):9–10.
Pallares, E. 1999. El borrego cimarrón: beneficio para los indígenas. *National Geographic* (spanish version, section Geografía America Latina) **3**(4):15–32.
Quiñonez, L., and M. Rodríguez. 1979. Isla Tiburón. *Bosques y Fauna* **2**(1):27–39.
SDNHM. 1996. An interview with George Lindsay. Unpublished transcription of a tape-recording conducted by Michael W. Hager. San Diego Natural History Museum, San Diego, California, 19 February 1976.
SEDUE. 1989. *Información básica sobre las Áreas Naturales Protegidas de México.* Subsecretaría de Ecología-SINAP. México, D.F.
Tobís, S.H. 1968. *Refugio de aves acuaticas migratorias Isla Rasa, B.C.* Direccion General de la Fauna Silvestre, Mexico, D.F.
Velarde, E. 1988. Baja's kingdom of the sea. *Animal Kingdom* **91**(4):24–31.
Velarde, E. 1993. Predation on nesting birds larids by Peregrine Falcons at Rasa Island, Gulf of California, Mexico. *Condor* **95**:706–708.
Velarde, E., and D.W. Anderson. 1994. Conservation and management of seabird islands in the Gulf of California. Setbacks and successes. In: D.N. Nettleship, J. Burger, and M. Gachfeld (eds), *Seabirds on Islands: Threats, Case Studies and Action Plans.* Birdlife Conservation Series No. 1, BirdLife International, Cambridge. pp. 229–243.
Velarde, E., D.W. Anderson, and S.B. Beebe. 1985. *Conservation of the Islands in a Desert Sea. Management and Planning Proposal for the Sea of Cortez and Its Islands.* University of California, Davis.
Velarde, E., M.S. Tordesillas, L. Vieyra, and R. Esquivel. 1994. Seabirds as indicators of important fish populations in the Gulf of California. *CALCOFI Reports* **35**:137–143.
Velázquez-Noguerón, V. 1969. *Aves acuaticas migratorias en isla Rasa. B.C.* Dirección General de la Fauna Silvestre, Mexico, D.F.
Vermeer, K., D.B. Irons, E. Velarde, and Y. Watanuki. 1993. Status, conservation, and management of nesting *Larus* gulls in the North Pacific. In: *The Status, Ecology, and Conservation of Marine Birds of the North Pacific.* Canadian Wildlife Service, Special Publication, Ottawa, pp. 131–139.
Vidal, N. 1967. Aportación al conocimiento de Isla Rasa, Baja California. Bachelor's dissertation, Universidad Nacional Autónoma de México, México, D.F.

Villa, B.R. 1983. Isla Rasa Paradigma. In: *Memorias del Simposio sobre Fauna Silvestre.* UNAM, Facultad de Medicina Veterinaria y Zootecnia, and Asociación de Acuarios y Zoológicos de México. México, D.F.; pp. 56–78.

Villa, B.R., M.A. Treviño, M. Herzig Z., M. Valdéz, G. Davis T., M. Manieu, and W. López-Forment. 1979. Informe de los trabajos de campo en Isla Rasa, Mar de Cortes, Baja California Norte durante la temporada de reproducción de las Aves Marinas, correspondiente a 1977. In: *Memorias del III Simposio Binacional sobre el Medio Ambiente del Golfo de California.* Instituto Nacional de Investigaciones Forestales, Publicación Especial no. 14, pp. 82–87.

Villa, B.R., M.A. Treviño, M. Herzig, M. Valdez, G. Davis, M. Manieux and W. López-Forment. 1980. Informe de los trabajos de campo en la Isla Rasa; Mar de Cortés, Baja California Norte, durante la temporada de reproducción de las aves marinas, correspondiente a 1977. *Calafia* (Journal of the Dirección de Extensión Universitaria, UABC) **4**(2): 25–30.

Walker, L.W. 1951. Sea birds of Isla Raza. *National Geographic* **99**:239–248.

Walker, L.W. 1965. Baja's island of birds. *Pacific Discovery* **18**:27–31.

Wiggins, I.L. 1962. Investigations in the natural history of Baja California. *Proceedings of the California Academy of Sciences* **30**(1):1–45.

PART IV

APPENDIXES

Appendix 1.1

New Measurements of Area and Distance for Islands in the Sea of Cortés

ROBERT W. MURPHY
FRANCISCO SANCHEZ-PIÑERO
GARY A. POLIS
ROLF L. AALBU

Studies of island biogeography frequently attempt to explain the numbers and composition of insular biota (e.g., MacArthur and Wilson 1963, 1967; Lack 1976; Abbott 1983). These investigations involve quantitative, statistical methods using the number of species as the dependent variable. The independent variables used in the correlative regression analyses may include island area (A), distance to nearest source fauna (D), habitat complexity or the number of plant species, elevation, and latitude. The predictive values from such analyses directly depend on the accuracy of the measurements of the independent variables.

Several investigations have examined species numbers on islands in the Sea of Cortés, as summarized in Case and Cody (1983). We measured A and D (to the nearest larger source fauna) for almost 80 islands in the Sea of Cortés clearly discernable from Earth Resources Observation Systems (EROS) satellite photographs. However, many small coastal islands were not measured, such as those at Bahía Coyote in Bahía Concepción, as well as all Pacific coast islands. In addition, we could not locate Isla Farallon on any photograph. Some islands that do not remain isolated at low tide, or that cannot be separated on EROS images, have either been combined in the table (e.g., Islas Los Coronados) or not measured for island area using satellite images (e.g., Isla San Luis Gonzaga). Nevertheless, many islands with faunal and floral records have been measured. The list includes islands associated with the peninsula of Baja California, both in the Pacific Ocean and the Gulf of California.

For measurements determined from EROS photographs, the error in repeated, direct measurements of D was ±0.2 km. Area, A, determined using a digitizing tablet, varied by less than 3% for the large islands, by about 5% for intermediate sized islands, and by as much as 60% for small islands. Estimates of A for extremely small islands (<0.2 km^2) should be considered tentative, unless specifically surveyed. Final estimates of A were determined by comparing digitized island shapes with aerial photographs, particularly those in Miller and Baxter (1977), to determine the best digitized replication of island shape, and by comparing relative island size among the smaller-sized islands. For statistical purposes, it should be noted that all measurements were made from the same scale (1 : 121,121), and thus all are scaled equally and differ proportionally.

On comparison, we found a surprising lack of correspondence with published values (e.g., Soulé and Sloan 1966; Case 1975; Gastil et al. 1983). Discrepancies in measurement of A and D between satellite and published measurements for some islands were as large as fourfold, and frequently reached twofold. In general, the new estimates of A were slightly smaller than those previously published. The discrepancies in measurements are not attributable to different scale maps, as evidenced by numerous shifts in rank order of size. The discrepancies probably result from the previous estimations of A and D being made from inaccurate navigational maps.

These new measurements may significantly affect the conclusions of previous biogeographic investigations. For example, Gastil et al. (1983) reported San Ildefonso and San Francisco islands as having equal areas (2.6 km^3). Surprisingly, the latter island supports almost three times as many lizard species as the former (Murphy and Ottley 1984), an observation previously attributed to island topography (Murphy 1983). Measurements from EROS satellite photographs, however, resulted in estimates of A of 0.6 km^2 for San Ildefonso and 1.5 km^2 for San Francisco, nearly a threefold difference. Thus, differences in A alone may adequately explain the variation in numbers of lizard species on these two islands as a quantification of habitat available to support stable populations of species; explanations of habitat complexity (Murphy 1983) are unnecessary.

Comments on the Table of Data

Completeness

We have attempted to list all islands in the region that have associated biological data. These include islands that have been remeasured, as well as those that have not. Some of the smaller islands could not be located on satellite images or in standard gazetteers. Thus, these islands have no coordinates. The problem is further complicated by the occurrence of numerous synonyms used for the same island. Our list does not include many unnamed rocks (e.g., those charted around Isla San Marcos), or named islets for which there are no biological data. We would appreciate receiving specific island coordinates obtained via Geographic Positioning Systems (GPS), as well as island areas and elevations determined through on-site surveys.

A cursory review of the table (A1.1-1) shows many missing data points. As far as we can determine, no reliable data exist for these missing values.

Island Synonyms

We have attempted to summarize all known synonyms of islands. In summarizing data, we have discovered many errors in attributing species to particular islands. Thus, our island-name synonyms reflect biological attributes as well as variances in geographic and map names. In table (A1.1), individual islands may still be listed more than once. For example, we give one listing for Isla de Altamura and another for Isla Altamura. The island coordinates indicate that these islands are very close to one another, but the separate listing by the U.S. Board on Geographic Names suggests the occurrence of two nearby islands.

Island Areas

Many islands have no estimated areas. This is particularly true for smaller islands and *rocas*. However, data are also missing for many larger islands, particularly those associated with the head of the gulf and in Pacific coast lagoons. In general, we have avoided making island area estimations from navigational maps.

Latitude and Longitude Coordinates

The latitude and longitude coordinates of most islands were taken from the U.S. Board on Geographic Names (USBGN). Our on-ground survey found no variance from those posted on the U.S.Geographic Survey web site (www-nmd.usgs.gov/www/gnis/index.html). Apart from these coordinates, the specific locations of islands vary from map to map and may be unreliable to within 5′–10′ or more. Most of the previously published island coordinates appear to have been taken from inaccurate maps, the same maps used to estimate island size and distance. For example, during a trip to the gulf islands in 2000, one of us (R.W.M.) used a Garman GPS3+ to navigate from San Esteban to Ángel de la Guarda in fog. This GPS unit, which shows maps, indicated that we were 3.5 km east of the island in the gulf when we were standing on Ángel de la Guarda. In our table of data, we unfortunately found it necessary to use some previously published coordinates due to the absence of more recently gathered data. Coordinates given to seconds were taken from the USBGN database, and these appear to be very accurate. Locations given in degrees and minutes were estimated from various maps or taken from the literature, and these are likely to be inaccurate to varying degrees. Among USBGN coordinates, some variation occurs. For example, the coordinates for Isla Salvatierra are listed as 30°03′00″ N, 114°29′00″ W, whereas the synonym Isla Lobos is listed as 30°03′00″ N, 114°31′00″ W.

Distance to Nearest Faunal or Floral Source

Estimates of distance for the major islands in the gulf are measured from EROS satellite images and are indicated as such in the table of data. Some estimates of distance have been taken from other published sources, and we caution against using these as accurate measures.

Elevations

Considerable variation occurs in published estimates of island elevations. We have made few direct measurements on our own, but rather repeated commonly published data.

Ocean Depths

Navigational charts differ considerably in ocean depth profiles. For studies of island biogeography, the most critical issues are (1) were the islands connected to the peninsula during the last (or older) glacial event, and (2) what is the sequence of isolation for landbridge islands? Isla Carmen has been listed as being both a deep-water island and a landbridge island, depending on which map is used to determine ocean depths. DNA sequence data suggest that Carmen is likely a deep-water island (chap. 8).

References

Abbott, I. 1983. The meaning of z in species/area regressions and the study of species turnover in island biogeography. *Oikos* **41**:385–390.

Case, T.J. 1975. Species numbers, density compensation, and colonizing ability of lizards on islands in the Gulf of California. *Ecology* **56**:3–18.

Case, T.J., and M.L. Cody. 1983. *Island Biogeography in the Sea of Cortéz.* University of California Press, Berkeley.

Gastil, R.G., J. Minch, and R.P. Phillips. 1983. The geology and ages of islands. In: T.J. Case and M.L. Cody (eds), *Island Biogeography in the Sea of Cortéz.* University of California Press, Berkeley, CA; pp. 13–25.

Lack, D. 1976. *Island Biology Illustrated by the Land Birds of Jamaica.* University of California Press, Berkeley.

Lewis, L., and P. Ebling. 1971. *Baja Sea Guide.* Miller Freeman Publications, San Francisco, CA.

MacArthur, R.H., and E.O. Wilson. 1963. An equilibrium theory of insular zoogeography. *Evolution* **17**:373–387.

MacArthur, R.H., and E.O. Wilson. 1967. *The Theory of Island Biogeography.* Princeton University Press, Princeton, NJ.

Miller T., and E. Baxter. 1977. *The Baja Book II.* Baja Trail Publications, Huntington Beach, CA.

Murphy, R.W. 1983. The reptiles: origins and evolution. In: T.J. Case and M.L. Cody (eds), *Island Biogeography in the Sea of Cortéz.* University of California Press, Berkeley, CA; pp. 130–158.

Murphy, R.W., and J. R. Ottley. 1984. Distribution of amphibians and reptiles on Islands in the Gulf of California. *Annals of Carnegie Museum* **53**:207–230.

Soulé, M., and A.J. Sloan. 1966. Biogeography and distribution of the reptiles and amphibians on islands in the Gulf of California, Mexico. *Transactions of the San Diego Society of Natural History* **14**:137–156.

Table 1.1-1 Island measurements

Name	State or Region	Island Group	Area (km²)	Latitude	Longitude	Distance (km)	D Measured to	Elevation (m)	Ocean Depth (m)
Pacific Islands									
Abaroa	BCS	Laguna San Ignacio		26°45′	113°13′				
Adelaida (Elide)	BCN	Rosarito		28°40′00″	114°17′00″	0.4	Peninsula	12[a]	
Adentro (El Zapato, Inner Islet)	PO	Guadalupe	14.58	28°51′00″	118°17′00″	<1	Guadalupe	227	
Afuera (El Toro, outer islet; Isla Exterior)	PO	Guadalupe	15.62	28°50′00″	118°16′00″	2	Guadalupe	206	
Alambre, El (Islote El Alambre)	BCS	Laguna Ojo de Liebre		27°45′	114°15′				
Arena (Ana)	BCS	Laguna San Ignacio		26°43′	113°12′				
Asuncion (Angulo Rock)	BCS	Bahía Asuncíon		27°07′00″	114°17′00″	1	Peninsula	50	
Cedros (Cerros)	BCN	Cedros	348	28°12′00	115°15′00	19[a]	Peninsula	1204[a]	
Cedros islet A	BCS	Cedros		28°07′	115°23′	<1	Cedros		
Cedros islet B	BCS	Cedros		28°07′	115°23′	<1	Cedros		
Cedros islet C	BCS	Cedros		28°07′	115°23′	<1	Cedros		
Cedros islet D	BCS	Cedros		28°07′	115°23′	<1	Cedros		
Cedros islet E	BCS	Cedros		28°07′	115°23′	<1	Cedros		
Cedros islet F	BCS	Cedros		28°07′	115°23′	<1	Cedros		
Chester islets (Chester rocks)	BCS	Punta Eugenia		27°52′	115°04′	0.9[a]	Peninsula	5.5[a]	
Choyita (Islote La Choyita)	BCS	Laguna Ojo de Liebre		27°39′	114°04′				
Clarión (Santa Rosa; San Bartolomé)	PO	Revillagigedo		18°22′00	114°44′00″	335	Mexico	335[a]	
La Concha	BCS	Laguna Ojo de Liebre		27°50′	114°13′				
Coronado islet (Roca Medio)	BCN	Los Coronados		32°25′	117°16′			31[a]	
Coronado del Centro (del Medio)	BCN	Los Coronados	2.5 total	32°25′00″	117°16′00″			77[a]	
Coronado del Norte	BCN	Los Coronados		32°26′00″	117°18′00″	13	Peninsula	142[a]	
Coronado del Sur	BCN	Los Coronados		32°24′00″	117°15′00″			205[a]	

(continued)

Table 1.1-1 (Continued)

Name	State or Region	Island Group	Area (km³)	Latitude	Longitude	Distance (km)	D Measured to	Elevation (m)	Ocean Depth (m)
Cresciente (Craciente)	BCS	Magdalena		24°23'00"	111°37'00"	<1	Peninsula	(low)	
Garzas	BCS	Laguna San Ignacio		26°56'	113°09'				
Guadalupe	PO	Guadalupe	255	29°00'00"	118°16'00"	232	Peninsula	1298	>365
La Islita	BCN	Guerrero Negro		28°06'	114°07'				
Los Morros	BCS	Bahía Tortugas		27°39'	114°53'				
Magdalena (Santa Magdalena)	BCS	Bahía Magdalena		24°55'00"	112°15'00"	1	Peninsula	415	
Mangrove	BCS	Bahía Magdalena		24°32'00"	111°49'00"				
Morro Hermoso	BCS	Tortugas		27°31'	114°43'				
Natividad	BCS	Natividad	7.2	27°52'00"	115°11'00"	9		150[a]	31[a]
Natividad North rock	BCS	Natividad		27°53'	115°13'				
Natividad-Roca Vela (Sail Rock)	BCS	Natividad		27°53'	115°10'			17[a]	
Negro	PO	Guadalupe		28°51'	118°17'				
North Rock	PO	Revillagigedo							
Pelicanos	BCS	Laguna San Ignacio		26°54'	113°09'				
La Piedra	BCS	Laguna Ojo de Liebre		27°43'	114°10'				
Piedra de San José	BCN	S of El Rosario							
Piedra Negra (Islote Piedra Negra)	BCS	Punta Eugenia		27°52'	115°03'				
Pirámide	PO	Revillagigedo		18°21'	114°43'	4	Clartión	73[a]	
Rocas Alijos (×3)	PO	Revillagigedo		24°57'	115°44'	245	Peninsula	34[a]	
Roca de Cuervo	PO	Revillagigedo		18°22'	114°44'	<1	Clarión	12[a]	
Roca Doble Pináculo	PO	Revillagigedo		18°46'	110°0'	<1	Socorro	8[a]	
Roca Elefante (Elefante)	PO	Guadalupe		29°07'	118°23'	<1	Guadalupe		
Roca Monumento	PO	Revillagigedo		18°22'	114°45'	<1	Clarión	61[a]	
Roca O'Neal	PO	Revillagigedo		18°45'	110°58'	1.6	Socorro	14[a]	
Roca Partida	PO	Revillagigedo		19°01'00"	112°02'00"	418	Peninsula	34[a]	
Roca Pináculo	PO	Revillagigedo		18°45'	110°02'	<1	Socorro		

Name	State	Location	Lat	Lon		Region	
Rocas del Vapor (Steamboat Rock)	PO	Guadalupe	20°06'	118°23"	<1	Guadalupe	29[a]
Rocas La Soledad (Piedra Santo Thomas)	BCN	Santo Thomás	ca. 31°30'	116°30'	2.3	Peninsula	6[a]
San Benedicto (Inocentes; Añublada)	PO	Revillagigedo	19°18'00"	110°49'00"	354	Peninsula	335[a]
San Benito del Este (Bonito E)	BCS	San Benitos	28°18'	115°34'	31	Cedros	128[a]
					66	Peninsula	
San Benito del Centro (Bonito C)	BCS	San Benitos	28°18'	115°34'	6.3 total		25[a]
San Benito del Oeste (Bonito W)	BCS	San Benitos	28°18'	115°33'			201[a]
San Jerónimo (San Geromino)	BCN	Bahía Rosario	29°47'00"	115°48'00"	0.4	Peninsula	40
San Martín	BCN	San Quintín	30°30'00"	116°07'00"	2.3	Peninsula	151[a]
San Roque	BCS	nr. Bahía Asuncion	27°08'	114°23'		Peninsula	15
Santa Margarita (Margarita; Santa Magdalena)	BCS	Magdalena	24°27'00"	111°50'00"		Peninsula	567
Santo Domingo	BCS	Magdalena					
Socorro (Islas de Santo Tomás)	PO	Revillagigedo	18°45'00"	110°58'00"	398	Peninsula	1040
Todos Santos (N)	BCN	Bahía de Todos Santos	31°49'	116°49'	8	Peninsula	17[a]
Todos Santos (S)	BCB	Bahía de Todos Santos	31°48'	116°47'	8	Peninsula	95[a]
Zacatoso	BCS	Laguna Ojo de Liebre	27°53'	114°13'	1.2 total		
Sea of Cortés							
Alcatraz	SON		28°49'00"	111°17'00"			
Alcatraz	SON		29°11'00"	113°35'00"			
Algadones, de los Algadones	SIN		24°52'00"	107°54'00"			
Almagre Chico (Almagres Chiquito)	SON	Bahía Guaymas	27°45'00"	110°38'00"	0.098[b]	Sonora	15[b] 5[c]
Almagre Grande (Almagres Grande)	SON	Bahía Guaymas	27°55'00"	110°54'00"	0.808[b]	Sonora	
			27°55'00"	110°54'00"	0.603[b]		25[b] 5[c]
Altamura, de Altamura	SIN		24°53'00"	108°08'00"		Tachichilte	
	SIN	Bahía Santa María	25°00'00"	108°12'00"	1		

(*continued*)

Table 1.1-1 (*Continued*)

Name	State or Region	Island Group	Area (km³)	Latitude	Longitude	Distance (km)	D Measured to	Elevation (m)	Ocean Depth (m)
Ángel de la Guarda (Angel de la Guardia)	BCN	Ángel de la Guarda	936.04[d]	29°20'00"	113°25'00"	12.12[d]	Peninsula	1316[c]	244
Ánimas, Las (Las Ánimas Sur)	BCS	San José	0.08[d]	25°05'46"	110°33'33"	11.88[d]	San José	27[c]	243
Arboleda	SON								
Ardilla	SON	Bahía Guaymas		27°55'00"	110°54'00"				
Ave	JAL								
Bahía Ánimas NE	BCN	Bahía de las Ánimas	0.004[b]	28°48'09"	113°20'53"[*h]	0.204[h]	Peninsula	10	10
Bahía Ánimas SW	BCN	Bahía de las Ánimas	0.004[b]	28°48'08"	113°20'59"[*h]	0.121[b]	Peninsula	10	10
Bajito, El	BCN	Colorado River		31°38'	114°45'				
Ballena	BCS	Espíritu Santo	0.46[d]	24°29'00"	110°25'00"	0.79[d]	Espíritu Santo	69[e]	12
Baredito	SIN								
Bargo (Barga)	BCS	Bahía Concepción		26°42'	111°56'		Peninsula	30[a]	18[c]
Basacori	SON	nr. Sinaloa		26°24'00"	109°15'00"				
Blanca	SON	Guaymas	0.051[b]	26°56'	110°59'	1.368[b]	Sonora	10[b]	9[c]
Blanca (Ánimas Bay)	BCN	Bahía de las Ánimas	0.029[b]	28°49'05"	113°21'23"	0.643[b]	Peninsula	12	10
Blanco (Roca Blanca, Islote Blanco)	BCS	nr. Danzante	0.030[b]	25°46'04"	111°16'58"[*h]	0.79[d]	Peninsula	20[b]	10
Blanco (Blanca, Concepción)	BCS	Bahía Concepción		26°43'	111°52'		Peninsula		18[c]
Bleditos	SIN	nr. Sonora		25°35'00"	108°59'00"				
Bledos	SIN	nr. Sonora		25°36'00"	108°59'00"				
Bota	BCN	Ventana	0.093[b]	29°01'	113°31'	0.73[d]	La Ventana	40	35
						2.85[d]	Peninsula		
Cabeza de Caballo	BCN	Cabeza	0.77[d]	28°58'00"	113°28'00"	2.24[d]	La Ventana	140	40
						2.00[d]	Peninsula	(68)[a]	
Cabeza de Natividad	JAL								

Name	State	Location		Lat	Lon	Region		
Calaveras	BCN	Calaveras	0.004[b]	29°01'	113°29'	Coronado Peninsula	17[b]	100
Candelero (Ventana, Leon Dormido)	SON	Bahía Guaymas						
Cardones	SIN	nr. Mazatlán		23°11'00"	106°25'00"	Partida N.		46[c]
Cardonosa Este (Cardonosa)	BCN	Partida N.	0.14[d]	28°53'	113°02'	Peninsula	479[c]	150
Carmen (El Carmen)	BCS	Carmen	143.03[d]	25°59'44"	111°08'36"	San José		61
Cayo	BCS	San José	0.02[d]	24°53'	110°37'	Peninsula		
Cerraja	BCN	Ventana	0.037[b]	29°00'	113°31'	Peninsula	12[b]	35
						La Ventana		
Cerralvo	BCS	Cerralvo	140.46[d]	24°15'00"	109°55'00"	Peninsula	767[c]	235
Cerro Blanco	BCS	Cape Region		22°52'00	109°53'00			
Cerro La Bufadora	BCS	Cape Region		22°52'00	109°53'00			
Chapetona (Rasa)	SON	Bahía Guaymas						
Cholla	BCS	Carmen	0.035[b]	26°01'	111°12'	Peninsula	6[a]	150
						Carmen		18[c]
						Coronados		
Cholludo (Lobos, La Foca, Roca Foca)	SON	S of Tiburón	0.02[a]	27°20'00"	110°36'00"	Dátil	100	
						Tiburón		
Chivos	SIN			23°11'00"	106°26'00"			
Ciari (Ciaris, Siari)	SON			26°58'00	109°56'00"			
Cocinas (Colorado)	JAL			19°33'	105°07'			
Dauto	SIN			24°30'00"	107°38'00"			
Consag (Roca Consag, Roca Consagracion)	BCN	San Felipe		31°06'	114°34'	Peninsula	30[a]	27[c]
Corm	BCN	San Pedro Mártir						
Coronadito	BCN	Coronado	0.072[b]	29°06'	113°32'	Coronado Peninsula	15	35
Coronado (Smith, Smiths)	BCN	Bahía de los Angeles	9.13[d]	29°04'00"	113°32'00"	Peninsula	465[c]	35
Coronados (Los Coronados)	BCS	Coronados	7.59[d]	26°07'00"	111°17'00"	Peninsula	283[c]	35

(*continued*)

Table 1.1-1 (Continued)

Name	State or Region	Island Group	Area (km^2)	Latitude	Longitude	Distance (km)	D Measured to	Elevation (m)	Ocean Depth (m)
Coronados islet SW (Santiago?)	BCS	Coronados	0.05^d	26°06'	111°19'	0.73^d	Coronados		9^c
						1.09^d	Peninsula		
Coyote (El Partido, El Coyote)	BCS	San José	0.01^d	24°50'	110°37'	1.15^d	San José	12^a	65
						9.27^d	Peninsula		
Coyote (El Coyote)	BCS	Bahía Concepción	0.262^b	26°43'	111°53'	0.746^b	Peninsula	110^b	6
Craveries	BCN	San Pedro Mártir							
Crestón	SIN	nr. Mazatlán		23°11'00"	106°27'00"				18^c
La Cueva	BCS	Bahía Concepción	0.017^d	26°44'	111°52'			25^b	
Cueva	SON			26°44'	111°52'	1.673^b	Sonora	106^c	30
Danzante	BCS		4.64^d	25°47'07"	111°14'59"	2.61^d	Peninsula	170^d	9^c
Dátil (Cholludo, Turners; Turner; Tornero)	SON	S of Tiburón	1.25^d	28°43'00"	112°19'00"	1.94^d	Tiburón		
Division (Navio; de la Vela)	BCN	Ángel de la Guarda		29°34'00"	113°34'00"	0.05	Ángel de la Guarda		3^c
La Doble (Algodones)	SON	Algodones							
Encantada East (Bird rock)	BCN	Encantadas	0.05^d	30°02'	114°27'	0.55^d	Encantada		46^c
Encantada (sometimes referred to as El Cholludo, but see above; Incantada)	BCN	Encantada	0.53^d	30°01'00"	114°30'00"	6.85^d	Peninsula	108^a	37^c
Encantada Grande (Salvatierra [in error], San Luis, San Luis Gonzaga [in error])	BCN	Encantadas	6.85^d	29°58'00"	114°26'00"	5.45^d	Peninsula	223^c	33
						5.76^d	Encantada		
de Enmedio (Blanca, Pastel)	SON	Bahía Guaymas		27°55'00"	110°59'00"				
Espíritu Santo	BCS	Espíritu Santo	87.55^d	24°30'00"	110°22'00"	6.15^d	Peninsula	576^c	12
Estanque (La Víbora, Pond)	BCN	Ángel de la Guarda	1.03^d	29°03'00"^{b}	113°07'00"	0.61^d	Ángel de la Guarda	122^a	<3
El Fierro	SIN	nr. SON		25°08'00"	108°18'00"	2.79^d	Peninsula	50	35
Flecha	BCN	La Ventana	0.129^b	29°00'	113°31'	0.61^d	La Ventana		

456

Name	State	Nearest		Lat	Lon		Region		
Las Galeras E	BCS	Monserrate	0.01	25°43'	111°03'	2.73[d]	Monserrat		30
Las Galeras W	BCS	Monserrat	0.01	25°43'	111°03'	2.73[d]	Monserrat		30
						0.30[d]	Las Galeras East		
Gallina	BCS	Espíritu Santo	0.05[d]	24°26'	110°23'	0.87[d]	Espíritu Santo		12
Gallo (El Gallo)	BCS	Bahía Concepción		26°53'	111°54'				9[c]
Gallo	BCS	Espíritu Santo	0.12[d]	24°28'00"	110°24'00"	0.46[d]	Espíritu Santo		12
Garrapata	SIN	nr. Sonora		25°10'00"	108°15'00"				
Gemelos E (Gemelitos)	BCN	Los Gemelos	0.047[b]	28°57'	113°28'	0.30[d]	Gemelitos W. Peninsula	20[b]	40
						0.79[d]			
Gemelos W (Gemelitos)	BCN	Los Gemelos	0.020[b]	28°57'	113°29'	0.85[d]	Peninsula	20[b]	40
del Giero	SIN	nr. Sonora		25°08'00"	108°23'00"				9[c]
Gore (Core)	BCN	Colorado River		31°44'00"	114°38'00"				37[c]
Granito (Granite)	BCN	Ángel de la Guarda	0.32[d]	29°35'00"	113°33'00"	1.39[d]	Ángel de la Guarda	85[a]	
Guapa (Guapo, Morro Tecomate)	BCS	Bahía Concepción		26°42'	111°52'			24[a]	18[c]
Habana	BCS	Mainland San Diego		25°08'00"	110°52'00"	2.1[a]	Peninsula		46[c]
Hermano del Norte	SIN								
Hermano del Sur	SIN								
Huerfanito, El	BCN	Encantadas	0.08[d]	30°07'00"	114°39'00"	0.48[d]	Peninsula	23[a]	27[c]
Huivuilay (Huivulay, Huivulai, Tóbara)	SON			27°03'00"	110°01'00"				
Iguana	SIN			24°27'00"	107°38'00"				
El Indio	SIN	nr. SON		25°41'00"	108°53'00"				
El Infiernito	SIN			24°29'00"	107°33'30"				
Isabel (San Isabela, Isabella, Gaviotas)	NAY	Tres Marías		21°51'00"	105°55'00"	32	Nayarit	85[a]	46
Las Islitas	BCS	nr. Danzante	0.03[d]	25°46'	111°17'	0.55[d]	Peninsula		18[c]
Islet E of Partida N	BCN	Partida N	0.14[a]	28°53'	113°01'	19.39[a]	Peninsula	15[a]	400
Jaltemba	JAL			21°03'00"	105°18'00"				
Jorobado	BCN	La Ventana	0.039[b]	29°01'	113°31'	0.36[d]	Flecha	25	35
						2.12[d]	Peninsula		

(continued)

Table 1.1-1 (*Continued*)

Name	State or Region	Island Group	Area (km³)	Latitude	Longitude	Distance (km)	D Measured to	Elevation (m)	Ocean Depth (m)
Lagartija	BCN	W of Is. Salsipuedes	0.05[d]	28°44′	112°59′	0.30[d]	Salsipuedes	15[a]	<46[c]
de la Lechuguilla (Lechuguilla)	SIN	nr. SON		25°43′00″	109°23′00″				
Leon (Santa Catalina)	SON	Bahía Guaymas							9[c]
La Liebre	BCS	Bahía Concepción		26°44′	111°54′	2.85[d]	Peninsula	8[b]	35
Llave	BCN	Ventana	0.022[b]	29°00′	113°31′	0.06[d]	La Ventana		
La Lobera (Islote La Lobera)	BCN			27°54′	112°45′				18[c]
Lobera	SON								
Lobos	SIN	nr. Mazatlán		23°13′00″	106°29′00″				
Lobos (Areo)	SON			27°52′	110°57′				
Lobos (Gaviota)	BCS	La Paz	0.14[d]	24°18′	110°21′	0.35[d]	Peninsula	4[a]	6
Los Islotes	BCS	Espíritu Santo	0.04[d]	24°36′	110°24′	0.75[d]	Partida Sur	15[a]	6
Luz	BCS								
Macapule (de Macapule, Racapule)	SIN			25°20′00″	108°38′00″	1	Sinaloa		
María Cleofas	Gulf	Tres Marías		21°16′00″	106°14′00″	87	Nayarit	402	
María Madre	Gulf	Tres Marías		21°35′00″	106°33′00″	101	Nayarit	616	
María Magdalena	Gulf	Tres Marías		21°25′00″	106°24′00″	96	Nayarit	366	
						16	María Cleofas		
						11	María Madre		
Mascocahui	SIN?	nr. Sonora		25°34′00″	109°00′00″				
Mejía	BCN	Ángel de la Guarda	2.26[d]	29°34′00″	113°35′00″	0.18[d]	Ángel de la Guarda	262[e]	9[c]
Melendres	SIN			24°54′00″	108°54′00″				
Del Mero	SIN			25°05′00″	108°11′00″				
Mestiza (Chuenque)	BCS	Mestiza	0.03[d]	25°50′	111°19′	0.30[d]	Peninsula	20	15
Miramar (El Muerto, Link)	BCN	Encantadas	1.33[d]	30°05′00″	114°32′00″[m]	3.39[d]	Peninsula	192[c]	35

Mitlán	BCN	Smith	0.156[b]	29°04'	113°31'	0.12[d] 2.00[d]	Coronado Peninsula	23[e] 35
Las Monas Norte	NAY	Isabella						
Las Monas Sur	NAY	Isabella						
Monserrat (Monserrate, Monserrato, Santa Cruz)	BCS	Monserrat	19.86[d]	25°41'00"	111°03'00"	13.70[d]	Peninsula	223[e] 158
Montague (Montagne)	BCN	Colorado River		31°45'00"	114°48'00"			9[c]
Morrito (Inglés, Barra El Morrito)	SON	Bahía Guaymas		27°55'00"	110°48'00"			2
Negrita	JAL							
Novilla	JAL							
Pájaros	SON	Bahía Guaymas	0.881[b]	27°54'00"	110°52'00"	1.042[b]	Sonora	110[b] 5[c]
Pájaros	SIN	nr. Mazatlán		24°49'00"	108°07'00"			
Pájaros	SIN	nr. Nayarit		23°15'00"	108°07'00"			
Palmito del Verde	SIN	nr. Mazatlán		22°40'00"	106°30'00"			
Palmito de la Virgen	SIN			23°00'00"	106°10'00"			
Pardo (Roca Sur, Los Candeleros Sur)	BCS	nr. Danzante	0.038[b]	25°44'	111°13"	0.48[d]	Peninsula	30[b] 20
Partida bird rock	BCN	Partida N						
Partida N (Cardonosa, Partida)	BCN	San Lorenzo Sur	1.36[d]	28°53'00"	113°04'00"	8.30[d] 17.88[d] 12.18[d]	Rasa Peninsula Ángel de la Guarda	122[e] 243
Partida	SON							
Partida Sur	BCS	Espíritu Santo	19.29[d]	26°11'00"	109°19'00"	0.17[d]	Espíritu Santo	335[a] 12
Passavera	JAL			24°34'00"	110°25'00"			
Pata	BCN	Ventana	0.136[b]	29°01'	113°31'	0.06[d] 2.55[d]	Bota Peninsula	40 35
Patos	SON	Patos	0.45[d]	29°17'00"	112°29'00"	7.45[d] 8.82[d]	Tiburón Sonora	15
Pelcáno (=Pelícano?)	SON			28°48'00"	111°58'00"			
El Pelícano (Alcatraz, Tassne)	SON	Colorado River		31°45'00"	114°37'00"	2.01	Sonora	165 5
Pelícano	SON	Bahía Kino		28°49'00"	112°00'00"	2.01	Sonora	165 9[c]
Pelon	NAY	Isabela						

(*continued*)

Table 1.1-1 (Continued)

Name	State or Region	Island Group	Area (km²)	Latitude	Longitude	Distance (km)	D Measured to	Elevation (m)	Ocean Depth (m)
Las Pelonas	SIN			24°36'38"	107°52'57"				
la Peña	NAY			21°02'00"	105°16'00"				
Peruano (Peruano Blanca)	SON		0.030[b]	26°55'	110°58'	0.843[b]	Sonora	15[b]	
Pescador (El Pescador)	BCN	Bahía Pescador	0.035[b]	28°55'05"	113°22'57"	0.516[b]	Peninsula	12[b]	35
de Piedra	SIN	Mazatlán		23°10'00"	106°20'00"				
de las Piedras	SIN			25°52'00"	109°25'00"				
El Piojo (Piojo)	BCN	Piojo	0.57[d]	29°02'00"	113°27'00"	2.61[d]	Coronado	38[d]	35
						6.30[d]	Peninsula		
						4.12[d]	La Ventana		
La Pitahaya (Pitahaya, Sin Nombre)	BCS	Bahía Concepción		26°45'	111°52'				18[c]
Pitahaya	SON	Bahía Guaymas	0.012[b]	26°45'	111°52'	0.537[b]	Sonora	30[b]	5[c]
Pitayosa	SON		0.094[b]	27°54'	110°53'	0.159[b]	Sonora	20[b]	
La Poma (Pomo, Encantada, Cantada)	BCN	Encantadas	0.22[d]	29°59'	114°52'	1.33[d]	Encantada Grande	145[d]	
Quevedo	SIN								
El Racito (Islote El Racito)	BCS			27°50'	112°44'				46[c]
Ramon	BCS	Bahía Concepción		26°45'	111°54'				9[c]
Rancho	SIN	nr. Sonora							
Rasa bird rock	BCN	San Lorenzo Sur		25°09'00"	108°21'00"				
Rasita									
Rasa (Raza, La Rasa, La Raza)	BCN	Midriff	0.68[d]	28°50'00"	113°00'00"	20.79[d]	Peninsula	31[e]	400
						0.58[d]	Salsipuedes		
						9.52[d]	Partida N.		
Rase (Raze, Raza)	SON		0.151	27°57'00"	111°02'00"	0.523	Sonora	10	
de Redo	SIN			24°38'36"	107°58'31"				
Requesón (El Requesón)	BCS	Bahía Concepción	0.134[b]	26°38'	111°50'	0.001[b]	Peninsula	20[b]	<2[a]
de la Risción	SIN			25°07'00"	108°18'00"				

(continued)

Name	State	Location	Area	Latitude	Longitude	Distance	Region		
Roca between Partida N and Ángel de la Guarda	BCN	Partida N Ángel de la Guarda Peninsula Division	0.05[d]	28°54'	113°02	0.61[d] 10.79[d] 19.58[d]		12.5[c]	27[c]
Roca Blanca (Piedra Blanca)	BCN			29°34'	113°33'	1.2			
Roca Gonlondrinas	NAY								7[c]
Roca Lobos (San Rafael)	BCS	San Marcos		27°09'	112°04'				
Roca Morena (Morena)	BCS	Santa Catalina		25°14'57"	110°55'17"				
Roca Negra	BCS	San José		25°15'00"	110°56'00"				
Roca Negra	NAY			19°46'00"	105°20'00"				
Roca Rasa	BCN		0.03[d]	28°50'	113°01	1.58[d] 19.82[d]	Rasa Peninsula	23[a]	46[c] 400
Roca San Marcial	BCS	S. of Monserrat		25°32'	111°10'	2.0	Peninsula	8[a]	18[c]
Roca Solitaria	BCS	S. of Monserrat		25°32'	111°14'	0.3	Peninsula	35[a]	18[c]
Roca Vela (nr. Mejía)	BCN	Ángel de la Guarda	0.03[d]	27°33'	113°37'	0.36[d]	Ángel de la Guarda	51[a]	46[c]
Rocallosa	BCN								
Saliaca	SIN								
Salsipuedes (Sal Si Puedes)	BCN	San Lorenzo Sur	1.16[d]	25°11'00" 28°44'00"	108°20'00" 112°59'00"	1.52[d]	San Lorenzo Norte Peninsula	114[e]	400
Salvatierra (Coloradito, Lobos, Los Lobos)	BCN	Encantadas	0.36[d]	30°03'00"	114°29'00"	19.21[d] 3.39[d]	Peninsula	169[a]	37[c]
San Andres	JAL					5.21[d]	Miramar Encantada		
San Aremar	BCN	Ventana	0.05[d]	29°01'	113°30'	2.97[d] 1.21[d] 3.27[d]	La Ventana Peninsula		27[c]
San Augustin	JAL								
San Basilio (×3) (Rocas de la Gaviota)	BCS	N of Loreto		26°22'	111°26'				5[c]
San Cosme (Roca San Cosme)	BCS	Peninsula-Monserrat	0.28[d]	25°35'	111°09'	1.45[d]	Peninsula	23[a]	46[c]

(continued)

Table 1.1-1 (Continued)

Name	State or Region	Island Group	Area (km³)	Latitude	Longitude	Distance (km)	D Measured to	Elevation (m)	Ocean Depth (m)
San Damian	BCS	nr. Monserrat	0.06[d]	25°35′	111°08′	2.62[d] 1.27[d]	Peninsula San Cosme	14[a]	18[a]
San Diego	BCS	San José	0.60[d]	25°12′55″	110°42′11″	6.36[d] 9.28[d]	Santa Cruz San José	220[c]	61
San Esteban	BCN	Midriff	40.72[d]	28°42′00″	112°36′00″	34.5 11.64[d] 16.85[d]	Peninsula Tiburón San Lorenzo Sur	431[c]	300
San Francisco	BCS	San José	4.49[d]	24°50′00″	110°35′00″	2.36[d] 7.39[d]	San José Peninsula	210[c]	63
San Ignacio Farallón (Farallon de San Ignacio, San Ygnacio Farallón)	SIN			25°27′00″	109°23′00″	1.5	Sinaloa		
San Ignacio (de San Ygnacio)	SIN			25°25′00″	108°54′00″				
San Ildefonso	BCS	San Ildefonso	1.33[d]	26°37′00″	111°26′00″	9.94[d]	Peninsula	117[c]	150
San Jorge (George)	SON	Bahía San Jorge		31°01′	113°15′	13.41	Sonora		18[c]
San Jorge islet-N	SON	Bahía San Jorge		31°01′	113°15′	13.41	Sonora		18[c]
San Jorge islet-S	SON	Bahía San Jorge		31°01′	113°15′	13.41	Sonora		18[c]
San José	BCS	San José	187.16[d]	25°00′13″	110°37′48″	4.16[d]	Peninsula	633[c]	61
San Juan Nepomuceno (San Juan Nepomuzeno, now a peninsula)	BCS	La Paz	1.38[d]	24°17′00″	110°22′00″	(0.30)[d]	Peninsula	24[a]	0[c]
San Juanito	Gulf	Tres Marías		21°43′00″	106°38′00″	109 3.2	Nayarit María Madre	46	366
San Lorenzo Norte (Las Ánimas N, Anima, Roca Blanca)	BCN	San Lorenzo	4.26[d]	28°42′00″	112°56′00″[b]	0.12[d]	San Lorenzo Sur	200[c]	400
San Lorenzo Sur (San Lorenzo)	BCN	San Lorenzo	33.03[d]	28°38′00″	112°51′00″[b]	16.36[d]	Peninsula	485[c]	400
San Lorenzo (La Paz)	BCS	Espíritu Santo		24°23′	110°18′				
San Luis (Algodones)	SON	Los Algodones		27°45′	110°38′				

Name	Code	Location		Lat	Lon		Region		
San Luis Gonzaga (Isla Willard)	BCN	Bahía San Luis Gonzaga	1.687	29°48'00"	114°23'00"	0.003	Peninsula	75	0
San Marcos	BCS	San Marcos	30.07^d	27°13'00"	112°06'00"	4.91^d	Peninsula	271	
San Nicolas	SON	Bahía Guaymas	0.019^b	27°56'	110°03'	0.510^b	Mainland	10^b	10
San Pedro	JAL								
San Pedro Mártir	BCN	San Pedro Mártir	2.90^d	28°22'	112°20'		Triburón	320^c	275
						39.09^d	Sonora		
						62.35^d	San Esteban		
						36.85^d	Lobos		
						35.82^d			
San Pedro Nolasco	SON	nr. Guaymas	3.45^d	27°58'00"	111°25'00"	14.61^d	Sonora	326^c	244
San Ramon	SON		0.045^b	26°45'	111°53'	5.696^b	Sonora	25^b	
San Vicente	SON			27°52'00"	110°51'00"				
San Vincente	SON	Bahía Guaymas		26°45'	111°52'	<0.15	Sonora		5^c
Santa Catalina (Catlán, Catalana, Catalano)	BCS	Santa Catalina	40.99^d	25°39'03"	110°49'03"^{h}	25.15^d	Peninsula	470^c	450
						20.73^d	Monserrat		
Santa Cruz	BCS	Santa Cruz	13.06^d	25°17'21"	110°43'48"^{h}	19.82^d	Peninsula	457^c	200
						16.91^d	San José		
Santa Ines (Santa Inéz)	BCS	Santa Ines	0.37^d	27°03'	111°54'	4.61^d	Peninsula	9^c	7
Santa María	SIN			25°38'00"	109°15'00"				
Satispac	BCN								
Sierra El Negro	SIN	nr. Sonora		25°30'00"	108°51'00"				
Tachichilte (de Tachichilte, Talchichilte, Tachichite, Tachichiltic)	SIN	Bahía Santa Maria		24°59'00"	108°04'00"	3	Sinaloa		
Tesobiáre	SIN			25°29'00"	108°49'00"				
Tecuacahui (A Tunosa)	SIN			25°35'00"	109°01'00"				
Tiburón (del Tiburón)	SON	Tiburón	1223.53^d	29°00'00"	112°25'00"	1.70^d	Sonora	1218^c	10
Tijeras (Las Tijeras, Roca Medio)	BCS	nr. Danzante	0.025^b	25°44'33"^{h}	111°13'31"^{h}	1.88^d	Peninsula	20^b	20
Timbabichi bird rock	BCS	Espíritu Santo							
Tío Ramón	SON	Bahía Guaymas		27°56'00"	110°52'00"				
Tóbari	SON			26°58'00"	109°57'00"				

(continued)

Table 1.1-1 (Continued)

Name	State or Region	Island Group	Area (km²)	Latitude	Longitude	Distance (km)	D Measured to	Elevation (m)	Ocean Depth (m)
Tortuga (de Tortuga)	BCS	Tortuga	11.36[d]	27°26'00"	111°52'00"	36.30[d]	Peninsula	309[c]	1200
						26.65[d]	San Marcos		
Tosacahui	SIN	Sinaloa		25°37'00"	109°01'00"				
Tres Marias islet A	Gulf	Tres Marias							
Tres Marietas (Las Tres Marietas)	NAY	Puerto Vallarta		20°43'	105°35'				
Las Tunas	SIN			24°51'00"	108°05'00"				
Tunosa	SIN								
Venado (La Venado, Algodones)	SON	Algodones	0.235	27°57'00"	111°07'00"	0.175	Sonora	25	
Venados	SIN								
La Ventana (Ventana)	BCN	La Ventana	1.41[d]	28°59'00"	113°30'00"	3.15[d]	Peninsula	120	35
de las Viejas	SON			26°42'00"	109°31'00"				
Vinorama	SIN			25°22'00"	108°46'00"				

Abbreviations for state or region: BCN, Estado de Baja California (northern state); BCS, Baja California Sur; JAL, Jalisco; NAY, Nayarit; PO, Pacific Ocean; SIN, Sinaloa; SON, Sonora

[a] Data from Lewis and Ebling (1971).
[b] Measured by island survey.
[c] Data from Sea of Cortez Charts (1998), Fish n Map Co. (Arvada, Colorado).
[d] Measured from EROS satellite photographs.
[e] Data from Soulé and Sloan (1966).

Appendix 4.1

Vascular Plants of the Gulf Islands

JON P. REBMAN
JOSÉ LUIS LEON DE LA LUZ
REID V. MORAN

This appendix summarizes the available records of vascular plant taxa on 20 major islands in the Sea of Cortés. In the previous edition of this book, the insular floras were reported with a total of 581 plant taxa (Moran 1983), but this new account includes 695 plant taxa. The majority of the new additions to this floristic appendix have been obtained by recent collecting efforts on the southern gulf islands conducted by José Luis Leon de la Luz and his assistants at the Centro de Investigaciones Biológicas del Noroeste and are deposited in the HCIB Herbarium in La Paz, Baja California Sur. However, there is no question that with more floristic surveys on the islands, especially during a good rainy season, the number of plant taxa will increase even more. Many of the islands have not yet been thoroughly explored due to a limited number of boat-landing areas and rugged terrain, so it is possible that even plant taxa new to science are yet to be discovered in this insular region.

Diversity and Composition of the Gulf Flora

At present, the compilation of the floras for the 20 major islands (see table A4.1-4) in the Gulf of California yields 695 plant taxa, representing 97 plant families, 331 genera, and 651 species (table A4.1-1). Of these, 11 taxa are ferns or fern allies, 1 is a gymnosperm, and 683 taxa are flowering plants. The most speciose plant families (table A4.1-2) represented in the insular flora are Asteraceae (80 species), Fabaceae (60), and Poaceae (52). The genus *Euphorbia* (*sensu lato*) is the most diversified genus on the gulf islands with 25 taxa (table A4.1-3), although many taxonomists split the genus and recognize *Chamaescyce* as a segregate genus. Two cactus genera are also well diversified in the insular flora. *Opuntia* (*sensu lato*; including *Cylindropuntia* and *Grusonia*) contains 13 taxa, and *Mammillaria* (not including the genus *Cochemiea*) has 11 taxa found on the gulf islands. With respect to plant endemism, the flora of the gulf islands is not unique. At present, only 28 endemic plant taxa are known to occur in the insular flora, representing only a 4% rate of endemism for the islands. For a more comprehensive account of plant endemism in the Gulf of California, see appendix 4.5.

Although detailed floristic analyses have not yet been published for every island in the gulf, some general observations on insular plant diversity can still be made. The most floristically diverse island in the gulf system, Tiburón, is also the largest island in the region, and it lies close to the Mexican mainland. This island has been recorded as having 298 plant species as compared with Ángel de la Guarda, the second largest island in the gulf, with only 199 species. However, Cerralvo (app. 4.2) has 232 plant taxa, but because it is the most southerly (tropical) island in the gulf and because it is topographically varied and receives a lot of summer rainfall, it supports a high diversity of plants.

The Floristic Lists

Table (A4.1-4) shows the vascular plants known from each of 20 principal islands of the gulf. This island list is the same as was originally selected by Reid Moran and used in the first edition of this book. Therefore, San Lorenzo includes not only the southern San Lorenzo proper, or San Lorenzo Sur, but also the smaller San Lorenzo Norte, also known as Las Animas, and Espíritu Santo includes Partida Island (Sur). It should also be noted that the floristic lists for some of the larger islands include plant records for the smaller islets surrounding each of them, such as for Tiburón, which includes Cholludo and Dátil islands; Ángel de la Guarda, which includes Estanque (Pond), Mejia, and Granito; and Espíritu Santo, which includes Partida, Ballena, Gallo, and Gallino. The floristic table distinguishes those plants represented by specimens (X) from those known only from sightings or observations (O). When an island is the type locality for a given taxon, this is indicated (T). Also, if there is a questionable report of a plant taxon on an island or an unverified specimen that may be misidentified, then it is shown by (?) in the table.

The new floristic lists in this appendix are compiled from the old table in the first edition of this book with more recent collections added by manually searching the SD Herbarium for deposited specimens from the gulf islands (e.g., duplicate specimens collected by T. Van Devender and R. Felger received in exchange from the ARIZ Herbarium, and recent insular collections made by J. Rebman). The table also includes voucher specimen data from the SD Herbarium electronic database and the HCIB Herbarium database in La Paz, Baja California Sur. During the development of this appendix, an electronic database was constructed that includes all of the specimen information for one voucher of each taxon from each island. It is hoped that this database will eventually be transformed into a searchable online resource. In this way, additions of plant taxa to each of the insular floras could be maintained in a more current and efficient manner.

The nomenclature used in table A4.1-4 has been updated using a variety of taxonomic monographs, revisions, and floristic treatments. For example, regional familial monographs and nomenclators finished since the first edition of this book were published for the Acanthaceae (Daniel 1997), Brassicaceae (Rollins 1993), Euphorbiaceae (Steinmann and Felger 1997), Malvaceae (Fryxell 1988), and Rubiaceae (Lorence 1999). Wiggins's (1980) flora of Baja California is always an important reference for the region, but due its age and incompleteness, other floristic manuals, family monographs, and synonymized checklists were also consulted (e.g., Gentry 1978; Gould and Moran 1981; Lenz 1992; Hickman 1993; Morin 1993, 1997; Kartesz 1994; Kubitzki 1998; Martin et al. 1998). In respect to the two most diverse plant families (Asteraceae and Fabaceae) in the region, various systematic treatments on groups, tribes, genera, and sections have been used, including, for the Asteraceae: Turner and Morris (1976), Whalen (1977), King and Robinson (1987), Smith (1989), and Schilling (1990), and for Fabaceae: Barneby (1977, 1991), Lavin (1988), Hernandez (1989), and Barneby and Grimes (1996). In respect to cacti, Craig (1989), Rebman (1995), and Lindsay et al. (1996) were used. Plus, the generic treatments for *Russelia* (Scrophulariaceae)(Carlson 1957) and the fern genus *Argyrochosma* (Windham 1987) were also consulted for taxonomic purposes.

Acknowledgments Many people have contributed to this appendix. At the SD Herbarium, both volunteers and staff manually searched the cabinets for island voucher specimens and helped to compile the data for table A4.1-4, including Judy Gibson, Joan Dowd, Karen Rich, and Annette Winner. Patty West and Gary Polis both contributed plant record information on islands in the gulf, especially near Bahía de los Ángeles. Dr. Norman Roberts facilitated an island trip to the central gulf islands, and Tom Murphy graciously donated his personal yacht for J.P.R. to obtain insular plant records. Also, Carolina Shepard Espinoza accompanied J.P.R. to the islands for field work in the central gulf area. Both Richard Felger and Tom Van Devender sent duplicate specimens of insular plant collections to the SD Herbarium and contributed information on the Sonoran islands of the gulf. At CIBNOR in La Paz, both Miguel Dominguez and Jose Juan Perez Navarro accompanied J.L.L.L. to the southern gulf islands, aided in plant collection, and helped process the specimens for the HCIB Herbarium.

References

Barneby, R. 1977. Daleae imagines. *Memoirs of the New York Botanical Garden* **27**:1–891.

Barneby, R. 1991. Sensitivae censitae; a description of the genus *Mimosa* Linnaeus (Mimosaceae) in the New World. *Memoirs of the New York Botanical Garden* **65**:1–835.

Barneby, R., and J. Grimes. 1996. Silk tree, Guanacaste, monkey's earring: a generic system for the synandrous Mimosaceae of the Americas; part 1. *Abarema, Albizia,* and allies. *Memoirs of the New York Botanical Garden* **74**:1–292.

Carlson, M. 1957. Monograph of the genus *Russelia* (Scrophulariaceae). *Fieldiana: Botany* **29**: 229–292.

Craig, R. 1989. *The Mammillaria Handbook.* Lofthouse Publications, West Yorkshire, UK.

Daniel, T. 1997. The Acanthaceae of California and the peninsula of Baja California. *Proceedings of the California Academy of Sciences* **49**:309–403.

Fryxell, P. 1988. Malvaceae of Mexico. *Systematic Botany Monographs* **25**:1–522.

Gentry, H. 1978. The agaves of Baja California. *Occasional Papers of the California Academy of Sciences* **130**:1–119.

Gould, F., and R. Moran. 1981. The grasses of Baja California. *San Diego Society of Natural History Memoir* **12**:1–140.

Hernandez, H. 1989. Systematics of *Zapoteca* (Leguminosae). *Annals of the Missouri Botanical Garden* **76**:781–862.

Hickman, J. 1993. *The Jepson Manual: Higher Plants of California.* University of California Press, Berkeley.

Kartesz, J. 1994. *A Synonymized Checklist of the Vascular Flora of the United States, Canada, and Greenland*, 2nd ed., vol. 1, Checklist. Timber Press, Portland, OR.

King, R., and H. Robinson. 1987. The genera of the Eupatorieae (Asteraceae). *Monographs in Systematic Botany, Missouri Botanical Garden* **22**:1–581.

Kubitzki, K. (ed). 1998. *The families and genera of vascular plants*, vol. 3. *Flowering Plants— Monocotyledons: Lilianae (except Orchidaceae).* Springer Press, Berlin.

Lavin, M. 1988. Systematics of *Coursetia* (Leguminosae-Papilionoideae). *Systematic Botany Monographs* **21**:1–167.

Lenz, L. 1992. *An Annotated Catalogue of the Plants of the Cape Region, Baja California Sur, Mexico.* The Cape Press.

Lindsay, G., et al. 1996. *The Genus* Ferocactus: *Taxonomy and Ecology; Exploration in the USA and Mexico.* Tireless Termites Press, Claremont, CA.

Lorence, D. 1999. A nomenclator of Mexican and Central American Rubiaceae. *Monographs in Systematic Botany from the Missouri Botanical Garden* **73**:1–171.

Martin, P., D. Yetman, M. Fishbein, P. Jenkins, T. Van Devender, and R. Wilson. 1998. *Gentry's Rio Mayo Plants.* The University of Arizona Press, Tucson.

Moran, R. 1983. Vascular plants of the Gulf islands. In T.J. Case and M. Cody (eds), *Island Biogeography in the Sea of Cortéz.* University of California Press, Berkeley; app.4.1, pp. 348–381.

Morin, N. (ed). 1993. *Flora of North America, North of Mexico, vol. 2. Pteridophytes and Gymnosperms.* Oxford University Press, New York.

Morin, N.(ed). 1997. *Flora of North America, North of Mexico, vol. 3. Magnoliophyta: Magnoliidae and Hamamelidae.* Oxford University Press, New York.

Rebman, J. P. 1995. Biosystematics of *Opuntia* subgenus *Cylindropuntia* (Cactaceae), the chollas of Lower California, Mexico. Ph.D. dissertation, Arizona State University, Tempe.

Rollins, R. 1993. *The Cruciferae of Continental North America.* Stanford University Press, Palo Alto, CA.

Schilling, E. 1990. Taxonomic revision of *Viguiera* subg. *Bahiopsis* (Asteraceae: Heliantheae). *Madroño* **37**:149–170.

Smith, E. 1989. A biosystematic study and revision of the genus *Coreocarpus* (Compositae). *Sytematic Botany* **14**:448–472.

Steinmann, V., and R. Felger. 1997. The Euphorbiaceae of Sonora, Mexico. *Aliso* **16**:1–71.

Turner, B., and M. Morris. 1976. Systematics of *Palafoxia* (Asteraceae: Helenieae). *Rhodora* **78**:567–628.

Whalen, M. 1977. Taxonomy of *Bebbia* (Compositae: Heliantheae). *Madroño* **24**:112–123.

Wiggins, I. 1980. *Flora of Baja California.* Stanford University Press, Palo Alto, CA.

Windham, M. 1987. *Argyrochosma*, a new genus of cheilanthoid ferns. *American Fern Journal* **77**:37–41.

Vascular Plants 469

Table A4.1-1 Taxonomic diversity of the insular flora

	Families	Genera	Species	Taxa
Pteridophytes	3	7	10	11
Gymnosperms	1	1	1	1
Angiosperms				
Dicots	81	280	566	604
Monocots	12	43	74	79
Totals	97	331	651	695

Table A4.1-2 Vascular plant families containing the largest number of taxa in the insular flora

Family	Genera	Species	Taxa
Asteraceae	38	80	91
Poaceae	28	52	54
Fabaceae	27	60	64
Euphorbiaceae	13	43	48
Cactaceae	10	41	43

Table A4.1-3 Genera containing the largest number of taxa represented on the gulf islands

Genus	Family	No. of Taxa in the Genus
Euphorbia (*sensu lato*)	Euphorbiaceae	25
Opuntia	Cactaceae	13
Mammillaria	Cactaceae	11
Boerhavia	Nyctaginaceae	10
Ambrosia	Asteraceae	9
Eriogonum	Polygonaceae	9
Lycium	Solanaceae	9
Cryptantha	Boraginaceae	8
Physalis	Solanaceae	8
Acacia	Fabaceae	7
Agave	Agavaceae	7
Aristida	Poaceae	7
Cuscuta	Cuscutaceae	7

470 Appendixes

Table A4.1-4 Vascular plants of the gulf islands

	Ángel de la Guarda	San Lorenzo	Tiburón	San Estéban	San Pedro Mártir	San Pedro Nolasco
Pteridophtes						
Lycopodiophyta						
Selaginellaceae						
Selaginella arizonica Maxon			x			
Polypodiophyta						
Marsileaceae						
Marsilea vestita Hooker & Grev. (=*M. fournieri*)						
Pteridaceae						
Argyrochosma peninsularis (Maxon & Weath.) Windham (=*Notholaena peninsularis*)						
Astrolepis cochisensis (Goodding) D.M. Benham & Windham =*Notholaena cochisensis*			x			
A. sinuata (Lag. ex Swartz) D.M. Benham & Windham ssp. *sinuata* (=*Notholaena sinuata* ssp. *sinuata*)			x			
Cheilanthes brandegeei D.C. Eaton	x					
C. wrightii Hook			x			
Notholaena californica D.C. Eaton	x		x			
N. lemmonii D.C. Eaton						x
Pentagramma triangularis (Kaulf.) Yatsk., Windham, & E. Wollenw. ssp. *maxonii* (Weath.) Yatsk., Windham, & E. Wollenw. (=*Pityrogramma triangularis* var. *maxonii*)						
P. triangularis ssp. *viscosa* (Nutt. ex D.C. Eaton) Yatsk., Windham, & E. Wollenw. (*Pityrogramma triangularis* var. *viscosa*)						
Gymnosperms						
Ephedraceae						
Ephedra aspera Engelm. ex S. Watson			x		x	
Angiosperms						
Magnoliopsida (Dicots)						
Acanthaceae						
Carlowrightia arizonica A. Gray (=*C. californica*; *C. californica* var. *pallida* [type from I. San Esteban])			x	x		
Dicliptera resupinata (Vahl) Juss. =*D. formosa*			x			
Elytraria imbricata (Vahl) Pers.			x			
Holographis virgata (Harv. ex Benth. & Hook.f.) T.F. Daniel ssp. *glandulifera* (Leonard & C.V. Morton) T.F. Daniel var. glandulifera (=*Berginia virgata* var. *glandulifera*)						
H. virgata ssp. *virgata*=*Berginia virgata* var *virgata*			x			
Justicia californica (Benth.) D.N. Gibson (=*Beleperone californica*)	x		x	x		
J. candicans (Nees) L.D. Benson(=*Jacobinia ovata* var. *ovata*)			x			
J. palmeri Rose						
Ruellia californica (Rose) I.M. Johnst. ssp. *californica*			x			

Vascular Plants 471

	Tortuga	San Marcos	San Ildefonso	Coronado	Carmen	Danzante	Monserrat	Santa Catalina	Santa Cruz	San Diego	San José	San Francisco	Espíritu Santo	Cerralvo
					x								x	
														x
				x										
					x		x	x	x	x			x	x
														x
x							x					x	x
							x					x	
				x										
		x		x		x	x	x	x	x		x	x
										x			x
		x		x	x					x		x	x
							x	x				x	x
												x	x
			x	x	x								

(continued)

472 Appendixes

Table A4.1-4 (Continued)

	Ángel de la Guarda	San Lorenzo	Tiburón	San Estéban	San Pedro Mártir	San Pedro Nolasco
R. californica ssp. *peninsularis* (Rose) T.F. Daniel (=*R. peninsularis*)						
Siphonoglossa longiflora (Torr.) A. Gray			x			
Achatocarpaceae						
Phaulothamnus spinescens A. Gray			x			x
Aizoaceae						
Sesuvium portulacastrum L.		x				
S. verrucosum Raf.	x	x	x			
Trianthema portulacastrum L.			x			
Amaranthaceae						
Amaranthus caudatus L.						
A. fimbriatus (Torr.) Benth.	x		x			x
A. lepturus S.F. Blake						
A. palmeri S. Watson						
A. watsonii Standl.	x	x	x	x		
Celosia floribunda A. Gray						
Dicraurus alternifolius (S. Watson) Uline & Bray						
Froelichia interrupta (L.) Moq.						
Iresine angustifolia Euphrasen						
Tidestromia lanuginosa (Nutt.) Standl.			x			
Anacardiaceae						
Cyrtocarpa edulis (Brandegee) Standl.						
Pachycormus discolor (Benth.) Coville var. *pubescens* (S. Watson) H.S. Gentry	x					
Rhus kearneyi Barkl. ssp. *borjaensis* Moran	x					
Apiaceae (Umbelliferae)						
Apiastrum angustifolium Nutt.	x					
Daucus pusillus Michx.			x			
Eryngium nasturtiifolium Juss. ex F. Delaroche						
Apocynaceae						
Haplophyton cimicidum A. DC. var. *crooksii* L.D. Benson			x			
Macrosiphonia hesperia I.M. Johnst.						
Vallesia glabra (Cav.) Link			x			
Aristolochiaceae						
Aristolochia watsonii Wooton & Standl. =*A. porphyrophylla*			x			
Asclepiadaceae						
Asclepias albicans S. Watson	x	x	x	x		
A. subulata Decne.			x			
Cynachum palmeri (S. Watson) S. F. Blake						
C. peninsulare S.F. Blake						
Marsdenia edulis S. Watson			x			
Matelea cordifolia (A. Gray) Woodson			x			
M. fruticosa (Brandegee) Woodson						
M. parvifolia (Torr.) Woodson	x					
M. pringlei (A. Gray) Woodson			x			
Metastelma arizonicum A. Gray (=*Cynanchum arizonicum*)			x			
M. californicum Benth.						

Vascular Plants 473

Tortuga	San Marcos	San Ildefonso	Coronado	Carmen	Danzante	Monserrat	Santa Catalina	Santa Cruz	San Diego	San José	San Francisco	Espíritu Santo	Cerralvo
									x			x	x
			x		x	x	x			x	x		x
			x	x	x	x	x			x	x		x
	x		x	x	x	o					x	x	x
x													
				x		x			x	x		x	x
												x	
		x	x		x	x				x		x	x
					x							x	x
					x			x					x
								x		x		x	x
								x				x	x
			o				x		x			x	x
			x				x		x			x	x
			x									x	
	x		T	x		x	x	x	x		x	x	
			x										
x	x		x	x	x	o				x	x		
													x
x	x		x	x		x	x x			x	x	x	x
			x	x				x					x
		x			x			x	x				x
						x							

(continued)

Table A4.1-4 (*Continued*)

	Ángel de la Guarda	San Lorenzo	Tiburón	San Estéban	San Pedro Mártir	San Pedro Nolasco
M. pringlei A. Gray *sensu lato*						x
Asteraceae (Compositae)						
Adenophyllum porophylloides (A. Gray) Strother	x					
Alvordia brandegeei A. Carter						
A. glomerata Brandegee var. *glomerata*						
A. glomerata var. *insularis* A. Carter (insular endemic)						
Ambrosia bryantii (Curran) Payne						
A. camphorata (Greene) Payne						
A. carduacea (Greene) Payne (=*Franseria arborescens*)						
A. chenopodiifolia (Benth.) Payne	x					
A. divaricata (Brandegee) Payne			x			
A. dumosa (A. Gray) Payne	x	x	x			
A. dumosa × *A. ilicifolia* hybrid	x					
A. ilicifolia (A. Gray) Payne	x	x	x	x		
A. magdalenae (Brandegee) Payne	x					
Baccharis						
B. salicifolia (Ruiz & Pavón) Pers. (=*B. glutinosa*)			x			
B. sarothroides A. Gray	o	x			x	
Bebbia atriplicifolia (A. Gray) Greene (=*B. juncea* var. *atriplicifolia*)						
B. juncea (Benth.) Greene var. *aspera* Greene	x	o	x	x		x
B. juncea var. *juncea*						
Brickellia brandegeei B.L. Rob.						
B. coulteri A. Gray			x	x		
B. glabrata (Rose) B.L. Rob.						
B. hastata Benth.						
Chromolaena sagittata (A. Gray) R.M. King & H. Rob. (=*Eupatorium sagittatum*)			x			
Coreocarpus dissectus (Benth.) S.F. Blake						
C. parthenioides Benth. var. *parthenioides*	x		x	x		
C. sanpedroensis E.B. Smith (endemic); *C. arizonicus* var. *filiformis* previously misapplied)						T
C. sonoranus Sherff			x			
Coulterella capitata Vasey & Rose						
Critonia peninsularis (Brandegee) R.M. King & H. Rob. (=*Eupatorium peninsulare*)						
Dyssodia concinna (A. Gray) B.L. Rob.			x			
D. speciosa A. Gray						
Encelia farinosa Torr. & A. Gray var. *phenicodonta* (S.F. Blake) I.M. Johnst.	x	x	x			
E. farinosa var. *radians* Brandegee ex S.F. Blake						
Filago californica Nutt.	x					
Gochnatia arborescens Brandegee						
Gutierrezia microcephala (DC.) A. Gray	x					
G. ramulosa (Greene) M.A. Lane =*Greenella ramulosa*						
Helianthus niveus (Benth.) *Brandegee* cf. ssp. *tephrodes* (A. Gray) Heiser			x			

	Tortuga	San Marcos	San Ildefonso	Coronado	Carmen	Danzante	Monserrat	Santa Catalina	Santa Cruz	San Diego	San José	San Francisco	Espíritu Santo	Cerralvo
				x				x						x
											x		x	
		x			T			x			x		x	
					x						x		x	
								x						
														x
					x									
			o											
					x	x				x	x		x	x
		x						x						
		x		x		x	x	x	x			x		
													x	
		x			T	x	x				x			x
														x
		x			x	x		x	x	x	x	x	x	x
		x		x							x		x	x
						x					x	x	x	
														x
												x	x	
	x	x				x		x			x			
					x						x			
													x	
														x
					x									

(continued)

Table A4.1-4 (*Continued*)

	Ángel de la Guarda	San Lorenzo	Tiburón	San Estéban	San Pedro Mártir	San Pedro Nolasco
Hofmeisteria crassifolia S. Watson						x
H. fasciculata (Benth.) Walp. var. *fasciculata*	x	x	x	x		
H. fasciculata var. *pubescens* (S. Watson) B.L. Robinson						
H. fasciculata var. *xanti* A. Gray						
H. filifolia I.M. Johnst. [Endemic]	T					
Hymenoclea monogyra A. Gray			x			
H. salsola A. Gray var. *pentalepis* (Rydb.) L.D. Benson (=*H. pentalepis*)	x		x			
Laennecia coulteri (A. Gray) Nesom (=*Conyza coulteri*)						
Machaeranthera arenaria (Benth.) Shinners (=*Haplopappus arenarius*)						
M. pinnatifida (Hook.) Shinners var. *gooddingii* (A. Nels.) B.L. Turner & R.L. Hartm. (=*Haplopappus gooddingii*)	x					
M. pinnatifida var. *incisifolia* (I.M. Johnst.) B.L. Turner & R.L. Hartm. (=*Haplopappus arenarius* var. *incisifolius* ; =*H. spinulosus* ssp. *incisifolius*; insular endemic?)		T	x	x		
M. pinnatifida var. *scabrella* (Greene) B.L. Turner & R.L. Hartm. (=*Haplopappus spinulosus* ssp. *scabrellus* ; ?=*H. arenosus* var. *rossii*; type from I. San Marcos)			x			
Malacothrix xanti A. Gray						
Malperia tenuis S. Watson (=*Hofmeisteria tenuis*)	x					
Palafoxia arida B.L. Turner & M.I. Morris var. *arida*			x			
P. linearis (Cav.) Lag. var. *glandulosa* B.L. Turner & M.I. Morris			x			
P. linearis var. *linearis* (=*P. leucophylla*; type from I. Carmen)						
Pectis angustifolia Torr.						
P. linifolia L.						
P. multiseta Benth. var. *ambigua* (Fernald) Keil (=*P. ambigua*)						
P. papposa Harv. & A. Gray			x			
P. rusbyi A. Gray (=*P. palmeri*)			x			
P. vollmeri Wiggins						
Pelucha trifida S. Watson	x		x	o	T	
Perityle aurea Rose	x	x	x	x		
P. californica Benth.						x
P. crassifolia Brandegee var. *crassifolia*						
P. crassifolia var. *robusta* (Rydb.) Everly						
P. emoryi Torr.	x	x	x	x	x	x
P. microglossa Benth.						
Peucephyllum schottii A. Gray (=*P. schottii* var. *latisetum*; type from I. San Marcos)	x	x		x		

	Tortuga	San Marcos	San Ildefonso	Coronado	Carmen	Danzante	Monserrat	Santa Catalina	Santa Cruz	San Diego	San José	San Francisco	Espíritu Santo	Cerralvo
	x													
		x	x	x	x	x	x	x	x	x	x	x		
													x	x
														x
													x	
				x	x			x	x	x	o	x	x	x
					x									
				x				x	x	x				
		x			x		x					x		
						x							x	x
				x										
				x	x		x					x		
				x	x					x				
														x
													x	
											x			
											x			
				x	x									
		x					x	x			x		x	
		x		x	x		x	x			x		x	x
		x		x			x	x			x		x	x
	x	x	x	x	x		x		x	x	x		x	x
		x												x

(continued)

478 Appendixes

Table A4.1-4 (*Continued*)

	Ángel de la Guarda	San Lorenzo	Tiburón	San Estéban	San Pedro Mártir	San Pedro Nolasco
Pleurocoronis laphamioides (Rose) R.M. King & H. Rob. var. *laphamioides* (=*Hofmeisteria laphamioides*; =*H. pluriseta* var. *laphamioides*)			x	x	x	T
P. laphamioides var. *pauciseta* (I.M. Johnst.) R.M. King (=*Hofmeisteria laphamioides* var. *pauciseta*; =*H. pluriseta* var. *pauciseta*)						T
Pluchea carolinensis (Jacq.) G. Don (=*P. symphytifolia*; =*P. odorata*)						
P. salicifolia (Mill.) S.F. Blake (=*P. adnata* var. *parvifolia*)						
Porophyllum confertum Greene var. *confertum*						
P. crassifolium S. Watson (=*P. leptophyllum*; type from I. Ángel de la Guarda)	x	x	x	x		
P. gracile Benth.	x		x	x		
P. ochroleucum Rydb.						
P. pausodynum Robins. & Greenm.						x
P. porfyreum Rose & Standl.						
Senecio mohavensis A. Gray.	x					
Trichoptilium incisum (A. Gray) A. Gray	x					
Trixis californica Kellogg	x	x	x	x	x	x
Verbesina palmeri S. Watson	x	x	x			
Vernonia triflosculosa Kunth var. *palmeri* (Rose) B.L. Turner (=*V. palmeri*)						
Viguiera chenopodina Greene (=*V. deltoidea* var. *chenopodina*)			x		x	
V. deltoidea A. Gray						
V. parishii Greene (=*V. deltoidea* var. *parishii*)			x	x		
V. tomentosa A. Gray						
V. triangularis M.E. Jones	x			x		x
Xylorhiza frutescens (S. Watson) Greene (=*Aster frutescens*)	x					
Xylothamnia diffusa (Benth.) G.L. Nesom (=*Ericameria diffusa*; =*Haplopappus sonoriensis*)	x	x	x			
Bataceae (Batidaceae)						
Batis maritima L.	x		x			
Bignoniaceae						
Tecoma stans (L.) Juss. ex Kunth						
Boraginaceae						
Antiphytum peninsulare (Rose) I.M. Johnst.						
Bourreria sonorae S. Watson						
Cordia curassavica (Jacq.) Roem. & Schult. ? (=*C. brevispicata*)						
C. parvifolia A. DC.			x			
Cryptantha angelica I.M. Johnst.	T		x			
C. angustifolia (Torr.) Greene	x		x	x		
C. echinosepala J.F. Macbr.						
C. fastigiata I.M. Johnst.	x	x	x	x		
C. grayi (Vasey & Rose) J.F. Macbr [var. undetermined]						

Vascular Plants 479

Tortuga	San Marcos	San Ildefonso	Coronado	Carmen	Danzante	Monserrat	Santa Catalina	Santa Cruz	San Diego	San José	San Francisco	Espíritu Santo	Cerralvo
x	x	x		x									
												x	
	x												
													x
	x		x	x	x	x						T	x
	x		x	x	x	x	x	x	x	x			x
												x	x
													x
x	x		x	x	x	x					x		
								x					
													x
x	x			x	x	x	x	x	x	x	x	x	x
								x				x	x
													x
									x				
	x	x	x	x							x	x	x
			x	x	x		x				x	x	
				x	x					x			x
	x												
				x	x	o	x	o	o	x	x	x	x
												x	x
										x			
	x		x	x							x		
			x		x								
	x			x	x								
												x	

(*continued*)

480 Appendixes

Table A4.1-4 (Continued)

	Ángel de la Guarda	San Lorenzo	Tiburón	San Estéban	San Pedro Mártir	San Pedro Nolasco
C. grayi var. *nesiotica* I.M. Johnst. [Insular endemic]						
C. holoptera (A. Gray) J.F. Macbr.						
C. maritima (Greene) Greene	x	x	x	x		
Heliotropium angiospermum Murr.						
H. curassavicum L.	?	x	x	x		
H. procumbens P. Mill.						
Pectocarya linearis (Ruiz & Pavón) DC. ssp. *ferocula* (I.M. Johnst.) Thorne	x					
Plagiobothrys jonesii A. Gray	x					
Tiquilia canescens (DC.) A. Richardson var. *canescens* (=*Coldenia canescens* var. *canescens*)			x			
T. cuspidata (I.M. Johnst.) A. Richardson (=*Coldenia cuspidata*)						
T. palmeri (A. Gray) A. Richardson (=*C. palmeri*)	x		x			
Tournefortia hartwegiana Steudel			x			
Brassicaceae (CRUCIFERAE)						
Caulanthus lasiophyllus (Hook. & Arn.) Payson	x					
Descurainia pinnata (Walter) Britton ssp. *halictorum* (Cockerell) Detling			x	x		
Draba cuneifolia Nutt. ex Torr. & A. Gray var. *integrifolia* S. Watson	x					
D. cuneifolia var. *sonorae* (Greene) Parish			x	x		
Dryopetalon crenatum (Brandegee) rollins var. *racemosum* Rollins (endemic)						
D. palmeri (S. Watson) O.E. Schulz						
D. purpureum Rollins						
Lepidium						
L. lasiocarpum Nutt. ex Torr. & A. Gray						
Lyrocarpa coulteri Hook. & Harv. var. *apiculata* Rollins						
L. coulteri var. *coulteri*			x			
L. linearifolia Rollins (insular endemic)	T			x		
L. xantii Brandegee						
Sibara angelorum (S. Watson) Greene (=*S. pectinata*)	x					
Sisymbrium irio L.						
Thysanocarpus laciniatus Nutt. ex Torr. & A. Gray						
Buddlejaceae (Loganiaceae)						
Buddleja corrugata M.E. Jones ssp. *corrugata*						
Burseraceae						
Bursera epinnata (Rose) Engler						
Bursera fagaroides (Kunth) Engler var. *elongata* McVaugh & Rzed. (=*B. odorata*)			x			
B. hindsiana (Benth.) Engler (=*B. epinnata*; =*B. rhoifolia*)	x		x	x		
B. laxiflora S. Watson			x			
B. microphylla A. Gray	x	o	x	x		x

Vascular Plants

Tortuga	San Marcos	San Ildefonso	Coronado	Carmen	Danzante	Monserrat	Santa Catalina	Santa Cruz	San Diego	San José	San Francisco	Espíritu Santo	Cerralvo
			x				x	x	x	x	T	x	x
x				x									
				x									
	x			x	x	x			x	x	x	x	
				x								x	
			x	x									
	T												
													T
				x	x					x		x	
										x			
										x			
	x					x				x		x	
						x							
						x						x	
										x			
				x									
								x					
						x							
													x
x	x	x	x	x	x	o	x	x	x	x	o	x	x
o	o		x	x	x	x	x	x	o	x	x	x	x

(*continued*)

482 Appendixes

Table A4.1-4 (*Continued*)

	Ángel de la Guarda	San Lorenzo	Tiburón	San Estéban	San Pedro Mártir	San Pedro Nolasco
Cactaceae						
Carnegiea gigantea (Engelm.) Britton & Rose			x			
Cochemiea poselgeri (Hildm.) Britton & Rose (=*Mammillaria poselgeri*)						
C. setispina (J.M. Coult.) Walton (=*Mammillaria setispina*)	x					
Echinocereus brandegeei (J. Coult.) K. Schum.						
E. grandis Britton & Rose (insular endemic)		x		T		
E. scopulorum Britton & Rose (=*E. pectinatus* var. *scopulorum*			x	x		
E. websterianus G. Lindsay (endemic)						T
Ferocactus diguetii (F.A.C. Weber) Britton & Rose var. *diguetii* (insular endemic)						
F. diguetii var. *carmenensis* G. Lindsay (endemic)						
F. johnstonianus Britton & Rose (endemic)	T					
F. peninsulae (Engelm. ex F.A.C. Weber) Britton & Rose var. *peninsulae*						
F. townsendianus Britton & Rose var. *townsendianus*						
F. wislizenii (Engelm.) Britton & Rose var. *tiburonensis* G. Lindsay (endemic)			T			
Lophocereus schottii (Engelm.) Britton & Rose			x	x		
Mammillaria			x	x		
M. albicans (Britton & Rose) Berger =*M. slevinii* (type from I. San Jose)						
M. cerralboa (Britton & Rose) Orcutt (endemic)						
M. dioica K. Brandegee (including *M. angelensis*)	x	x	x			
M. estebanensis G. Lindsay (endemic)				T		
M. evermanniana (Britton & Rose) Orcutt (endemic)						
M. fraileana (Britton & Rose) Boed.						
M. insularis H.E. Gates						
M. milleri (Britton & Rose) Boed. =*M. microcarpa*			x			
M. multidigitata G. Lindsay (endemic)						T
M. schumannii Hildm. =*Bartschella schumannii*						
M. tayloriorum Glass & Foster (endemic)						T
Opuntia subgenus *Opuntia*						x
Opuntia alcahes F.A.C. Weber var. *alcahes* =*O. brevispina* (type from I. Ballena, islet near I. Espíritu Santo)	x	x	x	x	x	x
O. bigelovii Engelm. var. *bigelovii*	x	o	x	x		
O. bravoana E. Baxter						
O. burrageana Britton & Rose						
O. cholla F.A.C. Weber	x	o	x	x	x	
O. fulgida Engelm. var. *fulgida*		o	x		o	x
O. fulgida var. *mamillata* (Schott ex Engelm.) J.M. Coult.			x	x		
O. invicta Brandegee						
O. leptocaulis DC.			x			
O. lindsayi Rebman (intermediate to *O. alcahes*)						

Vascular Plants   483

Tortuga	San Marcos	San Ildefonso	Coronado	Carmen	Danzante	Monserrat	Santa Catalina	Santa Cruz	San Diego	San José	San Francisco	Espíritu Santo	Cerralvo
		x	o	x	o	o	x	o	x	x	x	x	x
	o			x	x			x	x	x	x	x	x
			o		o	x	T	x	x				x
				T									
		x											
										x	x		x
o	x	o	x	o	o	o	x	o	x	x	x		x
x					o			T	x	x	x	x	
													T
x	o	x	o	x	x	x	x	o	x	x	x	x	x
							?					?	T
							x					x	
	x												
												x	
							x						
x		x		x	x	x	x	x	x	x	x	x	o
								x					
x													x
												x	
	o	o	x	x	x	o	x	x	x	x	x	x	x
	x			x									
							x	x					

(*continued*)

Table A4.1-4 (Continued)

	Ángel de la Guarda	San Lorenzo	Tiburón	San Estéban	San Pedro Mártir	San Pedro Nolasco
O. molesta Brandegee var. *molesta*						
O. tapona Engelm. (=*O. comonduensis* [J.M. Coult.]) Britton & Rose						
O. versicolor Engelm. ex J.M. Coult. (=*O. thurberi* ssp. *versicolor*)			x	x		
Pachycereus pringlei (S. Watson) Britton & Rose	x	o	x	x	x	x
Peniocereus johnstonii Britton & Rose						
P. striatus (Brandegee) Buxbaum (=*Neoevansia striata*)			x			
Stenocereus gummosus (Engelm.) A. Gibson & Horak (=*Machaerocereus gummosus*)	x	o	o	o		
S. thurberi (Engelm.) F. Buxbaum (=*Lemaireocereus thurberi*)	?		x	o	?	x
Campanulaceae						
Nemacladus glanduliferus Jepson *sensu lato*	x	x	x	x		
Capparaceae (Capparidaceae)						
Atamisquea emarginata Miers ex Hook. & Arn.	o		x	x		?
Cleome lutea Hook. ssp. *jonesii* J.F. Macbr.						
C. tenuis S. Watson			x			
Forchammeria watsonii Rose						
Caryophyllaceae						
Achyronychia cooperi Torr. & A. Gray	x	x	x			
Drymaria arenarioides Willd. ssp. *peninsularis* (S.F. Blake) J. Duke (=*D. johnstonii*; type from I. Espíritu Santo)						
D. debilis Brandegee (=*D. polystachya* var. *diffusa*; type from I. Carmen)						
D. glandulosa Presl (=*D. fendleri*)						
D. holosteoides Benth. var. *holosteoides*	x		x			
D. viscosa S. Watson						
Celastraceae						
Maytenus phyllanthoides Benth. (=*Tricerma phyllanthoides*)			x			
Schaefferia cuneifolia A. Gray						
Chenopodiaceae						
Allenrolfea occidentalis (S. Watson) Kuntze	x		x			
Atriplex barclayana (Benth.) D. Dietr. *sensu lato*	x	x	x	x		
A. canescens (Pursh) Nutt. ssp. *canescens*				o		
A. canescens ssp. *linearis* (S.Wats.) Hall & Clem. (=*A. linearis*)		o	x			
A. hymenelytra (Torr.) S. Watson	o					
A. polycarpa (Torr.) S. Watson	x	x	x	x		
Chenopodium murale L.	x		x			
Salicornia bigelovii Torr.	x		x			
S. subterminalis Parish (=*Arthrocnemum subterminale*)	x		x			
S. virginica L.	o		x			
Suaeda			x			
S. californica S. Watson	x		x	x		
S. moquinii (Torr.) Greene (=*S. ramosissima*; =*S. torreyana*)						

Tortuga	San Marcos	San Ildefonso	Coronado	Carmen	Danzante	Monserrat	Santa Catalina	Santa Cruz	San Diego	San José	San Francisco	Espíritu Santo	Cerralvo
x			x	x	x	o	o	o	o	x	x	x	o
x	o	o	o	x	o	o	x	o	o	x	o	x	x
						x				T			x
x	x		x	x	o	x				x	x	x	
x	o	o	o	x	o	x	x	o	x	x	x	x	x
o	o	o	o	o	o	o	o	o	o	x	x	x	x
x			o	x	o	o			x	x	x	x	
				x	x								
				x	x					x			
										x		x	x
	x												
												x	
	x		x	x		x	x	x				x	x
					x								
	x		x	x						x	x	x	
												x	
x	o		x	o	x		x			x	x	x	
										x		x	
			o	x		x				x	x	x	x
x	x	x	o	x	x	x	x	x	x	x	x	x	x
												x	
										x		x	
													x
			x	x	x					x	x	x	
			x							x	x	x	x
				x	x					x		x	
										x		x	

(continued)

Table A4.1-4 (Continued)

	Ángel de la Guarda	San Lorenzo	Tiburón	San Estéban	San Pedro Mártir	San Pedro Nolasco
Combretaceae						
Laguncularia racemosa (L.) C.F. Gaertn.			x			
Convolvulaceae						
Cressa truxillensis Kunth	x		x			
Evolvulus alsinoides (L.) L. var. *angustifolius* Torr. (=*E. alsinoides* var. *acapulcensis*)			x			
E. sericeus Sw.						
Ipomoea			x			
I. jicama Brandegee						
I. pes-caprae (L.) R. Br.						
Jacquemontia abutiloides Benth. var. *eastwoodiana* (I.M. Johnst.) Wiggins			x			
Merremia aurea (Kell.) O'Donnell						
Crassulaceae						
Crassula connata (Ruiz & Pavón) Berger (=*C. erecta*; =*Tillaea erecta*)	x					
Dudleya albiflora Rose						
D. arizonica Rose (=*D. pulverulenta* ssp. *arizonica*)	x					
D. nubigena (Brandegee) Britton & rose ssp. *cerralvensis* Moran [Endemic]						
D. nubigena ssp. *nubigena*						
Crossosomataceae						
Crossosoma bigelovii S. Watson			x			
Cucurbitaceae						
Echinopepon minimus (Kell.) S. Watson (including var. *peninsularis*)						
E. wrightii (A. Gray) S. Watson						
Ibervillea sonorae (S. Watson) Greene var. *peninsularis* (I.M. Johnst.) Wiggins (=*Maximowiczia sonorae*; =*M. sonorae* var. *peninsularis*)			x			
Vaseyanthus insularis (S. Watson) Rose var. *brandegeei* (Cogn.) I.M. Johnst. (=*V. brandegeei*)						
V. insularis var. *insularis* (=*V. insularis* var. *inermis*; =*V. insularis* var. *palmeri*)	x	x	x	x	T	x
Cuscutaceae						
Cuscuta			x			
C. corymbosa Ruiz & Pavón *sensu lato*	x			x		x
C. lepthantha Engelm. (=*C. leptantha* var. *palmeri*)	x		x			
C. macrocephala Schaff. & Yunck.						
C. odontolepis Engelm.						
C. pentagona Engelm. var. *pentagona* (=*C. campestris*)						
C. tuberculata Brandegee						
C. umbellata Kunth						
Ebenaceae						
Maba intricata (A. Gray) Hiern						
Euphorbiaceae						
Acalypha californica Benth.			x	x		
A. comonduana Millsp.			x			
Adelia virgata Brandegee						

Vascular Plants

	Tortuga	San Marcos	San Ildefonso	Coronado	Carmen	Danzante	Monserrat	Santa Catalina	Santa Cruz	San Diego	San José	San Francisco	Espíritu Santo	Cerralvo
					x	x					x	x	x	
						x					x	x		x
														x
														x
						x								
	x	x	T		x	x		x	x		x	x	x	x
						x					x		x	x
													x	
														T
													x	
		x									x		x	x
														x
						x		x	x	x	x	x	x	T
		x		x	x	x	x	x	x	x		x	x	x
	x	x		x		x		x	x	x	x		x	x
					x	x							x	x
						x								
					x									
						x					x			
											x			
						x								
					x									
											x			x
									x		x		x	x
											x		x	x
					x							x	x	x

(continued)

Table A4.1-4 (Continued)

	Ángel de la Guarda	San Lorenzo	Tiburón	San Esteban	San Pedro Mártir	San Pedro Nolasco
Andrachne microphylla (Lam.) Baill. (=*A. ciliatoglandulosa*)	x		x			
Bernardia viridis Millsp. (=*B. mexicana*)						x
Cnidoscolus cf. *maculatus* (Brandegee) Pax & K. Hoffm.						
C. palmeri (S. Watson) Rose			x			
Croton caboensis Croizat						
C. californicus Müll. Arg.			x			
C. magdalenae Millsp.			x			
C. sonorae Torr.			x			
Ditaxis brandegeei (Millsp.) Rose & Standley var. *brandegeei* (=*Argythamnia brandegeei* var. *brandegeei*)	x?					
D. brandegeei var. *intonsa* I.M. Johnst. (=*Argythamnia brandegeei* var. *intonsa*)						
D. lanceolata (Benth.) Pax & K. Hoffm. (=*Argythamnia lanceolata*)	x		x	x		
D. neomexicana (Müll. Arg.) A. Heller (=*Argythamnia neomexicana*)	x	x				
D. serrata (Torr.) A. Heller (=*Argythamnia serrata*)				x		
Euphorbia abramsiana L.C. Wheeler						
E. arizonica Engelm.						
E. bartolomaei Greene						
E. californica Benth.						
E. eriantha Benth.	x		x			
E. fendleri Torr. & A. Gray						
E. florida Engelm.			x			
E. incerta Brandegee						
E. leucophylla Benth. ssp. *comcaacorum* V. Steinm. & Felger			x			
E. leucophylla Benth. ssp. *leucophylla*						
E. magdalenae Benth.			x			x
E. micromera Boiss.						
E. misera Benth.	x		x	x		
E. pediculifera Engelm. var. *pediculifera*	x	x	x	x		
E. peninsularis I.M. Johnst.						
E. petrina S. Watson	x		x	x	T	
E. polycarpa Benth. var. *carmenensis* (Rose) Wheeler (insular endemic)						
E. polycarpa var. *johnstonii* Wheeler (insular endemic?)						
E. polycarpa var. *mejamia* Wheeler						
E. polycarpa var. *polycarpa* (including var. *hirtella* and var. *intermixta*)	x		x			
E. serpens Kunth						
E. setiloba Engelm. ex Torr.	x		x	x		
E. taluticola Wiggins						
E. tomentulosa S. Watson			x			
E. xanti Engelm. ex Boiss.						

Tortuga	San Marcos	San Ildefonso	Coronado	Carmen	Danzante	Monserrat	Santa Catalina	Santa Cruz	San Diego	San José	San Francisco	Espíritu Santo	Cerralvo
				x			x					x	x
						x		x				x	x
													x
				?	x			x			x		x
													x
			x			x							
	x			x	x	o	x	x	o	x	o	x	x
x?	x			x?	x								
			x						x				x
x	x		x	x	x	x	x	x		x	x	x	x
	x												
						x				x			
						x				x		x	
	x											x	x
												x	
										x			
													x
	x		x	x	x	o	x	x	x	x	x	x	x
			x								x	x	x
	x											x	
x	x		x	x	x	x				x		x	
						x							
x	x		x	T			x	x	x		x	x	
				x		T							
x	x		x	x					x	x	x	x	x
												x	
	x			x		x	x						x
			x	x		x						x	
x				x		x							

(continued)

Table A4.1-4 (Continued)

	Ángel de la Guarda	San Lorenzo	Tiburón	San Estéban	San Pedro Mártir	San Pedro Nolasco
Jatropha cinerea (Ortega) Müll. Arg.			x			
J. cuneata Wiggins & Rollins	x	o	x	x		x
Pedilanthus macrocarpus Benth.						x
Phyllanthus galleotianus Baill.						
Sebastiana bilocularis S. Watson (=*Sapium biloculare*)			x			
Tragia						
T. carteri R. Urtecho ined.						
T. jonesii Radcl.-Sm. & R. Govaerts			x			
Fabaceae (Leguminosae)						
Acacia constricta Benth.						
A. filicioides (Cav.) Trel.						
A. goldmanii (Britton & Rose) Wiggins						
A. greggii A. Gray var. *greggii*	x		x			
A. mcmurphyi Wiggins						
A. pacensis Rudd & A. Carter						
A. willardiana Rose			x			x
Aeschynomene nivea Brandegee						
Astragalus insularis Kell. var. *harwoodii* Munz & McBurney	x	x				
A. nuttallianus DC. var. *cedrosensis* M.E. Jones	x	x				
Caesalpinia californica (A. Gray) Standl.						
C. intricata (Brandegee) E.M. Fisher? (=*Hoffmanseggia intricata*)			x	x		
C. pannosa Brandegee						
C. placida Brandegee						
C. virgata E.M. Fisher (=*Hoffmanseggia microphylla*)	x					
Calliandra californica Benth.			x			
C. eriophylla Benth. var. *eriophylla*			x			
Cercidium floridum Benth. ex A. Gray ssp. *floridum*			x			
C. floridum ssp. *peninsulare* (Rose) A. Carter						
C. microphyllum (Torr.) Rose & I.M. Johnst.	x		x	x		
C. praecox (Ruiz & Pavón) Harms						
C. xsonorae Rose & I.M. Johnst.						
Coursetia caribaea (Jacq.) Lavin var. *caribaea* (=*Benthamantha brandegeei*; =*Cracca aletes*)						
C. glandulosa A. Gray			x			
Dalea mollis Benth.	x		x			
Desmanthus covillei (Britton & Rose) Wiggins ex Turner			x			
D. fruticosus Rose	x		x	x		
Desmodium procumbens (Mill.) A.S. Hitchc. var. *exiguum* (A. Gray) Schub.						
Ebenopsis confinis (Standl.) Barneby & Grimes (=*Pithecellobium confine*)			x			
Errazurizia megacarpa (S. Watson) I.M. Johnst.	x		x			
Erythrina flabelliformis Kearney						
Indigofera fruticosa Rose						

Vascular Plants 491

Tortuga	San Marcos	San Ildefonso	Coronado	Carmen	Danzante	Monserrat	Santa Catalina	Santa Cruz	San Diego	San José	San Francisco	Espíritu Santo	Cerralvo
				x	x		x		x			x	x
x	o		o	x	x	o	x	x	x	x	x	x	x
				x								x	x
												x	x
				x					x				x
				x									
												x	
							x		x			x	x
							x						
												x	
			x	x	x	o		o	o	x		x	x
	x											x	
									o			x	x
												x	
	x		x	x					o	x	x	x	
x	x		x	x	o	x	x		x			x	x
x	x	x		x	x	o			x				
x				x									x
x	x												
x	x		x	T	x	x	x	x	x	x	x	x	x
													x
	o				o	x	x	o	o			x	x
	x												
												x	x
													x

(continued)

Table A4.1-4 (Continued)

	Ángel de la Guarda	San Lorenzo	Tiburón	San Estéban	San Pedro Mártir	San Pedro Nolasco
I. nelsonii Rydb. (=*I. argentata*; =*I. nesophila* Lievens & Urbatsch nom. nov. in ed.)						
Lotus rigidus (Benth.) Greene	x					
L. salsuginosus Greene var. *brevivexillus* Ottley	x	x	x			
L. strigosus (Nutt.) Greene (=*L. tomentellus*)	x	x	x	x		
Lupinus arizonicus (S. Watson) S. Watson ssp. *lagunensis* (M.E. Jones) Christ. & Dunn			x	x		
L. arizonicus ssp. *setossimus* (C.P. Sm.) Christ. & Dunn	T	x		x		
Lysiloma candidum Brandegee						
L. divaricatum (Jacq.) J.F. Macbr. (=*L. microphyllum*)			x			
Macroptilium atropurpureum (DC.) Urban (=*Phaseolus atropurpureus*)						
Marina catalinae Barneby (endemic)						
M. maritima (Brandegee) Barneby (=*Dalea maritima*)						
M. oculata (Rydb.) Barneby (insular endemic)						
M. parryi (Torr. & A. Gray) Barneby (=*Dalea parryi*)	x		x	x		
M. vetula (Brandegee) Barneby (=*Dalea vetula*)			x			
Mimosa distachya Cav. var. *laxiflora* (Benth.) Barneby (=*M. brandegeei*; =*M. laxiflora*; =*M. purpurascens*)			x			
M. xanti A. Gray						
Olneya tesota A. Gray	x		x	x		
Phaseolus acutifolius A. Gray var. *acutifolius*						
P. acutifolius var. latifolius G.F. Freeman			x			
P. filiformis Benth.	x	x	x	x		
Prosopis						
P. articulata S. Watson						
P. glandulosa Torr. var. *torreyana* (L.D. Benson) M.C. Johnst. (=*P. juliflora*)			x	x		
P. velutina Wooton				x		
Psorothamnus emoryi (A. Gray) Rydb. var. *arenarius* (Brandegee) Barneby (=*P. emoryi* var. *arenaria*; =*Dalea tinctoria*)						
P. emoryi var. emoryi (=*Dalea emoryi*)	x	x	x			
P. spinosus (A. Gray) Barneby (=*Dalea spinosa*)			x			
Senna confinis (Greene) Irwin & Barneby (=*Cassia confinis*)	x		x	x		
S. covesii (A. Gray) Irwin & Barneby (=*Cassia covesii*)			x	x		
S. polyantha (Colladon) Irwin & Barneby (=*Cassia goldmanii*)						
Tephrosia palmeri S. Watson			x			
T. purpurea (L.) Pers.						
Zapoteca formosa (Kunth) H.M. Hern. ssp. *rosei* (Wiggins) H.M. Hern. (=*Calliandra schotti* ssp. *rosei*)			x			

Tortuga	San Marcos	San Ildefonso	Coronado	Carmen	Danzante	Monserrat	Santa Catalina	Santa Cruz	San Diego	San José	San Francisco	Espíritu Santo	Cerralvo
													T
		x					x	x					x
							x						x
	x	o	x	x	x	o		x		x	x	x	x
		x											x
							T						
												x	
								x				x	
x	x		x	x	x	x		x		x	x	x	T
x										x			
										x		x	
												x	
o	o		o	x	o	o	x	o	o	x	o	x	x
												x	
x	x	x	x	x	x	x	o			x		x	x
					o								
	x			x						x		x	
				x						x			
x						x	x			x	x		
						x							
	x		x	x		x		o	o	x	x	x	x
x						x	x	x		x	x	x	x
										x			
	x		x							x			x
		x								x			x

(continued)

Table A4.1-4 (*Continued*)

	Ángel de la Guarda	San Lorenzo	Tiburón	San Estéban	San Pedro Mártir	San Pedro Nolasco
Fouquieriaceae						
Fouquieria burragei Rose						
F. columnaris (Kell.) Curran (=*Idria columnaris*)	x					
F. diguetii (Van Tieghem) I.M. Johnst.	x	o	x	?		x
F. splendens Engelm.	x		x			
Frankeniaceae						
Frankenia palmeri S. Watson	x		x			
Hydrophyllaceae						
Eucrypta micrantha (Torr.) A. Heller			x			
Nama hispidum A. Gray			x			
Phacelia ambigua M.E. Jones var. *minutiflora* (J. Voss) Atwood (=*P. minutiflora*; =*P. crenulata* var. *minutiflora*)			x	x		
P. crenulata Torr. ex S. Watson			x	x		
P. pauciflora S. Watson	x					
P. pedicellata A. Gray	x		x	x		
P. scariosa Brandegee						
Pholistoma racemosum (Nutt. ex A. Gray) Const.	x					
Koeberliniaceae						
Koeberlinia spinosa Zucc.			x			
Krameriaceae						
Krameria erecta Willd. ex J.A. Schultes (=*K. parvifolia* var *imparata*)			x			
K. grayi Rose & Painter			x			
K. paucifolia Rose						
Lamiaceae (Labiatae)						
Hyptis emoryi Torr.	x	x	x	x		
H. decipiens M.E. Jones						
H. laniflora Benth. (=*H. laniflora* var. *insularis*; type from I. Espíritu Santo)						
Monardella lagunensis M.E. Jones ssp. *mediopeninsularis* Moran	x					
Salvia platycheila A. Gray (insular endemic)						
S. similis Brandegee						x
Loasaceae						
Eucnide aurea (A. Gray) H. Thompson & Ernst						
E. cordata Kell. ex Curran	x	x		x		
E. rupestris (Baill.) H. Thompson & Ernst.	x		x	x		x
Mentzelia adhaerens Benth.	x	x	x	x	x	x
M. hirsutissima S. Watson var. *stenophylla* (Urb. & Gilg.) I.M. Johnst.	T		x			
M. multiflora (Nutt.) A. Gray						
Petalonyx linearis Greene	x	x	x	x	x	
Loranthaceae						
Psittacanthus sonorae (S. Watson) Kuijt (=*Phrygilanthus sonorae*)						
Lythraceae						
Ammannia robusta Heer & Regel						
Malpighiaceae						
Callaeum macropterum (DC.) D.M. Johnson (=*Mascagnia macroptera*)			x			

Tortuga	San Marcos	San Ildefonso	Coronado	Carmen	Danzante	Monserrat	Santa Catalina	Santa Cruz	San Diego	San José	San Francisco	Espíritu Santo	Cerralvo
												x	x
x	x	x	o	o	o	o	x	x	o	x	x	x	x
x	x		x	x	x	x	x	x		x		x	
	x			x									
	x		x									x	
												x	
x	x			x			x				x	x	
												x	x
									x			x	x
			T	x	x	x	x						x
													x
	x	x		x	x		x	x		x			
x			o	x		x		o		x	x	x		x
	x											x	
x	x		x	x	x	x	x		o	x	x	x
									x				
x	x												
			x	x		x	x			x			x
												x	
			x	x	o								

(continued)

Table A4.1-4 (Continued)

	Ángel de la Guarda	San Lorenzo	Tiburón	San Esteban	San Pedro Mártir	San Pedro Nolasco
Galphimia angustifolia Benth. (=*G. brasiliensis* ssp. *angustifolia*; =*Thryallis angustifolia*)			x			x
Janusia californica Benth.			x			
J. gracilis A. Gray			x	x		
Malvaceae						
Abutilon californicum Benth.			x			
A. incanum (Link) Sweet			x			
A. xanti A. Gray						
Anoda crenatiflora Ortega						
A. palmata Fryxell						
Gossypium armourianum Kearney						
G. davidsonii Kellogg (=*G. klotzschianum* var. *davidsonii*)						x
G. harknessii Brandegee						
Herissantia crispa (L.) Brizicky			x			
Hibiscus biseptus S. Watson			x			
H. denudatus Benth.	x		x	x		
H. ribifolius A. Gray						
Horsfordia alata (S. Watson) A. Gray			x			
H. newberryi (S. Watson) A. Gray	x		x	x		
H. rotundifolia S. Watson			x			
Sida xanti A. Gray						
Sphaeralcea ambigua A. Gray ssp. *ambigua*			x			
S. ambigua ssp. *versicolor* Kearney	T			x		
S. axillaris S. Watson						
S. hainesii Brandegee	x	x			x	
Martyniaceae						
Proboscidea altheaefolia (Benth.) Decne.			x			
Molluginaceae						
Mollugo cerviana (L.) Ser.			x			
M. verticillata L.						x
Moraceae						
Ficus palmeri S. Watson (=*F. petiolaris* ssp. *palmeri*; =*F. brandegeei*)	x		x	x	T	x
Nyctaginaceae						
Abronia maritima Nutt. ex S. Watson	x	o	x			
Allionia incarnata L.	x		x			
Boerhavia						x
B. coccinea Mill.						
B. coulteri (Hook. f.) S. Watson			x	x		
B. erecta L.			x	x		
B. intermedia M.E. Jones (=*B. erecta* var. *intermedia*)						
B. maculata Standl.						
B. purpurascens A. Gray						
B. spicata Choisy						
B. Triquetra S. Watson	x					
B. wrightii A. Gray						
B. xantii S. Watson						
Commicarpus brandegeei Standl.						

Tortuga	San Marcos	San Ildefonso	Coronado	Carmen	Danzante	Monserrat	Santa Catalina	Santa Cruz	San Diego	San José	San Francisco	Espíritu Santo	Cerralvo
				x	x	x	x			x		x	
x	x		x	x	x	x	x			x		x	x
				x									
			x	x						x			x
x			x	x	x					x			x
			x										
										x		x	
	T												x
													x?
			x	x		x							x
	x		x	x		x	x						x
x			x		x					x			x
x	o		x	x	x	x	x	x	x	x	x	x	x
										x		x	
			x							x		x	x
	x												
				x						x	x	x	
										x	x		x
										x			
	x												
x	x			x	x	x				x			
										x	x		x
							x					x	
									x	x		x	
	x	x		x	x	x	x	x	x	x	x	x	x
				o		x	x		o	x	x	x	x
x	x		o	x	x	x		x		x	x		x
			x	x									x
				x		x	x						
					x	x	x		x			x	
	x			x		x	x			x		x	x
												x	x
										x			
										x			
		x											
				x						x			
										x			
										x			x

(continued)

Table A4.1-4 (*Continued*)

	Ángel de la Guarda	San Lorenzo	Tiburón	San Estéban	San Pedro Mártir	San Pedro Nolasco
C. scandens (L.) Standl.			x			
Mirabilis tenuiloba S. Watson	x	x		x		
Olacaceae						
Schoepfia californica Brandegee						
Oleaceae						
Forestiera shrevei Standl. (=*F. phillyreoides*)						
Onagraceae						
Camissonia californica (Nutt. ex Torr. & A. Gray) Raven (=*Oenothera leptocarpa*)	x		x			
C. cardiophylla (Torr.) Raven ssp. *cardiophylla* (=*Oenothera cardiophylla* ssp. *cardiophylla*)	x	x		x	x	
C. cardiophylla ssp. *cedrosensis* (Greene) Raven	x		x			
C. chamaenerioides (A. Gray) Raven (=*Oenothera chamaenerioides*)	x		x			
C. crassifolia (Greene) Raven			x			
Oenothera brandegeei (Munz) Raven)	x					
O. californica (S. Watson) S. Watson				x		
Oxalidaceae						
Oxalis nudiflora Moc. & Sesse						
Papaveraceae						
Argemone gracilenta Greene						
A. subintegrifolia G.B. Ownbey	x			x		
Eschscholzia minutiflora S. Watson	x		x			
Passifloraceae						
Passiflora arida (Mast. & Rose) Killip var. *arida*		x	x	x		
P. arida var. *cerralbensis* Killip						
P. foetida L.						
P. fruticosa Killip						
P. palmeri Rose	x	x	x	x		
Phytolaccaceae						
Stegnosperma halimifolium Benth.	x	x	x	x	x	
Plantaginaceae						
Plantago ovata Forsskal (=*P. insularis* var. *fastigiata*)	x		x			
Plumbaginaceae						
Plumbago scandens L.			x			
Polemoniaceae						
Gilia palmeri S. Watson	x					
G. stellata Heller	x					
Leptodactylon pungens (Torr.) Torr. ex Nutt.	x					
Polygonaceae						
Antigonon leptopus Hook. & Arn.						
Eriogonum angelense Moran [Endemic]	T					
E. austrinum (Stokes) Reveal	x					
E. elongatum Benth. *sensu lato*	x					
E. fasciculatum Benth. var. *flavoviride* Munz & I.M. Johnst.	x					
E. fasciculatum var. *polifolium* (Benth.) Torr. & A. Gray	x					

Vascular Plants

Tortuga	San Marcos	San Ildefonso	Coronado	Carmen	Danzante	Monserrat	Santa Catalina	Santa Cruz	San Diego	San José	San Francisco	Espíritu Santo	Cerralvo
	x			x						x			x
												x	
				x								x	
x	x												
													x
				x		x							
x	x		x			x				x	x		x
										x	x		T
x	x		x	T	x		x	x	x	x	x	x	
x	x		o	x	x	o	x	o		x	x		x
										x			
			o	o	x	x		o	o	x	o	x	x

(continued)

Table A4.1-4 (*Continued*)

	Ángel de la Guarda	San Lorenzo	Tiburón	San Estéban	San Pedro Mártir	San Pedro Nolasco
E. inflatum Torr. & Frém. var. *deflatum* I.M. Johnst.	x	x	x	x		
E. orcuttianum S. Watson	x					
E. thomasii Torr.	x					
E. wrightii Torr. ex Benth. *sensu lato*	x					
Portulacaceae						
Calandrinia maritima Nutt. (=*Cistanthe maritima*)	x					
Portulaca	x					
P. halimoides L. (=*P. parvula*)			x			
P. oleracea L. (=*P. retusa*)						
P. pilosa L. (=*P. californica*; =*P. mundula*)						
P. umbraticola Kunth ssp. *lanceolata* (Engelm.) Matthews & Ketron (=*P. lanceolata*)			x			
Talinum paniculatum (Jacq.) Gaertn.			x			
Primulaceae						
Samolus ebracteatus Kunth						
Resedaceae						
Oligomeris linifolia (Vahl) J.F. Macbr.	x	x	x	x		
Rhamnaceae						
Colubrina viridis M.E. Jones (=*C. glabra*)	x		x	x		x
Condalia globosa I.M. Johnst. var. *globosa*						
C. globosa var. *pubescens* I.M. Johnst.	x		x	T		
Karwinskia humboldtiana (Roem. & Schult.) Zucc.						
Ziziphus obtusifolia (Hook. ex Torr. & A. Gray) A. Gray var. *canescens* (A. Gray) M.C. Johnst. (=*Condalia lvcioides* var. *canescens*)	x		x	x		
Rhizophoraceae						
Rhizophora mangle L.			x			
Rubiaceae						
Chiococca petrina Wiggins			x			
Galium stellatum Kellogg var. *eremicum* Hilend & J. Howell	x	x				
Hedyotis brevipes (Rose) W.H. Lewis (=*Houstonia brevipes*)		x				
H. gracilenta (I.M. Johnst.) W.H. Lewis (=*Houstonia gracilenta*; endemic)						
H. mucronata Benth. (=*Houstonia mucronata*; =*Houstonia fruticosa*; type from Carmen)						
H. saxitilis W.H. Lewis (=*Houstonia australis*)						
Mitracarpus linearis Benth.						
M. schizangius DC.						
Randia capitata CD. (=*R. megacarpa*)						
R. thurberi S. Watson			x			
Rutaceae						
Amyris cf. *madrensis* S. Watson						
Esenbeckia flava Brandegee						
Thamnosa montana Torr. & Frém.	x					
Salicaceae						
Salix gooddingii Ball			x			

Vascular Plants

Tortuga	San Marcos	San Ildefonso	Coronado	Carmen	Danzante	Monserrat	Santa Catalina	Santa Cruz	San Diego	San José	San Francisco	Espíritu Santo	Cerralvo
x	x		x	x									
		x	x							x			x
													x
													x
	x												
x	x		x									x	
x	x		x	x	x	x	x	x		x		x	x
			x	o			x			x		x	x
										x		x	
										x		x	x
	x									x			x
			x	o						x		x	
	x		x	x	x	x	x		x	x	x		x
	x								T				
		x	x	x	x	x			x	x	x	x	x
									x				x
												x	
												x	
												x	o
										x			
						x				x	x		x

(continued)

Table A4.1-4 (*Continued*)

	Ángel de la Guarda	San Lorenzo	Tiburón	San Estéban	San Pedro Mártir	San Pedro Nolasco
Sapindaceae						
Cardiospermum corindum L.			x			
C. spinosum Radlk.						
C. tortuosum Benth.						
Dodonaea viscosa (L.) Jacq.			x			
Paullinia sonorensis S. Watson						
Sapotaceae						
Sideroxylon leucophyllum S. Watson	x			x		
S. occidentale (Hemsl.) T.D. Penn. (=*Bumelia occidentalis*)			x			
Scrophulariaceae						
Antirrhinum cyathiferum Benth.	x		x	x		x
A. kingii S. Watson var. *watsonii* (Vasey & Rose) Munz (=*Sairocarpus kingii*)	x	x	x			
Castilleja lanata A. Gray	x					
Conobea intermedia A. Gray						
C. polystachya (Brandegee) Minod						
Galvezia juncea (Benth.) Ball var. *pubescens* (Brandegee) I.M. Johnst. (=*G. juncea* var. *foliosa*)	x	x		x		x
Linaria texana Scheele (=*L. canadensis* var. *texana*)			x			
Mimulus floribundus Lindl.						
Mohavea confertiflora (Benth.) A. Heller	x					
Penstemon angelicus (I.M. Johnst.) Moran (=*P. clevelandii* ssp. *angelicus*; endemic)	T					
Russelia retrorsa Greene (*R. polyedra* misapplied)						
Stemodia durantifolia (L.) Sw.			x			
Simaroubaceae						
Castela peninsularis Rose						
C. polyandra Moran & Felger			x			
Simmondsiaceae						
Simmondsia chinensis (Link) Schneid.	x		x	x		x
Solanaceae						
Datura discolor Bernh.	x	x	x			
Lycium				x		
L. andersonii A. Gray (all vars. except *deserticola*)	x		x	x		
L. andersonii var. *deserticola* (C.L. Hitchc.) Jepson			x			
L. berlandieri Dunal cf. var. *peninsulare* (Brandegee) C.L. Hitchc.	x	x				
L. brevipes Benth. (=*L. vichii*)	x		x	x	x	
L. californicum Nutt. ex A. Gray	x		x			
L. carolinianum Walt.						
L. fremontii A. Gray			x			
L. megacarpum Wiggins	x			x		
L. parishii A. Gray						
Nicotiana clevelandii A. Gray			x			
N. glauca Graham						
N. greeneana Roem.						
N. obtusifolia M. Martens & Galeotti (=*N. trigonophylla*)	x	x	x	x	x	x

Vascular Plants

Tortuga	San Marcos	San Ildefonso	Coronado	Carmen	Danzante	Monserrat	Santa Catalina	Santa Cruz	San Diego	San José	San Francisco	Espíritu Santo	Cerralvo
	x		x	x	x	x	x	x		x		x	x
	x									x	x	x	x
	x		x								x	x	x
													x
										x	x		
x	x	x	x		x	o				x	x	x	x
	x	x		x	x					x		x	
												x	
												x	
	x			x		x						x	
													x
												x	
												x	x
	x						x	x	x	x	x	x	x
	o		x	x	x	o	x	x		x	x	x	
	x			x		x	o	x		o		x	x
							x		x				
x	x			x	x	x						x	x
						x				x			
x	x	x		x	x					x	x	x	x
										x		x	
										x			
										x			
										x			
	x					x							
										x			
										x			
x		x		x			x						

(continued)

Table A4.1-4 (*Continued*)

	Ángel de la Guarda	San Lorenzo	Tiburón	San Esteban	San Pedro Mártir	San Pedro Nolasco
Physalis angulata L.						
P. crassifolia Benth. var. *crassifolia* (=*P. greenei*)						
P. crassifolia var. *infundibularis* I.M. Johnst.	T	x		x		
P. crassifolia var. *versicolor* (Rydb.) Waterfall			x			
P. glabra Benth.						
P. leptophylla B.L. Rob. & Greenm.						
P. philadelphica Lam.						
P. pubescens L.			x			
Solanum hindsianum Benth.	x	x	x	x		
Sterculiaceae						
Ayenia compacta Rose	x		x	x		
A. filiformis S. Watson				x		
A. glabra S. Watson			x			
Hermannia palmeri Rose						
Melochia tomentosa L.			x			x
Waltheria indica L. (=*W. americana*)						
Tamaricaceae						
Tamarix aphylla (L.) H. Karst.						
T. ramosissima Ledeb.			x			
Theophrastaceae						
Jacquinia macrocarpa Cav. ssp. *pungens* (A. Gray) B. Stahl (=*J. pungens*)			x			
Ulmaceae						
Celtis pallida Torr.			x			
Urticaceae						
Parietaria hespera B.D. Hinton var. *hespera* (*P. floridana* misapplied)	x	x	x	x		x
Verbenaceae						
Aloysia barbata (Brandegee) Moldenke						
Avicennia germinans (L.) L.			x			
Citharexylum flabellifolium S. Watson						
Lantana hispida Kunth =*L. velutina*						
Lippia palmeri S. Watson			x			
Violaceae						
Hybanthus fruticulosus (Benth.) I.M. Johnst.			x			
H. verticillatus (Ortega) Baill.						
Viscaceae						
Phoradendron californicum Nutt.	x		x			
P. diguetianum Van Tieghem (=*P. eduardi*; type from Carmen; =*P tumidum*)			x			
Vitaceae						
Cissus trifoliata (L.) L.						
Zygophyllaceae						
Fagonia barclayana (Benth.) Rydb.						
F. densa I.M. Johnst.	x	x				
F. laevis Standl. =*F. californica*			x			
F. pachyacantha Rydb.	x		x			
F. palmeri Vasey & Rose			x			
F. villosa D.M. Porter						
Guaiacum coulteri A. Gray			x			

Vascular Plants 505

Tortuga	San Marcos	San Ildefonso	Coronado	Carmen	Danzante	Monserrat	Santa Catalina	Santa Cruz	San Diego	San José	San Francisco	Espíritu Santo	Cerralvo
	x			x	x				x	x	x	x	x
	x							x			x		
											x	x	
									x				
											x		x
	x												
x	o			x	x	x	x	x		x	x	x	x
	x		x	x	x	x				x			x
										x			
o	x		x	x	x			x	x	x	x	x	x
													x
													x
			x	x	x					x	x	x	x
						x							
	x		x	x		x				x		x	x
										x			x
										x			x
													x
				x	x	x	x	x	x	x		x	x
												x	
			x	x	x	x				x	x		
			x	x		x				x			
	x		x										
	x												
											x	x	

(continued)

Table A4.1-4 (*Continued*)

	Ángel de la Guarda	San Lorenzo	Tiburón	San Estéban	San Pedro Mártir	San Pedro Nolasco
Kallstroemia californica (S. Watson) Vail			x			
K. grandiflora Torr. ex A. Gray			x			
K. peninsularis D.M. Porter						
Larrea tridentata (Sessé & Moc. ex DC.) Coville	x		x			
Tribulus cistoides L.						
Viscainoa geniculata (Kell.) Greene var. *geniculata*	x	x	x	x	x	
Liliopsida (Monocots)						
Agavaceae						
Agave cerulata Trel. ssp. *cerulata*	x					
A. cerulata ssp. *dentiens* (Trel.) H.S. Gentry (=*A. dentiens*; endemic)				T		
A. cerulata ssp. *subcerulata* H.S. Gentry						
A. chrysoglossa I.M. Johnst.			x			T
A. sobria Brandegee ssp. *roseana* (Trel.) H.S. Gentry						
A. sobria ssp. *sobria*						
A. subsimplex Trel.			x			
Alliaceae						
Allium haematochiton S. Watson	x					
Amaryllidaceae						
Zephyranthes arenicola Brandegee						
Arecaceae (Palmae)						
Brahea armata S. Watson (=*Erythea armata*)	x					
B. cf. *brandegeei* (Purpurs) H.E. Moore (=*Erythea brandegeei*; introduced? extinct)						
Phoenix dactylifera L.						
Commelinaceae						
Commelina erecta L. var. *angustifolia* (Michx.) Fern.						
Gibasis heterophylla (Brandegee) Reveal & Hess (=*Tradescantia heterophylla*)						
Cyperaceae						
Cyperus dioicus I.M. Johnst.						
C. elegans L.			x			
C. esculentus L.						
C. squarrosus L. (=*C. aristatus*)					x	
Eleocharis geniculata (L.) Roem. & Schult.			x			
Nolinaceae						
Dasylirion wheeleri S. Watson			x			
Nolina bigelovii (Torr.) S. Watson	x					
Poaceae (Gramineae)						
Anthephora hermaphrodita (L.) Kuntze						
Aristida adscensionis L.	x	x	x	x	x	x
A. californica Thurber ex S. Watson var. *californica*	x	x	x			
A. californica var. *glabrata* (Vasey (=*A. glabrata*)						
A. orcuttiana Vasey (=*A. schiedeana*)						
A. purpurea Nutt. var. *nealleyi* (Vasey) K.W. Allred (=*A. glauca*)	x	x				

Tortuga	San Marcos	San Ildefonso	Coronado	Carmen	Danzante	Monserrat	Santa Catalina	Santa Cruz	San Diego	San José	San Francisco	Espíritu Santo	Cerralvo
			x	x	x				x	x			x
	x		x				x				x	x	x
x		x	o						x	x		x	
	x												
												T	
			x	x									
			x			x							
								x					
							o						
													x
												x	
			x			x							x
												x	
	x											x	x
x	x		x	x		x	x			x	x	x	x
	x		x										
												x	
												x	

(continued)

Table A4.1-4 (Continued)

	Ángel de la Guarda	San Lorenzo	Tiburón	San Esteban	San Pedro Mártir	San Pedro Nolasco
A. purpurea var. wrightii (Nash) K.W. Allred (=A. wrightii)			x			
A. ternipes Cav.			x			x
Arundo donax L.			x			
Bothriochloa barbinodis (Lag.) Herter						x
Bouteloua annua Swallen						
B. aristidoides (Kunth) Griseb.			x	x		x
B. barbata Lag.	x		x	x		
B. reflexa Swallan						
Brachiaria arizonica (Scribn. & Merr.) S.T. Blake (=Panicum arizonicum)			x	x		
B. fasciculata (Sw.) Parodi (=Panicum fasciculatum)						
Cathestecum erectum Vasey & Hack.			x			
Cenchrus palmeri Vasey	o	x	x	x		x
Chloris bandegeei (Vasey) Swallen				x		
C. crinita Lag. (=Trichloris crinita [Lag.] Parodi)						x
C. cf. verticillata Nutt.						
C. virgata Sw.						
Dactyloctenium aegyptium (L.) Willd.						
Digitaria californica (Benth.) Henr. (=Trichachne californica)			x	x		x
Distichlis palmeri (Vasey) Fassett ex I.M. Johnst.	x					
D. spicata (L.) Greene			x			
Enneapogon desvauxii Beauv.						
Eragrostis pectinacea (Michx.) Nees ex Steud. var. pectinacea (=E. diffusa)						x
E. viscosa (Retz.) Trin.						
Erioneuron pulchellum (Kunth) Tateoka	x	x	x	x		
Heteropogon contortus (L.) Roem. & Schult.	x	x	x	x		
Jouvea pilosa (J. Presl) Scribn.						
Lasiacis ruscifolia (Kunth) A.S. Hitchc. & Chase			x			
Leptochloa dubia (Kunth) Nees						
L. fascicularis (Lam.) A. Gray						
L. filiformis (Lam.) Beauv. (=L. mucronata)			x	x		x
Melica frutescens Scribn.	x					
Monanthochloe littoralis Engelm.	x		x			
Muhlenbergia brandegeei C. Reeder						
M. microsperma (DC.) Trin.	x	x	x	x	x	x
Panicum alatum Zuloaga & Morrone						
P. hirticaule C. Presl			x			
Phragmites australis (Cav.) Steud. (=P. communis)			x			
Setaria leucopila (Scribn. & Merr.) K. Schum.						
S. liebmannii Fourn.			x			
S. macrostachya Kunth						
S. palmeri Henr.						
Sporobolus contractus A.S. Hitchc.						
S. cryptandrus (Torr.) A. Gray			x			
S. patens Swallen			x			

	Tortuga	San Marcos	San Ildefonso	Coronado	Carmen	Danzante	Monserrat	Santa Catalina	Santa Cruz	San Diego	San José	San Francisco	Espíritu Santo	Cerralvo
														x
					x									
	x			x	x	x		x	x	x	x		x	
				x	x		x				x	x	x	
				x	x	x	x	x	x	x	x		x	
				x	x				x		x	x	x	x
	x				x								x	
		o		x	x	o	x				x		x	x
				x	x		x	x	x		x			x
									x					
	x									x			x	
		x												x
				x	x	x		x			x	x	x	x
										x				
													x	
		x			x	x	x	x		x	x	x	x	x
				x	x	x	x	x			x	x	x	x
				x							x		x	
				x	x	x		x		x				x
				x				x			x	x	x	x
							x						x	x
	x	x		x	x	x	x	x			x		x	x
				x	x						x			
		x	x				x			x			x	x
					x					x			x	x
				x							x		x	x
				x										
											x			
							x							

(continued)

Table A4.1-4 (*Continued*)

	Ángel de la Guarda	San Lorenzo	Tiburón	San Estéban	San Pedro Mártir	San Pedro Nolasco
S. pryamidatus (Lam.) A.S. Hitchc. (=*S. argutus*)						
S. virginicus (L.) Kunth						
Stipa speciosa Trin & Rupr. (=*Achnatherum speciosum*)	x					
Vulpia octoflora (Walt.) Rydb. var. *hirtella* (Piper) Henr.	x					
Ruppiaceae						
Ruppia maritima L.	x		x			
Themidaceae						
Bessera tenuiflora (Greene) Macbr. (*Behria tenuiflora*)						
Typhaceae						
Typha domingensis Pers.			x			
Zosteraceae						
Zostera marina L.			x			
Total taxa recorded	199	82	298	123	24	57

Tortuga	San Marcos	San Ildefonso	Coronado	Carmen	Danzante	Monserrat	Santa Catalina	Santa Cruz	San Diego	San José	San Francisco	Espíritu Santo	Cerralvo
			x	x						x			
				x	x	x	x			x			
												x	
												x	
79	142	34	127	195	128	127	122	100	75	219	109	249	232

Appendix 4.2

The Vascular Flora of Cerralvo Island

JOSÉ LUIS LEON DE LA LUZ
JON P. REBMAN

    Cerralvo is a mountainous island composed of granitic and volcanic rocks, approximately in equal portions. The granitic substrates make up the southern half of the island, and the volcanic rocks make up the northern half. The granite is part of the great batholith of lower Cretaceous age which underlies most of the peninsula and presumably the eastern part of the Sea of Cortés sea floor, since it appears intermittently on Cerralvo, Espíritu Santo, San José, San Diego, Santa Cruz, Catalina, and Granito (Gastil et al. 1978; Moore and Curray 1982). The volcanic deposits represent the southernmost outcrops of the Comondú Formation, which are the dominant rock types in the southern half of the peninsula. This formation consists of a mixture of Late Tertiary sandstone, conglomerates, agglomerates, tufas, and lavas. Some of the lowland areas of the island are underlain by Pleistocene marine sandstone (Hausback 1984). The probable origin of this island is by faulting and uplifting.

    Cerralvo lies relatively close to the peninsula, around 17 km along its western flank and 11 km at the southern point. In respect to other gulf islands, it is a high and narrow island about 26 km long and 7 km wide, with approximately 160 km^2 of surface area. The highest elevation on the island is 680 m, by altimeter, making this island the third highest in the gulf island system, after Tiburón and Ángel de la Guarda.

    As a result of its geographical position (24°09′N–24°22′N and 109°48′W–109°56′W), close to the Tropic of Cancer, and as a result of its high mountains, Cerralvo receives more rainfall than all of the other islands in the gulf. The majority of rainfall is from summer thunderstorms, which usually produce great amounts of water in a short time. However, smaller amounts of moisture can also come with winter precipitation (<10% of the annual total), or as fog, which is common in the uplands. As is the case for all gulf islands, there has never been a meteorological station on the island to record climatic data, but judging from the records at the nearest station on the peninsula (located around 30 km away and 550 m in elevation), total annual rainfall may reach 300–400 mm. However, due to the steep-slope topography of the island, most water runs off rapidly and into the sea, so its availability to terrestrial organisms is low.

Geomorphologically, the island is conformed mostly by the slopes of hills, small intermountain valleys, and arroyo beds along the canyons. These arroyos occasionally end in an alluvial flat, but mostly run directly into the sea. The coastal zone is made up mainly of narrow stony and sandy beaches; salt flats are scarce. No mangrove stands grow along the coastal fringe of the island due to the lack of bays and coves. Temporary ponds are sometimes formed in flat, rocky beds after heavy rains and can remain for some weeks. The elevations above 500 m represent less than 10% of the total surface area. The eastern side of the island is more precipitous than the western side, a pattern similar to that found on almost all of the southern gulf islands. It is believed that this is a consequence of the pulling effects of the southern peninsula, whose tectonic plate is moving faster than the central and northern segments. The more graded slope of the western side of the island has zigzagging canyons that reach hundreds of meters in elevation in only a few kilometers from the shore. These canyons run in an east–west direction, allowing the north and south slopes to receive different amounts of solar radiation. Thus, the microclimatic conditions from side to side are quite different in respect to temperature and soil moisture, which in turn selects for different plant species assemblages.

In reference to physiognomy, the dominant plant species of the lowland slopes are composed of cacti such as *Pachycereus pringlei, Stenocereus gummosus,* and *Ferocactus diguetii* var. *diguetii*; legumes like *Olneya tesota* and *Aeschynomene nivea*; and other common species such as *Bursera microphylla, Fouquieria diguetii,* and *Colubrina viridis*. In the arroyo beds, stands of *Lysiloma candidum, Olneya tesota, Cercidium floridum* ssp. *peninsulare,* and *Jacquemontia abutiloides* are typical. The few but extensive alluvial plains are dominated by *Ebenopsis confinis, Castela peninsularis, Phaulothamnus spinescens,* and *Olneya tesota*. These four spinescent species grow in dense, shrubby populations that make it difficult to travel across this land form. The narrow zone of coastal vegetation is composed of species such as *Atriplex barclayana, Perityle* spp., *Sesuvium portulacastrum, Vaseyanthus insularis, Abronia maritima,* and *Jouvea pilosa*. *Hofmeisteria fasciculata* var. *xantii* is a common occupant on most rocky cliffs. The upland portions of the island contain plant species such as *Randia capitata, Dodonaea viscosa, Esenbeckia flava, Karwinskia humboldtiana, Erythrina flabelliformis, Gochnatia arborescens,* and *Cordia curassavica,* as well as the herbaceous *Commelina erecta*. Most of these species represent more modern colonizing attempts in the wetter habitats of the island by taxa from dry tropical areas or are remnants of a dry tropical forest that grew there in ancient times, such as can be found in the mountains of the nearby cape region on the peninsula.

To analyze the vegetation, we sampled a plot of 1000 m^2 in a belt transect of 5×200 m on a north-facing slope at 110 m in elevation. Sampling was done by measuring the height and two diameters of the canopy coverage for each woody individual taller than 0.2 m. From these field data, we calculated the mean height, mean and total canopy coverage, and the number of individuals for each species. Each species was classified into one of three life forms based on its growth habit: trees, bushes, and perennials. A general ranking table (table A4.2-1) was created according to the relative importance values for each species, which was calculated from the sum of the relative values of these three variables divided by 3 (Brower and Zar 1977). Shrub vegetation is the dominant life form in this analysis because on the north-facing slopes the accumulated canopy is higher than the sampled area. However, because the plants tend to be clumped, some bare soils are also exposed. Only a few species are quantitatively dominant. This trend seems to be characteristic of most of the gulf islands.

Cerralvo ranks third in the Sea of Cortés in respect to plant endemism with three island-specific endemic taxa (although two taxa have undocumented populations in other areas) and six insular endemics. Formerly, it was thought to be the island with the highest rate of endemism. However, many of the presumed endemic taxa have recently been found on the adjacent peninsula or on other islands. For example, *Marina oculata* is now known to occur on Espíritu Santo, and *Dudleya nubigena* ssp. *cerralvensis* and *Passiflora arida* var. *cerralbensis* have been seen on the cliffs and bajadas of the east cape region. *Ferocactus diguetii* var. *diguetii*, a taxon which can be physiognomically important in the landscape, is shared between Cerralvo and the distant Santa Catalina Island (where it is one of the most characteristic plants), but it is also found occasionally on Coronado, Danzante, Monserrat, San Diego, and Santa Cruz. *Mammillaria evermanniana* may also have a similar pattern; it has been observed on Espíritu Santo and Santa Catalina, but has not yet been documented. *Ibervillea sonorae* var. *peninsularis*, a perennial climbing vine, was originally described by Johnston (1924) as a variety for Cerralvo, but later he also recognized populations of the southern cape region in the same taxon. The only strict endemic taxon that has not yet been reported off of the island seems to be *Mammillaria cerralboa*, which is one of the tallest members of this genus. Another interesting taxon is *Indigofera nelsonii*, which was once described as the endemic *I. argentata* by Johnston (1924), who also reported a small population of *Acacia filicioides*, a species that grows in southern Mexico, but we have not yet found this population. On the slopes of the canyons above 200 m in elevation grow stands of copalquín (*Pachycormus discolor* var. *pubescens*), representing the southern limit of distribution of this typically mid-peninsular, endemic genus. Curiously, not one representative of the genus *Agave* has yet been encountered on Cerralvo, although the environment seems suitable for some of the Cape Region species.

Cerralvo is not permanently inhabited by humans, but it is believed to have been by indigenous peoples some centuries ago. However, local fishermen do irregularly occupy camps along the shore during some seasons of the year. Currently, there is a high impact on the vegetation of the island, including the scarce soil, which can be attributed to the grazing effects of goats. No other island in the gulf has such a high density of feral goats. We estimate that the current goat population is more than 1000 animals. Fishermen regularly hunt them for food, but the rate of births still seems to be higher than the number of deaths. Recently, the common peninsular hare, *Lepus californicus*, was introduced into the southern part of the island, but it is difficult to ascertain whether it will become a pernicious influence on the natural balance of the vegetation.

The flora of the island presented here is based on the work of the authors (eight trips between 1995 and 1998) and Moran's compilation of vascular plants of the gulf islands, cited as "SD" (presented in Case and Cody 1983). Moran previously reported 143 plant taxa for the island as a product of several explorations between 1952 and 1976, but we now have registered 232 plant taxa (see table A4.2-2), which greatly increases the known flora for the island. The current listing includes 89 new plant records that we found on CIBNOR expeditions. Thirty-four taxa reported by Moran have not been found again, although 109 of the same taxa have been found. If the fieldwork in both eras has recorded most of the flora, then the floristic discrepancies could be interpreted as a result of local extinctions and colonization from the flora of the peninsula.

In respect to floristic composition, the Asteraceae (28 taxa), Euphorbiaceae (23), Fabaceae (22), Cactaceae (16), and Poaceae (16) are the families with the greatest species diversity on the island. The plants in these five plant families make up 45% of the total taxa represented on the island. Only 8 genera have more than 2 taxa: *Euphorbia* (7), *Perityle* (5), *Opuntia* (4), *Porophyllum* (4), *Mammillaria* (3), *Boerhaavia* (3), *Heydiotis* (3), *Bursera* (3), and *Physalis* (3); 124 genera are monospecific (54%). Considering the life forms of the various listed plant taxa, perennial herbs and bushes are the most common on the island (84 and 65 taxa, respectively), followed by annuals (44), trees (19), vines (16), and parasites (3). Analyzing the distribution of each taxon, we find that 94 of these are restricted to the peninsula, 58 to the Sonoran Desert, 31 to Tropical America, 14 to the Gulf of California coasts and islands, and the remaining taxa are more widely distributed in North America, Mexico, or are cosmopolitan.

References

Brower, J.E., and J.H. Zar. 1977. *Field and Laboratory Methods for General Ecology.* Wm. C. Brown, Dubuque, IA.

Case, T.J., and M.L. Cody. 1983. *Island Biogeography in the Sea of Cortéz.* University of California Press, Berkeley.

Gastil, G., G.J. Morgan, and D. Krummenacher. 1978. Mesozoic history of Peninsular California and related areas east of the Gulf of California. In: D.G. Howell and K.A. McDougall (eds), *Mesozoic Paleogeography of the Western United States.* Society of Economic Paleontologists and Mineralogists, Pacific Coast Paleogeography Symposium 2; pp. 107–115.

Hammond, E.H. 1954. *A Geomorphic Study of the Cape Region of Baja California.* University of California Publications in Geography. University of California Press, Berkeley.

Hausback, B.P. 1984. Cenozoic volcanic and tectonic evolution of Baja California Sur, Mexico. In: V.A. Frizzel (ed), *Geology of the Baja California Peninsula,* Pacific Section. Society of Economic Paleontologists and Mineralogists 39:219–236.

Johnston, I.M. 1924. Expedition of the California Academy of Sciences to the Gulf of California in 1921: The botany (the vascular plants). *Proceedings of the California Academy of Science,* 4 ser. **12**:951–1218.

Moore, D.G., and R.J. Curray. 1982. Geologic and tectonic history of the Gulf of California. In: R.J. Curray and D.G. Moore, (eds) *Initial Reports of the Deep Sea Drilling Project,* vol. 64. U.S. Government Printing Office, Washington, DC; 1279–1294.

Table A4.2-1 Results of sampling a 1000-m² site on a slope on Cerralvo

Growth Form	Species[a]	No. of Individuals	Height Average (m)	Canopy Average (m²)	Coverage Sum (m²)	IVI (%)
Bush	*Jatropha cuneata*	113	1.07	1.91	215.37	18.643
Bush	*Stenocereus gummosus*	41	1.22	7.63	313.17	14.512
Bush	*Bursera microphylla*	50	1.56	5.14	257.12	14.140
Tree	*Olneya tesota*	26	2.45	8.23	213.98	11.465
Bush	*Solanum hindsianum*	18	1.44	1.41	25.44	4.086
Bush	*Opuntia cholla*	25	0.96	0.55	13.94	3.926
Tree	*Lysiloma candidum*	1	3.01	12.37	12.37	3.707
Bush	*Euphorbia magdalenae*	20	0.81	0.86	17.21	3.352
Tree	*Pachycereus pringlei*	7	2.23	0.58	4.08	3.228
Bush	*Fouquieria diguetii*	4	1.85	7.04	28.18	3.221
Bush	*Stenocereus thurberi*	2	2.51	4.47	8.95	3.166
Bush	*Desmanthus fruticosus*	3	1.86	0.66	2.01	2.370
Perennial	*Melochia tomentosa*	7	1.31	1.11	7.72	2.327
Bush	*Aeschynomene nivea*	4	1.62	1.46	5.84	2.321
Bush	*Colubrina viridis*	2	1.61	2.78	5.57	2.093
Perennial	*Pedilanthus macrocarpus*	2	1.51	1.75	3.51	1.925
Perennial	*Porophyllum sp.*	6	0.98	0.14	0.87	1.682
Bush	*Bourreria sonorae*	1	1.41	1.63	1.63	1.662
Perennial	*Echinocereus brandegeei*	2	1.05	3.25	6.51	1.525
Perennial	*Cochemiea poselgeri*	2	0.41	0.37	0.75	0.653
	Total	336			1144.29	100

[a]Species ranked according their importance value index (IVI) obtained from the sum of the relative values for number of individuals, height average, and sum of canopy coverage.

Table A4.2-2 Vascular flora of Cerralvo

	Growth Form	Distribution	Habitat	Abundance	Flowering Phenology	Collection
Pteridophytes						
Pteridaceae						
Argyrochosma peninsularis (Maxon & Weath.) Windham	Per	BC	Sl, Nf	U	SUM	HCIB
Notholaena californica D. C. Eaton	Per	BC	Sl, Nf	U	SUM	HCIB, SD
N. lemmonii D. C. Eaton	Per	BC	Sl, Nf	U	SUM	HCIB, SD
Pentagramma triangularis (Kaulf.) Yatsk., Windham & E. Wollenw. ssp. *maxonii* (Kaulf.) Yatsk., Windham & E. Wollenw.	Per	BC	Sl, Nf	U	SUM	HCIB
Angiosperms Magnoliopsida (dicots)						
Acanthaceae						
Carlowrightia arizonica A. Gray	Per	SD +	Bj, Sl	C	SPR-WIN	HCIB
Dicliptera resupinata (Vahl) Juss.	Per	SD +	Bj, Sl	C	SPR-WIN	HCIB, SD
Holographis virgata (Harv. ex Benth. & Hook f.) T.F. Daniel ssp. *glandulifera* (Leonard & C.V. Morton) T.F. Daniel var. *glandulifera*	Bus	BC	Bj, Sl	L	SUM-AUT	HCIB
Justicia californica (Benth) D. Gibson	Per	SD	Bj, Sl	C	SUM-AUT	HCIB
J. palmeri Rose	Per	BC	Bj, Sl	O	SUM-AUT	SD
Ruellia californica (Rose) I.M. Johnst. ssp. *peninsularis* (Rose) T.F. Daniel	Bus	BC	Bj, Sl	C	SUM-WIN	HCIB
Achatocarpacea						
Phaulothamnus spinescens A. Gray	Per	Mex	Sl, Bj	O	SUM	HCIB, SD
Aizoaceae						
Sesuvium portulacastrum L.	Per	TA	Sh	C	SUM-WIN	SD
Trianthema portulacastrum L.	Ann	SD +	Ar, Sl, Bj	O	SUM	HCIB, SD
Amaranthaceae						
Amaranthus fimbriatus (Torr.) Benth.	Ann	SD +	Ar, Bj	O	SUM	HCIB
A watsonii Standl.	Ann	SD ±	Ar, Sl, Bj	L	SUM	SD
Celosia floribunda A. Gray	Bus	BC	Ar, Bj	L	SPR-SUM	HCIB, SD
Dicraurus alternifolius (S. Watson) Uline & Bray	Bus	SD +	Ar	O	SUM	HCIB
Froelichia interrupta (L.) Moq.	Ann	TA	Ar, Bj, Sl	O	SPR-SUM	HCIB, SD
Iresine angustifolia Euphrasén	Per	TA	Ar, Bj, Sl	O	SPR	HCIB

(*continued*)

Table A4.2-2 (*Continued*)

	Growth Form	Distribution	Habitat	Abundance	Flowering Phenology	Collection
Anacardiaceae						
Cyrtocarpa edulis (Brandegee) Standl.	Tre	BC	Ar, Bj, Sl	C	SPR-SUM	HCIB, SD
Pachycormus discolor (Benth.) Coville var. *pubescens* (S. Watson) H.S. Gentry	Tre	BC	Sl, Nf	L	SPR	HCIB
Apocynaceae						
Macrosiphonia hesperia I.M. Johnst.	Bus	GC	Sl	L	SUM	HCIB
Asclepiadaceae						
Asclepias subulata Decne.	Per	SD+	Ar, Bj, Sl	O	SPR-WIN	HCIB, SD
Cynanchum palmeri (S. Watson) S.F. Blake	Vin	BC	Ar, Bj, Sl	C	SUM, WIN	HCIB, SD
Matelea cordifolia (A. Gray) Woodson	Vin	SD	Ar, Bj, Sl	U	SUM, AUT	SD
M. pringlei (A. Gray) Woodson	Vin	BC	Bj, Sl	U	SUM, AUT	SD
Metastelma pringlei A. Gray sens lat.	Vin	BC	Sl	R	SUM	SD
Asteraceae (Compositae)						
Ambrosia carduacea (Greene) Payne	Per	SD+	Sl	L	SUM, AUT	SD
Bebbia atriplicifolia (A. Gray) Greene	Per	SD	Ar, Sl	C	SPR	HCIB, SD
Brickellia glabrata (Rose) B.L. Rob.	Ann	BC	Ar, Sl	U	SUM	HCIB
B. hastata Benth.	Per	BC	Ar, Sl	U	SUM	HCIB
Coreocarpus dissectus (Benth.) S.F. Blake	Ann	BC	Ar, Bj, Sl	L	WIN	SD
C. parthenioides Benth. var. *parthenioides*	Ann	SD	Ar, Bj, Sl	A	WIN	HCIB, SD
Critonia peninsularis (Brandegee) R.M. King & H. Rob.	Bus	BC	Sl	U	SUM	HCIB, SD
Gochnatia arborescens Brandegee	Tre	BC	Bj, Sl	L	SPR	HCIB, SD
Hofmeisteria fasciculata (Benth.) Walp. var. *xantii* A. Gray	Per	GC	Sh	L	WIN	HCIB, SD
Hymenoclea monogyra A. Gray	Bus, Eve	SD	Ar	L	SUM	SD
Machaeranthera arenaria (Benth.) Shinners	Per	BC	Ar, Bj	C	SPR	HCIB
Malacothrix xantii A. Gray	Ann	BC	Ar, Bj, Sl	U	SPR	HCIB, SD
Pectis linifolia L.	Ann	BC	Bj, Ar	L	SUM	HCIB
Perityle californica Benth.	Ann	BC	Ar, Bj, Cl, Sh	C	WIN	HCIB
P. crassifolia Brandegee var. *crassifolia*	Per	SD	Ar, Bj, Cl, Sh	C	WIN	HCIB, SD
P. crassifolia Brandegee var. *robusta* (Rydb.) Everly	Per	SD	Ar, Bj, Cl, Sh	C	WIN	HCIB, SD
P. emoryi Torr.	Ann	SD	Ar, Bj, Cl, Sh	C	WIN	HCIB

Species	Form	Region	Habitat	Abund	Season	Locality
P. microglossa Benth.	Ann	SD	Ar, Bj, Cl, Sh	U	WIN	HCIB, SD
Pluchea salicifolia (Mill.) S.F. Blake	Per	BC	Ar, Bj	U	SUM, AUT	SD
Porophyllum confertum Greene var. confertum	Bus	BC	Cl, Sl	L	SUM- AUT	HCIB
P. gracile Benth.	Per	SD	Ar, Bj, Sl	U	SUM-AUT	HCIB
P. ochroleucum Rydb.	Bus	BC	Ar, Bj, Sl	L	SUM-AUT	HCIB
P. porfyreum Rose & Standl.	Bus	BC	Ar, Bj, Sl	U	SUM-AUT	HCIB
Vernonia triflosculosa Kunth var. palmeri (Rose) B.L. Turner	Ann	BC	Bj, Sl	U	SUM	HCIB
Viguiera chenopodina Greene	Bus	BC	Bj, Sl	C	SUM-AUT	HCIB
V. deltoidea A. Gray	Bus	BC	Bj, Sl	C	SUM-AUT	HCIB, SD
V. tomentosa A. Gray	Bus	BC	Bj, Nf, Sl	U	SUM-AUT	HCIB
Xylothamnia diffusa (Benth.) G.L. Nesom	Bus	BC+	Ar	L	AUT	SD
Bignoniaceae						
Tecoma stans (L.) Juss. ex Kunth	Bus	TA	Sl, Ar	O	SUM-WIN	HCIB
Boraginaceae						
Bourreria sonorae S. Watson	Bus	Sd	Ar, Bj, Sl	C	AUT	HCIB, SD
Cordia curassavica (Jacq.) Roem. & Schult.	Bus	TA	Sl	U	SPR-AUT	SD
Cryptantha grayi Macbr. var. nesiotica I.M. Johnst.	Ann	GC	Ar, Bj, Fl	C	SUM	HCIB
Burseraceae						
Bursera epinnata (Rose) Engler	Tre	BC	Ar, Bj, Sl	O	SUM	HCIB
B. hindsiana (Benth.) Engler	Ar	GC	Ar, Bj, Sl	C	SUM	HCIB, SD
B. microphylla A. Gray	Ar	BC+	Bj, Sl	A	SUM	HCIB, SD
Cactaceae						
Cochemiea poselgeri (Hildm.) Britton & Rose	Per, Suc	BC	Bj, Sl	O	SPR	HCIB, SD
Echinocereus brandegeei (J.M. Coult.) K. Schum.	Per, Suc	BC	Bj, Sl	O	SUM	HCIB
Ferocactus diguetii (F.A.C. Weber) Britton & Rose var. diguetii	Bus, Suc	GC	Bj, Sl	A	SPR-SUM	HCIB, SD
Ferocactus townsendianus Britton & Rose var. townsendianus	Bus, Suc	BC	Bj, Sl	L	SPR-SUM	HCIB
Lophocereus schottii (Engelm.) Britton & Rose	Tre, Suc	BC	Bj, Sl	U	SPR, SUM	HCIB, SD
Mammillaria cerralboa (Britton & Rose) Orcutt	Per, Suc	End	Bj, Sl	L	SUM	HCIB, SD
M. dioica K. Brandegee	Per, Suc	BC	Sl	U	SUM	SD
M. evermanniana (Britton & Rose) Orcutt	Per, Suc	End	Bj, Sl	U	SUM	HCIB, SD
Opuntia alcahes F.A.C. Weber var. alcahes	Bus, Suc	GC	Bj, Sl	C	SPR	HCIB, SD
Opuntia bravoana E. Baxter	Bus, Suc	BC	Sl	U	SPR	SD
O. cholla F.A.C. Weber	Bus, Suc	SD	Bj, Sl	A	SPR-AUT	HCIB, SD

(continued)

Table A4.2-2 (*Continued*)

	Growth Form	Distribution	Habitat	Abundance	Flowering Phenology	Collection
O. tapona Engelm.	Bus, Suc	BC	Sl	L	SPR	SD
Pachycereus pringlei (S. Watson) Britton & Rose	Tre, Suc	SD	Bj, Sl	A	SUM	HCIB, SD
Peniocereus johnstonii Britton & Rose	Per, Suc	GC	Ar, Bj	R	SPR	SD
Stenocereus gummosus (Engelm.) A. Gibson & Horak	Bus, Suc	SD	Bj, Sl	A	SUM	HCIB, SD
S. thurberi (Engelm.) F. Buxbaum.	Tre, Suc	SD	Bj, Sl	C	SPR-SUM	HCIB, SD
Capparaceae (Capparidaceae)						
Forchammeria watsonii Rose	Tre, Eve	BC	BC +	U	SPR	HCIB, SD
Caryophyllaceae						
Drymaria debilis Brandegee	Per	BC	Ar, Bj, Sl	L	SUM-AUT	HCIB, SD
Chenopodiaceae						
Allenrolfea occidentalis (S. Watson) Kuntze	Bus, Suc	NA	Sf	L	SUM, AUT	HCIB, SD
Atriplex barclayana (Benth.) D. Dietr. *sensu. lat.*	Per, Eve	SD	Fl, Sf	L	SPR-SUM	HCIB, SD
Chenopodium murale L.	Ann	TA	Bj	U	SUM	HCIB
Salicornia virginica L.	Per, Suc	NA	Sf	L	SPR-SUM	HCIB
Convolvulaceae						
Cressa truxillensis Kunth	Per	TA	Du, Sh, Sf	L	SPR, SUM	SD
Evolvulus alsinioides (L.) L. var. *angustifolius* Torr.	Per	TA	Ar, Nf, Sl	C	SUM	HCIB
E. sericeus Sw.	Per	TA	Ar, Nf, Sl	C	SUM	HCIB
Ipomoea pes-caprae (L.) R. Br.	Vin	TA	Du, Sh	U	SUM	HCIB, SD
Jacquemontia abutiloides Benth. var. *eastwoodiana* (I.M. Johnst.) Wiggins	Per	BC	Ar, Bj	A	SUM-WIN	HCIB, SD
Merremia aurea (Kell.) O'Donnell	Vin	BC	Bj, Sl	L	SPR-WIN	HCIB, SD
Crassulaceae						
Dudleya rubigena (Brandegee) Britton & Rose ssp. *cerralvenis* Moran	Per, Suc	End	Nf, Sl	R	SPR	HCIB, SD
Cucurbitaceae						
Echinopepon minimus (Kell.) S. Watson	Vin	SD	Ar, Sl	L	SUM-AUT	HCIB, SD
E. wrightii (A. Gray) S. Watson	Vin	SD	Ar, Bj, Sl	U	SUM-AUT	HCIB
Ibervillea sonorae (S. Watson) Greene var. *peninsularis* (I.M. Johnst.) Wiggins	Vin	End	Sl, Bj	A	SUM-AUT	HCIB, SD

V. insularis (S. Watson) Rose var. brandegeei (Cogn.) I.M. Johnst.	Vin	GC	Ar, Sl	L	SUM-AUT	HCIB, SD
Vaseyanthus insularis (S. Watson) Rose var. insularis	Vin	GC	Ar, Sl	L	SUM-AUT	HCIB, SD
Cuscutaceae						
Cuscuta corymbosa Ruiz & Pavon	Par	TA	Fl	L	SUM	HCIB, SD
Ebenaceae	Per	BC	Ar, Sl	L	SUM	HCIB, SD
Maba intricata (A. Gray) Hiern.						
Euphorbiaceae						
Acalypha californica Benth.	Bus	SD	Sl	U	SPR-AUT	HCIB, SD
A. comonduana Millsp.	Bus	BC	Sl, Bj	L	SPR-SUM	HCIB, SD
Adelia virgata Brandegee	Bus	BC	Ar, Sl	L	SUM-AUT	HCIB
Andrachne microphylla (Lam.) Baill.	Ann	BC	Ar, Sl	U	AUT-SPR	HCIB
Bernardia viridis Millsp.	Bus	TA	Sl	U	SUM	HCIB, SD
Cnidoscolus cf. maculatus (Brandegee) Pax & K. Hoffm.	Per	SD+	Ar, Bj	U	SPR-SUM	HCIB
C. palmeri (S. Watson) Rose	Per	GC	Sl	U	SPR	HCIB, SD
Croton caboensis Croizat	Per	BC	Sl	L	SUM	HCIB
C. magdalenae Millsp.	Bus	BC	Sl	L	SPR-SUM	HCIB, SD
Ditaxis brandegeei (Millsp.) Rose & Standl. var. intonsa I.M. Johnst.	Per	BC	Sl, Bj	L	SPR, SUM	SD
D. lanceolata (Benth.) Pax & K. Hoffm.	Per	SD+	Ar, Sl	L	SPR-SUM	HCIB
Euphorbia californica Benth. var. californica	Bus	SD+	Sl	L	AUT- SPR	HCIB
E. fendleri Torr. & A. Gray	Per	SD	Sl	U	SPR-SUM	HCIB
E. leucophylla Benth. ssp. leucophylla	Per	GC	Du	L	SUM-WIN	HCIB, SD
E. magdalenae Benth.	Bus	BC	Bj, Sl	L	WIN-SUM	HCIB, SD
E. polycarpa Benth. var. mejamia Wheeler	Ann	BC	Ar, Sl, Du	L	SPR, SUM	SD
E. polycarpa Benth. var. polycarpa	Ann	BC	Ar, Sl, Du	L	SPR, SUM	SD
E. setiloba Engelm. ex Torr.	Per	SD+	Sl	L	SPR-AUT	HCIB, SD
Jatropha cinerea (Ortega) Muell.-Arg.	Bus	Mex	Sl, Bj	L	SUM	HCIB
J. cuneata Wiggins & Rollins	Bus	SD	Ar, Fl, Sl	C	SUM	HCIB, SD
Pedilanthus macrocarpus Benth.	Bus	SD, Mex	Bj, Sl	O	SPR-WIN	HCIB
Phyllanthus galleotianus Baill.	Per	Mex	Sl	L	SUM-WIN	HCIB
Tragia cf. amblyodonta (Muell.-Arg.) Pax & K. Hoffm.	Vin	SD+	Sl	L	SUM	HCIB, SD

(continued)

Table A4.2-2 (Continued)

	Growth Form	Distribution	Habitat	Abundance	Flowering Phenology	Collection
Fabaceae (Leguminosae)						
Acacia filicioides (Cav.) Trel.	Bus	TA	Sl	L	SPR	HCIB
A. goldmanii (Britton & Rose) Wiggins	Bus	BC	Sl, Bj	L	SPR	HCIB, SD
Aeschynomene nivea Brandegee	Bus	BC	Sl	L	WIN, SPR	HCIB, SD
Caesalpinia pannosa Brandegee	Bus	BC	Ar, Sl	L	SUM, AUT	HCIB, SD
Cercidium floridum Benth. ex A. Gray ssp. peninsulare (Rose) A. Carter	Tre	BC	Ar	L	SPR	HCIB, SD
Coursetia carbaea (Jacq.) Lavin var. caribaea	Ann	Mex	Sl, Bj	L	SUM, AUT	HCIB
Desmanthus fruticosus Rose	Bus	BC	Ar, Bj, Sl	C	SUM, AUT	HCIB, SD
Desmodium procumbens (Mill.) A.S. Hitchc. var. exiguum (A. Gray) Schub.	Ann	TA	Ar	L	SUM, AUT	HCIB
Ebenopsis confinis (Standl.) Barneby & Grimes	Tre	BC	Ar, Bj, Sh	C	SPR	HCIB, SD
Erthrina flabelliformis Kearney	Tre	Mex	Ar, Sl	U	SPR, SUM	HCIB, SD
Indigofera fruticosa Rose	Bus	BC	Sl, Bj	C	SUM	HCIB
I. nelsonii Rydb.	Bus	End	Ar	C	SUM	HCIB, SD
Lupinus arizonicus (S. Watson) S. Watson sensu. lat.	Ann	SD	Ar, Bj	L	WIN	SD
Lysiloma candidum Brandegee	Tre	BC	Ar	L	SPR	HCIB, SD
Macroptilium atropurpureum (DC.) Urban	Per	TA	Ar, Bj, Sl	U	SPR, SUM	HCIB
Marina oculata (Rydb.) Barneby	Per	End	Ar	U	SUM, AUT	HCIB, SD
Olneya tesota A. Gray	Tre	SD	Ar, Sl, Bj, Sh	C	SPR	HCIB, SD
Phaseolus filiformis Benth.	Ann	SD	Ar, Sl, Bj, Sh	C	SUM, AUT	HCIB
Senna confinis (Greene) Irwin & Barneby	Ann	BC	Ar, Sl	U	SUM-WIN	HCIB, SD
S. covesii (A. Gray) Irwin & Barneby	Per	NA	Ar, Bj, Sl	U	SPR, SUM	HCIB
Tephrosia palmeri S. Watson	Per	BC	Ar, Du	U	SUM	SD
T. purpurea (L.) Pers.	Per	SD	Ar, Sl, Sh	U	SUM	HCIB, SD
Fouquieriaceae						
Fouquieria burragei Rose	Bus	GC	Sl	L	SUM	HCIB
F. diguetii (Van Tieghem) I.M. Johnst.	Bus	BC	Bj, Sl	A	SPR-WIN	HCIB, SD
Lamiaceae (Labiatae)						
Hyptis decipiens M.E. Jones	Bus	SD	Ar, Sl	U	SPR-SUM	HCIB

H. laniflora Benth.	Bus	SD	Ar, Sl	L	SPR-SUM	HCIB, SD
Salvia platycheila A. Gray	Bus	GC	Ar, Bj, Sl	L	AUT, WIN	HCIB
S. similis Brandegee	Per	BC	Sl, Nf	C	SUM, AUT	HCIB, SD
Loasaceae						
Eucnide cordata Kell. ex Curran	Per	BC	Ar, Du	C	SPR-AUT	HCIB, SD
Mentzelia adhaerens Benth.	Ann	BC	Ar, Sl	C	SUM, AUT	HCIB, SD
Loranthaceae						
Psittacanthus sonorae (S. Watson) Kuijt	Par	BC	Sl	U	SPR, SUM	SD
Malpighiaceae						
Janusia californica Benth.	Vin	BC	Sl, Bj	U	SPR, SUM	SD
Malvaceae						
Abutilon californicum Benth.	Bus	SD+	Ar, Bj, Sl	L	SUM, AUT	HCIB, SD
A. incanum (Link) Sweet	Per	Mex	Ar, Bj, Sl	L	SUM, AUT	HCIB, SD
Anoda palmata Fryxell	Ann	BC	Ar, Bj, Sl	L	SUM	HCIB
Gossypium davidsonii Kellogg	Bus	SD+	Ar, Bj, Sl	L	SUM	HCIB
G. harknessii Brandegee	Bus	GC	Ar, Bj, Sl	L	SUM, AUT	HCIB, SD
Herissantia crispa (L.) Brizicky	Per	TA	Ar, Bj, Sl	L	SPR-WIN	HCIB
Hibiscus denudatus Benth.	Per	SD	Ar, Bj, Sl	L	SPR-AUT	HCIB, SD
Horsfordia alata (S. Watson) A. Gray	Per	NA	Ar, Sl	L	SPR-AUT	HCIB, SD
Sida xanti A. Gray	Ann	BC+	Sl	L	SPR-AUT	HCIB, SD
Martyniaceae						
Proboscidea altheaefolia (Benth.) Decne.	Per	SD+	Ar, Du, Sh	L	SPR, SUM	HCIB, SD
Moraceae						
Ficus palmeri S. Watson	Tre	BC	Ar, Sl	L	SUM, AUT	HCIB, SD
Nyctaginaceae						
Abronia maritima Nutt. ex S. Watson	Per	GC	Du, Sh	L	SPR-WIN	HCIB, SD
Allionia incarnata L.	Ann	SD	Ar, Du, Sl, Bj	L	SUM	SD
Boerhavia coccinea Mill.	Per	SD+	Ar, Sl	C	SUM	HCIB
B. intermedia M.E. Jones	Ann	SD+	Ar, Sl	C	SUM	HCIB
B. maculata Standl.	Ann	SD+	Ar, Sl	C	SUM	HCIB, SD
Commicarpus brandegeei Standl.	Per	BC	Ar, Sl	L	AUT, WIN	HCIB, SD
C. scandens (L.) Standl.	Per	TA	Ar, Sl	L	SUM, AUT	HCIB

(*continued*)

Table A4.2-2 (Continued)

	Growth Form	Distribution	Habitat	Abundance	Flowering Phenology	Collection
Oxalidaceae						
Oxalis nudiflora Moc. & Sesse	Ann	Mex	Ar	L	SUM	HCIB
Passifloraceae						
Passiflora arida (Mast. & Rose) Killip var. *cerralbensis* Killip	Vin	End	Ar, Sl	L	SUM	HCIB, SD
Phytolaccaceae						
Stegnosperma halimifolium Benth.	Bus, Eve	BC	Ar, Sh, Du	L	SPR-AUT	HCIB, SD
Polygonaceae						
Antigonon leptopus Hook. & Arn.	Vin	BC +	Ar, Bj, Sl	C	SUM	HCIB, SD
Portulacaceae						
Portulaca pilosa L.	Ann	SD +	Ar, Du	L	SUM	SD
Portulaca umbraticola Kunth ssp. *lanceolata* (Engelm.) Matthews & Ketron	Ann	TA	Ar, Du, Sh	L	SUM	HCIB
Talinum paniculatum (Jacq.) Gaertn.	Per	TA	Ar, Sl	L	SPR, SUM	HCIB
Rhamnaceae						
Colubrina viridis M.E. Jones	Tre	SD	Sl, Bj	U	SUM	HCIB, SD
Condalia globosa I.M. Johnst. var. *globosa* I.M. Johnst.	Bus	SD	Sl, Bj	U	SUM	HCIB, SD
Karwinskia humboldtiana (Roem. & Sch.) Zucc.	Tre	SD	Sl, Bj	U	SUM	HCIB, SD
Ziziphus obtusifolia (Hook ex Torr. & A. Gray) A. Gray var. *canescens* (A. Gray) M.C. Johnst.	Bus	BC	Sl, Bj	U	SPR	SD
Rubiaceae						
Heydiotis brevipes (Rose) W.H. Lewis	Ann	GC	Sl, Bj, Ar	U	SUM, AUT	HCIB, SD
H. mucronata Benth.	Per	GC	Du, Sh	C	SUM, AUT	HCIB, SD
H. saxitilis W. H. Lewis	Per	BC	Sl, Bj, Ar	C	SUM, AUT	HCIB
Randia capitata DC.	Bus	BC	Sl	U	SUM, AUT	SD
Rutaceae						
Esenbeckia flava Brandegee	Tre	BC	Sl	U	AUT	SD
Sapindaceae						
Cardiospermum corindum L.	Vin	TA	Ar, Bj, Sl	C	SUM	HCIB, SD
C. tortuosum Benth.	Bus	BC	Ar, Sl, Du	U	SUM	HCIB, SD
Dodonaea viscosa (L.) Jacq.	Bus, Eve	TA	Sl	L	AUT	SD

Scrophulariaceae						
Antirrhinum cyathiferum Benth.	Ann	BC	Ar	L	SUM, AUT	SD
Linaria texana Scheele	Ann	BC	Ar, Sl, Bj	L	SUM, AUT	SD
Russelia retrorsa Greene	Per	Mex	Sl	U	SPR, SUM	HCIB
Simaroubaceae						
Castela peninsularis Rose	Bus	GC	Sl, Bj	U	SPR	HCIB, SD
Solanaceae						
Datura discolor Bernh.	Ann	TA	Ar, Bj	U	SPR-AUT	HCIB, SD
Lycium andersonii A. Gray	Bus	NA	Bj	U	WIN	SD
L. brevipes Benth.	Bus	NA	Bj, Sl	U	WIN	HCIB, SD
Physalis angulata L.	Per	TA	Ar,	U	SUM	HCIB
Physalis crassifolia Benth. var. crassifolia	Per	SD	Ar, Bj	U	SUM	HCIB, SD
P. leptophylla B.L. Rob. & Greenm.	Per	SD	Ar, Bj	U	SUM	HCIB
Solanum hindsianum Benth.	Bus	SD	Ar, Sl, Fl	C	SUM	HCIB, SD
Sterculiaceae						
Ayenia compacta Rose	Per	BC	Sl	U	SUM	HCIB
Hermannia palmeri Rose	Per	BC	Ar, Sl	U	WIN-SUM	HCIB, SD
Melochia tomentosa L.	Per	TA	Ar, Sl	C	SPR, SUM	HCIB, SD
Waltheria indica L.	Bus	TA	Ar, Sl, Bj	U	SPR, SUM	HCIB
Tamaricaceae						
Tamarix aphylla (L.) H. Karst.	Bus, Eve	Cos	Ar	R	SPR, SUM	HCIB
Verbenaceae						
Aloysia barbata (Brandegee) Moldenke	Bus	BC	Ar	U	SPR	HCIB, SD
Lantana hispida Kunth	Bus	TA	Sl, Bj	U	SPR, SUM	HCIB
Violaceae						
Hybanthus fruticulosus (Benth.) I.M. Johnst.	Per	BC +	Sl, Bj	U	SPR	HCIB
H. verticillatus (Ortega) Baill.	Per	NA	Sl, Bj	U	WIN	HCIB
Viscaceae						
Phoradendron diguetianum Van Tieghem	Par	BC	Sl	L	SPR	SD
Zygophyllaceae						
Kallstroemia californica (S. Watson) Vail	Ann	NA	Ar, Sl, Bj	U	SUM	HCIB
K. peninsularis D.M. Porter	Ann	BC	Ar, Sl, Bj	L	SUM	HCIB

(continued)

Table A4.2-2 (Continued)

	Growth Form	Distribution	Habitat	Abundance	Flowering Phenology	Collection
Liliopsida (monocots)						
Commelinaceae						
Commelina erecta L. var. *angustifolia* (Michx.) Fern.	Per	TA	Ar	L	SUM	HCIB
Cyperaceae						
Cyperus dioicus I.M. Johnst.	Per	BC	Ar, Sl	L	SUM	HCIB, SD
C. squarrosus L.	Ann	Cos	Ar, Sl	L	SUM	HCIB
Poaceae (gramineae)						
Aristida adscensionis L. sensu. lat.	Ann	Cos	Ar, Bj, Sl	C	WIN-SUM	HCIB, SD
A. ternipes Cav.	Per	Cos	Ar, Bj, Sl	C	WIN-SUM	HCIB
Brachiaria arizonica (Scribn. & Marr.) S.T. Blake	Ann	Mex	Ar, Bj, Sl	L	WIN	HCIB
Cenchrus palmeri Vasey	Ann	SD	Ar, Du, Fl	L	SUM	HCIB
Chloris brandegeei (Vasey) Swallen	Per	BC	Ar, Bj, Sl	L	SUM	HCIB
Digitaria californica (Benth.) Henr.	Per	BC	Ar	U	SUM	SD
Distichlis palmeri (Vasey) Fassett ex. I.M. Johnst.	Per	SD	Du, Sh	L	SPR	HCIB
Heteropogon contortus (L.) Roem & Schult.	Per	TA	Sl	L	SUM	HCIB, SD
Jouvea pilosa (J. Presl) Scribn.	Per	TA	Du, Sh	L	SPR-SUM	HCIB, SD
Leptochloa fascicularis (Lam.) Beauv.	Per	TA	Sl	L	SUM	HCIB
Monanthochloe littoralis Engelm.	Per	Mex	Sf, Sh	L	SUM-AUT	HCIB
Muhlenbergia brandegeei C. Reeder	Ann	BC	Sl	L	SUM	HCIB
M. microsperma (DC.) Trin.	Ann	TA	Ar, Bj, Sl	U	SUM	HCIB
Phragmites australis (Cav.) Steud.	Bus, Eve	Cos	Ar	L	SPR	HCIB
Setaria liebmannii Fourn.	Ann	Cos	Sl	L	SUM, AUT	HCIB
S. palmeri Henr.	Per	BC	Sl	L	SUM	HCIB

Explanation of flora. Growth form: Ann = annual or ephemeral, Bus = bush or shrub, Tre = tree, Vin = vine, Per = ± herbaceous perennial, Suc = succulent; Dec = deciduous, Eve = evergreen. Distribution: NA = North America; BC = Baja California; Cos = cosmopolitan; End = endemic; GC = islands and shores of the Gulf of California; SD = Sonora Desert; TA = Tropical America. Habitat: Ar = arroyo beds; Sl = rocky slope; Bj = bajada or outwash slope; Cl = cliff; Du = dunes; Fl = flat or coastal plain; Nf = north-facing slope; Sf = salt flat; Sh = shore. Abundance (designated subjectively based on field work and observations): R = rare; U = uncommon; O = occasional; C = common; A = abundant; L = localized. Flowering months (based on little information specific to the island): SPR = spring; SUM = summer; AUT = autumn; WIN = winter. Collections and references: SD = Ivan Murray Johnston and Reid V. Moran; HCIB = CIBNOR expeditions.

Appendix 4.3

Plants on Some Small Gulf Islands

These lists follow closely those of appendix 4.3 in the earlier version of the *Biogeography*, compiled by R. Moran. The nomenclature has been updated by J. Rebman, following names and authorities given in appendix 4.1.

Table A4.3-1 Plants on small gulf islands

Fam	Species	Alcatraz	Cholludo	Datil	Granito	Las Ánimas	La Víbora	Mejía	Partida Norte	Patos	Rasa	Salsipuedes	San Luis	Santa Ines
Acanthaceae	Carlowrightia arizonica			X										
	Holographis virgata virgata	X	X											
	Justicia californica			X										
Achatocarpaceae	Phaulothamnus spinescens	X	X	X										
Aizoaceae	Sesuvium verrucosum	X									X			
	Trianthema portulacastrum	X				X								
Amaranthaceae	Amaranthus fimbriatus	X		X						X				
	A. watsonii	X		X					X			X		
Anacardiaceae	Pachycornus discolor pubescens							X						X
Asclepiadaceae	Asclepias albicans			X				X						
	Matelea pringlei			X										
Asteraceae	Ambrosia divaricata			X										
	A. ilicifolia							X						
	Bebbia juncea juncea			X										
	Coreocarpus parthenioides parthenioides			X										
	Encelia farinosa	X		X										
	E. f. phenicodonta		X	X										
	Hofmeisteria fasciculata fasciculata		X											
	H. filifolia							X						
	Palafoxia arida	X												
	Pectis rusbyi		X											

Family	Species	1	2	3	4	5	6	7	8	9	10
	Perityle crassifolia robusta					X					
	P. emoryi	X	X								
	Pleurocoronis laphamioides	X	X								
	Porophyllum crassifolium		X	X							
	P. gracile	X								X	
	Senecio mohavensis	X	X	X							
	Trixis californica	X	X	X							
	Viguiera chenopodina			X		X					
	Xylorhiza frutescens	X	X	X							
	Xylothamnia diffusa	X	X	X							
Bataceae	*Batis maritima*			X			X				
Boraginaceae	*Cryptantha angelica*			X	X						
	C. angustifolia			X	X						
	C. fastigiata	X									
	C. grayi cryptochaeta								X		
	C. maritima pilosa	X	X	X		X					
Brassicaceae	*Tiquilia canescens*	X	X								
	Descurainia pinnata halictorum	X		X							
	Draba cuneifolia sonorae	X		X							
	Lyrocarpa coulteri coulteri				X						
	Sibara angelorum	X	X								
Burseraceae	*Bursera hindsiana*	X	X		X						
	B. microphylla	X	X		X						X
Cactaceae	*Carnegiea gigantea*	X		X		X					
	Cochemiea poselgeri			X							
	Lophocereus schottii	X	X		X	X					
	Mammillaria albicans		X	X	X	X					
	M. dioica		X	X							X

(continued)

Table A4.3-1 (Continued)

Fam	Species	Alcatraz	Cholludo	Datil	Granito	Las Ánimas	La Víbora	Mejía	Partida Norte	Patos	Rasa	Salsipuedes	San Luis	Santa Ines
	M. milleri	X												
	Opuntia bigelovii	X	X				X					X		
	O. cf. burrageana						X	X	X			X		
	O. cholla				X									
	O. fulgida fulgida	X					X		X	X	X			
	O. fulgida mamillata	X	X	X										
	O. leptocaulis			X										
	O. engelmannii	X												
	O. versicolor			X										
	Pachycereus pringlei	X	X	X	X	X	X	X	X	X	X	X		
	Peniocereus striatus			X						X	X	X		
	Stenocereus gummosus		X	X		X	X			X	X	X		
	S. thurberi	X	X	X		X								X
Campanulaceae	Nemacladus glanduliferus													
Capparaceae	Atamisquea emarginata		X											
Caryophyllaceae	Achyronychia cooperi							X						
Celastraceae	Maytenus phyllanthoides	X												
Chenopodiaceae	Allenrolfea occidentalis	X			X		X							
	Atriplex barclayana	X	X			X	X	X	X	X	X	X	X	X
	A. polycarpa			X							X			
	Chenopodium murale	X						X			X			
	Salicornia subterminalis													
	Suaeda moquinii	X		X										

Convolvulaceae	Cressa truxillensis		X							X
Cucurbitaceae	Vaseyanthus insularis insularis							X	X	X
Cuscutaceae	Cuscuta corymbosa stylosa		X	X						
	C. leptantha		X		X					
Euphorbiaceae	Acalypha californica		X							
	Cnidoscolus palmeri		X							
	Ditaxis lanceolata		X							
	D. serrata		X		X					
	Euphorbia sp.			X						
	Euphorbia misera		X							
	E. pediculifera		X							
	E. petrina	X								
	E. polycarpa polycarpa		X		X		X			
	Jatropha cuneata		X		X					
	Tragia aff. glandulifera		X							
Fabaceae	Caesalpinia intricata		X							
	C. virgata		X		X					
	Calliandra californica	X	X							
	Dalea mollis	X								
	Desmanthus fruticosus		X		X					
	Ebenopsis confinis		X							
	Lotus strigosus		X		X					
	Marina parryi		X							
	Olneya tesota	X	X							
	Phaseolus filiformis		X					X		
	Prosopis glandulosa torreyana	X	X		X					
	Senna covesii		X							

(continued)

Table A4.3-1 (Continued)

Fam	Species	Alcatraz	Cholludo	Datil	Granito	Las Ánimas	La Víbora	Mejía	Partida Norte	Patos	Rasa	Salsipuedes	San Luis	Santa Ines
Foquieriaceae	Fouquieria diguetii?						X				X			
	F. splendens		X											
Frankeniaceae	Frankenia palmeri						X							
Hydrophyllaceae	Phacelia pedicellata			X				X						
Lamiaceae	Hyptis emoryi			X				X						
Loasaceae	Eucnide cordata												X	
	E. rupestris			X										
	Mentzelia adhaerens			X								X		
	M. hirsutissima							X						
Malpighiaceae	Janusia californica			X										
	J. gracilis			X										
Malvaceae	Abutilon incanum			X										
	Herissantia crispa			X										
	Hibiscus denudatus			X										
	Sphaeralcea ambigua ambigua			X										
	S. orcuttii	X												
Moraceae	Ficus palmeri	X	X	X								X		
Nyctaginaceae	Abronia maritima	X												
	Allionia incarnata			X		X								
	Boerhavia erecta			X										
Onagraceae	Camissonia cardiophylla							X						
Passifloraceae	Passiflora palmeri			X										
Phytolocaceae	Stegnosperma halimifolium		X	X										
Plantaginaceae	Plantago ovata							X						
Polemoniaceae	Gilia palmeri						X							
Polygonaceae	Eriogonum austrinum						X							
	E. galioides												X	

Family	Species								
Portulacaceae	Portulaca californica					X			
Resedaceae	Oligomeris linifolia		X						
Rhamnaceae	Colubrina viridis	X							
	Condalia globosa pubescens		X						
	Ziziphus obtusifolia canescens		X						
Sapindaceae	Cardiospermum corindum		X						
Scrophulariaceae	Antirrhinum cyanthiferum	X	X		X				
	A. kingii		X						
	Mohavea confertiflora				X				
Simmondsiaceae	Simmondsia chinensis		X X						
Solanaceae	Datura discolor	X		X	X				
	Lycium sp.				X				
	Lycium andersonii andersonii								
	L. brevipes	X	X		X			X	
	Nicotiana obtusifolia	X	X		X X			X X	
	Solanum hindsianum		X						
Sterculiaceae	Melochia tomentosa		X						
Urticaceae	Parietaria hespera		X						
Verbenaceae	Lippia palmeri		X						
	Fagonia laevis		X						
Zygophyllaceae	Kallstroemia californica		X						
	Larrea tridentata								
	Viscainoa geniculata geniculata	X	X		X				X
Agavaceae	Agave subsimplex		X		X				

(continued)

Table A4.3-1 (Continued)

Fam	Species	Alcatraz	Cholludo	Datíl	Granito	Las Ánimas	La Víbora	Mejía	Partida Norte	Patos	Rasa	Salsipuedes	San Luis	Santa Ines
Agavaceae	Agave subsimplex		X											
Poaceae	Aristida adscensionis		X	X			X							
	Bouteloua aristidoides	X		X										
	B. barbata	X							X	X				
	Brachiaria cf. arizonica		X	X										
	Cenchrus palmeri	X												
	Digitaria californica		X	X										
	Erioneuron pulchellum			X										
	Monanthochloe littoralis	X					X							
	Muhlenbergia microsperma		X								X			
	Setaria macrostachya			X										
	Sporobolus virginicus	X												
Zosteraceae	Zostera marina	X												
	Total species	43	28	99	4	11	26	32	16	10	14	15	10	5

Appendix 4.4

Plants of Small Islands in Bahía de Los Ángeles

PATRICIA A. WEST
JON P. REBMAN
GARY A. POLIS
L. DAVID HUMPHREY
RICHARD S. FELGER

The 14 small islands listed in this appendix are all in the vicinity of Bahía de Los Ángeles. These floral lists are the result of ongoing ecological and biogeographical studies by a research group headed by the late Gary Polis. The list is dedicated to him. Nomenclature follows appendix 4.1. We deposited voucher specimens at the San Diego Natural History Herbarium (SDNHM) and duplicates at the Universidad Autónoma de Baja California in Ensenada. Islands in the list are arranged by decreasing size (Polis et al. 1997). The corresponding appendix in the previous edition of this book contained the plants of 13 other small islands in the eastern mid-gulf region (Case and Cody 1983). The current list does not repeat the information for those 13 islands because new information is not available for those islands. One known addition to the previous list for the 13 islands is the invasive, non-native bufflegrass [*Pennisetum ciliare* (L.) Link] on Alcatraz in Kino Bay (West and Nabhan in press).

References

Case, T.J., and M.L. Cody. 1983. *Island Biogeography in the Sea of Cortéz*. University of California Press, Berkeley.
Polis, G.A., S.D. Hurd, C.T. Jackson, and F. Sanchez-Piñero. 1997. El Niño effects on the dynamics and control of a terrestrial island ecosystem in the Gulf of California. *Ecology* **78**:1884–1897.
West, P.A., and G.P. Nabhan. In press. Invasive plants: their occurrence and possible impact on the Central Gulf Coast of Sonora and the midriff islands of the Sea of the Cortés. In: B. Tellman (ed), *Invasive Exotic Species in the Sonoran Region*. University of Arizona Press, Tucson.

Appendix

Table A4.4-1 Floras of small islands in Bahía de Los Ángeles

	Smith	Ventana	Cabeza de Caballo	Piojo	Mitlan	Pata	Flecha	Bota	Coronadito	Gemelito East	Jorobado	Cerraja	Llave	Gemelito West
Gymnosperms														
Ephedraceae														
Ephedra aspera Engelm. ex. S. Watson		x												
Angiosperms														
Magnoliopsida (Dicots)														
Aizoaceae														
Mesembryanthemum crystallinum L.	x	x	x		x			x				x	x	x
Sesuvium portulacastrum L.	x	x												
Amaranthaceae														
Amaranthus watsonii Standl.	x			x		x	x			x	x	x	x	x
Asclepiadaceae														
Asclepias albicans S. Watson	x	x	x	x	x	x								
Asteraceae (Compositae)														
Ambrosia ilicifolia (A. Gray) Payne	x													
Bebbia juncea (Benth.) Greene var. *juncea*	x				x									
Coreocarpus parthenioides Benth. var. *parthenioides*	x	x	x	x	x	x	x	x	x			x		
Encelia farinosa Torr. & A. Gray var. *phenicodonta* (S. F. Blake) I.M. Johnst.	x	x	x	x	x	x			x					x
Hofmeisteria fasciculata (Benth.) Walp. var. *fasciculata*	x	x			x	x			x			x		
Malacothrix glabrata A. Gray		x						x						
Perityle emoryi Torr.	x	x	x	x	x	x	x	x	x	x		x	x	x
Peucephyllum schottii A. Gray	x	x		x										
Pleurocoronis pluriseta (A. Gray) R. King & H. Robinson	x		x											
Porophyllum crassifolium S. Watson	x	x												
P. gracile Benth.	x													
Trixis californica Kellogg	x	x			x									
Xylorhiza frutescens (S. Watson) Greene	x	x	x	x	x				x					
Bataceae (Batidaceae)														
Batis maritima L.	x	x												
Boraginaceae														
Cryptantha fastigiata I.M. Johnst.		x			x			x						
C. maritima (Greene) Greene var. *maritima*					x					x				
C. maritima var. *pilosa* I.M. Johnst.								x				x		
C. racemosa (S. Watson) Greene	x													
Burseraceae														
Bursera hindsiana (Benth.) Engler						x								
B. microphylla A. Gray	x	x				x								
Cactaceae														
Echinocereus				x								x	x	
Ferocactus gatesii G. Lindsay	x	x				x	x	x				x	x	

Table A4.4-1 (*Continued*)

	Smith	Ventana	Cabeza de Caballo	Piojo	Mitlan	Pata	Flecha	Bota	Coronadito	Gemelito East	Jorobado	Cerraja	Llave	Gemelito West
Mammillaria	x			x	x		x	x		x			x	
Mammillaria cf. *dioica* K. Brandegee		x	x			x								
Lophocereus schottii (Engelm.) Britton & Rose var. *schottii*	x		x				x	x	x					
Opuntia alcahes F.A.C. Weber var. *alcahes*	x	x	x	x	x	x	x	x		x	x	x	x	
O. bigelovii Engelm. var. *bigelovii*	x	x	x	x		x	x	x	x			x		
O. cholla F. A. C. Weber														x
O. molesta Brandegee var. *molesta*		x		x								x		
Pachycereus pringlei (S. Watson) Britton & Rose	x	x	x	x	x	x	x	x	x	x	x	x	x	x
Stenocereus gummosus (Engelm.) A. Gibson & Horak	x		x	x			x	x				x		
Campanulaceae														
Nemacladus glandulifera Jepson sens. lat.	x													
Caryophyllaceae														
Achyronychia cooperi Torr. & A. Gray	x	x	x						x					
Drymaria holosteoides Benth. var. *holosteoides*		x	x											
Chenopodiaceae														
Allenrolfea occidentalis (S. Watson) Kuntze	x	x			x	x		x						
Atriplex barclayana (Benth.) D. Dietr. sens. lat.	x	x	x	x	x	x	x	x	x	x	x	x	x	x
A. hymenelytra (Torr.) S. Watson	x	x												
A. polycarpa (Torr.) S. Watson	x				x	x								
Chenopodium murale L.									x		x	x	x	x
Salicornia subterminalis Parish	x	x												
Suaeda moquinii (Torr.) Greene	x	x												
Convolvulaceae														
Cressa truxillensis Kunth	x													
Cucurbitaceae														
Vaseyanthus insularis (S. Watson) Rose	x	x	x	x	x					x				
Euphorbiaceae														
Ditaxis serrata (Torr.) A. Heller	x	x	x	x	x	x		x						
Euphorbia eriantha Benth.	x													
E. misera Benth.			x											
E. pediculifera Engelm. cf. var. *pediculifera*		x				x								
E. polycarpa Benth.	x													
Stillingia linearis S. Watson	x													
Fabaceae (Leguminosae)														
Astragalus nuttallianus DC. var. *cedrosensis* M.E. Jones	x		x									x		
Caesalpinia virgata E.M. Fisher	x	x												
Dalea mollis Benth.	x					x								

(*continued*)

Table A4.4-1 (Continued)

	Smith	Ventana	Cabeza de Caballo	Piojo	Mitlan	Pata	Flecha	Bota	Coronadito	Gemelito East	Jorobado	Cerraja	Llave	Gemelito West
Lotus strigosus (Nutt.) Greene	x	x	x	x	x			x	x			x		
Lupinus arizonicus (S. Watson) S. Watson	x							x						
Phaseolus filiformis Benth.	x	x	x	x	x	x								
Psorothamnus emoryi (A. Gray) Rydb.	x	x												
Fouquieriaceae														
Fouquieria diguetii (Van tieghem) I.M. Johnst.	x	x	x	x		x								
F. splendens Engelm.	x	x	x	x		x	x	x						
Frankeniaceae														
Frankenia palmeri S. Watson	x	x				x								
Hydrophyllaceae														
Phacelia pedicellata A. Gray	x													
Loasaceae														
Eucnide rupestris (Baill.) H. Thompson & Ernst.	x	x			x	x			x	x	x	x		
Mentzelia adhaerens Benth.	x	x			x	x			x					
M. hirsutissima S. Watson var. *stenophylla* (Urb. & Gilg.) I.M. Johnst.									x					
Mentzelia				x										
Petalonyx linearis Greene		x			x				x					
Malvaceae														
Hibiscus denudatus Benth.		x												
Nyctaginaceae														
Allionia incarnata L.		x				x								
Mirabilis tenuiloba S. Watson	x													
Onagraceae														
Cammisonia californica (Nutt.) Raven			x											
C. cardiophylla (Torr.) Raven ssp. *cedrosensis* (Greene) Raven	x	x	x	x	x	x			x	x				x
Phytolaccaceae														
Stegnosperma halimifolium Benth.	x	x												
Plantaginaceae														
Plantago ovata Forsskal	x	x	x	x	x		x					x		
Polygonaceae														
Eriogonum austrinum (Stokes) Reveal	x													
E. inflatum Torr. & Frém var. *deflatum* I.M. Johnst.	x													
Portulacaceae														
Calandrinia maritima Nutt.	x													
Resedaceae														
Oligomeris linifolia (Vahl) J.F. Macbr.		x			x	x			x			x		
Rhizophoraceae														
Rhizophora mangle L.	x													
Scrophulariaceae														
Antirrhinum cyathiferum Benth.	x	x	x		x	x		x						

Table A4.4-1 (Continued)

	Smith	Ventana	Cabeza de Caballo	Piojo	Mitlan	Pata	Flecha	Bota	Coronadito	Gemelito East	Jorobado	Cerraja	Llave	Gemelito West
Mohavea confertiflora (Benth.) A. Heller	x	x			x	x		x						
Simmondsiaceae														
Simmondsia chinensis (Link) Schneid.	x													
Solanaceae														
Datura discolor Bernh.		x	x											
Lycium				x									x	
Lycium andersonii A. Gray		x												
L. brevipes Benth.		x			x				x		x	x	x	x
L. fremontii A. Gray		x				x								
Nicotiana obtusifolia M. Martens & Galeotti	x	x	x	x	x	x	x	x	x	x				x
Physalis crassifolia Benth.	x	x	x			x								
Zygophyllaceae														
Fagonia pachyacantha Rydb.	x	x	x			x		x						
Viscainoa geniculata (Kell.) Greene var. *geniculata*	x	x	x	x	x	x	x	x	x	x				
Liliopsida (monocots)														
Agavaceae														
Agave cerulata Trel. ssp. *cerulata*		x	x											
Poaceae (gramineae)														
Aristida adscensionis L.	x	x	x	x	x	x	x	x	x				x	
A. californica Thurb. ex. S. Watson var. *californica*	x	x												
Bouteloua barbata Lag.	x		x											
Monanthochloe littoralis Engelm.	x	x												
Muhlenbergia microsperma (DC.) Trin.		x				x	x							

Appendix 4.5

Plants Endemic to the Gulf Islands

JON P. REBMAN

In the previous edition of this book, an appendix on plant endemism reported 18 endemic taxa for the insular flora of the Sea of Cortés. However, the present floristic analysis recognizes 28 insular endemic taxa (table A4.5-1). The differences are in the publication of new species since 1983: *Coreocarpus sanpedroensis* (Smith 1989), the lumping of *Opuntia brevispina* into *O. alcahes* var. *alcahes* based on biosystematic research of *Opuntia* subgenus *Cylindropuntia* (Rebman 1995), and the recognition of other taxa (e.g., *Alvordia glomerata* var. *insularis, Dryopetalon crenatum* var. *racemosum, Dudleya nubigena* ssp. *cerralvensis, Euphorbia polycarpa* var. *johnstonii, Euphorbia polycarpa* var. *carmenensis, Ferocactus wislizenii* var. *tiburonensis, Hedyotis gracilenta,* and *Mammillaria evermanniana*) not included in the first edition.

The known insular flora contains 695 plant taxa (see app. 4.1), of which 28 taxa are endemic (table A4.5-1), yielding a 4.0% rate of endemism for the islands of the Sea of Cortés. According to Wiggins (1980), the Lower California flora has an endemism percentage of approximately 23.2%. There are significant differences in the rate of endemism of the insular floras as compared to the flora of the entire Lower California region. Not only are the percentages of endemism for all plant groups drastically different (4.0% vs. 23.2%), but the family representation of endemics is quite different as well. For example, the Cactaceae is the plant family with the most endemics on the gulf islands. It far exceeds all other plant families with insular endemism by having 11 endemics as compared to the next highest, the Asteraceae, with only 4 endemic taxa (table A4.5-2). For the Cactaceae, that yields a 25.6% endemism (table A4.5-3) for the gulf islands, which is considerably lower than the 72.3% which has been estimated for all of the cacti in Lower California (Rebman et al. 1999). In respect to the Asteraceae, the endemism percentage for the insular flora is only 4.4% versus

30.7%, which was estimated for the entire Lower California region by Wiggins (1980). It is not known why the floristic endemism percentages are so low for the islands. However, it can be postulated that because most of the islands are relatively near to the main landmasses and may have been connected as recently as the Pleistocene and because they lack a great deal of habitat diversity, they may have not had the opportunity for much isolation or adaptive radiation to evolve new species. The cacti may have evolved so many endemics in respect to other plant families represented on the islands due to their rather limited dispersal mechanisms and ability to significantly radiate in arid environments.

The genera with the highest number of endemic taxa in the insular flora are *Mammillaria* and *Ferocactus*, both in the Cactaceae. The genus *Mammillaria* contains five endemic species found on the islands and other taxa such as *M. angelensis*, which is currently lumped into *M. dioica*, but probably deserves taxonomic recognition at least at the varietal level. The genus *Ferocactus* is represented by two endemic species and four endemic taxa on the islands. However, some taxa such as *F. gatesii* restricted to the islands of Bahía de Los Ángeles are still of questionable taxonomic status; *F. gatesii* is currently recognized as a variant of *F. cylindraceus*. Many of the cacti of the insular flora need to be studied more thoroughly and may eventually be designated as endemic taxa. For example, some chollas (*Opuntia* subgenus *Cylindropuntia*) from Cerralvo and Ángel de la Guarda are currently lumped into *O. alcahes* var. *alcahes*, but they do exhibit some morphological differences from the rest of the taxon. However, until chromosome analyses and better reproductive material can be obtained from the island plants, they are best treated within the extremely variable *O. alcahes* complex. Other potential endemics that do not correspond to any known insular taxa include nopals/prickly pears (*Opuntia* subgenus *Opuntia*) from San Pedro Nolasco and Santa Catalina, which need better biosystematic investigation to determine their affinities and taxonomic designations.

The island with the most endemic plant taxa represented is Ángel de la Guarda, which is home to four species restricted to the island and one insular endemic that is known from one other island. Other islands with at least five insular endemics present include Carmen, Cerralvo, San Diego, San Esteban, Santa Catalina, and Santa Cruz (table A4.5-4).

It should be noted that Cerralvo and San José need to be more extensively surveyed after the summer rains. Both of these islands are good candidates for the discovery of new endemic species due to their size, habitat diversity, and annual rainfall.

Acknowledgments Richard Felger provided information on the current status of a few cacti in the state of Sonora. José Luis Leon de la Luz kindly reviewed this appendix, gave invaluable information on endemics of the southern Gulf islands, and provided the current status on a few insular plant species also found in the cape region.

References

Gentry, H. 1978. The agaves of Baja California. *Occasional Papers of the California Academy of Sciences* **130**:1–119.

Rebman, J. P. 1995. Biosystematics of *Opuntia* subgenus *Cylindropuntia* (Cactaceae), the chollas of Lower California, Mexico. Ph.D. dissertation, Arizona State University, Tempe.

Rebman, J., M. Resendiz, and J. Delgadillo. 1999. Diversity and documentation for the Cactaceae of Lower California, Mexico. *Cactáceas y Suculentas Mexicanas* **44**:20–26.

Smith, E. 1989. A biosystematic study and revision of the genus *Coreocarpus* (Compositae). *Sytematic Botany* **14**:448–472.

Wiggins, I. 1980. *Flora of Baja California*. Stanford University Press, Palo Alto, CA.

Table A4.5-1 Endemic plant taxa of the insular flora

Endemic Taxon	Family	Island(s)
Agave cerulata Trel. ssp. *dentiens* (Trel.) H.S. Gentry	Agavaceae	San Estéban
Alvordia glomerata Brandegee var. *insularis* A. Carter	Asteraceae	Carmen (type), San José, San Marcos, Santa Catalina
Coreocarpus sanpedroensis E.B. Smith	Asteraceae	San Pedro Nolasco
Cryptantha grayi (Vasey & Rose) J.F. Macbr. var. *nesiotica* I.M. Johnston	Boraginaceae	Cerralvo, Coronado, Espíritu Santo, San Diego, San Francisco (type), San José, Santa Catalina, Santa Cruz
Dryopetalon crenatum (Brandegee) Rollins var. *racemosum* Rollins	Brassicaceae	Espíritu Santo
Dudleya nubigena (Brandegee) Britton & Rose ssp. *cerralvensis* Moran[a]	Crassulaceae	Cerralvo
Echinocereus grandis Britton & Rose	Cactaceae	San Estéban (type), San Lorenzo
Echinocereus webstarianus G.E. Linds.	Cactaceae	San Pedro Nolasco
Eriogonum angelense Moran	Polygonaceae	Ángel de la Guarda
Euphorbia polycarpa Benth. var. *carmenensis* (Rose) L.C. Wheeler	Euphorbiaceae	Carmen (type), Coronado, Espíritu Santo, San Diego, San Marcos, Santa Catalina, Santa Cruz, Tortuga
Euphorbia polycarpa Benth. var. *johnstonii* L.C. Wheeler	Euphorbiaceae	Carmen, Monserrat (type)
Ferocactus diguetii (F.A.C. Weber) Britton & rose var. *carmenensis* G.E. Linds.	Cactaceae	Carmen
Ferocactus diguetii (F.A.C. Weber) Britton & Rose var. *diguetii*	Cactaceae	Cerralvo, Coronado, Danzante, Monserrat, San Diego, Santa Catalina (type), Santa Cruz
Ferocactus johnstonianus Britton & Rose	Cactaceae	Ángel de la Guarda
Ferocactus wislizenii (Engelm.) Britton & Rose var. *tiburonensis* G.E. Linds.[b]	Cactaceae	Tiburón
Hedyotis gracilenta (I.M. Johnst.) W.H. Lewis	Rubiaceae	San Diego (type), San Marcos
Hofmeisteria filifolia I.M. Johnst.	Asteraceae	Ángel de la Guarda
Lyrocarpa linearifolia Rollins	Brassicaceae	Ángel de la Guarda (type), San Estéban

Table A4.5-1 (*Continued*)

Endemic Taxon	Family	Island(s)
Machaeranthera pinnatifida (Hook.) Shinners var. *incisifolia* (I.M. Johnst.) B.L. Turner & R.L. Hartm.	Asteraceae	Coronado, San Diego, San Estéban, San Lorenzo (type), Santa Catalina, Santa Cruz, Tiburón
Mammillaria cerralboa (Britton & Rose) Orcutt	Cactaceae	Cerralvo
Mammillaria estebanensis G.E. Linds.	Cactaceae	San Estéban
Mammillaria evermanniana (Britton & Rose) Orcutt	Cactaceae	Cerralvo
Mammillaria multidigitata G.E. Linds.	Cactaceae	San Pedro Nolasco
Mammillaria tayloriorum Glass & R.C. Foster	Cactaceae	San Pedro Nolasco
Marina cataliniae Barneby	Fabaceae	Santa Catalina
Marina oculata (Rydb.) Barneby	Fabaceae	Cerralvo (type), Espíritu Santo, Santa Cruz
Penstemon angelicus (I.M. Johnst.) Moran	Scrophulariaceae	Ángel de la Guarda
Salvia platycheila A. Gray	Lamiaceae	Carmen (type), Cerralvo, Danzante, Monserrat, Santa Catalina, Santa Cruz

[a]Reported to occur on the east side of the cape region of Baja California Sur by José Luis Leon de la Luz (pers. commun., 1999), but needs verification.
[b]Reported to occur on mainland in Sonora by Richard Felger (pers. commun., 1999), but needs verification.

Table A4.5-2 Number of insular endemics represented by plant family

Family	No. of endemic taxa
Cactaceae	11
Asteraceae	4
Brassicaceae	2
Euphorbiaceae	2
Fabaceae	2
Agavaceae	1
Boraginaceae	1
Crassulaceae	1
Lamiaceae	1
Polygonaceae	1
Rubiaceae	1
Scrophulariaceae	1
Total	28

Table A4.5-3 Insular endemism as compared to endemism in Lower California, Mexico

Family	No. of Insular Endemic Taxa	No. of Insular Taxa Represented	% Insular Endemism	% Endemism in Lower California
Cactaceae	11	43	25.6	72.3
Asteraceae	4	91	4.4	30.7
Brassicaceae	2	15	13.3	15.7
Euphorbiaceae	2	48	4.2	30.3
Fabaceae	2	64	3.1	30.5
Agavaceae	1	7	14.3	90.0
Boraginaceae	1	21	4.8	16.9
Crassulaceae	1	5	20.0	42.3
Lamiaceae	1	6	16.7	30.6
Polygonaceae	1	10	10.0	36.5
Rubiaceae	1	10	10.0	57.1
Scrophulariaceae	1	13	7.7	19.3

Values for endemism in Lower California are based on Wiggins (1980).

Table A4.5-4 Number of island-specific endemics and insular endemics on each island

Island	No. of Island Endemics	Total No. of Insular Endemics
Ángel de la Guarda	4	5
San Pedro Nolasco	4	4
Cerralvo	3	7
San Estéban	2	5
Santa Catalina	1	7
Carmen	1	5
Espíritu Santo	1	4
Tiburón	1	2
Santa Cruz	0	6
San Diego	0	5
Coronado	0	4
Monserrat	0	3
San Marcos	0	3
Danzante	0	2
San José	0	2
San Lorenzo	0	2
San Francisco	0	1
Tortuga	0	1

Appendix 5.1

Checklist of the Ants of the Gulf of California Islands

PHILIP S. WARD
APRIL M. BOULTON

Subfamily	Species	Northern Miramar	Coronadito	Coronado	Mitlán	Piojo	Jorobado	Pata	Bota	Flecha	Llave	Cerrajal	La Ventana	Cabeza de Caballa	Los Gemelitos [Est]	Los Gemelitos [Oeste]	El Pescador
Dolichoderinae	*Dorymyrmex bicolor*																
	Dorymyrmex insanus			X	X	X		X	X				X	X		X	
	Dorymyrmex sp. BCA-1													X			
	Forelius mccooki	X															
	Forelius prunosus			X	X								X				
	Forelius sp. BCA-1			X	X												
	Forelius sp. BCA-2												X				
Ecitoninae	*Neivamyrmex leonardi*																
	Neivamyrmex nigrescens																
	Neivamyrmex opacithorax																
Formicinae	*Camponotus festinatus*		X	X	X								X	X			
	Camponotus mina											X					
	Camponotus ochreatus												X				
	Myrmecocystus kennedyi																
	Myrmecocystus mimicus																
	Myrmecocystus navajo																
	Myrmecocystus nequazcatl																
	Myrmecocystus placodops													X			
	Myrmecocystus semirufus																
	Paratrechina cf. *terricola*																
	Paratrechina bruesii			X					X				X		X		
	Paratrechina longicornis																
Myrmicinae	*Acromyrmex versicolor*																
	Aphaenogaster boulderensis																
	Aphaenogaster megommata			X	X		X	X	X				X	X			
	Crematogaster arizonensis												X				
	Crematogaster californica (s.l.)																
	Crematogaster depilis												X	X			
	Crematogaster rossi												X	X			
	Cyphomyrmex flavidus																X

Species	1	2	3	4	5	6	7	8	9	10	11	12	13	14	15	16
Cyphomyrmex wheeleri												X				
Cyphomyrmex sp. BCA-1			X													
Leptothorax obliquicanthus																
Leptothorax politus																
Leptothorax sp. BCA-1												X				
Leptothorax sp. BCA-6																
Leptothorax sp. BCA-7																
Messor julianus			X													
Messor pergandei																
Monomorium ergatogyna																
Pheidole clydei			X													
Pheidole hyatti																
Pheidole rhea			X				X				X	X	X			
Pheidole sp. nr. *californica*																
Pheidole sp. nr. *vistana*							X									
Pheidole sciophila												X				
Pheidole spadonia																
Pheidole tucsonica																
Pheidole vistana																
Pheidole yaqui		X			X		X					X				
Pheidole sp. BCA-2																
Pheidole sp. BCA-4			X		X	X	X	X	X	X	X					
Pheidole sp. BCA-5			X				X							X		
Pogonomyrmex californicus	X													X		
Pogonomyrmex imberbiculus		X														
Pogonomyrmex laevinodis				X												
Pogonomyrmex pima																
Pogonomyrmex rugosus																
Pogonomyrmex tenuispinus					X											
Solenopsis molesta			X	X									X			
Solenopsis xyloni		X	X	X	X	X	X	X	X	X	X	X	X		X	
Tetramorium spinosum																
Ponerinae																
Odontomachus cf. *brunneus*																
Pseudomyrmecinae																
Pseudomyrmex apache							X								X	
Pseudomyrmex major							X								X	
Pseudomyrmex pallidus													X			X
Total number of species	1	5	16	9	5	3	7	6	2	2	4	14	10	2	4	2

Family	Species	Bahía de Los Ángeles								Midriff							Sonora	
		Bahía Animas Blanca	Bahía Animas Est	Bahía Animas Oeste	Mejía	Ángel de la Guarda	Estanque	Partida	Islote east of Partida	Raza	Salsipuedes	Las Ánimas	San Lorenzo	San Esteban	San Pedro Mártir	Patos	Tiburón
Dolichoderinae	*Dorymyrmex bicolor*					X											
	Dorymyrmex insanus																
	Dorymyrmex sp. BCA-1																X
	Forelius mccooki		X														X
	Forelius pruinosus					X							X				X
	Forelius sp. BCA-1					X								X			X
	Forelius sp. BCA-2					X											X
Ecitoninae	*Neivamyrmex leonardi*																X
	Neivamyrmex nigrescens																X
	Neivamyrmex opacithorax																X
Formicinae	*Camponotus festinatus*					X											X
	Camponotus mina		X			X											
	Camponotus ochreatus																
	Myrmecocystus kennedyi																X
	Myrmecocystus mimicus																X
	Myrmecocystus navajo																X
	Myrmecocystus nequazcatl																X
	Myrmecocystus placodops													X			
	Myrmecocystus semirufus					X											
	Paratrechina cf. *terricola*					X											X
	Paratrechina bruesii																
	Paratrechina longicornis																
Myrmicinae	*Acromyrmex versicolor*																X
	Aphaenogaster boulderensis					X									X		
	Aphaenogaster megommata	X	X	X													
	Crematogaster arizonensis											X					
	Crematogaster californica (s.l.)					X											X
	Crematogaster depilis	X															
	Crematogaster rossi					X											X
	Cyphomyrmex flavidus					X											X

548

Species	1	2	3	4	5	6	7	8	9	10	11	12	13	14	Total
Cyphomyrmex wheeleri												X			
Cyphomyrmex sp. BCA-1												X		X	
Leptothorax obliquicanthus												X		X	
Leptothorax politus															
Leptothorax sp. BCA-1												X			
Leptothorax sp. BCA-6															
Leptothorax sp. BCA-7															
Messor julianus															
Messor pergandei				X								X			
Monomorium ergatogyna				X							X	X			
Pheidole clydei				X											
Pheidole hyatti												X			
Pheidole rhea															
Pheidole sp. nr. *californica*				X					X			X			
Pheidole sp. nr. *vistana*				X								X			
Pheidole sciophila	X														
Pheidole spadonia															
Pheidole tucsonica															
Pheidole vistana				X				X				X		X	
Pheidole yaqui				X								X			
Pheidole sp. BCA-2				X								X			
Pheidole sp. BCA-4				X								X			
Pheidole sp. BCA-5															
Pogonomyrmex californicus			X									X			
Pogonomyrmex imberbiculus												X			
Pogonomyrmex laevinodis															
Pogonomyrmex pima															
Pogonomyrmex rugosus												X		X	
Pogonomyrmex tenuispinus															
Solenopsis molesta															
Solenopsis xyloni	X	X		X	X	X		X	X	X	X	X	X	X	
Tetramorium spinosum				X										X	

Ponerinae

Species	1	2	3	4	5	6	7	8	9	10	11	12	13	14	Total
Odontomachus cf. *brunneus*														X	

Pseudomyrmecinae

Species	1	2	3	4	5	6	7	8	9	10	11	12	13	14	Total
Pseudomyrmex apache				X											
Pseudomyrmex major														X	
Pseudomyrmex pallidus														X	
Total number of species	4	4	1	25	1	2	1	3	3	3	3	10	4	1	37

Family	Species	Sonora							Southern								
		Alcatraz	San Pedro Nolasco	Santa Inés	Ildefonso	Carmen	Mestiza	Islote Blanco	Danzante	Las Tijeras	Pardo	Santa Cruz	San Diego	San José	San Francisco	Partida Sur	Espiritu Santo
Dolichoderinae	*Dorymyrmex bicolor*	X															
	Dorymyrmex insanus																
	Dorymyrmex sp. BCA-1										X						
	Forelius mccooki																
	Forelius pruinosus					X											
	Forelius sp. BCA-1													X			
	Forelius sp. BCA-2																
Ecitoninae	*Neivamyrmex leonardi*																
	Neivamyrmex nigrescens													X			
	Neivamyrmex opacithorax																
Formicinae	*Camponotus festinatus*					X			X	X	X	X	X	X	X		
	Camponotus mina								X	X	X	X		X		X	
	Camponotus ochreatus					X											
	Myrmecocystus kennedyi							X									
	Myrmecocystus mimicus							X									
	Myrmecocystus navajo																
	Myrmecocystus nequazcatl																
	Myrmecocystus placodops																
	Myrmecocystus semirufus																
	Paratrechina cf. *terricola*										X						
	Paratrechina bruesii					X											
	Paratrechina longicornis						X										
Myrmicinae	*Acromyrmex versicolor*																
	Aphaenogaster boulderensis		X														
	Aphaenogaster megommata					X	X					X					
	Crematogaster arizonensis					X											
	Crematogaster californica (s.l.)							X									
	Crematogaster depilis					X			X		X						
	Crematogaster rossi						X							X			
	Cyphomyrmex flavidus														X		

	Species	C1	C2	C3	C4	C5	C6	C7	C8	C9	C10	C11	C12	C13	C14	C15
	Cyphomyrmex wheeleri					X										
	Cyphomyrmex sp. BCA-1															
	Leptothorax obliquicanthus															
	Leptothorax politus															
	Leptothorax sp. BCA-1					X	X									
	Leptothorax sp. BCA-6					X	X			X						
	Leptothorax sp. BCA-7							X								
	Messor julianus															
	Messor pergandei															
	Monomorium ergatogyna					X	X									
	Pheidole clydei															
	Pheidole hyatti			X												
	Pheidole rhea															
	Pheidole sp. nr. californica					X	X	X		X	X X					
	Pheidole sp. nr. vistana															
	Pheidole sciophila					X										
	Pheidole spadonia															
	Pheidole tucsonica							X						X		
	Pheidole vistana					X	X									
	Pheidole yaqui															
	Pheidole sp. BCA-2									X						
	Pheidole sp. BCA-4											X				
	Pheidole sp. BCA-5					X	X									
	Pogonomyrmex californicus	X														
	Pogonomyrmex imberbiculus												X			X
	Pogonomyrmex laevinodis															
	Pogonomyrmex pima															
	Pogonomyrmex rugosus															
	Pogonomyrmex tenuispinus								X							X
	Solenopsis molesta								X							
	Solenopsis xyloni				X	X	X	X	X	X			X	X		
	Tetramorium spinosum					X			X							
Ponerinae	Odontomachus cf. brunneus			X	X											
Pseudomyrmecinae	Pseudomyrmex apache															
	Pseudomyrmex major															
	Pseudomyrmex pallidus															
	Total number of species	2	2	1	1	14	8	7	14	5	10	3	2	7	3	2

Distribution Summaries

Family	Species	Southern Cerralvo	Baja California Islands	Baja California Sur Islands	Sonoran Islands	Baja California	Baja California Sur	Sonora
Dolichoderinae	*Dorymyrmex bicolor*			X		X	X	X
	Dorymyrmex insanus					X	X	X
	Dorymyrmex sp. BCA-1		X		X	X	X	X
	Forelius mccooki		X		X	X	X	X
	Forelius pruinosus		X	X	X	X	X	X
	Forelius sp. BCA-1		X			X		
	Forelius sp. BCA-2		X			X	X	
Ecitoninae	*Neivamyrmex leonardi*				X	X	X	X
	Neivamyrmex nigrescens			X	X	X	X	X
	Neivamyrmex opacithorax				X	X	X	X
Formicinae	*Camponotus festinatus*	X	X	X	X	X	X	X
	Camponotus mina		X	X	X	X	X	X
	Camponotus ochreatus			X		X	X	X
	Myrmecocystus kennedyi		X			X		X
	Myrmecocystus mimicus				X	X		X
	Myrmecocystus navajo				X	X		X
	Myrmecocystus nequazcatl				X			X
	Myrmecocystus placodops				X			
	Myrmecocystus semirufus		X			X		
	Paratrechina cf. *terricola*		X		X	X		X
	Paratrechina bruesii			X			X	
	Paratrechina longicornis		X	X		X	X	X
Myrmicinae	*Acromyrmex versicolor*				X	X		X
	Aphaenogaster boulderensis		X	X	X	X	X	X
	Aphaenogaster megommata		X		X	X	X	X
	Crematogaster arizonensis	X	X	X		X	X	X
	Crematogaster californica (s.l.)	X		X	X	X	X	X
	Crematogaster depilis		X	X		X	X	X
	Crematogaster rossi		X	X		X	X	X

Species	1	2	3	4	5	6	7
Cyphomyrmex flavidus						X	X
Cyphomyrmex wheeleri							
Cyphomyrmex sp. BCA-1		X				X	X
Leptothorax obliquicanthus				X		X	X
Leptothorax politus				X			
Leptothorax sp. BCA-1			X			X	X
Leptothorax sp. BCA-6			X			X	X
Leptothorax sp. BCA-7			X			X	
Messor julianus		X					
Messor pergandei		X		X		X	X
Monomorium ergatogyna		X		X		X	X
Pheidole clydei		X	X			X	
Pheidole hyatti		X		X			
Pheidole rhea		X		X		X	X
Pheidole sp. nr. californica		X		X		X	X
Pheidole sp. nr. vistana		X	X	X		X	X
Pheidole sciophila		X		X			
Pheidole spadonia		X		X		X	X
Pheidole tucsonica		X		X		X	X
Pheidole vistana		X		X		X	X
Pheidole yaqui		X	X	X		X	X
Pheidole sp. BCA-2		X				X	X
Pheidole sp. BCA-4		X	X			X	X
Pheidole sp. BCA-5		X	X	X		X	X
Pogonomyrmex californicus	X	X		X			
Pogonomyrmex imberbiculus		X		X		X	X
Pogonomyrmex laevinodis			X			X	
Pogonomyrmex pima		X		X			
Pogonomyrmex rugosus		X		X		X	
Pogonomyrmex tenuispinus		X		X		X	X
Solenopsis molesta		X		X		X	X
Solenopsis xyloni	X	X		X		X	X
Tetramorium spinosum		X		X		X	X
Ponerinae							
Odontomachus cf. brunneus				X			X
Pseudomyrmecinae							
Pseudomyrmex apache		X		X		X	X
Pseudomyrmex major				X		X	X
Pseudomyrmex pallidus		X		X		X	X
Total number of species	5	37	28	42	52	46	49

Appendix 6.1

Tenebrionid Beetle Species on Islands in the Sea of Cortés and the Pacific

The island group to which the island belongs is shown in parentheses under the island name. The numbers next to the island indicate sampling category (1, occasional collecting; 2, low-confidence survey; 3, high-confidence survey). The letters next to species indicate endemic status (A, single-island endemic; B, multiple-island endemic; C, island-group endemic).

Sea of Cortés

Ángel de la Guarda
(Ángel de la Guarda) 1
1 *Asbolus mexicana* (Horn)
2 *Batuliodes confluens* Blaisdell
3 *Eleodes discinctus* Blaisdell
4 *Hylocrinus oblongulus* Casey
5 *Nocibiotes* n. sp. 1
6 *Phaleria latus* Blaisdell
7 *Phaleria pilifera* LeConte
8 *Steriphanus subopacus* (Horn)
9 *Tonibius sulcatus* (LeConte)
10 *Triphalopsis californicus* Doyen
11 *Triphalopsis partida* Blaisdell

Ballena
(Espíritu Santo) 1
1 *Orthostibia frontalis* Blaisdell B

Blanca (Bahía de las Ánimas)
(Blanca) 2
1 *Argoporis apicalis* Blaisdell
2 *Cryptadius tarsalis* Blaisdell

Blanco
(Blanco) 3
1 *Argoporis apicalis* Blaisdell
2 *Conibius opacus* (LeConte)
3 *Cryptoglossa asperata* Horn
4 New Genus 2 n. sp. B
5 *Triphalopsis partida* Blaisdell

Bota
(Ventana) 3
1 *Argoporis apicalis* Blaisdell
2 *Craniotus* n. sp. 1
3 *Cryptadius tarsalis* Blaisdell
4 *Eschatomoxys* n. sp. 1 B
5 *Microschatia championi* Horn
6 New Genus 1 n. sp. 1
7 *Phaleria pilifera* LeConte
8 *Tonibius sulcatus* (LeConte)
9 *Triphalopsis californicus* Doyen

Cabeza de Caballo
(Cabeza de Caballo) 3
1 *Argoporis apicalis* Blaisdell
2 *Batuliodes confluens* Blaisdell
3 *Craniotus* n. sp. 1
4 *Cryptoglossa spiculifera* (LeConte)
5 *Eschatomoxys* n. sp. 1   B
6 *Eusattus dubius* Doyen
7 *Microschatia championi* Horn
8 New Genus 1 n. sp. 1
9 *Stibia sparsa* Blaisdell
10 *Tonibius sulcatus* (LeConte)
11 *Triphalopsis californicus* Doyen

Calaveras
(Calaveras) 3
1 *Eschatomoxys* n. sp. 1   B

Cardonosa (Islet E of Partida N,
Cardonosa E)
(Partida N) 2
1 *Argoporis apicalis* Blaisdell
2 *Asidina parallela (LeConte)*
3 *Batuliodes confluens* Blaisdell
4 *Cryptadius tarsalis* Blaisdell
5 *Eusattus dubius* Doyen
6 *Steriphanus subopacus* (Horn)
7 *Steriphanus tardus* Blaisdell   B
8 *Stibia sparsa* Blaisdell

Carmen
(Carmen) 2
1 *Argoporis apicalis* Blaisdell
2 *Asbolus mexicana* (Horn)
3 *Asidina parallela* (LeConte)
4 *Batuliodes confluens* Blaisdell
5 *Cerenopus concolor* LeConte
6 *Conibius opacus* (LeConte)
7 *Cryptadius tarsalis* Blaisdell
8 *Cryptoglossa asperata* Horn
9 *Cryptoglossa spiculifera* (LeConte)
10 *Eleodes eschscholtzi* Solier
11 *Eleodes insularis* Linell
12 *Eleodes loretensis* Blaisdell
13 *Emmenides obsoletus* Blaisdell   A
14 *Hylocrinus insularis* Blaisdell
15 *Hypogena marginata* (LeConte)
16 *Latheticus prosopis* Chittenden
17 *Metoponium abnorme* Casey
18 *Nocibiotes rossi* (Blaisdell)
19 *Telabis latipennis* Blaisdell
20 *Trimytis obtusa* Horn
21 *Triphalopsis partida* Blaisdell

Cayo
(San José) 1
1 *Emmenides subdescalceatus* Blaisdell   B
2 *Stibia fallaciosa* (Blaisdell)

Cerraja
(Ventana) 3
1 *Argoporis apicalis* Blaisdell
2 *Batuliodes confluens* Blaisdell
3 *Craniotus* n. sp. 1
4 *Cryptadius tarsalis* Blaisdell
5 *Microschatia championi* Horn
6 *Tonibius sulcatus* (LeConte)
7 *Triphalopsis californicus* Doyen

Cerralvo
(Cerralvo) 1
1 *Asidina* n. sp. 1   A
2 *Conibius opacus* (LeConte)
3 *Cryptoglossa asperata* Horn
4 *Cryptoglossa seriata* (LeConte)
5 *Cryptoglossa spiculifera* (LeConte)
6 *Eleodes eschscholtzi* Solier
7 *Emmenides apicalis* Blaisdell   A
8 *Eusattus ceralboensis* Doyen   A
9 *Eusattus erosus* (Horn)
10 *Eusattus laevis* (LeConte)
11 *Heterasida bifurca* (LeConte)
12 *Heterasida conivens* (LeConte)
13 *Hypogena marginata* (LeConte)
14 *Stibia fallaciosa* (Blaisdell)
15 *Stibia puncticollis* Horn
16 *Telabis lunulata* Blaisdell
17 *Trimytis ceralboensis* Blaisdell   A
18 *Trimytis obtusa* Horn
19 *Triphalopsis partida* Blaisdell
20 *Triphalus subcylindricus* Blaisdell   B

Cholla
(Carmen) 1
1 *Triphalopsis partida* Blaisdell

Coronadito
(Smith) 3
1 *Argoporis apicalis* Blaisdell
2 *Batuliodes confluens* Blaisdell
3 *Cheirodes californica* (Horn)
4 *Craniotus* n. sp. 1
5 *Cryptoglossa spiculifera* (LeConte)
6 *Eschatomoxys* n. sp. 1   B
7 *Triphalopsis californicus* Doyen

Coronado
(Coronados) 1
1 *Cryptadius tarsalis* Blaisdell
2 *Cryptoglossa asperata* Horn
3 *Cryptoglossa spiculifera* (LeConte)

Coyote (Bahía Concepción)
(Bahía Concepción) 1
1 *Cryptoglossa asperata* Horn

Coyote (El Partido, El Coyote)
(San José) 1
1 *Cryptoglossa spiculifera* (LeConte)
2 *Triphalus subcylindricus* Blaisdell B

Danzante
(Danzante) 2
1 *Araeoschizus aalbui* Papp
2 *Batuliodes confluens* Blaisdell
3 *Chilometopon cribricolle* Blaisdell
4 *Conibius opacus* (LeConte)
5 *Cryptoglossa asperata* Horn
6 *Cryptoglossa spiculifera* (LeConte)
7 New Genus 2 n. sp. B
8 *Nocibiotes rossi* (Blaisdell)
9 *Triphalopsis partida* Blaisdell
10 *Triphalus impressifrons* Blaisdell
11 *Triphalus subcylindricus* Blaisdell B
12 *Typhlusechus* n. sp. B

Espíritu Santo
(Espíritu Santo) 1
1 *Argoporis cribratus* LeConte
2 *Argoporis niger* Aalbu et al
3 *Cerenopus concolor* LeConte
4 *Conibius ventralis* Blaisdell
5 *Cryptadius tarsalis* Blaisdell
6 *Cryptoglossa asperata* Horn
7 *Cryptoglossa spiculifera* (LeConte)
8 *Emmenides subdescalceatus* Blaisdell B
9 *Eusattus erosus* (Blaisdell)
10 *Eusattus pallidus* Doyen
11 *Orthostibia frontalis* Blaisdell B
12 *Telabis* sp.
13 *Trimytis obtusa* Horn

Estanque (Pond, La Vibora)
(Ángel de la Guarda) 1
1 *Argoporis apicalis* Blaisdell
2 *Batuliodes confluens* Blaisdell
3 *Tonibius sulcatus* (LeConte)
4 *Triphalopsis partida* Blaisdell

Flecha
(Ventana) 3
1 *Argoporis apicalis* Blaisdell
2 *Batuliodes confluens* Blaisdell
3 *Craniotus* n. sp. 1
4 *Microschatia championi* Horn
5 *Tonibius sulcatus* (LeConte)
6 *Triphalopsis californicus* Doyen

Gallina
(Espíritu Santo) 1
1 *Cryptoglossa asperata* Horn

Gemelitos E. (Gemelos E.)
(Gemelitos) 3
1 *Argoporis apicalis* Blaisdell
2 *Batuliodes confluens* Blaisdell
3 *Stibia sparsa* Blaisdell

Gemelitos O. (Gemelos W.)
(Gemelitos) 3
1 *Argoporis apicalis* Blaisdell
2 *Cryptoglossa spiculifera* (LeConte)
3 *Eschatomoxys* n. sp. 1 B
4 *Stibia sparsa* Blaisdell

Granito (Granite)
(Ángel de la Guarda) 1
1 *Phaleria pilifera* LeConte

Islote Bahía de las Ánimas NE
(Bahía de las Ánimas) 2
1 *Argoporis apicalis* Blaisdell
2 *Cryptadius tarsalis* Blaisdell
3 *Tonibius sulcatus* (LeConte)
4 *Triphalopsis californicus* Doyen

Islote Bahía de las Ánimas SW
(Bahía de las Ánimas) 2
1 *Argoporis apicalis* Blaisdell
2 *Cryptadius tarsalis* Blaisdell
3 *Steriphanus durus* Blaisdell
4 *Stibia sparsa* Blaisdell
5 *Triphalopsis californicus* Doyen

Jorobado
(Ventana) 3
1 *Argoporis apicalis* Blaisdell
2 *Craniotus* n. sp. 1
3 *Cryptadius tarsalis* Blaisdell
4 *Triphalopsis californicus* Doyen

Las Ánimas N (San Lorenzo N)
(San Lorenzo) 2
1 *Argoporis apicalis* Blaisdell
2 *Conibius opacus* (LeConte)
3 *Cryptoglossa spiculifera* (LeConte)
4 *Eschatomoxys* n. sp. 2 A
5 *Steriphanus subopacus* (Horn)
6 *Stibia sparsa* Blaisdell
7 *Triphalopsis partida* Blaisdell

Las Ánimas S
(San José) 1
1 *Orthostibia frontalis* Blaisdell B

**Las Galeras East
(Monserrate)** 1
1 *Cryptoglossa asperata* Horn

**Las Galeras West
(Monserrate)** 1
1 *Cryptoglossa asperata* Horn
2 *Stibia fallaciosa* (Blaisdell)

**Llave
(Ventana)** 3
1 *Argoporis apicalis* Blaisdell
2 *Batuliodes confluens* Blaisdell
3 *Craniotus* n. sp. 1
4 *Microschatia championi* Horn
5 *Triphalopsis californicus* Doyen

**Mejía
(Ángel de la Guarda)** 1
1 *Asbolus mexicana* Horn
2 *Batuliodes confluens* Blaisdell
3 *Phaleria latus* Blaisdell
4 *Phaleria pilifera* LeConte
5 *Triphalopsis californicus* Doyen
6 *Triphalopsis partida* Blaisdell

**Mestiza (Chuenque)
(Mestiza)** 2
1 *Conibius opacus* (LeConte)
2 *Triphalopsis partida* Blaisdell
3 *Typhlusechus* n. sp. B

**Mitlan
(Smith)** 3
1 *Craniotus* n. sp. 1
2 *Cryptadius tarsalis* Blaisdell
3 *Eschatomoxys* n. sp. 1 B
4 New Genus 1 n. sp. 1
5 *Tonibius sulcatus* (LeConte)
6 *Triphalopsis californicus* Doyen

**Monserrat (Monserrate)
(Monserrat)** 1
1 *Batuliodes confluens* Blaisdell
2 *Cerenopus concolor* LeConte
3 *Conibius opacus* (LeConte)
4 *Cryptoglossa asperata* Horn
5 *Cryptoglossa spiculifera* (LeConte)
6 *Eleodes eschscholtzi* Solier
7 *Eleodes insularis* Linell
8 *Stibia fallaciosa* (Blaisdell)
9 *Stibia puncticollis* Horn

**Pardo (Roca Sur; Candeleros Sur)
(Los Candeleros)** 3
1 *Conibius opacus* (LeConte)
2 *Cryptoglossa asperata* Horn
3 *Eusattus erosus* (Blaisdell)
4 New Genus 2 n. sp. B

**Partida N (Cardonosa)
(Partida N)** 2
1 *Argoporis apicalis* Blaisdell
2 *Asbolus mexicana* (Horn)
3 *Asidina parallela* (LeConte)
4 *Batuliodes confluens* Blaisdell
5 *Cryptadius tarsalis* Blaisdell
6 *Steriphanus subopacus* (Horn)
7 *Stibia sparsa* Blaisdell
8 *Tonibius sulcatus* (LeConte)
9 *Triphalopsis partida* Blaisdell

**Partida S
(Espíritu Santo)** 1
1 *Argoporis cribratus* LeConte
2 *Argoporis niger* Aalbu et al
3 *Asbolus mexicana* (Horn)
4 *Conibius opacus* (LeConte)
5 *Cryptoglossa asperata* Horn
6 *Cryptoglossa spiculifera* (LeConte)
7 *Eusattus erosus* (Blaisdell)
8 *Orthostibia frontalis* Blaisdell B

**Pata
(Ventana)** 3
1 *Argoporis apicalis* Blaisdell
2 *Craniotus* n. sp. 1
3 *Cryptadius tarsalis* Blaisdell
4 *Microschatia championi* Horn
5 *Tonibius sulcatus* (LeConte)
6 *Triphalopsis californicus* Doyen

**Patos
(Patos)** 1
1 *Argoporis apicalis* Blaisdell
2 *Cryptoglossa variolosa* Horn
3 *Eleodes caudatus* (Horn)
4 *Steriphanus subopacus* (Horn)
5 *Triorophus laevis* LeConte
6 *Triphalopsis partida* Blaisdell

**Pelicano (Alcatraz, Tassne)
(Colorado)** 1
1 *Hypogena marginata* (LeConte)
2 *Telabis punctulata* (LeConte)

Pescador (El Pescador)
(Pescador)   2
1 *Argoporis apicalis* Blaisdell
2 *Batuliodes confluens* Blaisdell
3 *Cryptadius tarsalis* Blaisdell
4 *Eusattus dubius* Doyen
5 *Steriphanus durus* Blaisdell

Piojo
(Piojo)   3
1 *Argoporis apicalis* Blaisdell
2 *Microschatia championi* Horn
3 *Tonibius sulcatus* (LeConte)
4 *Triphalopsis californicus* Doyen

Rasa
(Rasa)   2
1 *Argoporis apicalis* Blaisdell
2 *Batuliodes confluens* Blaisdell
3 *Cryptadius tarsalis* Blaisdell
4 *Tonibius sulcatus* (LeConte)
5 *Triphalopsis partida* Blaisdell

Salsipuedes
(San Lorenzo)   2
1 *Argoporis apicalis* Blaisdell
2 *Cryptoglossa spiculifera* (LeConte)
3 *Steriphanus subopacus* (Horn)
4 *Stibia sparsa* Blaisdell
5 *Tonibius sulcatus* (LeConte)

San Diego
(San José)   1
1 *Conibius opacus* (LeConte)
2 *Cryptoglossa asperata* Horn
3 *Emmenides subdescalceatus* Blaisdell   B
4 *Triphalus subcylindricus* Blaisdell   B

San Esteban
(San Esteban)   1
1 *Argoporis estebanensis* Berry   A
2 *Craniotus marecortezi* in litteris   A
3 *Eleodes caudatus* (Horn)
4 *Eupsophulus castaneus* (Horn)
5 *Hymenorus thoracicus* Fall
6 *Steriphanus estebani* Blaisdell   B
7 *Steriphanus subopacus* (Horn)
8 *Steriphanus tardus* Blaisdell   B

San Francisco
(San José)   1
1 *Asidina parallela* (LeConte)
2 *Cerenopus concolor* LeConte
3 *Conibius opacus* (LeConte)
4 *Cryptadius tarsalis* Blaisdell
5 *Cryptoglossa asperata* Horn
6 *Eleodes mexicanus* Blaisdell
7 *Eusattus franciscanus* Doyen   B
8 *Eusattus pallidus* Doyen
9 *Stictodera pinguis* (LeConte)
10 *Telabis* sp.
11 *Triphalopsis partida* Blaisdell
12 *Triphalus cribricollis* Horn
13 *Triphalus subcylindricus* Blaisdell   B

San Ildefonso
(San Ildefonso)   1
1 *Conibius opacus* (LeConte)
2 *Cryptoglossa asperata* Horn
3 *Emmenides subdescalceatus* Blaisdell   B
4 *Stibia cribrata* Blaisdell   B

San Jorge (George)
(Northern)   1
1 *Triorophus laevis* LeConte

San José
(San José)   1
1 *Adelonia filiformis* (Castelnau)
2 *Cerenopus concolor* (LeConte)
3 *Conibius opacus* (LeConte)
4 *Cryptadius tarsalis* Blaisdell
5 *Cryptoglossa asperata* Horn
6 *Eleodes insularis* Linell
7 *Eleodes mexicanus* Blaisdell
8 *Eusattus franciscanus* Doyen   B
9 *Eusattus pallidus* Doyen
10 *Hypogena marginata* (LeConte)
11 *Platydema nigratum* (Motschulsky)
12 *Rhipidandrus peninsularis* (Horn)
13 *Telabis* sp.
14 *Triphalus cribricollis* Horn

San Lorenzo (San Lorenzo Sur)
(San Lorenzo)   1
1 *Argoporis apicalis* Blaisdell
2 *Conibius opacus* (LeConte)
3 *Cryptoglossa spiculifera* (LeConte)
4 *Doliopinus cucujinus* Horn
5 *Hymenorus* sp.
6 *Steriphanus subopacus* (Horn)
7 *Stibia sparsa* Blaisdell
8 *Tonibius sulcatus* (LeConte)
9 *Triphalopsis partida* Blaisdell

San Luis Gonzaga (Isla Willard)
(Bahía San Luis Gonzaga) 1
1 *Phaleria latus* Blaisdell

San Marcos
(San Marcos) 1
1 *Cryptoglossa asperata* Horn
2 *Cryptoglossa spiculifera* (LeConte)
3 *Edrotes ventricosus* LeConte
4 New Genus 1 n. sp. 3 A
5 *Triphalopsis partida* Blaisdell

San Pedro Mártir
(San Pedro Mártir) 1
1 *Argoporis aequalis* Blaisdell B
2 *Asidina parallela* (LeConte)
3 *Batuliodes confluens* Blaisdell
4 *Conibius opacus* (LeConte)
5 *Cryptoglossa asperata* Horn
6 *Eusattus ciliatoides* Doyen
7 *Hylocrinus longulus* (LeConte)
8 *Metoponium bicolor* Horn
9 *Steriphanus estebani* Blaisdell B

San Pedro Nolasco
(San Pedro Nolasco) 1
1 *Argoporis aequalis* Blaisdell B
2 *Argoporis alutacea* Casey
3 *Asidina parallela* (LeConte)
4 *Conibius opacus* (LeConte)
5 *Steriphanus subopacus* (Horn)

Santa Catalina (Catlan; Catalana; Catalano)
(Santa Catalina) 1
1 *Asidina catalinae* Blaisdell A
2 *Conibius opacus* (LeConte)
3 *Cryptadius tarsalis* Blaisdell
4 *Cryptoglossa asperata* Horn
5 *Emmenides catalinae* Blaisdell
6 *Eusattus catalinensis* Doyen A
7 *Eusattus erosus* (Blaisdell)
8 *Stibia granulata* Blaisdell
9 *Telabis* sp.

Santa Cruz
(Santa Cruz) 1
1 *Conibius opacus* (LeConte)
2 *Eleodes insularis* Linell
3 *Eusattus erosus* (Blaisdell)
4 *Telabis* sp.
5 *Triphalus cribricollis* Horn

Santa Inés
(Santa Inés) 1
1 *Asidina parallela* (LeConte)
2 *Eusattus araneosus* Blaisdell A
3 *Stibia sparsa* Blaisdell
4 *Telabis hirtipes* Blaisdell
5 *Triphalopsis partida* Blaisdell

Smith (Coronado)
(Smith) 3
1 *Argoporis apicalis* Blaisdell
2 *Batuliodes confluens* Blaisdell
3 *Craniotus* n. sp. 1
4 *Cryptadius tarsalis* Blaisdell
5 *Cryptoglossa spiculifera* (LeConte)
6 *Edrotes ventricosus* LeConte
7 *Eleodes discinctus* Blaisdell
8 *Eleodes mexicanus* Blaisdell
9 *Eschatomoxys* n. sp. 1 B
10 *Eusattus difficilis* LeConte
11 *Eusattus dubius* Doyen
12 *Hymenorus* sp.
13 *Microschatia championi* Horn
14 New Genus 1 n. sp. 1
15 *Nocibiotes* n. sp. 1
16 *Phaleria latus* Blaisdell
17 *Stibia sparsa* Blaisdell
18 *Telabis serrata* (LeConte)
19 *Tonibius sulcatus* (LeConte)
20 *Triphalopsis californicus* Doyen

Tiburón
(Tiburón) 1
1 *Cryptoglossa muricata* LeConte
2 *Cryptoglossa variolosa* Horn
3 *Metopoloba pruinosa* Casey
4 New Genus 1 n. sp. 2 A
5 *Notibius puberulus* LeConte
6 *Phaleria pilifera* LeConte
7 *Steriphanus subopacus* (Horn)
8 *Triphalopsis partida* Blaisdell
9 *Zophobas sublaevis* Horn

Tijeras (Las Tijeras; Roca Medio)
(Los Candeleros) 3
1 *Argoporis apicalis* Blaisdell
2 *Conibius opacus* (LeConte)
3 *Stibia fallaciosa* (Blaisdell)
4 *Triphalopsis partida* Blaisdell

Tortuga
(Tortuga) 1
1 *Conibius opacus* (LeConte)
2 *Stibia sparsa* Blaisdell

Ventana (La Ventana)
(Ventana)   3
1 *Argoporis apicalis* Blaisdell
2 *Batuliodes confluens* Blaisdell
3 *Craniotus* n. sp. 1
4 *Cryptadius tarsalis* Blaisdell
5 *Eschatomoxys* n. sp. 1   B
6 *Eusattus dubius* Doyen
7 *Microschatia championi* Horn
8 New Genus 1 n. sp. 1
9 *Nocibiotes* n. sp. 1
10 *Phaleria pilifera* LeConte
11 *Stibia sparsa* Blaisdell
12 *Tonibius sulcatus* (LeConte)
13 *Triphalopsis californicus* Doyen

Pacific

Adentro (El Toro)
(Guadalupe)   1
1 *Coelotaxis punctulatus* Horn   C

Afuera (Zapato)
(Guadalupe)   1
1 *Coelotaxis punctulatus* Horn   C

Asunción (Ángulo Rock)
(Bahía Asunción)   1
1 *Argoporis impressa* Blaisdell   B
2 *Cryptadius inflatus* Thomas
3 *Eleodes adumbratus* Blaisdell   B
4 *Eleodes discinctus* Blaisdell
5 *Eleodes femoratus* LeConte
6 *Eleodes moestus* Blaisdell
7 *Eleodes ursus* Triplehorn
8 *Stibia puncticollis* Horn

Cedros
(Cedros)   1
1 *Argoporis impressa* Blaisdell   B
2 *Cerenopus concolor* LeConte
3 *Coelocnemis slevini* Blaisdell   A
4 *Cryptoglossa spiculifera* Blaisdell
5 *Eleodes discinctus* Blaisdell
6 *Eleodes femoratus* LeConte
7 *Eleodes insularis* Linell
8 *Eleodes ursus* Triplehorn
9 *Eusattus cedrosensis* Doyen   A
10 *Eusattus costatus* (Horn)
11 *Eusattus erosus* (Horn)
12 *Gnatocerus angelicus* Blaisdell
13 *Microschatia cedrosensis* Brown   A
14 *Stibia puncticollis* Horn
15 *Tonibius sulcatus* (LeConte)

Clarión (Santa Rosa)
(Revillagigedo)   1
1 *Blapstinus pacificus* Aalbu et al   A

Coronado del Medio
(Los Coronados)   1
1 *Blapstinus brevicollis* LeConte
2 *Eleodes acuticaudus* LeConte
3 *Stibia puncticollis* Horn

Guadalupe
(Guadalupe)   1
1 *Coelotaxis punctulatus* Horn   C
2 *Conibius guadelupensis* Casey   A
3 *Helops crockeri* Blaisdell   A
4 *Helops guadelupensis* Casey   A

Magdalena (Santa Magdalena)
(Magdalena)   1
1 *Eleodes mexicanus* Blaisdell
2 *Tonibiastes costipennis* (Horn)

Natividad
(Natividad)   1
1 *Argoporis impressa* Blaisdell   B
2 *Cryptadius inflatus* Thomas
3 *Eleodes discinctus* Blaisdell
4 *Stibia puncticollis* Horn
5 *Tonibius sulcatus* (LeConte)

San Benito del Centro (middle)
(San Benitos)   1
1 *Argoporis impressa* Blaisdell   B
2 *Eleodes adumbratus* Blaisdell   B
3 *Helops benitensis* Blaisdell   A
4 *Stibia fallaciosa* (Blaisdell)
5 *Stibia williamsi* Blaisdell   C

San Benito del Este (east)
(San Benitos)   1
1 *Argoporis impressa* Blaisdell   B
2 *Eleodes adumbratus* Blaisdell   B
3 *Stibia sparsa* Blaisdell
4 *Stibia williamsi* Blaisdell   C

San Benito del Oeste (west)
(San Benitos)   1
1 *Argoporis impressa* Blaisdell   B
2 *Eleodes adumbratus* Blaisdell   B
3 *Eleodes discinctus* Blaisdell
4 *Stibia williamsi* Blaisdell   C

San Gerónimo
(Bahía Rosario) 1
1 *Coelus maritimus* Casey
2 *Eleodes moestus* Blaisdell
3 *Micromes ovipennis* (Horn)
4 *Stibia puncticollis* Horn

San Martín
(San Quintin) 1
1 *Apsena pubescens* Blaisdell
2 *Coniontis keiferi* Blaisdell A
3 *Eleodes moestus* Blaisdell
4 *Eleodes nigropilosus* LeConte
5 *Eleodes sanmartinensis* Blaisdell A
6 *Eusattus difficilis* LeConte
7 *Micromes ovipennis* (Horn)
8 *Stibia puncticollis* Blaisdell
9 *Stibia puncticollis* Horn

San Roque
(Bahía Asunción) 1
1 *Argoporis impressa* Blaisdell B
2 *Cryptadius inflatus* Thomas
3 *Eleodes moestus* Blaisdell

Santa Margarita
(Magdalena) 1
1 *Argoporis inconstans* Horn B
2 *Cryptoglossa spiculifera* (LeConte)
3 *Eleodes discinctus* Blaisdell
4 *Eleodes femoratus* LeConte
5 *Eleodes insularis* Linell
6 *Eusattus costatus* (Horn)
7 *Tonibiastes costipennis* (Horn)

Socorro
(Revillagigedo) 1
1 *Blapstinus faulkneri* Aalbu et al A

Todos Santos (N)
(Bahía de Todos Santos) 1
1 *Eleodes carbonarius* Blaisdell
2 *Eleodes nigropilosus* LeConte
3 *Hypogena tricornis* (Castelnau)
4 *Microschatia inaequalis* (LeConte)

Appendix 6.2

List of Tenebrionidae on Islands in the Sea of Cortés and the Pacific

The numbers after species names indicate level of adaptation to arid conditions (1, most highly adapted; 2, highly adapted; 3, least adapted).

Sea of Cortés

Alleculinae
Alleculini
   1 *Hymenorus thoracicus* Fall   3
   2 *Hymenorus* sp.   3

Bolitophaginae
Bolitophagini
   3 *Rhipidandrus peninsularis* (Horn)   3

Diaperinae
Diaperini
   4 *Doliopinus cucujinus* Horn   3
   5 *Platydema nigratum* (Motschulsky)   3
Phaleriini
   6 *Phaleria pilifera* LeConte   3
   7 *Phaleria latus* Blaisdell   3

Lagriinae
Belopini
   8 *Adelonia filiformis* (Castelnau)   3

Opatrinae
Melanimimi
   9 *Cheirodes californica* (Horn)   2
Opatrini
   10 *Conibius ventralis* Blaisdell   2
   11 *Conibius opacus* (LeConte)   2
   12 *Nocibiotes* n. sp. 1   2
   13 *Nocibiotes rossi* (Blaisdell)   2
   14 *Notibius puberulus* LeConte   2
   15 *Tonibius sulcatus* (LeConte)   2

Pimeliinae
Anepsiini
   16 *Batuliodes confluens* Blaisdell   1
Asidini
   17 *Asidina parallela* (LeConte)   1
   18 *Asidina* n. sp. 1   1
   19 *Asidina catalinae* Blaisdell   1
   20 *Craniotus* n. sp. 1   1
   21 *Craniotus marecortezi* in litteris   1
   22 *Heterasida bifurca* (LeConte)   1
   23 *Heterasida conivens* (LeConte)   1
   24 *Microschatia championi* Horn   1

Coniontini
25 *Eusattus ciliatoides* Doyen 1
26 *Eusattus laevis* (LeConte) 1
27 *Eusattus franciscanus* Doyen 1
28 *Eusattus erosus* (Horn) 1
29 *Eusattus difficilis* LeConte 1
30 *Eusattus pallidus* Doyen 1
31 *Eusattus ceralboensis* Doyen 1
32 *Eusattus catalinensis* Doyen 1
33 *Eusattus araneosus* Blaisdell 1
34 *Eusattus dubius* Doyen 1
Cryptoglossini
35 *Asbolus mexicana* (Horn) 1
36 *Cryptoglossa asperata* Horn 1
37 *Cryptoglossa muricata* LeConte 1
38 *Cryptoglossa seriata* (LeConte) 1
39 *Cryptoglossa spiculifera* (LeConte) 1
40 *Cryptoglossa variiolosa* Horn 1
Epitragini
41 *Metopoloba pruinosa* Casey 3
Eurymetopini
42 *Chilometopon cribricolle* Blaisdell 1
43 *Cryptadius tarsalis* Blaisdell 1
44 *Edrotes ventricosus* LeConte 1
45 *Emmenides obsoletus* Blaisdell 1
46 *Emmenides catalinae* Blaisdell 1
47 *Emmenides subdescalceatus* Blaisdell 1
48 *Emmenides apicalis* Blaisdell 1
49 *Eschatomoxys* n. sp. 1 1
50 *Eschatomoxys* n. sp. 2 1
51 *Hylocrinus insularis* Blaisdell 1
52 *Hylocrinus oblongulus* Casey 1
53 *Hylocrinus longulus* (LeConte) 1
54 *Metoponium bicolor* Horn 1
55 *Metoponium abnorme* Casey 1
56 New Genus 1 n. sp. 1 1
57 New Genus 1 n. sp. 2 1
58 New Genus 1 n. sp. 3 1
59 New Genus 2 n. sp. 1
60 *Orthostibia frontalis* Blaisdell 1
61 *Steriphanus estebani* Blaisdell 1
62 *Steriphanus subopacus* (Horn) 1
63 *Steriphanus durus* Blaisdell 1
64 *Steriphanus tardus* Blaisdell 1
65 *Stibia sparsa* Blaisdell 1
66 *Stibia cribrata* Blaisdell 1
67 *Stibia fallaciosa* (Blaisdell) 1
68 *Stibia granulata* Blaisdell 1
69 *Stibia puncticollis* Horn 1
70 *Stictodera pinguis* (LeConte) 1
71 *Telabis hirtipes* Blaisdell 1
72 *Telabis latipennis* Blaisdell 1
73 *Telabis serrata* (LeConte) 1
74 *Telabis punctulata* (LeConte) 1
75 *Telabis* sp. 1
76 *Telabis lunulata* Blaisdell 1

77 *Trimytis ceralboensis* Blaisdell 1
78 *Trimytis obtusa* Horn 1
79 *Triorophus laevis* LeConte 1
80 *Triphalopsis californicus* Doyen 1
81 *Triphalopsis partida* Blaisdell 1
82 *Triphalus subcylindricus* Blaisdell 1
83 *Triphalus cribricollis* Horn 1
Stenosini
84 *Araeoschizus aalbui* Papp 1
85 *Typhlusechus* n. sp. 1
Vacronini
86 *Eupsophulus castaneus* (Horn) 3

Tenebrioninae
Cerenopini
87 *Argoporis alutacea* Casey 2
88 *Argoporis aequalis* Blaisdell 2
89 *Argoporis apicalis* Blaisdell 2
90 *Argoporis niger* Aalbu et al 2
91 *Argoporis estebanensis* Berry 2
92 *Argoporis cribratus* LeConte 2
93 *Cerenopus concolor* LeConte 2
Eleodini
94 *Eleodes loretensis* Blaisdell 2
95 *Eleodes mexicanus* Blaisdell 2
96 *Eleodes discinctus* Blaisdell 2
97 *Eleodes eschscholtzi* Solier 2
98 *Eleodes insularis* Linell 2
99 *Eleodes caudatus* (Horn) 2
Tenebrionini
100 *Zophobas sublaevis* Horn 2
Triboliini
101 *Hypogena marginata* (LeConte) 3
102 *Latheticus prosopis* Chittenden 3

Pacific

Coelometopinae
Coelometopini
103 *Coelocnemis slevini* Blaisdell 2

Diaperinae
Diaperini
104 *Gnatocerus angelicus* Blaisdell 3

Opatrinae
Opatrini
105 *Blapstinus faulkneri* Aalbu et al 2
106 *Blapstinus pacificus* Aalbu et al 2
107 *Conibius guadelupensis* Casey 2
108 *Tonibiastes costipennis* (Horn) 2
109 *Tonibius sulcatus* (LeConte) 2

Pimeliinae
Asidini
110 *Microschatia inaequalis* (LeConte) 1
111 *Microschatia cedrosensis* Brown 1
Coniontini
112 *Coelotaxis punctulatus* Horn 1
113 *Coelus maritimus* Casey 1
114 *Coniontis keiferi* Blaisdell 1
115 *Eusattus erosus* (Horn) 1
116 *Eusattus difficilis* LeConte 1
117 *Eusattus cedrosensis* Doyen 1
118 *Eusattus costatus* (Horn) 1
Cryptoglossini
119 *Cryptoglossa spiculifera* (LeConte) 1
Eurymetopini
120 *Cryptadius inflatus* Thomas 1
121 *Micromes ovipennis* (Horn) 1
122 *Stibia puncticollis* Horn 1
123 *Stibia williamsi* Blaisdell 1
124 *Stibia fallaciosa* (Blaisdell) 1
125 *Stibia sparsa* Blaisdell 1

Tenebrioninae
Cerenopini
126 *Argoporis impressa* Blaisdell 2
127 *Argoporis inconstans* Horn 2
128 *Cerenopus concolor* LeConte 2
Eleodini
129 *Eleodes nigropilosus* LeConte 2
130 *Eleodes mexicanus* Blaisdell 2
131 *Eleodes sanmartinensis* Blaisdell 2
132 *Eleodes acuticaudus* LeConte 2
133 *Eleodes moestus* Blaisdell 2
134 *Eleodes insularis* Linell 2
135 *Eleodes femoratus* LeConte 2
136 *Eleodes discinctus* Blaisdell 2
137 *Eleodes carbonarius* Blaisdell 2
138 *Eleodes adumbratus* Blaisdell 2
139 *Eleodes ursus* Triplehorn 2
Eulabini
140 *Apsena pubescens* Blaisdell 2
Helopini
141 *Helops benitensis* Blaisdell 2
142 *Helops crockeri* Blaisdell 2
143 *Helops guadalupensis* Casey 2
Triboliini
144 *Hypogena tricorinis* (Castelnau) 3

Appendix 7.1

Location of Ichthyocide Collections of Rocky-Shore Fishes in the Sea of Cortés

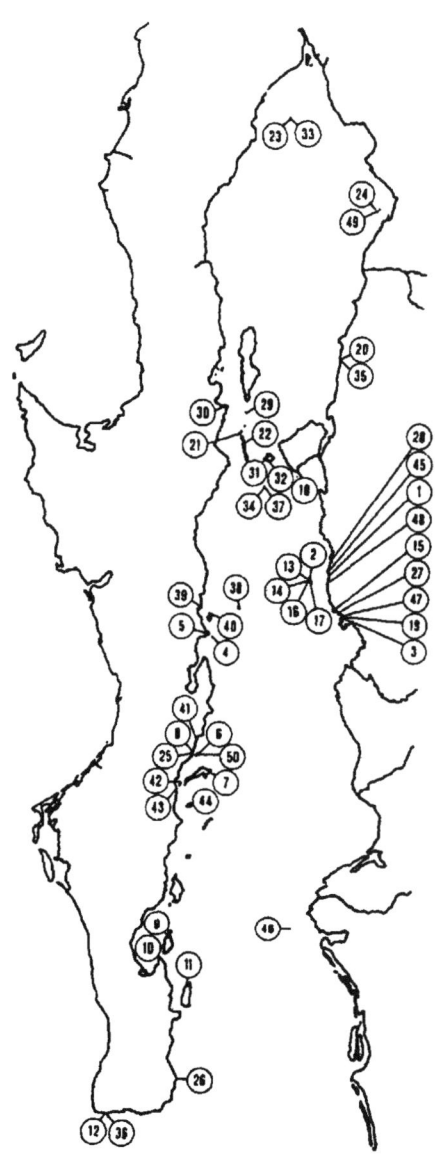

Map numbers correspond to collection numbers in appendixes 7.2–7.4.

Appendix 7.2

List of Data of Island and Mainland Fish Collection Stations in the Gulf of California (1973–1976)

Sample No.	Locality	Date	UA No.	Lat. (°N)	Sea Temp. (°C)	S.D.	S	N	Biomass	H'_n	H'_b
1.	Punta San Pedro	7/05/75	75-91	28.05	29.0	—	12	115	137.5	1.997	1.405
2.	San Pedro Nolasco	7/23/74	74-28	27.95	27.2	8	19	276	512.5	2.271	1.678
3.	San Carlos	7/24/74	74-29	27.97	—	—	10	159	231.6	1.505	1.690
4.	Santa Inez	7/28/74	74-32	27.05	29.0	4.6	13	247	349.5	2.123	2.070
5.	Punta Chivato	7/28/74	74-33	27.10	26.1	2.0	8	83	355.9	1.391	1.053
6.	Coronado	7/31/74	74-34	26.13	28.1	6.4	28	770	845.1	2.260	2.072
7.	Carmen	8/01/74	74-35	26.02	28.8	7.0	22	408	388.6	2.391	2.040
8.	Loreto	8/01/74	74-36	26.12	30.5	—	18	412	221.8	1.880	1.874
9.	Espíritu Santo	8/05/74	74-37	24.42	28.3	—	24	489	315.0	2.253	1.998
10.	La Paz	8/06/74	74-38	24.35	28.3	—	18	392	798.2	2.076	1.469
11.	Cerralvo	8/08/74	74-40	24.32	28.5	10.0	20	248	114.8	1.950	2.048
12.	Cabo San Lucas	8/09/74	74-41	22.87	26.7	5.0	20	363	768.9	2.237	1.485
13.	San Pedro Nolasco	10/19/74	74-57	27.95	26.0	6.1	26	847	1039.6	2.275	2.149
14.	San Pedro Nolasco	3/27/75	75-32	27.95	18.0	7.0	19	134	201.5	2.462	1.780
15.	Venado	3/28/75	75-33	27.98	18.0	—	16	151	221.6	2.052	1.577
16.	San Pedro Nolasco	7/04/75	75-38	27.95	27.0	12.0	22	508	218.0	2.169	2.315
17.	San Pedro Nolasco	7/01/75	75-37	27.95	27.0	12.0	21	514	747.1	2.290	1.939
18.	Tiburón	6/27/75	75-35	28.78	29.0	4.0	15	125	529.9	2.193	1.322
19.	Punta San Antonio	3/26/75	75-31	27.95	16.0	5.0	11	46	68.2	2.053	1.433
20.	South of Libertad	6/22/75	75-34	29.70	25.0	10.0	10	114	114.4	1.805	1.452
21.	San Lorenzo	7/28/75	75-74	28.67	26.0	10.0	22	428	1225.5	2.463	2.110
22.	San Lorenzo	7/28/75	75-75	28.67	27.0	8.5	13	370	432.7	1.813	1.790
23.	Roca Consag	10/10/75	75-64	31.12	28.0	4.8	10	226	377.4	1.331	1.369

Sample No.	Locality	Date	UA No.	Lat. (°N)	Sea Temp. (°C)	S.D.	S	N	Biomass	H'_n	H'_b
24.	San Jorge	10/11/75	74-66	31.03	28.8	6.0	15	225	1088.5	1.977	1.460
25.	Loreto	6/16/73	73-73	26.00	21.0	5.1	20	172	433.3	2.322	1.757
26.	Cabo Pulmo	6/23/72	73-80	23.45	26.7	5.0	21	234	287.4	2.092	1.817
27.	Venado	10/13/73	73-100	27.95	22.2	—	20	318	489.2	2.032	1.374
28.	Punta San Pedro Area	6/30/75	75-36	28.05	28.0	12.0	25	309	802.5	2.420	1.829
29.	Partida Norte	7/29/75	75-76	28.87	25.0	5.5	17	300	485.4	2.102	2.132
30.	Punta Qué Malo	7/30/75	75-77	28.95	29.0	5.5	13	419	153.9	0.963	1.180
31.	San Esteban	7/31/75	75-78	28.67	27.0	9.0	12	259	229.1	1.740	1.600
32.	San Esteban	7/31/75	75-79	28.67	27.0	—	17	485	508.8	1.777	1.702
33.	Roca Consag	7/09/73	73-89	31.12	25.6	1.8	8	69	54.3	1.004	1.156
34.	San Pedro Mártir	7/27/75	75-73	28.37	29.0	6.1	21	575	354.4	2.026	1.929
35.	Punta San Pedro Area	6/22/75	75-51	28.08	25.0	4.0	14	385	1081.3	1.516	1.618
36.	Cabo San Lucas	6/28/73	73-83	22.87	21.0	—	14	43	53.5	1.866	1.349
37.	San Pedro Mártir	6/30/76	76-13	28.37	28.5	20.0	21	257	344.7	2.489	2.196
38.	Tortuga	7/01/76	76-14	27.45	29.0	15.0	24	338	584.7	2.616	2.015
39.	Santa Rosalia	7/02/76	76-15	27.33	25.0	8.0	19	152	435.2	2.562	1.879
40.	San Marcos	7/02/76	76-16	27.33	26.0	5.0	24	541	1510.5	2.722	1.778
41.	Loreto Area	7/20/76	76-24	26.20	29.0	—	21	362	304.2	2.160	1.878
42.	Loreta Area	7/25/76	76-26	25.73	29.5	6.0	31	431	985.3	2.796	1.612
43.	Danzante	7/27/76	76-31	25.80	29.7	15.0	21	755	461.9	2.036	2.132
44.	Monserrate	7/28/76	76-32	25.70	29.3	5.5	27	1138	650.5	2.333	1.916
45.	Punta San Pedro	7/02/75	75-121	28.08	29.0	—	16	193	434.9	2.043	1.410
46.	Farallón	3/15/76	76-02	25.50	18.0	6.3	13	195	1117.0	1.713	0.522
47.	San Carlos	7/15/76	76-22	27.95	30.0	—	16	162	196.3	1.879	1.581
48.	Bahía San Pedro	10/19/74	74-58	28.03	26.0	3.0	18	439	311.3	1.684	1.852
49.	San Jorge	10/11/75	75-65	31.05	27.8	4.8	11	141	133.2	1.337	1.377
50.	Coronado	7/22/76	76-25	26.12	28.5	15.0	23	1017	790.9	2.209	2.039

Abbreviations: UA No., University of Arizona Fish Collection catalogue number; Lat., latitude given in decimal degrees; Sea temp., surface sea temperature taken during collection; S.D., secchi disc reading in meters; S, total species in collection; N, total number of individuals in collection; biomass, wet preserved weight in grams of total collection; H'_n, Shannon-Weiner species diversity index using numbers of individuals; H'_b, Shannon-Weiner species diversity index using weight in grams per species lot.

Appendix 7.3

Physical Data of Mainland Areas Sampled in the Gulf of California

Area	No. of Collections	Sample Numbers[a]	Rocky Coastline (km)[b]	Mean Monthly Minimum Temp. (°C)[c]	Mean Monthly Maximum Temp. (°C)[c]	Volume of Water Within 20 km (km^3)[d]
South of Libertad	2	20, 35	20	16	30	43
Punta Qué Malo	1	30	98	14	28	334
Guaymas vicinity	9	28, 45, 1, 48, 15, 27, 47, 19, 3	45	16	30	73–132
Santa Rosalia	1	39	98	16	31	460
Punta Chivato	1	5	5	17	31	470
North of Loreto	3	41, 8, 25	98	18	31	478
South of Loreto	1	42	98	18	31	478
La Paz	1	10	27	19	31	483
Cabo Pulmo	1	26	5	20	30	490
Cabo San Lucas	2	12, 36	10	21	29	483

[a]Sample numbers shown in appendix 7.1 and listed in appendix 7.2.
[b]Km length of major rocky coastline contiguous with a collection or collections.
[c]Mean monthly minimum and maximum sea-surface temperatures from Robinson (1973).
[d]Volume of water within a 20-km radius of collection sites; calculated from areas between major depth isoclines measured with a planimeter from Fisher et al. (1964).

Appendix 7.4

Physical Data of Islands Sampled in the Gulf of California

Name	No. of Collections	Collection Numbers[a]	Area (km^2)[b]	Perimeter (km)[c]	Distance to Mainland (km)[d]	Mean Monthly Minimum Temp. (°C)[e]	Mean Monthly Maximum Temp. (°C)[e]	Volume of Water Within 20 km (km^3)[f]
Roca Consag	2	23, 33	0.6	2.8	31	15	30	33
San Jorge	2	24, 49	2.4	5.5	13	16	31	25
Partida Norte	1	29	1.2	3.9	17	15	29	446
Tiburón	1	18	1,208.0	123.2	2	15	30	82
San Lorenzo	2	21, 22	7.5	9.7	18	14	29	538
San Esteban	2	31, 32	43.0	23.2	37	15	29	447
San Pedro Mártir	2	34, 37	1.9	4.9	48	15	30	550
San Pedro Nolasco	5	2, 13, 14, 16, 17	3.2	6.3	10	16	30	360
Tortuga	1	38	6.3	8.9	37	16	31	467
San Marcos	1	40	32.0	20.1	5	16	31	464
Santa Ines	1	4	0.4	2.2	2	17	31	470
Coronado	2	6, 50	8.5	10.3	2	18	31	477
Carmen	1	7	151.0	43.6	6	18	31	477
Danzante	1	43	4.9	7.8	1	17	31	477
Monserrate	1	44	19.4	15.6	13	18	31	479
Farallón	1	46	0.7	3.0	20	18	30	479
Espíritu Santo	1	9	99.0	35.3	6	18	31	483
Cerralvo	1	11	160.0	44.8	11	19	31	479

[a] Sample numbers shown in appendix 7.1 and listed in appendix 7.2.
[b] Island area as in table 2.2 or as measured with planimeter by M. Gilligan.
[c] Perimeter of island calculated from area assuming circular islands.
[d] Distance to nearest mainland as in table 2.2 or measured by M. Gilligan.
[e] Mean monthly minimum and maximum sea surface temperature from Robinson (1973).
[f] Volume of water within a 20-km radius of collection sites; calculated from areas between major depth isoclines measured with planimeter from Fisher et al. (1964).

Appendix 7.5

List of Rocky-Shore Fishes Used in the Analyses, Including Their Residency Status, Biogeographic Affinity, Type of Egg, and Food Habits

Family	Genus Species	Residency[a]	Affinity[b]	Egg Type[c]	Food Habits[d]
Muraenidae (morays)	*Gymnothorax castaneus* (Jordan & Gilbert 1882)	2°	P	P	b,f
	G. panamensis (Steindachner 1876)	2°	P	P	b
	Muraena lentiginosa (Jenyns 1843)	2°	P	P	b
	Uropterygius necturus[e] (Jordan & Gilbert 1882)	2°	P	P	b
	U. sp. (undescribed species)	2°	?	P	b
Bythitidae (brotulas)	*Ogilbia* sp. (undescribed species)	1°	—	V	b
	Oligopus diagrammus (Heller & Snodgrass 1903)	1°	P	?	b
Scorpaenidae (scorpionfishes)	*Scorpaena mystes* (Jordan & Starks 1895)	2°	P	P	b,f
	Scorpaenodes xyris (Jordan & Gilbert 1882)	2°	P	P	b
Grammistidae (Soapfishes)	*Rypticus bicolor* (Valenciennes 1846)	2°	P	P	b
Apogonidae (cardinalfishes)	*Apogon retrosella* (Gill 1863)	1°	M	d	z,b
Sciaenidae (drums)	*Pareques viola* (Gilbert 1898)	2°	P	P	b
Cirrhitidae (hawkfishes)	*Cirrhitichthys oxycephalus* (Bleeker 1855)	2°	I	P	b
	Cirrhitus rivulatus (Valenciennes 1855)	2°	P	P	b

Family	Genus Species	Residency[a]	Affinity[b]	Egg Type[c]	Food Habits[d]
Blenniidae (combtooth blennies)	*Entomacrodus chiostictus* (Jordan & Gilbert 1883)	1°	P	d	h
	Hypsoblennius brevipinnis (Günther 1861)	1°	P	d	h
	H. gentilis (Girard 1854)	1°	N	d	h
	Ophioblennius steindachneri (Jordan & Evermann 1898)	1°	P	d	h
	Plagiotremus azaleus (Jorday & Bollman 1890)	1°	P	d	o
Tripterygiidae (triplefin blennies)	*Crocodilichthys gracilis* (Allen & Robertson 1991)	1°	G	d	b
	Axoclinus carminalis (Jordan & Gilbert 1882)	1°	G	d	b
	Axoclinus nigricaudus (Allen & Robertson 1991)	1°	G	d	b
	Enneanectes sexmaculatus (Fowler 1944)	1°	P	d	b
	Enneanectus reticulatus (Allen & Robertson 1991)	1°	G	d	b
Clinidae[f] (clinid blennies)	*Exerpes asper* (Jenkins & Evermann 1889)	1°	G	d	z
	Labrisomus striatus (Hubbs 1953)	1°	P	d	b
	L. xanti (Gill 1860)	1°	M	d	b
	Malacoctenus gigas (Springer 1959)	1°	G	d	b
	M. hubbsi (Springer 1959)	1°	P	d	b
	M. margaritae (Fowler 1944)	1°	P	d	b
	M. tetranemus (Cope 1877)	1°	P	d	b
	Paraclinus beebei (Hubbs 1952)	1°	P	d	b
	P. mexicanus (Gilbert 1904)	1°	P	d	b
	P. sini (Hubbs 1952)	1°	M	d	b
	Starksia spinipenis (Al-Uthman 1960)	1°	P	d	b
	Stathmonotus sinuscalifornici (Chaubanaud 1942)	1°	G	d	b
	Xenomedea rhodopyga (Rosenblatt & Taylor 1971)	1°	G	d	b
Chaenopsidae (tubeblennies)	*Acanthemblemaria balanorum* (Brock 1940)	1°	P	d	z
	A. crockeri (Beebe & Tee-Van 1938)	1°	G	d	z
	A. macrospilus (Brock 1940)	1°	P	d	z
	Chaenopsis alepidota (Gilbert 1890)	1°	M	d	z
	Coralliozetus angelica (Böhlke & Mead (1957)	1°	P	d	z
	C. micropes (Beebe & Tee-Van 1938)	1°	G	d	z
	C. rosenblatti (Stephens 1963)	1°	G	d	z
	Ekemblemaria myersi (Stephens 1963)	1°	P	d	z
	Protemblemaria bicirris (Hildebrand 1946)	1°	P	d	z
Gobiidae (gobies)	*Aruma histrio* (Jordan 1884)	1°	G	d	b
	Barbulifer pantherinus (Pellegrin 1901)	1°	G	d	b

Family	Genus Species	Residency[a]	Affinity[b]	Egg Type[c]	Food Habits[d]
	B. sp. (undescribed species)[g]	1°	?	d	b
	Bathygobius ramosus (Ginsburg 1947)	1°	P	d	b
	Chriolepis zebra (Ginsburg 1938)	1°	G	d	b
	Coryphopterus urospilus (Ginsburg 1938)	1°	P	d	b
	Elacatinus digueti (Pellegrin 1901)	1°	P	d	o
	E. punticulatus (Ginsburg 1938)	1°	P	d	b
	E. sp. (undescribed species)	1°	P	d	o
	Gobiosoma chiquita (Jenkins & Evermann 1889)	1°	G	d	b
	Gobulus hancocki (Ginsburg 1938)	1°	P	d	b
	Gymneleotris seminudus (Günther 1864)	1°	P	d	b
	Lythrypnus dalli (Gilbert 1890)	1°	N	d	b
	L. pulchellus (Ginsburg 1938)	1°	G	d	b
	Pyconomma semisquamatum (Rutter 1904)	1°	G	d	b
Gobiescocidae (clingfishes)	*Arcos erythrops* (Jordan & Gilbert 1882)	1°	M	d	b
	Gobiesox adustus (Jordan & Gilbert 1882)	1°	P	d	b
	G. pinniger (Gilbert 1890)	1°	G	d	b
	Pherallodiscus funebris (Gilbert 1890)	1°	G	d	b
	Tomicodon boehlkei (Briggs 1955)	1°	G	d	b
	T. eos (Jordan & Gilbert 1882)	1°	G	d	b
	T. humeralis (Gilbert 1890)	1°	G	d	b

[a] Residency: 1°, primary resident; 2°, secondary resident.
[b] Affinity: G, Gulf of California endemic; M, restricted to Mexican waters; N, northern (San Diegan) affinities; P, Panamic, range extends to Panama and sometimes farther south; I, Indo-west Pacific immigrant.
[c] Eggs: p, pelagic; d, demersal; v, viviparous.
[d] Food habits: h, herbivorous, grazes on benthic algae; b, benthic predator, feeds on small benthic invertebrates such as crustaceans, mollusks, worms, etc.; f, piscivorous, feeds on fishes; z, planktivorous, feeds on zooplankton; o, other (feeds on ectoparasites or mucus of other fishes).
[e] Synonym of *Uropterygius macrocephalus* (Bleeker 1865).
[f] Labrisomidae (labrisomid blennies).
[g] Now described as *Barbulifer mexicanus*; Hoese and Larson 1985.

Appendix 8.1

Updated mtDNA Phylogeny for *Sauromalus* and Implications for the Evolution of Gigantism

KENNETH PETREN

TED J. CASE

Populations of the large, herbivorous chuckwalla lizard (*Sauromalus*) differ substantially with regard to body size throughout their range (Case 1976, 1982; Tracy 1999). Among populations of *S. obesus*, body size varies from approximately 180 g to 380 g, or 175 mm to 218 mm UDL (upper decile snout-vent length). Size differences are partly determined by vegetation density, which is closely linked to winter rainfall (Case 1976). On islands in the Sea of Cortés, two endemic species attain substantially larger body sizes: *S. hispidus* on Ángel de la Guarda island (fig. A8.1-1) reaches about 1400 g, 219 mm UDL, and *S. varius* on San Esteban reaches 1900 g, 301 mm UDL (Case 1982).

A central question is whether the gigantic species diverged early from the rest of the genus, as proposed by Shaw (1945). This would imply that the gigantic chuckwallas may have simply retained large body size from ancestors shared with other large-bodied iguanine genera (*Iguana, Conolophus, Amblyrhynchus, Cyclura, Ctenosaura*). Alternatively, large body size may have evolved recently in response to the island environment (Case 1982).

Earlier, we tested Shaw's (1945) hypothesis by reconstructing a phylogeny of *Sauromalus* using mtDNA (cytochrome b) sequence variation (Petren and Case 1997). Results showed that the island gigantics diverged from other *Sauromalus* on the nearby Baja Peninsula after most of the other populations across the entire range were established. Furthermore, the two gigantic species were firmly supported to be sister taxa, an unexpected result given the extensive phenotypic divergence between them. The resulting phylogenetic topology supported the conclusion that body size had undergone two major changes: the ancestor of all chuckwallas became substantially smaller than the closest iguanine relative (*Iguana*), and the ancestor of the gigantics evolved large size again after becoming isolated on a deep water island.

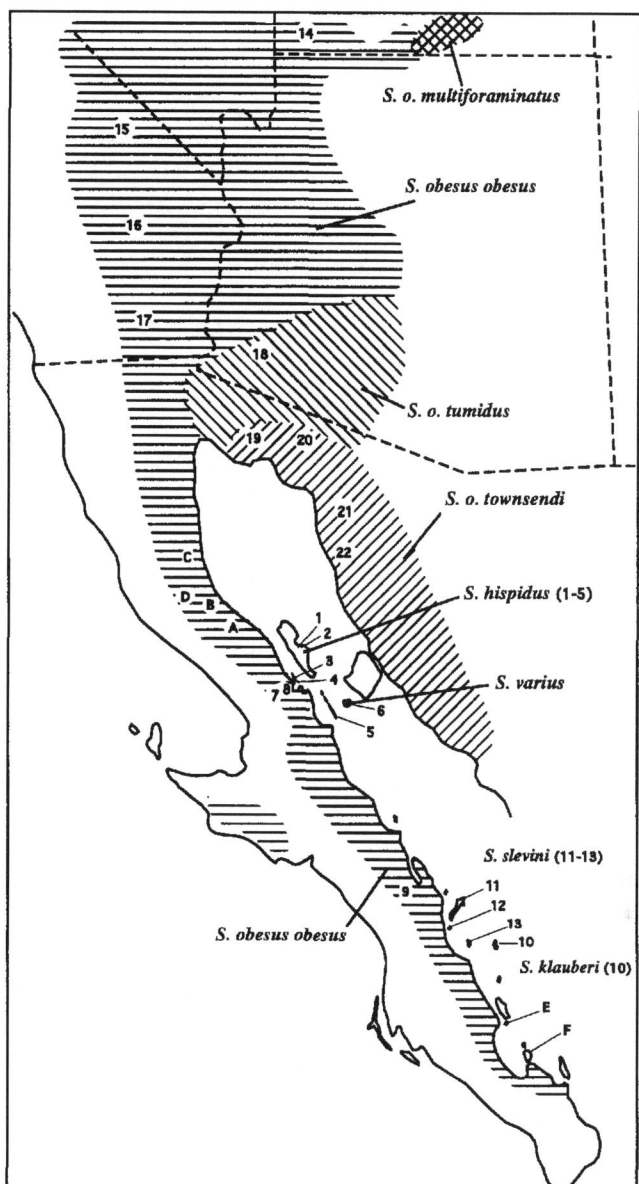

Figure A8.1-1 Distribution of *Sauromalus* species and geographic locations of sampled populations (Shaw 1945; Petren and Case 1997). Numbered locations correspond to Petren and Case (1997). Six individuals from new populations were added to previous analyses. Individuals A–D were sampled from the northern Baja California peninsula; E and F were taken from individuals on the islands of San Francisco and Espíritu Santo, respectively.

Our sampling of populations has been extended further to reveal the patterns of genetic differentiation in *Sauromalus*. Humans have historically moved gigantic forms to other islands to provide a source of food (see chaps. 8, 15). For instance, *S. hispidus* has been introduced to islands near the Baja peninsula (fig. A8.1-1). Sampling was extended beyond historically human-occupied regions to rule out recent gene flow. Populations on the southern islands of San Francisco and Espíritu Santo (formerly referred to as subspecies *S. ater ater*) extend our sampling to the southern edge of the genus's range.

To reconstruct a molecular phylogenetic hypothesis, a total of 902 bp of the cytochrome *b* gene was sequenced (Petren and Case 1997). The *Iguana iguana* sequence was used as the outgroup, as it is most closely related to *Sauromalus*. The mean Jukes-Cantor distance to all *Sauromalus* cytochrome *b* haplotypes is 13.3% for *Iguana*, 15.0% for *Ctenosaura*, and 15.5% for *Conolophus* (Petren and Case 1997). Using different outgroups (e.g., *Ctenosaura* or *Conolophus*) did not change our results. Unweighted maximum parsimony, neighbor-joining, and quartet puzzling maximum likelihood were used to reconstruct trees (Strimmer and von Haeseler 1996; Swofford 1999). All methods agree with respect to the divisions of the four major clades, as well as the branching order among them (fig. A8.1-2). The resulting topology is similar to the previous one (Petren and Case 1997) and does not alter our original conclusions regarding body size evolution in *Sauromalus*. The northern peninsular populations cluster with other nearby populations, whereas the southern island populations cluster as expected with the southern peninsular and island populations. Four main clades are well supported by bootstrap analysis: the northern/eastern, the southern, the north Baja peninsular, and the island gigantics. Roughly, these same geographic divisions for mainland forms were also found by Hollingsworth (1998) based on morphological characters, although his phylogeny is different (see also app. 8.2).

Within the northern clade, the more isolated populations from the eastern side of the Sea of Cortés form a well-supported clade corresponding to their classification as *S. o. townsendi*. In contrast, the more northern subspecies *S. o. tumidus* is not clearly distinct from *S. o. obesus*. The widely distributed *S. o. obesus* shows large genetic distances among populations, reaching an extreme with the genetically distinct northern and southern Baja California clades. The result is that based on cytochrome *b*, *S. obesus* is a paraphyletic taxon with regard to other *Sauromalus* species.

The distinct northern Baja California clade suggests that there may have been two distinct vicariant boundaries in the peninsular region. However, among *Sauromalus*, genetic distance correlates strongly with geographic distance (Mantel test; $r = .77$; permuted $p < .005$, approximated $p < .0001$). Thus, caution must be used in interpreting molecular genetic discontinuities in regions that have not been extensively sampled. Nevertheless, the possible existence of a genetic discontinuity in the northern peninsular region agrees with patterns found in other taxa (see chap. 8).

The pattern of isolation by distance is evident even at reduced spatial scales, as nearby populations contain substantially different haplotypes (Lamb et al. 1992; Petren and Case 1997). Previous taxonomic classifications of species and subspecies based on phenotype are reflected in the molecular data primarily where taxa are isolated geographically, as in the eastern and southern clades. Mitochondrial DNA is maternally inherited; thus female migration and gene flow appear to be limited. We hypothesize that this may be a consequence of the strong preference of mainland

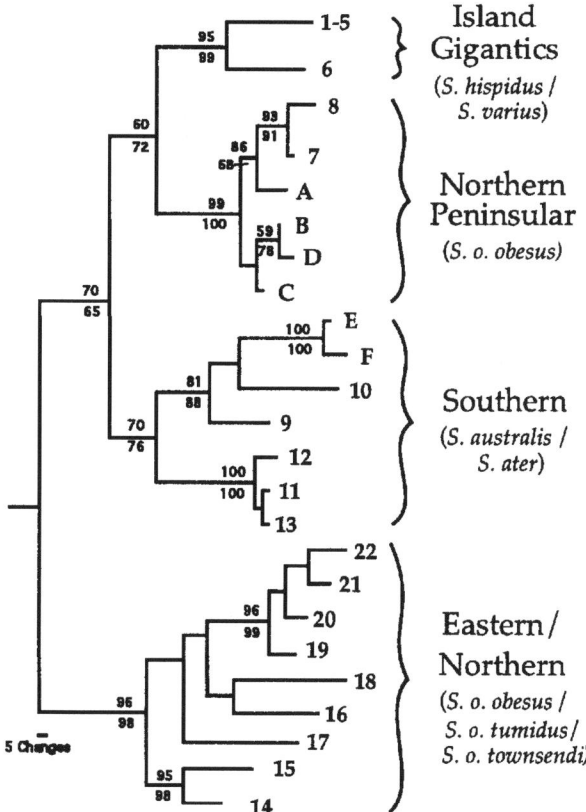

Figure A8.1-2 A phylogeny of *Sauromalus* based on Petren and Case (1997) and newly sampled populations (A–F). A total of 902 bp of the cytochrome *b* gene was sequenced. *Iguana iguana* was used as the outgroup. One of 16 most parsimonious trees is shown, with branch lengths proportional to inferred numbers of base changes. The percentage representations in 1000 bootstrap pseudoreplicates is given for maximum parsimony (above node) and neighbor-joining (below node). Discrepancies among the 16 maximum parsimony trees, the neighbor-joining tree, and the quartet puzzling tree are confined to rearrangements within the southern and northern/eastern clades. The composition of the four main clades and their relative arrangement was consistent among methods. These relationships remain unchanged from the original analysis, except for the addition of new taxa. Note the well-supported *S. slevini* clade on the islands of Carmen (11), Danzante (12), and Monserrat (13). Hollingsworth's (1998) phylogeny based on morphological characters does not include Danzante in this group, but rather within *S. obesus* (*ater*). Sequences are deposited in Genbank under accession numbers AF020223-AF020259.

chuckwallas for rocky outcrops and their reduced tendency to cross sandy arroyos, where they may be vulnerable to predators. This islandlike population structure reduces gene flow and enhances divergence rates, both from drift and selection. Similar color pattern "classes" on the mainland have discontinuous ranges (Hollingsworth 1998).

Based on the phylogenetic topology, we previously concluded that gigantism evolved in response to the island environment. The timing of events supports this interpretation. The island of Ángel de la Guarda is between 1 and 3 Ma old and was apparently formed by block-fault movement of a large landmass away from the Baja peninsula (Gastil et al. 1983; Carreño, chap. 2, this volume). Mitochondrial DNA sequence of the endemic *S. hispidus* from this island is approximately 2% divergent from chuckwallas found in the region from where the island originated. This is in line (or if anything, less divergent) with the widely used estimated rate of mtDNA mutation of 2% per million years. The much greater sequence divergence among mainland forms (up to 7%) strongly suggests that these lineages were established before the island gigantic species arose.

An alternative interpretation is that all mainland and island populations were once large bodied, and all of the mainland forms evolved smaller size (Shaw 1945; Hollingsworth 1998). This is unlikely because of the large number of independent evolutionary events that must be invoked: each mainland lineage must have responded dramatically and in parallel to some selective force. The selective factors responsible would have had to change dramatically and in concert less than 1 million years ago across the entire range of the genus on the mainland.

The selective factors involved in the evolution of large size in *S. hispidus* and *S. varius* are reviewed in chapter 9. We have proposed a model (Petren and Case 1997) whereby predation pressure from a diverse suite of mainland predators has acted to constrain the body size of mainland chuckwallas soon after their divergence from *Iguana* relatives. Large size may also be favored because of variation in the food supply (Case 1979). Both islands experience long periods of drought that can last many years. Larger lizards can store more energy during periods of greater resource availability and therefore are able to withstand longer periods without adequate resources. However, Case (chap. 9) finds that the largest individuals on Ángel suffer disproportionately higher mortality rates during drought years than smaller and presumably younger individuals.

Iguanines show substantial variation not only in body size, but also in sexual dimorphism and behavioral traits such as territoriality. For this reason they have been characterized as a model system for the study of the evolution of these traits (Burghardt and Rand 1982; Carothers 1994). The gigantic chuckwallas have relaxed territorial tendencies and show reduced sexual dimorphism in body size when compared to mainland forms. The selective factors shaping these traits may also be linked to predation and food availability. Manipulative field experiments would be of great value for understanding the interplay of the factors affecting body size in *Sauromalus*.

References

Burghardt, G. M., and A. S. Rand. 1982. *Iguanas of the World*. Noyes Publications, Park Ridge, NJ.

Carothers, J. H. 1984. Sexual selection and sexual dimorphism in some herbivorous lizards. *American Naturalist* **124**:244–254.

Case, T. J. 1976. Body size differences between populations of the chuckwalla, *Sauromalus obesus*. *Ecology* **57**:313–323.

Case, T. J. 1979. Optimal body size and an animal's diet. *Acta Biotheoretica* **28**:54–69.

Case, T. J. 1982. Ecology and evolution of insular gigantic chuckwallas, *Sauromalus hispidus* and *Sauromalus varius*. In: G. M. Burghardt and A. S. Rand (eds.), *Iguanas of the World*. Noyes Publications, Park Ridge, NJ; pp. 184–212.

Gastil, G., J. Minch and R. P. Phillips 1983. The geology and ages of the islands. In: T. J. Case and M. L. Cody (eds.), *Island Biogeography in the Sea of Cortez*. University of California Press, Berkeley; pp. 13–25.

Hollingsworth, B. D. 1998. The systematics of chuckwallas (*Sauromalus*) with phylogenetic analysis of other iguanid lizards. *Herpetological Monographs* **12**:38–191.

Lamb, T., T. R. Jones, and J. C. Avise. 1992. Phylogenetic histories of representative herpetofauna of the southwestern U.S.: Mitochondrial DNA variation in the desert iguana (*Dipsosaurus dorsalis*) and the chuckwalla (*Sauromalus obesus*). *Journal of Evolutionary Biology* **5**:465–480.

Petren, K., and T. J. Case. 1997. A phylogenetic analysis of body size evolution and biogeography in chuckwallas (*Sauromalus*) and other iguanines. *Evolution* **51**:206–219.

Shaw, C. E. 1945. The chuckwallas, genus *Sauromalus*. *Transactions of the San Diego Society of Natural History* **10**:269–306.

Stimmer, K. N., and A. von Haeseler. 1996. Quartet puzzling: a quartet maximum likelihood method for reconstructing tree topologies. *Molecular Biology and Evolution* **14**: 210–211.

Swofford, D. L. 1999. PAUP*. *Phylogenetic Analysis Using Parsimony* (*and Other Methods), Version 4.0b. Sinauer Associates, Sunderland, MA.

Tracy, C. R. 1999. Differences in body size among chuckwalla (*Sauromalus obesus*) populations. *Ecology* **80**:259–271.

Appendix 8.2

Distributional Checklist of Nonavian Reptiles and Amphibians on the Islands in the Sea of Cortés

ROBERT W. MURPHY

GUSTAVO AGUIRRE-LÉON

Biologists depend on checklists for many reasons, ranging from evolutionary studies to conservation and management of species and areas to compilations in popular guidebooks. Many uses are critically dependent on names and the information that they carry, especially when the taxonomy reflects genealogical relationships (Hull 1964; Hennig 1966; Wiley 1981). For example, in this volume alone, two chapters depend on such information for formulating hypotheses of origins and relationships and studies of evolutionary ecology. Consequently, we provide an updated checklist to the amphibians and reptiles on the islands in the Sea of Cortés. In part, this makes the volume complete, but we also clarify the taxonomic relationships of many lineages and species from a phylogenetic perspective. The distribution records are based largely on Grismer (1999b). We add the common kingsnake, *Lampropeltis getula* (documentary photos on file in the Royal Ontario Museum) and night snake *Hypsiglena torquata* (specimen deposited in the herpetological collections of the Royal Ontario Museum) to the herpetofauna of Santa Cruz. We also note that the leaf-toed gecko *Phyllodactylus tuberculosus* occurs on San Ignacio Farralon (Dixon 1966; Murphy 1983c; Murphy and Ottley 1984). Like Grismer, we have avoided referring to subspecies in this checklist (but see below).

Our taxonomy differs significantly for many groups of species, particularly for lizards. Some of the taxonomic controversy arises from different interpretations of Frost and Hillis's (1990) paper on species concepts as applied to herpetology. They recommend recognizing allopatric populations as species so long as (1) the lineage can be discretely diagnosed either anatomically or biochemically (i.e., demonstration that the independent lineages are on their own, nonephemeral evolutionary trajectory), and (2) monophyly is maintained in the taxonomy for all taxa (see also Frost and Kluge 1994; Graybeal 1995).

Grismer (1999a) required discrete anatomical diagnosis for species recognition; in other words, there could be no overlap in at least one characteristic, and all insular populations that could be anatomically diagnosed were elevated to the status of species. He ignored all biochemical evidence. His rejection of molecular data is a problem because it precludes recognition of cryptic species as can be discovered using biochemical characters or behavior. For example, using anatomical characters only, Grismer could not recognize *Hyla chrysoscelis* and *H. versicolor* as separate species because they can only be separated based on calls, ploidy levels, and, usually, distribution (Hillis et al. 1986). Similarly, Grismer does not recognize divergent biochemical lineages on islands in the Sea of Cortés (chap. 8).

The anatomy-only operational criterion does not require history and may not reflect such. In and of itself this is not disturbing. We simply accept that mistakes will be made, and that these will be corrected if phylogenies require such (Frost and Hillis 1990). Although Grismer (1999a) claimed to apply the evolutionary species concept according to Wiley (1978, 1981) and Frost and Hillis (1990) to the herpetofauna on islands in the Sea of Cortés, he has not. History must be faithfully reflected in the taxonomy. Exceptions are not allowed.

Violation of this fundamental premise of monophyly is clear in Grismer's recognition of species of side-blotched lizards. Upton and Murphy (1997) and Grismer's student Hollingsworth (1999) showed that side-blotched lizards on Ángel de la Guarda, San Esteban, San Pedro Mártir, Santa Catalina, and Carmen and Danzante (and associated satellite islands) could be biochemically diagnosed (see chap. 8, fig. 8.12). They are genetically divergent. These insular lineages are basal in cladograms relative to those on landbridge islands. Thus, the insular lineages were isolated before those on landbridge islands. Consequently, older lineages on deep-water islands must be accorded species status in order to recognize Grismer's three landbridge island species *Uta lowei, U. encantadae,* and *U. tumidarostra* on Miramar (=El Muerto), Encantada (and Islotes Blancos), and Coloradito, respectively. Grismer's species cannot be recognized without all older insular populations being accorded equal, species-level status (Upton and Murphy 1997; chap. 8, this volume). Furthermore, to maintain Grismer's species, we must also describe most other lineages on landbridge islands. Consequently, we not only reject recognition of Grismer's three species of *Uta*, we recognize *U. antiqua* and list the two side-blotched lizard lineages as undescribed species in order to maintain recognition of *U. squamata* from Santa Catalina and *U. palmeri* on San Pedro Mártir. Species descriptions are in preparation.

Grismer's use of grades (extent of anatomical divergence, and not phylogeny) is not limited to side-blotched lizards. The western whiptail lizards, *Cnemidophorus tigris*, and their associated species have identical problems. Whereas Grismer (1999a,b) relegated *C. estebanensis* to synonymy with *C. tigris*, cladogenic relationships (fig. 8.13) either require its recognition or the synonymization of *C. martyris* and *C. canus* with *C. tigris*. Thus, we recognize *C. estebanensis*. Grismer does not list the chuckwalla on Danzante as being *Sauromalus slevini*, although the DNA sequence data unambiguously demonstrate that it is a member of this lineage (app. 8.1). We include it in this taxon as being on Danzante.

We are also concerned about nomenclatorial stability. Unfortunately, for most species groups there is no phylogenetic hypothesis upon which to base the validity of Grismer's alpha taxonomic changes. However, patterns of relationships repeat because of similar histories, as evidenced by cladograms based on DNA sequence data (chaps. 8, 12). Therefore, we predict that other inappropriate changes have been made in more speciose groups, such as leaf-toed geckos (*Phyllodactylus*). Our checklist maintains previously recognized species on deep-water islands pending phylogenetic studies. In doing so, we hope to stabilize the nomenclature until the required evaluations are completed.

Based on the phylogenetic work of Kluge (1993), we have recognized the genus *Charina*, as opposed to *Lichanura*, for rosy boas. Following the International Code for Zoological Nomenclature (ICZN), Article 82.1, we use the name *Sauromalus obesus*, rather than *S. ater*, for the peninsular and landbridge island chuckwallas pending the outcome of a petition filed with the ICZN to conserve the former, more commonly used name (Montanucci et al. in press). We also recognize the name *Phyllodactylus partidus*, rather than *P. partitus*, following Article 24.2.3 of the Code. The generic name *Sator* has been retained due to extensive DNA sequence evidence that its inclusion in the genus *Sceloporus* results in a paraphyletic taxonomy (chap. 8). The inclusion of *Sator* into *Sceloporus* would also necessitate the synonymization of *Petrosaurus* and *Urosaurus* into *Sceloporus*. (The genus *Sceloporus* may remain paraphyletic with respect to some tropical Mexican species. A taxonomic correction of this problem is beyond the scope of this checklist.)

Some other taxonomic differences relate to a misunderstanding of the utility of molecular allozyme data. As noted in chapter 8, Grismer has ignored the significance of overall genetic similarity, which can provide indirect evidence of the absence of gene flow, although precise values cannot be used as arbitrators for recognizing species. Low levels of similarity (high divergence) translate to fixed allelic differences among populations. In turn, fixed allelic differences provide refutable hypotheses that gene flow is not occurring, at least within the framework of Mendelian genetics. Accordingly, some of our taxonomy is based on the largely unpublished character allozyme data summarized in Murphy (1983a) as well as our more recent unpublished data. This is especially evident in the alpha taxonomy of leaf-toed geckos (*Phyllodactylus*) and desert spiny lizards (*Sceloporus zosteromus* group), as was the taxonomy of Murphy (1983a) and Murphy and Ottley (1984). For example, in leaf-toed geckos, genus *Phyllodactylus*, Murphy's (1983a) data indicated that 7–9 (actually 8) of the 27 surveyed loci exhibited fixed allelic differences for *P. xanti* in the Cape Region and those north of there. Because the populations were not interbreeding, Murphy (1983a) recognized *P. nocticolus* as the northern species and *P. xanti* in the Cape Region. In contrast, Grismer (1994), based on similarity of scale characters (Dixon 1964), subsequently concluded that *P. xanti* occurred throughout the Peninsular Ranges from the Cape Region to San Gorgonio Pass. Similarly, in spiny lizards, *Sceloporus*, our allozyme data have no heterozygotes at five fixed loci in sympatric populations of *Sceloporus zosteromus* in the Cape Region, and *S. monserratensis* which occurs from the Cape Region northward to the mid-peninsula. Our checklist reflects these data.

When possible, our taxonomy faithfully reflects genealogical hypotheses. Where explicit phylogenetic hypotheses are not available, we have retained the species-level taxonomy given in Murphy and Ottley (1984). Whereas the previously noted differences center on the application of phylogenetic theory to taxonomy, and Grismer's summary on rejection of biochemical data, other issues deal with less precise aesthetics, or rather individual ideas about what alpha taxonomy should reflect to maximize its utility.

Populations on islands in the Sea of Cortés are the natural equivalents to bottles of *Drosophila*. Some populations have been altered by human intervention, and gene flow may occur via overwater colonization, as gene flow occurs among escaped *Drosophila*. The selection pressures vary among populations depending on effective population size, climate, and other variables. Bottlenecks in population size may accelerate adaptive change and quickly fix mutations in only a few generations (Waddington 1961). After all, these attributes are responsible for the considerable attention paid to the region. We believe that it is no more desirable to name all 370+ herpetological lineages distributed on 68+ islands in the Sea of Cortés as species than it is to name all of the estimated 180,000 strains of *Drosophila melanogaster* (fruit fly diversity estimate from Kathy Matthews, pers. commun., 2000 Bloomington *Drosophila* Stock Center, Indiana University).

If all insular herpetological lineages were recognized as species (undoubtedly, all can be diagnosed using microsatellites), then the alpha taxonomy will not reflect genealogical relationships, and new hierarchical levels will be required, such as recognition of new genera or subgenera (formalized species groups). We find this unappealing. This problem is particularly acute for ephemeral lineages on landbridge islands, both in the Sea of Cortés and elsewhere. Landbridge islands were frequently connected to a larger faunal source (Emslie 1998). Oxygen isotope records in marine sediments have documented at least 21 glacial cycles during the past 2.3 million years (van Donk 1976). These were rapid (10^4-10^5 years for the last interglacial; Cronin 1987) and glaciation occurred for 94% of the Pleistocene (Van Devender and Burgess 1985). Interglacial events, including our current condition, were short in duration. Apparently such events occurred without genetic consequences, at least following the last glacial event (Murphy 1983a,b; chap. 8, this volume) (i.e., unencumbered interbreeding continued after reunification). Therefore, we believe alpha taxonomic species names should be maintained for peninsular and insular populations of diagnosable lineages on landbridge islands. The only exceptions might include (1) preventing a paraphyletic taxonomy, and (2) demonstration of perpetual isolation (e.g., that glacial cycles have ceased). No data suggest that glacial cycles have stopped. All predict continuance. Therefore, we do not recognize any endemic species on landbridge islands that have a demonstrable recent genetic association with nearby peninsular regions. We do recognize endemic species on Carmen and Danzante because multiple cladograms suggest that they are not landbridge islands (chap. 8), as originally indicated for Carmen by Soulé and Sloan (1966).

Although we do not recognize endemic ephemeral species on landbridge islands, we recognize that anatomical divergence can occur very quickly, particularly on small islands. Subspecies names seem appropriate for such ephemeral lineages for ease of transferring information within a Linnean alpha taxonomic system. Frost and Hillis (1990, p. 94) stated that the "use of the subspecies category would necessitate the hypothesis that the subspecies *will* be subsumed into a larger species in the future, not *could* potentially be subsumed; they would necessarily be seen as temporary isolated parts of the larger species" (their emphasis). Given the weight of evidence, we must assume that landbridge islands in the Sea of Cortés are ephemeral and, thus, that peripheral isolates on these islands will be subsumed in the future. We agree with the caveat that monophyly of all lineages be maintained, even for subspecies. Therefore, many insular subspecies will defy recognition, including Grismer's three species of side-blotched lizards, unless all older populations are accorded formal alpha taxonomic recognition, at least at the subspecific level. Age of isolation can be estimated using channel depths (Wilcox 1978; Upton and Murphy 1997) and, given certain unpalatable assumptions, by the extent of genetic differentiation.

Our checklist is given in two appendixes. Appendix 8.2 lists the herpetofauna on major islands in the Sea of Cortés—islands with a relatively large number of species, or with oceanic (non-landbridge, noncontinental) origins. Appendix 8.3 provides additional data for relatively minor islands; these are landbridge islands either of the mainland or larger adjacent islands. Our tables also give the documented or presumed origins of the insular faunas, estimates of the percentage of overwater colonists, and levels of endemism calculated from the tables, which is dependent on our taxonomy. We are certain that levels of endemism on the deep-water islands will increase as knowledge accumulates, both anatomically and biochemically.

References

Cronin, T. 1987. Quaternary sea-level studies n the eastern United States of America: a methodological perspective. In: M.J. Tooley and I Shennan (eds.), *Sea-Level Changes*. Institute of British Geographers Special Publication 20. Basil Blackwell, Oxford; pp. 225–248.

Dixon, J.R. 1964. The systematics and distribution of lizards of the genus *Phyllodactylus* in North and Central America. New Mexico State University Research Center Scientific Bulletin 64, Albuquerque.

Dixon, J.R. 1966. Speciation and systematics of the gekkonid lizard genus *Phyllodactylus* of the islands of the Gulf of California. *Proceedings of the California Academy of Sciences* **33**:415–452.

Emslie S.D. 1998. Avian community, climate, and sea level changes in the Plio-Pleistocene of the Florida Peninsula. *Ornithological Monographs 50*.

Frost, D.R. and Hillis, D.M. 1990. Species in concept and practice: herpetological applications. *Herpetologica* **46**:87–104.

Frost, D.R., and Kluge, A.G. 1994. A consideration of epistemology in systematic biology, with special reference to species. *Cladistics* **10**:259–294.

Graybeal, A. 1995. Naming species. *Systematic Biology* **44**:237–250.

Grismer, L.L. 1994. The evolutionary and ecological biogeography of the herpetofauna of Baja California and the Sea of Cortes, Mexico. Dissertation, Loma Linda University, Loma Linda, CA.

Grismer, L.L. 1999a. An evolutionary classification of reptiles on islands in the Gulf of California, México. *Herpetologica* **55**:446–469.

Grismer, L.L. 1999b. Checklist of amphibians and reptiles on islands in the Gulf of California. *Bulletin of the Southern California Academy of Sciences* **98**:45–56.

Hennig, W. 1966. *Phylogenetic Systematics*. University of Illinois Press, Urbana.

Hillis, D.M., Collins, J.T. and Bogart, J.P. 1986. Distribution of diploid and tetraploid species of gray tree frogs (*Hyla chrysoscelis* and *Hyla versicolor*) in Kansas. *The American Midland Naturalist* **117**:214–217.

Hollingsworth, B.D. 1999. The molecular systematics of the side-blotched lizards (Iguania: Phrynosomatidae: *Uta*). Ph.D. dissertation, Loma Linda University, Loma Linda, CA.

Hull, D.L. 1964. Consistency and monophyly. *Systematic Zoology* **36**:381–386.

Kluge, A.G. 1993. *Calabaria* and the phylogeny of erycine snakes. *Zoological Journal of the Linnean Society* **107**:293–351.

Montanucci, R.R., Smith, H.M., Adler, K., Auth, D.L., Axtell, R.W., Case, T.J., Chiszar, D., Collins, J.T., Conant, R., Murphy, R.W., Petren, K. and Stebbins, R.C. 2001. *Sauromalus obesus* (Baird) 1858 (Reptilia, Lacertilia): proposed precedence of the specific name over that of *Sauromalus ater* Duméril, 1856. *Bulletin of Zoological Nomenclature* **58**:37–40.

Murphy, R.W. 1983a. Paleobiogeography and genetic differentiation of the Baja California herpetofauna. *Occasional Papers of the California Academy of Sciences* **137**:1–48.

Murphy, R.W. 1983b. The reptiles: origins and evolution. In: T.J. Case and M.L. Cody (eds), *Island Biogeography in the Sea of Cortéz*. University of California Press, Berkeley; pp. 130–158.

Murphy, R.W. 1983c. A distributional checklist of the amphibians and reptiles on the islands in the Sea of Cortez. In: T.J. Case and M.L. Cody (eds), *Island Biogeography in the Sea of Cortez*. University of California Press, Berkeley; pp. 429–437.

Murphy, R.W. and Ottley, J.R. 1984. Distribution of amphibians and reptiles on Islands in the Gulf of California. *Annals of Carnegie Museum* **53**:207–230.

Soulé, M. and Sloan, A.J. 1966. Biogeography and distribution of the reptiles and amphibians on islands in the Gulf of California, Mexico. *Transactions of the San Diego Society of Natural History* **14**:137–156.

Upton, D.E. and Murphy, R.W. 1997. Phylogeny of the side-blotched lizards (Phrynosomatidae: *Uta*) based on mtDNA sequences: Support for a midpeninsular seaway in Baja California. *Molecular Phylogenetics and Evolution* **8**:104–113.

Van Devender, T.R., and Burgess, T.L. 1985. Late Pleistocene woodlands in the Bolson de Mapimi: a refugium for Chihuahuan Desert biota? *Quaternary Research* **24**:346–353.

van Donk, J. 1976. O18 record of the Atlantic Ocean for the entire Pleistocene Epoch. *Geological Society of America Memoirs* **145**:147–163.

Waddington, C.H. 1961. Genetic assimilation. *Advances in Genetics* **10**:257–293.

Wilcox, B.A. 1978. Supersaturated island faunas: a species-age relationship for lizards on post-Pleistocene land-bridge islands. *Science* **199**:996–998.

Wiley, E.O. 1978. The evolutionary species concept reconsidered. *Systematic Zoology* **27**:17–26.

Wiley, E.O. 1981. Convex groups and consistent classifications. *Systematic Botany* **6**:346–358.

Wiley, E.O. 1981. *Phylogenetics: The Theory and Practice of Phylogenetic Systematics*. John Wiley and Sons, New York.

Appendix 8.3

Distribution of Nonavian Reptiles and Amphibians on Major Islands in the Sea of Cortés

Species	Tiburón	Mejía	Ángel de la Guarda	Pond	Partida Norte	La Rasa	Salsipuedes	San Lorenzo Norte	San Lorenzo Sur	San Esteban	San Pedro Mártir	San Pedro Nolasco	Tortuga	San Marcos	Coronados	Carmen	Danzante	Monserrat	Santa Catalina	Santa Cruz	San Diego	San José	San Francisco	Espíritu Santo-Partida Sur	Cerralvo
Amphibia Anura																									
Family Bufonidae																									
Bufo punctatus	N																							C	C
Family Pelobatidae																									
Scaphiopus couchii	N																							C	C
Chelonia																									
Family Emydidae																									
Gopherus agassizii	N																								
Reptilia, Squamata																									
Family Iguanidae																									
Subfamily Crotaphytinae																									
Crotaphytus dickersonae	N																								
Crotaphytus insularis		E																							
Gambelia wislizenii	N																								
Subfamily Iguaninae																									
Ctenosaura conspicuosa										M															
Ctenosaura hemilopha														C		C	C	C				C		C	C
Ctenosaura nolascensis												S													
Dipsosaurus catalinensis																			E						
Dipsosaurus dorsalis			C											C							C	C	C	C	C
Sauromalus obesus	N							R-1	R-1												C	C	C	C	
Sauromalus hispidus		R	R												R	R	R						C		
Sauromalus klauberi			R														R	R							
Sauromalus slevini																R	R	E		C					
Sauromalus varius										R															

Species	Tiburón	Mejía	Ángel de la Guarda	Pond	Partida Norte	La Rasa	Salsipedes	San Lorenzo Norte	San Lorenzo Sur	San Esteban	San Pedro Mártir	San Pedro Nolasco	Tortuga	San Marcos	Coronados	Carmen	Danzante	Monserrat	Santa Catalina	Santa Cruz	San Diego	San José	San Francisco	Espíritu Santo-Partida Sur	Cerralvo
Family Phrynosomatinae																									
Callisaurus draconoides	N		E											C	C	C						C	C	C	C
Callisaurus splendidus																	C								
Petrosaurus mearnsi																									
Petrosaurus repens																				M	M			C	S
Petrosaurus slevini		R	R																						
Petrosaurus thalassinus	N																								
Phrynosoma solare	N																								
Sator angustus																			E						
Sator grandaevus																		C				C			
Sceloporus clarkii	N																								
Sceloporus hunsakeri																						C	C	C	
Sceloporus lineatulus																									
Sceloporus magister	N											N													
Sceloporus monserratensis																		E							
Sceloporus orcutti														C	C	C						C	C		
Sceloporus zosteromus														C	C	C	C	C				C	C	C	
Urosaurus nigricaudus															C	C	C	C							
Urosaurus ornatus	N																								
Uta antiqua						R		R	R																
Uta nolascensis												S													
Uta palmeri											E														
Uta squamata																									
Uta stansburiana																			E						
Uta sp. #1	N	R	R	R	R	R							C	C	C	C	C	C				C	C	C	C

588

Species												
Uta sp. #2					E							
Family Eublepharidae												
Coleonyx switaki	N						C					
Coleonyx variegatus		C					C	C			C	C
Family Gekkonidae												
Phyllodactylus angelensis		R										
Phyllodactylus bugastrolepis		R										
Phyllodactylus homolepidurus		R						E				
Phyllodactylus nocticolus	C			C		N	C	C			C	C
Phyllodactylus partidus		N		C				C				
Phyllodactylus santacruzensis				C					E			
Phyllodactylus tinklei		E		C								
Phyllodactylus unctus											C	C
Family Teiidae												
***hyperythrus* group**												
Cnemidophorus carmenensis							E					
Cnemidophorus ceralbensis								C				
Cnemidophorus hyperythrus						S	C		E			
***tigris* group**												
Cnemidophorus bacatus						S						
Cnemidophorus canus				M	M	M						
Cnemidophorus catalinensis		R										
Cnemidophorus dickersonae		R										
Cnemidophorus estebanensis		R				S			E			
Cnemidophorus martyris						S						
Cnemidophorus tigris	N						C	C			C	C
Snakes (serpentes)												
Family Leptotyphlopidae												
Leptotyphlops humilis							C	C				C
Family Boidae												
Charina trivirgata	N	C	C				C					C

Species	Tiburón	Mejía	Ángel de la Guarda	Pond	Partida Norte	La Raza	Salsipuedes	San Lorenzo Norte	San Lorenzo Sur	San Esteban	San Pedro Mártir	San Pedro Nolasco	Tortuga	San Marcos
Family Colubridae														
Bogertophis rosaliae	N													C
Chilomeniscus cinctus	N													
Chilomeniscus savagei														
Chilomeniscus stramineus														
Eridiphas slevini	N	C	C											
Hypsiglena torquata					N/C			C	C	N/C	N/C		C	C
Lampropeltis catalinensis														C
Lampropeltis getula			C				C	C	C	N/C	N/C	N	C	
Masticophis aurigulus														
Masticophis bilineatus	N													C
Masticophis flagellum	N									S				
Masticophis slevini			C											C
Phyllorhynchus decurtatus														
Pituophis catenifer	N													
Rhinocheilus etheridgei														C
Salvadora hexalepis	N													
Sonora semiannulata														
Tantilla planiceps	N													C
Trimorphodon biscutatus														C

	1	2	3	4	5	6	7	8	9	10	11	12	13	14	15	16	17	18	19	20	21	22	23
Family Elapidae																							
Micruroides euryxanthus	N																						
Family Viperidae																							
Crotalus angelensis		E																					
Crotalus atrox	N									N													
Crotalus catalinensis																		E					
Crotalus cerastes	N												C		C						C		
Crotalus enyo													C	C								C	C
Crotalus estebanensis									S														
Crotalus lorenzoensis									E														
Crotalus mitchellii	N					C									C	C					C	C	C
Crotalus molossus			C	C																			
Crotalus ruber	N												C		C	C					C		
Crotalus tigris																							
Crotalus tortugensis												S							S				
Crotalus sp																							
Totals	29	6	15	5	4	2	5	7	9	4	6	5	22	15	16	16	13	10	6	3	22	10	20
% Native endemic species	0	30	53	80	50	100	40	57	67	50	50	20	0	0	13	6	0	80	50	33	0	0	20
% Species with peninsular origin	3	100	100	75 (?)	100	100	80	71	56	25	0	80	100	100	100	100	100	100	60	67	100	100	90

C, on the peninsula of Baja California; E, of peninsular origin and endemic to one island; R, or peninsular origin and endemic to multiple islands; N, mainland Mexico taxon and origin; S, mainland Mexico origin and endemic to one island; M, mainland origin and endemic to multiple islands; I, probably introduced by Seri Indians.

Appendix 8.4

Distribution of Nonavian Reptiles on Minor Islands in the Sea of Cortés

Species	Sonora and Sinaloa					N. Baja			Midriff			Bahía de los Angeles						Bahía Concepción			Baja California Sur					Cape Region		
	Alcatraz	Cholludo	Dátil	Patos	San Ignacio Farallon	Encantada Grande	Miramar	Willard	Cardonosa Este	Granito	Roca Lobos	Cabeza de Caballo	Flecha	La Ventana	Mitlán	Piojo	Coronado	Santa Ines (S)	Cayo	El Coyote	San Ildefonso	Pardo	Rocas San Cosme	El Pardito	Las Animas	Ballena	Gallina	Gallo
Chelonia																												
Family Emydidae																												
Gopherus agassizii			?																									
Reptilia, Squamata																												
Family Iguanidae																												
Subfamily Iguaninae																												
Ctenosaura conspicuosa	M-I																											
Dipsosaurus dorsalis[a]	N					C																						
Sauromalus obesus	R-I							C																				
Sauromalus hispidus	R-I											R-I	R-I	R-I	R-I													
Sauromalus varius[b]										R	R-I																	
Subfamily Phrynosomatinae																												
Callisurus draconoides				N			C											C										
Petrosaurus mearnsi																	C											
Sceloporus hunsakeri																					C							
Sceloporus orcutti																		C			C							
Sceloporus zosteromus												C	C	C	C	C	C		C	C	C	C	C	C	C	C	C	C
Urosaurus nigricaudus[c]																										C	C	C
Uta stansburiana[b]	N			N	N	C	C	C	R	C	C	C	C	C	C	C	C		C	C	C	C	C	C	C	C	C	C
Uta sp. 1																												
Family Eublepharidae																												
Coleonyx variegatus																												

(continued)

	Sonora and Sinaloa					N. Baja			Midriff			Bahía de los Angeles						Bahía Concepción			Baja California Sur					Cape Region		
Species	Alcatraz	Cholludo	Dátil	Patos	San Ignacio Farallón	Encantada Grande	Miramar	Willard	Cardonosa Este	Granito	Roca Lobos	Cabeza de Caballo	Flecha	La Ventana	Mitlán	Piojo	Coronado	Santa Ines (S)	Cayo	El Coyote	San Ildefonso	Pardo	Rocas San Cosme	El Pardito	Las Animas	Ballena	Gallina	Gallo
Family Gekkonidae																												
Phyllodactylus nocticolus[d]	C	C	C				C							C		C	C		C	C	C	C		C	C		C	C
Phyllodactylus partidus									R																			
Phyllodactylus unctus					C																							
Phyllodactylus tuberculosus																										C	C	C
Family Teiidae																												
Cnemidophorus dickersonae									R																			
Cnemidophorus tigris																	C						C					
Snakes (Serpentes)																												
Family Colubridae																												
Hypsiglena torquata																	C				C							
Masticophis flagellum			N				C					C				C	C				C	C						
Trimorphodon biscutatus							C																					
Family Viperidae																												
Crotalus atrox			N									C										C						
Crotalus enyo																												
Crotalus mitchellii																												
Totals	5	2	4–5	2	1	3	5	2	3	2	2	3	2	3	2	4	7	2	2	3	4	4	3	2	2	5	2	5

C, on the peninsula of Baja California; E, of peninsular origin and endemic to one island; R, or peninsular origin and endemic to multiple islands; N, mainland Mexico taxon and origin; S, mainland Mexico origin and endemic to one island; M, mainland origin and endemic to multiple islands; I, probably introduced by Seri Indians.

[a] Also occurs on Isla Santiago.
[b] Also occurs on el Requesón, Gaviota, Islitas, San Damian, and Tijeras.
[c] Also occurs on Islas Bota, Cerraja, Coloradito, Encantada, Lagartija, Las Galeras (two islands), Pata, and Islotes Blancos.
[d] Also occurs on Isla Moscas.

Appendix 10.1

Bird Census Data from Ten Mainland and Peninsular Sites in Similar Sonoran Desert Habitat

Species	Organpipe (32°11'N, 112°50'W)	Puertecitos (30°30'N, 114°47'W)	Rcho. Arenoso (30°00'N, 115°12'W)	Cataviña (29°48'N, 114°45'W)	El Arco (28°12'N, 113°12'W)	Casas Blancas (28°21'N, 113°11'W)	San Juan Rd. (24°31'N, 110°46'W)	La Paz (24°0'N, 110°35'W)	Cd. Guzmán (19°45'N, 103°20'W)	Huatabampo (25°58'N, 109°30'W)	Hermosillo (29°17'N, 111°3'W)	Tucson (32°19'N, 110°47'W)
1. Banded Quail *Philortyx fasciatus*									X			
2. California Quail *Callipepla californica*		X	X	X	X	X	X					
3. Gambel's Quail *Callipepla gambelii*	X	X								X	X	X
4. White-winged Dove *Zenaida asiatica*	X	X	X	X	X	X	X	X	X	X	X	X
5. Mourning Dove *Zenaida macroura*	X	X	X	X	X	X		X	X	X	X	X
6. Inca Dove *Columbina inca*									X			
7. Common Ground Dove *Columbina passerina*									X	X	X	
8. Greater Roadrunner *Geococcyx californianus*	X	X		X		X			X	X	X	X
9. Poorwill *Phalaenoptilus nuttallii*	X			X	X	X	X				X	
10. Lesser Nighthawk *Chordeiles acutipennis*		X		X						X		
11. White-throated Swift *Aeronautes saxatalis*	X	X	X	X			X	X			X	X
12. Black-chinned Hummingbird *Archilochus alexandri*	X	X										
13. Broad-billed Hummingbird *Cynanthus latirostris*									X			X

Species										
13. Costa's Hummingbird *Calypte costae*	x	x	x	x	x	x		x	x	x
14. Xantus' Hummingbird *Basilinna xantusii*					x	x		x		
15. Northern (Gilded) Flicker *Colaptes (chrysoides) cafer*	x	x	x	x	x	x		x	x	x
16. Gila Woodpecker *Melanerpes uropygialis*	x	x	x	x	x	x		x	x	x
17. Ladderbacked Woodpecker *Picoides scalaris*	x	x	x	x	x	x	x	x	x	x
18. Ash-throated Flycatcher *Myiarchus cinerascens*	x	x	x	x	x	x	x	x	x	x
19. Purple Martin *Progne subis*			x							x
20. Violet-green Swallow *Tachycineta thalassina*		x		x	x		x			
21. Western Scrub-jay *Aphelocoma californica*				x	x	x	x	x		
22. Common Raven *Corvus corax*	x	x	x	x	x	x	x	x	x	x
23. Verdin *Auriparus flaviceps*	x	x	x	x	x	x	x	x	x	x
24. Cactus Wren *Campylorhynchus brunneicapillus*	x	x	x	x	x	x	x	x	x	x
25. Rufous-naped Wren *Campylorhynchus rufinucha*								x		
26. Canyon Wren *Catherpes mexicanus*	x		x	x	x	x	x		x	
27. Rock Wren *Salpinctes obsoletus*	x	x		x	x	x				
28. Bewick Wren *Thryomanes bewickii*		x	x	x		x				
29. Northern Mockingbird *Mimus polyglottos*	x	x	x	x	x	x	x	x	x	x

(continued)

Species	Organpipe (32°11'N, 112°50'W)	Puertecitos (30°30'N, 114°47'W)	Rcho. Arenoso (30°01'N, 115°12'W)	Cataviña (29°48'N, 114°45'W)	El Arco (28°12'N, 113°12'W)	Casas Blancas (28°21'N, 113°11'W)	San Juan Rd. (24°31'N, 110°46'W)	La Paz (24°0'N, 110°35'W)	Cd. Guzmán (19°46'N, 103°20'W)	Huatabampo (25°58'N, 109°30'W)	Hermosillo (29°17'N, 111°3'W)	Tucson (32°19'N, 110°47'W)
30. Bendire's Thrasher *Toxostoma bendirei*		X										X
31. Curve-billed Thrasher *Toxostoma curvirostre*	X									X	X	X
32. Gray Thrasher *Toxostoma cinereum*			X	X	X	X	X	X				
33. Black-tailed Gnatcatcher *Polioptila melanura*	X	X		X	X	X				X	X	X
34. California Gnatcatcher *Polioptila californica*			X	X	X	X	X	X				
35. Black-capped Gnatcatcher *Polioptila nigriceps*									X			
36. Phainopepla *Phainopepla nitens*	X	X	X	X	X	X	X	X				X
37. Loggerhead Shrike *Lanius ludovicianus*	X	X	X	X	X	X						X
38. Lucy's Warbler *Vermivora luciae*	X											X
39. Hooded Oriole *Icterus cucculatus*	X	X	X	X	X	X		X			X	X
40. Scott's Oriole *Icterus parisorum*	X	X	X	X		X						X
41. Streak-backed Oriole *Icterus pustulatus*									X		X	

Species												
42. Brown-headed Cowbird *Molothrus ater*		X										
43. Red Cardinal *Cardinalis cardinalis*	X			X	X	X		X		X	X	X
44. Pyrrhuloxia *Cardinalis sinuata*	X				X		X			X	X	X
45. House Finch *Carpodacus mexicanus*	X	X	X	X	X	X		X		X	X	X
46. Lesser Goldfinch *Carduelis psaltria*	X	X	X					X				X
47. Brown Towhee *Pipilo fuscus*	X	X	X	X	X	X		X		X	X	X
48. Rufous-winged Sparrow *Aimophila carpalis*										X		
49. Rusty Sparrow *Aimophila rufescens*								X				
50. Stripe-headed Sparrow *Aimophila ruficauda*									X			
51. Black-throated Sparrow *Amphispiza bilineata*	X	X	X	X	X		X				X	X
Total species censused	31	28	26	30	25	29	23	22	21	21	25	29

Appendix 10.2

Analysis of Bird Species Counts in Twelve Mainland and Peninsular Sites

All sites (see app. 10.1) are of Sonoran Desert vegetation, except for Cd. Guzmán (Jal), which is a succulent desert scrub of similar physiognomy. The figures along the diagonal are the numbers of species, S, recorded at each site over three to six visits during the breeding season. The figures below the diagonal are the numbers of species in common, C, at each pair of sites, and the figures above the diagonal are measures of species turnover between sites, or γ-diversity, using the formula $100 [1 - C*(S_1 + S_2)/2\ S_1\ S_2)]$.

	Organ-pipe	Puerte-citos	Rcho. Arenoso	Cataviñá	El Arco	Casas Blancas	San Juan Rd	La Paz	Cd. Guzmán	Huatabampo	Hermosillo	Tuscon
Organpipe (OP)	31	10.7	23.1	23.3	16.0	17.2	21.7	18.2	47.6	14.3	8.0	6.9
Puertecitos (PT)	25	28	19.2	16.7	24.0	17.2	30.4	18.2	52.4	19.0	24.0	17.2
Rcho. Arenoso (RA)	20	21	26	20.0	20.0	24.1	30.4	13.6	57.1	38.1	36.0	31.0
Cataviñá (CA)	23	25	24	30	8.0	6.9	17.4	13.6	52.4	9.5	20.0	27.6
El Arco (EL)	21	19	20	23	25	13.8	13.0	9.1	57.1	28.6	24.0	34.5
Casas Blancas (CB)	24	24	22	27	25	29	13.0	9.1	57.1	28.6	20.0	27.6
San Juan Rd (SJ)	18	16	16	19	20	20	23	13.6	66.7	42.9	32.0	44.8
La Paz (LP)	18	18	19	19	20	20	19	22	57.1	33.3	36.0	41.4
Cd. Guzmán (CG)	11	10	9	10	9	9	7	9	21	47.6	52.0	62.0
Huatabampo (HU)	18	17	13	19	15	15	12	14	11	21	20.0	37.9
Hermosillo (HE)	23	19	16	20	19	20	17	16	12	20	25	24.1
Tuscon (TU)	27	24	20	21	19	21	16	17	11	18	22	29

Appendix 10.3

Factors Affecting γ-Diversity (Species Turnover) among the Birds of Desert Sites in Southwestern North America

The sites compared (first column) are those of appendix 10.2. Species turnover is treated as the dependent variable in simple and multiple regression analyses with the remaining variables independent, and all significant contributors to the regression (see text).

Intersite Comparison	Species Turnover	Intersite Distance (km)	Proportion Overwater Separation	Size (km) of Water Gap	% Difference in Summer Rainfall
OP-PT	10.71	376	0.62	233	22
OP-AR	23.08	349	0.38	133	36
OP-CA	23.33	334	0.67	224	32
OP-EL	16.00	432	0.41	177	22
OP-CB	17.24	424	0.41	174	19
OP-SJ	21.74	1312	0.48	630	0
OP-LP	18.18	1456	0.53	722	0
OP-CD	47.62	1624	0.00	0	23
OP-HU	14.29	1040	0.00	0	16
OP-HE	8.00	568	0.00	0	19
OP-TU	6.90	288	0.00	0	12
PT-AR	19.23	107	0.00	0	14
PT-CA	16.67	117	0.00	0	10
PT-EL	24.00	296	0.00	0	0
PT-CB	17.24	290	0.00	0	3
PT-SJ	30.44	1264	0.13	164	22
PT-LP	18.18	1368	0.11	150	22

Intersite Comparison	Species Turnover	Intersite Distance (km)	Proportion Overwater Separation	Size (km) of Water Gap	% Difference in Summer Rainfall
PT-CD	52.38	1605	0.43	690	45
PT-HU	19.05	1072	0.51	547	38
PT-HE	24.00	616	0.55	339	41
PT-TU	17.24	648	0.35	227	34
AR-CA	20.00	77	0.00	0	4
AR-EL	20.00	282	0.00	0	14
AR-CB	24.14	301	0.00	0	17
AR-SJ	30.44	693	0.00	0	36
AR-LP	13.64	795	0.00	0	36
AR-CD	57.14	1619	0.46	745	59
AR-HU	38.10	466	0.52	242	52
AR-HE	36.00	403	0.37	149	55
AR-TU	31.03	494	0.32	158	48
CA-EL	8.00	203	0.00	0	10
CA-CB	6.90	203	0.00	0	13
CA-SJ	17.39	640	0.00	0	32
CA-LP	13.64	733	0.00	0	32
CA-CD	52.38	1538	0.57	876	55
CA-HU	9.52	344	0.69	237	48
CA-HE	20.00	330	0.40	132	51
CA-TU	27.59	461	0.32	147	44
EL-CB	13.79	32	0.00	0	3
EL-SJ	13.04	437	0.00	0	22
EL-LP	9.09	523	0.00	0	22
EL-CD	57.14	1352	0.55	744	45
EL-HU	28.57	218	0.72	157	38
EL-HE	24.00	222	0.50	111	41
EL-TU	34.48	499	0.27	135	34
CB-SJ	13.04	432	0.00	0	19
CB-LP	9.09	514	0.00	0	19
CB-CD	57.14	1328	0.72	956	42
CB-HU	28.57	198	0.83	165	35
CB-HE	20.00	208	0.51	106	38
CB-TU	27.59	499	0.27	135	31
SJ-LP	13.64	120	0.00	0	0
SJ-CD	66.67	931	0.86	801	23
SJ-HU	42.86	432	0.76	328	16
SJ-HE	32.00	816	0.62	506	19
SJ-TU	44.83	1352	0.38	514	12
LP-CD	57.14	843	0.63	531	23
LP-HU	33.33	512	0.89	456	16
LP-HE	36.00	904	0.70	633	19
LP-TU	41.38	1440	0.45	648	12
CD-TU	47.62	1173	0.00	0	7
CD-HE	52.00	1280	0.00	0	4
CD-TU	62.07	1547	0.00	0	11
HU-HE	20.00	488	0.00	0	3
HU-TU	37.93	984	0.00	0	4
HE-TU	24.14	528	0.00	0	7

Appendix 10.4

Bird Census Data from Eleven Mainland and Peninsular Sites in Similar Thorn-Scrub Habitats

Nine mainland sites are listed from north to south; they are similar in vegetation structure to the last two sites (in the cape region of Baja California), although species composition in the cape thorn scrub is much more similar in bird species composition to adjacent Sonora Desert sites.

Species	Chiricahua Mts., AZ (31°55'N, 109°08'W)	Nácori Chico, SON (29°44'N, 109°24'W)	Moctezuma SON (29°39'N, 109°33'W)	Ures SON (29°30'N, 110°16'W)	Mocuzari SON (27°14'N, 109°13'W)	Rio Cuchujaqui SON (27°03'N, 108°34'W)	Mazatlan SIN (23°24'N, 106°19'W)	Tonoya SIN (20°24'N, 103°33'W)	Estepec JAL (20°16'N, 103°33'W)	San Bártolo BCS (23°43'N, 109°47'W)	San Matías (23°32'N, 110°02'W)
1. Gambel's Quail *Callipepla gambelii*	X										
2. California Quail *Callipepla californica*											X
3. Elegant Quail *Callipepla douglasii*		X		X	X	X	X				
4. Banded Quail *Philortyx fasciatus*									X		
5. White-winged Dove *Zenaida asiatica*		X	X	X	X		X	X	X	X	X
6. Mourning Dove *Zenaida macroura*	X	X	X	X	X	X		X	X	X	X
7. White-tipped Dove *Leptotila verrauxii*					X	X					
8. Common Ground Dove *Columbina passerina*		X		X	X	X	X	X	X	X	X
9. Inca Dove *Columbina inca*			X					X	X		
10. Blue-rumped Parrotlet *Forpus cyanopygius*					X						
11. White-fronted Parrot *Amazona albifrons*					X	X	X				
12. Yellow-billed Cuckoo *Coccyzx americanus*	X	X		X	X	X				X	
13. Mangrove Cuckoo *Coccyzx minor*							X				

(continued)

Species	Chiricahua Mts., AZ (31°55'N, 109°08'W)	Nácori Chico, SON (29°44'N, 109°24'W)	Moctezuma SON (29°39'N, 109°33'W)	Ures SON (29°30'N, 110°16'W)	Mocuzari SON (27°14'N, 109°13'W)	Rio Cuchujaqui SON (27°03'N, 108°34'W)	Mazatlan SIN (23°24'N, 106°19'W)	Tonoya SIN (20°24'N, 103°33'W)	Estepec JAL (20°16'N, 103°33'W)	San Bártolo BCS (23°43'N, 109°47'W)	San Matías (23°32'N, 110°02'W)
14. Groove-billed Ani *Crotophaga sulcirostris*	X	X									
15. Greater Roadrunner *Geococcyx californianus*		X	X	X	X	X	X				X
16. White-throated Swift *Aeronautes saxatalis*	X				X	X		X			
17. Black-chinned Hummingbird *Archilochus alexandri*	X										
18. Broad-billed Hummingbird *Cynanthus latirostris*		X	X	X	X	X	X	X	X		
19. Violet-crowned Hummingbird *Amazilia violiceps*		X			X						
20. Plain-capped Starthroat *Heliomaster constantii*					X	X					
21. Costa's Hummingbird *Calypte costae*		X		X	X		X			X	X
22. Xantus' Hummingbird *Basilinna xantusii*										X	X
23. Lucifer Hummingbird *Calothorax lucifer*								X	X		
24. Elegant Trogon *Trogon elegans*					X	X					
25. Russet-crowned Motmot *Momotus mexicanus*					X	X		X	X		
26. Ivory-billed Woodcreeper *Xiphorhynchus flavogaster*								X	X		

	1	2	3	4	5	6	7	8	9	10
27. Northern (Gilded) Flicker *Colaptes (chrysoides) cafer*	X	X	X	X	X		X		X	X
28. Gila Woodpecker *Melanerpes uropygialis*	X	X		X			X		X	X
29. Golden-cheeked Woodpecker *Centurus chrysogenys*				X	X		X			
30. Ladderbacked Woodpecker *Picoides scalaris*	X	X	X	X	X		X		X	X
31. Tropical Kingbird *Tyrannus melancholicus*			X							
32. Thick-billed Kingbird *Tyrannus crassirostris*		X	X	X			X		X	
33. Cassin's Kingbird *Tyrannus vociferans*	X		X							
34. Sulfur-bellied Flycatcher *Myiodynastes luteiventris*		X	X		X					
35. Ash-throated Flycatcher *Myiarchus cinerascens*	X	X	X	X	X		X		X	X
36. Nutting's Flycatcher *Myiarchus nuttingi*		X	X		X					
37. Common Raven *Corvus corax*	X	X	X		X		X		X	X
38. Mexican Crow *Corvus imparatus*			X		X					
39. Magpie Jay *Calocitta formosa*			X		X					
40. Purplish-backed Jay *Cyanocorax beecheii*								X		
41. Western Scrub-jay *Aphelocoma californica*									X	X
42. Verdin *Auriparus flaviceps*	X		X		X		X			X
43. Cactus Wren *Campylorhynchus brunneicapillus*	X		X		X		X		X	X

(continued)

Species	Chiricahua Mts., AZ (31°55'N, 109°08'W)	Nácori Chico, SON (29°44'N, 109°24'W)	Moctezuma SON (29°39'N, 109°33'W)	Ures SON (29°30'N, 110°16'W)	Mocuzari SON (27°14'N, 109°13'W)	Río Cuchujaqui SON (27°03'N, 108°34'W)	Mazatlán SIN (23°24'N, 106°19'W)	Tonoya SIN (20°24'N, 103°33'W)	Estepec JAL (20°16'N, 103°33'W)	San Bártolo BCS (23°43'N, 109°47'W)	San Matías (23°32'N, 110°02'W)
44. Canyon Wren *Catherpes mexicanus*					X	X					
45. Sinaloa Wren *Thryothorus sinaloa*							X				
46. Happy Wren *Thryothorus felix*					X	X	X	X	X		
47. Northern Mockingbird *Mimus polyglottos*		X	X	X				X	X		
48. Curve-billed Thrasher *Toxostoma curvirostre*		X	X	X	X		X	X	X		
49. Gray Thrasher *Toxostoma cinereum*										X	X
50. Crissal Thrasher *Toxostoma crissale*	X										
51. Bendire's Thrasher *Toxostoma bendirei*	X										
52. Black-tailed Gnatcatcher *Polioptila melanura*	X		X	X							
53. California Gnatcatcher *Polioptila californica*										X	
54. Black-capped Gnatcatcher *Polioptila nigriceps*		X			X		X	X	X		
55. Blue-gray gnatcatcher *Polioptila caerulea*										X	X
56. Phainopepla *Phainopepla nitens*		X		X							X

57. Loggerhead Shrike *Lanius ludovicianus*	X													
58. Bell's Vireo *Vireo belli*			X											
59. Solitary Vireo *Vireo solitarius*					X									
60. Lucy's Warbler *Vermivora luciae*	X													
61. Rufous-capped Warbler *Basileuterus rufifrons*							X							
62. Great-tailed Grackle *Cassidix mexicanus*									X					
63. Hooded Oriole *Icterus cucculatus*				X								X		
64. Streak-backed Oriole *Icterus pustulatus*							X			X		X		
65. Black-vented Oriole *Icterus wagleri*							X				X			
66. Brown-headed Cowbird *Molothrus ater*	X					X	X		X				X	
67. Red-eyed Cowbird *Molothrus aeneus*						X	X		X					
68. Red-winged Blackbird *Agelaius phoeniceus*						X	X			X				
69. Greyish Saltator *Saltator coerulescens*									X	X				
70. Red Cardinal *Cardinalis cardinalis*	X					X	X		X	X				X
71. Pyrrhuloxia *Cardinalis sinuatus*	X					X	X		X					X
72. Yellow Grosbeak *Pheuticus chrysopeplus*						X			X	X				
73. Blue Grosbeak *Guiraca caerulea*	X					X	X		X			X		

(continued)

Species	Chiricahua Mts., AZ (31°55'N, 109°08'W)	Nácori Chico, SON (29°44'N, 109°24'W)	Moctezuma SON (29°39'N, 109°33'W)	Ures SON (29°30'N, 110°16'W)	Mocuzari SON (27°14'N, 109°13'W)	Rio Cuchujaqui SON (27°03'N, 108°34'W)	Mazatlan SIN (23°24'N, 106°19'W)	Tonoya SIN (20°24'N, 103°33'W)	Estepec JAL (20°16'N, 103°33'W)	San Bártolo BCS (23°43'N, 109°47'W)	San Matías (23°32'N, 110°02'W)
74. Lazuli Bunting *Passerina amoena*	X										
75. Varied Bunting *Passerina versicolor*		X	X	X	X	X	X			X	X
76. White-collared Seedeater *Sporophila torqueola*							X				
77. Ruddy-breasted Seedeater *Sporophila minuta*								X			
78. Blue-black Grassquit *Volatinia jacarina*							X				
79. Brown Towhee *Pipilo fuscus*	X	X	X	X	X	X	X			X	X
80. Rufous-winged Sparrow *Aimophila carpalis*		X	X	X	X						
81. Stripe-headed Sparrow *Aimophila ruficauda*			X						X		
82. Black-chested Sparrow *Aimophila humeralis*								X			
83. Rusty-crowned Ground-sparrow *Melozone kieneri*									X		
84. Black-throated Sparrow *Amphispiza bilineata*	X									X	X
85. Lesser Goldfinch *Carduelis psaltria*									X		X
86. House Finch *Carpodacus mexicanus*	X	X		X						X	X
Total species censused	26	30	25	27	41	31	26	26	26	20	26

Appendix 10.5

Explanation of Island Numbers and References

In appendixes 10.6–10.10, island numbers correspond to the list shown in table A10.5-1. Numbered references within the appendices are to the following publications: 1. Banks (1963a); 2. Banks (1964); 3. Townsend (1923); 4. Grinnell (1928); 5. Brewster (1902); 6. Lamb (1924); 7. Banks (1969); 8. Van Rossem (1932); 9. Vaughn and Schwartz (1980); 10. Thayer and Bangs (1909); 11. Maillard (1923); 12. Boswall and Barrett (1978); 13. Henny and Anderson (1979); 14. Wauer (1978); 15. T. J. Case (unpublished data); 16. Van Rossem (1945b); 17. K. Wilde (pers. commun.); 18. George (1987a); 19. George (1987b). See chapter 10 for full references.

Table A10.5-1 Island Numbers

Ángel de la Guarda	1	Los Islotes	16	San Esteban	31
Animas Norte	2	Lobos	17	San Francisco	32
Animas Sur	3	Montserrat	18	San Ildefonso	33
Ballena	4	Montague	19	San Jorge	34
Cardonosa	5	Partida Norte	20	San José	35
Carmen	6	Partida Sur	21	San Lorenzo	36
Sta. Catalina	7	Patos	22	San Luis	37
Cerralvo	8	Pelícano	23	San Marcos	38
Coronado	9	Pichilingue	24	San Pedro Mártir	39
Danzante	10	Rasa	25	Salsipuedes	40
Encantada	11	Roca Partida	26	Tiburón	41
Estanque	12	Roca Rasa	27	Tortuga	42
Espíritu Santo	13	Roca Salsipuedes	28	Turners	43
Gallo	14	Santa Cruz	29		
Granito	15	San Diego	30		

Appendix 10.6

Distribution of Shoreline Species over Islands

Species listed in the table are those that breed within the Gulf of California, but breeding has not been verified, nor is it expected, on all islands on which the species have been recorded. Most records of American Oystercatcher likely represent breeding birds, but most shoreline species nest only on small islands or in specific habitats such as mangrove thickets or on sandy beaches. A majority of the tabulated records pertain to feeding activity of wide-ranging species whose nearest nesting sites are generally unknown.

Species		Island Number										
		1	2	3	4	5	6	7	8	9	10	11
Least Bittern	*Ixobrychus exilis*											
Great Blue Heron	*Ardea herodias*	XYZ	XY				XY	Y1	Y4	Y1	X1	Z
Great Egret	*Egretta alba*										9	
Snowy Egret	*Egretta thula*	X	Y			Y	XY			Y	X	
Reddish Egret	*Egretta rufescens*	11										Z
Green Heron	*Butorides virescens*							Y				
Black-crowned Night-heron	*Nycticorax nycticorax*						Z	Y				
Yellow-crowned Night-heron	*Nycticorax violaceus*							Y				
White Ibis	*Eudocimus albus*											
Clapper Rail	*Rallus longirostris*											
Wilson's Plover	*Charadrius wilsonia*						Y6		Y			
American Oystercatcher	*Haematopus palliatus*	YZ3	Y			Y	YZ4	Y3	YZ4			Z
Belted Kingfisher	*Ceryle alcyon*	X1					XY		Y			
Number of species		5	3	0	0	2	6	5	4	2	3	3

Species		Island Number										
		12	13	14	15	16	17	18	19	20	21	22
Least Bittern	*Ixobrychus exilis*						Z					
Great Blue Heron	*Ardea herodias*	Y	XYZ	Y	Y	Y		XY		Z	XY	Y
Great Egret	*Egretta alba*										X	
Snowy Egret	*Egretta thula*	Y	XY		Y	Y	Z			Y	X	
Reddish Egret	*Egretta rufescens*		YZ					Y				
Green Heron	*Butorides virescens*		X				Z					
Black-crowned Night-heron	*Nycticorax nycticorax*											
Yellow-crowned Night-heron	*Nycticorax violaceus*		Z									Z
White Ibis	*Eudocimus albus*											
Clapper Rail	*Rallus longirostris*		Z4				Z					
Wilson's Plover	*Charadrius wilsonia*		YZ			Y		Y				
American Oystercatcher	*Haematopus palliatus*	YZ	YZ4	Y		Y	Z	Y17	Z	YZ4		
Belted Kingfisher	*Ceryle alcyon*		Y					X		Y		
Number of species		3	9	2	2	4	5	5	1	4	3	2

Species		Island Number										
		23	24	25	26	27	28	29	30	31	32	33
Least Bittern	*Ixobrychus exilis*											
Great Blue Heron	*Ardea herodias*			Y1		Y		Y		Y	X	Y3
Great Egret	*Egretta alba*			12								
Snowy Egret	*Egretta thula*			Y	Y		Y			XY	XZ	Y
Reddish Egret	*Egretta rufescens*			Y								
Green Heron	*Butorides virescens*		Z									
Black-crowned Night-heron	*Nycticorax nycticorax*			Y								
Yellow-crowned Night-heron	*Nycticorax violaceus*	Z		Y								
White Ibis	*Eudocimus albus*											
Clapper Rail	*Rallus longirostris*											
Wilson's Plover	*Charadrius wilsonia*										Z	
American Oystercatcher	*Haematopus palliatus*	Z		XYZ	Y		Y	X		XYZ	XZ	YZ
Belted Kingfisher	*Ceryle alcyon*		Z	Y				Y		Y		Y1
Number of species		2	2	8	2	1	2	3	0	4	4	4

Species		\multicolumn{10}{c}{Island Number}	
		34	35	36	37	38	39	40	41	42	43	No. of islands
Least Bittern	*Ixobrychus exilis*											1
Great Blue Heron	*Ardea herodias*		YZ1	XY	Z	XY		Y	Y12	Y		30
Great Egret	*Egretta alba*		1									4
Snowy Egret	*Egretta thula*		YZ4			Y		Y	YZ3	Y		24
Reddish Egret	*Egretta rufescens*		YZ3		Z				Y			8
Green Heron	*Butorides virescens*		Z3									5
Black-crowned Night-heron	*Nycticorax nycticorax*		YZ3									4
Yellow-crowned Night-heron	*Nycticorax violaceus*		YZ		Z				Z		Z	9
White Ibis	*Eudocimus albus*		YZ4									1
Clapper Rail	*Rallus longirostris*		Z4									3
Wilson's Plover	*Charadrius wilsonia*		YZ						Y	Y		9
American Oystercatcher	*Haematopus palliatus*		YZ3	YZ		Y		Y	YZ			28
Belted Kingfisher	*Ceryle alcyon*		XY	Y		Y	Y			Y		16
Number of species		0	12	3	3	4	1	3	6	4	1	

Records are marked as X: MLC surveys; Y: EV surveys; Z: records of Adolfo Navarro et al., *Atlas de las Aves de Mexico* (in prep.).

Appendix 10.7

Raptors, Owls, and Goatsuckers

	Island Number											
Species	1	2	3	4	5	6	7	8	9	10	11	12
Resident species												
Turkey Vulture *Cathartes aura*[a]	XYZ					XY	Y9	XY	Y1	X		
Osprey *Pandion haliaetus*		XY		Y	Y	XY	Y1	XY	Y1	X1		Y
Bald Eagle *Haliaeetus leucocephalus*												
Harris' Hawk *Parabuteo unicinctus*[a]	Y							Y				
Red-tailed Hawk *Buteo jamaicensis*[a]	XY	XY				XY	Y9	XY	1,18	X9		
Golden Eagle *Aquila chrysaetos*[a]		Z										
Prairie Falcon *Falco mexicanus*	Y											
Peregrine Falcon *Falco peregrinus*	Y	Y			Y	Y	Y	Y	Y15	1		
American Kestrel *Falco sparverius*	X	Y			Y	XYZ	Y	XY				
Barn Owl *Tyto alba*												
Western Screech Owl *Otus asio*												
Great Horned Owl *Bubo virginianus*	XYZ	Y				17						
Elf Owl *Micrathene whitneyi*												
Burrowing Owl *Athene cunicularia*[b]	XY					Y		Y	1			
Lesser Nighthawk *Chordeiles acutipennis*						9		Y				
Poorwill *Phalaenoptilus nuttallii*[b]						9	9	Y1				
Nonbreeding species												
Northern Harrier *Circus cyaneus*												
Sharp-shinned hawk *Accipiter striatus*												
Cooper's Hawk *Accipiter cooperii*												
Zone-tailed Hawk *Buteo albonotatus*												
Swainson's Hawk *Buteo swainsoni*												
Long-eared Owl *Asio otus*	Y											
Resident species	8	6	1	1	3	9	6	8	5	4	0	1
Total no. of species	9	6	1	1	3	9	6	8	5	4	0	1

Records are marked as follows: X, MLC field data; Y, EV field data; Z, Adolfo Navarro et al., Atlas de las Aves de México (in prep.). Numbered references as listed in appendix 10.5.
[a]Larger raptor species, other than Osprey and Peregrine Falcon, likely nonbreeding on smaller islands; boldface indicates confirmed breeding.
[b]Some island records may be of wintering rather than breeding birds. [c]Mainland census sites are those listed in appendix 10.2. [d]Also *Caracara plancus*. [e]Also *Coragyps atratus*.
[f]These species are breeding at the site.

		Island Number											
Species		13	14	15	16	17	18	19	20	21	22	23	24
Resident species													
Turkey Vulture	*Cathartes aura*[a]	XY			Y		XY			XY			
Osprey	*Pandion haliaetus*	XY	Y	Y			XY		XYZ	XY			
Bald Eagle	*Haliaeetus leucocephalus*	4											
Harris' Hawk	*Parabuteo unicinctus*[a]	Y											
Red-tailed Hawk	*Buteo jamaicensis*[a]	XY					XY			Y			
Golden Eagle	*Aquila chrysaetos*[a]												
Prairie Falcon	*Falco mexicanus*												
Peregrine Falcon	*Falco peregrinus*	Y		Y	Y		XY		XY				
American Kestrel	*Falco sparverius*	XY					17		Y				
Barn Owl	*Tyto alba*						9		Y				
Western Screech Owl	*Otus asio*												
Great Horned Owl	*Bubo virginianus*	XY								X			
Elf Owl	*Micrathene whitneyi*						Y						
Burrowing Owl	*Athene cunicularia*[b]												
Lesser Nighthawk	*Chordeiles acutipennis*						17						
Poorwill	*Phalaenoptilus nuttallii*[b]						9						
Nonbreeding species													
Northern Harrier	*Circus cyaneus*												
Sharp-shinned hawk	*Accipiter striatus*	Y											
Cooper's Hawk	*Accipiter cooperii*	Y							Y				
Zone-tailed Hawk	*Buteo albonotatus*												
Swainson's Hawk	*Buteo swainsoni*												
Long-eared Owl	*Asio otus*												
Resident species		8	0	2	2	0	9	0	4	4	0	0	0
Total no. of species		10	0	2	2	0	9	0	5	4	0	0	0

(continued)

Species		25	26	27	28	29	30	31	32	33	34	35	36
Resident species													
Turkey Vulture	*Cathartes aura*[a]	Y				X		Y	XY			Y	
Osprey	*Pandion haliaetus*	XY		Y	Y	XY	X	XY	X	Y1		XY	XY
Bald Eagle	*Haliaeetus leucocephalus*								4				
Harris' Hawk	*Parabuteo unicinctus*[a]											YZ	
Red-tailed Hawk	*Buteo jamaicensis*[a]	Y				XY		XY	XY			XY	XY
Golden Eagle	*Aquila chrysaetos*[a]												
Prairie Falcon	*Falco mexicanus*												
Peregrine Falcon	*Falco peregrinus*	YZ	Y			Y1		Y		Y6		XY	Y
American Kestrel	*Falco sparverius*	Y						X				XY	Y
Barn Owl	*Tyto alba*	Y											Y
Western Screech Owl	*Otus asio*								1			X	
Great Horned Owl	*Bubo virginianus*											X	
Elf Owl	*Micrathene whitneyi*												
Burrowing Owl	*Athene cunicularia*[b]	XY											
Lesser Nighthawk	*Chordeiles acutipennis*							Y16		Y15		YZ	
Poorwill	*Phalaenoptilus nuttallii*[b]										Z	YZ	
Nonbreeding species													
Northern Harrier	*Circus cyaneus*												
Sharp-shinned hawk	*Accipiter striatus*												
Cooper's Hawk	*Accipiter cooperii*	Y											
Zone-tailed Hawk	*Buteo albonotatus*												
Swainson's Hawk	*Buteo swainsoni*												
Long-eared Owl	*Asio otus*												
Resident species		7	1	1	1	4	1	6	5	3	1	10	5
Total no. of species		8	1	1	1	4	1	6	5	3	1	10	5

		Island Number							No. of Islands	Mainland Census Sites[c]											
Species		37	38	39	40	41	42	43		OP	PT	RA	CA	EI[d]	CB	SJ	LP	CG	HU[e]	HE[e]	TU
Resident species																						
Turkey Vulture	*Cathartes aura*[a]		Y1		Y	Y8			17	X	X	X	X	X	X	X	X	X	X	X	X
Osprey	*Pandion haliaetus*	Z11	Y1		XY	Y12	Y1	Z	31												
Bald Eagle	*Haliaeetus leucocephalus*[a]								2												
Harris' Hawk	*Parabuteo unicinctus*[a]								4				X	X	X						
Red-tailed Hawk	*Buteo jamaicensis*[a]		Y1		Y	Y8	1		21	X			X	X	X	X	X	X	X	X	X
Golden Eagle	*Aquila chrysaetos*[a]					14,16			2												
Prairie Falcon	*Falco mexicanus*					Y			2						X						
Peregrine Falcon	*Falco peregrinus*	4	Y	XYZ	Y1	Y16	Y1		27												
American Kestrel	*Falco sparverius*		15	X	Y	Y8			17	X	X		X	X	X	X	X	X	X	X	X
Barn Owl	*Tyto alba*					Y	1		7												
Western Screech Owl	*Otus asio*					12			2		X					X					
Great Horned Owl	*Bubo virginianus*			X		12			8	X	X		X	X	X	X					
Elf Owl	*Micrathene whitneyi*					12	Y		3	X	X								X		X
Burrowing Owl	*Athene cunicularia*[b]			XY	Y	Y15			9	X	X										
Lesser Nighthawk	*Chordeiles acutipennis*					Y			5	X									X	X	X
Poorwill	*Phalaenoptilus nuttallii*[b]					YZ12			7		X		X	X	X	X			X	X	X
Nonbreeding species																						
Northern Harrier	*Circus cyaneus*					16			1												
Sharp-shinned hawk	*Accipiter striatus*					16			2												
Cooper's Hawk	*Accipiter cooperii*					16			4												
Zone-tailed Hawk	*Buteo albonotatus*				Y	Y			2			X[f]	X[f]				X				
Swainson's Hawk	*Buteo swainsoni*				Y	Y			3												
Long-eared Owl	*Asio otus*					16			1												
Resident species		2	5	5	6	14	5	1		6	7	7	7	7	7	6	4	3	5	7	5
*Total no. of species		2	5	5	8	20	5	1		6	8	7	7	7	7	6	4	4	5	7	5

Appendix 10.8

Land Bird Records from the Northern Islands

Northern islands (from San Marcos north) are ranked left to right in descending area. This appendix lists all recorded bird species that breed in the area, and either are known to breed on the islands indicated or are possible breeders.

| | | Island Number | | | | | |
Species		41	1	31	33	38	42
Gambel's Quail	*Callipepla gambelii*	1,8,12,Yz					
White-winged Dove	*Zenaida asiatica*	1,3,8,12,Y	XY	XY		15	
Mourning Dove	*Zenaida macroura*	3,8,12,Yz	XY	X	X	Y	
Common Ground Dove	*Columbina passerina*	12,Y					
White-throated Swift	*Aeronautes saxatalis*			X		X	
Costa's Hummingbird	*Calypte costae*	1,12,Yz	XY	XY	XY	1,15,Yz	1,Y
Northern Flicker	*Colaptes auratus*	3,12					
Gila Woodpecker	*Melanerpes uropygialis*	1,3,12,Yz		Y			
Ladderbacked Woodpecker	*Picoides scalaris*	z	XY	XYz	XY		Y
Ash-throated Flycatcher	*Myiarchus cinerascens*	3,Yz	XYz	XYz	XY	1,15,Y	1
Nutting's Flycatcher	*Myiarchus nuttingii*	y					
Brown-crested Flycatcher	*Myiarchus tyrannulus*	y					y
Purple Martin	*Progne subis*	12,z					
Violet-green Swallow	*Tachycineta thalassina*	12,Y	Y	XY	X	Y	Y
Common Raven	*Corvus corax*	1,3,12,Y	XY	XY	Xyz	1,15,Y	Y
Verdin	*Auriparus flaviceps*	1,3,12,Yz	XYz	XYz	XY	1,15,Y	Yz
Rock Wren	*Salpinctes obsoletus*	1,Y	X	XY	XY	1	Y
Canyon Wren	*Catherpes mexicanus*	12,yz	X				
Cactus Wren	*Campylorhynchus brunneicapillus*	3,yz		Y			
Blue-gray Gnatcatcher	*Polioptila caerulea*		y	7			
Black-tailed Gnatcatcher	*Polioptila melanura*	1,3,Yz	Z	XY		y	
California Gnatcatcher	*Polioptila californica*		XY		Y	1,15,Y	
Northern Mockingbird	*Mimus polyglottos*	1,3,12,yz	XY	XYz	X	15,y	y
Bendire's Thrasher	*Toxostoma bendirei*	3,y					
Curve-billed Thrasher	*Toxostoma curvirostre*	12,y		XY			
Phainopepla	*Phainopepla nitens*	1,3,12,yz	Xy	XY	y		
Loggerhead Shrike	*Lanius ludovicianus*	3,Y	XYz	15,Y	XY		
Yellow Warbler	*Dendroica petechia*	y					
Red Cardinal	*Cardinalis cardinalis*	1,3,12,Yz					
Pyrrhuloxia	*Cardinalis sinuatus*	12					
Varied Bunting	*Passerina versicolor*						
Brown Towhee	*Pipilo fuscus*	3,12,yz					
Black-throated Sparrow	*Amphispiza bilineata*	1,12,Yz	XY	XYz	Y	1,15,Y	1,Yz

(continued)

Species		Island Number						
		37	2	39	20	43	23	40
Gambel's Quail	*Callipepla gambelii*							
White-winged Dove	*Zenaida asiatica*							
Mourning Dove	*Zenaida macroura*		X	X	X			X
Common Ground Dove	*Columbina passerina*							
White-throated Swift	*Aeronautes saxatalis*				y			
Costa's Hummingbird	*Calypte costae*	z	Y			z		
Northern Flicker	*Colaptes auratus*							
Gila Woodpecker	*Melanerpes uropygialis*							
Ladderbacked Woodpecker	*Picoides scalaris*		XY		X			
Ash-throated Flycatcher	*Myiarchus cinerascens*		XY		z			
Nutting's Flycatcher	*Myiarchus nuttingii*							
Brown-crested Flycatcher	*Myiarchus tyrannulus*							
Purple Martin	*Progne subis*							
Violet-green Swallow	*Tachycineta thalassina*		Y		XY			Y
Common Raven	*Corvus corax*	z	XY	XY	XY			XY
Verdin	*Auriparus flaviceps*		Y					
Rock Wren	*Salpinctes obsoletus*	z	XY	Xz	XY		Yz	Xyz
Canyon Wren	*Catherpes mexicanus*		y		y			
Cactus Wren	*Campylorhynchus brunneicapillus*		Y	Y				
Blue-gray Gnatcatcher	*Polioptila caerulea*							
Black-tailed Gnatcatcher	*Polioptila melanura*							
California Gnatcatcher	*Polioptila californica*				z			
Northern Mockingbird	*Mimus polyglottos*		X	X	X			
Bendire's Thrasher	*Toxostoma bendirei*							
Curve-billed Thrasher	*Toxostoma curvirostre*				X			
Phainopepla	*Phainopepla nitens*							
Loggerhead Shrike	*Lanius ludovicianus*			XY	Y	Y		
Yellow Warbler	*Dendroica petechia*							
Red Cardinal	*Cardinalis cardinalis*							
Pyrrhuloxia	*Cardinalis sinuatus*							
Varied Bunting	*Passerina versicolor*							
Brown Towhee	*Pipilo fuscus*							
Black-throated Sparrow	*Amphispiza bilineata*	1,Yz		Y		z	z	

12	25	11	22	15	5	26	28	27	No. of Islands
									1
									4
									9
									1
									3
									9
									1
									2
									7
									8
									1
									2
									1
	Y								10
yz	12,XY			y	y	y		y	17
Y		Y							8
	Y	z							14
									4
									4
									2
									4
									4
									9
									1
									3
									4
Y									8
	y								2
									1
									1
	y								1
	y								2
									9

(*continued*)

		Island Number								
Species		41	1	31	33	38	42	37	2	39
Great-tailed Grackle	*Quiscalus mexicanus*	y								
Brown-headed Cowbird	*Molothrus ater*	12,y	y							
Hooded Oriole	*Icterus cucculatus*	12,y					y			
Scott's Oriole	*Icterus parisorum*	3,7	X	X						
House Finch	*Carpodacus mexicanus*	1,3,12,Yz	XY	XYz	XY	1,15,z	1,Y		1	Xz
House Sparrow	*Passer domesticus*					Y				
Number of species		34	21	20	15	14	12	3	14	8

Numbered references as listed in appendix 10.5. Records are marked as follows: X; MLC field data; Y, y, EV field data, where uppercase denotes confirmed breeding, lowercase observation only; z, museum and collection records of Adolfo Navarro et al., *Atlas de las Aves de México* (in prep.).

					Island Number								No. of Islands
20	43	23	40	12	25	11	22	15	5	26	28	27	
													1
					y								3
													2
													3
Xyz					Y								10
					y								2
13	2	1	4	3	9	1	0	1	1	1	0	1	

Appendix 10.9

Land Birds Records from Southern Islands

Southern islands (south of San Marcos) are ranked left to right in descending area. This appendix lists all recorded bird species that breed in the area, and either are known to breed on the islands indicated or are possible breeders.

Species		35	6	8	13	7	21	18	29	9	10	32	33	30	4	14	3	16	44	No. of Islands
California Quail	*Callipepla californica*	16,Y	2z																	2
White-winged Dove	*Zenaida asiatica*	XYz	XY	XY	XY	1,9,Y	X	XY	1	19										9
Mourning Dove	*Zenaida macroura*	16	X	X	X	1,Y						1								7
Common Ground Dove	*Columbina passerina*		Y	Y				Y												3
White-throated Swift	*Aeronautes saxatalis*	XY	19,X		y			19,X			X	X								7
Xantus' Hummingbird	*Basilinna xantusi*	XYz		1,Yz	y															3
Costa's Hummingbird	*Calypte costae*	XYz	XYz	XYz	XY	9,Y	XY	XY	X	1,Y	19,X	X								11
Gila Woodpecker	*Melanerpes uropygialis*	XYz	XY	XYz	XYz	9,Y	X	XYz												7
Ladderbacked Woodpecker	*Picoides scalaris*	XYz	XYz	XYz	XY	1,9,Yz	XY	XY	Xz	19,Y	X	Xz	1			X				13
Ash-throated Flycatcher	*Myiarchus cinerascens*	XYz	XY	XYz	XYz	1,9,Y	X	XYz	X	1,19	19,X	X								11
Black Phoebe	*Sayornis nigricans*	z	yz		y															3
Vermillion Flycatcher	*Pyrocephalus rubinus*		y																	1
Purple Martin	*Progne subis*	Y						Xy			X							Y		4

(*continued*)

Species		35	6	8	13	7	21	18	29	9	10	32	33	30	4	14	3	16	44	No. of Islands
Violet-green Swallow	*Tachycineta thalassina*	Y	Y	XY	XYz		X			15,19								y		7
Common Raven	*Corvus corax*	XYz	XYz	XY	XY	1,9,Y	XY	XYz	XY	1,y	19,X	Xz	1,6,YZ	X		Xy	y			15
Verdin	*Auriparus flaviceps*	XYz	XY	XYz	XYz	1,9,Yz	XY	XY	XY	1,Y	X	Xz	Y							12
Rock Wren	*Salpinctes obsoletus*	XY	XY	Xz	XY	1,Y	X	X	X	1,YZ	X	X	6,YZ	X			Y			14
Canyon Wren	*Catherpes mexicanus*	Xy		Xz	XYz		Xy						4							5
Cactus Wren	*Campylorhynchus brunneicapillus*	y												z						2
Blue-gray Gnatcatcher	*Polioptila caerulea*	1,Yz	3,4,Yz	3,4,Yz	1,Yz	3	Y	1,Yz		1,19	Xz									9
Black-tailed Gnatcatcher	*Polioptila melanura*	z			y	y														3
California Gnatcatcher	*Polioptila californica*	XYz	X	X	XYz	3,9,Y	X	X	X	1	X	X								11
Northern Mockingbird	*Mimus polyglottos*	1,16,Y	Xy	XYz	Xyz	11,y	y	Xy	Y	15,y	X						y			11
Gray Thrasher	*Toxostoma cinereum*	XYz		1,Y		y														2
Loggerhead Shrike	*Lanius ludovicianus*	XY	XY	Y	XY	Y		19,Y	1,3	19										8
Yellow Warbler	*Dendroica petechia*	Yz			Xyz		X													3

Species		Island Number																			No. of Islands
		35	6	8	13	7	21	18	29	9	10	32	33	30	4	14	3	16	44	
Belding's Yellowthroat	*Geothlypis beldingi*	yz																		1
Red Cardinal	*Cardinalis cardinalis*		XYz	XYz	XY	X	1,9,Yz	X	XY	X	15,Y	19,X	1							11
Varied Bunting	*Passerina versicolor*							Y												1
Black-throated Sparrow	*Amphispiza bilineata*		XYz	XYz	XY	XYz	1,3,4,9,Yz	XY	XYz	Xyz	1,4,Y	Xz	Xz	Y	X		X			X	15
Hooded Oriole	*Icterus cucculatus*		XY	3,Y	XY	y														5
Scott's Oriole	*Icterus parisorum*			Y	Y															2
House Finch	*Carpodacus mexicanus*		XYz	XYz	XY	XYz	1,3,9,Yz	XY	XY	XY	1,4,Y	Xz	Xz	1	1		X				14
Lesser Goldfinch	*Carduelis psaltria*				Xy			X												2
	Number of species	1	29	24	23	25	19	17	20	13	16	14	12	7	5	0	4	3	2	1	

Numbered references as listed in appendix 10.5. Records are marked as follows: X, MLC field data; Y, y, EV field data, where uppercase denotes confirmed breeding, lowercase observation only; z, CONABIO database of museum and collection records, assembled by Adolfo Navarro and Hesiquio Benitez.

Appendix 10.10

Wintering, Migrant, and other Land Birds of Casual Occurrence

Species		Status	1	2	3	4	5	6	7	8	9	10	11	12	13
Common Snipe	*Gallinago gallinago*	P		x											
Black Swift	*Cypseloides niger*	P													
Vaux' Swift	*Chaetura vauxi*	O	z												
Black-chinned Hummingbird	*Archilochus alexandri*	L	y												
Rufous Hummingbird	*Selasphorus rufus*	P	x					y							
Allen's Hummingbird	*Selasphorus sasin*	P													
Tropical Kingbird	*Tyrannus melancholicus*	O													
Cassin's Kingbird	*Tyrannus vociferans*	W													
Western Wood-pewee	*Contopus sordidulus*	P		x											
Pacific Flycatcher	*Empidonax difficilis*	W						y							
Dusky Flycatcher	*Empidonax oberholseri*	O													
Willow Flycatcher	*Empidonax traillii*	P													
Gray Flycatcher	*Empidonax wrightii*	W						xyz		yz					4yz
Say's Phoebe	*Sayornis saya*	W	x					y							
Black Phoebe	*Sayornis nigricans*	P						yz							yz
Nutting's Flycatcher	*Myiarchus nuttingi*	W?													
Brown-crested Flycatcher	*Myiarchus tyrannulus*	W?													
Vermillion Flycatcher	*Pyrocephalus rubinus*	P	15												
Horned Lark	*Eremophila alpestris*	W						3							
Barn Swallow	*Hirundo rustica*	P													
Cliff Swallow	*Petrochelidon pyrrhonota*	P													
N. Rough-winged Swallow	*Stelgidopteryx serripennis*	O								yz					
Sedge Wren	*Cistothorus platensis*	W						yz							
Bewick's Wren	*Thryomanes bewickii*	W	x					x							
House Wren	*Troglodytes aedon*	W													
Sage Thrasher	*Oreoscoptes montanus*	W	1												
Bendire's Thrasher	*Toxostoma bendirei*	W		x											
Hermit Thrush	*Catharus guttatus*	P	x							yz					x
Swainson's Thrush	*Catharus ustulatus*	P													

(*continued*)

Species		Status	1	2	3	4	5	6	7	8	9	10	11	12	13
Townsend's Solitaire	*Myadestes townsendi*	P						y							
Tropical Gnatcatcher	*Polioptila plumbea*	O													
Ruby-crowned Kinglet	*Regulus calendula*	P												17	
Water Pipit	*Anthus rubescens*	W	y					3z							
Bell's Vireo	*Vireo bellii*	W						3							
Hutton's Vireo	*Vireo huttoni*	P	x												
Gray Vireo	*Vireo vicinior*	W	x					x		x					
Yellow Warbler	*Dendroica petechia*	W?													z
Yellow-rumped Warbler	*Dendroica coronata*	W	xyz					y	y						y
Magnolia Warbler	*Dendroica magnolia*	P													
Black-throated Gray Warbler	*Dendroica nigrescens*	P													
Townsend's Warbler	*Dendroica townsendi*	P													
Common Yellowthroat	*Geothlypis trichas*	W						xy							
MacGillivray's Warbler	*Oporornis tolmiei*	P	x						y						
Northern Waterthrush	*Seiurus novaboracensis*	W													
Tennessee Warbler	*Vermivora peregrina*	P													
Nashville Warbler	*Vermivora ruficapilla*	P													
Orange-crowned Warbler	*Vermivora celata*	W	x					xy	y	y					yz
Wilson's Warbler	*Wilsonia pusilla*	P	x												
Western Tanager	*Piranga ludoviciana*	P	x												
Botteri's Sparrow	*Aimophila botterii*	L													
Cassin's Sparrow	*Aimophila cassinii*	L													
Rufous-crowned Sparrow	*Aimophila ruficeps*	O													
Lark Bunting	*Calamospiza melanocorys*	W	3,4					3y							
Lark Sparrow	*Chondestes grammacus*	W													y
Lincoln's Sparrow	*Melospiza lincolnii*	W	x												
Song Sparrow	*Melospiza melodia*	L	y												
Savannah Sparrow	*Passerculus sandwichensis*	W	x	x				5z							xz
Lazuli Bunting	*Passerina amoena*	P													

Varied Bunting	*Passerina versicolor*	O?							
Black-headed Grosbeak	*Pheucticus melanocephalus*	P							
Green-tailed Towhee	*Pipilo chlorurus*	W	xy			xy	y	x	1	x	16
Chipping Sparrow	*Spizella passerina*	W	x	1						xyz
Clay-colored Sparrow	*Spizella pallida*	W				z				y
Brewer's Sparrow	*Spizella breweri*	W	xy	x		xz				z
White-throated Sparrow	*Zonotrichia albicollis*	O					y			3z
White-crowned Sparrow	*Zonotrichia leucophrys*	W	xy	xy		y	y			
Brewer's Blackbird	*Euphagus cyanocephalus*	W	y							y
Northern Oriole	*Icterus galbula*	P								
Western Meadowlark	*Sturnella neglecta*	W	y							
Total nonbreeding land birds			26	7	0	21	5	7	1	1	0	1	14
No. of wintering species			17	5	0	18	3	6	1	1	0	1	13
No. of Passage species			9	2	0	3	2	1	0	0	0	0	1

Numbered references as in appendix 10.5. Records are marked as follows: x, MLC field data; y, EV field data; z, Adolfo Navarro et al., *Atlas de las Aves de México* (in prep.). Status symbols. L, local migrant; O, occasional, vagrant; P, passage migrant; W, wintering species.

Island Number

Species		14	15	16	17	18	19	20	21	22	23	24	25	26	27	28	29	30
Common Snipe	*Gallinago gallinago*																	
Black Swift	*Cypseloides niger*																	
Vaux' Swift	*Chaetura vauxi*																	
Black-chinned Hummingbird	*Archilochus alexandri*							y										
Rufous Hummingbird	*Selasphorus rufus*																	
Allen's Hummingbird	*Selasphorus sasin*												y					
Tropical Kingbird	*Tyrannus melancholicus*																	
Cassin's Kingbird	*Tyrannus vociferans*							x										
Western Wood-pewee	*Contopus sordidulus*																	
Pacific Flycatcher	*Empidonax difficilis*					x												
Dusky Flycatcher	*Empidonax oberholseri*							z	x								4	
Willow Flycatcher	*Empidonax traillii*																	
Gray Flycatcher	*Empidonax wrightii*																4	
Say's Phoebe	*Sayornis saya*									yz								
Black Phoebe	*Sayornis nigricans*																	
Nutting's Flycatcher	*Myiarchus nuttingi*																	
Brown-crested Flycatcher	*Myiarchus tyrannulus*																	
Vermillion Flycatcher	*Pyrocephalus rubinus*																	
Horned Lark	*Eremophila alpestris*																	
Barn Swallow	*Hirundo rustica*																	
Cliff Swallow	*Petrochelidon pyrrhonota*																	
N. Rough-winged Swallow	*Stelgidopteryx serripennis*																	
Sedge Wren	*Cistothorus platensis*																	
Bewick's Wren	*Thryomanes bewickii*												y					
House Wren	*Troglodytes aedon*																	
Sage Thrasher	*Oreoscoptes montanus*																	
Bendire's Thrasher	*Toxostoma bendirei*							x										
Hermit Thrush	*Catharus guttatus*																	

Common Name	Scientific Name	C1	C2	C3	C4	C5	C6	C7
Swainson's Thrush	*Catharus ustulatus*							3z
Townsend's Solitaire	*Myadestes townsendi*							
Tropical Gnatcatcher	*Polioptila plumbea*							
Ruby-crowned Kinglet	*Regulus calendula*							
Water Pipit	*Anthus rubescens*						yz	
Bell's Vireo	*Vireo bellii*							
Hutton's Vireo	*Vireo huttoni*							
Gray Vireo	*Vireo vicinior*							
Yellow Warbler	*Dendroica petechia*					y		
Yellow-rumped Warbler	*Dendroica coronata*	x				y		
Magnolia Warbler	*Dendroica magnolia*					y		
Black-throated Gray Warbler	*Dendroica nigrescens*	z						
Townsend's Warbler	*Dendroica townsendi*					y		
Common Yellowthroat	*Geothlypis trichas*							
MacGillivray's Warbler	*Oporornis tolmiei*							
Northern Waterthrush	*Seiurus novaboracensis*							
Tennessee Warbler	*Vermivora peregrina*							
Nashville Warbler	*Vermivora ruficapilla*							
Orange-crowned Warbler	*Vermivora celata*		y					1
Wilson's Warbler	*Wilsonia pusilla*	16						
Western Tanager	*Piranga ludoviciana*		y					
Botteri's Sparrow	*Aimophila botterii*							
Cassin's Sparrow	*Aimophila cassinii*							
Rufous-crowned Sparrow	*Aimophila ruficeps*							
Lark Bunting	*Calamospiza melanocorys*							
Lark Sparrow	*Chondestes grammacus*		x					
Lincoln's Sparrow	*Melospiza lincolnii*							
Song Sparrow	*Melospiza melodia*							
Savannah Sparrow	*Passerculus sandwichensis*	z		z	z			y
Lazuli Bunting	*Passerina amoena*							
Varied Bunting	*Passerina versicolor*	y	y					
Black-headed Grosbeak	*Pheucticus melanocephalus*	y	y					
Green-tailed Towhee	*Pipilo chlorurus*	x	x					

Species		14	15	16	17	18	19	20	21	22	23	24	25	26	27	28	29	30
Chipping Sparrow	*Spizella passerina*												y					
Clay-colored Sparrow	*Spizella pallida*																	
Brewer's Sparrow	*Spizella breweri*					y							yz				y	
White-throated Sparrow	*Zonotrichia albicollis*																	
White-crowned Sparrow	*Zonotrichia leucophrys*							xy					y					
Brewer's Blackbird	*Euphagus cyanocephalus*												y					
Northern Oriole	*Icterus galbula*																	
Western Meadowlark	*Sturnella neglecta*																	
Total nonbreeding land birds		0	0	0	0	6	0	8	5	1	1	1	14	0	0	0	5	0
No. of wintering species		0	0	0	0	4	0	6	4	1	1	1	10	0	0	0	3	0
No. of Passage species		0	0	0	0	2	0	2	1	0	0	0	4	0	0	0	2	0

Numbered references as in appendix 10.5. Records are marked as follows: x, MLC field data; y, EV field data; z, Adolfo Navarro et al., *Atlas de las Aves de México* (in prep.). Status symbols. L, local migrant; O, occasional, vagrant; P, passage migrant; W, wintering species.

| | | \multicolumn{13}{c|}{Island Number} | No. of |
Species		31	32	33	34	35	36	37	38	39	40	41	42	43	Islands
Common Snipe	*Gallinago gallinago*														1
Black Swift	*Cypseloides niger*					y						y			2
Vaux' Swift	*Chaetura vauxi*														1
Black-chinned Hummingbird	*Archilochus alexandri*	y					y		y			y	y		8
Rufous Hummingbird	*Selasphorus rufus*	xz				yz	x					12			5
Allen's Hummingbird	*Selasphorus sasin*	xy					xy								4
Tropical Kingbird	*Tyrannus melancholicus*			yz		z									2
Cassin's Kingbird	*Tyrannus vociferans*	x		yz						x					3
Western Wood-pewee	*Contopus sordidulus*					4									2
Pacific Flycatcher	*Empidonax difficilis*	x				3yz				x		12z			8
Dusky Flycatcher	*Empidonax oberholseri*					z									2
Willow Flycatcher	*Empidonax traillii*					y						16yz			2
Gray Flycatcher	*Empidonax wrightii*					yz						12			6
Say's Phoebe	*Sayornis saya*					yz				x		yz			6
Black Phoebe	*Sayornis nigricans*					4z									3
Nutting's Flycatcher	*Myiarchus nuttingi*											y			1
Brown-crested Flycatcher	*Myiarchus tyrannulus*											yz	yz		2
Vermillion Flycatcher	*Pyrocephalus rubinus*	x													2
Horned Lark	*Eremophila alpestris*														1
Barn Swallow	*Hirundo rustica*					y						12z			2
Cliff Swallow	*Petrochelidon pyrrhonota*											12			1
N. Rough-winged Swallow	*Stelgidopteryx serripennis*											yz			2
Sedge Wren	*Cistothorus platensis*											z			2
Bewick's Wren	*Thryomanes bewickii*														2
House Wren	*Troglodytes aedon*					yz									4
Sage Thrasher	*Oreoscoptes montanus*	y										yz	yz		3
Bendire's Thrasher	*Toxostoma bendirei*		z								x	z			4
Hermit Thrush	*Catharus guttatus*	x					x					z			4

(continued)

Species		31	32	33	34	35	36	37	38	39	40	41	42	43	No. of Islands
Swainson's Thrush	*Catharus ustulatus*	x													1
Townsend's Solitaire	*Myadestes townsendi*														1
Tropical Gnatcatcher	*Polioptila plumbea*														1
Ruby-crowned Kinglet	*Regulus calendula*						y					16			2
Water Pipit	*Anthus rubescens*											1z			5
Bell's Vireo	*Vireo bellii*														2
Hutton's Vireo	*Vireo huttoni*	x													2
Gray Vireo	*Vireo vicinior*	xy					x					12			6
Yellow Warbler	*Dendroica petechia*					yz						y			4
Yellow-rumped Warbler	*Dendroica coronata*	x				y				x		12y			10
Magnolia Warbler	*Dendroica magnolia*														1
Black-throated Gray Warbler	*Dendroica nigrescens*	x										12			4
Townsend's Warbler	*Dendroica townsendi*						15		y						2
Common Yellowthroat	*Geothlypis trichas*					z									2
MacGillivray's Warbler	*Oporornis tolmiei*	x				y	x					12			6
Northern Waterthrush	*Seiurus novaboracensis*	15													1
Tennessee Warbler	*Vermivora peregrina*								y						1
Nashville Warbler	*Vermivora ruficapilla*								y			12			3
Orange-crowned Warbler	*Vermivora celata*	x		1		yz	1yz		1			12	1		14
Wilson's Warbler	*Wilsonia pusilla*	x					15yz			x		12y			6
Western Tanager	*Piranga ludoviciana*	y		1			15								4
Botteri's Sparrow	*Aimophila botterii*											z			1
Cassin's Sparrow	*Aimophila cassinii*											12			1
Rufous-crowned Sparrow	*Aimophila ruficeps*					z									1
Lark Bunting	*Calamospiza melanocorys*											16z			4
Lark Sparrow	*Chondestes grammacus*			1											2

Species	Scientific name									
Lincoln's Sparrow	*Melospiza lincolnii*	x		z			x	16z		5
Song Sparrow	*Melospiza melodia*	y						y		3
Savannah Sparrow	*Passerculus sandwichensis*	16	z	4z	z		x	16z		15
Lazuli Bunting	*Passerina amoena*							3,12		1
Varied Bunting	*Passerina versicolor*									2
Black-headed Grosbeak	*Pheucticus melanocephalus*	x		16				12	y	6
Green-tailed Towhee	*Pipilo chlorurus*	xy		16y	y	1	x	1y		15
Chipping Sparrow	*Spizella passerina*	x								5
Clay-colored Sparrow	*Spizella pallida*		yz							3
Brewer's Sparrow	*Spizella breweri*	x		yz	x		x	16yz		15
White-throated Sparrow	*Zonotrichia albicollis*							z		1
White-crowned Sparrow	*Zonotrichia leucophrys*	x	1	z	xy	15	x	16yz		15
Brewer's Blackbird	*Euphagus cyanocephalus*				z					3
Northern Oriole	*Icterus galbula*	x						12y		2
Western Meadowlark	*Sturnella neglecta*							y		2
Total nonbreeding land birds		27	7	23	15	7	8	41	5	0
No. of wintering species		12	5	12	6	3	6	21	3	0
No. of Passage species		15	2	11	9	4	2	20	2	0

Numbered references as in appendix 10.5. Records are marked as follows: x, MLC field data; y, EV field data; z, Adolfo Navarro et al., *Atlas de las Aves de México* (in prep.). Status symbols. L, local migrant; O, occasional, vagrant; P, passage migrant; W, wintering species.

Appendix 10.11

Breeding Bird Densities Recorded at Sonoran Desert Sites on Peninsula, Mainland, and Gulf Islands

		Density		Species or Family[b]															Σ Density
Site[a]	No. of species	Pairs/ha	Individuals/h	Amphispiza bilineata	Carpodacus mexicanus	Picoides scalaris	Troglodytidae	Auriparus flaviceps	Myiarchus cinerascens	Trochilidae	Mimus polyglottos	Polioptila spp.	Cardinalis cardinalis	Melanerpes uropygialis	Toxostoma spp.	Colaptes auratus	Cardinalis sinuatus	Pipilo fuscus	
Organpipe	31	12.4	90.8	2	13	2	9.5	5	3	2	0.5	3	2	7	3	3	1	2	58
Tucson	29	7.7	65.0	0.2	4.1	0	5.8	5.9	3.4	2.3	0.5	1.6	3.2	2	6.8	2.3	2.5	1.1	41.7
Puertecitos	28	7.6	57.1	6.5	4.6	0.8	2.1	7.8	2	1.8	0.3	2.9	0.5	1.9	1	1.3		0.7	34.2
San Juan Road	23	7.1	69.4	2.4	2.8	2.4	8.8	8.8	3.2	9.6	0	8	3.2	4.8	0.8	1.6	0.5	2.4	59.3
La Paz	22	10.9	[81.6]	0	3.8	2	5.5	6.6	6.5	12	1.6	1.8	2.5	6.4	6.6	3	0	4	62.3
Hermosillo	25	9.4	[70.4]	6.5	2.2	0.8	4.7	7.3	5.7	5	2.2	3	5	1.3	1.6	2.3	2.9	0.7	51.2
Huatabampo	21	11.6	74.1	0	1.2	2.4	7.2	10.8	3.6	4	1.5	2.4	3.6	4.8	3.6	1.2	1.2	0	47.5
Mainland Mean	25.6	9.5	72.6	2.5	4.5	1.5	6.2	7.5	3.9	5.2	0.9	3.2	2.9	4.0	3.3	2.1	1.4	1.6	50.6
SE	1.4	0.8	4.2	1.1	1.5	0.4	1.0	0.7	0.6	1.5	0.3	0.8	0.5	0.9	1.0	0.3	0.4	0.5	10.2
2 SD	7.6	4.2	25.2	5.8	15.6	3.6	10	7.6	6.4	16	3.2	8.8	5.6	9.2	10	3.2	4.8	5.6	
San Jose	29		55.0	4.5	5.5	1.4	0.5	7.5	3.1	7.5	0.7	5.5	4.1	3.8	1.5				44.9
Carmen	24		49.8	10	6.7	0.7	1.3	6.3	4.8	7		3.3	1.1	1.3					43.2
Monserrat	20		40.2	15	3.4	2.1	0.5	4.4	2.3	2.6	1.3	1.6	1.8	2.6					37.6
Santa Cruz	13		19.5	2.3	2.9	1.1	0.2	6.3	1.5	2.9	0	0.5	0.3						18
Danzante	14		39.0	12	2	1	3	8	2	5	1	2	3						39
San Francisco	12		26.7	6.9	0	0.7	7.2	2.6	0.9	1.7		2.4	0						22.4
San Diego	5		14.0	1.7	0		3.4												5.1
Gallo	4		5.1	4	6	2													12

Bracketed values are projected from census data in pr/ha rather than measured directly.

[a]First seven sites are mainland; last eight are southern islands in decreasing size.

[b]Species known present on island but not seen in survey are scored 0; those absent from site or island left blank. Bird species are listed in the rank order of table 10.2.

Appendix 12.1

Native Terrestrial Mammals Recorded from Islands in the Sea of Cortés

Taxonomic designations follow Lawlor (1983) as updated by Wilson and Reeder (1993) (see also changes proposed by Riddle et al. 2000a, b; Hafner et al. 2001).

Islands are mapped in figures 1.1–1.3. Landbridge islands are set in roman type, oceanic ones in italics. Endemic species are denoted by boldface. Parenthetical numbers identify islands in figure 12.7.

Species	Willard (1)	Mejía (2)	Granito (3)	Angel de la Guarda (4)	Smith (Coronado) (5)	Salsipuedes (6)	San Lorenzo Norte (7)	San Lorenzo Sur (8)	San Esteban (9)	Tiburón (10)	Turner (Dátil) (11)	San Pedro Nolasco (12)	Tortuga (13)	San Marcos (14)	Coronados (15)	Carmen (16)	Danzante (17)	Monserrat (18)	Santa Catalina (19)	Santa Cruz (20)	San Diego (21)	San José (22)	San Francisco (23)	Espíritu Santo (24)	Cerralvo (25)	Santa Margarita (26)	Magdalena (27)	Asunción (28)	San Roque (29)	Natividad (30)	Cedros (31)	San Gerónimo (32)	San Martín (33)	Todos Santos (34)
Notiosorex crawfordi																																		
Lepus alleni tiburonensis										X																								
Lepus californicus magdalenae																																	X	
Lepus californicus sheldoni																X																		
Lepus insularis																								X										
Sylvilagus bachmanni cerrosensis																															X			
Sylvilagus mansuetas																						X												
Ammospermophilus insularis																								X										
Ammospermophilus leucurus														I																				
Spermophilus tereticaudus										X																								
Spermophilus variegatus grammurus										X																								
Thomomys bottae magdalenae																											X							
Dipodomys merriami insularis																						X												
Dipodomys merriami margaritae																										X								
Dipodomys merriami mitchelli										X																								
Chaetodipus arenarius albulus																											X							
Chaetodipus arenarius ammophilus																										X								
Chaetodipus arenarius siccus																									X									

Species	Willard (1)	Mejía (2)	Granito (3)	Angel de la Guarda (4)	Smith (Coronado) (5)	Salsipuedes (6)	San Lorenzo Norte (7)	San Lorenzo Sur (8)	San Esteban (9)	Tiburón (10)	Turner (Dátil) (11)	San Pedro Nolasco (12)	Tortuga (13)	San Marcos (14)	Coronados (15)	Carmen (16)	Danzante (17)	Monserrat (18)	Santa Catalina (19)	Santa Cruz (20)	San Diego (21)	San José (22)	San Francisco (23)	Espíritu Santo (24)	Cerralvo (25)	Santa Margarita (26)	Magdalena (27)	Asunción (28)	San Roque (29)	Natividad (30)	Cedros (31)	San Gerónimo (32)	San Martín (33)	Todos Santos (34)
Chaetodipus baileyi					X													E																
Chaetodipus baileyi formicatus										X																								
Chaetodipus baileyi insularis																															X			
Chaetodipus fallax anthonyi										X																								
Chaetodipus intermedius											X																							
Chaetodipus intermedius minimus																																		
Chaetodipus penicillatus seri										X																								
Chaetodipus spinatus bryanti																						X												
Chaetodipus spinatus evermanni		X																																
Chaetodipus spinatus guardiae				X																														
Chaetodipus spinatus lambi																								X										
Chaetodipus spinatus latijugularis																							X											
Chaetodipus spinatus lorenzi							X	X																										
Chaetodipus spinatus marcosensis														X																				
Chaetodipus spinatus magdalenae																											X							

644

Species	1	2	3	4	5	6	7	8
Chaetodipus spinatus margaritae						X		
Chaetodipus spinatus occultus								
Chaetodipus spinatus pullus			X					E
Chaetodipus spinatus seorsus		X						
Neotoma albigula seri		X		X	X			
Neotoma albigula varia				X				
Neotoma anthonyi			E					
Neotoma bryanti		X			X			
Neotoma bunkeri				X				
Neotoma lepida abbreviata								
Neotoma lepida insularis				X				
Neotoma lepida latirostra				X				
Neotoma lepida marcosensis			X		X			
Neotoma lepida nudicauda			X		X			
Neotoma lepida perpallida								
Neotoma lepida pretiosa				X		X X	E	
Neotoma lepida vicina						X		
Neotoma martinensis			X					
Peromyscus boylii glasselli				X				
Peromyscus caniceps[a]								
Peromyscus crinitus pallidissimus	X							
Peromyscus dickeyi[b]			X					
Peromyscus eremicus avius						X		
Peromyscus eremicus cedrosensis								X

Species	Willard (1)	Mejía (2)	Granito (3)	Angel de la Guarda (4)	Smith (Coronado) (5)	Salsipuedes (6)	San Lorenzo Norte (7)	San Lorenzo Sur (8)	San Esteban (9)	Tiburón (10)	Turner (Dátil) (11)	San Pedro Nolasco (12)	Tortuga (13)	San Marcos (14)	Coronados (15)	Carmen (16)	Danzante (17)	Monserrat (18)	Santa Catalina (19)	Santa Cruz (20)	San Diego (21)	San José (22)	San Francisco (23)	Espíritu Santo (24)	Cerralvo (25)	Santa Margarita (26)	Magdalena (27)	Asunción (28)	San Roque (29)	Natividad (30)	Cedros (31)	San Gerónimo (32)	San Martín (33)	Todos Santos (34)
Peromyscus eremicus cinereus																						X												
Peromyscus eremicus collatus																								X										
Peromyscus eremicus insularis																										X	X							
Peromyscus eremicus polypolius											X																							
Peromyscus eremicus tiburonensis										X																								
Peromyscus eva carmeni																X																		
Peromyscus guardia guardia				X																														
Peromyscus guardia harbitsoni			E																															
Peromyscus guardia mejiae		E																																
Peromyscus interparietalus interparietalus[c]								X																										
Peromyscus intermedius lorenzi[c]							X																											
Peromyscus intermedius ryckmani[c]						X																												
Peromyscus maniculatus																											X							

Species									
Peromyscus maniculatus cineritius							E		
Peromyscus maniculatus dorsalis								X	
Peromyscus maniculatus dubius									X
Peromyscus maniculatus exiguus								X	
Peromyscus maniculatus geronimensis								X	
Peromyscus maniculatus hueyi	X								
Peromyscus maniculatus magdalenae							X		
Peromyscus maniculatus margaritae							X		
Peromyscus pembertoni				E					
Peromyscus pseudocrinitus					X				
Peromyscus sejugis						X X			
Peromyscus slevini						X			
Peromyscus stephani[d]		?							
Canis latrans		X							
Canis latrans jamesi			X						
Canis latrans peninsulae			X					X	
Urocyon cinereoargenteus			X						
Bassariscus astutus flavus									
Bassariscus astutus insulicula								X	
Bassariscus astutus saxicola								X	
Odocoileus hemionus								X	

Species	Islands																																		
	Willard (1)	Mejia (2)	Granito (3)	Angel de la Guarda (4)	Smith (Coronado) (5)	*Salsipuedes* (6)	*San Lorenzo Norte* (7)	*San Lorenzo Sur* (8)	*San Esteban* (9)	Tiburón (10)	Turner (Dátil) (11)	*San Pedro Nolasco* (12)	*Tortuga* (13)	San Marcos (14)	Coronados (15)	Carmen (16)	Danzante (17)	*Monserrat* (18)	*Santa Catalina* (19)	*Santa Cruz* (20)	*San Diego* (21)	*San José* (22)	San Francisco (23)	Espíritu Santo (24)	*Cerralvo* (25)	Santa Margarita (26)	Magdalena (27)	Asunción (28)	San Roque (29)	Natividad (30)	Cedros (31)	San Gerónimo (32)	San Martín (33)	Todos Santos (34)
Odocoileus hemionus cerrosensis																															X			
Odocoileus hemionus peninsulae																						X												
Odocoileus hemionus sheldoni										X																								
Ovis canadensis										X																								
Total species	2	2	1	3	2	1	2	2	2	14	3	2	1	3	3	4	2	2	2	1	1	7	2	6	2	7	9	1	1	1	5	1	3	2
Total endemic species	0	0	1	1	1	1	1	1	1	0	0	0	0	0	2	0	1	1	1	1	1	1	0	0	0	0	0	0	0	0	1	0	1	1

E, probably extinct; I, probably introduced; ?, probably transient. Sources include Hall (1981), Lawlor (1983), and M. A. Bogan (pers. commun. 1999).

[a] Probably *Peromyscus eva caniceps* (Hafner et al. 2001).
[b] Probably *Peromyscus merriami dickeyi* (Hafner et al. 2001).
[c] Considered to be populations of *Peromyscus eremicus* by Hafner et al. (2001).
[d] Probably *Peromyscus boylii stephani* (Hafner et al. 2001).

Appendix 12.2

Records of Occurrence and Likely Origins of Terrestrial Mammals from Islands in the Sea of Cortés

Island populations are organized by clades (species or species groups) from the list of species in appendix 12.2. Islands are arranged from north to south in their respective groupings and are mapped in figures 1.1–1.3. Landbridge islands are set in roman type, oceanic ones in italics.

Species/Species Group	Pacific Ocean									Sea of Cortés																					Mainland Islands			
	Todos Santos	San Martín	San Gerónimo	Cedros	Natividad	San Roque	Asunción	Magdalena	Santa Margarita	Willard	Mejía	Granito	Angel de la Guarda	Smith (Coronado)	Salsipuedes	San Lorenzo Norte	San Lorenzo Sur	Tortuga	San Marcos	Coronados	Carmen	Danzante	Monserrat	Santa Catalina	Santa Cruz	San Diego	San José	San Francisco	Espíritu Santo	Cerralvo	San Esteban	Tiburón	Turner (Dátil)	San Pedro Nolasco
Notiosorex crawfordi	P																																	
Lepus alleni																																M		
Lepus californicus[a]				W	W																								W			M		
Sylvilagus bachmanni[b]				P																														
Ammospermophilus leucurus[c]																			P								P					M		
Spermophilus tereticaudus																																		
Spermophilus variegatus								P																							M			
Thomomys bottae									W																		W					W		
Dipodomys merriami																																		
Chaetodipus arenarius								P	P														W							P				
Chaetodipus baileyi														W																		M	M	
Chaetodipus fallax				P																														
Chaetodipus intermedius													P			P	P		P	P	P	P												
Chaetodipus penicillatus																											P					M	M	
Chaetodipus spinatus													P														P	P	P					M
Neotoma albigula																																M	M	
Neotoma lepida[d]	P	P											P														P	P	P			M	M	
Peromyscus boylii[e]																															M			M

650

Peromyscus crinitus														P										
Peromyscus eremicus[f]					W					W	W W W	W W W		W					W	W W			W W	
Peromyscus eva[g]																								
Peromyscus maniculatus[h]	P	P	P		P P P P P				P	P		P			P		P P P							M
Peromyscus merriami[i]												M													
Canis latrans					W																			W	
Urocyon cinereoargenteus																								W	
Bassariscus astutus					W														W	W				W	
Odocoileus hemionus					W														W					W	
Ovis canadensis																								W	
Total species	2	3	1	5	1 1 1 8 7	2	2	1	3	2	1	2	2	2	1	2	3	4	2	2	1 1 1 7	2	6	2	1 14 3 2

M, affinities with Mexican mainland relatives; P, affinities with peninsular (Baja California) relatives; W, affinities with mainland relatives widespread on both sides of Sea Cortés.

[a] Includes *Lepus insularis*.
[b] Includes *Sylvilagus mansuetas*.
[c] Includes *Ammospermophilus insularis*.
[d] Includes *Neotoma anthonyi, N. bryanti, N. bunkeri,* and *N. martinensis*.
[e] Includes *Peromyscus stephani*.
[f] Includes *Peromyscus guardia, P. interparietalis,* and *P. pseudocrinitus* (but see Riddle et al. 2000a).
[g] Includes *Peromyscus caniceps* (but see Riddle et al. 2000a).
[h] Includes *Peromyscus sejugis* and *P. slevini*; the latter may be closest related to *P. melanophrys* (Carleton and Lawlor, in prep.).
[i] Includes *Peromyscus dickeyi* and *P. pembertoni*.

Appendix 12.3

Distribution of Bats on Islands in the Sea of Cortés

Pacific islands not included. Santa Margarita is the only Pacific island from which a bat species (*Pipistrellus hesperus*) has been recorded. Landbridge islands are set in roman type, oceanic ones in italics.

Species	San Jorge	Encantada	Willard	Granito	Angel de la Guarda	Pond (Estanque)	Pescadora (BLA)	Cordonosa (BLA)	Partida Norte	Alcatraz	San Esteban	Salsipuedes	Patos	Tiburón	Turner (Dátil)	San Pedro Nolasco	Blanca	San Ildefonso	Coronados	Carmen	Monserrat	Santa Catalina	San José	Cayo	Los Islotes	Espíritu Santo	Ballena	Islet S end Bahía Rosario	Cerralvo
Macrotus californicus														X						X			X			X			
Leptonycteris curosoae					X									X						X			X						
Antrozous pallidus											X		X							X		X							
Corynorhinus townsendii														X								X	X			X			
Eptesicus fuscus														X									X						X
Lasiurus cinereus															X														
Lasiurus xanthinus																										X			
Myotis californicus														X															
Myotis vivesi	X	X	X	X	X	X	X	X	X	X		X	X	X		X		X	X		X	X		X	X	X	X	X	X
Pipistrellus hesperus														X							X	X	X			X			X
Nyctinomops femorosaccus																										X			
Tadarida brasiliensis																	X												
Total species	1	1	1	1	2	1	1	1	1	1	1	1	2	9	1	1	2	1	1	3	2	3	5	1	1	6	1	1	3

BLA, Bahía de los Ángeles. For sources, see text.

Index

f indicates figure; *t* indicates table

α-component of diversity (α-diversity), 85–86, 100, 273
Accession numbers, 113
Acronyms, list, 427*t*
Agavaceae (family Agavaceae, agaves and yuccas), 97–98
Alarcón, Hernando de, 7, 388
Alcatraz (Tassne or Pelíano)
   Sauromalus on, 411–413
   size, 383
Allozymes (reptile), 191–192
Anchovies, northern. *See Engraulis mordax* (northern anchovies)
Ángel de la Guarda
   location and geology, 3, 5*f*, 23, 25
   size, 383
Ángel de la Guarda island group
   reptiles, 205–207
   tectonics, 205, 206*f*
Annual net primary productivity (ANPP), 363, 365, 372
Ants
   Acromyrmex versicolor (leaf-cutting ants), 118, 123

   arboreal (twig-nesting) ants, 118, 124
   collections, 113–114, 116–117
   Cyphomyrmex spp., 117, 118
   data for islands, 115*t*
   elevation, relation to species number, 118–120
   fungus-growing ants (tribe Attini), 117, 118
   granivorous (seed harvesting) ants, 117, 118, 123–124
   ground-nesting ants, 117, 118
   isolation, relation to ant species number, 118–120, 124
   Leptothorax spp., 117, 118, 124
   Myrmecocystus spp., 118
   myrmicine ants, 124
   Neivamyrmex spp. (army ants), 117, 118, 123, 124
   non-native species, 117, 118
   Pheidole spp., 117, 118, 124
   pitfall trapping, 113, 114
   plant diversity, relation to ant species number, 120–121, 122–123, 124. *See also* Habitat heterogeneity

Ants (*continued*)
Pogonomyrmex californicus, 117, 118
ponerine ants, 117, 118, 124
sea birds, relation to ant species number, 120–123, 124, 125
Solenopsis xyloni, 118
species-accumulation curves, 114, 116*f*
species-area curve, 118–120
species associations, 123, 124*t*, 125–126
species in the Sea of Cortés, 113, 543*t*–561*t*
from spider middens, 113, 114
Asteraceae (family Asteraceae, formerly Compositae)
achenes, dispersal of, 107–108
Ambrosieae tribe, 105–106
Astereae tribe, 106
distribution, by growth form, 103–104
distribution, factors affecting, 107
diversity, effects of island size, 104
Eupatorieae tribe, 106–107
Helenieae tribe, 107
Heliantheae tribe, 104–105
Mutisieae tribe, 104, 105
Perityle spp. (rock-daisies), 107, 108*f*
Senecioneae tribe, 105
Asunción, location, 4*f*
Avifauna. *See* Birds
Axelrod, D., 74–75

β-component of diversity (β-diversity), 85–86, 100, 166–167, 273
Baegert, Johann Jakob, S.J., *Observations in Lower California*, 9
Bahía de las Ánimas islands, 336*t*
Bahía de Loreto, 363, 364*f*, 436–438
Bahía de Los Ángeles (BLA) islands
food resources, 337–338
location and characteristics, 336*t*, 337, 363, 364*f*, 422*f*
plants, 534*t*–537*t*
Bahre, C. J., 391, 408
Baja California
formation of peninsula, 14, 18, 29–30, 74
paleostratigraphy, 183, 193–195
Ballena, location, 6*f*
Barco, Miguel del, S.J., *The Natural History of Baja California*, 8
Baroclonic circulation, 52
Bathymetry of the Gulf of California, 42
Bats (Chiroptera). *See* Chiroptera (bats)

Bechtel, Kenneth, 12, 417, 423, 424
Beier, E., 52
Beltrán, Enrique, 424, 427
Belvedere Scientific Fund, 12, 424
Bennett, C. F., 414
Binomial taxonomy, 182
Biosphere reserves, 428, 429, 432
Birds
α-diversity, 273–274
Amphispiza bilineata (black-throated sparrow), 273, 281, 293, 296, 308–309
Aphelocoma californica (western scrub-jay), 277
Baja peninsula species, 276–277
Basilinna xantusii (Xantus hummingbird), 271, 276, 279*f*
β-diversity, 273
bird densities, 303–309
Callipepla californica (California quail), 273, 276–277, 286
Callipepla gambelii (Gambel's quail), 273, 286
Campylorhynchus brunneicapillus (*C. brunneinucha*, cactus wren), 273, 277, 287
cape region endemic species, 275–276
Cardinalis cardinalis (red cardinal), 280, 281, 286
Cardinalis sinuata (pyrrhuloxia), 280–281, 286
Carpodacus mexicanus (*house finch*) (western scrub-jay), 272, 279, 281, 296
on Cerralvo, 281
densities, factors affecting, 305–306, 307
density compensation, 303–305, 307–309
differentiation by climate or vegetation zone, 277–281
distribution, 272–282, 292–294, 300–301, 603*t*–607*t*, 612*t*–618*t*
diversity, 273–275
drainage basin area effects on species numbers, 297–300
endemism, 272, 275–276, 281–282
Falco peregrinus (peregrine falcon), 283, 284, 315
γ-diversity (species turnover), 273–274, 275, 608*t*–611*t*
habitat types, effects on species numbers, 296–300, 305
hummingbirds, breeding ranges, 279*f*
incidence functions, 285*f*, 295*f*–296*f*, 297*f*, 298*f*, 302*f*

island endemics, 281–282
Junco bairdi (junco), 276
Larus heermanni (Heermann's gulls) (*See Larus heermanni* (Heermann's gulls))
Larus livens (yellow-footed gull), 313, 315, 399
migrant species on the gulf islands, 301–303, 304f, 638t–647t
nestedness of species data, 286–287, 288t–291t, 292
northern and southern islands, differences, 292–294
Oceanodroma melania (black storm-petrel), 313, 397
Oceanodroma microsoma (least storm-petrel), 313, 397
owls, 284–285
Pelecanus occidentalis californicus (California brown pelicans), 55, 313, 322, 323, 399
Philortyx fasciatus (banded quail), 273
Picoides scalaris (*Dendrocopus scalaris*, ladderback woodpecker), 281
Pipilo fuscus (brown towhee), 281, 286
Polioptila californica (California gnatcatcher), 287
Polioptila melanura (black-tailed gnatcatcher), 277, 287
productivity, effects of, 305–306
quail, breeding ranges, 280f
raptors, 283–285, 623t–627t
reproductive effort, 239
Salpinctes obsoletus (rock wren), 273, 286, 287, 296
on San Estéban, 281
shoreline birds on the islands, 282–283, 620t–622t
small land birds on northern islands, 285–292, 628t–633t
small land birds on southern islands, 290t–291t, 292, 634t–637t
Sonoran desert species, 272–275, 603t–607t
species-area relations, 284f, 294–296, 298f, 304f
species turnover (species turnover), 273–274, 608t–611t
Sterna elegans (elegant terns), 313, 322, 384, 421f
Sula spp. (boobies), 263, 313, 322–323, 384

Synthliboramphus craveri (Craveri's murrelet), 313
thorn-scrub species, 275, 612t–618t
on Tiburón, 281, 286
Toxostoma cinereum (gray thrasher), 276–277, 278f
Toxostoma curvirostre (curve-billed thrasher), 276–277, 278f, 281
Toxostoma lecontei (Le Conte's thrasher), 191, 272, 278f
watershed areas, effects on species numbers, 297–300
wintering species, 301–303
See also Larus heermanni (Heermann's gulls); Sea birds
Body-size trends
mammals, 352–354
reptiles, 238–239, 240–241, 243–245, 246–261
rocky-shore fishes, 167–169, 173
Bourillón, Luis A., 424, 426
Bowen, Tom, *Seri Prehistory—The Archaeology of the Central Coast of Sonora, Mexico; Unknown Island: Seri Indians, Europeans, and San Esteban Island in the Gulf of California*, 388
Browne, J. Ross, *Resources of the Pacific Slope*, 10
Búrquez, A., 410

Cacti (family Cactaceae)
Carnegiea gigantea (saguaro), 93, 409
characteristics, 92
classification difficulties, 92
cultural dispersal 408–411, 92
Echinocereus spp. (hedgehog cacti), 95
Ferocactus spp. (barrel cacti), 10, 95, 423f
Fouquieria columnaaaris (boojum or cirio), 411
hybridization, 92–93
incidence functions, 84f
Lophocereus schottii (senita or garambullo), 93–94, 315
Mammillaria and *Cochemiea* spp. (nipple cacti), 95
morphology variations, 92, 94–95, 96
Opuntia spp. (chollas and prickly pears), 95–96, 315, 410
Pachycereus pringlei (cardon), 64, 93–94, 315, 409, 422f
Stenocereus spp., 93, 409–410

Cacti (family Cactaceae) (*continued*)
  Stenocereus thurberi (organpipe), 93, 409
  vegetative reproduction, 92
California Academy of Sciences, *Sea of Cortez and Baja California*, 12, 423
Cape region
  origin of cape region block, 74–75, 194, 195f
  phytogeographic zone, 71–72
  plant distributions in, 79
Carmen, 27, 213, 437–438
Case, T. J.
  on extinction rates, 227
  on lizard species number, 222
  on niche overlap, 236, 245
Case, T. J., and M. L. Cody, 182, 226
Cedros, location, 4f
Cerralvo
  birds, 281
  location and geology, 6f, 21t, 25t
  plant distributions in, 79
  reptiles, 209, 213
Cetaceans, and El Niño, 55–56
Chaetodipus (pocket mice)
  abundance near shore and inland, 339, 340f
  competition, 341–342, 374
  diet, 335, 338, 352–353
  distribution, 327, 335–336, 338, 343
  dwarfism, 352–354
  effects of El Niño, 340
  granivorous diet, 328, 335, 338–339, 368, 374
Chiroptera (bats)
  diet, 328
  distribution, 333–334, 660t–661t
  Leptonycteris curasoae (long-nosed bat), 328, 333
  Macrotus californicus (California leaf-nosed bat), 328
  Myotis vivesi (fish-eating bat, fishing bat), 327–328, 397
Chlorophyll *a* concentrations in the Gulf of California, 45–46
Chuckwalla lizard. *See Sauromalus* (chuckwalla lizards)
Circulation in the Gulf of California, 52–54
Cladistic methodology, 182
Cladogenesis, patterns of relationships, 192–193

Cladograms
  Cnemidophorus hyperythrus (orangethroat whiptail lizards), 202f
  Cnemidophorus tigris (western whiptail lizards), 186, 187f, 201f, 205
  Crotalus (rattlesnakes), 202f
  iguanid genera, 190f
  interpreting, 184–186, 201–203
  Lampropeltis zonata (California mountain kingsnakes), 189f
  Pituophis catenifer (Pacific gopher snakes), 188f
  Urosaurus (brush lizards), 192f
  Uta (side-blotched lizards), 187f, 198f
Clavijero, Francisco Javier, S.J., *History de la California*, 8
Climate in Gulf of California
  climate of gulf islands, 64–68
  effect of mountains, 42–43
Climax communities in patch reefs communities, 155, 175
Cnemidophorus spp. (whiptail lizards)
  body-size trends, 243–245
  character displacement, 243
  cladograms, 187f, 200, 201f, 202f
  clutch-size trends, 245–246
  dispersal, 197
  DNA sequences, 186–188, 200
  niche overlap, 245
  origins, 197
  taxa, 196
Cochimí (Comondú), 387, 388–389
Coloradito (Los Lobos, Lobos, Salvatierra), geology, 22
Consag, Fernando, S.J., 8, 388
Conservation
  financial issues, 439–440
  in the lower gulf, 436–439
  in the Midriff islands, 432–436
  in the northern Sea of Cortés, 430–432
  timeline, 425t–426t
Coronados (Coronado, Smith)
  conservation issues, 436–438
  location and geology, 6f, 25, 422f
  size, 383
Cortés, Hernán, 7
Crassulacean acid metabolism (CAM), 92, 338–339
Crotalus (rattlesnakes)
  body-size trends, 259–260
  cladograms, 202f

DNA sequences, 188, 200
  target prey defensive behavior, 260–261, 262t
  taxa and distribution, 196, 205, 207, 208–209
Cultural dispersal
  cacti dispersal, 408–411
  and equilibrium theory of island biodiversity, 407, 410–411
  exotic plant species on gulf islands, 79–80, 398
  island ethno-biogeography, 408
  mammals introduction by humans, 329, 348–351, 397
  marine resources, use by introduced feral populations, 376
  reptile dispersal, 237, 411–413
  Seri introduction of plants and animals, 394
  stowaway organisms, 413–414

Danzante, 6f, 28, 213
Darwin, Charles, 181
Deer mice. See *Peromyscus* (deer mice)
Defensive behavior of prey species, 260–261, 262t
Demersal eggs of rocky-shore fishes, 154
Dendographs of rocky-shore fishes, 162, 163f
Density compensation
  in birds, 303–305, 307–309, 337
  in lizards, 262–263
  in *Peromyscus*, 337
Diamond, J. M., 83, 233
Diaz, Antero, 434, 435
Diguet, Leon, 10–11
DNA sequences
  advantages of, in cladistic evaluation, 184
  Callisaurus draconoides (Zebra-tailed lizards), 191
  Chilomeniscus (sand snakes), 191
  Cnemidophorus hyperythrus (orangethroat whiptail lizards), 186–188, 200
  Cnemidophorus tigris (western whiptail lizards), 186, 200
  Crotalus (rattlesnakes), 188, 200
  Eumeces (skinks), 189
  iguanid genera, 189–191
  Lampropeltis zonata (California mountain kingsnakes), 188
  Pituophis catenifer (Pacific gopher snakes), 188

Sauromalus (chuckwallas), 186, 200
Toxostoma lecontei (Le Conte's thrasher), 191
Uma notata (Colorado Desert fringe-toed lizards), 186
Uta (side-blotched lizards), 186, 199
See also Mitochondrial DNA (mtDNA)
Dunham, A. E. et al., on *Uta* body size, 240
Dwarfism, 352–354

Eastern Pacific Barrier (EPB), 169
Echeverría, Luis, 427
Ecotourism, 12, 399–400, 434, 435
Egg collecting, 390, 397–398
Ekman transport, 45
El Cholludo, geology, 22
El Huerfanito, geology, 22
El Muerto, location, 4f
El Niño-Southern Oscillation (ENSO)
  causes, 315–316
  effects on arthropods, 340, 372–373
  effects on birds, cetaceans, and fish, 55–56
  effects on circulation in gulf, 54–55
  effects on phytoplankton, 42, 55
  effects on plant productivity, 335, 339–340, 372
  effects on reptiles, 248f, 254, 256, 258, 263–264
  effects on rodent populations, 340
  effects on water masses, 55
  ocean currents, 316
  and oceanic productivity, 316
  pulse-reserve dynamics in desert communities, 373
  and reproductive success of Heermann's Gulls, 318–323
  sea level pressure (SLP) anomalies, 316
  Southern Oscillation Index (SOI), 317, 318f
Emlen, J. T., 309
Encantada, location, 4f
Endemic plant species on gulf islands, 73–74, 538t–542t
Engraulis mordax (northern anchovies), 55, 56, 316, 392
Equilibrium theory of island biogeography
  aquatic studies, 155
  biogeographic regions in Gulf of California, 159, 162–163

Equilibrium theory of (*continued*)
  colonization potentials of marine fishes, 155
  compared to historical biogeography, 222
  cultural dispersal of plants, 407, 410–411
  deterministic nature of colonization on islands in Sea of Cortés, 210
  and deterministic processes (predictability), 155–156, 175, 177, 201–203
  distribution patterns of rocky-shore fishes, 155–156, 169–170, 175–177
  effect of species turnover, 155, 175
  effects of predation, 155
  immigration *vs.* extinction rates, 114
  reef-fish recolonization after defaunation, 155, 175
  reptile immigration and extinction rates, 226–229
  rocky-shore fishes colonization and extinction, 154, 170
  and Sale's lottery hypothesis, 177
  sea bird immigration and extinction rates, 124–125
  species-area curve for Polynesian ants, 112
  and stochastic processes, 155–156, 170, 175, 201–203
  z-values, significance, 344, 347
  See also MacArthur, H. R.; Walker, B. W.
Espiritu Santo
  conservation issuse, 438–439
  location and geology, 6f, 28
Estanque (Pond), 26, 383
Evaporation from the Gulf of California, 46
Exotic plants on gulf islands, 79–80, 398
Exploration, biological, of the Sea of Cortés, 8–12
Exploration, physical and political, of the Sea of Cortés, 6–8
Extinctions and extinction rates
  inferred, 226–229
  and island biogeographic theory, 114, 202–203, 210
  mammals, 227, 348–350
  reptiles, 210–214, 226–229
  rocky-shore fishes, 170
  sea birds, 124–125
  variation over time, 228

Farallón, location, 4f
Felger, R. S., 69, 393, 409, 410
Fisheries, overexploitation of, 5, 430–432

Fishing, commercial, 391–392, 400
Fishing camps, environmental impacts, 398–399
Food chains in planktonic communities, 41, 172
Food webs
  diagram, 372f
  plant productivity, 365
  sea bird effects, 366–367
  species interactions, 374
  subsidies from marine resources, 374, 376
Formation of islands in Sea of Cortés, 14–15, 19–28
Fossil deposits, 28–33
Fronts between water masses, 41, 48

γ-component of diversity (γ-diversity), 85–86, 273
Gastil, G. et al., *The Geology and Ages of Islands*, 33–34
Gentry, Howard Scott, *Agaves of the Sonoran Desert*, 11
George, T. L., 309
Gigantism, 246–248, 352–354, 582*t*–587*t*
Global Environmental Facility (GEF), 430
Global Environment Fund (GEF), 439–440
Goldman, Edward A., *Plant Records of an Expedition to Lower California*, 11
Gomez Pompa, Arturo, 431
Grismer, L. L., 199
Guano
  effects on ANPP, 365
  effects on ants, 125
  effects on plants, 125, 146, 369, 375–376
  effects on soil, 314, 369
  effects on tenebrionid beetle abundance, 146–147, 370–371
  mining, 389–390, 396–397
Guaymas, connection to plant distributions on gulf islands, 78
Gulf of California. *See* Sea of Cortés
Gyres in the Gulf of California, 52–53

Habitat heterogeneity
  effect on tenebrionid beetle species richness, 144, 149
  relation to plant diversity, 121, 125
Heat exchange in the Gulf of California, 46
Heermann's gulls. *See Larus heermanni* (Heermann's gulls)

Herpetofauna, associations
  East-West split, 196
  peninsular archipelago, 196–197, 214
  transgulfian vicariance, 195–196
Herpetology, topics investigated in Sea of Cortés, 181–182
Hews, D. K., 241
Hinds, Richard Brinsley, *The Botany of H.M.S. Sulphur*, 10
Historical biogeography, 222, 329–334, 346
Hollingsworth, B. D., 199
Horn's index, 162, 163*f*
Horton's laws, 300
Huey, Laurence M., 11, 12
Hurricanes (*chubascos*), 385
Hutchinson, G. E., 369

Incidence functions, 83–85
Island biogeography theory, 112, 114, 124
  See also Equilibrium theory of island biogeography
Island ethno-biogeography, 408
Islands, in Sea of Cortés. *See* Midriff; individual islands
Islands, origins
  central gulf (Midriff) islands, 23–27
  northern gulf islands, 19–23
  southern gulf islands, 27–28
Islas del Golfo de California, 424
Isolation factor (ISOL), 90
Isotopic stage 5 (I.S.5), 19, 22, 27, 34n
Isthmus of La Paz, 30–31, 330, 332

Krutch, Joseph Wood, 417

Landbridge islands
  colonization by mammals, 329, 335, 354
  distribution of reptiles, 209–214, 212*t*, 223
  landbridge effect on mammal distributions, 223
  landbridge effect on plant distributions, 75–76
  in north and south Sea of Cortés, 211*t*
  relictual mammals, 329
  species-area relationships, 344, 345*f*
  supersaturation of lizard fauna, 213, 226
  tendency for higher species richness, 90
La Niña, 316
La Paz
  history, 7
  Isthmus of, 30–31, 330, 332

La Paz fault, 30–31
Larus heermanni (Heermann's gulls)
  banding, 316–317
  body mass, 318–319, 320–321, 322*f*
  breeding success, 317, 319–321
  clutch size and egg mass, 319, 321, 323
  diet, 316
  and El Niño, 55–56, 318–323
  nesting, 313–314, 316
  nesting success, 317, 319–321
  photo, 420*f*, 421*f*
  population fluctuations, 315
  reproductive success, 317
  See also Rasa
Las Ánimas, 6*f*, 25–26, 383
Lawlor, T. E., 223, 351–352
Ley General de Equilibrio Ecológico y la Protección al Ambiente, 428–429
Lindbergh, Charles, 417, 423–424
Lindblad, Sven Olof (*Sea Bird*), 431
Lindsay, George, 417, 423, 434
Lizards
  Callisaurus draconoides (Zebra-tailed lizards), DNA sequences, 191
  El Niño effects, 248*f*, 254, 256, 258, 263–264
  Eumeces (skinks), 189
  iguanid genera, 189–191
  incidence functions, 233, 234*f*
  marine resources, 263–264, 368
  Petrosaurus (rock lizards), 191, 258–259
  Phyllodactylus spp. (geckos), 196, 207, 315, 411, 418*f*
  sand lizards, 182
  Sceloporus (desert spiny lizards), 191, 196, 197, 213
  sea bird effects, 263
  snout-vent length (SVL) as a measure of size, 240
  Uma notata (Colorado Desert fringe-toed lizards), 186
  Urosaurus (brush lizards), 191–192
  See also Cnemidophorus spp. (whiptail lizards); *Sator*; *Sauromalus* (chuckwalla lizards); *Uta* (side-blotched lizards)
Lomolino, M. V., 346, 354
Longinos Martínes, José, 9
López Mateos, Adolfo, 5, 427
López Portillo, José, 5
Los Angeles Bay. *See* Bahía de Los Ángeles (BLA) islands

Los Cabos block, 30–31
Lowe, C. H., 69, 411

MAB. *See* United Nations' Man and the Biosphere Program (MAB)
MacArthur, H. R., 154, 199, 345–346. *See also* Equilibrium theory of island biogeography
MacArthur-Wilson equilibrium. *See* Equilibrium theory of island biogeography
Magdalena, location, 4f
Mammals
  Bassariscus astutus (ring-tailed cat), 335, 384, 419f
  body size variation, 352–354
  Canis latrans (coyote), 350, 368, 384
  colonizing abilities, 223, 334–335
  competition, 341–342
  dispersal abilities, 223, 329, 332–333, 342
  distribution on islands of the sea of Cortés, 326–328, 650t–659t
  extinctions, 348–350
  faunal relationships, 332–334
  feral cats and dogs, 348, 350, 397
  habitat quality and diversity, 335–337
  introduction by humans, 329, 348–351, 397
  mainland species, shifts in distribution, 333
  melanism, 351
  Odocoileus hemionus sheldoni (desert mule deer), 384, 395, 432
  origins and evolution of insular fauna, 328–334
  Ovis canadensis mexicana (bighorn sheep), 384, 395, 432–434
  Phocoena sinus (vaquita porpoise), 430–431
  resource availability, 337–340
  species-area relationships, 344–346
  species assemblies on landbridge and oceanic islands, 346–347
  variation, 351–354
  Zalophus californianus (California sea lions), 390, 419f
  See also Chaetodipus (pocket mice); Chiroptera (bats); Rodents (order Rodentia)
Marine environments, comparison of islands and mainlands, 155, 170–172
Marine resources
  consumption by *Peromyscus*, 338–339, 368, 374

effect on arthropods, 145–146, 366, 367–368, 369–371
effect on land birds, 368
effect on lizard densities, 263–264, 368
effect on population stability, 374–375
effect on species distribution and diversity, 375
island perimeter/area (P/A) ratios, 145–146, 366
sea bird effects, 146, 366–367, 368–371
shoreline effects, 365–368
shoreline scavengers and predators, 366, 368, 369
use by introduced feral populations, 376
windblown sea foam and spray, 366
See also Guano
McGee, W. J., 393
Mediterranean-climatic vegetation, 70f, 71, 76, 81–82
Mejía, 23, 383
Microphyllous desert vegetation, 68–69
Midriff
  distribution of mainland and peninsular plants, 77–78
  location, 5f, 364f, 383
  plant and vertebrate species introduced by humans, 407–415
  prehistoric archaeology, 386–387, 407
  See also Islands; Sea of Cortés
Miramar (El Muerto or Link), geology, 22
Mitochondrial DNA (mtDNA), 184, 330–332, 582t–587t. *See also* DNA sequences
Monserrat
  fauna, 205
  location and geology, 6f, 28, 203
Montague, location, 4f
Moran, Reid, 12
Moser, M. B., 393, 409, 410
MtDNA. *See* DNA sequences; Mitochondrial DNA (mtDNA)
Murphy, R.W.
  on evolution of herpetofauna, 183, 199, 330–331, 332–333
  on patterns of extinction, 202–203, 210

Natividad, location, 4f
Nature Conservancy, 424
Nelson, Edward W., *Lower California and its Natural Resources*, 11

Nestedness of species data
  birds, 286–287, 288t–291t, 292
  reptiles, 231–232

Oceanic islands in the Sea of Cortés
  colonization by mammals, 329, 335, 354
  colonization by reptiles, 204–206
  lizard distributions, 212t, 222
  in north and south, 211t
  reptile distributions, 212t, 229–238
Ostracoda, distribution and diversity, 31–32

Packrat middens, 183
Paleobiogeography of gulf reptiles
  allozymes, 191–192
  alternative methods, 183
  cladogenesis patterns, 192–193
  cladograms, interpreting, 184–186, 201–203
  DNA Sequences, 186–191
  east-west split, 196
  peninsular archipelago, 196–197, 214
  transgulfian vicariance, 195–196
Paleoecology, 183
Paleogeography, 193–195, 202–203, 206f, 210–214
Paleostratigraphy, origins of Baja peninsula, 183, 193–195
Palmer, Edward, 10
Pangas, 392
Parque Marino Nacional Bahía de Loreto, 436
Partida Sur (Partida), location and geology, 6f, 28
Patos
  guano mining, 389, 397
  size, 383
Pearl hunting, 10–11, 388
Pearl oyster (*Pteria sterna*), 388
Pelagic eggs and larvae of rocky-shore fishes, 154
Peninsular archipelago, 196–197, 214
Peromyscus (deer mice)
  abundance near shore and inland, 339, 340f
  colonization abilities, 332, 342
  competition, 341–342, 374
  density compensation, 337
  distribution, 327, 335–336, 338, 342–343
  effects of El Niño, 340, 341f
  gigantism, 352–354
  marine resources in diet, 338–339, 368, 374
  omnivorous diet, 327, 335, 338–339, 352–353, 368
  Peromyscus eremicus avius, 332–333
  relationships of species, 331f
Phylogenies, discrepancies between molecular and anatomical systems, 184
Phytogeographic regions
  definition, 69
  distribution of vegetation types, 69–72, 70f
Phytoplankton in the Gulf of California, 45
Phytoplankton in the Sea of Cortés, 41
Pigment (chlorophyll a) in the Gulf of California, 45–46
Plant distributions on gulf islands
  cape region, 79
  Cerralvo, 79, 511–513, 514t–524t
  connection to Guaymas, 78
  distribution across Midriff chain, 77–78
  endemic species, 73–74, 538t–542t
  exotics and recent invasions, 79–80, 398
  geographic and climate factors, 72–73
  landbridge islands, 75–76, 90
  links to Mojave Desert and northwestern gulf coast, 77
  phytogeographic regions, 69–72
  plant species on gulf islands, 72, 80, 463t–509t, 525t–537t
  relicts from southern mainland in cape region and Cerralvo, 79
  relicts of Mediterranean-climate vegetation, 76–77
  Sonoran Desert history, impact of, 74–75
  species-area curves, 80–83, 90
Plant diversity
  adaptations to island life, 91–92
  diversity components, 85–86, 100
  effects of area on diversity, 80–83, 90, 104
  on landbridge islands, 90
  species turnover from one island to another (SPTURN), 86–89, 91
Plant productivity
  and consumer abundance, 365
  effects of El Niño, 335, 339–340
  effects of guano, 146
  and food web structure, 365
Plants
  Acacia spp., 99, 100–101

Plants (*continued*)
  Atriplex spp. (saltbush), 85, 315
  Avicennia germinans (black mangrove), 98
  Burseraceae (family Burseraceae), 101–102, 103*f*
  Bursera spp. (elephant trees), 408
  Cercidium spp. (palo verdes), 99
  exotic species, 79–80, 398
  Ficus palmeri (fig), 98
  incidence functions, 84
  Larrea tridentata (creosote bush), 69, 84*f*, 85
  leaf morphology of *Acacia* spp., 100, 100*f*
  Lysiloma spp., 99
  Olneya tesota (ironwood), 99
  Pachycormus discolor (elephant tree), 10, 98
  Prosopis spp. (mesquites), 99
  Rhizophora mangle (red mangrove), 64, 98
  trees, non-leguminous, 98–99
  vegetation, types of, 68–69, 70*f*–71*f*
  See also Agavaceae (family Agavaceae, agaves and yuccas); Asteraceae (family Asteraceae, formerly Compositae); Cacti (family Cactaceae); Vegetation, types of
Plate tectonics
  origin of cape region block, 74–75, 194, 195*f*
  origin of Gulf of California, 16–18, 193–195
Pocket mice. *See Chaetodipus* (pocket mice)
Polis, G. A. et al., 373
Pómez (El Pomo, El Puma), geology, 22
Precipitation in Gulf of California
  mean annual precipitation, 43, 44*f*, 65–68, 363, 385
  seasonal distribution of rainfall, 43, 44*f*, 67*f*, 68, 385
  variability, 365
Predation
  effect on tenebrionid beetle abundance, 147–148, 149
  effects of low predation on *Sauromalus*, 257–258
  and equilibrium theory of island biogeography, 155
Propagules, 169–170, 172, 176
Protected areas, 428–430
Proto-gulf, 18, 34, 194
Pteria sterna (pearl oyster), 388
Puerto Peñasco, 158, 159*f*

Rainfall. *See* Precipitation in Gulf of California
Ramirez Ruiz, Jesús, 426, 435
Rasa
  conservation efforts, 434–435
  designation as migratory waterfowl preserve, 5, 424, 434
  eradication of rats and mice, 315, 397, 426, 435
  guano mining, 390, 396, 397
  Heermann's gulls (*Larus heermanni*) nesting on, 313–323, 384
  island environment, 313, 314–315
  location and geology, 26, 314
  oceanic productivity, 315–316
  plants and animals, 315
  sea bird population, changes in, 434–435
  size, 383
  Sterna elegans (elegant terns) nesting on, 313, 322, 384
Rattlesnakes. *See Crotalus* (rattlesnakes)
Recolonization after defaunation
  in reef-fish communities, 155, 175
Reeder, T. W., on iguanid DNA sequences, 189–190
Refugia, Pleistocene, 334
Regional diversity, 80
Reptiles
  body-size trends, 238–239, 240–241, 243–245, 246–261
  Chelonia mydas agassizi (black sea turtle), 384, 390, 391–392
  clutch-size trends, 239–240, 241–243, 245–246, 247
  cultural dispersal, 237, 411–413
  distribution on islands in the Sea of Cortés, 204*t*, 229–238, 588*t*–592*t*, 594*t*–602*t*
  effects of glaciation in the Pleistocene, 183
  El Niño effects, 248*f*, 254, 256, 258, 263–264
  endemic species, 230*t*
  immigration and extinction rates, 226–229
  incidence functions, 232–236
  marine resources, 263–264, 366, 368
  mesophilic species, ranges of, 183
  nestedness, 231–232
  occurrence, 233–234, 235*f*
  population densities, 261–264
  relationship between colonization success and persistence, 224–226

reptiles with ancestors in Baja California or Sonora, 230f
sea bird effects, 263
species associations, 236–238
species number, factors related to, 222–229
See also *Crotalus* (rattlesnakes); Lizards; Snakes
Reséndiz, Antonio, 435
Residuals (RESID), 90
Richman, A. et al., 226, 227
Ricketts, Edward F. (Doc), 11, 326
Roca Consagrada, location and geology, 4f, 22
Rocky-shore fishes
  α-diversity, 172
  Acanthemblemaria crockeri, 164, 165t, 168
  Axoclinus spp., 164, 165t, 168, 174
  β-diversity, 158
  barriers to distribution, 158, 169
  biogeographic regions in Gulf of California, 159, 162–163
  biomass H' diversity, 157–158, 161, 574t–575t
  body-size trends, 167–169, 173
  collection sites, 156, 157f, 573f, 574t–575t
  comparison of island and mainland sites, 164–166, 172–174
  Coralliozetus micropes, 164, 165f
  Crocodilichthys gracilis, 164f, 165t, 168
  demersal eggs, effects on dispersal, 154, 160, 169
  dendographs of, 162, 163f
  dispersal mechanisms, 154, 155, 169, 176
  distribution, 155, 159–161
  effects of predation, 155, 174, 176
  effects of vertical relief and substrate, 158, 172, 174
  endemism, in gulf fishes, 159–161
  endemism, in Pacific fishes, 169
  Enneanectes reticulatus, 164, 165t
  equilibrium theory of island biogeography, 169–170, 175–177
  habitat differences of juveniles and adults, 155, 174, 176
  ichthyocide collections, 156, 573f
  island-mainland differences in rocky-shore fish communities, 172–174
  island-mainland environmental differences, 170–172
  latitudinal trends, 158, 160, 164
  Malacoctenus hubbsi, 164f, 165t, 168
  most abundant species, 163–164, 165t, 173
  numerical H' diversity, 157–158, 161, 574t–575t
  pelagic eggs and larvae, 154, 160, 169–170
  physical data of islands sampled, 577t–578t
  physical data of mainland areas sampled, 576t
  planktonic early life stages, 170, 172, 173, 176
  primary residents, 159–160, 169
  propagules, 169–170, 172, 176
  sea temperature and species diversity, 161
  secondary residents, 159–160
  short-lived larvae, 154
  species-area curves, 166, 167f
  species diversity, 157–161, 574t–575t
  species sampled in gulf, 579t–581t
  species turnover, 166–167, 173
  Tomicodon boehlkei, 164f, 165t, 167–168
  use of tidal pools as refuges, 158, 172
  visual censuses, 158, 166, 167f, 174
  Xenomedea rhodopyga, 164, 165t, 168–169
  zoogeographical affinities, 160t
Rodents (order Rodentia)
  Ammospermophilus (antelope ground squirrels), 350–351
  diet and resources, 337–341
  distribution, 327
  effects of El Niño, 340
  Lepus alleni (white-sided jackrabbit), 384
  Lepus californicus (jackrabbits), 336–337
  mtDNA studies, 330
  Mus musculus (house mouse), 315, 397, 435
  Neotoma lepida (woodrats), 336–337, 352, 354
  Rattus rattus (black rat), 315, 397, 435
  See also *Chaetodipus* (pocket mice); *Peromyscus* (deer mice)

Sale, P. F., 170, 177
Salinas de Gortari, Carlos, 432
Salsipuedes, 25–26, 383
San Diego, 6f, 28

San Esteban
  birds, 281
  cardon cacti, 422f
  reptiles, 207
  size, 383
San Francisco, location, 6f
San Gorgonio Filter Barrier, 196
San Gorgonio Pass, 196–197
San Ildefonso, geology, 27
San Jerónimo, location, 4f
San José, location and geology, 6f, 28
San Lorenzo
  geology, 25–26
San Luis (Encantada Grande), location and geology, 4f, 22
San Luis Gonzága (Willard)
  fauna, 327
  location and geology, 4f, 23
San Marcos, location and geology, 3, 4f, 27
San Martín, location, 4f
San Pedro Mártir
  booby colonies, 384
  guano mining, 390, 396, 397
  location and geology, 4f, 207
  rats, 397
  reptiles, 207
  size, 383
San Pedro Nolasco, location and geology, 4f, 27
San Roque, location, 4f
Santa Catalina, 28, 208, 213, 423f
Santa Cruz
  history, 7
  location and geology, 6f, 28
  reptiles, 208–209
Santa Inés, location, 4f
Santa Margarita, location, 4f
Santa Rosalía, 390
Sarcocaulescent desert vegetation, 68–69
Sardinops caeruleus (Pacific sardines), 316, 392
Sator
  aggressiveness, 237
  endemic species, 231, 233
  species associations, 233, 236–237
  taxa and distribution, 196, 208, 213, 332–333
Sauromalus (chuckwalla lizards)
  clutch size trends, 247
  cultural dispersal by Seri, 237–238, 411–413
  distribution, 205
  DNA sequences, 186, 200
  effects of drought, 252–255, 256
  El Niño effects, 248f, 254, 256, 258, 263–264
  gigantics, 246–258
  mtDNA phylogeny, 582t–587t
  photo, 418f
  population densities, 256
  predation, 257–258
  Sauromalus hispidus (Ángel de la Guarda block), 248–255
  Sauromalus varius (San Esteban), 248f, 249f, 250f, 251f, 255–256
  size trends, 246–247
Savage, J. M., on evolution of herpetofauna, 183, 198–199
Schulte, J. A. I. et al., on iguanid DNA sequences, 189–191
Scientific field research, environmental effects, 399
Sea-bird egg collecting, 390, 397–398
Sea birds
  effect on plant diversity, 122–123, 125
  and El Niño, 55–56
  as marine resources in food webs, 146, 366–367, 368–371
  relation to ant species number, 120–123, 124, 125
  See also Guano; *Larus heermanni* (Heermann's Gulls)
Sea-level changes, 226, 330, 332
Sea lion hunting, 389, 390
Sea of Cortés (el *Mar de Cortés*, Gulf of California)
  appearance and surface features of islands, 63–64
  bathymetry, 42
  biogeographic regions in gulf, 159, 162–163
  geographic parameters of islands in the Gulf of California, 24t–25t
  island age, using fossil deposits to determine, 28–33, 30t
  island names and measurements, 3, 445t–462t
  origin and age, 15–19, 329–330
  origins and geology of islands, 19–28, 20t–21t, 33, 385
  thermohaline circulation, 42
  tidal mixing, 41, 42

Sea turtle hunting, 384, 390, 391–392
Secretaría de Agricultura y Recursos Hidráulicos (SAHR, Secretariat of Agriculture and Water Resources), 432
Seeman, Berthold, 10
Seri (Comcaác)
  diet, 391, 393–395
  early European contacts, 388
  fishing, 391–392, 395
  handicrafts, 392
  hunting, 395
  introduction of chuckwallas on gulf islands, 205, 411–413
  introduction of spiny-tailed iguana to San Esteban, 394, 411, 413
  oral history, 387, 409, 411, 412–413
  placental burials, 409
  prehistoric archaeology, 386–387, 407, 414
  prehistory, 387–388, 414
  and Tiburón, 395, 426–428, 432–433
  transplantation and translocation of cacti, 408–411
Shark fishing, 390, 391
Sheridan, T. E., 395
Shreve, F., 68–69, 80
Shrimp fishing, 391, 392, 431
Sloan, A. J., 199
Small-island effect, 81, 344, 345f
Smith, G. B., on reef-fish communities, 175
Snakes
  body-size trends, 259–261
  Chilomeniscus (sand snakes), DNA sequences, 191
  distribution, 224–225
  incidence functions, 233, 235f
  kingsnakes, distribution pattern on islands, 225–226
  Lampropeltis zonata (California mountain kingsnakes), 188, 189f, 196–197
  Pituophis catenifer (Pacific gopher snakes), 188
  Rhinocheilus etheridgei (Cerralvo long-nose snake), 209
  See also Crotalus (rattlesnakes)
Snout-vent length (SVL) in lizards, 240
Sonoran Desert
  history, 74–75
  map, 70f
  sources of plant species, 75
Soulé, Michael, 12, 199, 240

Southern Oscillation Index (SOI), 317, 318f
Species-area curves, 80–83
Species number, related factors
  in ants, 118–125
  in birds, 296–300
  in reptiles, 222–229
Species turnover
  definition, 166–167
  and equilibrium theory of island biogeography, 155
Species turnover between islands (SPTURN)
  effect of island size on flora species turnover, 87–89, 91
  relationship to distance, 86, 87f, 91, 274
Stable isotope analysis (SIA), 338, 339f
Stapp, P. et al., 337, 338, 339f
Steinbeck, John, The Log of the *Sea of Cortez*, 11, 326
Stratigraphic history of Baja California, 183
Subsidence, 19, 22f
Substrate (soil), effects on vegetation and plant distribution, 68, 337, 363
Supersaturation, 213, 226, 329
Systematics, 182

Temperatures (air) in Gulf of California, 43, 44f, 66f, 67
Temperatures (sea-surface), 158, 159f, 161, 177, 385–386
Tenebrionidae (tenebrionid beetles)
  adaptations to desert conditions, 129, 131, 148
  Argoporis spp., 134, 135, 138, 140, 148
  categorization by tolerance of desert conditions, 131, 135, 570t–572t
  characteristics, 129, 130f
  classification by habitat use, 135
  common genera, 134
  Cryptoglossa spp., 134, 135, 148
  different habitats for adults and larvae, 135
  disharmony of island fauna with mainland, 138
  distribution on Pacific islands, 568t–569t
  distribution on Sea of Cortés islands, 562t–568t
  diversity on Baja California peninsula, 134
  diversity on Pacific islands, 134
  diversity on Sea of Cortés islands, 134
  endemic species, 133, 135–137
  endemism, related to isolation and island area, 136–137, 149

668 Index

Tenebrionidae (*continued*)
  guano, effects of, 146–147
  habitat heterogeneity, effect on species richness, 144, 149
  island area, effect on species richness, 142, 143*f*, 144*t*, 149
  island size, effect on species occurrence, 139–140
  isolation and species occurrence, 140–141
  isolation and species richness, 142–144, 145*f*
  methods of collecting, 132
  plant species richness, effect on beetle species richness, 144, 149
  predation, effect on beetle abundance, 147–148, 149
  resistance to immersion in sea water, 131
  resource availability, effect on beetle abundance, 145–147
  sea birds, effects of, 144–145, 146–147, 148*f*, 149, 369–371
  similarity of mainland and island faunas, 134, 137, 148
  species abundance on islands, 135
  species associations and competition, 141–142, 149
  species found in Sea of Cortés, 570*t*–571*t*
  species found in the Pacific, 571*t*–572*t*
  substrate type, effect of, 138–139, 149
  transition temperatures, 131
  Triphalopsis spp., 135, 138, 148
  winged forms, 131, 135
Tershy, B. R. et al., 396
Tiburón
  birds, 281, 286
  designation as protected area, 426, 432
  location, geology, 3, 5*f*, 26
  Odocoileus hemionus sheldoni (desert mule deer), 384, 395, 432
  Ovis canadensis mexicana (bighorn sheep), 384, 395, 432–434
  potable water, 383
  Seri, 395, 426–428, 432–433
  size, 383
Tiburón Plain Ware ("egg shell") pottery, 387
Tides, 41, 42, 49–51, 158, 171–172
Todos Santos, location, 4*f*
Tortuga
  location and geology, 3, 4*f*, 27
  reptiles, 204–205, 229

Totoaba macdonaldi (totoaba or totuava), 384, 390–391, 430–431
Transgulfian vicariance of reptile species, 195–196
Transition temperatures (beetles), 131
Trophic flow. *See* Food webs

United Nations Conference on Environment and Development (UNCED, the Rio Summit), 428
United Nations' Man and the Biosphere Program (MAB), 428, 429
Universidad Nacional Autónoma de México (National University of Mexico, UNAM), 424, 433
Uplift, related to origins of islands, 19, 22*f*
Upton, D. E., 330–331
Upwelling in the Gulf of California, 41–42, 45, 46
Uta (side-blotched lizards)
  body-size trends, 240–241
  cladograms, 187*f*, 198*f*
  clutch-size trends, 241–243
  diet, 240–241
  distribution, 197, 205, 233, 315
  DNA sequences, 186, 196, 199, 331
  negative association with *Sator*, 233, 236, 237

Vegetation, types of, 68–69, 70*f*–71*f*
Velarde, Enriqueta, 424
Venegas, Miguel, S.J., *A Natural and Civil History of California*, 8
Ventana (La Ventana), geology, 25
Vicariance biogeography, 183, 186, 195–196
Vicariant plant species, 77, 80, 85
Vidal, N., 314
Villa, Bernardo, 424, 430, 434
Vizcaíno, Sebastián, 7, 388
Vizcaíno Seaway, 29, 195, 197, 210, 330–331
Voznesenskii, I. G., 9

Walker, B. W., 159, 162–163
Walker, Lewis Wayne, 424, 434
Wallace, Alfred, *Island Life*, 181
Water loss in insects, causes of, 129
Water masses, 46–49
Whiptail lizards. *See Cnemidophorus* spp. (whiptail lizards)

Wiggins, Ira L., *Flora of the Sonoran Desert; Flora of Baja California*, 11
Wilcox, B. A., 226, 228, 241
Wilson, E. O., 112, 154, 199, 346. *See also* Equilibrium theory of island biogeography
Winds and upwelling in the Gulf of California, 41–42, 44–46, 317

Xántus, John (János), 10, 271
Xerophilic species, distribution of, 183

Yetman, D. A., 410

z-values, significance, 344, 347

Printed in the USA/Agawam, MA
June 8, 2010

541735.008